Environmental and Natural Resource Economics

Frank A. Ward

New Mexico State University

PEARSON
Prentice Hall

Upper Saddle River, New Jersey
Columbus, Ohio

Library of Congress Cataloging-in-Publication Data
Ward, Frank A.
Environmental and natural resource economics / Frank A. Ward.
p. cm.
Includes bibliographical references and index.
ISBN 0-13-113163-X (casebound)
1. Environmental economics. 2. Environmental policy. 3. Natural resources. I. Title.
HC79.E5W358 2006
333.7—dc22
2005043147

Executive Editor: Debbie Yarnell
Assistant Editor: Maria Rego
Production Editor: Alexandrina Benedicto Wolf
Production Coordination: Carlisle Publishers Services
Design Coordinator: Diane Ernsberger
Cover Designer: Thomas Mack
Cover image: Thomas Moran, U.S. Department of Interior Museum
Production Manager: Matt Ottenweller
Marketing Manager: Jimmy Stephens

This book was set in Times by Carlisle Communications, Ltd. It was printed and bound by Hamilton Printing. The cover was printed by Phoenix Color Corp.

Pearson Education Ltd.
Pearson Education Singapore Pte. Ltd.
Pearson Education Canada, Ltd.
Pearson Education—Japan
Pearson Education Australia Pty. Limited
Pearson Education North Asia Ltd.
Pearson Educación de Mexico, S.A. de C.V.
Pearson Education Malaysia Pte. Ltd.

PEARSON
Prentice
Hall

10 9 8 7 6 5 4 3 2 1
ISBN 0-13-113163-X

Preface

Few modern issues have so dramatically influenced consumer decisions, personal lifestyles, corporate planning, and public policy as environmental and natural resource problems. Since 1970, we have seen changes in product design and packaging, corporate investments, personal career and lifestyle choices, government tax policies, and emerging technologies—to name but a few—because public and private decisions are now shaped by environmental and natural resource issues.

With this book, I hope to bring you close to today's exciting policy issues and introduce you to the theories and methods of analysis that economists use to approach these issues. This book shows how economic tools, such as benefit-cost analysis, can be used to understand a wide range of real environmental and natural resource problems. These tools can help design practical policy solutions to environmental problems. They also can contribute to public debate about complicated and emotional environmental issues by providing a universal, rigorous, and testable framework.

This book is intended for students who have completed one semester of microeconomic principles, but it also will engage students with a wider economics background. It provides an introduction to environmental and natural resource issues and describes economic theories and methods used by experts working in the field.

Chapters begin with a vignette that motivates and introduces the economic theories, methods of analysis, and policy questions, followed by clearly-stated goals. Relevant economic principles are introduced as a resource for posing questions, testing and applying theories, understanding policy debates, and posing future questions. Footnotes are included to cover advanced or specialized treatment and to encourage students to pursue their own reading or research. Current methods of analysis, data sources, and ongoing studies are presented. Policy debates are analyzed along with unanswered questions posed by those debates. References to the current literature are open and transparent.

Graphs, tables, and numerical examples are included to reinforce the principles presented in the text. No mathematics beyond simple algebra is required; all mathematics beyond algebra (e.g., calculus) is relegated to footnotes and the appendixes. Most chapters have several numbered examples that present current real-life examples of research and policy applications of economic principles. End-of-chapter questions provide an opportunity to apply the material through class discussion, personal reflection, and exam preparation. The appendixes include advanced quantitative methods of analysis.

The topics and structure of this book encourage its use for various courses. All of the chapters could find a place in an upper-division environmental economics course. Some chapters could be a source for graduate students or those wishing to pursue more specialized research interests. There may be more material in this book than can be covered well in one semester. The additional information opens up options for selecting topics that best cover the interests, expectations, and backgrounds of instructors and students. For example, a one-semester course in environmental economics could be organized around Chapters 1–7 and 14–19. If a brief introduction to natural resource economics is desired, Chapter 8 and/or Chapter 12 could be added. A one-term course in natural resource economics could cover Chapters 1–6 and 8–13. An introduction to environmental economics could be accomplished by adding Chapter 7 and/or Chapter 16.

This book focuses on environmental and natural resource policy and presents a considerable amount of economic theory, methods, and data to understand policy solutions. Although this is an economics book, it is not limited to economics. Economics is not a stand-alone discipline; rather, other disciplines complement the ability of economics to sharpen our understanding of policy issues. The disciplines of biology, the physical sciences, law, history, political science, and philosophy are introduced where their insights enhance the power of economics in confronting policy problems. In some cases, combining other disciplines with economics presents contrasting views of a problem, focuses a policy debate, or suggests competing hypotheses in the search for policy solutions.

This book is optimistic. It presumes that past gains in environmental protection and natural resource management will be maintained and that future gains are both desirable and possible. There is little future in pessimism. Students are eager to understand how economic concepts, like markets, can explain and address environmental and natural resource problems. The text is pro-environment, and it also emphasizes the role of profit incentives in promoting the cooperation of business in protecting the environment and ensuring adequacy of natural resource supplies.

ACKNOWLEDGMENTS

This book has evolved over my years of college teaching, beginning in 1990. I owe a deep debt to the thousands of students who have patiently sat through my classes when they had so many other interesting things to occupy their time. Without searching their eyes, watching their faces, seeing their reactions, listening to their questions, and reading their essays and papers, I could not have written this book. If this book has meaning, relevance, or utility, my students at New Mexico State University can take much of the credit. The original inspiration for this book began with classes I took at Colorado State University from Ken Nobe, Lee Gray, Bob Young, John McKean, and many others.

A book like this depends on the published and unpublished contributions of the hundreds of scholars whose creative work has pushed back the frontiers of knowledge and methods while advancing understanding of environmental issues. This book makes few original contributions. Its source material comes from many classic economics books and papers dating back to the 1930s and draws on several hundred journal articles and other papers written since 1995, as well as hundreds of World Wide Web citations posted since 2000.

I would like to thank the following reviewers for their valuable feedback: Chris Azevedo, Central Missouri State University; Stacey Brook, University of Sioux Falls; Andrew Keeler, University of Georgia; Leah Greden Matthews, University of North Carolina-Asheville; Mark Phillips, Moorehead State University; and Jeffrey R. Williams, Kansas State University

Finally I thank my wife Erin, whose loving support has made this book possible.

Frank A. Ward

Additional Resources

Online Supplements to Accompany the Text

An online Instructor's Manual and TestGen are also available to instructors through the Ward catalog page at *www.prenhall.com*. Instructors can search for a text by author, title, ISBN, or by selecting the appropriate discipline from the pull down menu at the top of the catalog home page. To access supplementary materials online, instructors need to request an instructor access code. Go to *www.prenhall.com*, click the **Instructor Resource Center** link, and then click **Register Today** for an instructor access code. Within 48 hours after registering you will receive a confirming e-mail including an instructor access code. Once you have received your code, go to the site and log on for full instructions on downloading the materials you wish to use.

Agriculture Supersite

This site is a free online resource center for both students and instructors in the Agricultural field. Located at *http://www.prenhall.com/agsite*, students will find additional study questions, job search links, photo galleries, PowerPoints, *The New York Times* eThemes archive, and other agricultural-related links.

Instructors will find a complete listing of Prentice Hall's agriculture titles, as well as instructor supplements supporting Prentice Hall Agriculture textbooks available for immediate download. Please contact your Prentice Hall sales representative for password information.

The New York Times Themes of the Times for Agriculture

Taken directly from the pages of *The New York Times,* these carefully edited collections of articles offer students insight into the hottest issues facing the industry today. These free supplements can be packaged along with the text.

Agribooks: A Custom Publishing Program for Agriculture

Just can't find the textbook that fits *your* class? Here is your chance to create your own ideal book by mixing and matching chapters from Prentice Hall's agriculture textbooks. Up to 20% of your custom book can be your own writing or come from outside sources. Visit us at: *http://www.prenhall.com/agribooks*.

SafariX: Textbooks Online

SafariX Textbooks Online™ is an exciting new choice for students looking to save money. As an alternative to purchasing the print textbook, students can subscribe to the same content online and save up to 50% off the suggested list price of the print text. With a SafariX WebBook, students can search the text, make notes online, print out reading assignments that incorporate lecture notes, and bookmark important passages for later reviews. For more information, visit *www.safarix.com*.

Contents

CHAPTER 1 Approaching Economics 1

Introduction 1
Natural Resource and Environmental Economics 2
Natural Resource Issues 2
The Cedars of Lebanon 2
The Anasazi and Drought 2
Water: California's Ultimate Resource 3
Environmental Issues 3
Drinking Water 4
Ozone Depletion 5
Toxic Wastes 5
Air Pollution 6
Why Study Environmental Economics? 6
Informing Environmental Debates 6
Empowerment 7
Reality Checks 7
Improved Communication 7
Focused Scientific Debates 8
Focused Policy Debates 8
Importance of Economic Analysis 9
Brundtland Report 9
Chinese Government's Population Report 9
U.S. EPA's Acid Rain Program 10
Economics and Environmental Policy 10
Setting the Environmental Policy Agenda 10
Design of Environmental Policy 11
Implementation of Environmental Policy 11
Evaluation of Environmental Policy 12
Summary 12
Review Questions 13
References 13

CHAPTER 2 Economic Thinking 14

Introduction 14
Hard Choices in Policy Issues 15
Timber versus Endangered Species 15
Affordable Energy versus Functioning Ecosystems 15
Clean Air versus Affordable Cars 16
Endangered Wolves versus Livestock and Wildlife 16
Endangered Fish versus Food 17
Power versus Recreation 18
Mining versus Public Lands Recreation 18
Riparian Ecosystems versus Livestock 18
Controlling Global Warming: Current Costs versus Future Benefits 19
Living Space versus Human Rights 19
Energy versus Wildlife 20
Short-Run Human Benefits versus Long-Run Environmental Damages 20
Choices Displace Opportunities 20
The Market Mechanism and Choices 22
What Things Are Produced 22
How Things Are Produced 23
For Whom Things Are Produced 23
Who Controls the Market? 23
The Invisible Hand of the Market 24
The Market, the Environment, and Natural Resources 24
How Markets Settle Environmental Management Questions 24
When Markets Are Best at Sustaining the Environment 25
Limits of the Market System 25

Goals and Means of Environmental Policies 26
Ethics versus Externalities 27
Benefits of Environmental Protection 27
The Costs of Environmental Protection 27
How Much Protection? 28
Setting Protection Goals: Benefit-Cost Analysis 29
Achieving Protection Goals: Cost-Effectiveness Analysis 29
Environmental Policy Goals 30
Human-Centered Goals 30
Nature-Centered Goals 30
Nature- versus Human-Centered Goals 32
Summary 33
Review Questions 33
References 34

CHAPTER 3 Economic Theory 35

Introduction 35
Competitive Markets 36
Demand 37
Law of Demand 37
Demand Curves 38
Market Demand 38
Change in Quantity Demanded 40
Change in Demand 40
Supply 42
Law of Supply 42
Supply Curves 43
Change in Quantity Supplied 44
Change in Supply 44
Equilibrium Demand and Supply 46
Surpluses 46
Shortages 48
Equilibrium 48
Market Price and Quantity 48
Price as a Rationing Method 49
Changes in Supply, Demand, and Equilibrium 49
Other Things Being Equal 52
Accomplishing Environmental Goals: Markets versus Governments 54
Markets 55
Governments 63
Summary 65
Review Questions 65
Problems 67
References 67

CHAPTER 4 Institutional Breakdown 69

Introduction 69
Efficient Institutions 70
Market Failures 70
Externalities 71
Public Goods 75
Common Property 79
Government Failures 86
Rent Seeking 87
Rational Ignorance 87
Empire Building 88
Abuses of Economic Analysis 88
Summary 88
Review Questions 89
Problems 90
References 90

CHAPTER 5 Decision Support for Environmental Policy 92

Introduction 92
Overview of Benefit-Cost Analysis 93
Why Measure Benefits and Costs? 93
What Benefit-Cost Analysis Is 93
What Benefit-Cost Analysis Does 93
How Benefit-Cost Analysis Is Used 94
Economic Principles of Benefit-Cost Analysis 95
Goals 95
Scope of Analysis 97
Incremental Analysis 97
Ten Steps to Net Present Value 101
Step 1: Scoping 101
Step 2: Benefits per Unit 101
Step 3: Costs per Unit 102
Step 4: Output Quantity 102
Step 5: Input Quantity 102
Step 6: Benefits by Period 103
Step 7: Costs by Period 103
Step 8: Net Benefits by Period 104
Step 9: Net Present Value 104
Step 10: Decision Rules 108
Measuring Benefits 108
Marginal and Nonmarginal Changes 109
Mathematical Methods 110
Alternative Cost Saved 110
User Surveys 111

Measuring Costs 111
Underused Resources 112
Opportunity Cost of Unpriced Resources 112
Treatment of Time 112
Issues Affecting Both Benefits and Costs 113
Secondary Effects 113
Employment Impacts 114
Inflation 115
Risk and Uncertainty 115
Alternatives to Benefit-Cost Analysis 115
Cost-Effectiveness Analysis 115
Environmental Impact Analysis 119
Advantages and Limits of Benefit-Cost Analysis 119
Advantages 119
Limits 120
Resolution 121
Summary 122
Review Questions 122
Problems 123
References 125

CHAPTER 6 The Discount Rate 127

Introduction 127
Using the Discount Rate to Find Net Present Value 128
Tables of Discounted Values 130
Application to Private Investments 134
Application to Environmental Policy 135
Search for a Discount Rate 136
Monetary Cost of Government Borrowing 137
Social Opportunity Cost of Capital 137
Social Rate of Time Preference 139
Summary 146
Review Questions 147
Problems 147
References 148

CHAPTER 7 Valuing the Environment 150

Introduction 150
Relevance of Environmental Values 151
Uses of Information on Economic Values of the Environment 151
Monetary Measures of Environmental Values 151
Individual Values and Social Welfare 152
Valuation of Environmental Systems 152
Expenditure Function 153
Compensating and Equivalent Variation 155
Consumer Surplus Approximation 158
Individual Demand for Environmental Improvement 160
Valuing Price and Quality Changes 162
Aggregation and Social Welfare 162
Methods for Measuring Environmental Values 166
Market Prices 166
Productivity 168
Hedonic Pricing 169
Travel Cost Model 171
Avoided Costs 175
Contingent Valuation 176
Benefit Transfer 182
Summary 184
Review Questions 185
Problems 185
References 187

CHAPTER 8 Forests 189

Introduction 189
Forest Management 190
Single-Use Forest Management: Timber 190
Multiple-Use Forest Management 200
Trade-Off Analysis 204
The Equimarginal Principle 207
Linking Benefits and Costs to Decisions 209
Distribution of Benefits and Costs 210
U.S. Forest Legislation 211
Multiple Use Sustained Yield Act 211
Wilderness Act 211
National Environmental Policy Act 212
National Forest Management Act 212
Designs for Efficient, Equitable, and Sustainable Forestry 212
Green Certification 213
Conservation Easements 214
Extractive Reserves 215
Community Management 216
Ecotourism 217
Mitigation of Global Warming 218
Gender, Property, and Health 218
Stewardship Incentives 220
Food Security 220
Sustainable Management and Property Rights 221

Summary 221
Review Questions 222
Problems 222
References 223

CHAPTER 9 Rangelands 225

Introduction 225
The Rangeland Resource 226
Scarcity and Competition 226
Desirable Outcomes 227
Efficiency 227
Equity 228
Sustainability 228
Unregulated Market Outcomes 228
Economically Optimum Stocking Rate 230
Economic Value of Forage 238
Economic Value of Range Improvements 239
Grazing and Forage Dynamics 240
Profit Motive and Overgrazing 240
Multiple Uses for Rangelands 241
Designs for Efficient, Equitable, and Sustainable Rangelands 241
Market Incentives 242
The Right Prices 243
Regulations 244
Institutional Development 244
Infrastructure Development 245
U.S. Grazing Fees 245
U.S. Legislation on Rangelands 251
Emerging Issues 252
Summary 253
Review Questions 254
Problems 254
References 256

CHAPTER 10 Water 258

Introduction 258
The Water Resource 259
Use Patterns 263
Management 264
Desirable Outcomes 264
Human Rights 265
Efficiency 265
Equity and Sustainability 269
Actual Outcomes 270
Water Laws 270
Supply Management 271
Demand Management 274
Regulations 279
Institutional Designs for Economically Efficient Water Use 281
Marginal Cost Pricing 281
Two-Tiered Pricing 285
Integrated Demand/Supply Management 287
Transboundary Management of River Basins 287
Virtual Water 288
Managing Links Between Poverty and Water 289
Decentralized Management 289
Politically Workable Marginal Cost Pricing 290
Stakeholder Participation 291
Summary 292
Review Questions 293
Problems 293
References 294

CHAPTER 11 Food 296

Introduction 296
Food Production and Distribution 297
Food Production Mechanisms 297
Food Distribution Problems 297
Desirable Outcomes 300
Market Outcomes 301
Complexity of Farm Decisions 301
Farmers' Goals 301
Product-Product Relations 302
Factor-Factor Relations 304
Factor-Product Relations 306
Technical Change 308
Output-Price Relations 309
Impact of Agriculture on the Environment 312
Institutional Design for Improving Food Policy Outcomes 315
Short-Term Measure: Emergency Relief 316
Intermediate-Term Measures 316
Long-Term Measures 323
Summary 328
Review Questions 329
Problems 329
References 330

CHAPTER 12 Fisheries 333

Introduction 333
Biological Model of a Fishery 335
Economically Efficient Model of a Fishery 337

Market Allocations 339
Sole Owner 339
Joint Owners: Open Access 340
Institutional Designs for Real-World Fisheries 342
Market Solutions: Fish Farming and Fish Ranching 342
Open Access Regulations 344
Limited Entry Regulations: Economic Analysis 347
Expanded Territorial Waters 351
Enforcement Issues 351
Summary 353
Review Questions 354
Problems 354
References 355

CHAPTER 13 Energy 356

Introduction 356
The Energy Resource 357
Measuring Energy Use 357
Energy and Poverty 358
Need for an Energy Policy 358
Energy Policy Objectives 359
Market Allocations of Energy 360
Economic Concepts 360
Energy Sources 369
The OPEC Cartel 372
Market Outlooks for the Future 377
Institutional Designs for Renewable Energy Sources 380
Marginal Cost Pricing 381
Environmental Adders 383
Energy and Pollution Taxes 383
Promotion of Renewable Energy 384
Standards 387
Green Power 388
Voluntary Action 388
Conservation 389
Summary 392
Review Questions 393
References 393

CHAPTER 14 Population 396

Introduction 396
Population and the Environment 397
Structural Analysis 398
Forecasting 398
Policy Evaluation 398
Population and Demography 399
Macroeconomic Forces 400
Effect of Economic Growth on Population Growth 401
Effect of Population Growth on Economic Growth 403
Microeconomic Forces: Economic Theory of Fertility 407
Demand for Children 407
Cost of Children 408
Emerging Questions 413
Summary 415
Review Questions 415
Problems 416
References 416

CHAPTER 15 Climate Change 418

Introduction 418
Scientific Evidence of Climate Change 419
Economic Analysis 421
Role of Technological Change 421
Economic Impacts 422
Analytical Framework 423
Benefits of Controlling Greenhouse Gases 425
Costs of Controlling Greenhouse Gases 425
Benefit-Cost Analysis 426
Policy Options for Dealing with Climate Changes 427
Adjustment 428
Mitigation 428
Prevention 428
Planet Insurance 433
International Agreements on Global Warming 434
United Nations Framework (1992) 434
Berlin (1995) 435
Geneva (1996) 437
Kyoto (1997) 437
Buenos Aires (1998) 439
Marrakesh (2001) 439
Summary 442
Review Questions 443
Problems 443
References 444

CHAPTER 16 Environmental Quality and Pollution 445

Introduction 445
Physical Basis of Environmental Pollution 446
Absorptive Capacity: Stock versus Flow 446
Spatial Scale: Local versus Regional versus Global 447

Origin: Point versus Nonpoint Sources 447
Occurrence: Continuous versus Episodic 447
Desirable Outcome Criteria 448
Economic Efficiency 448
Equity in Pollution Control 459
Environmental Sustainability 460
Unregulated Market Outcomes 461
Institutional Designs for Controlling Pollution 461
Decentralization 461
Command-and-Control Approach 466
Harnessing Market Incentives 473
Enforcement 482
Summary 486
Review Questions 486
Problems 487
References 489

CHAPTER 17 Environmental Risk 491

Introduction 491
Nature of Risk 492
Complexity 492
Causality 493
Voluntary versus Involuntary 493
Perceived versus Objective 494
Desirable Outcomes 495
Economic Efficiency 496
Equity 499
Sustainability 500
Unregulated Market Outcomes 500
Choices under Risk 501
Environmental Risk Insurance 501
Insurability and the Market 504
Institutional Designs for Coping with Environmental Risks 504
Risk Assessment 505
Risk Management 507
Risk Management Institutions 518
Summary 523
Review Questions 523
Problem 524
References 524

CHAPTER 18 Environmental Justice 526

Introduction 526
What Is Environmental Justice? 527
Measuring Environmental Justice 529
Economic Theory and Policy Analysis 530
Applications to Achieve Environmental Justice 534
Benefits and Costs of Controlling Environmental Degradation 534
Quantifying Environmental Justice 538
Evidence of Environmental Justice 540
Summary 543
Review Questions 544
Problem 544
References 545

CHAPTER 19 Sustainable Development 546

Introduction 546
Deciding What to Sustain 547
Reasons for Sustainability 547
Definitions of Sustainability 548
Measuring Progress of Sustainability 553
Conditions for Sustainability 554
Instruments for Implementing Sustainability 555
Finding Right Prices 555
The Hartwick-Solow Rule 555
Critical Natural Capital 556
Safe Minimum Standard 557
Avoiding Irreversibility 558
Steady-State Principles 559
Policy Applications 560
Institutional Design 561
Global Actions 566
Summary 568
Review Questions 568
Problems 569
References 569

Appendix A Supply and Demand Mathematics 570

Appendix B Total Benefit Marginal Benefit, and Consumer Surplus 573

Appendix C Use of Programming Methods to Value Public Actions 575

Appendix D Improving Cost-Effectiveness of Environmental Programs 576

Appendix E Preferences, Demand, and Benefits 577

Appendix F Incentive Compatibility in Contingent Valuation Studies 579

Appendix G Critiques of Contingent Valuation 581

Appendix H Infinite Time Horizons and Forest Rotation Analysis 588

Appendix I Dynamic Efficiency for a Depletable Resource 591

Glossary 593

Index 604

Chapter 1

APPROACHING ECONOMICS

Introduction

History teaches us that human welfare depends on maintaining reliable access to natural resources. No culture has survived long without making investments to secure a sustained supply of these resources. Because natural resources are vital to our communities and to the survival of our culture, they are considered one of the most significant disciplines of study. Economists study natural resources to determine their adequacy in meeting human needs. They examine policies that impact natural resource supplies, allocations, and use. This field of study is called *natural resource economics*, and it is one of the oldest fields in economics.

A much newer history also recognizes the importance of the natural environment. Since the mid-twentieth century, we have learned the price paid by failing to take care of the natural environment: environmental damages produced by major oil spills, dying forests and lakes, nuclear accidents, holes in the ozone layer, greenhouse gasses, and global climate change. Because of the high price of environmental mistakes, all cultures are discovering the importance of taking care of the environment. *Environmental economics* is the application of economic principles to protect the environment.

Guarding against environmental and natural resource mistakes in the future requires asking and answering several kinds of questions:

- Does the earth have a limited carrying capacity?
- If these limits exist, how do we measure them?
- Does current economic activity exceed that capacity?
- How do economic systems deal with natural resource scarcities?
- How can political systems be designed to draw on the strengths of the economic system and avoid its weaknesses?
- How can institutions be designed to provide acceptable living standards for current generations without damaging standards for future generations?
- How should institutions be designed to avoid natural resource and environmental problems that have not yet occurred?

This book will help you learn techniques that can be used to analyze and answer each of these questions. Application of these techniques will help find ways to ensure the adequacy of our natural resource supplies and to take care of the environment for the future.

Outline

Natural Resource and Environmental Economics 2

Natural Resource Issues 2

Environmental Issues 3

Why Study Environmental Economics? 6

Importance of Economic Analysis 9

Economics and Environmental Policy 10

Objectives

After studying this chapter, you should be able to

- Describe the disciplines of environmental and natural resource economics
- Identify several natural resource and environmental issues
- Explain what can be gained by the study of natural resource and environmental economics
- Discuss the importance of economic analysis
- Describe how economic principles contribute in setting the environmental agenda and in designing, implementing, and evaluating policies

NATURAL RESOURCE AND ENVIRONMENTAL ECONOMICS

The environment is an asset that provides a variety of beneficial services to people. Because it is a special asset that provides the life support system we all need, it is important to avoid irreversible depreciation of its value. The environment also provides us with raw materials that can be converted into useful products through application of energy to production processes. Typically, these raw materials and energy are returned to the environment as waste.

Economists make a distinction between natural resource economics and environmental economics. Natural resource economics analyzes policy questions about natural resource *quantities,* for example, water, forests, fisheries, agriculture, minerals, energy, and land. Environmental economics deals with natural resource *qualities,* for example, acid rain, ozone depletion, safe drinking water, toxic wastes, air pollution, soil degradation, and climate change. (See Pearce, 2002, for a review of the history of environmental economics.)

Today, as in the past, the economic welfare of people and nations depends on maintaining reliable access to natural resources. It also depends greatly on securing and maintaining a clean environment. This book shows how understanding the discipline of *economics* is crucial in gaining insight into policies that sustain that economic welfare.

NATURAL RESOURCE ISSUES

The adequacy of natural resources has been a concern throughout history. The following three examples come from different time periods and places, but they tell a similar story. People devote considerable time and energy into maintaining their access to natural resources. In this book you will see how economics offers insights into the best way to maintain that access and provides techniques for analyzing natural resource issues.

The Cedars of Lebanon

Forests have always provided important benefits for humanity. Six thousand years ago, cedar trees from Lebanon were prized for their beauty, fragrance, and utility in construction. By the time of the philosopher Plato, the Greek people had harvested most of their forests and were importing wood, mostly from what is now Lebanon. The reputation of Lebanese cedars had been established in the Persian empire, which incorporated the wood in the construction of its famous navy vessels of the fifth century B.C. As the desire for cedar grew, the supply of the wood shrunk. Eventually, unsustainable harvesting of the cedar forests caused them to disappear and the landscape of Lebanon changed. Without the protection of the forest, the soil washed away and plant nutrients were lost. Mountains were left bare and vulnerable to erosion. Today, much of the region remains without forests despite the area's climate and topography, which are favorable to the growth of trees.

Why is this story about cedar trees important? It tells us that thousands of years ago nations made major economic investments to ensure access to a reliable timber supply. It also tells us that the economic damages of deforestation can be high, sometimes irreversible. Living forests absorb carbon dioxide (CO_2) in the atmosphere and reduce the harmful effects of *global warming.* Because of decisions made in ancient times, much of temperate Europe has lost much of its forests (Williams, 2000; Zagas et al., 1998). Today, this region of the world can do little sequestering of CO_2 toward mitigating the global warming problem.

Economic principles can be a significant factor in sustaining or failing to sustain the environment and natural resources. These principles will be described, developed, and applied throughout this book to real environmental issues.

The Anasazi and Drought

In the American Southwest, the Anasazi people settled near what is now Mesa Verde National Park early in the first century. They lived in caves carved into the sides of tall mountain cliffs. They began building these caves about 450 A.D. and survived in the region for centuries. During the late thirteenth century, however, the cave dwellings were abandoned. One explanation for this sudden disappearance is based on modern tree-ring analysis, which indicates the presence of a severe drought from 1276 to 1300. It appears that at first the Anasazi attempted to escape the drought by shifting locations and building a variety of water-collection structures and devices. Despite their efforts, though, they were gone from Mesa Verde by the year 1300. It is fair to guess that although the Anasazi were able to survive in a barren land and dry conditions for nearly a thousand years, the extreme drought of 1276 to 1300 overwhelmed their ingenuity and ability to adapt. Many mysteries remain about the Anasazi's disappearance, their structures, and the people themselves, but their behavior indicates they understood the importance of capturing and controlling water. This book will show how economic principles can be used in making decisions about how to cope with drought and other natural resource shortages.

Water: California's Ultimate Resource

Californians covet and mine water today with as much energy and passion as miners searched for gold in 1849. The economic worth of water to California is illustrated by the necessity to import it to make possible the state's status as the world's fifth largest economy. In the mid-nineteenth century Californians discovered that the loss of control over water limited their ability to live safely and plan for the state's future.

The climate in California ranges from flood to drought: Downpours occur from November through March, but little precipitation occurs from April to October. Although much of California is bone dry in the summer, the state has experienced huge costs and suffering in coping with devastating floods.[1] In the 1930s the U.S. federal government came to the rescue. President Franklin Roosevelt issued an executive allocation of $4.2 million under the Emergency Relief Appropriations Act for construction of the Central Valley Project (CVP), a program that would control floods and bring irrigation water to California's Central Valley. Throughout the late 1940s and the 1950s, the federal government authorized expansions of the project, with economic feasibility the only criterion. By the end of the 1960s, the project was a conglomeration of federal and state projects funded under a variety of government banners.

Despite the social, environmental, and political controversy that has surrounded the CVP, it remains an impressive accomplishment. The CVP contains three quarters of the irrigated land in California and one sixth of the irrigated land in the United States. Four of the top five U.S. counties in gross farm receipts are served by the CVP. Between 1950 and 2000, the CVP ranked first among reclamation projects in value of flood damage prevented *and* for agricultural output.

What does this simplified history of California's complex water problem tell us? First, it reminds us that in dry places people must incur costs to maintain steady access to water. More significantly, it tells us that human suffering can be reduced considerably and human prosperity can grow significantly when *economic principles* are properly applied in making decisions about natural resources. The chapters of this book will show how to apply economic principles to a range of diverse and challenging natural resource issues.

■ Environmental Issues

Since the mid-twentieth century, protecting the environment has grown in importance from a concern first expressed by the world's wealthiest nations to a rallying cry of nations and cultures worldwide. Today, environmental issues, such as the following, attract the attention of all nations:

- Providing access to safe drinking water
- Understanding and preventing ozone depletion

[1] Throughout the nineteenth century, Sacramento experienced repeated flooding. The city faced no effective means of control until someone thought of building huge upstream reservoirs, at the federal government's expense. For an excellent early history of northern California flooding and struggles to control it, see the article by Bishofberger (1975). Also see a more recent history of water projects in the western United States by Reisner (1993).

- Reducing toxic wastes and enforcing proper disposal
- Controlling air pollution

The field of economics has an immense influence on defining environmental debates, shaping the outcomes, and contributing to policy decisions. Economics is a rigorous discipline: Its hypotheses are testable by reliable statistical means, and it can prescribe instruments for implementing policy. These qualities give economics a special appeal, as its findings can be translated into policy through a wide range of instruments such as taxes, subsidies, and investment programs. The chapters in this book develop and apply economic principles to a range of global environmental issues.

Drinking Water

Safe drinking water is inaccessible for 1.1 billion people in the world. Approximately 400 children younger than the age of 5 are estimated to die per hour as a result of drinking poor-quality water (Gadgil, 1998).[2] At any time, about half the population of the developing world suffers from one or more *preventable* debilitating or fatal disease associated with a poor water supply and poor sanitation: diarrhea, schistosomiasis, and trachoma. Longer lives, reduced infant mortality, and better health, productivity, and material well-being are associated with good-quality drinking water. Most developing countries rank low on these quality of life indicators compared with industrialized nations. The World Health Organization states that infectious diseases caused by pathogenic bacteria, viruses, and protozoa or parasites are the most common and widespread health risks associated with poor drinking water. The World Bank estimates that more than 3 million children younger than age 5 die each year from diarrheal illnesses contracted through poor-quality drinking water.

A lack of understanding of economics is both a cause of this age-old problem and a barrier to its solutions. Heated debates on the best way to deal with the distribution of safe drinking water revolve largely around the economics of water purification, transmission, and distribution. One view of how to deal with the problem is summarized by the following statement: "Some (safe water) for all, rather than more for some" (United Nations, 2004, p. 1). This principle has a strong appeal to our sense of fairness. Carrying out the principle requires the following four steps (Gadgil, 1998):

1. Estimate the cost of extending service at some known level (e.g., 100,000 gallons per week of drinkable water) to the population that lacks safe drinking water.
2. Find the money from the national government or from an external source.
3. Spend the money to deliver the services, with little practical hope of recovering the costs through user charges, since the poor cannot afford to pay the charges.
4. Return to step 1 and repeat the analysis for a higher known level of service (e.g., 500,000 gallons per week).

The problem with this approach is that it usually works badly. What results is a water delivery system that soaks up much taxpayer money to deliver cheap, piped water to the lucky few and that leaves many poor people without clean water or at the mercy of expensive water vendors, who might charge 10 times the price of piped water. Hundreds of millions of dollars of annual government subsidies are required to keep the water utilities from going broke. The justification for the high level of government subsidy is the claim of weak ability of the poor to pay for services. The problem with applying this economic theory is that heavily subsidized services lead to little or no service expansion and poor quality of service.

Since 1991, a vigorous debate on this issue has challenged old views on the best way to supply safe drinking water to the world's poor. A new and more ambitious economic theory is based on the discovery that even very poor people are willing to pay a high percentage of their incomes

---

[2] World leaders met in Johannesburg in 2002 at the World Summit on Sustainable Development to focus on worldwide sustainability of water. Governments committed themselves to providing clean water to at least half a billion more people by 2105 and sanitation services to at least 1.2 billion. For details, see United Nations (2002).

for safe drinking water.[3] This issue is unlikely to be resolved soon because of the moral conflict in asking poor people to spend a large percentage of their incomes for safe drinking water. Nevertheless, one thing remains clear: Development, testing, and application of economic theories will provide major contributions to debates on how best to deal with the problem of unsafe drinking water as well as other environmental problems. The chapters of this book show how economics can help resolve these debates.

Ozone Depletion

The *ozone layer* is found in the stratosphere between 3 and 30 miles above ground. Ozone molecules have three oxygen atoms instead of the normal two. That atmospheric layer protects us from the harmful effects of certain solar ultraviolet rays. Increases in levels of radiation that are normally absorbed by the ozone layer can result in increases in human skin cancers, immune system disorders, and eye disorders and may affect plants and animals.

Chloroflurocarbons (CFCs) were developed in the early 1930s and later used in a variety of industrial, commercial, and household applications, such as air conditioning. Because they were chemically inert, cheap, and effective, demand for CFCs grew. However, a 1985 discovery of ozone depletion over Antarctica prompted scientific investigations into ozone depletion processes, along with a search for policies that could reduce ozone depletion effectively, cheaply, and fairly (Columbia University, 2001). CFCs were generally believed to be a major contributing factor to the depletion process. The Montreal Protocol on Substances That Deplete the Ozone Layer was adopted in 1987. Its provisions were strengthened through amendments adopted in London (1990), Copenhagen (1992), Vienna (1995), Montreal (1997), and Beijing (1999). The protocol aims to reduce and eventually eliminate the emissions of man-made, ozone-depleting substances; nations ratifying parts of the protocol are listed at United Nations Ozone Secretariat, 2004.

Economic impacts have been a significant factor in designing CFC reduction policies. One focus of the economic debate has been over the costs and benefits[4] of reducing CFCs. The costs of the relevant policy options vary, often by substantial magnitudes, so it is important to find control policies that are affordable and effective. CFCs were banned in the United States in the 1990s, and elsewhere major policy proposals for regulating CFCs are being scrutinized.

The *distribution* of costs and benefits from an environmental policy is part of the global environmental debate. People and nations who pay the greatest costs for reduced ozone depletion are not necessarily those who receive the greatest benefits. Even though some CFC reduction policies might increase overall social welfare in the sense that the benefits outweigh the costs, the distribution of those benefits and costs must be seen as fair for the policies to be politically acceptable. Poorer nations, for example, may suffer reductions in quality of life after CFC reduction policies are implemented, whereas citizens of wealthier nations may feel little impact on their lifestyles. Therefore, designing effective policies to deal with the ozone hole requires careful attention to those who gain and those who lose. The chapters in this book emphasize the role of distribution of gains and losses in the design of environmental policies.

Toxic Wastes

The Comprehensive Environmental Response, Compensation, and Liability Act (CERCLA), commonly known as Superfund, was enacted by the U.S. Congress in 1980. (In 2003 China and Japan passed similar laws.) This U.S. law created a tax on the chemical and petroleum industries

---

[3] Guiding principles supporting this new economic theory emerged at the 1992 United Nations International Conference on Water and Sanitation in Dublin. The application of the new theory focuses on providing at full cost the level of private services that people want and are willing to pay for (United Nations, 2000).

[4] You will see many references in this book to costs and benefits associated with a proposed or actual policy or program. Unless stated otherwise, the terms refer to *incremental* or *marginal* costs and benefits produced by the program. Marginal costs and benefits are additions to existing totals and not absolute totals. Economics almost always focuses on *additions* to costs or benefits produced by a proposed program.

and provided federal authority to respond directly to releases or threatened releases of hazardous substances that may endanger public health or the environment. Over a period of five years, $1.6 billion was collected in taxes, and the revenues went to a fund for cleaning up abandoned hazardous waste sites. The success of the cleanup program rests in large part on the ability of the Environmental Protection Agency (EPA) to convince major corporations to share in the costs of this multibillion dollar effort. Since 1990, EPA's enforcement strategy has evolved from accommodation to enforcement. There also has been a considerable increase in the application of the fund's most powerful enforcement tools. Continuing to vigorously enforce this cleanup program depends on two factors: the willingness of society to accept the large costs of confronting powerful special interests, and the assumption that the benefits of the program are larger than the costs. The chapters in this book describe the importance of program benefits and costs in designing environmental policies.

Air Pollution

Economists long have argued that economic incentives can reduce *pollution* more cheaply than rigid regulatory actions, but this theory has seen limited application. Southern California sometimes experiences the worst air pollution in the United States. To address this problem, state and local authorities have developed the most complex set of regulatory approaches in the world. Rising costs and increasing resistance to further regulation, however, have prompted California to adopt a new approach, the Regional Clean Air Incentives Market (RECLAIM) program. For details of the program, see South Coast Air Quality Management District, 2004. The program is expected to reduce the costs of air pollution control about 50% over a decade compared to the costs of direct regulation by allowing air polluters to trade marketable *emission permits.*

Economists widely support two innovative pollution control strategies: marketable emission permits (emissions trading) and emission charges, both of which are described in detail in Chapter 17. The predominant method of control has been that of setting standards and penalizing noncompliance, but occasionally marketable emission permits or emission charges make up a part of pollution control programs. We will likely see a growing use of these strategies in the future. For that reason, the role of markets in solving today's environmental problems is a major focus of this book.

■ Why Study Environmental Economics?

The natural resource and environmental issues analyzed so far have hinted at an answer to the question, What can you gain by studying environmental economics? The answer lies in the fact that as individuals we use economic principles every day to make better personal decisions (e.g., what to buy, what to recycle, what career to choose). Communities (e.g., nations) use the same principles to make political and environmental decisions. Let us consider several reasons for studying environmental economics.

Informing Environmental Debates

Most of us have seen, read, or participated in heated environmental debates. Many of these debates are complex, emotional, and reach few conclusions. You might be asking yourself how the study of economics can address the wide range of complex and passionate issues that surround the natural environment. The answer lies in the realm of economic principles. Used properly, economic principles create a common, well-understood framework for addressing environmental issues. With these principles and economic tools, you can organize information to approach objectively environmental policy debates that most people are only equipped to approach emotionally.

Those who value the wise use of the environment can use economic principles to organize information that supports a sustainable future. After you study environmental economics, you can better use economic principles and methods of analysis in your search for solutions to complex environmental issues. Material in this book will help you show all sides of debates the economic consequences of their proposed policies.

Empowerment

Knowledge of environmental economics can pave the way to a successful career through personal empowerment. Political leaders around the world face environmental issues daily. Presidents, prime ministers, kings, and other political leaders have studied economics in college more than any other field of study. Economics empowers you by supplying a framework to better understand choices made by individuals and communities. For example, oil is the common economic interest of the Organization of Petroleum Exporting Countries (OPEC). A shared economic interest explains why 11 countries bind themselves together to control oil production. Likewise, members of the Nature Conservancy, a group dedicated to preserving rare natural environments, are empowered by banding together to purchase undeveloped land.

Economics helps you understand the motivations of people and communities. For example, many economies worldwide rely heavily on cheap access to natural resources such as grazing land, water, timber, minerals, wildlife, and outdoor recreation. Using economic principles, you can understand why proposals to increase the price or reduce access to any of these resources will be unpopular in the affected economies. You can understand why farmers, ranchers, mining and timber companies, fishing-boat owners, canneries, and resort owners will fight vigorously any proposals to remove these subsidies. Why? Because it is in their economic self-interest to do so.

Economics helps you understand motivations behind environmental debates. Economics is based on accepted theories and facts, not on moral or aesthetic principles. Economics gives you the language and analytical framework to approach objectively important environmental issues. By learning that profits are the lifeblood of manufacturers, you can predict why proposals to restrict pollution will be fought vigorously by the affected industries and why certain industries will resist pleas for polluters to act ethically. Private firms survive by maximizing profits, which requires that they minimize costs. Economics not only explains why a business that pollutes resists regulations to limit pollution, it teaches why business is not unethical for doing so—it simply wants to stay in business. Much of the content in this book emphasizes how understanding economics gives you insight into and tolerance of environmental policy debates driven by strong economic interests.

Reality Checks

Experience shows that the practical success of an environmental proposal often depends on its economic impact. Accordingly, economics offers a reality check for such environmental proposals. An environmental organization, for example, may propose a program requiring manufacturers to install a new and expensive device, *X*, on automobiles to reduce emissions by *Y* units. Economics teaches you to ask two questions: First, is *X* the cheapest way to lower emissions by *Y*? Second, if *X* is cheapest, does reducing emissions by *Y* produce a greater benefit than the cost? As you will learn in this book, if the answer to either question is no, you should doubt that the proposal is in society's best interest.

Improved Communication

All nations, cultures, and religions understand and use economic principles, and those principles supply a common language for communicating widely divergent interests and viewpoints. Posing economic questions when debating environmental policy issues can help resolve environmental problems fairly and in a universally acceptable manner. A major goal of this book is to help you communicate economic principles more effectively.

Focused Scientific Debates

Economics can contribute to the resolution of scientific debates even if the debates do not start with economics. Consider several examples. One of the longest rivers in America, the Rio Grande, dries up in summer when farmers use the river's water to irrigate crops. In the 1990s biologists discovered the river's low summer flows could drive into extinction a tiny fish called the silvery minnow. Biologists believe the minnow needs somewhere in the range of 10,000 to 30,000 acre-feet of water per year to keep it from going extinct. (An acre-foot covers an acre a foot deep, which is about a third of a million gallons.) Economics teaches us to ask how important is it to narrow that range? If taking anything between 10,000 and 30,000 acre-feet out of other water uses results in no lost income, then gaining the extra precision is of little economic value. Economics helps identify the relevance of this scientific debate.

Similarly, a Mexico City environmentalist might propose converting the city's diesel-powered buses into propane-powered vehicles to reduce air pollution. After a class in environmental economics, you would likely pose two questions: (1) Does the cost of the conversion exceed the benefits gained in improved health? (2) Could the money be better spent on other measures, such as enforcing existing air pollution standards?

The technical sciences typically ask *if* a technology or proposal works and *how well* it works. By contrast, economics attempts to translate the technical scientist's answer into the level and distribution of the costs and benefits produced by implementing the proposed course of action. Policymakers are keenly interested in costs and benefits, because that information tells them what society can afford to pay for the decision as well as who suffers and by how much. If the proposal does not work, then the costs certainly exceed the benefits. Even if the proposal works well technically, it may not be in society's best interest because it might cost more to carry out than it produces in human benefit. Economics can sharpen and focus environmental scientific debates, a fact demonstrated throughout this book.

Focused Policy Debates

Economics helps focus emotional environmental policy debates. Typically, environmental debates focus on losses (gains) in the physical quantity of an environmental resource needed to pay for gains (losses) in material standards of living. For example, should we give up a piece of wilderness for better access to and more production of oil, or should we preserve the wilderness for future generations to enjoy in its natural state?

The debate over access to the U.S. Arctic National Wildlife Refuge is a good example. Containing anywhere from an estimated 6 to 16 billion barrels of oil, this refuge may hold the largest undeveloped onshore reserves in North America. For Alaskans, the refuge debate is an old one, but for the first time since the 1989 *Exxon Valdez* oil spill, supporters of development see a real chance to open the refuge to lucrative oil exploration. Oil supporters argue that the industry has refined its practices over 30 years of Arctic operations and that the reserves can be developed with minimal impact to the tundra and wildlife. Opponents say that pipelines and production would permanently change the tundra, lead to oil spills, and threaten caribou and other wildlife. Furthermore, any development would alter the wild character of the entire refuge. In the village of Kaktovik, Alaska, refuge oil would mean jobs, tax revenues, and perhaps some of the oil itself. For Alaskans, drilling in the refuge would mean more high-paying jobs and a new injection of dollars to the state treasury, as much as $400 million a year.

The antidevelopment fight is old and vigorous. Environmentalists have supported protection of the refuge area for nearly 50 years, and in 1960 it was set aside as a wildlife range. As companies discovered oil, the refuge has risen in profile to become a symbol of the U.S. environmental debate. Perhaps more than any other factor, the *idea* of this sprawling, undeveloped land drives opposition to development of the refuge. For millions of people around the world, there is value in simply knowing that the refuge is there, preserved and untouched. Regardless of how this debate plays out, economics teaches us to focus on cost and benefit. What is the benefit of leaving the wilderness untouched compared to the benefit of the additional oil? As demonstrated throughout this book, economics helps focus such debates by using well-established principles that can be communicated using a language everybody understands.

Importance of Economic Analysis

This book *supplies* the economic principles needed to understand important environmental policy issues for the twenty-first century. It also *applies* those principles to issues people find most interesting. The issues that excite people's curiosity—global warming, tropical deforestation, gender equity, natural resource depletion, food production adequacy, air pollution, environmental justice, and the like—are also important in the environmental policy arena. The theoretical subjects people find dry (e.g., memorization of dates, price and quantity indexes, utility theory, and the like) are typically less important to environmental policy.

The following three examples illustrate the importance of understanding economics to contribute to environmental policy debates. The examples show how economics sharpens our understanding of causes, effects, and policy solutions to various environmental problems.

Brundtland Report

A report titled "Our Common Future" (Brundtland, 1987) was published by the World Commission on Environment and Development in 1987. It is called the Brundtland report after the chair of the commission, then prime minister of Norway, Mrs. Gro Harlem Brundtland, and it is one of the path-breaking environmental documents of the twentieth century. After the report was published, the commission decided to reexamine the world's critical environment and development problems, formulate realistic proposals to solve them, and create a global agenda for change. It was to work within the principle of environmentally sustainable development, which it defined as "development that meets the needs of the present without compromising the ability of future generations to meet their own needs" (p. 54). The report issues a call to action, for all nations to find solutions to the *tragedy of the commons*, in which the lack of property rights damaged incentives to take care of environmental and natural resources (see Chapter 4).

The Brundtland report approaches environmental and development issues as a worldwide challenge to be solved by collective action, rather than by the pursuit of personal or national self-interest. It examines population and human resources, food security, species and ecosystems, energy, industry, and cities. Important results and shared visions have come from this report, including major international agreements like the Montreal and Kyoto protocols. This book is designed to help you consider critically the important contributions of this and other far-reaching reports.

Chinese Government's Population Report

On August 23, 1995, the Chinese government issued its long-awaited white paper titled "Family Planning in China" (Chinese Government, 1995). In the paper, the need for population programs is given a clearly articulated *economic* rationale. The report places special emphasis on population-to-resource ratios and on limits imposed by scarce capacity for food production. The government explicitly lays out the goal to control rapid population growth. According to the report, that success is "necessary to bring about sustained economic growth and sustainable development, satisfying the daily increasing material and cultural demand of the whole people, and guaranteeing the fundamental and long-term interests of the current generation and their posterity."

The paper lays out an impressive set of goals with immense environmental and human consequences. After reading this book, you will be able to understand *what* China wants to do and *why;* then you can reach your own conclusions about whether China is succeeding.

U.S. EPA's Acid Rain Program

The goal of the U.S. EPA's Acid Rain Program is to produce environmental and human health benefits through reductions in emissions of sulfur dioxide (SO_2) and nitrogen oxides (NO_x), which are believed to be the main causes of *acid rain.* The program is designed to achieve this goal at minimum cost to society. The first phase of the program began in 1995 and affected 110 mostly coal-burning electric utility plants located in 21 states. Later, an additional 182 units joined the first phase. Emissions data indicate that 1995 SO_2 emissions at these units nationwide were reduced by nearly 40% below their required level. The program's second phase, starting in 2000, tightened annual emissions limits on larger, higher-emitting plants as well as restricted smaller, cleaner plants fired by coal, oil, and gas.

When you read the main features of this innovative program (U.S. EPA, 2002), you will encounter many technical terms. This important report would be hard to understand without knowledge of environmental economics.

■ Economics and Environmental Policy

Disciplined study gives you the tools to understand, predict, control, and improve human welfare. Many questions facing the world today require such study; here are a few examples: How do price and profit incentives facing timber companies affect the pace and extent of tropical deforestation? How can economic progress of poor farming nations be improved? What kinds of water policies in dry places improve living standards and distribute the net benefits of those improvements fairly? What are the benefits and costs of a proposed wastewater treatment plant and how will they be distributed? What factors affect the decision of a poor Mexican brick maker to invest in pollution control measures to reduce the thick smoke belching from his kiln that pollutes his home city? What are the benefits and costs of a policy that requires employers to invest in health and safety in the workplace if the extra costs of the safety measures cause the employer to lay off a sizable percent of the workforce? What does it cost to keep a wild trout stream flowing freely if the water that stays in that stream could be used to develop agriculture to feed a hungry world?

This book challenges you to understand the economic factors underlying environmental policy issues and debates. To succeed, you must first cultivate an ability to objectively see things as they are, regardless of your personal biases. Why is this so hard to do? Because consideration of both economic and environmental issues is influenced by human passions. *Hoping* that noble human behavior produces the right environmental impact will lead to disappointments. *Wishing* that the business world will respond to motivations other than profit is a poor way to identify which financial incentives will reduce factory air pollution. *Imagining* that homeowners will reduce their water consumption during a drought because of their commitment to civic duty is a poor way to find out what motivates people to reduce water use.

Environmental economists do not always agree on what makes for a good policy. One economist may be in favor of keeping a "warm and fuzzy" endangered species from going extinct at all costs. Another may vote to use the same resources to save five "cold and slippery" species, and that resource reallocation may drive the warm and fuzzy species into extinction. A third economist may think that the same resources should not go into species protection at all, but should be used for entirely different programs, such as early education for children. Yet another economist may believe that the market system will take care of the problem on its own within a few years, and that tax dollars that would have saved the species should stay in consumers' pockets. This book will help you distinguish between facts and values in the formulation of environmental policies.

Setting the Environmental Policy Agenda

This book's first aim is to show how economic principles can be used to help set the environmental agenda, which means deciding which environmental policy issues are the most important.

It is tough to reach consensus on an environmental agenda when you have to deal with issues as diverse as reducing radon, guaranteeing safe drinking water, cleaning toxic wastes, preserving biodiversity, controlling pesticides, protecting food supplies, and deciding the best way to manage depletable forests, fisheries, and groundwater. How should policymakers assess, prioritize, and evaluate problems faced by the people of their constituency?

Many factors could serve as standards for choosing environmental issues to place at the top of the agenda. Some people favor reduced potential danger to society as a way to decide which issues need attention. Others favor political popularity. Still others look for a moral imperative of some kind, that is, defining the most important issues based on what is fair and just. Whatever the standards, economics always plays a role. The incremental net benefits to society produced by addressing the issue successfully, or to state it differently, the net costs of failing to address the issue, will always be an important factor in deciding which environmental issues receive attention. This book will give you a deeper understanding of those costs and benefits.

Design of Environmental Policy

This book's second aim is to show how economic principles can be used in the design of fair, efficient, and sustainable policies. Environmental policy design means looking for workable ways to address a recognized environmental issue or problem. Policy design involves three steps: (1) Review various possible configurations of the policy. (2) Develop information about each possible configuration. (3) Select the best configuration for implementation. Economics plays an important part in all three steps. Each step, if done improperly, could produce major errors with high economic costs, environmental damages, or both. Because economic incentives play such an important part in business decisions, economics plays an important role in designing policies to regulate those businesses. Many people believe that environmental policies have failed to deliver sufficient returns to the community in part because the policies were poorly designed from the beginning. The material in this book is designed to help you design policies that increase returns to the community.

Implementation of Environmental Policy

This book's third aim is to show how economic principles and methods of analysis can be used in the implementation of fair, efficient, and sustainable policies. Environmental policy implementation happens when the policy is formalized into laws, regulations, and other actions that put the policy into place. Typically, this is done by government agencies. Implementation puts the policy on the road by assigning staff and by mobilizing money and other supporting resources; that is, implementation carries out the policy and monitors its effectiveness.

There are few clear pathways for success, and the problems of implementation raise many questions. For example, is it best to try out a new policy on a limited scale to test whether the design and implementation arrangements will work? Will the stakeholders respond to the incentives and disincentives as anticipated?[5] Limited testing may be essential when the regulators and the regulated have little experience with the policy or when there are major risks involved if the

---

[5] Businesses and consumers respond to incentives. Marginal changes in costs or benefits motivate people to respond. The decision to choose one alternative over another occurs when that alternative's marginal benefit exceeds its marginal cost. Because people make decisions by comparing costs and benefits, their behavior changes when the costs or benefits change. Public policies change the costs or benefits that people face, and therefore change their behavior. For example, a tax on gasoline encourages people to drive more fuel-efficient cars and take public transportation. When policymakers overlook how policies affect incentives, they often end up with unintended results. There is evidence that people respond significantly to incentives even in situations where we do not usually expect that to occur. Some psychologists have discovered in experiments that when you hand a person an unexpectedly hot cup of coffee, she usually drops the cup if she believes it to be cheap but manages to hang on if she thinks the cup is expensive.

policy fails to deal with the problem it was designed to solve. Testing also is essential if politically powerful special interests may be affected.

Implementation is neither easy nor cheap. Mistakes are made, and we try to learn from them. Mistakes are costly both economically and environmentally. Policies that overprotect the environment may produce unacceptable income and job losses, and these losses may fall heavily on poor people who cannot withstand the losses. Policies that underprotect the environment cannot be sustained. This book will help you understand more about the sustainability of environmental policies.

Evaluation of Environmental Policy

The fourth aim of this book is to show how to use economic thinking to evaluate environmental policies. Environmental policy evaluation is the review and analysis of information related to the policy's previously described implementation. How well has the policy performed and how strongly have the mechanisms for carrying out the policy performed? Sometimes, after a policy is implemented it is forgotten. Evaluation is often one of the weakest links in environmental policy analysis, especially in developing countries. Evaluation is expensive, so the budget, resources, and will are often lacking. In addition, environmental policies often are evaluated based on their process and not their output. Unless evaluations are based on predetermined environmental *objectives,* people who do the evaluations are forced to base evaluations on mere inputs or processes, such as number of staff trained and money spent on vehicles, offices, and equipment.

Policy evaluation must start with a set of indicators of performance, but typically these indicators do not exist, or are hard or expensive to find. For example, performance indicators of a program of water quality improvement could be a measurable improvement in human health, greater productivity in the workplace, more water-based recreation fishing days at affected streams, and the like. Performance indicators of an air quality policy could be reduced number of respiratory illnesses in the affected population or increased recreational use of natural resource facilities that are improved by the program. Such indicators are scarce and/or expensive to secure.

Once performance indicator data are found (if they are), somebody must interpret them in terms of (1) the impacts of the policy on environmental quality, (2) the sustainability of natural resource use, or (3) economically important factors such as tax revenues, income, and jobs. Typically, the impacts are measured in number of reduced polluted days, increases in water quality, and the like. That is, the impacts are not measured in ways that allow the benefits to be easily compared with the costs of the program, and data must be translated through a variety of valuation methods. Thus, it is difficult and expensive to evaluate environmental policies in terms of the extent to which they met their objectives. Nevertheless, economic costs and benefits of proposed policies play a major role in the method chosen to evaluate environmental policies as well as in what evaluations are produced. This book will help you understand more about those costs and benefits.

■ Summary

Natural resource economists use economic principles to determine the adequacy of those resources in meeting human needs. Environmental economics is the application of economic principles to protect the environment.

Historically, people and nations have incurred large political and economic costs to ensure continued access to a clean environment and abundant natural resources. Natural resources and the environment are depletable, and nations incur high costs to guard against that depletion. Communities have discovered that it is good policy to avoid making poor decisions about the environment and natural resources.

As fields of study, environmental economics and natural resource economics aim to understand the distribution of natural resource use over time and how that use is changed by technology, individual preferences, and social institutions. They also aim to understand how economic forces change that distribution and to formulate standards for identifying the distribution that best serves society's needs. Understanding these aspects of the environment and natural resources is a noble aim in itself. However, the larger objective is to use the knowledge gained to improve human welfare. The aims of this text are to use this knowledge to help set the agenda and to design, implement, and evaluate public and private actions dealing with the environment and natural resources. If these goals are achieved, your time spent studying the methods of economics as applied to the environmental and natural resources will be justified.

■ Review Questions

1. Cite at least four historical or current examples that show how communities have suffered economically by making poor natural resource or environmental decisions.

2. Describe the economic forces that led to the loss of the cedars of Lebanon. What lessons learned from their loss can be applied to the future?

3. Why did the Anasazi people disappear and what does this tell us about the economic benefits of policies that guarantee people access to a reliable water supply?

4. Why is water so important to dry places like California? Why were the economic benefits from California's Central Valley Project so high for agriculture and for the control of flooding?

5. Explain how economics can provide insights into the following environmental problems: safe drinking water, ozone depletion, toxic wastes, and air pollution.

6. Why does economics empower people to better contribute to environmental policy debates that are driven by economic interests?

7. Explain the meaning of environmental proposals and give an example of how economics provides reality checks for those proposals.

8. Give an example of how economics can help improve communication in environmental debates.

9. Explain the difference between an environmental policy that works technically and one that pays economically.

10. Describe how economics focuses environmental policy debates.

11. How can understanding economics empower you to better approach environmental debates?

12. What does it mean to *set the agenda* for environmental and natural resource policy? How can economic principles help set that agenda?

13. What role can economics play in designing environmental and natural resource policy?

14. In what ways can economics help to more effectively implement environmental and natural resource policies?

15. Describe how economics can contribute to the process of policy evaluation.

■ References

Bishofberger, T. E. 1975. Early flood control in the California Central Valley. *Journal of the West.* 14 (3): 85–94.

Brundtland, G. H. 1987. *Our Common Future: The World Commission on Environment and Development.* Oxford, UK: Oxford University Press.

Chinese Government. 1995. *Family Planning in China.* Available from *http://www.china.org.cn/e-white/familypanning/* [Accessed October 13, 2004].

Columbia University Center for International Earth Science Information Network. 2001. *Ozone depletion and global environmental change.* Available from *http://www.ciesin.org/TG/OZ/oz-home.html* [Accessed July 30, 2003].

Gadgil, A. 1998. Drinking water in developing countries. *Annual Review of Energy and the Environment.* 23: 253–86.

Pearce, D. 2002. An intellectual history of environmental economics. *Annual Review of Energy and the Environment.* 27: 57–81.

Reisner, M. 1993. *Cadillac Desert: The American West and Its Disappearing Water.* New York: Penguin Books.

South Coast Air Quality Management District. 2004. *Web page.* Available from *http://www.aqmd.gov/* [Accessed October 13, 2004].

U.S. Environmental Protection Agency. 2002. *Acid rain program.* Available from *http://www.epa.gov/airmarkets/trading/index.html* [Accessed July 30, 2003].

United Nations. 2000. *The Dublin statement on water and sustainable development.* Available from *http://www.unesco.org/science/waterday2000/dublin.htm* [Accessed October 13, 2004].

———. 2002. *The Johannesburg summit test: What will change?* Available from *http://www.johannesburgsummit.org/* [Accessed October 13, 2004].

———. 2004. *Water.* Available from *http://www.unesco.org/water/wwap/milestones/index.shtml* [Accessed October 13, 2004].

United Nations Ozone Secretariat. 2004. *Status of ratification.* Available from *http://www.unep.ch/ozone/Treaties_and_Ratification/2C_ratificationTable.asp* [Accessed October 13, 2004].

Williams, M. 2000. Dark ages and dark areas: Global deforestation in the deep past. *Journal of Historical Geography.* 26 (1): 28–46.

Zagas, T., A. Hatzistathis, T. Tsitsoni, and P. Ghanatsas. 1998. Degradation of Mediterranean forest ecosystems and silvicultural measures for their restoration. *East-West Series in Economics, Business, and the Environment.* 1 (2): 53–60.

Chapter 2

ECONOMIC THINKING

Outline

Hard Choices in Policy Issues 15
Choices Displace Opportunities 20
The Market Mechanism and Choices 22
Goals and Means of Environmental Policies 26
Environmental Policy Goals 30

Objectives

After studying this chapter, you should be able to

- Illustrate hard choices presented by environmental and natural resource policies
- Show how policy choices displace opportunities
- Explain the role of the market mechanism in making the choices of what, how, and for whom
- Describe how economics offers important insights into the goals and means of environmental policies
- Identify several environmental policy goals

Introduction

Economics is about choices. Environmental and natural resources—land, water, wildlife, and unpolluted air—may be (1) developed and used by the current human generation, (2) preserved for use by later generations, (3) managed using any combination of current use and preservation, or (4) left untouched. For example, ancient redwood trees in northern California can be (1) cut down now to produce lumber for houses or fences, (2) left untouched for later harvest, (3) harvested in phases and replanted over time, or (4) designated as wilderness that is protected from current or future timber harvest. During one time period, a natural resource may be faced with pressures from more than one type of use. Moreover, people from one generation may have a different idea about how best to use a natural resource than those from another generation. Economics deals with the problem of how best to protect the environment and manage natural resources to produce sustained benefit to people in current and future generations.

■ HARD CHOICES IN POLICY ISSUES

Environmental and natural resource economics is important in making choices when no single choice is best for all people in all generations. Governments have always tried to design institutions—including laws, policies, and regulations—to help make better choices in meeting society's objectives. Environmental and natural resource managers are expected to improve the general welfare by choosing wisely among environmental protection and natural resource management proposals. Adam Smith (1776), in his innovative book *The Wealth of Nations,* was the first to show that *markets* are powerful and efficient institutions for allocating resources to maximize human benefits for current and future generations. This theme has played an important role in economics ever since. The following examples illustrate a few of the hard choices we face today in dealing with the environment and natural resources.

Timber versus Endangered Species

In the American and Canadian Pacific Northwest, less than 5% of land originally containing ancient forests, or *old growth,* remains today. These trees are an exceptionally valuable source of high-grade timber as well as high-paying logging jobs. The same trees, however, are a valuable home for northern spotted owls, a species so near extinction that in 1990 it was listed in the United States as endangered (U.S. Fish and Wildlife Service, 2004). Because the supply of old-growth forest is scarce, it is essential to make the right decisions in allocating this natural resource if present and future generations of people are to receive maximum sustained benefit. The choice is made more difficult by the mandates of the U.S. Endangered Species Act (U.S. Congress, 1973) and other laws,[1] which have emerged just as many environmental and natural resources have grown more scarce and competition for them has intensified. Underlying the issue of logs versus owls is a more fundamental economic issue of measuring the value associated with each use.

Figure 2–1 shows that when we are at point *A,* we are producing many commercial logs but protecting the habitat of few spotted owls. Ensuring that spotted owls survive as a species requires that high-quality, old-growth forests remain intact. This moves us to point *B,* where the benefit is the value of avoiding species extinction. The cost is the value of timber lost. Choosing correctly in this difficult situation depends in part on whether the benefits gained from avoiding species extinction outweigh the costs of lost timber.[2]

Affordable Energy versus Functioning Ecosystems

How much more are you willing to pay for fuel to ensure that ecosystems are being protected? The U.S. Congress has passed laws requiring oil companies to build double-hulled tankers by 2015 so that accidents do not cause an ecological catastrophe, such as the *Exxon Valdez* spill in Alaska's Prince William Sound. But that policy would raise the price of gasoline, since most of the added costs of double-hulled tankers are ultimately paid for by gasoline consumers. The benefit of the policy is the value of the ecological catastrophe prevented; the cost is the higher price of gasoline. So an important question facing policymakers is whether the benefits of requiring double-hulled tankers exceed the cost.

---

[1] Canada faces similar conflicts between old-growth forests and spotted owls (CBC News, 2004). Loomis and Gonzales Caban (1997) analyze conflicts surrounding spotted owl habitat in California and Oregon. Montgomery and Brown (1992) analyze the cost of various increments of protection for the northern spotted owl in the U.S. Pacific Northwest.

[2] It is difficult to measure the benefits of protecting an endangered species from going extinct. In the United States, the Endangered Species Act and its amendments (U.S. Congress, 1973) have assigned only a minor role to efforts at valuing species preservation.

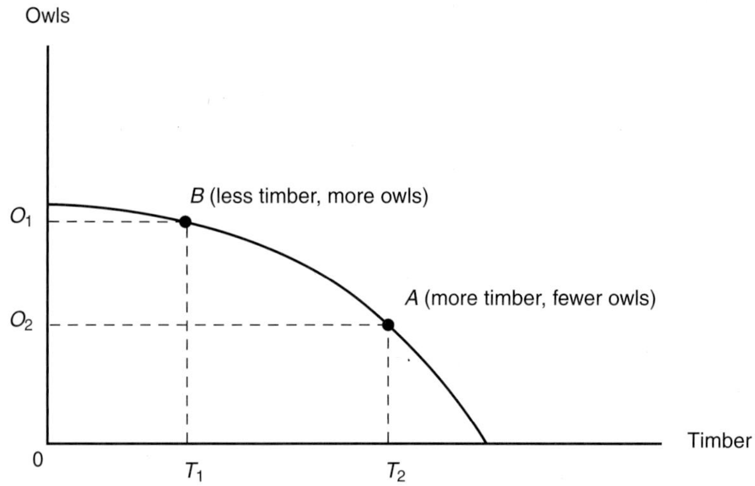

Figure 2–1
Tradeoff between timber and spotted owls

Clean Air versus Affordable Cars

People in many of the world's largest cities breathe polluted air daily. This pollution can be reduced and the air cleaned, but the costs are high. In the United States the Clean Air Act of 1970 and later amendments set limits on air emissions. Achieving objectives of air pollution control established by the U.S. Congress in the 1990 Clean Air Act amendments will require an estimated $25 to $50 billion each year to be spent by both the public and private sectors. Prior to the 1990 amendments, the United States regulated fewer than 20 pollutants. Now the U.S. Clean Air Act regulates more than 380 pollutants at considerable cost to industry. The steel industry is particularly hard hit by the added rules, which require the industry to spend billions of dollars during the next 10 years, a cost that adds substantially to operating costs and to the costs of products such as new cars. Air pollution control laws will have similar economic effects in other industrialized countries. The debate over clean air and the cost of steel might be stated as follows: Are the benefits of cleaner air greater than the cost of higher priced cars?

Endangered Wolves versus Livestock and Wildlife

What are you willing to pay in higher beef and lamb prices and smaller deer and elk herds to restore wolves to their native habitat? Recent reintroductions of wolves in Yellowstone National Park and in Arizona and New Mexico may result in lower sustainable numbers of wildlife. Wolves also are known to eat domestic cattle and sheep, therefore, costs for cattle and sheep ranchers will increase. So reintroduction of wolves may reduce beef and lamb supplies in our grocery stores, prompting an increase in the price we all pay for beef and lamb. A similar situation in the Scottish Highlands is described by Bullock, Elston, and Chalmers (1998).

Figure 2–2 shows the impact of reintroducing wolves in a sheep-grazing region. For awhile, the price of lamb stays at the preintroduction equilibrium level, P_A. Over time, however, a number of lambs will be eaten by hungry wolves and never make it to the slaughterhouse; this is shown by a leftward shift in supply (S) from point A to C at the going price P_A. With fewer lambs to sell, less lamb meat is available to meat packers; that is, quantity supplied falls from Q_A to Q_C. Eventually, consumers find there is less lamb meat available in groceries, meat markets, and restaurants. As meat buyers scramble to buy shrinking supplies of lamb, retail sellers raise the price of lamb because they know they can do it without losing buyers. The higher price sends a signal to everybody in the lamb supply chain. It signals ranchers to expand their sheep herds at the expense of other ranching activities. It signals packing houses to pack more lamb. It sends a message to grocery stores to buy and stock more lamb. The market system, through the power of price and incentives for profit, rewards suppliers the most who can respond fastest to the rising price.

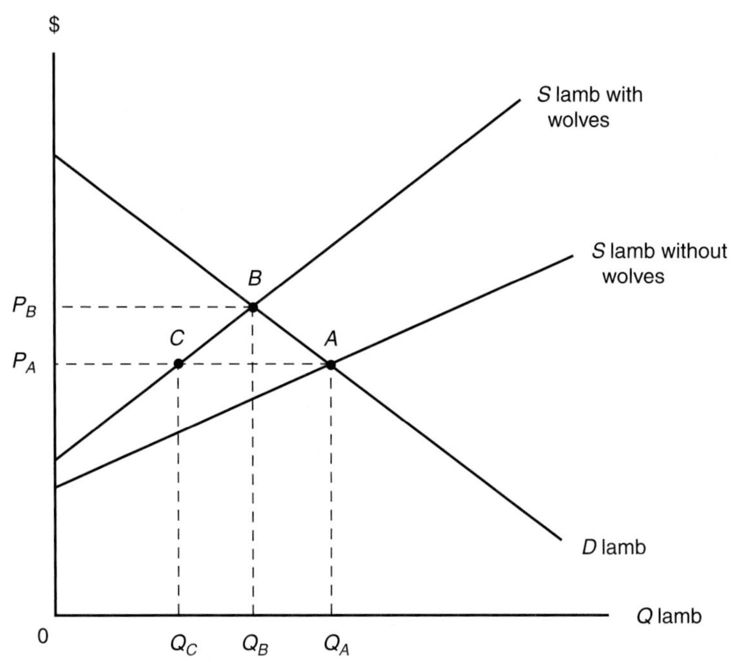

Figure 2–2
Reduction in lamb supply caused by increased reintroductions of wolves

Figure 2–2 shows all this as a movement up and to the right (to the northeast) along the left-hand supply curve. Rising prices and increased lamb production continue in the face of new profit opportunities until a new equilibrium is reached at point *B*. At that point, the supply (*S*) and demand (*D*) for lamb are again equal and the forces that bid up the price after the wolves were introduced are dissipated.

Ranchers lucky enough to live away from the reintroduced wolves, for example, get a higher price for their existing herds and may be able to respond more quickly to the market trend. Producers who lose lambs to wolves may suffer. Although these producers receive a higher price for their remaining lambs, their losses also increase. Buyers of lamb are worse off: They pay a higher price no matter what. For economists and policymakers this issue poses an important question: Are the benefits of wolves restored to their native habitats greater than the costs of increased beef and lamb prices?

Endangered Fish versus Food

The history of the Klamath Basin conflict in northern California and southern Oregon date to the mid-nineteenth century. In 1864 the Klamath, Modoc, and Yahooskin Native Americans signed a treaty with the federal government guaranteeing them perpetual hunting and fishing rights along Upper Klamath Lake. In the early twentieth century Congress authorized construction of the Klamath Project, one of the Bureau of Reclamation's (BOR) first irrigation projects and designed to promote crop irrigation and food production. A year after the Klamath Project opened, President Theodore Roosevelt created the nation's first waterfowl refuge on Lower Klamath Lake. Then, in the 1930s and 1940s, the U.S. government deeded the Klamath Project land to homesteaders and war veterans with the promise of irrigation water for years to come. In 1988 two fish, the Lost River sucker and short-nosed sucker, were listed as endangered species under the U.S. Endangered Species Act.

Things remained quiet until 1992, when the U.S. Fish and Wildlife Service set a minimum water level in Upper Klamath Lake to maintain water quality for the suckers. Unfortunately, Upper Klamath Lake is also the area's primary irrigation reservoir. Droughts followed in 1994, 2000, 2001, and 2002. A flash point occurred in 2001 when the BOR shut off Klamath Project irrigation water to 1,400 farms, encompassing 90% of Klamath Basin's farmland.

A U.S. federal court denied a request to restore the irrigation water, making a strong economic argument when it said the need to protect endangered fish outweighed the economic needs of farmers. The federal government then offered farmers low-interest loans and other emergency aid. Many farmers sold their lands to the American Land Conservancy, which planned to restore farmlands to marshland status. A new twist in this tale occurred in early 2002, when the BOR released a draft of a biological assessment that would return irrigation water to farmers. Environmental groups quickly criticized the assessment, charging that the study was based on flawed science.

The Klamath Basin conflict continues. None of the interests—tribes, farmers, fish, environmentalists, wildlife or government agencies—is winning the battle. A series of court cases has determined that Klamath Basin water will be allocated as follows: first, to meet the needs of endangered species; second, for Native American tribes and the wildlife they depend on; third, for farmers in the Klamath Project; and, fourth, for wildlife refuges that depend on farm runoff for much of their water. The price of saving the endangered species will be higher food prices and lost farm income, both of which result from increased costs of producing food. Do you think that the benefits of the endangered species saved outweigh the costs of higher food prices and lost farm income?

Power versus Recreation

Glen Canyon Dam, which backs up the Colorado river on the Arizona–Utah border, was completed in the early 1960s, mostly to produce low-cost hydroelectric power. For electric power production, the best time to release water over the dam's turbines is on hot summer afternoons, when power is expensive as people use more for cooling and air conditioning. However, for environmental benefits, the best time to release flows over the dam is in March so as to reproduce natural springtime runoff. At this time of year hydroelectric power is at its lowest value.

Here is the hard choice. By releasing water from the dam in March for environmental benefits, Phoenix and Las Vegas lose some high-valued electricity in the summer that must be made up by burning coal, a fuel that is economically costly for electric utilities and environmentally costly for the public. By holding the high flows back until summer, all the March environmental benefits are lost. While the U.S. taxpayer struggles with these and other choices, the field of economics provides some guidance by asking this question: Do the benefits of a more natural seasonal flow in the Colorado River exceed the cost of higher power prices?

Mining versus Public Lands Recreation

Mining activities of various kinds can conflict with outdoor recreation and other environmental assets. In the late 1970s a conflict emerged in southern Utah between coal mining interests and recreational users of nearby Bryce Canyon National Park. Recreational users claimed that a proposed coal field that would supply coal to a nearby electric utility would damage outdoor recreation at the southern end of the park. Yet, protecting the scenic vista for the park's visitors came at a cost of coal not supplied, which resulted in higher power prices. An economic study published at the time found that air emissions from burning coal would reduce visibility, thereby considerably reducing the value associated with recreational use of the park (Haspel and Johnson, 1982). Later, the hard choice was made to scrap the coal mining project because it was believed that the benefits of the environmental asset saved were larger than the cost of higher power prices that resulted from saving that asset.

Riparian Ecosystems versus Livestock

In dry places like the world's deserts, riparian areas (near streams and rivers) attract a diverse assortment of plants and animals, many of which would not survive without the water. Domestic livestock that are allowed to graze in these areas compete with wildlife and other species, such as the endangered southwestern willow flycatcher of North America. Ensuring the survival of the flycatcher may require removing cattle from a large percentage of stream-side grasslands, thereby raising the cost of producing livestock for affected ranchers. Are the benefits that result from guaranteeing survival of the flycatcher and other species greater than the cost of higher beef prices and lost ranching incomes resulting from enforcing that guarantee?

Controlling Global Warming: Current Costs versus Future Benefits

Since the late twentieth century, average global temperatures have increased. Many scientists believe that higher average temperatures are related to growth in emissions of CO_2 and other greenhouse gasses (see Chapter 15; Dasgupta, 1996; Lin and Chang, 1996). Policymakers have a keen interest in the CO_2 problem because current and future policies affect the amount by which atmospheric CO_2 builds up. Different control policies will produce different levels of CO_2 accumulation in the years ahead. Total volume of CO_2 emissions from all sources around the world is huge, about 7 billion tons per year (Schelling, 1992), and the costs of controlling these emissions are estimated to be large, which will raise the costs of production for electric utilities and other activities that produce greenhouse gasses. The climate change problem is unique because it requires that actions be taken in which large costs are incurred long before benefits can be realized or measured with confidence. The stakes are high in this policy debate, involving potential investments and/or economic impacts of hundreds of billions of dollars a year worldwide.

If we chose to invest in these environmental measures, the largest costs are likely to be incurred by industrialized countries. Doing so will produce the largest benefits to future residents of the world. Industrialized nations are faced with the hard choice of deciding how large an economic cost they are prepared to incur now to reduce greenhouse gas emissions in the future, and for benefits that are shared by people and ecosystems worldwide.

Living Space versus Human Rights

Larger populations are generally believed to put pressure on the environment and natural resource base (see Chapter 14). Yet, many couples want large families. Social scientists and political leaders believe that the coercive power of the state may be the only way to reduce family sizes. China's one-child policy illustrates the power of the state to intervene in private family planning when the state believes society as a whole will benefit from a reduced family size.

In 2000 China housed 20% of the world's population, some 1.2 billion people. China's one-child policy, in which couples cannot have more than one child without heavy penalties, was first introduced in the early 1980s. At the time, China was feeding 1.2 billion people, but only 5% of its territory was producing food. By comparison, the United States cultivated 20% of its total land area, but has a much smaller population (having nine times China's per capita farmland). China's farmers were faced with feeding a continuing growing population squeezed onto a shrinking land base, producing falling living standards for everyone.

China's one-child policy has had several impacts, some of which were unplanned. Chinese culture places a high value on male children to continue family bloodlines. In some families, girl babies are abandoned or killed to make space for a boy. This cultural preference for boys skews the ratio of boys to girls, sometimes approaching 1.4 to 1. In some areas, more than 90% of abandoned babies are girls. In China children are the sole support for their parents as the parents grow old, so the one-child policy may place pressure on the government for publicly funded retirement pensions. High ratios of men to women will force many men into bachelorhood and may contribute to social tensions and a greater emphasis on the military. Today, many Chinese children receive sole attention from two parents and four grandparents. They are criticized by some as living the lives of "spoiled little emperors," excelling individually but unable to work as part of a team.

China's one-child policy and its implementation illustrates the hard policy choices that governments of crowded communities face: What culturally compatible measures can control population, accommodate government budget constraints, and maintain basic human rights?

Energy versus Wildlife

The Arctic National Wildlife Refuge comprises 19 million acres in northeast Alaska. It is administered by the U.S. Fish and Wildlife Service within the Department of Interior. A portion of the refuge, some 1.5 million acres along the northern coastal plain, has been found to contain large reserves of oil and gas. The U.S. Geological Survey reports this federal land could contain as much economically recoverable oil as the huge field at Alaska's Prudhoe Bay, now estimated at 11 to 13 billion barrels. At the same time, the refuge, especially its coastal plain, is home to a variety of plants and animals. The presence of caribou, polar bears, grizzly bears, wolves, migratory birds, and other species in a nearly undisturbed state makes it one of America's premier wildlife areas.

The conflict between energy potential and nearly pristine nature creates a series of hard choices for the American people. Should the U.S. Congress open up the refuge for oil and gas development or should this ecosystem be permanently closed to energy development, thus increasing future costs and prices of energy? Are the benefits of preserving the environment larger than the costs of higher energy prices required by that preservation? What factors should determine whether to open the area to development? If the area is opened, how can damages be avoided, minimized, or reduced? To what extent should Congress legislate special management of the area if it permits development, and to what extent should federal agencies manage the area under existing law?

Short-Run Human Benefits versus Long-Run Environmental Damages

Short-run human benefits sometimes produce long-run damages to the environment. Consider the case of the drug Taxol (produced by Bristol Myers-Squibb), which was found to be effective in treating several cancers, including ovarian, lung, and breast cancer. Taxol research continues today and medical interest in the drug has led to conflicts. Large quantities of the drug are needed to conduct the necessary clinical trials. The natural source of Taxol, the Pacific yew tree, is an environmentally protected species and one of the slowest growing trees in the world. Isolation of the chemical compound in Taxol, which is contained in the bark of the Pacific yew, involves killing the tree. Even then, the quantities available are painfully small. It takes six 100-year-old trees to provide enough Taxol to treat just one patient. Propagation of the Pacific yew from seed is slow, uncertain, and expensive.

The U.S. government faced a hard choice in trading the present benefits of saving lives against long-run environmental consequences of destroying the protected trees. The U.S. Pacific Yew Act of 1992 (expired in 1997) increased awareness of the need to manage federal lands with ancient yew trees for human benefit. The law became null and void when a sufficient supply of the trees was found available on nonfederal lands. Some environmental projects face equally tough choices between short-term human gain (benefit) and long-term environmental loss (cost): groundwater protection, global climate change, radioactive waste disposal, mineral depletion, and loss of biodiversity.[3] The field of economics provides methods for evaluating these tough choices and supplies important quantitative tools for comparing and analyzing the impact of environmental and natural resource decisions.

■ Choices Displace Opportunities

Decisions about community policies are often difficult to make because most social or economic gains entail costs. Socrates said the unexamined life is not worth living. Nowhere is the need for examination more true than with community choices regarding environmental protection and nat-

---

[3] Castaneda (1999) describes trade-offs in Chile between economic benefit and environmental sustainability.

ural resource management. Almost without exception, in decisions regarding the use of natural resources, certain opportunities are opened while others are displaced. The opportunity gained is the benefit; the opportunity displaced is the cost. Economics helps us think about which choices should be made so that human benefits exceed human costs by the greatest amount possible.

All societies invest a large amount of time, energy, and money in the design of institutions (see Lele and Norgaard, 1996). Institutions include collective rules, practices, or organizations for governing individuals' behavior. The way a society goes about solving its economic problems tells us a lot about the way it will approach its environmental protection and natural resource problems. All societies face three basic economic problems.

First, all societies, from the smallest village to the largest complex urban state, either accidently or intentionally, must determine *what* commodities will be produced and what quantities of each. For example, will we use our forest lands to produce timber and support livestock without consideration for outdoor recreation, water use, and habitat (living area) for wildlife? Or will we use those forest lands to produce smaller quantities of timber and livestock, but provide opportunities as well for outdoor recreation? Will we allocate our rivers only for the production of electric power? Or will we use rivers to produce smaller amounts of a greater number of commodities? Should we let our rivers run wild to the ocean and take the inevitable floods and droughts that result? Or will we take taxpayer resources (money, land, and water) out of private consumption and allocate them to building dams to reduce the risk of floods and minimize the impact of droughts?

Second, all societies need to determine *how* they will produce their goods. By whom and with what resources will products be made and with what technologies? For example, will businesses use environmentally friendly technologies for producing paper in a way that keeps pulp from being dumped in trout streams but raises the cost of a ream of paper? Or will businesses use environmentally insensitive technology that reduces the price of a ream of paper but deposits wastes into streams and rivers? Will houses be produced at low cost with lumber harvested cheaply from big trees that displace spotted owls? Or will houses be produced with alternative, but often more expensive, resources such as bricks, stone, and adobe, while preserving the habitats of spotted owls?

Third, all societies must decide *for whom* goods will be produced. Who will enjoy the benefits of the community's production of commodities? In other words, who gets how much income to buy produced commodities? Will forest-product workers receive high wages and live in expensive houses in unpolluted environments? Or will society choose to pay low wages to its daily workers but reserve high profits and a clean environment for sawmill owners? Will laws be established that ignore protection of the environment, thereby depressing salaries for environmental scientists and increasing incomes for miners and lumberjacks? Or will we pass laws that require environmentally friendly technologies that increase salaries for biologists but reduce incomes for miners and lumberjacks?

Not only do all societies face the three problems of what, how, and for whom, but when environmental and natural resources are scarce, societies must make the right decisions for solving these problems. The cost of making mistakes can be high. For example, in desert climates like Los Angeles, where water is scarce, it is essential to establish and implement good plans for finding, storing, saving, and using water. Making the right decision is especially important when it is difficult for the environment to recover from poor decisions. An example of such a decision is the use of a nonrenewable resource, such as a nonrechargeable underground aquifer that may be tapped to supply a city's water. Similarly, the cost of a failed effort to save a species from extinction is especially high when the species is forever lost. Where protein is scarce, which often occurs in Africa, and in situations of war or famine, policies that trade livestock ranching with other activities, such as tourism, must be thought through carefully. The wrong policy could force people to go hungry, hoard food, even starve.

Economics studies the choices people make in allocating scarce resources among competing demands. Economic principles provide insights into human behavior and suggest how to alter human behavior in a way that promotes the welfare of current and future generations. Economics helps solve the problems of what, how, and for whom.

Figure 2–3
Increase in demand for organically grown grapes

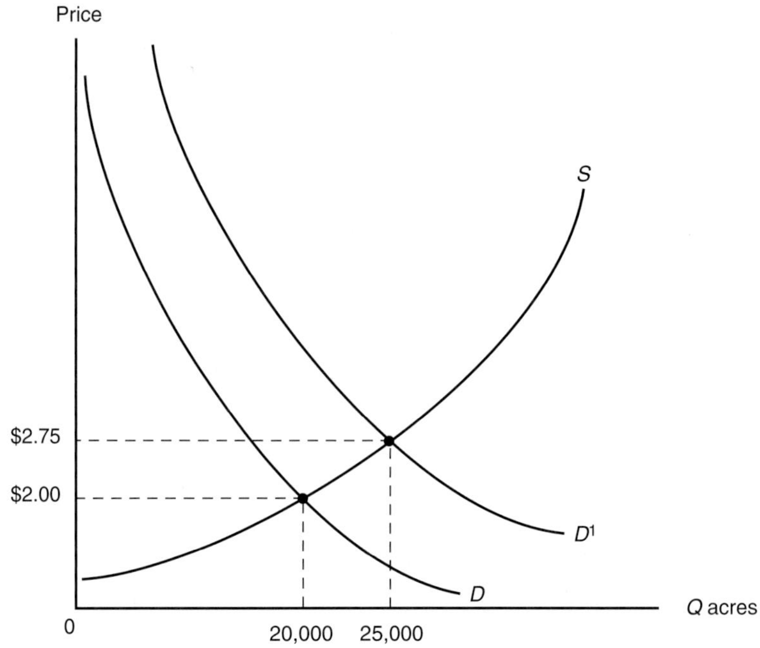

■ The Market Mechanism and Choices

The market is more than a place where goods are traded such as the Tokyo Stock Exchange or the local livestock or automobile auction. The market system is actually a *process* by which buyers and sellers of any good or service—including electric power, water, livestock, timber, or northern spotted owls—interact to determine its scarcity (quantity) and value (price). The market system is one mechanism to address the three problems of what, how, and for whom.

What Things Are Produced

How does the market accommodate people's desire for more of something (economists call this increases in demand)? If people want to buy more of something, such as organically grown grapes, a surge of new orders will be placed for the grapes. We can show this graphically, illustrated in Figure 2–3, as an increase in demand from D to D^1. Demand temporarily exceeds supply at the going price of $2.00 per pound. As buyers scramble to buy more organic grapes, sellers raise the price because they know can get away with it without losing customers. The higher price sends a signal that lures farmers to develop more organic soils, packers to pack more organic grapes, and grocery stores to buy and stock more organic grapes. And so the story goes. The market process, through the price and profit signal, rewards people who respond to buyers' desires for more grapes.

The market process also moves in the opposite direction. If people want fewer *conventionally* grown grapes, fewer orders will be placed for them. Sellers, anxious to get rid of their unsold conventional grapes rather than have them rot on the shelves, lower the price. Producers and sellers will allocate resources (e.g., land, pesticides, water, and shelf space) normally assigned to conventional grapes, for something else.

This example shows the way in which consumer's wishes—transmitted through the market mechanism—attract resources to produce *what* people want. Business firms are enticed into producing goods that are in high demand by the potential to earn greater profits. Remember, profit is the sum of what is left when a business subtracts its total cost from its total revenue. Profit is the main reason businesses exist. It is the main measure of a business's success. The lack of profit is the main measure of a business's failure. If consumers want less of something like conventionally grown grapes, prices will fall, and producers will leave behind those areas of low profit rather than risk losses.

How Things Are Produced

What determines *how* things are produced in market economy? What determines why one firm applies a small amount of labor to work large plots of land and another hires a large number of laborers to work smaller parcels of land? How things are produced in a market economy (the technologies that firms choose) is determined by the competition among producers, who must minimize costs to stay in business. Businesses compete for a limited number of customers and the limited dollars those customers have to spend. Why do businesses need to minimize costs? A business that fails to minimize its costs will need to charge higher prices to its customers than a competing business that successfully minimizes costs. A business that fails to minimize costs will lose business to competitors who succeed in minimizing costs.

Producers maximize profits by minimizing their costs of supplying their chosen level of output. To keep costs at a minimum, producers reduce the use of expensive inputs and make greater use of cheap inputs. Agriculture provides a good example of how this idea works in different parts of the world. In the United States land is plentiful and cheap, labor is scarce, and wages are high. For U.S. farmers, the cheap land and costly labor rewards the use of labor-saving technologies for the production of cotton, wheat, livestock, as well as organic grapes. Where labor is abundant and land is scare, such as parts of Africa and Asia, farmers find they can maximize their profits and minimize costs in producing cotton, wheat, and livestock with labor-intensive technology.

For Whom Things Are Produced

In a market economy, who gets how much of the goods produced? Goods are produced *for whom*—for people with *income* to buy them. The distribution of income in a market economy is determined by who owns productive resources—such as land, natural resources, labor, and capital—and how valuable those resources are in meeting the desires of consumers.

In a market economy, your income is measured by the price of each productive resource you own, times the quantity you offer to the market. What determines the income earned by our growers of organic grapes? As you would guess, the price of organic grapes plays a major part. If consumers want more organic grapes, they buy more, so the price increases and the grower of the grapes gets richer. The grape farmers who can meet the extra demand at the least cost will get even richer yet. In a market system, income increases for people who can respond most efficiently to changes in consumer buying patterns.

Who Controls the Market?

No one person controls the market. The three rulers of the marketplace are consumers, technology, and profits. Consumers are directed by their natural or learned preferences; they vote with their dollars. Television marketers, for example, try to influence preferences so that consumers want to buy what they have to sell. Growers of organically grown grapes, for example, may try to convince you through advertising that if you buy their product, you will be sophisticated, better looking, and well loved.

Technology sets limits on what can be produced and at what cost. In 1849, California gold miners probably wanted to talk to their families thousands of miles away. Miners would have gladly paid in gold for the opportunity to speak in person to their family members. However, there were no telephones in 1849. Until the telephone was invented, the phone lines stretched, and the cable laid, miners could not talk to their family members, no matter how high the demand. Demand must be backed by the technology to carry it out. When consumers "vote" in favor of wanting more of a good or service, firms are lured into these business opportunities by the potential to earn greater profits. Similarly, firms abandon business areas to avoid reduced profits when consumers want fewer of their goods and services.

The Invisible Hand of the Market

In *The Wealth of Nations,* Adam Smith emphasized the importance of self-interest as a human motivator, a fact that has led some people to call economics the "dismal science." Smith wrote that every individual uses his resources to produce the greatest value to himself. He neither intends to promote the public interest, nor knows how much he is promoting it. He intends only to maximize his own security and his own gain. Smith goes on to introduce an idea that has become a cornerstone of microeconomic theory: In selfishly pursuing our own personal interest, we are led, as if by an invisible hand, to accomplish the best good for the larger society. In other words, Smith says that producers who attempt to maximize their own profits, consumers who buy when and where prices are the lowest, and workers who work where their wages are highest, are doing a great service to society. According to Smith, individuals pursuing their own self-interests improve social welfare even if they have no intention of doing so.

In a market-driven society, the lure of profit encourages businesses to produce what people want to buy, motivates businesses to produce using least cost methods, and rewards businesses if they conserve on scarce resources. The profit signal tells businesses how and when to take action. But businesses get rich only if they are successful at solving the market's three problems of what, how, and for whom. In short, the profit motive, combined with free market competition for customers, ensures that products are made at minimum cost and with maximum choice. And we are all consumers.

The Market, the Environment, and Natural Resources

Oil companies, foresters, ranchers, political leaders, environmentalists, and others have asked whether the profit motive, as powerful as it is, can be harnessed to help protect the environment. The answer to the question is important, because it tells us how much or how little government is needed to protect environmental resources. The answer depends on whether or not there are *property rights* (i.e., rights to own, use, buy, and trade natural resources). Where property rights to natural resources are well defined, the profit motive helps protect private forests, private grazing lands, and private water supplies. The following examples explain why.

High timber prices encourage forest owners to plant trees so they can have something to harvest for the future (see Chapter 8). High cattle prices encourage ranchers to expand their herds so they can have beef to sell for the future, and to nurture land with enough grass on it to support future herds. In places where wildlife can be privately owned, high tourist expenditures encourage landowners to invest in habitat for watchable wildlife. Where water rights are secure and are marketable, high prices encourage owners of water rights to use their water carefully and/or save it for future use or sale. In the early 1990s California established an emergency water bank to minimize damages from an ongoing and severe drought. The state allowed high payments to farmers, if the farmers agreed to deposit their unused water into the bank for resale to cities. The California Water Banking Law included specific legal assurances that depositing water into the bank would not change a farmer's permanent water rights. Because the law increased a farmer's certainty over possibly insecure water rights, the law proved successful.

How Markets Settle Environmental Management Questions

By and large, capitalistic countries allow free markets to determine environmental and natural resource allocations between the present and the future. There is little chance that this practice will be abandoned. Several U.S. states tax coal production, for example, but not for the purpose of saving coal for the future. They tax coal and other minerals to raise revenues and make up for the loss of the mineral. Deciding how much copper to extract over time is left to the mining company.

Likewise, production and distribution of food between present and future generations is largely in the hands of private farmers.

The market settles environmental protection and natural resource management issues by renewing, producing, and selling those resources over time to maximize the stream of income over time to owners. A private forest is a good example. The opportunity for current and future profit tells the private forester when to plant saplings, when to harvest old trees, and when and how long to wait for little trees to grow into big ones. Will these profit-maximizing decisions lead to depletion of forests? If the private-forest owner has secure property rights, the answer is a resounding no. Private-forest owners have a strong financial incentive to take care of their trees. Cutting down all trees and selling them today without replanting means a forester will have nothing to sell in future years. (This assumes no alternative uses of the land; if building homes on land exceeds the land's value for trees, private foresters will clear the land for houses.)

The profit system can help provide for future national and global timber needs. If a private-forest owner expects the population to grow and people to keep buying wood products, she can plan for high future demand and high future prices. Under these circumstances, a private forester has every incentive to plant more trees now, wait for growth, and harvest fewer trees today to save more for the future when prices are higher. In a free market, personal self-interest acts like an invisible hand, in this case, providing for the wood product needs of future generations.

When Markets Are Best at Sustaining the Environment

The question of when markets are best at sustaining the environment has been widely debated for years. The answer depends on how you define the phrase *best at sustaining* the environment. Let us say for now that we are willing to define *best at sustaining* to mean maximizing long-term total benefits, regardless of how those benefits are distributed among different people. This is what we mean by *economic efficiency*. If economic efficiency is what we care about, we can say markets do the job best in situations when all potential resource users can effectively voice their demands in a monetary form. Again, take forests as an example. If forests were auctioned so that private bidders could effectively voice their demands for all forest products—including spotted owls, timber, livestock, water, and recreation—then the market system would make it profitable for businesses to produce the best mix of timber, spotted owls, and other forest products.

This analysis leads to the following general principle: When all resource users are able to voice their demands monetarily, the price of resources is determined by current and anticipated future demands and current and anticipated future supplies. Under these conditions, anybody who decides to use a resource at a particular time is faced with a price equal to the value of other uses that are displaced. Natural resources are then priced at their real cost.

Limits of the Market System

There are major limits of relying on the market system to settle questions of environmental protection and natural resource management.[4] Many publicly owned resources, such as clean air, are unpriced. The market system has little incentive to take care of its clean air because it is unprofitable to do so. The next paragraphs present three examples of how the market fails to protect the environment. The first example of the market system failing to protect an *underpriced resource* involves the price of coal used to produce electric power in the western United States. In an unregulated market, burning coal to produce electric power causes air pollution, which is

---

[4] Markets can also be unfair because they are impartial. There are no guarantees that markets will produce a fair distribution of income, power, or wealth. Left alone, a market economy may create inequalities that grow larger as time passes. Both the economic successes and failures of individuals and businesses can be cumulative—the successful tend to succeed again while the unsuccessful may remain unsuccessful.

transported by prevailing winds; acid rain results, which is thought to damage forest and lakes in the northeast United States and in Canada. The cost of damages to these resources is not included in the price of the coal-fired electricity charged to customers in Phoenix and Las Vegas. Electricity buyers in those cities, therefore are not charged a price to pay for the full cost of their power.

Spillover benefits provide a second example of why free markets sometimes fail to protect the environment. Spillover benefits are those that fall into the lap of someone as a direct by-product of an economic trade between a buyer and seller. Spillover benefits are important because the benefits accruing to third parties cannot be sold for profit. Even if benefits to society exceed the costs of the environmental investment, a business will not make the investment if it cannot charge its customers the full price. Consider the pulp mill owner who deliberates on whether to spend money on controlling pollution to reduce discharges into a trout stream. Even if the benefits to society exceed the costs to the mill owner, the owner is still likely to decide against the investment because he cannot make money on selling the cleaner water to the many anglers who fish there. So what happens in an unregulated market system? The mill owner fails to invest in pollution controls and the trout stream remains polluted.

Spillover costs are a third example of the failure of the market system to protect the environment. Spillover costs are those shifted onto somebody as a by-product of an economic trade between a buyer and seller. If the traders are not charged for costs spilled over onto the rest of society, they ignore those costs, even when damages to the environment are severe. Consider a logging company that clear cuts a patch of forest and is not charged for the resulting silt in the downstream river. The effect of the muddy stream is of no concern to the company or its customers (the traders) if they are not charged for the downstream damages. But the costs are important to the downstream users of water, downstream users are the ones who live with the dirty water or pay to clean it up. This example shows how the presence of spillover costs dulls the potential for market incentives to promote efficient decisions.

Spillover benefits and spillover costs are both called *externalities* and are excluded from market prices. Businesses are unlikely to spend money on environmental resource protection when they cannot charge their customers for that cost. Put differently, underpriced resources are inadequately protected by the market system. (This issue is discussed in greater detail in Chapter 4.)

Goals and Means of Environmental Policies

Two questions are widely debated when developing environmental policies. The first, *What is the goal?,* seeks to determine the desirable degree of environmental protection. Consider the following three examples:

- How clean is clean enough? Is a 7-million ton per year reduction in sulphur dioxide emissions adequate? Should it be only 5 million tons? Perhaps it should be 10 million tons.
- Suppose we want to protect a certain endangered species from going extinct in the wild. Do we want to know with 80% confidence that the species will survive in the wild? Or is 95% confidence enough? Or do we want 100% confidence?
- Annual global carbon emissions of *greenhouse gasses* are about 7 billion tons. They are believed to contribute to global warming. Reducing those emissions will reduce the rate of increase in the earth's temperature. Should those emissions be unchanged at 7 billion tons per year, reduced by 10% per year, or reduced by some other amount—perhaps scaled back to 1990 levels.

The second question, *What are the means?,* seeks to find the best methods to achieve the goal. Once a goal has been established, usually there are many methods by which to achieve it:

- Is the best public policy for reducing sulfur dioxide emissions by 8 million tons a tax, a legal liability plan, command-and-control regulations, or an emissions trading program?

- Is the best way to have 99% confidence in the species survival to establish a breeding program, to reduce timber harvests, or to set aside its current habitat as a nature preserve?
- If carbon emissions are to be reduced to 1990 levels, should all nations establish a uniform carbon tax? Or would it be cheaper to plant adequate numbers of trees to absorb all the unwanted carbon?

Economics sharpens the way we think about determining the *goals* themselves (question 1) as well as the *means* for accomplishing the goals (question 2).

Ethics versus Externalities

Global warming is believed to be caused by greenhouse gasses (see Chapter 15). Think about the economic causes of global warming. You might ask, "Why don't profit-making firms simply reduce greenhouse gas emissions on their own? Are they unethical?" Many environmentalists believe that people in private industry are ethically impure. Unfortunately, this belief misdiagnoses the problem, and even if the diagnosis were correct, it would not help solve the problem. After some research, we find that private firms in a market economy may indeed wish to limit greenhouse gas emissions, but paying the costs of that control may drive them into bankruptcy. Businesses avoid controlling greenhouse gas emissions because the environmental benefits of that control do not appear in the firms' profit-and-loss statements. Yes, they might get some economic benefit from better public relations. However, most of firms' damages to the environment, as well as most of the benefits of controlling those damages, are received by somebody outside the firms. For this reason, economics teaches us to think of greenhouse gas emissions and other forms of environmental pollution as a problem of externalities.

Benefits of Environmental Protection

Answering questions about goals requires knowledge about the benefits of environmental protection. In concept, the benefits of environmental protection are the value of the environmental damages that are avoided. The benefits are not merely what social philosophers or newspaper editors say society needs. Economic benefits of environmental protection are equal to the sum of the actual sacrifices that people are prepared to make—what they are willing to pay (see Chapter 7).

Economics teaches us that the benefits of environmental protection are measured as the willingness to pay for it or willingness to accept compensation to take the place of environmental protection. We measure those benefits by finding out what people are willing to give up to have a clean environment, but how can we possibly measure that? Since the late 1940s, through much detective work, cost, and effort, economists have developed a series of methods for determining the economic value of environmental protection. Some of these methods have unusual names: the household production method, hedonic property method, the travel cost method, contingent behavior method, and contingent valuation method. None is perfect. Year by year, advances in the development and application of these methods reveal more about the benefits of environmental protection in rigorous, testable, economic terms.

The Costs of Environmental Protection

If you polled people about the economic costs of environmental protection, most would respond in terms of what shows up in the budgets of government enforcers, typically the monetary costs of enforcement. For the most part, however, those budget numbers are a tiny part of the costs, somewhere from 3% to 5% or, occasionally, up to 10% of the total real costs. Most of the costs of environmental protection are the private sector costs for new capital equipment required to comply with environmental regulations. These make up 80% to 90% of the real cost of environmental protection, and these costs are passed on to consumers through higher prices.

Figure 2–4 Total cost of greenhouse gas mitigation (GHG)

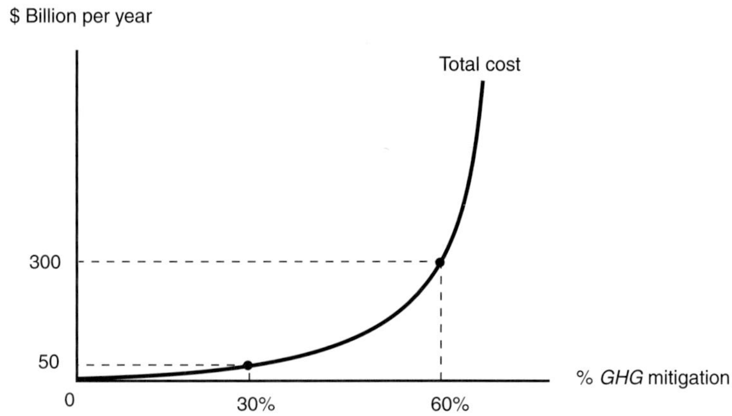

Figure 2–4 shows that when seeking ways to clean up the environment, it is always best to begin with the least expensive environmental cleanup. As we move to stricter and stricter environmental standards, we discover that not only do the total costs of environmental protection increase, but the incremental (marginal) costs of environmental protection increase. For example, the cost to move from a 10- to a 12-million ton reduction of sulfer dioxide for acid rain prevention in the United States has been estimated to be five times greater than the incremental cost to go from an 8- to 10-million ton reduction. Figure 2–4 shows rapidly rising marginal costs for controlling greenhouse gas emissions, and marginal costs matter most in setting environmental policy.[5]

How Much Protection?

Policymakers often want to know how much environmental protection should occur. This question is especially important when the cost of achieving added protection beyond current levels is high. One way to answer this question comes from thinking about the added benefit versus the added cost of additional protection that is, marginal analysis. Simpson (1998) summarizes the importance of marginal analysis. *Marginal benefit* is the added benefit of one more unit of environmental protection. *Marginal cost* is the added cost. If marginal costs are increasing with stricter environmental standards and if marginal benefits are decreasing, it means that at some intermediate level the marginal benefit of additional cleaning will equal the marginal cost. This is shown in Figure 2–5. At that level, the difference between total benefits and total costs is at a maximum. (This difference is sometimes referred to as *net benefit,* as described in Chapter 5.) When this difference is maximized, we have achieved what is called economic efficiency (see Chapter 3). Information on economic efficiency allows us to use economics when we face this important public policy question: How much environmental protection is enough? To answer this question, we must determine how much protection is optimal in achieving a level of protection that produces benefits that exceed costs by the most.

Every day we plan and carry out personal environmental policy decisions; for example, how often to vacuum the floor. Your living room rug may not be perfectly clean. You probably don't devote all your energy and time into cleaning your rug because you have other uses for those resources. You stop cleaning your rug at the point where the marginal benefit of further cleaning equals the marginal cost of other opportunities foregone because of that cleaning. But for the manufacturer of computer chips, the story is different. The principles are the same, but the result is different. The plant manager makes sure that the floor of the chip assembly room has a much higher standard of cleanliness than a living room rug. Is that because the plant manager is lazy at

[5] Total costs also matter. Costanza et al. (1997) describe total values of the world's ecosystem services, values which would be the costs of losing them all.

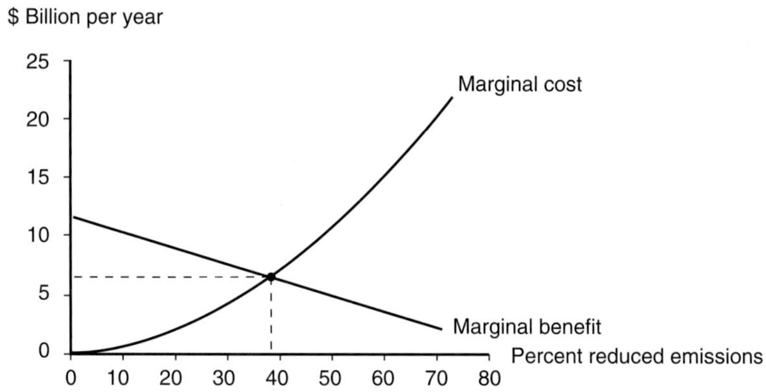

Figure 2–5 Marginal cost and marginal benefits of emissions mitigation

home, but not at work? No. The manager uses the same principle in both places. The marginal benefit of controlling computer chip pollution equals the avoided loss of a batch of dirty chips. This avoided loss is much greater than the loss of an unclean living room.

Setting Protection Goals: Benefit-Cost Analysis

People often balance added benefits against added costs.

The use of this idea, marginal benefits versus marginal costs, is called economic efficiency (see Chapters 3 and 4). It is based on the observation that people at home or at business do things that maximize their net benefits. The efficiency standard involves much more than saying benefits should exceed costs; it says that the difference between benefits and costs should be as large as possible. We may design or discover an environmental program that has benefits greater than costs. If we are flexible, however, we might find that by relaxing or tightening the environmental standard, the *difference* between benefits and costs could be increased.

Efficiency is not the only important environmental standard. Another is equity: How are these benefits and costs distributed? An environmental policy may be efficient in terms of maximizing net benefits in total, but the policy may have socially unacceptable consequences for particular income classes or geographic areas. Most of us would find it unacceptable that poor people pay all the environmental costs of living near a toxic waste dump (see Chapter 18). This leads us to the question of the means.

Achieving Protection Goals: Cost-Effectiveness Analysis

Once we have chosen how much environmental protection is enough, we must decide the best way to achieve that goal. Here, again, economic analysis is a sharp and powerful tool to help identify the least-cost method. If we decide our goal is a 25% reduction in greenhouse gas emissions, we can try to carry out that goal as cheaply as possible. The task amounts to seeking the most cost-effective means and uses a method called cost-effectiveness analysis (see Chapter 5).

Two kinds of environmental policy instruments have predominated. *Command-and-control approaches* have been used in most parts of the world for environmental protection. In these approaches governments tell firms *how much* to protect the environment and *how to carry out* the protection. Command-and-control approaches tend to be costly because they lock in existing technologies, and block incentives that could release creative urges and impulses to develop new technological solutions. Since the early 1990s, more flexible approaches have received considerable attention. Typically, these are market-based approaches. The idea behind them is to harness the power of market forces to protect the environment. The challenge for regulators, policymakers, political leaders, polluters, and environmentalists is to select the most cost-effective policy instrument for each environmental problem.

■ Environmental Policy Goals

Economics offers a powerful framework for *designing, applying,* and *evaluating* environmental and natural resource decisions. Despite the power of this framework, it cannot be applied without a clear statement of the goals to be accomplished. For that reason, several possible environmental policy goals are described next.

Human-Centered Goals

In Western nations, human-centered philosophical traditions have played a major role in environmental and natural resource policies. Three such traditions are: the egalitarian, elitist, and utilitarian.

Egalitarian. Egalitarianism, represented in the writings of early philosophers such as Immanuel Kant (1724–1804) and more modern writers such as John Rawls (1921–2002), is guided by the ethical standard that the excellence of a society is measured by its worst-off person. Environmental policy A is preferred to policy B, according to that philosophy, only to the extent that policy A improves the satisfaction of the worst-off or most vulnerable members of society by a greater amount than does policy B. As described in Chapter 18, placing toxic dumps in poor people's neighborhoods fails the test of that ethical system.

Elitist. Elitism, represented by the writings of the German philosopher Friedrich Nietzsche (1844–1900), states that the welfare of society can be best accomplished by investing in society's most talented people. According to this philosophy, policies would be implemented that improved opportunities for people who have the greatest potential for contributing to society. An example of an elitist environmental policy is the use of public resources to preserve access to unspoiled nature for people who, by virtue of their education, talents, or sensitivities, could gain the most from it. The acceptance of elitist philosophy by public policymakers, even in the modern Western democracies, is more common than might be imagined. For example, most flagship universities in the United States and western Europe limit admission to students with the best academic-performance records. This policy is based, at least in part, on the philosophy that investing scarce resources in elite students contributes more to society's future leadership, commerce, and the arts than democratic admission policies.

Utilitarian. The utilitarian ethic is represented by the writings of English political philosophers Jeremy Bentham (1748–1832) and John Stuart Mill (1806–1873). A utilitarian ethical system requires that public policy result in the greatest satisfaction (utility) and/or least displaced satisfaction for the people affected. For a utilitarian environmental policy to be ethically correct, the utility gained by that action should exceed the utility lost when summed over all affected members of society. Basing policy actions on this philosophy would support any program that promoted economic benefits exceeding costs to the affected population. Historically, the strong influence of utilitarianism in nineteenth-century Britain led to the introduction of progressive social and safety laws. Economic efficiency, the single most widely studied policy goal studied by economists, is based on utilitarian philosophy.

Nature-Centered Goals

A nature-centered goal is based on the idea that environmental and natural resource policies should serve the welfare of nature. Four such goals are the land ethic, deep ecology, animal rights, and ecofeminism.

The Land Ethic. The land ethic was formulated by the famous wildlife ecologist Aldo Leopold (1886–1948).[6] Leopold developed his land ethic over a lifetime of observation and interpretation. Leopold's significance in formulating environmental and natural resource policy goals lies in his unique attempt to devise an ethic from a foundation of ecological science.

Leopold is best known as the author of *A Sand County Almanac* (1949), a volume of philosophical ecological essays dealing with the connection between people and land. He states his land ethic in the following way: "A thing is right when it tends to preserve the integrity, stability, and beauty of the biotic community. It is wrong when it tends otherwise" (pp. 224–225).

Four fundamental principles compose Leopold's land ethic:

1. Land (an ecosystem) is a system of interdependent parts that are best regarded as a community, not a commodity.
2. People are members and not masters of the land community.
3. The whole informs the part (i.e., we can only understand and appreciate our place in nature and the place of our fellow living creatures by way of understanding the whole).
4. As stewards of the land, our duty is to preserve the integrity, stability, and beauty of the land.

Leopold (1949) made this observation about the problem of natural resource depletion: "We abuse land because we regard it as a commodity belonging to us. When we see land as a community to which we belong, we may begin to use it with love and respect." Leopold saw that a reversal of natural resource depletion would not occur without a change in attitudes, which would require that "a land ethic changes the role of *homo sapiens* from conqueror of the land-community to plain member and citizen of it. It implies respect for his fellow-members, and also respect for the community as such" (1949, p. 20).

Deep Ecology. The foundations of a philosophy known as deep ecology, an outgrowth of Leopold's land ethic, were laid in 1973 by Arne Naess, professor of philosophy at Oslo University in Norway (Naess, 1973). In asking searching questions about the moral standing of human and nonhuman life, he identified eight principles that have since been since called deep ecology:

1. All life has value in itself, independent of its utility to people.
2. Richness and diversity contribute to life's well-being and have inherent value.
3. Humans have no right to reduce this richness and diversity except to satisfy vital needs responsibly.
4. The impact of people on nature is excessive and getting worse.
5. Human lifestyles and population are key parts of this impact.
6. The diversity of life, including human culture, can flourish only with reduced human impact.
7. Essential ideological, political, economic, and technological structures must therefore change.
8. People who accept the previous principles have a moral obligation to help implement needed changes and to do so peacefully and democratically.

Although most of us would agree that these are noble and lofty principles, they are not yet the foundation of any nation's environmental policy. It may take considerable increases in education and per capita income, as well as a reduction in personal struggles with nature, before national decision makers base environmental policies on these principles. If they are accepted, environmental and natural resource policies will look very different from policies of today.

[6] Leopold received his Master of Forestry from Yale University in 1909. After that he worked for 15 years in the U.S. Forest Service in Arizona, New Mexico, and Wisconsin. In 1928 he left the forest service and contracted his talents to conduct wildlife game surveys. In 1933 he was appointed professor of game management in the agricultural economics department at the University of Wisconsin, Madison, where he taught until he died in 1948.

Animal Rights. An animal rights philosophy maintains that at least some nonhumans have rights that function similar to human rights. This philosophy guarantees that relevant animal interests are protected and may not be sacrificed for the benefit of people, even if the animals whose interests are at stake are exploited humanely. The animal rights philosophy rejects animal exploitation and the regulation of that exploitation; instead, it calls for its abolition. The philosophy rejects the treatment of animals as means to human ends or as the property of people. According to this philosophy, animals have moral standing that must be respected regardless of the consequences to humans. Animal rights philosophers reject the use of animals in experiments or for human consumption, but not simply because these activities cause animals to suffer. Such uses, in the view of animal rights philosophers violate the basic obligations of justice that people owe animals.

Ecofeminism. Ecofeminism is a label attached to a worldview that sees many people living unsatisfying lives because of male domination of women, the poor, and the environment.[7] This worldview, only partly a nature-centered view, attempts to understand how Western philosophical, economic, cultural, and religious systems have historically identified women, the poor, the underrepresented, and nature as objects to be controlled and manipulated. Some ecofeminists draw a connection between ecological concerns and justice, by which people who are concerned with a lack of justice may take an interest in ecological issues (see Chapter 18). Kirk (1997) examines the relationship between ecofeminism and environmental justice.

Several kinds of behaviors, seen or experienced, may motivate people to embrace ecofeminism: rich people dumping their trash in poor neighborhoods, women in third-world villages walking great distances daily to find safe drinking water for their families, and native cultures pushed from forested land that was cleared for industrialization. Destruction of the land, water, and air may be highly correlated with the oppression of people who must live in polluted environments. Although no single ecofeminist philosophy exists, one may emerge that calls for policies that treat all living beings as equals, whose needs and interests shape the design of economic, political, and religious systems.

Nature- versus Human-Centered Goals

Some people see nature from an anthropocentric, or people-oriented, point of view. This viewpoint gives higher standing to humans, human needs, and human satisfaction and lower standing to nonhuman needs. It assumes that the nonhuman part of the environment is an instrument to be used by people as they see fit. The environment is viewed as a resource defined in terms of the benefits it provides to people. The nonhuman world is viewed as having instrumental value to people only.

The biocentric view sees all natural organisms, species, and ecosystems as equally important. Although the biocentric view recognizes that human desires and human values are important, it sees them as a part of a larger natural scene. It believes living nature in various forms has moral standing on its own, independent of its instrumental value to people. Some have characterized the biocentric view as the essence of environmentalism.

Two people with an identical mix of anthropocentric and biocentric values may make very different choices depending on the cost and constraints they face.[8] For example, suppose two people, a European and an African, agree that their children outrank elephants morally and that elephants have moral standing over ivory chess sets. Despite this agreement on values, the European denounces as immoral hunting elephants for their ivory tusks to make ivory chess sets, and the African defends the practice because the ivory revenues produced by his hunting lets him feed and clothe his children. Although both people have identical values, they face widely different costs of implementing those values. For the European, a ban on elephant hunting means no ivory chess set, but a nearly perfect substitute—wood chess set—can be bought in place of the ivory

---

[7] See, for example, Agarwal (2001) for analysis of village environmental programs in Nepal and India.

[8] Spash (1997) examines the role of ethical belief systems on individual attitudes toward the environment.

set with little compromise or cost to personal values. For the African, a ban on elephant hunting takes away his means to support his children.

The long history of human struggle with disease, hunger, wildfire, drought, floods, plagues, storms, earthquakes, and the like may have produced a human-centered view that sees wild nature as a threat to life. People may see this threat as most (least) dangerous when their resources for living and working with an unpredictable nature are most scarce (abundant). Since the mid-twentieth century, in Western industrialized nations, the fear of uncontrolled nature and the perceived need to control it has been reduced. In industrialized nations people with high incomes who believe they have secure futures are often insulated against the unpredictable side of nature. This insulation may give them the luxury of seeing the beautiful side of nature and embracing biocentric views. People who are least insulated from the threats of nature may hold the most anthropocentric views. Held values are complex, however; those most insulated from nature (e.g., people living in big cities) may also have strongly anthropocentric views. The anthropocentric view is associated by some with Jewish-Christian beliefs; indeed, one passage of the Bible's book of Genesis[9] encourages the rule of people over nature.

Summary

Economics gives us insights into understanding human behavior. Through those insights, economics helps us design institutions for managing natural resources and environmental quality. By influencing behavior through economic incentives, we can better meet the objectives of sustaining the needs of both current and future generations. Economics also studies the choices we make in allocating scarce resources among competing demands. Difficult choices are part of most environmental and natural resource policy questions.

The market system is one mechanism to address the three problems of what, how, and for whom. Consumers' wishes, transmitted through the market mechanism, attract resources to produce *what* people want. *How* things are produced in a market economy (technologies that are selected) is determined by competition among producers who must minimize costs to stay in business. The distribution of income *for whom* in a market economy is determined by who owns productive resources like land, natural resources, labor, and capital. That distribution of income is also determined by how valuable those resources are in meeting the desires of consumers.

Two questions are widely debated in environmental policy circles. The first, *what is the goal?*, asks what is the desirable degree of environmental protection? Benefit-cost analysis is one way to set environmental protection goals. The second question, *what are the means?*, asks what are the best methods to achieve the goal? Once a goal has been established there are usually many methods by which to achieve it. Cost-effectiveness analysis is one way to achieve environmental protection goals.

Environmental and natural resource policy goals can be based on a nature-centered view of the world as well as on a human-centered view. The human-centered philosophies of egalitarianism, elitism, and utilitarianism have been important in western countries. Nature-centered goals include land ethic, deep ecology, animal rights, and ecofeminism. The goal of economic efficiency is based on utilitarian philosophy of the greatest good for the greatest number of people. This philosophy has had widespread support among both economists and policy-makers in most countries throughout much of the twentieth century.

Review Questions

1. Why is it important to make good decisions about environmental protection and natural resource management when those resources are scarce?

2. Explain what the market mechanism is and describe how the market solves the problems of what, how, and for whom.

3. Why is it important to know if profit incentives can protect the environment and natural resources?

4. What is the major limitation of the market mechanism in regard to environmental protection and natural resource management?

5. Explain the meaning of the *goal* of environmental protection and natural resource management. Use an example to demonstrate how economics can help set such a goal.

6. Explain the meaning of this question: What are the best *means* to achieve the goals of environmental protection and natural

[9] Chapter 1, Verse 28: "And God blessed them, and God said unto them, Be fruitful, and multiply, and replenish the earth, and subdue it: and have dominion over the fish of the sea, and over the fowl of the air, and over every living thing that moveth upon the earth." There are many interpretations of this statement. Some say that rule and dominion over the earth is that of a steward or a caretaker, not a reckless exploiter.

resource management? How are the means for achieving the goals different from the goals themselves?

7. Explain what is meant by the *benefits* of environmental protection. How are they different from the *costs* of environmental protection?

8. How can a measurement of costs and benefits be used to set the desired level of environmental protection?

9. Why are environmental and natural resource policy goals important to state before formulating those policies?

10. Explain the egalitarian philosophy and apply it to an environmental policy.

11. Explain the elitist philosophy and apply it to environmental policy.

12. Describe the utilitarian ethic and explain its importance to environmental and natural resource policies.

13. Describe nature-centered philosophy.

14. What are the ethical principles of Leopold's land ethic? What kinds of environmental policies would be favored by people who support the land ethic?

15. Summarize the deep ecology philosophy and describe the kinds of policies it calls for.

16. Explain the animal rights ethic and describe its likely future importance in environmental policy.

17. What is ecofeminism? Describe its call for action.

18. Describe the conflicts likely to occur between nature-centered and human-centered goals.

19. Show how two people with identical preferences for "people versus nature" can choose opposite outcomes where resources are scarce.

References

Agarwal, B. 2001. Participatory exclusions, community forestry, and gender: An analysis for South Asia and a conceptual framework. *World Development.* 29 (10): 1623–1648.

Bullock, C. H., D. A. Elston, and N. A. Chalmers. 1998. An application of economic choice experiments to a traditional land use—deer hunting and landscape change in the Scottish Highlands. *Journal of Environmental Management.* 52 (4): 335–351.

Castaneda, B. E. 1999. An index of sustainable economic welfare (ISEW) for Chile. *Ecological Economics.* 28 (2): 231–244.

CBC News. 2004 (March). *Ottawa asked to save B.C.'s spotted owls.* Available from *http://www.cbc.ca/stories/2004/03/02/canada/spottedowl_040302* [Accessed October 14, 2004].

Costanza, R., R. dArge, R. deGroot, S. Farber, M. Grasso, B. Hannon, K. Limburg, S. Naeem, R. V. O'Neill, J. Paruelo, R. G. Raskin, P. Sutton, and M. vandenBelt. 1997. The value of the world's ecosystem services and natural capital. *Nature.* 387 (6630): 253–260.

Dasgupta, P. 1996. The economics of the environment. *Environment and Development Economics.* 1 (4): 387–428.

Haspel, A., and F. R. Johnson. 1982. Multiple destination trip bias in recreation benefit estimation. *Land Economics.* 58 (3): 364–372.

Kirk, G. 1997. Ecofeminism and environmental justice: Bridges across gender, race, and class. *Frontiers—A Journal of Women's Justice.* 18 (2): 2–20.

Lele, S., and R. B. Norgaard. 1996. Sustainability and the scientist's burden. *Conservation Biology.* 10 (2): 354–365.

Leopold, A. 1949. *A Sand County Almanac and Sketches Here and There.* New York: Oxford University Press.

Lin, S. J., and T. C. Chang. 1996. Decomposition of SO_2, NOx and CO_2 emissions from energy use of major economic sectors in Taiwan. *Energy Journal.* 17 (1): 1–17.

Loomis, J. B., and A. Gonzalez Caban. 1997. Comparing the economic value of reducing fire risk to spotted owl habitat in California and Oregon. *Forest Science.* 43 (4): 473–482.

Montgomery, C., and G. M. Brown. 1992. Economics of species preservation: The spotted owl case. *Contemporary Policy Issues.* 10: 1–12.

Naess, A. 1973. The shallow and the deep, long-range ecology movement: A summary. *Inquiry.* 16: 95–100.

Schelling, T. 1992. Some economics of global warming. *The American Economic Review.* 82 (1): 1–14.

Simpson, R. D. 1998. Economic analysis and ecosystems: Some concepts and issues. *Ecological Applications.* 8 (2): 342–349.

Smith, A. 1776. *An Inquiry into the Nature and Causes of the Wealth of Nations.* Edinburgh, Scotland.

Spash, C. L. 1997. Ethics and environmental attitudes with implications for economic valuation. *Journal of Environmental Management.* 50 (4): 403–416.

U.S. Congress. 1973. *The endangered species act.* Available from *http://endangered.fws.gov/esa.html* [Accessed October 14, 2004].

U.S. Fish and Wildlife Service. 2004. *Northern spotted owl, Strix occidentalis caurina.* Available from *http://arcata.fws.gov/es/birds/ns_owl.html* [Accessed October 14, 2004].

Chapter 3

ECONOMIC THEORY

Introduction

Sometime around 1740 Adam Smith entered Glasgow University to study moral philosophy. In his mid-20s he began wondering about what it takes to increase a community's wealth. In 1776, Smith moved to London where he published his famous book *An Inquiry into the Nature and Causes of the Wealth of Nations.* In that book, he coined the expression "the invisible hand," which he used to demonstrate how *individual* self-interest guides the most efficient use of resources for a *nation,* with increased *public* welfare coming as an *unintended* by-product.

Smith convincingly explained how individual self-interest was an underlying force in the economy and how competition served as its regulator. (For a recent examination of the role of self-interest in environmental policies, see Gintis, 2000). He set forth the lesson that all economists have been taught since: Unregulated markets, motivated by self-interest, can serve society's material needs very well indeed. Through his book, Smith was influential in creating today's discipline of economics and serving as an intellectual and political foundation for many nations' commercial policies. In relation to this textbook, Adam Smith's writings raise an important question: Do unregulated market forces tend to damage natural resources and the environment? Or can market forces be harnessed to protect natural resources and the environment? Answers to these questions are debated by environmentalists, economists, landowners, and political leaders worldwide.

Economic theory provides the foundation for understanding an economic approach to the environment, so that we can better interpret the facts we see. Obtaining facts is important, but simply having them is only the beginning. Economic theory provides the standard that lets us interpret the facts. Economic theory provides the source of questions (maintained hypotheses) and tells us how to interpret answers to the questions (results of hypothesis tests). This chapter develops the concepts economists use to understand today's environmental and natural resource problems and to develop policy solutions.

Outline

Competitive Markets 36

Demand 37

Supply 42

Equilibrium Demand and Supply 46

Accomplishing Environmental Goals: Markets versus Governments 54

Objectives

After studying this chapter, you should be able to

- Explain how competitive markets channel consumer preferences into productive action
- Describe the factors that influence demand
- Describe the factors that influence supply
- Explain how demand and supply work together to determine market equilibrium
- Compare the performance of markets and governments as institutions for managing society's environmental resources

■ Competitive Markets[1]

Chapter 2 introduced the three problems every economy faces:

- *What* will be produced out of the variety of potential goods and services and in what quantities?
- *How* will the economy combine its various productive resources—land, labor, capital—to produce each good?
- *For whom* shall goods be produced; that is, how shall the economy's total output be distributed among the different people who contributed to its production?

In a market economy, consumers are like voters, using their money like votes to buy what they want when they want it. A major task is to analyze how the spending of money turns consumers' desires into productive action in a market system.

In 1797 the first survey on the state of the poor in England found that up to 75% of a poor family's income was spent on food, mostly bread and potatoes (Eden, 1797). For most of us, things are better. Suppose today that we are rich enough to afford fresh fruit every day if we want it. If we would otherwise be eating mostly bread and potatoes, as in eighteenth-century England, how does our wish to substitute fresh fruit for bread and potatoes get turned into action? Do we call local politicians and ask them to tell farmers and ranchers to grow less wheat and plant fewer potatoes, while encouraging them to plant more fruit orchards? How many and what kinds of government policies do politicians decide are needed to get farmers to transfer land from wheat and potato production to fruit orchards? And how do we communicate to the fruit growers that consumers should get the kind of fruits they want? Actually, there is little sense in pursuing this line of questioning because things do not work this way in a market economy. Here is what actually happens.

As incomes rise, consumers begin to buy less bread and more fruit. As fruit starts disappearing from the shelves and bread grows stale, sellers raise the price of scarce fruit and lower the price of bread. Wheat farmers begin to experience losses and fruit growers increase their profits. Fruit pickers find they can demand higher wages; meanwhile, grain workers from the farm to the flour mill start to leave their jobs for better pay elsewhere. Over time, increased fruit prices attract increased production of more fruit.

As people's needs, desires, and incomes change, as technical methods of production change, as supplies of natural resources and the capacity to protect the environment change, changes occur in the prices and quantities of commodities and services sold. A market economy advances standards of living and enforces order through a system of prices that signals producers and consumers to act in their own self-interest. An understanding of how supply and demand works "like an invisible hand" to produce what people want through an efficient combination of resources is an essential insight into how the world of markets works.

A market is an institution in which buyers and sellers come together to trade goods, services, or resources. Many forms of markets exist: the local car dealer, a favorite grocery store, a software store, a farmer's fruit stand. The Tokyo, London, or New York stock exchanges are markets where buyers and sellers of stocks and bonds communicate to trade simultaneously from all over the world. Local auction houses attract prospective buyers and sellers of used appliances, art, livestock, and jewelry. Northern Colorado and southern New Mexico have active water markets in which thirsty cities buy water from water-rich farmers so that both can survive the latest drought. Each year, thousands of acre-feet are traded in this market through private, voluntary transactions. Any institution that links prospective buyers with motivated sellers is a market. Some markets are local, while others are regional, national, or global. Some markets are up-close and personal, involving face-to-face contact; in other markets buyer and seller never meet.

This chapter focuses on markets made up of large numbers of independently acting buyers and sellers of homogeneous products. Think of these as the large, impersonal markets (e.g., a na-

[1] Much of this chapter contains review material, which can be skipped by those of you who retained your principles of microeconomics. For another review, see Caviglia-Harris, 2003.

tional grain exchange, a commodity market, or a market for livestock) in which the price is determined by the joint decisions of sellers and buyers. They are *not* the markets in which a small number of influential producers set prices.

DEMAND

Demand is desire backed by purchasing power. There is no demand when a consumer has a great desire for a product but lacks income to buy it. There also is no demand when a consumer who has a large income has no desire for the product. Demand is summarized by a schedule that shows the amount of a product a consumer will buy at each possible price during a given time period. Demand is not a single point but an entire schedule. It shows how quantities of a product that will be bought change according to the price, other things being equal. The quantity demanded does not stand alone. Knowing the quantity demanded requires knowing the price. Table 3–1 is a hypothetical demand schedule for a single consumer purchasing organically grown grapes.

The table shows the relation between possible prices for organic grapes and the quantity of grapes a consumer will buy at each possible price. If the price of grapes is $5 per pound, the consumer will buy 10 pounds per month; if it is $4, the consumer will buy 20 pounds per month; and so forth. The table does not say which of the five possible prices for grapes will actually prevail. Actual price depends on demand interacting with supply. Demand is just a statement of a buyer's conditional plans; it summarizes buying plans, plans that are flexible enough to vary according to price. A demand schedule is like a grocery list that says buy *X* if the price is $5, but buy *Y* (more than *X*) if the price is $4.

A demand schedule summarizes buying plans for particular period such as a day, a month, or a year. Stating that a consumer will buy 10 pounds of grapes at $5 per pound gives little information. Stating that a consumer will buy 10 pounds of grapes *per month* at $5 per pound reveals something useful. We need to state a specific time period before we know whether the demand for a product is high, low, growing, or shrinking.

Law of Demand

All else being equal,[2] as price falls, the quantity demanded by consumers will rise, and as price increases, the quantity demanded will fall. In other words, there is an inverse relationship between price and quantity demanded. This inverse relationship is called the *law of demand.* Remember the assumption about all else being equal. Many influences other than the price of the product in question affect the amount the consumer will buy. For example, the quantity of grapes bought depends

Table 3–1 An individual buyer's demand for organic grapes

Price per pound (dollars)	Quantity demanded per month (pounds)
5	10
4	20
3	35
2	55
1	80

[2] The Latin phrase for "holding everything else constant" is *ceteris paribus.* The term is commonly used as a way to isolate important relations when conducting economic analyses, such as analysis of supply or demand. Holding everything else fixed (*ceteris paribus*) makes it possible to identify the cause-and-effect relation between two factors, such as price and quantity supplied. In fact, *ceteris paribus* goes beyond economics; it is a central feature of the scientific method, in which mental or actual experiments are conducted to piece together important puzzles. The use of *ceteris paribus* in economics makes it possible to isolate the effect one factor has on another in the formulation and testing of hypotheses and in the conduct of policy analysis based on those hypotheses.

not only on the price of grapes, but also on the prices of other fruits. The law of demand in this case says, if the price of grapes rises, fewer grapes will be bought, other things being equal. However, if the price of grapes and the price of all other competing (and complementary) fruits increase by some amount, say $1 per pound, consumers might buy more, less, or the same amount of grapes. There are three good reasons why we see an inverse relationship between price and quantity demanded.

First, the law of demand squares with our intuition. People usually buy more of a product at a low price than at a high price. Because the price must be paid out of a consumer's income, price is a barrier that restrains consumers from buying. The higher that barrier, the less of a product consumers buy; the lower the barrier, the more they buy. When you see a business having a clearance sale, you are seeing that it believes in the law of demand.

Second, in any given time period, a buyer of a good or service receives less benefit or satisfaction (utility) from each *added* unit of the product consumed. The first bottle of water produces a considerable amount of benefit, but the second produces less added benefit. The third bottle produces even less added benefit than did the second. Economists say that consumption is subject to diminishing marginal (added) utility. Because each increased unit of a particular good returns less and less *marginal* utility, consumers will voluntarily buy more units only if the price of those units is progressively reduced. (Remember, buying more of a good in the marketplace takes spendable income away from other consumption.)

Third, the law of demand can be explained by income and substitution effects. Imagine that sellers move the price of all goods one decimal point to the left (i.e., everything costs only 10% of its original price). Anything originally priced at $10 is marked down to $1. When all prices fall by a factor of 10, with unchanged money income, you can buy 10 times more than before: Your standard of living has increased by a factor of 10. The gain in real income produced by a falling price is called the *income effect.* Even when the price of a single good falls by a known amount, its income effect is measured by the increased purchases of the good produced by the consumer's increased real income. A higher price has the opposite effect. The *substitution effect* is the second reaction to a change in price. When the price of a good falls, consumers have the incentive to substitute that product for another whose price has not fallen. The substitution effect of a price decline is measured by consumers' increased purchases of the lower-priced good, even if the real income gain is removed. The product whose price falls becomes a better buy compared to other products.

For example, a reduced price for grapes increases the purchasing power of buyers' incomes, enabling them to buy more grapes (the income effect). At the reduced price, grapes are also relatively more attractive and buyers tend to purchase them instead of peaches, apricots, and strawberries (the substitution effect). The income and substitution effects jointly act on a budget, enabling and encouraging consumers to buy more of any product whose price falls.

Demand Curves

Economists represent a product's inverse relationship between price and quantity demanded on a simple graph. Quantity demanded is typically shown on the horizontal axis while price is measured on the vertical axis. Figure 3–1 shows a plot of the five price-quantity points shown in Table 3–1; it also connects those points with a curve, *D.* The curve (schedule), *D,* is called a *demand curve.* It slopes downward, showing the law of demand: Consumers buy more of a product or resource as the price falls. Figure 3–1 and Table 3–1 are two ways of showing the same facts. Of course, those facts can also be summarized in words.

Market Demand

Normally, markets consist of many individual buyers. Accordingly, an analysis for an individual demand can be used to determine the sum of those demands, called market demand. You can determine market demand by simply adding up quantities demanded by all consumers at each of the various potential prices. For example, Table 3–2 shows a market consisting of only two buyers. You can use the table to find the total quantity demanded by those two people at each

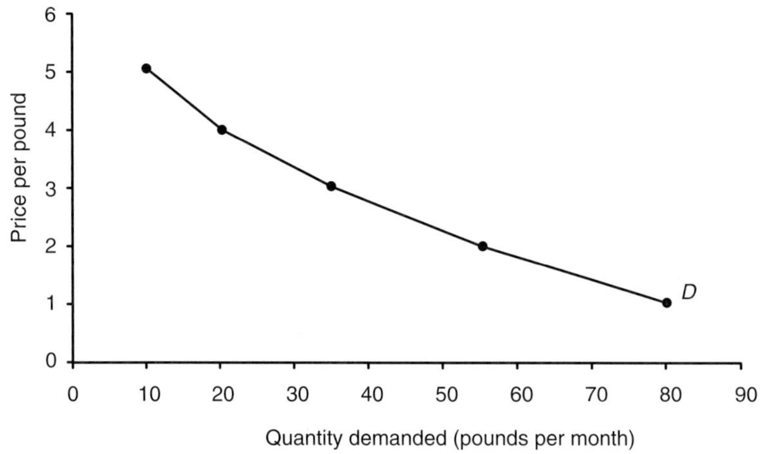

Figure 3–1 Individual buyer's demand for organic grapes

Table 3–2 Market demand for organic grapes, two buyers

Price per pound (dollars)	Quantity demanded per month (pounds)				Quantity demanded per month (pounds)
	First buyer		Second buyer		
5	10	+	12	=	22
4	20	+	23	=	43
3	35	+	39	=	74
2	55	+	60	=	115
1	80	+	87	=	167

price. Figure 3–2 shows what's called the horizontal summation method for finding market demand: At each price you sum all individual quantities demanded by the two participants ($D_1 + D_2$) to compute total market demand at that price (D); you then plot the price and total quantity demanded as a single point for the market demand curve. The term *horizontal summation* comes from the fact that total market demand is the simple (horizontal) addition of each of the two individual demand curves for any given price.

Real markets typically are made up of many more than two product buyers. To avoid thousands of horizontal additions, economists somtimes think of a typical (representative) buyer who buys a known amount of a product at each possible price. Based on that concept, you simply multiply the typical buyer's quantities by the total number of buyers in the market to compute the market demand. This is the method used to find the curve D_1 in Figure 3–3 for a market with 400 grape buyers whose demand is that shown in Table 3–1. Table 3–3 shows the calculations.

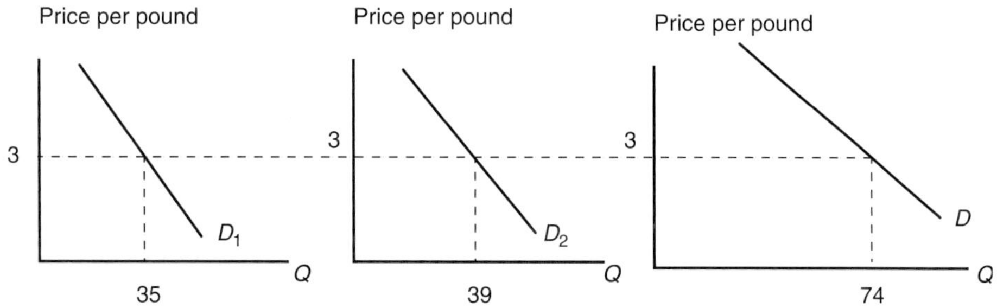

Figure 3–2 Market demand for organic grapes, two buyers

Figure 3–3 Changes in the demand for organic grapes

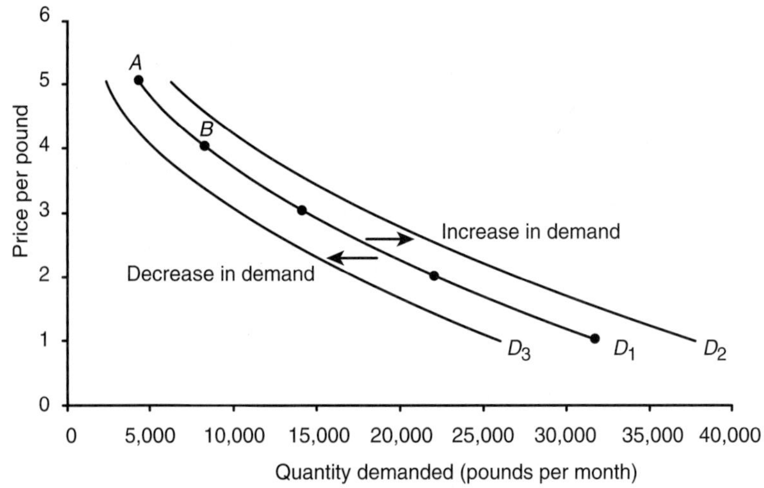

Table 3–3 Market demand for organic grapes, 400 buyers

1 Price per pound (dollars)	*2* Quantity demanded per month, single buyer (pounds)		*3* Number of buyers in the market		*4* Total quantity demanded per month (pounds)
5	10	×	400	=	4,000
4	20	×	400	=	8,000
3	35	×	400	=	14,000
2	55	×	400	=	22,000
1	80	×	400	=	32,000

Change in Quantity Demanded

Remember that demand is a schedule, so a change in quantity demanded is a movement along that schedule from one point to another, that is, from one price-quantity combination to another on that unchanged schedule. The change in quantity demanded comes from a decrease or increase in the product whose price we care about.[3] In Table 3–3, for example, a reduced price of grapes from $5 to $4 will increase the quantity of grapes demanded from 4,000 to 8,000 pounds per month. In Figure 3–3, the movement from point *A* to point *B* on curve D_1 represents a change in quantity demanded. The entire curve D_1 itself stays fixed, as it has not moved at all.

Change in Demand

A change in quantity demanded is quite different from a change in demand. A change in quantity demanded is a movement from one point on a fixed demand schedule to another such point in the face of a price change. A *change in demand* is a shift of the entire demand curve to the right (increased demand) or to the left (decreased demand). The demand curve D_1 in Figure 3–3 highlights the importance of price in the amount of a product purchased, but other factors have a major influence on consumers' purchases. For instance, your purchases of winter coats are influenced by their price, but no matter what the price is, you will buy few of them on hot summer afternoons.

---

[3] Berbel and Gomez-Limon (2000) and Gomez-Limon and Berbel (2000) show how price influences the quantity of water demanded by irrigation farmers in Spain.

These nonprice factors, called *demand determinants* (predictors), are assumed fixed when a graph like Figure 3–3 shows a demand curve like D_1.

Demand determinants are the other things we hold equal when graphically expressing the relation between quantity demanded and price. When any demand determinant changes, more or less of the product will be bought even if its price is held constant. For example, consumers buy more air conditioners per week when the outdoor temperature is 100°F than when it is a cool 50°F, even if the price of air conditioners stays the same. When a demand determinant changes the entire demand curve, the demand curve typically shifts either right or left. For this reason, "demand shifters" is a label commonly given to demand determinants.

The five demand determinants discussed next are number of buyers, consumers' tastes and preferences, consumers' incomes, prices of related goods, and consumer expectations about future prices and incomes. A change in one or more of these determinants of demand will change the demand data (the demand schedule) in Table 3–3 and, therefore, the location of the demand curve in Figure 3–3. A change in the demand schedule, or graphically a shift in the demand curve, is called a change in demand.

If consumers want to buy more grapes at each possible price than is reflected in column 4 in Table 3–3, that increase in demand is shown as a shift of the demand curve to the right, say, from D_1, to D_2. Conversely, a decrease in demand occurs when consumers buy fewer grapes at each possible price than is indicated in column 4 of Table 3–3. The leftward shift of the demand curve from D_1 to D_3 in Figure 3–3 shows such a situation. Now consider how changes in each determinant affect demand.

Size of Market. An increase in the size of the market (number of buyers in a market) increases demand. Reducing the number of buyers decreases demand. For example, increased public awareness of the importance of a clean and healthful environment has greatly increased the size of the market for green-certified wood products; therefore, the demand for these products has increased in recent years. (Green-certified wood products are made from timber harvested from sustainably managed forests. See Chapters 8 and 19.)

Tastes and Preferences. Changes in consumers' tastes and preferences that make a product more desirable increase the demand for that product at any price. Demand will increase and the demand curve shifts to the right. A good example is the recent increased demand for bottled water driven by changes in tastes and preferences. Interest in health and fitness has led people to buy and drink bottled water. Producers of bottled water often use sport and fitness images in their advertisements. In addition, the occasional, well-publicized incidents of drinking water pollution increase preferences for bottled water by reducing preferences for substitutes like tap water. A reduction in consumer preferences will reduce demand and shift a product's demand curve to the left.

Income. Income exerts a major influence on demand. For most products, an increase in income increases demand. Consumers typically buy more computer software, new homes, and organic foods as their incomes increase. Similarly, the demand for the same products falls as consumers' incomes decline. Products whose demand increases with growing income are called normal goods.

Not all products are normal goods. Some are inferior goods, defined as those goods for which demand falls with increased consumer incomes. As consumers' incomes rise beyond a certain level, the demand for old cars, low-rent apartments, previously worn clothing, potatoes, and low-cost cuts of meat decreases. Consumers substitute higher quality products for the goods their former incomes constrained them from buying.

Prices of Related Goods. A change in the price of a related good may either increase or decrease the demand for a product, depending on whether the related good is a substitute or a complement. A substitute good is one that can be substituted for another good; a complementary good is one that is used with another good. Fish and chicken are examples of substitute goods. When the price of fish rises, perhaps because of reduced fish populations (see Chapter 12), consumers buy less fish, increasing their demand for the substitute chicken. Similarly, as the price of fish falls, perhaps due to increased fish harvests, consumers buy more fish, reducing their demand for the substitute chicken. When two products are substitutes, the price of one and the demand for the substitute move in the same direction.

Gasoline (petrol) and motor oil are complementary goods. These are goods that are consumed together, so their demands are closely intertwined. If the price of gasoline increases and, as a result, consumers take fewer and shorter trips in their cars, then demand for motor oil is reduced. Because cars require both gasoline and oil, these two goods are jointly demanded and they are complements. So it is with bread and butter, tennis balls and tennis rackets, bottled water and running shorts. For two products that are complements, the price of one and the demand for the other move in opposite directions. Most goods are independent; that is, they are neither substitutes nor complements for each other. A price change in one has no appreciable influence on the demand for the other (e.g., bread and tennis rackets, tennis balls and television sets, sweet potatoes and bottled water). For two independent goods, a change in the price of one has no effect on the other's demand.

Expectations. Consumers' expectations about a product's future supply, demand, or price can have a major impact on the product's current demand. Changes in these expectations often shift current demand. For example, consumers who expect future gasoline shortages because of anticipated stricter environmental regulations for oil drilling will buy more gasoline now to beat the expected price rise. This increases the current demand for gasoline. If oil buyers expect an oil-producing country to be at war, thus reducing its future supplies of oil, these buyers may reason that the future price of oil will increase. So they may stock up on oil now to beat the anticipated future price hike. For similar reasons, expectations of falling prices for a commodity may encourage consumers to delay current purchases in anticipation of future lower prices. This reduces current demand.

A change in expectations regarding future income may prompt consumers to adjust their current spending patterns.[4] For example, fourth-year college students may buy a new car in anticipation of a lucrative professional career looming around the corner, or workers who fear losing their current high-paying jobs may reduce their demand for various luxury goods to increase their savings.

In summary, an increase in demand—the decision by consumers to buy larger quantities of a product at each possible price—may be caused by several factors: size of the market, change in tastes and preferences, change in income, change in price of related goods, and change in expectations. Table 3–4 summarizes the determinants of demand. The size of the market is the only factor that does not shift the *individual's* demand curve. It shifts the market demand curve because it adds more curves to the horizontal summation. The rest of the factors shift the market demand curve because each one shifts individual demand curves.

■ SUPPLY

Supply is a schedule or curve showing the quantities of a product that producers are willing and able to make available for sale at each of a series of possible prices during a specific period. Table 3–5 is a hypothetical supply schedule for a single producer (grower) of organic grapes. It shows the quantities of grapes that will be supplied at various prices, other things being equal.

Law of Supply

Table 3–5 shows that a positive or direct relationship prevails between price and quantity supplied. As price rises, the quantity supplied rises; as price falls, the quantity supplied falls. This re-

[4] Milton Friedman (1976) and Franco Modigliani (1985) won the Nobel Prize in economics for explaining current demand patterns by future expectations. See Royal Swedish Academy of Sciences (2004 a, b).

Table 3–4 Demand shifters: Factors that shift the demand curve

Determinant	Example
Change in buyer numbers	A decline in birthrate reduces buyers' demand for disposable diapers.
Change in buyer tastes	Physical fitness grows in popularity, increasing the demand for hiking boots and fad diets.
Change in income	Increases in per capita income raise the demand for normal goods like fine wines, air travel, and steak, while reducing the demand for inferior goods like cheap wine, intercity buses, and hamburger.
Change in prices of related goods	A reduced price of plastic packaging for soft drinks reduces the market price of recycled glass (substitute good). A reduced price of plastic packaging for soft drinks increases the quantity of soft drinks purchased (complementary good).
Change in expectations	Heavy winter and spring rains in California create an expectation of higher future prices of organic grapes, which increases today's demand for organic grapes.

Table 3–5 An individual producer's supply of organic grapes

Price per pound (dollars)	Quantity supplied per week (pounds)
5	60
4	50
3	35
2	20
1	5

lationship is called the *law of supply*. A supply schedule tells us that firms will produce and offer for sale more of their product at a high price than at a low price. Why do producers act this way?

Price is an obstacle from the view of the consumer who is on the paying end. The higher the price, the less the consumer will buy. But the supplier is on the receiving end of the product's price. To a supplier, price represents revenue, which serves as an incentive to produce and sell a product. The higher the price, the greater this incentive and the greater the quantity supplied. Consider a farmer who can shift resources among alternative products. As price moves up, as shown in Table 3–5, the farmer finds it profitable to take land out of wheat, oats, and soybean production and put it into grapes. The higher grape prices enable the farmer to cover the increased costs associated with more intensive cultivation and the use of more seed, labor, and water. The overall result is more grapes produced.

Now consider a manufacturer. Beyond some quantity of production, a manufacturer usually encounters increasing costs per added unit of output. Certain productive resources—in particular, a firm's plant and machinery—cannot be expanded quickly, so the firm uses more of other resources, such as labor, to produce more output. As time passes, however, the existing plant becomes increasingly crowded and congested. As a result, each added worker produces less added output, and the cost of successive units of output rises accordingly. The firm will not produce the more costly units unless it receives a higher price for them. Again, price and quantity supplied are directly related.

Supply Curves

As with demand, it is convenient to represent supply graphically. In Figure 3–4, curve S_1 is a *picture* of the market supply *numbers* given in Table 3–6. Those numbers assume there are 400

Figure 3–4
Changes in the supply of organic grapes

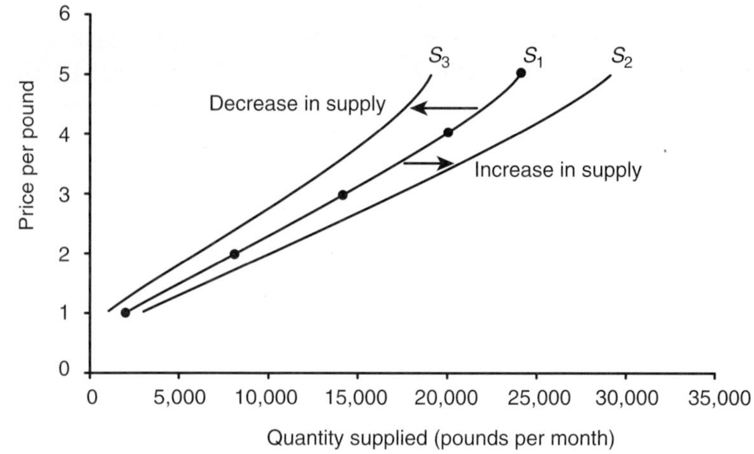

Table 3–6 Market supply of organic grapes, 400 producers

1 Price per pound (dollars)	*2* Quantity supplied per month, single producer (pounds)		*3* Number of sellers in the market		*4* Quantity supplied per month (pounds)
5	60	×	400	=	24,000
4	50	×	400	=	20,000
3	35	×	400	=	14,000
2	20	×	400	=	8,000
1	5	×	400	=	2,000

suppliers in the market, each willing and able to supply grapes according to the numbers in Table 3–5 shown for a single individual supplier. The *market* supply curve comes from horizontally adding the *individual* supply curves of all producers in the market. Note that the axes in Figure 3–4 are the same as those used in the graph of market demand (Figure 3–3), except for the change from "quantity demanded" to "quantity supplied" on the horizontal axis.

Change in Quantity Supplied

A change in the quantity supplied is a movement from one point to another on a fixed supply curve. The cause of that movement on the fixed supply curve is a change in the price of the specific product being considered. In Table 3–6 and Figure 3–4, a decline in the price of grapes from $4 to $3 decreases the quantity of grapes supplied per month from 20,000 to 14,000 pounds. That reduction in quantity supplied comes from a price reduction that reduces producers' incentives to supply to the market the quantity of product they supply at $4. That same change in quantity supplied is also shown by a movement along the supply curve S_1 in Figure 3–4, down and to the left. The complete supply schedule itself is a schedule of prices and quantities, and this schedule does not change when price changes.

Change in Supply

In building a single supply curve, the influence of price on quantity supplied to the marketplace is the central focus. Of course, many factors other than price influence what producers supply.[5]

[5] The supply curve is drawn on the assumption that these other things are fixed. If one of them does change, a change in supply will occur, meaning that the entire supply curve will shift.

Six determinants of supply are number of sellers in the market, resource prices, technology, taxes and subsidies, prices of other goods, and price expectations. A change in any one of these supply determinants, sometimes called supply shifters, shifts the supply curve for a product either left or right. A shift to the left, as from S_1 to S_3 in Figure 3–4, indicates a decrease in supply: Producers supply smaller quantities of the product at any given price. A shift to the right, as from S_1 to S_2, shows a supply increase: Producers supply more output to the market at each possible price.

The distinction between a change in supply and a change in quantity supplied parallels the distinction between a change in demand and a change in quantity demanded. Because supply is a schedule or curve, a change in supply means a change in the entire schedule and a shift of the entire curve. An increase in supply shifts the curve to the right; a decrease in supply shifts it to the left. The cause of a change in supply is a change in one or more of the determinants of supply. Let us consider how changes in each of the determinants affect supply. The key idea is that costs are a major factor underlying supply curves; anything that affects costs (other than changes in output itself) usually shifts the supply curve.

Seller Numbers. *Ceteris paribus,* larger numbers of producers will supply larger amounts of a product to the market, shifting the entire supply curve to the right. As more firms enter any given industry, the supply curve shifts to the right. Likewise, as firms leave an industry, the supply curve shifts to the left. For example, in the mid-1990s, the U.S. federal government imposed timber harvesting restrictions to protect the habitat of the northern spotted owl on federal forests in the Pacific Northwest. Those restrictions considerably reduced the number of timber harvesters and led to reduced timber supplies available for processing by the forest products industry. These restrictions effectively shifted the supply of timber to the left and produced a marked decline in the market supply of timber. See Murray and Wear (1998) for details.

Resource Prices. A producer's output comes from inputs, called resources. Resources are inputs like land, labor, capital, water, and a quality environment. Higher prices for resources raise production costs and, if the product price does not change, reduce profits. Those falling profits provide incentives for firms to reduce their supply of output at a given product price. For example, stumpage (e.g., uncut timber standing in the woods) is needed to produce lumber. An increase in the price of stumpage caused by nearby forest fires reduces the supply of lumber and, ultimately, reduces the supply of homes, wood furniture, and other wood products that use this lumber as an input. In contrast, reduced resource prices lower production costs, raise profits, and increase incentives to add supply to the market. So when a resource price falls, firms that use this resource in their production process supply greater output at any given product price.

Technology. Improvements in technology (knowledge applied to production) permit firms to produce a given quantity of output with fewer or cheaper resources. This reduces production costs and increases supply for a given product price. For example, improvements in the speed of microprocessors has reduced the cost of building computers. Therefore, computer suppliers offer larger and more powerful computers than previously at any given computer price; the supply of computers continues to increase.

Another example is improvements in technology of solar photovoltaics (PV), the direct conversation of solar energy into direct current electricity (see Chapter 13). PV systems first became recognized in the 1950s for powering space satellites. After the oil crisis in the early 1970s, PV systems were developed further for a range of applications. PV systems need no fossil fuel supplies and give off no pollution. At first the costs of PV systems were high, but with improved technology prices have steadily fallen. The stage has been reached when PV systems are both technically and economically suitable for many applications, particularly those involving relatively small amounts of power in remote locations, where the cost of operating and maintaining a conventionally powered system is high. Since 1990, PV systems have been installed in developing countries to supply power for water pumping, refrigeration, lighting, electrification, and communications. The supply of PV-generated electric power continues to increase.

Taxes and Subsidies. Producers typically treat taxes as costs of production. An increase in a sales, property, or excise tax increases production costs and reduces supply. Subsidies are the opposite of taxes. If a government subsidizes the production of a good, that government effectively reduces producers' costs and increases supply. The impact of taxes on supply have major implications for environmental policies. Governments can reduce the supply of products that pollute by taxing these products (see Chapter 16); this is an important policy lever government has at its disposal to reduce pollution.

Prices of Other Goods. Most business firms supply more than a single product. And even those that supply only one product can potentially switch to another product if that other product becomes more lucrative. Firms that produce a particular product, say, ordinary lumber, can sometimes use their plant and equipment to produce alternative products, say, green-certified lumber (see Chapter 8, and Little, 2002). An increased price of green-certified lumber might entice producers to switch production from ordinary lumber to green-certified lumber in order to increase profits. This substitution results in a decline in the supply of ordinary lumber. Alternatively, when the prices of ordinary lumber rise relative to the price of green-certified lumber, producers of those goods may decide to produce more ordinary lumber to increase their supply. Thus, the price of related goods has an important influence on the production of environmentally sensitive products.

Price Expectations. Changed expectations in regard to the future price of a product is likely to affect the producer's present willingness to supply that product. For example, producers of conventional cotton who expect the future price of naturally colored cotton[6] to increase might withhold some of their current cotton production from the market to prepare their soils to grow naturally colored cotton. This action would decrease the current supply of conventional cotton. Similarly, if people expect the price of green-certified lumber to increase, the supply of conventional lumber might decrease in the near term, raising current prices for conventional lumber prices. Table 3–7 summarizes supply determinants.

Equilibrium Demand and Supply

Now we are ready to bring together supply and demand to see how household buying decisions and business selling decisions interact. The outcome of that interaction is the observed price and quantity of the product bought and sold. Columns 1 and 2 of Table 3–8 reproduce the quantity and price from the market supply of grapes of Table 3–6. Columns 2 and 3 reproduce the price and quantity from the market demand for grapes of Table 3–3. This analysis assumes that the number of sellers and buyers is large enough that no single one has enough market power to set the price from its own actions.

Surpluses

Table 3–8 is simplified considerably by limiting the analysis to five possible prices. A compelling question is which one of the five will prevail as the observed market price for grapes. One way to find the answer is by experimenting with several prices. If we begin with a price of $5 per pound, we see immediately that this cannot be the actual market price. Why not? Because at that

[6] Many of today's cotton garments are made from cotton grown with chemical pesticides, bleached, then colored with chemical dyes. Naturally colored cotton is one ecofriendly alternative to conventionally grown cotton. It is grown without bleaching or harmful dyes. Unlike dyed material, the color of the fabrics made from colored cotton sometimes deepens with washing. Colored cotton is naturally pest and disease tolerant, which makes it easier to grow organically, and it is suitable for chemically sensitive people. Colored cotton provides new yarn and fabric design potentials. Without the need for abundant water or energy sources, new mill sites have many more choices for location. Recent growing demands for colored cotton has encouraged companies like Esprit and Levi's to launch green lines of cotton outerwear using dye-free, unbleached, organic green cotton. Colored cotton is now being grown in the United States, Europe, and Australia.

Table 3–7 Supply determinants: Factors that shift the supply curve

Determinant	Example
Change in seller numbers[a]	An increase in the number of Internet service providers increases the supply of such services; formation of women's professional volleyball leagues increases the supply of women's professional volleyball games.
Change in resource prices	A decrease in the price of microchips increases the supply of computers; an increase in the price of crude oil reduces the supply of gasoline.
Change in technology	Development of more effective wireless technology increases the supply of cell phones.
Changes in taxes and subsidies	An increase in the excise tax on cigarettes reduces the supply of cigarettes; a decline in subsidies to state universities reduces the supply of higher education.
Change in prices of other goods	An increase in the price of cucumbers decreases the supply of watermelons.
Change in expectations	An expectation of a substantial rise in future log prices decreases the supply of logs today.

[a] A change in seller numbers does not shift individual supply curves. Rather, it shifts the market supply curve because it adds more curves to the horizontal summation. The rest of the factors in this table shift the market supply curve because they shift the individual supply curves.

Table 3–8 Market supply of and demand for organic grapes

1 Total quantity supplied per month (pounds)	*2* Price per pound (dollars)	*3* Total quantity demanded per month	*4* Surplus (+) or shortage (−)	*5* Effect on price
24,000	5	4,000	+20,000	↓
20,000	4	8,000	+12,000	↓
14,000	3	14,000	0	none
8,000	2	22,000	−14,000	↑
2,000	1	32,000	−30,000	↑

price, producers are willing to produce and offer for sale 24,000 pounds of grapes per month, but buyers are willing to purchase only 4,000 pounds. The $5 price encourages producers to offer many grapes for sale, but discourages consumers from buying. The result is a 20,000 pound surplus, or an excess supply of grapes. This surplus, shown in column 4 of Table 3–8, is the excess quantity supplied over quantity demanded at $5. Grape producers find themselves with 20,000 pounds of unsold grape output that is rotting in the fields, warehouses, and grocery stores. A high price of $5 for grapes, even if it lasted for a short time, cannot last for long. Large surpluses of grapes encourage competing grape sellers to lower the price to encourage buyers to purchase the surplus, rather than to let the unsold grapes spoil.

Suppose the price drifts downward to $4. This reduced price has two effects on human behavior: It encourages buyers to buy more grapes and it encourages grape producers to offer fewer grapes for sale. The grape surplus falls to 12,000 pounds. However, because there is still a surplus, competition among sellers exerts more downward pressure on price. We can conclude that the grape prices of $5 and $4 cannot last, because they are too high to match the desires of buyers and sellers. The market price of grapes must fall below $4.[7]

[7] We described a market adjustment in which price rises when there is excess demand and falls when there is excess supply. This is called Walrasian adjustment, named after the Italian economist Leon Walras. In another process called Marshallian adjustment, named after the British economist Alfred Marshall, quantity rises when demand price exceeds supply price and falls when supply price exceeds demand price.

Shortages

Now start low by experimenting with $1 as the potential market price of grapes. Column 4 of Table 3–8 shows the result of this experiment: At the $1 price, quantity supplied falls short of quantity demanded by 30,000 units. Grape buyers would surely love to buy grapes at that price. But there are shortages. Why?

The $1 price *discourages* producers from committing resources to grape production and *encourages* consumers to attempt to buy more than is available. The result is a 30,000 pound shortage of grapes. The $1 price is not sustainable: it cannot last as the market price. A few lucky buyers will get some cheap grapes at that price, but many will not. Many buyers who want to buy grapes at this price will stand in long lines and complain bitterly, but get no grapes. Shortages will occur, and those shortages promote the following action: Consumers begin offering a higher price than $1 to secure some of the available output to avoid standing in long lines. Competition among these buyers for scarce output drives up the price to an amount exceeding $1.

Competition among buyers who bid for available supplies boosts the price to $2. Is that price high enough? This higher price will reduce, but will not eliminate, the grape shortage. At $2, grape producers are financially rewarded when they put more resources into grape production. Some grape buyers who were willing to pay $1 per pound are no longer willing to buy grapes at $2. The bad news is that a shortage of 14,000 pounds still occurs at a $2 price. The good news is that the shortage puts upward pressure on the market price that helps eliminate the shortage.

Equilibrium

Review of each experiment leads us to reject every price but $3 per pound. Only at that price does the quantity of grapes that producers are willing to produce match the quantity that consumers are prepared to buy. There is neither shortage nor surplus at the price of $3 per pound. Lacking both a shortage and surplus, there is no reason for the price to change. The price where quantity demanded equals quantity supplied is called the market-clearing or *market equilibrium* price. Equilibrium means "in balance." At $3, quantity supplied and quantity demanded are in balance at a quantity of 14,000 pounds. Table 3–8 shows that $3 is the only equilibrium price of grapes under these supply and demand conditions.

Market Price and Quantity

The price of grapes, or of any other product traded in competitive markets, is established where the supply decisions of producers and the demand decisions of buyers are compatible. Compatible decisions occur only at the equilibrium price (in this case, $3) and equilibrium quantity (in this case, 14,000 pounds). At any higher price, suppliers will offer more for sale than consumers want to buy, which produces a surplus. At any price below the equilibrium, consumers are prepared to buy more than producers offer for sale, which produces a shortage. Such differences between the supply intentions of sellers and demand intentions of buyers prompt price changes that bring the two sets of intentions closer, and then equal, to each other.

An analysis of supply and demand using graphs leads to identical conclusions. Figure 3–5 shows superimposed market supply and demand schedules on a single graph. In that graph, quantity demanded and quantity supplied are both on the same horizontal axis.

Graphs are popular for several reasons: They simplify a complex reality, they are realistic, they are powerful, they are graceful, they are elegant, they are useful, and they provide tremendous insight into the process and outcome of human behavior. Like any other picture, a graph can be worth a thousand words. On a graph, identifying the intersection of a product's supply and demand curves is a fast way to find the market equilibrium. The graph of Figure 3–5 shows that the equilibrium is $3 per pound (price) and 14,000 pounds (quantity). At any price above equilibrium, quantity supplied exceeds quantity demanded. The resulting surplus causes price reductions by sellers who are eager to rid themselves of their surplus rather than face unsold inventories. The

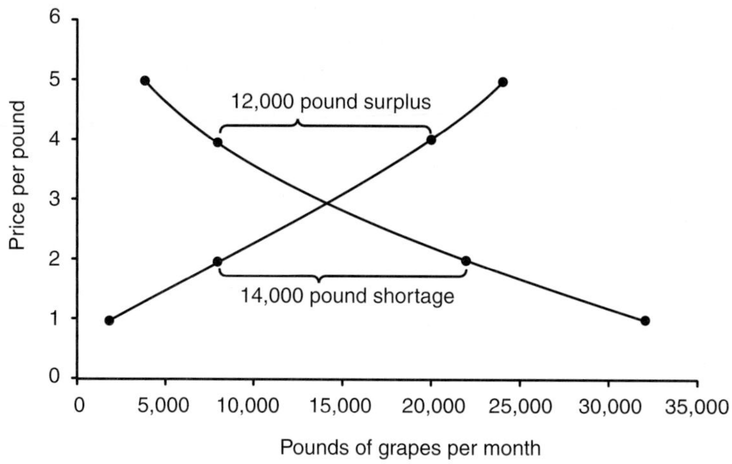

Figure 3–5
Equilibrium price and quantity of organic grapes

falling price causes fewer pounds of grapes to be offered for sale and at the same time encourages consumers to buy more. These adjustment forces press the market toward equilibrium.

Any price that is less than equilibrium produces market forces that create a shortage, because demand exceeds supply. At such a low price, buyers stand in lines to get the product and begin offering a higher price. That action exerts upward pressure on the price, moving it toward equilibrium. The rising price causes producers to raise their quantity supplied and encourages some buyers to leave the market. This process eliminates the shortage, and the market regains equilibrium. See Example 3–1.

Price as a Rationing Method

Rationing means distributing a scarce resource or product. The synchronization of decisions by buyers and sellers that creates equilibrium is called the rationing function of prices. Prices are indicators of *scarcity.* A higher price occurs when scarcity increases. A higher price rations goods to those who are willing to pay more. The price increase means that some who bought at the low price will now buy less or none at all.

There are many rationing methods that do not involve prices (called non-price rationing methods). Goods can be rationed on a first-come, first-served basis, through a person's access to political influence, through physical force, by coupons, by lotteries, and by cultural or religious traditions (e.g., first priority goes to people who belong to a particular religion). Scarcity means that it is not possible for everyone to have everything they want. When goods are scarce, there is always some method of rationing. Rationing by a freely functioning price system is efficient, because goods are allocated to people who are willing to pay the most and are supplied by those who are able to produce and sell at the lowest price. A good is allocated to its highest valued use and is produced with the cheapest production methods.

The market equilibrium price rations a good to its highest-valued uses. For the example shown in Figure 3–5, the equilibrium price of $3 clears the market; that is, it leaves no unsold surplus for sellers and no haunting shortage for potential buyers. The combination of freely made decisions by individual buyers and sellers sets this market-clearing price. All buyers who are willing and able to pay $3 for a pound of grapes will have them; all buyers who are not prepared to pay $3 will go without grapes. Similarly, every producer who is prepared to supply grapes at $3 a pound will sell them; producers who seek more than $3 will not find buyers. See Example 3–2.

Changes in Supply, Demand, and Equilibrium

Now we know what determines price and output, but prices and quantities rarely stay the same for long. Remember that demand can change because of fluctuations in consumer preferences, consumer incomes, adjusted consumer expectations, or changes in prices of related goods.

Example 3–1
Applying Demand and Supply: Ecolabeling

Ecolabeling encourages production and use of products that have reduced impacts on the environment. Market forces are harnessed as a force for environmental improvements. The European Union (EU) is developing a common ecolabeling system for its 25 member states (see Swedish Environmental Protection Agency, 2000). The EU ecolabel, the flower, is a symbol that can be used by goods and services that have a lower environmental impact than competing products that perform the same function (see U.K. Department for Environment, Food, and Rural Affairs, 2004). By setting strict ecological standards for product groups, the scheme enables consumers to make informed decisions. By making purchasing choices clearer for environmentally friendly consumers, the program financially rewards producers who design products with reduced environmental impacts.

Where consumer demand is high for environmentally friendly products, those bearing the EU flower will have a competitive economic advantage. Building environmentally friendly products is rarely cheap, which means that shifting production from ordinary products to environmentally sensitive ones will shift the supply curve up and to the left. So rewarding environmentally friendly business with consumer demand that shifts out and to the right is an important way to motivate and encourage businesses to produce goods that meet high environmental standards. EU ecolabeling covers a wide range of products such as washing machines, dishwashers, refrigerators, light bulbs, vacuum cleaners, televisions, personal computers, paper products, textiles, shoes, mattresses, laundry detergents, floor coverings, paints, fertilizers, and hotels.

If you are a producer, why would you want the ecolabel on your product? Because it would enhance your product's appeal and make to stand out in the market. That makes your demand curve shift to the right to compensate for your increased costs of doing business. This rightward demand shift rewards your "green behavior" by increasing your profitability and sales.

Example 3–2
Applying Demand and Supply: Cap-and-Trade Programs

Cap-and-trade programs for emissions harness market forces to control pollution. They reduce pollution by setting a permanent cap on emissions, then allowing trading within that cap. Cap-and-trade programs ensure that environmental goals are met, while providing private firms an alternative to the installation of costly pollution control technologies. In the United States they were first used nationally by the Environmental Protection Agency (EPA) in its acid rain program to reduce sulfur dioxide (SO_2). They were then used by several northeastern U.S. states to reduce nitrogen oxides (NOx), to reduce SO_2 and NOx in southern California and to reduce smog-forming volatile organic compounds in Chicago. Before trading can occur, EPA requires careful monitoring and reporting standards. This monitoring is required to ensure that each allowance traded represents 1 ton of emissions, regardless of its source.

By harnessing the power of the market, both EPA's acid rain program and the northeastern NOx program have reduced emissions faster than would have occured with conventional command-and-control approaches. As of the year 2000, the acid rain program reduced SO_2 emissions by 6 million tons per year from 1980 levels, and the cost has been 75% below original projections by industry. EPA forecasts that by the program's full implementation date of 2010, emissions from power plants will be half of 1980 levels, thereby improving the environmental quality of lakes and streams damaged by acid rain and delivering more than $50 billion per year in health benefits to Americans. Many nations around the world are considering similar programs.

Source: U.S. Environmental Protection Agency. 2004. *Acid rain.* Available from *http://www.epa.gov/airmarkets/acidrain/* [Accessed October 18, 2004].

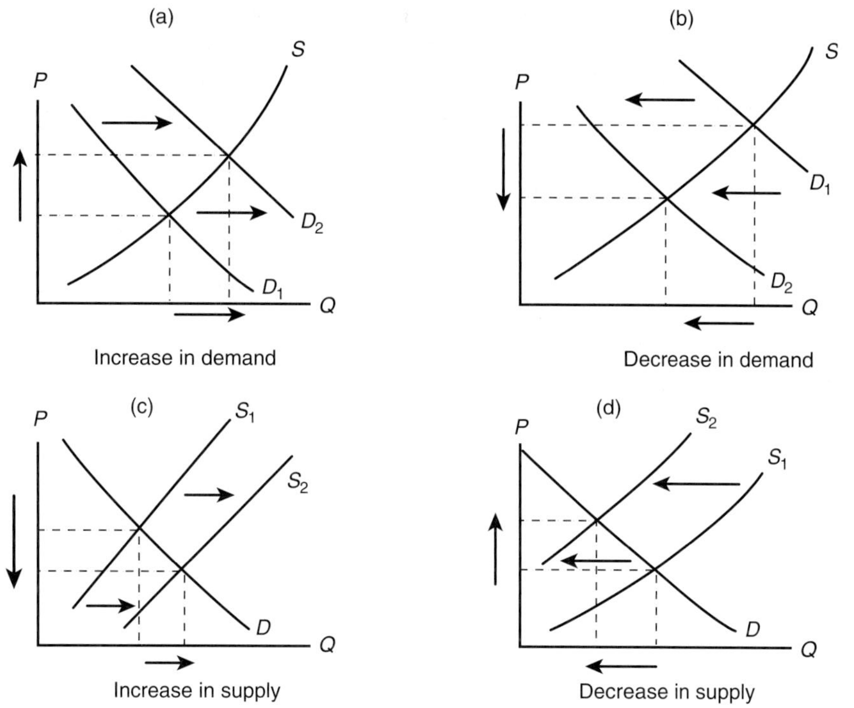

Figure 3–6 Changes in demand and supply of organic grapes

Supply can change because of adjustments in resource prices, new technology, or changed taxation. An important question is what effects will such changes in supply and demand have on equilibrium price and quantity?

Figure 3–6a demonstrates the power of graphic analysis (compared to tables of numbers) by showing a situation in which supply is constant but demand increases. It permits us to easily compare old and new equilibrium points. For example, suppose a television news story reports on a health scare in connection with pesticides applied to conventional grapes, so the demand for organically grown grapes increases. After demand increases, shortages occur at the old price. Rather than suffer shortages, buyers bid up prices. The intersection of supply and the new demand curve occurs at a higher price and a higher quantity. The increase in demand raises equilibrium price and quantity. For similar reasons, Figure 3–6b shows that a decrease in demand reduces both equilibrium price and equilibrium quantity.

Figure 3–6c shows a case where demand for grapes is unchanged but supply increases. Perhaps a new technology is discovered for growing organic grapes more cheaply, or a more productive soil type is discovered or developed, or a more efficient grape-harvesting method is developed. For whatever reason, at the old price there is more supply and the same demand. The surpluses that result encourage producers to offer a lower price rather than accumulate unsold inventories. The reduced price encourages a higher quantity demanded. This process continues until a new equilibrium occurs at a higher quantity and a lower price. An increase in supply lowers equilibrium price and raises equilibrium quantity. Figure 3–6d shows the what happens when supply decreases: Shortages occur and consumers bid up prices. Ultimately, equilibrium price rises and equilibrium quantity falls.

Many kinds of environmental policies affect both demand and supply simultaneously. For example, a policy of newly set-aside habitats for an endangered species combined with growth in per capita income would reduce the supply of timber from the newly protected area while increasing demand for wood products. When both supply and demand change, the effect is a combination of each individual effect in isolation. Four combinations are possible:

1. A *reduced* supply and an *increased* demand each increases price on its own. You would expect their combined effect to increase equilibrium price by more than either can do in isolation, and you are right. What about the impact on equilibrium quantity? The combined effect on equilibrium quantity is not known for certain (indeterminate). The effect on equilibrium quantity depends on the comparative sizes of the supply and demand changes (comparing the change of one to the other at a given equilibrium price). If the supply decrease is greater than the demand increase, equilibrium quantity decreases. However, if the demand increase exceeds the supply decrease, equilibrium quantity increases. If the demand and supply changes are equal, there is no change in equilibrium quantity; the result is an unchanged volume of output and a large price increase.

2. Imagine a mix of environmental policies that increases a product's supply while reducing its demand. What effect will this have on equilibrium price? Each change, in isolation, exerts downward pressure on price, so when both changes occur together, the net result is a price reduction greater than that resulting from either change alone. (The net result on price is the algebraic sum of the two changes in isolation.) What about the impact on equilibrium quantity? In this case, the effects of the changes in supply and demand are offsetting: The increase in supply raises equilibrium quantity, but the decrease in demand lowers it. So the net effect on quantity is indeterminate. The final change in quantity depends on the comparative sizes of the changes in supply and demand. Where the supply increase exceeds the demand decrease, equilibrium quantity increases. However, if the demand decrease exceeds the supply increase, equilibrium quantity falls. For the special case where the two are equal, equilibrium quantity is unchanged.

3. Consider the effects of reducing both supply and demand. If the output of a product imposes major pollution damages on the environment, policymakers may search for policies to limit equilibrium output by operating on both sides of the market (i.e., they may attempt to reduce both demand and supply). What is the effect of a policy that decreases both supply and demand? If the decrease in supply is greater than the decrease in demand, equilibrium price rises. If the opposite is true, equilibrium price falls. However, since decreases in supply and demand each reduce equilibrium quantity on their own, equilibrium quantity falls by more than each reduction accomplishes by itself.

4. Think of the effects of increasing both supply and demand.[8] Any policy that increases both supply and demand simultaneously will produce a considerable increase in output. Remember that an increase in supply or demand by itself raises equilibrium quantity. Equilibrium quantity will increase by an amount greater than that caused by either change in isolation. In this case, the impact on equilibrium price is indeterminate. Recall that a supply increase reduces equilibrium price, whereas a demand increase raises it. If the supply growth exceeds the demand growth, equilibrium price falls. If the opposite occurs, the equilibrium price will rise. If the two are equal, price stays the same. See Example 3–3.

Table 3–9 summarizes the four cases just described. To understand them fully, draw demand and supply diagrams for each case to produce effects shown in the table. (If you are mathematically inclined, see Appendix A.)

Other Things Being Equal

The downward-sloping demand curves tell us that price and quantity demanded are inversely related, other things being equal. The upward-sloping supply curves say that price and quantity supplied are directly related, other things being equal. It is easy to forget the assumption about other things being equal. If you forget it, you might encounter situations that appear to produce upward-sloping demands or downward-sloping supplies. For example, suppose bottled

[8] The outputs of some products help the environment. Examples include various kinds of pollution control equipment, alternative energy sources such as wind or solar power, and various kinds of organic or pesticide-free agriculture. Some policymakers will take an interest in designing policies that increase the outputs of these products without resorting to costly subsidies or regulations.

Example 3–3
Applying Demand and Supply: Africa's Ivory Trade

The African elephant, which symbolizes the spirit of Africa, has been declining in numbers since the 1970s. Elephant population declines were first seen in Kenya and Uganda, and later Kenya banned ivory exports. That ban, by shifting the legal supply up and to the left, raised the price of ivory. Later, in the early 1990s, some African countries recognized that a small ivory trade would produce needed revenue for cash-starved wildlife conservation programs. In 1999 a one-time auction of ivory stock from Botswana, Namibia, and Zimbabwe to buyers in Japan was allowed, provided that Japan did not reexport the ivory. The sale raised $5 million but led to a vigorous debate about whether additional state-supervised ivory sales should occur to support wildlife conservation programs.

Despite the formal international ban, the market for ivory in several countries continues to grow. For some financially struggling African governments, it seems unwise to uphold a total international ban under which a strong illegal market flourishes. These same governments could legally sell or tax sales of a quota of ivory and use the profits to invest in local communities and wildlife programs. Any single ivory-producing country has a strong economic incentive to issue hunting permits and/or tax hunters when the price of ivory increases to take advantage of other countries' ban on ivory trade.

The economic laws of supply and demand are a driving force behind the illegal ivory trade: Wherever there is a high and growing demand (demand lies to the right), increased prices will attract growing quantities supplied. Regulations that limit ivory supply will shift the legal supply to the left, considerably increasing the price and economic return price of illegal ivory production methods like poaching that contribute to that supply. How can the laws of supply and demand be harnessed to protect elephants? Their protection may require focusing on the demand side of the market; that is, measures could be examined for reducing people's demands for ivory. For example, develop a new supply of ivory substitutes and/or reduce the price of ivory substitutes. But will people look hard for those substitutes when the status of ivory grows in the face of its increased scarcity? As greater numbers of countries ban the ivory trade, the price of ivory will continue to grow, thereby providing substantial financial rewards for people and countries who continue the trade illegally.

Source: Lesniewski, S. 2002. *Ivory trade.* Available from *http://www.cowac.org/studyguideaugzero3sdc1.html* [Accessed October 18, 2004].

Table 3–9 Effects of changes in both supply and demand

Change in demand	Change in supply	Effect on equilibrium price	Effect on equilibrium quantity
Increase	Decrease	Increase	Indeterminate
Decrease	Increase	Decrease	Indeterminate
Decrease	Decrease	Indeterminate	Decrease
Increase	Increase	Indeterminate	Increase

water producers sell 1 million bottles of water at $3 a bottle in year 1; 2 million bottles at $4 in year 2; and 3 million at $6 a bottle in year 3. These facts tell you that price and quantity purchased are moving up together, which looks like a violation of the law of demand, but further consideration shows there's no conflict whatsoever. Remember, the law of demand is based on the assumption of other things being equal. In fact, the assumption of other things being equal has been violated during the three years of water sales. Simply put, demand has shifted out over all those three years, perhaps because of growing tastes and increased income, as Figure 3–6a shows. That figure shows that a change in demand causes a change in quantity supplied.

Similarly, poor harvests shift supply to the left and produce rising prices and reduced quantities at the same time. This produces price quantity combinations that appear to refute the law of supply, but the assumption of other things being equal has been violated. A supply that shifts to the left is not holding other things equal that affect supply, and it will produce rising prices combined with falling quantities, as shown in Figure 3–6d. Note that Figure 3–6d shows that a change in supply causes a change in quantity demanded, whereas Figure 3–6a shows that a change in demand causes a change in quantity supplied.

■ Accomplishing Environmental Goals: Markets versus Governments

The two main social institutions for meeting environmental goals are markets and governments. Depending on our cultural background, we may place greater faith in markets or in governments for dealing with environmental problems. Regardless of our preference, both markets and governments are involved around the world. Before we examine these two institutions, let us consider the meaning of *social institutions.*

Social institutions are collectively agreed-upon rules that constrain *individual* behavior (Commons, 1931) for the purpose of increasing *social* welfare. They are designed to improve total social welfare by constraining individual actions that would otherwise reduce that welfare. Social institutions are society's rules that structure political, economic, and social interaction. Social institutions consist of informal rules (e.g., sanctions, taboos, customs, values, beliefs, traditions, and codes of conduct) and formal rules (e.g., constitutions, laws, and property rights). Furthermore, social institutions contain enforcement mechanisms, such as courts and the power of public opinion, to ensure compliance with the rules. Groups of people develop institutions to create order and reduce uncertainty. Institutions form the incentive structure of a society. Because incentives have such an important effect on economic output, they can be designed to increase economic benefits by more than their costs.[9] (The social institution of property is discussed in Chapter 4.)

Like others constraints in life, institutional constraints are meant to be respected. However, like other constrains in life, violations occur. When a society designs its institutions efficiently, it requires violators to pay a cost (two examples are fines and jails). The cost to an individual for violating an institutional constraint is the punishment society believes is appropriate for the violation. Efficient punishments will be set by society to make the punishment of violation equal to the social costs[10] produced by the behavior that violates the constraint. In an efficient legal system, violations that cause small damages to society (small costs) will have small punishments. Where government rules and regulations are established to encourage environmental compliance, efficiently designed fines for breaking those laws are proportional to the environmental damages caused. Larger fines are imposed for larger environmental damages.

If efficient behavior is desired, violations that cause society to suffer large costs will have more severe punishments. Thus, the individual who violates an accepted institutional constraint will see the punishment as the price of the violation. If that price is equal to the opportunity cost to society of violating the institution, then the institution is efficient. Under an efficient institution, the individual is faced with sanctions that require the individual to face the real cost of her actions. Thus, an efficient price of violating any constraint is much like an efficient price of producing and buying any commodity: This price signals to the individual the net opportunity cost, measured as displaced benefits incurred by the rest of society resulting from her actions.

Social institutions may seem complex, so consider a simple analogy: A traffic stop sign is familiar to most of us. It shares many of the characteristics of a social institution: It is a jointly

---

[9] Douglass North won the 1993 Nobel Prize in economics for a lifetime of insightful contributions to the study of social institutions. According to his autobiography, North felt the tools of neoclassical economic theory (analysis of demand, supply, and markets based on the assumption of income and profit maximization) were not up to the job of explaining the kind of fundamental social change he had seen. He was looking for a view of institutions that accounted for why they produced results that failed to produce economic growth. Later he abandoned the assumption that institutions were efficient and attempted to explain why inefficient rules would exist and last. For more, see Royal Swedish Academy of Sciences (2004c).

[10] Pearce (1998) summarizes the application of cost-benefit analysis to environmental institutions in the United Kingdom. Also see Chapter 5.

agreed-upon constraint that limits each person's behavior. Cities typically place stop signs on side streets that intersect busy streets. By constraining each driver's behavior, a stop sign imposes costs. The sign also benefits drivers on the busy street who are able to drive more efficiently. The following example shows that analysis of costs and benefits plays a role in locating stop signs as well as in designing of institutions. Drivers of cars know that stop signs promote social order and efficiency, yet we each perceive a private welfare gain by ignoring a stop sign.[11] As a driver who acts in his or her own self-interest, an individual may lobby the city council to remove the side street's stop sign that slows the entrance to a boulevard. In a moment of bold self-interest, the driver might even ask the council to rotate the sign 90 degrees so that it faces boulevard traffic, thereby giving the driver the safest and easiest access to the boulevard. However, the city council is elected by all voters and has seen many appeals motivated by self-interest. It denies the individual's request.

The city council knows that, like a market price, the stop sign is a constraint that restricts an individual's actions and signals the costs imposed on society by violating that constraint. If the council removes the stop sign, it increases the individual's liberties. An efficient-minded council calculates the costs and benefits of the proposed removal, as well as the fairness of the distribution of those costs and benefits among the affected population. It requires little deliberation to reject the proposal. The council knows that by accepting the proposal the total value of liberties gained falls short of the total value of liberties lost. Eliminating one stop sign removes a constraint, but the benefits of the individual's added freedom are small compared with the costs to society. Unrestricted freedom for every person brings chaos to all. Fines for running a stop sign or for violating institutions signal the costs imposed on society by the individuals who violate it.

Markets

One social institution that has an enormous impact on natural and environmental resources is the market system. In terms of economic significance, the market system has an important influence on the environment and natural resources. Two aspects of markets are perfect competition and economic efficiency.

Competitive Supply: Where Price Equals Marginal Cost. Perfect competition occurs when no individual firm can affect the market price. Each firm is a price-taker, which means that it must sell at the going market price. Perfect competition occurs when there is a large number of small firms, each producing an identical product and each too small to affect the market price by increasing or reducing output. Figure 3–7 shows that under conditions of perfect competition, each firm faces a completely horizontal demand curve equal to the same market price (*P*) at all levels of output. In fact, even large changes in output by an individual firm have no effect on market price. Therefore, each firm faces an infinite *price elasticity of demand.*[12] Because of this, the elasticity of demand is much larger to a single competitive firm than to the entire industry. Changes in the firm's output have no effect on the firm's price, so its total revenue—price times output—is proportional to total output (shown in Figure 3–7 as the dotted line *TR*).

Marginal cost is defined as the addition to total cost that comes from producing an additional unit of output. It is the extra cost from one more unit produced, not the total cost of all units produced.

With the concepts of perfect competition and marginal cost defined, we can ask this question: How does the competitive firm decide on the amount of output it will supply? Presume the firm wants to maximize its profits, defined as revenues minus costs. To maximize the difference

---

[11] Marglin (1963) described the dilemma posed by our public desire to produce good social institutions like speed limits and our private wish to bribe police who stop us for speeding when we violate that institution.

[12] The price elasticity of demand is defined as the percentage change in quantity produced by a price change divided by the percentage change in price that produced the change: $n = \%\Delta Q/\%\Delta P$. A 10% change in output of a single firm has no effect on its price. A 10% change in output of an industry can have an enormous effect on its price.

Figure 3–7
Product price under perfect competition

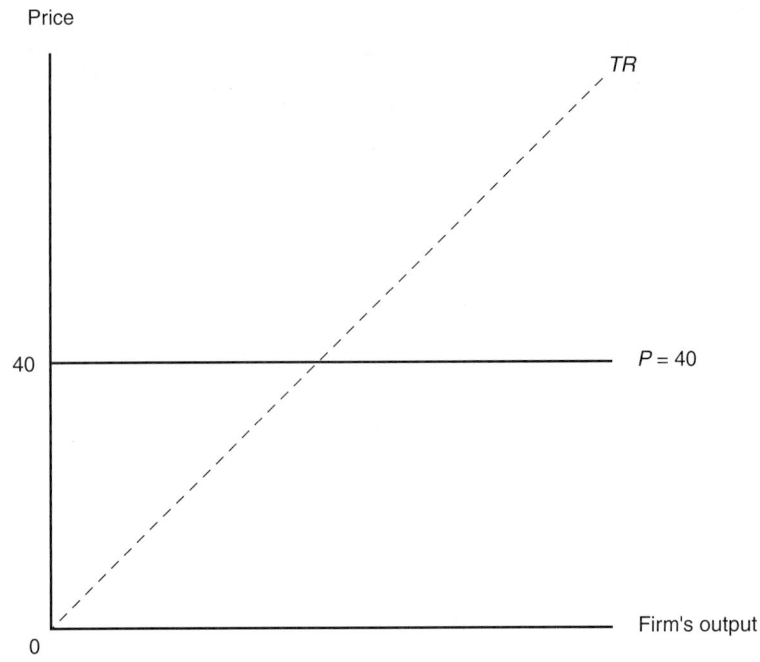

between revenues and costs, the firm's chosen output level will be influenced by both revenues and costs. For example, no rational firm will supply teak computer tables at $1 apiece, because those revenues will not cover the cost of the tables. Likewise, if teak computer tables sold for $1 million apiece, there would be a rush of competitors starting up new computer table enterprises. Still, within that broad range of price and output, the firm's best decision is not obvious.

Given that a single, perfect competitor faces a flat demand curve, which means it can produce and sell any level of output it wishes at the going price, what should it do? A profit-maximizing perfect competitor chooses the quantity of output to supply by setting its output level so that the marginal cost of the last unit of output equals its price. Why? Because the firm wishes to maximize the total profit it can earn. The firm increases total profit by increasing output as long as the added revenue brought in from one more unit output exceeds the added (marginal) cost.

Total profit reaches a maximum when there is no longer added profit to be earned by selling added output; the last unit produced and sold by the perfect competitor is at that point exactly in balance from the view of added revenue and added cost. What is the firm's added revenue from one more unit output? It is the constant price per unit. What is the firm's added cost? It is its marginal cost.

To illustrate with numbers, Table 3–10 shows the costs and revenue for a hypothetical vendor who sells drinking water at a price of $40 per tank. The firm is now producing 5,000 units, and it is considering *expanding* output to 5,001. The cost of expanding output from 5,000 to 5,001 is $44 (marginal cost column). The added revenue fetched from the 5,001st unit is $40. Should the firm increase output to 5,001? Doing so increases cost by $44 and increases revenue by $40, so the firm loses money on the 5,001st unit. Marginal cost (*MC*) is greater than price (*P*).

This story tells us that increasing output is a bad economic decision when marginal cost is larger than price. Now consider the consequences of *reducing* output. Should the firm produce and sell one unit less than 5,000? By reducing output from 5,000 to 4,999, it would save the marginal cost of the 5,000th unit, shown in the marginal cost column to be $36; but being a perfect competitor, it would lose $40 in revenue. Therefore, by reducing output, the firm would discover that the lost revenue is greater than reduced cost. That means $MC < P$. So it does not pay to reduce output whenever price equals marginal cost.

What does all this mean? The profit-maximizing output of a vendor's drinking water (tank of water sold per unit time) is precisely where *price equals marginal cost.* At the output level

Table 3–10 Cost concepts

1 Quantity	2 Fixed cost	3 Variable cost	4 Total cost	5 Marginal cost
←		dollars		→
0	40,000	0	40,000	–
1,000	40,000	24,000	64,000	24
2,000	40,000	52,000	92,000	28
3,000	40,000	84,000	124,000	32
4,000	40,000	120,000	160,000	36
5,000	40,000	160,000	200,000	40
6,000	40,000	204,000	244,000	44
7,000	40,000	252,000	292,000	48
8,000	40,000	304,000	344,000	52

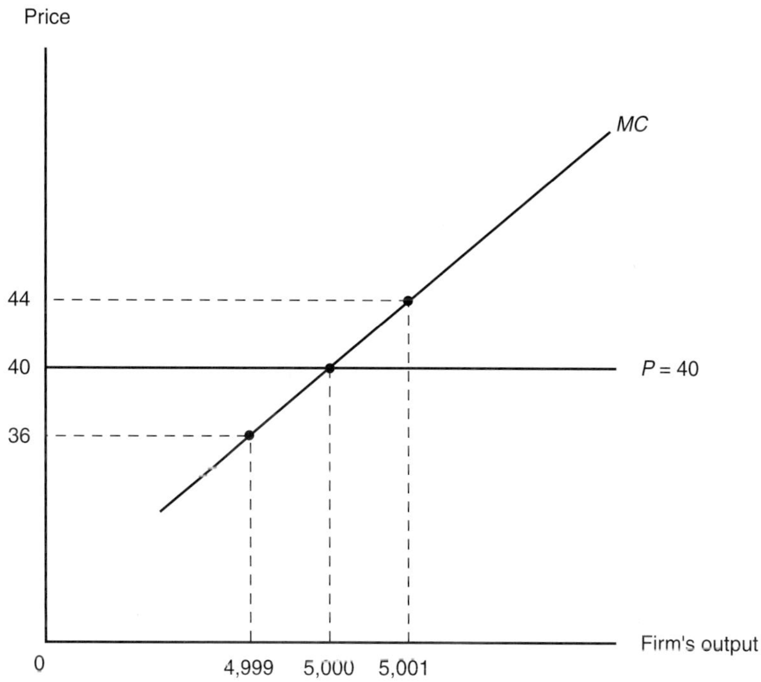

Figure 3–8 Price and marginal cost

where $P = MC$, any larger output will result in $MC > P$, which reduces the maximum profit. Similarly, when $P = MC$, any reduction in output also reduces profits because the reduced revenue (P) exceeds the reduction in cost (MC). Either expanding or contracting output is bad economics for the profit maximizer when $P = MC$. Sticking with the status quo, when price equals marginal cost, makes the most profit possible for the firm. So competitive firms follow the rule $P = MC$, as shown in Figure 3–8.

You now have a rule that lets you peer inside the head of a profit-maximizing perfect competitor and use that rule to predict its behavior. Knowledge of that rule lets you do powerful things, even if you are not a perfect competitor yourself. For example, if you want to predict the output of a firm that operates in a competitive industry, you can expect to make a good prediction if you say it will produce a level of output for which its price equals its marginal cost. Maybe you have been asked to predict the amount of new piped water resulting from a policy that would reduce the cost of supplying water. You can predict that this policy will increase the industry's

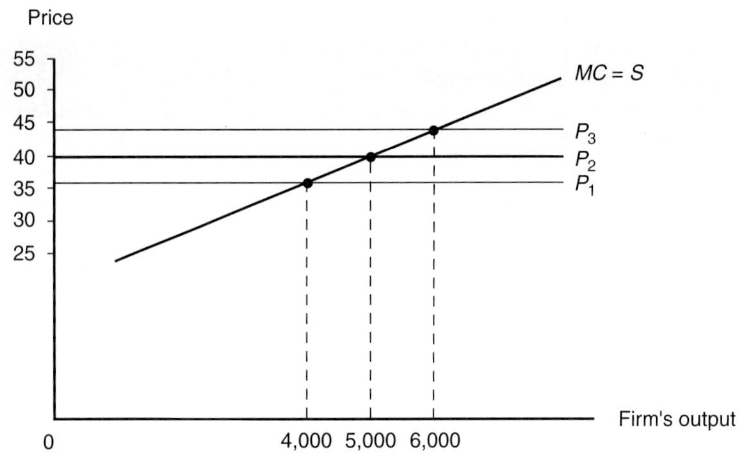

Figure 3–9 Price, marginal cost, and supply

water output: Because it is an industry of profit maximizers, each firm will adjust its output upward so that its lower[13] marginal cost again equals the existing price.

The Competitive Firm's Supply Curve. We have seen that a perfectly competitive firm makes the maximum profit when it expands output whenever $MC < P$. It also reduces output whenever $MC > P$. So the competitive firm's maximum-profit behavior comes when it follows this simple rule:

$$P = MC$$

This equation means that the competitive firm's supply curve is described by its *MC* curve, as shown in Figure 3–9. Why is this? Table 3–10 shows that at the low price of $36, the firm maximizes profits by supplying 4,000 units (also shown in Figure 3–9). At the intermediate price of $40, the firm maximizes profits by supplying 5,000 units, and at the high price of $44 it maximizes profits by supplying 6,000 units. As the price increases, the firm responds by re-adjusting its profit-maximizing output level by moving to a higher level on its marginal cost curve. This means that the firm's marginal cost curve and its competitive supply curve[14] are one and the same thing.

Economic Efficiency of Perfect Competition. How well does the institution of competitive markets perform the aim of meeting society's goals? We know that the institution of competitive markets is an excellent system for maximizing private profits. But is the institution of competitive markets a poor way to satisfy *society's* economic needs? Or should it receive high marks as a coordinator of economic activity? After all, the real purpose of economic activity is the opportunity for the human consumption it provides.

Finding the answer to the first question is not easy. We begin by considering the meaning of economic efficiency, which turns out to be a central yardstick used for evaluating the performance of a social institution for coordinating economic activity. We then examine how competitive markets perform in an economy with many consumers, goods, and firms.

---

[13] Marginal cost, though, must be rising. If marginal cost falls with increased output, then profits are *minimized* when $P = MC$.

[14] In fact, the firm only continues to stay in business and keep producing if price exceeds average variable costs, so the competitive firm's supply curve is that part of its rising marginal cost curve that lies above its average variable cost curve. At the critically low market price, in which the firm barely recovers its variable cost, it is close to shutting down. Below that price, it shuts down and produces nothing. At prices above average variable cost, the firm will expand output along its marginal cost curve for its fixed plant size. For any of these higher prices it will be receiving some positive net revenue for paying back its fixed costs. But in the long run, all costs must be paid for, including fixed costs. In the long run, the firm's supply curve is that part of its marginal cost that lies above its average total cost.

Economic Efficiency. An allocation of resources is defined as economically efficient if it is organized in a way to provide people the largest possible value of output for the economy's available resources and technology. *Economic efficiency occurs when no possible reorganization could make anyone better off without making at least one person worse off.* Under conditions of economic efficiency, so much human benefit is already squeezed from available resources and technologies that no possible change from the current situation that improves one person's welfare can do it without cost to someone else. At this point there is no slack left in the already efficient system. This concept may be the most pivotal concept in all of microeconomics. *Economic efficiency occurs when available resources and technology produce the greatest possible monetary value of human benefits.*

The opposite of economic efficiency is economic inefficiency. Economic inefficiency occurs when some resources are wasted in the sense that they could be reallocated (reorganized) to improve the total monetary value of human welfare. A change from the status quo will produce an improvement in economic efficiency if the monetary value of the gains exceeds the value of losses. The gains of the gainers are calculated based on what they would be willing (prepared) to pay to secure their net gains. The losses of the losers are calculated as what they would be willing (prepared) to accept as full compensation for their losses.[15] Choosing a more economically efficient alternative typically leaves some members of society worse off, because the monetary value of losses to losers is rarely compensated. Losers typically live with the consequences.

Using the yardstick of willingness-to-pay to define economic efficiency makes the value judgment that the existing distribution of income and wealth is morally acceptable. This is so because the willingness to pay for the gains depends critically on the existing distribution of income and wealth (who earns how much and who owns what). For any individual, the willingness to pay for any proposed action depends on that person's ability to pay (income and wealth).

The management of publicly owned lands provides an illustration of the concept of economic efficiency. Suppose a well-organized group of hunters is willing to pay $100 per acre per year to set aside certain productive lands for wildlife habitat and that this habitat is the highest-valued use of that piece of land. Suppose also that current institutions lock these same lands into cattle grazing that is worth only $4 per acre per year. Finally, suppose these institutions prohibit ranchers from renting the land to the hunters for the wildlife habitat. These facts let us say that the current pattern of land use is economically inefficient, for the institution effectively takes the land from higher-valued uses and locks it into lower valued ones. To state it more positively, any action that removes this institutional barrier increases economic efficiency.

For a second example, turn to water. Many institutional barriers in dry places block maximum beneficial use of water, thereby forcing limited supplies of the resource to fall short of their power to satisfy greater human needs. Suppose farmers are earning an income of $50 per acre-foot of water by using a given amount of water for irrigation. Suppose too that a nearby city would gladly pay $300 per acre-foot for the same water to supply its thirsty urban customers. Many water laws in dry places around the world discourage economic efficiency in water use by prohibiting farmers from renting, leasing, or selling water to cities. These laws actually hurt the farmers by prohibiting the development of certain markets for transferring water. These laws also hurt the cities by taking away a potential source of water. Laws that block the highest-valued benefits produced by water also block economic efficiency. So we begin to see that the maximum benefit produced by a scarce resource is really the same thing as its economic efficiency. See Example 3–4.

---

[15] This is the definition of Potential Pareto Efficiency. If you accept this definition, you make the strong moral assumption that an acceptable yardstick for combining values across people is willingness to pay. If so, then a gain produced by a decision for which person A is willing to pay $10 dollars to secure exactly balances a loss from the same decision that person B is willing to pay $10 dollars to avoid. In fact, you may believe it is morally unacceptable for any policy to make a rich person better off and a poor person worse off. Pareto Efficiency is named after the famous Italian economist, Vilfredo Pareto (1848–1923). He laid the foundation of modern welfare economics with a concept known as Actual Pareto Efficiency. According to this principle, an efficient allocation of a society's resources is not attained so long as it is possible to make at least one individual better off while keeping others equally well off. Potential Pareto Efficiency was later invented as a way of using Pareto's brilliant idea, while recognizing that few actual policies can be implemented without making somebody worse off.

Example 3–4
Economic Efficiency of the Clean Air Act

Since the U.S. Clean Air Act was enacted in 1970, many people have asked whether the health and environmental benefits of air pollution control justify the costs incurred by industry, taxpayers, and consumers. People affected by those costs, particularly affected industries and their customers, have vigorously debated the usefulness of the added costs those regulations impose. Industry, for example, cites inability to compete in the marketplace with competitors from less environmentally regulated countries. These complaints raise a legitimate question: How do the overall welfare, ecological, health, and economic benefits produced by the Clean Air Act programs compare with their costs? The question is one of economic efficiency: Do the total benefits, to whomsoever they may accrue, exceed the costs?

The U.S. Congress wanted to know the answer to that question. So Congress added to the 1990 Clean Air Act Amendments a stipulation that benefits and costs be estimated and presented, and in 1997 the EPA produced a report that estimated benefits and costs. How did the EPA do this? It compared the nation's benefits and costs *with* the Act compared to the equivalent benefits and costs that would have occured *without* the Act. What did the EPA find? First, the benefits of the Clean Air Act and its control programs considerably exceeded costs. Second, most of the economic benefits of the Clean Air Act programs came from reducing two especially damaging pollutants: lead and particulate matter.

When used properly, efficiency analysis sheds light on important effects of changes in policy and helps set priorities for findings needed information. Still, as you will see in Chapter 5, the proper use of efficiency analysis requires that adequate levels of time, resources, and money be made available to permit careful, exhaustive, and scientifically defensible data and data interpretation.

Source: Environmental Protection Agency. 1997. *The benefits and costs of the Clean Air Act, 1970–1990.* Washington, D.C. EPA.

Cooperation of Marginal Cost and Marginal Benefit. Now that we know what perfect competition does and what economic efficiency means, we can ask how well an economy of perfectly competitive firms scores in terms of promoting economic efficiency. Let us begin the analysis by considering the supply and demand for a single commodity, wheat.

We assume that the *suppliers* (farmers) who produce the wheat incur costs to do so and that each unit of each resource required to grow the wheat (e.g., land, water, labor, and equipment) requires the farmer to pay a price. Because wheat production takes place on fixed land plots, we know from the law of diminishing marginal returns that each extra application of the variable resources to the fixed land brings less and less additional wheat. Recall that this law says *additions* to output produced *by one more* unit of input must ultimately fall at suitably high levels of input.

For the wheat *demanders* (consumers) we also see the impacts of diminishing marginal returns: Each extra unit of wheat brings diminished marginal benefit, since more wheat offered to the wheat market is only bought if the price is reduced (see Appendix B). Figure 3–10 shows the outcome for this single competitive market. When we add up the supply (*MC*) curves for all wheat farmers in the market, we have the upward-sloping *MC* curve, which is also the wheat industry's supply curve. For any given price on the vertical axis, the quantity it intersects on the *MC* curve is the quantity of wheat supplied by the industry from individual profit maximizers. The *MC* curve is labeled *S* to indicate it is also the industry's supply curve. The figure also shows the downward-sloping demand curve for wheat consumers. This curve is the horizontal summation of all wheat buyers' marginal benefit (*MB*) curves for wheat; it is labeled *D* to indicate it is the industry's demand curve.

The point at which the *S* and *D* curves intersect in Figure 3–10 shows the competitive equilibrium for wheat. At point *C* wheat suppliers produce just what wheat demanders will buy. At this point a rather remarkable thing occurs: *Each farmer is pressing his scarce resources up to the point where the diminishing marginal benefit of consuming wheat intersects the increasing marginal cost of growing wheat.*

Point *C* holds two essential messages: It is the competitive equilibrium, and it is economically efficient. At point *C* each wheat buyer's total wheat-consuming benefits are maximized at

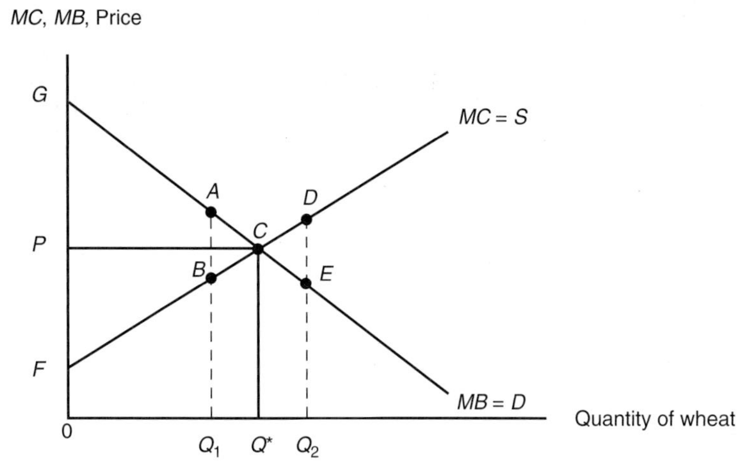

Figure 3–10 Consumer and producer surplus

the going price of wheat.[16] At that point the marginal benefit of more wheat equals the price, which in turn equals the marginal cost of added wheat.

For the consumer, the price equals marginal benefit of wheat consumption.[17] Each wheat consumer, who is free to choose how much wheat to buy at the going price, chooses to buy a quantity of wheat for which price equals the marginal benefit ($P = MB$). As a result of each buyer making this personal optimization decision, every buyer gains a marginal benefit from the last unit of wheat consumed equal to the price.

For the producer, the price equals the marginal cost of added wheat production. Since each wheat producer is a profit maximizer, each buys wheat inputs and supplies wheat up to the point where the price of wheat equals the marginal cost of the last unit of wheat supplied. So price equals the marginal cost of wheat-producing resources. The marginal costs of those resources consists of the losses from bidding those resources away from other consumption and putting them into producing the last unit of wheat ($P = MC$).

We can combine consumer and producer behavior into a larger story of social behavior. We know that point *C* is where the marginal benefit of the last unit of wheat consumed equals the marginal cost of the last units produced ($MB = MC$). So the benefits gained from the last unit of wheat consumed are equal to the benefits lost (cost) by taking wheat-producing resources out of other uses and putting them into wheat. It is precisely this condition—that the *incremental* gain to society from the last unit of wheat consumed equals the *incremental* costs to society of that last unit of wheat produced—which tells us that a competitive equilibrium is economically efficient.

Consumer and Producer Surplus. Figure 3–10 shows another way of measuring the economic efficiency of competitive equilibrium at point *C*. The top triangle in the figure, *PGC*, is called *consumer surplus*. For each unit of wheat output, consumer surplus is measured as the difference between the marginal benefit of that unit of output and the price. At the level of output defined by point *C*, total consumer surplus is *PGC*.

[16] This does not suggest that everybody is as well-off as possible; they aren't. Although each is maximizing her welfare subject to income and price constraints, some could be better off under a different situation. For example, if price is lower, supply is reduced and there are wheat shortages. But those lucky wheat buyers who can find the wheat at the lower price are better off. Also, if you can afford no wheat at all at the equilibrium price, you do not receive maximum welfare. The need to increase welfare of poor people is behind government subsidies for basic foods.

[17] White, Bennett, and Hayes (2001) show some of the limitations of the willingness-to-pay approach for valuing the human benefits produced by policies that conserve nature. Zizzo (2003) describes the relation between marginal benefits and the evolution of the formation of individual preferences that underlie those expressions of benefit.

The bottom triangle in the figure, *PCF,* is called *producer surplus.* It is the difference between what resource owners who supply inputs for wheat production receive in payment and the minimum they would accept in pay to supply their services. It is the excess of their actual receipts over the minimum at which they would be willing to supply their output. *PCF* is an important measure of social benefit of wheat production. At point *C* the sum of consumer and producer surplus from wheat production is maximized. Now, let us see why any level of wheat output reduces economic efficiency if it is not at point *C*. Suppose wheat producers expect a higher price than they receive, like *D,* and they supply Q_2 units of wheat. At any level of output larger than the *C* level, the marginal cost of wheat is larger than its marginal benefit: Wheat producers put more expensive resources into wheat production than wheat consumers gain from the added wheat consumption. The little triangular area to the right of *C, CDE,* is the loss of consumer and producer surplus from producing Q_2 units of wheat in comparison to producing a quantity of wheat equal to Q^*.

Similarly, suppose wheat producers mistakenly expect a low price of wheat, like *A,* and produce only Q_1 units. At any level like Q_1, smaller than Q^* units, the marginal benefit of added wheat is larger than its marginal cost. Wheat producers put lower-valued resources into wheat production than wheat consumers gain from extra wheat consumption. The triangular area to the left of *C, ABC,* is the money value of lost opportunity for gaining added consumer-plus–producer-surplus when producers mistakenly produce only Q_1 of wheat.

Interdependent Markets. The real world is not limited to a single wheat market; there are hundreds of thousands of interdependent markets operating simultaneously. One market may be an office worker who types letters to communicate important news to customers, another may be a logger who cuts trees to build homes and furniture, and another market may be a forester who plants new trees for future homes. Can an economy consisting of hundreds of thousands of perfectly competitive firms still achieve the desired social goal of economic efficiency? The answer to the question is yes, if the following three conditions are met:

1. Everyone's demand curve must equal their true marginal benefits (see Appendix B). This means we should not apply the theory when people avoid buying things because they lack information on the existence of products and services they want and can afford.
2. There must be no externalities. If the production of cheap coal-fired electric power emits sulfur dioxide into the air that damages a downwind forest for which the power buyers are not charged, then the competitive market charges too low a price. In this case the power price equals the marginal cost of production, but it ignores the cost of the damaged forest (see Chapter 4).
3. There can be no *monopoly* power (a monopolist is the sole seller in a market and therefore has nobody directly competing for customers). When there are monopoly sellers, they maximize profit not by setting price to marginal cost, but by setting marginal revenue to marginal cost. Price (marginal benefit) exceeds marginal cost and too little is produced for economic efficiency to occur. More resources should be put into production for efficiency to result, but the monopolist will not do it.

Bonnieux (2003) and Fromm (2000) show how economic efficiency fails to occur in markets for water when all three conditions do not occur. Van Den Bergh, Ferrer-i-Carbonell, and Munda (2000) examine limits of markets in promoting economic efficiency for environmental policy.

There may be a group of industries for which these three conditions do not occur. Certain types of food, clothing, and shelter may fall into this category. If consumer information is widely available and cheap, and many firms compete for customers, competition *may* produce economic efficiency. *Therefore, policies that permit or even encourage each firm to act in its own self-interest by maximizing profit at a level of output where price equals its marginal cost can serve society's best interest in terms of economic efficiency.*

Marginal Cost Pricing. Perfectly competitive firms choose to produce an amount of output so that the price over which they have no control equals the marginal cost, over which they *do* have control. Marginal cost pricing is an institution that describes the behavior of any perfectly competitive firm. Shepherd (2003) summarizes recent challenges facing marginal cost pricing for roads in Europe, and Suarez and Guerra (2000) examine its potential for mitigating global climate change. However, marginal cost pricing is much more interesting than merely describing the behavior of profit-maximizing competitive firms.

Marginal cost pricing has everything to do with designing an institution that produces maximum human benefit from scarce resources. The price of each good must be equal to its marginal cost for the economy to be getting the greatest economic efficiency (i.e., producing the highest valued output from its resources and technical knowledge). Only when each producer of output sets its marginal cost equal to every other producer's marginal cost (e.g., when each marginal cost is set to a common price) can all sources be producing output at minimum total cost. Under that condition, the economy will be squeezing the maximum value of output from its resources for human benefit. See Example 3–5.

The social efficiency produced by price equaling marginal cost is as valid for a capitalistic country as it is for a socialist one. For example, China for many years has spent too much and obtained too little from its power production. For this reason, Chinese planners study marginal cost pricing. Giant firms' corporate leaders in New York and London and directors of regulated utilities study marginal cost pricing. People far removed from the private sector—national forest managers, military planners, organizers of centrally planned economies—study marginal cost pricing.

Because of its valuable economic efficiency property, marginal cost pricing has been used as a measure for defining, understanding, and seeking economic efficiency. It is even used to help identify inefficiency in public and private organizations. Even if perfectly competitive markets never existed, the leaders responsible for public welfare can accomplish great service to society from understanding, defining, and learning about the principle of marginal cost pricing.

Governments

The market institution is attractive because it is based on freedom of choice by the individual who makes decisions by buying and selling in the marketplace. However, for many parts of our lives we depend on cooperation with others to meet our needs. It is important, therefore, that our governments (collective-choice institutions) make the kinds of choices that will improve and sustain the welfare of individual members when markets fail. We need look no further than depletion of the ozone layer, the greenhouse effect, and the history of war[18] to see that the need for efficient collective action is no guarantee that such action will occur (see DeBlonde, 2000).

Greater efficiency of collective action in serving the needs of a group's members is likely if the size of the group is small, if the members are similar in their shared values, and if the net benefits from group membership are large and fairly distributed. However, collective action itself does not guarantee fairness. Women, children, and other vulnerable groups, for example, may have little voice, in the collective decision-making process, but they still may be asked to contribute their resources to support that collective action.

Collective Choice Process. Like the market system, the collective choice process (political system) has its rules and its players. Political decisions occur according to a set of rules. In addition to the voting system, these rules include the basic constitution or long-established customs.

---

[18] Winston Churchill (1948) explained the tragic story of how weak international organizations that failed to stand up to the Nazi threat contributed to World War II, a weakness that cost millions of lives on all sides. He emphasized the futility of appeasement and the need for better collective institutions for confronting dictators. Churchill's theme is now considered a self-evident truth in Western society.

Example 3–5
Pricing the Environment at Marginal Cost

Many European cities suffer major congestion, noise, and air pollution due to high and growing use of cars, as well as considerable scarcity of space and highways. This situation exists despite a threefold to fivefold higher price of gasoline in Europe than in the United States. Since the 1970s, economists and transport planners have reached growing consensus that pricing measures are a potential way to limit the demand for road use, especially in the most crowded European cities.

Strong economic efficiency arguments have been advanced to support a transport policy based on marginal cost pricing of roads to reduce crowding, noise, and air pollution. Such pricing confronts people who make the decision to drive a car (as opposed to using substitute transport) to pay the full cost of a public service that was previously provided for free. Despite the fact that total benefits would probably outweigh costs (potential efficiency gains), these proposals are typically unpopular with citizens and their elected representatives. A successful pricing scheme was introduced in London in 2002 and it reduced car traffic by 30%. The marginal cost of one more car entering an already crowded road can be quite high, since it slows down all other cars noticeably. When a road is not crowded, the marginal cost of one more car is essentially zero.

A common criticism of marginal cost pricing of roads is that it is unfair to the poor, as these are precisely the people who can least afford the extra cost of driving imposed by a congestion tax. These would be the people most often "screened" off the road. In Norway, Fridstrom et al. (2000) and Niskanan et al. (2001) analyzed this screening hypothesis by developing a transportation model in which the economic efficiency as well as the income distribution effects of various road pricing schemes were tested. Their studies were based on a traffic network model for the greater Oslo area. The authors found that road pricing had adverse income distribution effects. However, creative policies could avoid some of these undesirable consequences if all or part of the road receipts were redistributed to the households progressively, for example, as an income tax rebate.

In a democracy, decisions are made by elected representatives. These politicians perform a function similar to that of individual firms in the marketplace. Elected representatives interpret the public's demand for collective goods and look for efficient ways of supplying those goods. The theory of *public choice*, developed by James Buchanan (Winner of the 1986 Nobel Prize in economics), makes the assumption that politicians are presumed to behave as if to maximize their probability of reelection.[19] That is, the politicians are assumed to act to maximize votes in much the same way as firms are assumed to maximize profits. The collective choice process, therefore, differs in a major way from the market institution in that the major players in a democratic system are vote maximizers and not profit maximizers. One way to measure the excellence of a political system is the extent to which its process promotes efficient and fair decisions.

Collective Choice and the Environment. Legislation and the resources to support its enforcement are major government actions that influence the environment and natural resources. Environmental legislation governs the relationship people have with the environment and provides needed controls to ensure compliance. Legislation consists of the laws, ordinances, rules, and regulations passed by government. Efficient legislation will reflect the will of the people by producing results for which the benefits exceed costs. For example, when the U.S. Flood Control Act was enacted in the 1930s, it required that benefits of flood control projects exceed the costs. Since then, numerous environmental and natural resources laws, executive orders, and agency actions in the United States and around the world have been initiated that assign major importance to economic efficiency. Properly applied, analysis of economic efficiency has considerable power to improve the environment for human welfare.

[19] Hewitt and Brown (2000) examine motivations that influence government environmental agencies. Judge (2002) analyzes the relation between individual property rights and public environmental policy.

■ SUMMARY

A market is an institution that brings sellers and buyers together to trade certain goods, services, or resources. Some markets are personal, involving face-to-face contact; others are impersonal—buyer and seller never meet.

A demand schedule shows the willingness of buyers in a particular period to buy a given product at each of several possible prices. The law of demand says that consumers are willing to buy more of any product as the price is reduced. The relationship between price and quantity demanded is negative, other things being equal. Demand is graphed as a downward-sloping curve. Market demand curves are found by adding horizontally individual demand curves of all consumers in a market.

The demand for a product has determinants: number of buyers in the market, consumer tastes, incomes, prices of related goods, and price expectations. A shift in the market demand occurs when any of these determinants change. Demand increases by shifting to the right; it decreases when it shifts to the left. Economists distinguish a change in demand from a change in quantity demanded. A change in quantity demanded is a movement from one location to another on a fixed demand curve brought about by a change in product price.

Supply is a schedule or curve that shows the various amounts of a product that producers will offer to the market at each possible price during a known period. The law of supply says that producers will offer more of a product at a high price than at a low price, if other things remain equal. According to this law there is a positive relation between price and quantity supplied, which means that the graph of supply is upward sloping—a higher price draws forth greater amounts of supply. A product's market supply curve is the horizontal sum of all its individual producers' supply curves.

A supply schedule has several determinants; number of sellers in the market, resource prices, production techniques, taxes or subsidies, price of related goods, and price expectations. Changes in one or more of these determinants shift a product's supply curve. An increase in supply means the supply curve shifts to the right; a decrease means it shifts it to the left. Economists distinguish a change in supply from a change in quantity supplied. A change in price causes a change in the quantity supplied but not a change in supply. A change in quantity supplied is shown as a movement from one point to another point on an unchanged supply curve.

Forces of market demand and supply adjust price until quantity demanded equals quantity supplied. The price that equalizes quantity demanded and supplied is called the equilibrium price. Graphically, that price is established at the intersection of the supply and demand curves. The quantity produced where supply and demand intersect is the equilibrium quantity.

Price eliminates potential surpluses and shortages by coordinating selling and buying decisions. The synchronization of decisions by buyers and sellers that creates equilibrium is called the rationing function of prices. Prices are indicators of relative scarcity; they ration goods to those people who are willing to pay the most. When goods are scarce, their rationing function of price sets priorities by allocating available supplies. An efficient price system allocates goods to people who are willing to pay the most for them and allocates output to producers who can sell at the lowest price. The good is allocated to its highest-valued use and is produced with the least cost methods.

Both demand and supply can change. Changes in either changes the equilibrium price and quantity. Increased demand raises equilibrium quantity and equilibrium price; decreased demand reduces equilibrium quantity and equilibrium price. Increased supply increases equilibrium output and reduces equilibrium price; decreased supply lowers equilibrium quantity and raises equilibrium price. Simultaneous changes in demand and supply affect equilibrium price and quantity. These effects depend on the direction of the changes and on their relative sizes.

The market mechanism is a remarkable institution for promoting economic efficiency where industries operate under conditions of perfect competition. Under perfect competition, each firm sets the price equal to its marginal cost of supply, and each consumer sets the market price equal to his or her own marginal benefit of consumption. Perfectly competitive markets have the potential to promote economic efficiency of environmental and natural resources. Any institution such as a market will promote economic efficiency in the use of environmental and natural resources when every user of the resources faces a price equal to the real cost of his or her actions. Collective-choice institutions (governments) also influence economic efficiency in the use of environmental and natural resources.

■ REVIEW QUESTIONS

1. Why does a demand curve for ecolabeled seafood slope downward? What happens to the demand curve for ecolabeled seafood when each of the five determinants of demand changes? Distinguish between a change in demand and a change in the quantity demanded.

2. What effect will each of the following have on demand for ecolabeled seafood?

a. The price of nonlabeled seafood increases.

b. Consumer income increases and ecolabeled seafood is a normal good.

c. Ecolabeled seafood becomes fashionable.

d. Consumers believe that the price of ecolabeled seafood will rise in the future.

e. The price of seafood sauce falls.

f. Domestic import tariffs on ecolabeled seafood are eliminated.

3. Why does the supply curve for organically produced fruit slope upward? What happens to the supply curve for organically produced fruit when each of the six determinants of supply changes? Distinguish between a change in supply and a change in the quantity supplied, noting the cause(s) of each.

4. What effect will each of the following have on supply of organically produced fruit?

a. A technological advance occurs in the methods of producing organic fruit.

b. The number of firms in the organic fruit industry increases.

c. The price of resources needed to produce organic fruit decreases.

d. Producers expect that the price of organic fruit will increase in the future.

e. A decline occurs in the price of organic vegetables, a production substitute for organic fruit.

f. Government passes a law subsidizing production of organic fruit.

g. Government establishes more stringent standards for growers of organic fruit.

5. Consider these two statements: In the organic grape market, demand often exceeds supply but supply often exceeds demand. The market price of organic grapes rises and falls in response to changes in supply and demand. In which of these statements are the terms *supply* and *demand* used correctly? Explain.

6. Suppose total annual demand and supply of organic grapes are as shown in the following table.

Pounds demanded[a]	Price per pound (dollars)	Pounds supplied[a]	Surplus (+) or shortage (−)
1,100	1.60	700	
1,000	1.80	800	
900	2.00	900	
700	2.20	1,000	
600	2.40	1,100	
600	2.60	1,200	

[a]*Hundreds of thousands.*

a. What is the equilibrium price? What is the equilibrium quantity? Fill in the surplus/shortage column and use it to explain why your answers are correct.

b. Graph the demand for wheat and the supply of organic grapes. Be sure to label the axes of your graph correctly. Label equilibrium price *P* and equilibrium quantity *Q*.

c. Why will $2.20 not be the equilibrium price in this market? Why not $1.80? Do you agree with this statement: Surpluses drive prices up; shortages drive them down.

d. Suppose the government establishes a maximum legal (ceiling) price of $1.60 for organic grapes. Explain carefully the effects of this ceiling price. Show your answer graphically. What might motivate the government to establish a ceiling price?

7. Each of the following changes in supply and/or demand affect equilibrium price and equilibrium quantity in a competitive market. Explain what that impact will be; namely, do price and quantity rise, fall, or stay the same? Or is there no clear answer because they depend on the size of the shifts? Use supply and demand diagrams to support your answers.

a. Supply increases and demand is unchanged.

b. Demand increases and supply increases.

c. Demand increases and supply is constant.

d. Supply decreases and demand is unchanged.

e. Demand decreases and supply is constant.

f. Supply is unchanged and demand decreases.

g. Demand decreases and supply decreases.

h. Demand increases and supply decreases.

8. Explain what the following statement means; state whether you agree or disagree and explain why or why not: Prices regulate markets by ensuring that production and consumption are not out of line with each other.

9. Explain the following statement: Even though water prices produce little revenue, they should be retained because of the rationing function they perform.

10. Use a pair of market supply and demand diagrams to show how a government price support of ecolabeled wood products will affect the price and quantity of both ecolabeled wood products and ordinary wood products.

11. Explain these statements: A comparison of the two equilibrium positions in Figure 3–6a shows that consumers buy a larger quantity at a higher price. This evidence refutes the law of demand.

12. Suppose you receive $25 per ton at a recycling center for your old newspapers. However, the local landfill charges you a disposal fee of $40 per ton to accept your old kitchen appliances. Use demand and supply diagrams to show both markets. What economic factors might cause you to get paid for old papers but pay to dispose of old appliances?

13. Suppose the demand for a product is summarized by the equation $Qd = 10 - 0.2P$ and supply by the equation $Qs = 2 + 0.2P$ where Qd and Qs, are quantity demanded and quantity supplied, respectively, and P is price. Remember that price automatically adjusts to make quantity demanded equal to quantity supplied, that is, $Qs = Qd$. Solve these two equations to determine equilibrium price. Then determine equilibrium quantity. Graph the two equations to verify your answers. (For advanced analysis, see Appendix A.)

14. Why does the profit-maximizing firm set its marginal cost equal to price?

15. Explain why the competitive firm's supply curve is its marginal cost curve.

16. Explain why perfect competition can promote economic efficiency. Be sure to define both terms.

PROBLEMS

1. Suppose that in a competitive market for organically grown grapes, demand is given by the equation $P = 600 - Q$, and supply is given by the equation $P = 160 + Q$, where P is price (dollars per unit) and Q is quantity (units) of grapes.

a. Compute the market equilibrium level of P and Q.
b. Compute the total net economic welfare from grapes (consumer and producer surplus).

2. A single firm that sells in this market (organic grapes) has the following cost schedules:

$$MC = 160 + 10\,Q \quad \text{(marginal cost)}$$
$$AC = 160 + 5\,Q \quad \text{(average cost)}$$
$$TC = 160\,Q + 5\,Q^2 \quad \text{(total cost)}$$

a. Compute that firm's profit-maximizing level of grape output.
b. Compute the profits the firm receives at the profit-maximizing level of output. Hint: profits equal total revenue minus total cost.

3. Suppose government enacts a policy that increases market demand for grapes by $10 per unit.

a. Recalculate the market equilibrium P and Q and total net economic welfare.
b. Recalculate the firm's profit-maximizing grape output and the firm's profits.
c. Based on your answers, would producers of grapes support this policy? Why or why not?

REFERENCES

Berbel, J., and J. A. Gomez-Limon. 2000. The impact of water-pricing policy in Spain: An analysis of three irrigated areas. *Agricultural Water Management.* 43 (2): 219–238.

Bonnieux, F. 2003. Using the principles of environmental economics to manage water resources. *Houille Blanche-Revue Internationale de L Eau.* (1): 55–59.

Caviglia-Harris, J. L. 2003. Introducing undergraduates to economics in an interdisciplinary setting. *Journal of Economic Education.* 34 (3): 195–203.

Churchill, W. S. 1948. *The Second World War: The Gathering Storm.* Boston: Houghton Mifflin.

Commons, J. R. 1931. Institutional economics. *American Economic Review.* (21): 648–657.

DeBlonde, M. K. 2000. Environmental economics: The meaning of an "objective" policy science. *Environmental Values.* 9 (2): 235–248.

Eden, F. M. 1797. *The State of the Poor: A History of the Labouring Classes in England, with Parochial Reports.* New York: Benjamin Blom.

European Union. 2004. *Europa.* Available from *http://europa.eu.int/* [Accessed October 18, 2004].

Fridstrom, L., H. Minken, P. Moilanen, S. Shepherd, and A. Vold. 2000. Economic and equity effects of marginal cost pricing in transport: Case studies from three European cities. Government Institute for Economic Research. Available from *http://data.vatt.fi/afford/reports/deliverable-2a.pdf* [Accessed February 12, 2005].

Fromm, O. 2000. Ecological structure and functions of biodiversity as elements of its total economic value. *Environmental and Resource Economics.* 16 (3): 303–328.

Gintis, H. 2000. Beyond *Homo economicus:* Evidence from experimental economics. *Ecological Economics.* 35 (3): 311–322.

Gomez-Limon, J. A., and J. Berbel. 2000. Multicriteria analysis of derived water demand functions: A Spanish case study. *Agricultural Systems.* 63 (1): 49–72.

Hewitt, J. A., and D. K. Brown. 2000. Agency costs in environmental not-for-profits. *Public Choice.* 103 (1–2): 163–183.

Judge, R. P. 2002. Restoring the commons: Toward a new interpretation of Locke's theory of property. *Land Economics.* 78 (3): 331–338.

Little, J. B. 2002. Can green-certified lumber make it? *High Country News.* Available from *http://www.hcn.org/servlets/hcn.Article?article_id=11301* [Accessed October 18, 2004].

Marglin, S. A. 1963. The social rate of discount and the optimal rate of investment. *Quarterly Journal of Economics.* 77 (1): 95–111.

Murray, B. C., and D. N. Wear. 1998. Federal timber restrictions and interregional arbitrage in U.S. lumber. *Land Economics.* 74 (1): 76–91.

Niskanen, E., N. Adler, J. Berechman, J. K. Brueckner, D. Milne, C. Nash, S. Proost, and E. Verhoef. 2001. Implementation of marginal cost pricing in transport – integrated conceptual and applied model analysis: MC-ICAM approach to pricing in transport. European Comission, 5th Framework Programme. Available from *http://www.strafica.fi/mcicam/handouts/deliverables/del1.pdf* [Accessed February 12, 2005].

Pearce, D. 1998. Cost-benefit analysis and environmental policy. *Oxford Review of Economic Policy.* 14 (4): 84–100.

Royal Swedish Academy of Sciences. 2004a. This year's economics prize awarded for pioneering studies of saving and of financial markets. Available from *http://nobelprize.org/economics/laureates/1985/press.html* [Accessed October 18, 2004].

———. 2004b. This year's economics prize to an American. Available from *http://nobelprize.org/economics/laureates/1976/press.html* [Accessed October 18, 2004].

———. 2004c. This year's prizewinners are leading figures within the field of the "new economic history." Available from *http://nobelprize.org/economics/laureates/1993/press.html* [Accessed October 18, 2004].

Shepherd, S.P. 2003. Towards marginal cost pricing: A comparison of alternative pricing systems. *Transportation.* 30 (4): 411–433.

Smith, A. 1776. *An Inquiry into the Nature and Causes of the Wealth of Nations.* Edinburgh, Scotland.

Suarez, M. L. H., and S. M. G. Guerra. 2000. Neoclasic economic theory and environmental policy instruments. *Interciencia.* 25 (2): 102–110.

Swedish Environmental Protection Agency. 2000. *Ecolabelling reduces impacts.* Available from *http://www.environ.se/enviroreport/envirorep2/enviro2_8.htm* [Accessed October 18, 2004].

UK Department for Environment, Food, and Rural Affairs (DEFRA). 2004. *Consumer products: The European ecolabel.* Available from *http://www.defra.gov.uk/environment/consumerprod/ecolabel/* [Accessed October 18, 2004].

Van Den Bergh, J. C. J. M., A. Ferrer-i-Carbonell, and G. Munda. 2000. Alternative models of individual behaviour and implications for environmental policy. *Ecological Economics.* 32 (1): 43–61.

White, P. C. L., A. C. Bennett, and E. J. V. Hayes. 2001. The use of willingness-to-pay approaches in mammal conservation. *Mammal Review.* 31 (2): 151–167.

Zizzo, D. J. 2003. Empirical evidence on interdependent preferences: Nature or nurture? *Cambridge Journal of Economics.* 27 (6): 867–880.

Chapter 4

INSTITUTIONAL BREAKDOWN

Introduction

In February 1968 Garett Hardin, professor of ecology at the University of California at Santa Barbara (see UCSB, 2004) wrote and sent this paper to *Science* magazine: "The Tragedy of the Commons." In it he made these conclusions: In a crowded world, an unmanaged commons (i.e., one resource owned and managed by a group of people) cannot possibly work. Some kind of government coercion of individual action is required for the group's sustained benefit. Hardin was sure that *group coercion of individual action* was the only way to sustain a crowded common property resource in the long run. The now famous and highly readable paper (Hardin, 1968) was published in the December 13, 1968, issue of *Science*.[1]

An unmanaged commons in an uncrowded world causes few difficulties. As soon as crowding starts, however, an unmanaged commons breaks down. Each person, seeking his own self-interest, is trapped into mistreating the commons. An unmanaged commons fosters the rule of "help yourself." Another rule is needed, and there are two possibilities, both of which can work:

1. Set up a community manager—someone with authority to make everybody obey the group rules. This is a managed commons.
2. Divide up the territory and assign it to individuals, making each owner also the manager of her own private property.

The overriding hypothesis of the previous chapter was simple: Results of individual actions, each motivated by self-interest, are good for the group. People who accept this hypothesis recognize the power of the market institution to improve society's welfare through its impact on individuals' economic incentives. The current chapter is based on Hardin's hypothesis, which contradicts Adam Smith's conclusions. Hardin's hypothesis is that results of individual actions, each motivated by self-interest, may damage the group.

Outline

Efficient Institutions 70
Market Failures 70
Government Failures 86

Objectives

After studying this chapter, you should be able to

- Identify the most important standards of an efficient institution
- Explain how market failures can be related to externalities, public goods, and common property
- Describe potential policy remedies to deal with market failures
- Define four types of government failures that hinder conservation

[1] Years after his paper was published, Hardin recognized that coercion is the definition of any law passed in a democracy (UCSB, 2004). Every democratically passed law is one by which we mutually agree to coerce each other. The majority coerces the minority.

■ Efficient Institutions

Institutions are rules that a society designs to shape its individual members' behavior to improve the welfare of the group. For a one-person society, institutional design is the simple process of finding behavior patterns that best meet the one person's objectives in light of available resources and technologies.

Suppose an economy consists of one person, Robinson Crusoe. He designs institutions that produce the greatest possible satisfaction for himself through use of his own means. The institutions he designs help him allocate his labor among productive activities so that the marginal benefits are equal in each activity. He faces all the costs and receives all the benefits of his own decisions. By picking coconuts from a tree, he faces all the costs of construction of a grass hut displaced by that picking, because picking coconuts diverts time from the shelter he could have built for that night. Likewise, since the benefits of the coconuts picked accrue entirely to him, he consumes the fruits of his own labor. As long as the marginal benefit from one more coconut picked exceeds the marginal cost of lost shelter, he continues to pick more coconuts. Facing all the costs means that he pays a price for all his coconut consumption equal to the value of shelter lost (nights spent sleeping in the rain). In a one-person economy there is no need to design institutions that signal the costs of each action. As long as the person has good information on the consequences of each activity, he has no need for institutions to signal that cost to him. There is no need to write laws or establish government bureaucracies that tell him the costs of all his activities.

Once a society consists of two or more people however, no person is likely to be self-sufficient, and the act of production is partly separated from consumption. Each person begins to specialize in the production of activities for which he or she is well suited and starts to trade away his or her unused production for other people's excess production. In a complex society, there arises a need for institutions that signal to each individual the cost his or her activities impose on others. Among other things, the institutions should inform each person of the costs and benefits of each actual and potential action.

If coconuts are cheap, an efficient price signal sends out the good news. The information produced by that signal indicates the low cost of picking coconuts and encourages everybody to consume many. When coconuts become expensive, as the nearby trees get picked clean and coconuts can be picked only by people going greater distances, efficient institutions send out the news warning of this impending scarcity to encourage people to reduce their coconut consumption. For with the rising price of coconuts, eating them displaces dangerously large numbers of other enterprises, such as the building of grass huts. In this situation there is a clear danger of failing to signal the high cost of coconuts; namely, people will keep on working and living as if coconuts were cheap. The lack of an institutional alarm to signal the need to reallocate scarce resources from picking coconuts to building grass huts causes people to consume coconuts at a dangerously high cost: People feel safe but they ignore the approaching rainy season, a breakdown that risks causing them to build too few protective huts. What does this story tell us about standards that an efficient institution should possess? It says that as long as society's institutions permit each person to reap the full rewards of his or her own actions, as well as signal the real cost to other people of those actions, institutions will succeed in producing efficiency.

■ Market Failures

The market institution has the potential to be an efficient and impressive mechanism for issuing signals that promote economic efficiency. As long as market institutions permit each person to reap the full rewards of his or her own actions, as well as signal the real cost imposed on others produced by those actions, markets will help obtain the most human benefit from society's resources. Nevertheless, promoting economic efficiency is a heavy burden for the market institution, and markets cannot always bear the full burden. When the burden is too heavy, markets may suffer one of three kinds of *market failures:* externalities, public goods, and common property.

Externalities

An externality occurs when a market transaction between a buyer and seller directly affects a third party (a bystander), and that third party receives no compensation for the impacts produced by the transaction. The term *externality* is clumsy, but it is based on the idea that when the outcome of a two-party market transaction affects a bystander without compensation, that outcome is "external" to the market. Because it is external to the market, society loses its protection of the market's efficiency.

Negative externalities take the form of a cost imposed on society in which the third party, who bears the cost, fails to receive compensation. The classic example is the impact of a factory that discharges wastes into a stream, which damages downstream water quality. Positive externalities are seen as a benefit received by society in which the person who provides the benefit fails to receive compensation from the third party who gets the benefit. A good example is the household that installs a low-flush toilet, which results in more water being available for another household.

Negative Externalities. Suppose firm *A* could receive a $10 revenue per ream of top-grade paper it manufactures and estimates its marginal cost at $7. From *A*'s view, production of the paper is good economics—it brings in $3 of marginal profit per ream. But suppose each ream of *A*'s paper production dumps chemical pollutants into a stream, which costs $5 of pollution control expenses to people living downstream, costs that occur *without compensation.* The marginal cost to *A* may only be $7 per ream, but the marginal cost to society overall is $12 ($7 of private costs and $5 of social costs). Since society's marginal cost ($12) exceeds the marginal benefit *A* receives ($10), paper production is inefficient. By inefficient we mean that the $3 private gain *A* receives is less than the $5 loss imposed on others (for the definition of Potential Pareto Efficiency, see Chapter 3). Even though *A* has a private incentive to produce paper, society is worse off if it does.

Firm *A* might be selfless and put society's interest ahead of its own. Can we trust *A* to sacrifice the $3 private gain it could receive (legally) to avoid hurting others? Maybe. History shows a few examples of private individuals who sacrificed personal gain for the good of the larger group, but such behavior is not common. The incentive to maximize net personal benefit is strong and often wins out over the maximization of society's net benefit. So more commonly, firm *A* acts to maximize net personal benefit more often than to maximize society's net benefit. So, when we see negative externalities, what do we expect from *A* most of the time? Firm *A overproduces* the product in question.

Can you think of other examples? Consider how we dispose of trash. When we dump trash in a stream or toss wrappers out a car window, we are behaving like firm *A*. We minimize inconvenience to ourselves, but make others pay. When we abuse others, we create external costs for others. In these examples allowing each individual the freedom to act in an unrestrained manner damages others.

You probably see the point. Giving each individual the freedom to act as he or she pleases works well if he or she is the only one affected or pays the full cost (social plus private). Maximum individual freedom is economically efficient (good for society) if that freedom imposes no cost on others. When the actions of one individual negatively impact others without full compensation for those costs, private self-interest *conflicts* with social interest and maximum individual freedom of choice is no longer economically efficient. This is why we want economically efficient laws (enforced, not just written) as well as efficient prices, even though neither comes free. Both efficient laws and efficient prices restrict our personal freedoms. They take away our freedoms to pollute, play music too loudly, and abuse people in other ways. Even though both restrict our individual freedoms, well-written laws and well-designed prices increase social efficiency. They maximize society's total consumer and producer surplus. Recall our discussion in Chapter 3.

Figure 4–1 is an example of a negative externality for the case of air pollution emitted by the burning of coal to produce electric power. When there is a negative externality, the marginal social (opportunity) cost of producing electricity (MSC_n) consists of two pieces: (1) the marginal cost (MC) to the utility of producing the power, equal to Q_cC, and (2) the marginal external cost

Figure 4-1 Impact of an externality

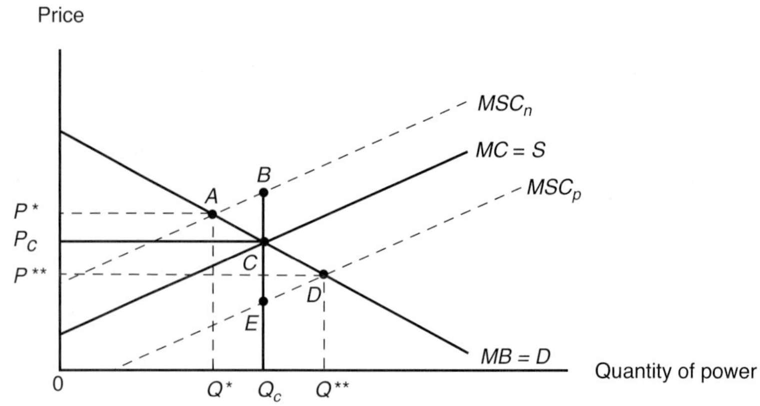

to all bystanders affected by the air pollution, equal to *BC*. Because of the negative externality, the marginal social cost of power exceeds the marginal private cost paid by the utility.

The economically efficient amount of power production occurs where the area between the demand and real marginal cost is maximized, which occurs at point *A*. It is efficient because at that point the total surplus, the value of power to consumers minus the real cost of producing it, is maximized. This point of efficiency occurs where the MSC_n curve intersects the demand curve (*MB*) at an efficient price of P^*. As long as the negative externality produced by the pollution exists, the market supply curve does not reflect the true cost of producing power, and the power market will continue producing where market demand intersects market supply at the point *C*.[2] At point *C*, the market produces Q_c minus Q^* more power than is economically efficient.

Society suffers a welfare loss produced by this negative externality. The welfare loss is measured as the amount by which marginal social cost exceeds marginal benefit over all the units of excess power produced by the market. This total loss, equal to area *ABC*, is called the *deadweight loss* caused by the externality. It is the area between the marginal social cost curve and the demand curve, between the actual level of output (Q_c) and the efficient level (Q^*).

Positive Externalities. Some spillovers create benefits rather than costs for others. If you make the environmentally correct decision to install a low-flush toilet in your house, somebody else gets the water you save. If you put landscape lighting (e.g., floodlights) in your front yard, your house is better lit at night, but your light also illuminates your neighbors' yard, thereby increasing their safety.[3] Will you install efficient quantities of low-flush toilets and put in efficient quantities of landscape lighting? If you act only out of self-interest and ignore the benefits that spill over to your neighbors, you will perceive low-flush toilets and landscape lighting to be less valuable than they really are, so you are less willing to buy them.

Suppose the landscape lighting creates a marginal benefit of \$10 to you and \$15 more to your neighbors for a total of \$25. Suppose also that there is a \$13 marginal cost to install the light. Will you put in the light based on your own self-interest? Remember the \$25 marginal benefit to society (\$10 to you and \$15 to your neighbors) more than pays for your \$13 marginal cost. So, putting in the lighting is economically efficient to society; however, it is not privately profitable (efficient) to you. Your \$13 marginal cost exceeds your \$10 marginal benefit. Installing the light

---

[2] Dasgupta (2000) provides an example of four kinds of negative social externalities associated with fertility behavior.

[3] Floodlights can also produce a negative externality if your neighbors are backyard astronomers, because the floodlights increase light pollution. Floodlights also disrupt nocturnal wildlife. For negative externalities produced by floodlights, see International Dark Sky Association (2004) and Sky and Telescope (2004). Diakoulaki et al. (2001) calculate the value of positive externalities resulting from investments in solar water heaters. Leggett and Bockstael (2000) analyze positive externalities produced by water quality improvements in the U.S. Chesapeake Bay.

may create \$12 of net benefit to your neighbors, but it creates a \$3 net loss to you. Will you sacrifice for the good of the neighborhood in this case?

What is the lesson taught by external benefits? Society *underproduces* the good action in question. Inefficiently small quantities are produced. Figure 4–1 shows that the economically efficient output level, where the marginal social cost curve, MSC_p, intersects the demand curve at point *D*, at output level Q^{**} and at a price equal to P^{**}. This output level is higher than the competitive market level, equal to Q_c.

The Coase Theorem. Ronald Coase, a British-born University of Chicago economist, earned fame for the theorem that bears his name and was awarded a Nobel Prize in economics (Nobel Foundation, 2004b). The *Coase theorem* states that if there are clear property rights, low bargaining costs, perfect competition, perfect information, and an absence of wealth and income effects, resources will be used efficiently regardless of who owns them. The existence of these conditions harnesses the power of the market to manage externalities.

Virtually all economists before Coase accepted Pigou's idea (discussed later in this chapter) that if, say, a rancher's cattle destroyed a neighboring farmer's crops, the government should intervene by stopping the rancher from stocking more cattle, or should at least charge the rancher a tax for damaging the farmer. Otherwise, the cattle would continue to destroy crops because the rancher had no incentive to stop that destruction. But Coase challenged that view. He showed that even if the rancher had no legal liability to compensate the farmer for destroyed crops, if *transactions costs* (i.e., the costs of resources necessary to transfer, establish, and maintain property rights) were zero, the farmer and rancher could come to a mutually beneficial agreement in which the farmer paid the rancher to reduce his numbers of stocked cattle. Coase believed this would happen if the damage to the farmer from added cattle stock exceeded the rancher's net returns on these cattle. If the rancher's net income on a marginal stocked steer is \$2, the rancher will accept any amount over \$2 as compensation to avoid stocking the added steer. That is, the rancher can make more income by not stocking than by stocking. If the steer does \$3 worth of crop damage, the farmer will be willing to pay the rancher up to \$3 to avoid stocking the steer. A mutually beneficial bargain would be struck with no need for government intervention into private market affairs. A considerable amount of resources can be saved on lawyers, litigation, and government regulations when society's resources can be allocated under a system of mutually beneficial negotiation.

Example of the Coase Theorem. Suppose a factory moves into a streamside area and begins dumping pollutants into the water. An existing group of sport anglers who fish the stream receive benefits (consumer surplus) by catching fish, but now the pollution reduces their catch. Suppose the pollution reduces the fishers' benefits by \$1,000 a year.[4] Suppose further that if the factory uses pollution-cleanup equipment priced at \$500 a year, the fishers' benefits increase by \$1,000 a year to prepollution levels. There are two possible ways to assign property rights to the lake:

1. **Right to be free of pollution:** Under this system of rights, the factory cannot pollute the lake, for it would violate the fishers' property rights. The factory could reach an agreement with the fishers to continue polluting the lake and compensate them a minimum of \$1,000 a year for lost benefits. However, it would be cheaper for the factory to buy the \$500 pollution-cleaning equipment. The result is that the market system encourages the factory to buy the equipment and let the fishers use a clean lake.

[4] The term *benefits* is more deceptive than it seems. Two ways of measuring benefits are (1) the willingness to pay in money for the service and (2) the willingness to accept monetary payment in lieu of the service. If there are no income or wealth effects, the two are equal. But if the demand for the service increases with higher income (a positive income effect), the willingness to pay is smaller than the willingness to accept payment. For more details on the two measures of benefit, see Chapters 5 and 7.

2. **Right to pollute:** If the fishers want to keep fishing, they have two options. They can suffer the $1,000 in lost benefits, or they can pay for fishing rights on the stream by paying the factory $500 to buy the equipment. The result is that the market encourages the fishers to buy the equipment so they can fish in a clean lake.

This simple example illustrates the power and scope of the Coase theorem. The pollution-cleaning equipment gets purchased no matter who owns the lake, and the fishers fish in a clean lake. Furthermore, economic efficiency results no matter who owns the property: The benefits of the clean lake ($1,000) exceed the cost of the equipment ($500).

Suppose the facts are different and that the fishers lose only $200 a year in benefits from a polluted lake, instead of $1,000. The pollution-cleaning equipment still costs $500, but the cost of tolerating the pollution (lost fishing benefits) is only $200. If we value economically efficient decisions, we would hope that pollution would continue, because the cost of the cleanup ($500) is higher than the smaller cost of the pollution ($200). Again, consider each property right scenario:

1. **Right to be free of pollution:** If the factory wants to continue dumping waste into the lake, it can either buy the $500 in equipment or pay the fishers a minimum of $200 in pollution damages. The result is that the factory pollutes the lake.
2. **Right to pollute:** If the fishers want to keep fishing, they can either buy the $500 cleaning equipment for the factory or suffer a $200 loss as the factory continues to pollute without the equipment. Result: The factory pollutes the lake because the $200 loss to the fishers is less than the $500 loss they would suffer from buying the equipment.

These results illustrate the application of the Coase theorem. The pollution equipment remains unsold no matter who owns the lake and the fishers fish in a polluted lake, but economic efficiency still occurs. Remember that the Coase theorem neither guarantees nor hopes for an elimination of the externality. The theorem asserts that once property rights are defined, the unregulated market will find the least cost (most economically efficient) method of coping with externalities.

The findings of the Coase theorem can be summarized as follows: (1) If there are no barriers to bargaining, assigning property rights to somebody results in the efficient decision in which total economic benefits are maximized. (2) Economic efficiency is achieved regardless of who is assigned the property rights. (3) The distribution of property rights does affect who receives how many benefits. Whichever party has property rights receives compensation by the party without them.

Figure 4–2 illustrates the main prediction of the Coase theorem. Quantity of emissions are measured on the horizontal axis, ranging from a minimum of zero to a maximum of 10 units. The

Figure 4–2 Coase theorem

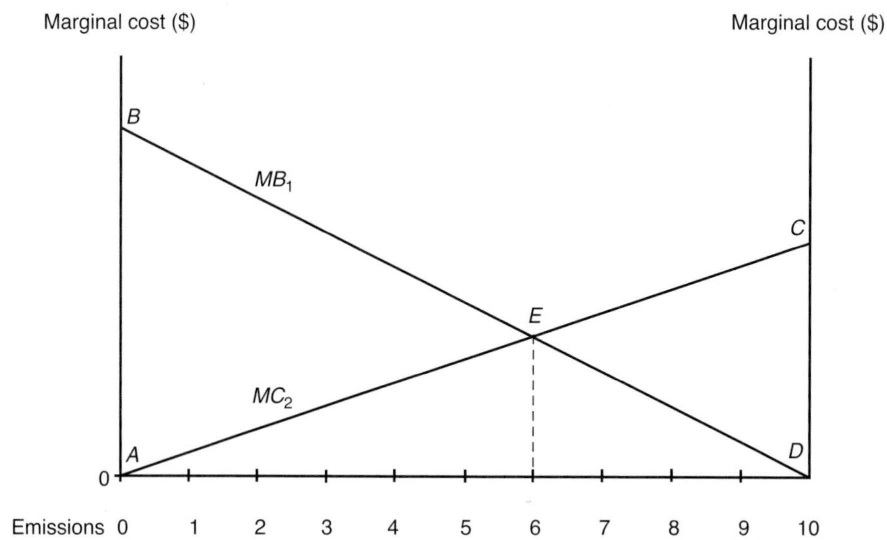

downward-sloping line shows the marginal benefit of abatement costs not incurred by the emitter, MB_1, which starts at point *B* and declines to zero at the maximum (uncontrolled) emission level of 10 units. People who suffer damages from the emissions, receivers, incur rising marginal costs from those emissions, MC_2, with increased emissions. If property rights to a clean environment are assigned to receivers, then the initial allocation of emissions will be zero (the far left side of the graph). However, bargaining between the emitter and the receiver will continue until the efficient (higher) level of pollution of 6 units is reached at point *E*.

If the legal system assigns a property right to emit to the emitter, the prebargaining allocation of emissions occurs at 10 units, shown as the far right side of the horizontal axis. At that uncontrolled level of emissions, the marginal benefit of the first unit of reduced emissions (from 10 units to 9), the savings in marginal costs suffered by the receiver equals *C*. The marginal abatement cost of that first unit controlled, assumed equal to zero, occurs at point *D*. In this case, the receiver bargains with the emitter to continue reducing emissions until point *E* is reached. The figure shows the Coase prediction, namely, that the same final point is reached regardless of who has the property rights. (For a detailed analysis of efficient pollution control policy, see Chapter 16.)

Limits of the Coase Theorem. For economic efficiency to occur, the two sides must be able to bargain successfully. Detractors of the Coase theorem cite three reasons that bargaining attempts fail even though their benefits exceed their costs:

1. If transactions costs are too high, it may be too expensive for the two sides to meet. For example, if there are many fishers and only one polluting firm, the cost of contacting the fishers and discovering their benefits of reduced pollution may be too high.
2. If either side lacks information about the costs or benefits of limiting pollution, an inefficient result could occur. It is hard to know how much to offer to the other side and to come to a bargain if you are ignorant of how the polluting activity affects the other side.
3. If either party engages in strategic behavior, an economically efficient agreement may be blocked. If one player, for example, demands the maximum possible return and will not back down from that position, it might not be possible to reach an efficient agreement.[5]

Public Goods

Understanding public goods requires knowing something about property rights.[6] A property right is a kind of human right. It is exclusive authority to decide how a resource is used, whether it is owned by private individuals or by government. Society affirms the uses chosen by the holder of the property right based on the government's authority to enforce those rights. If the government owns the resource, the people who decide on its use must manage the resource under the rules established by the community.

The three basic elements of *private* property are (1) exclusive rights to use a resource, (2) exclusive rights to the services of a resource (e.g., to lease a resource for income), and (3) rights to

---

[5] However here is how a purist might defend the efficiency attributes of the Coase theorem in face of each qualification: (1) Transactions costs: These are a real cost like any other cost. As long as they are lower than the benefits to be gained by the bargain, incentives exist to bring the bargainers together. For example, if the benefits of the bargain are $500 but are spread over 10 fishers, there are up to $500 in incentives to be had by an enterprising facilitator who consummates the bargain. (2) Information: One good way to find out information about the other side's benefits of the bargain is to offer up to the maximum benefits the bargain would give you and see if the other party agrees. If the party agrees, you are both better off and efficiency has been reached. If the party does not agree, then the benefits to your offer are evidently less the benefits lost by accepting the offer. (3) Strategic behavior: If one player demands maximum benefit for the bargain and will not budge, and if this is more than the other player is willing to pay, then the bargain should be blocked because it is economically inefficient.

[6] This discussion of property rights is based on Alchain (2002).

Example 4–1
Property Rights and Resource Depletion at the Frontier

Frontiers are remote, unsettled places. Landowners do not enjoy the clear presence of government and legal enforcement to support their ownership claims. The first settler on the frontier finds that even though the law may support her land claim, the frontier's remoteness makes it hard to prove. The limited presence of a legal infrastructure creates an opportunity for future settlers to contest the settler's claim and even attempt to evict her.

Defining and enforcing property rights at the frontier is expensive. When the owner of a site decides to exploit a natural resource—pasture, fish, water, forest, hunting ground, oil field—she must decide on the intensity of exploitation of the resource and on the level of spending needed to define and enforce her property rights. When later contestants at the site are expensive to exclude, the original settler may increase the speed and intensity of the resource's current exploitation in order to reduce spending on excluding future encroachers.

Hotte analyzes three historical case studies where this happened: (1) In the Brazilian Amazon frontier, huge distances to government centers, and low commercial value of the land made the costs of registering land titles greater than its benefits. (2) Similar events occurred in Honduras during a cotton and cattle boom in the 1950s. The owner of a large hacienda fenced an area inside of which small farmers had been building homes for years. Unable to prove ownership, the settlers were forcibly removed. (3) In the 1980s, the threat of a land reform in Paraguay led to widespread deforestation by landowners for fear that their forested areas would be declared unproductive and be taken over by somebody else.

What common thread ties these diverse cases together? It is the cost or difficulty for the land's original occupier to prove ownership rights. That lack of security at the frontier encourages short-term, quick payoff, unsustainable exploitation actions. Any policy aimed at increasing the value of land's output in order to promote conservation should be combined with better government support for protecting property rights.

Sources: Hotte, L. 2001. Conflicts over property rights and natural resource exploitation at the frontier. *Journal of Development Economics.* (66): 1–21. Campbell, J. R. 2004. Ethnic minorities and development: A prospective look at the situation of African pastoralists and hunter-gatherers. *Ethnicities.* 4 (1): 5–26.

transfer a resource to a new owner at mutually agreeable terms. Under a system of private property rights, the market values of property reflect the preferences and demands of those members of society who can put the property to use as a resource. See Example 4–1.

No matter who the owner is, the use of a private resource is influenced by what others believe is the most valuable use. An owner who chooses to use a resource for other than its highest-valued use gives up that use and the price others would pay for the resource's use. What this means is the market value of private property is based on the highest private value to which anybody could potentially put it. An important function served by the institution of private property is that it eliminates competition for control of economic resources through nonpeaceful means. Well-defined and secure rights promote peaceful transfer of resources among owners.

Public goods were described by Nobel laureate Paul Samuelson in two famous articles (Samuelson, 1954, 1955) that have dramatically altered the way economists look at government. Samuelson discovered this special characteristic of a public good: Once it has been designed, built, supplied, and made available for use, many people can simultaneously use it without limiting the opportunity of others to do the same.

The flood protection offered by a dam is a public good. Once money has been set aside, land bought, engineers hired, and the river dammed, all the people who live downstream of the dam and who might otherwise get flooded *simultaneously* enjoy the flood protection services. One person's consumption of the dam's flood protection services does not displace another's. These days, with fast and cheap communication, an inventor's new idea posted on the Web is a public good. Once the idea has been accepted, everybody with Web access can enjoy it without displacing other's benefits.

Table 4–1 Classification of economic goods

		Rival: Yes	Rival: No
Excludable	Yes	**Private Goods** Electric power Bottled water Backpacks	**Club Goods**[a] Pay television Public lands recreation Uncongested library
	No	**Common Property** Shared aquifers Oil deposits Shared grazing lands	**Public Goods** Common knowledge Flood control Lighthouse

[a] Sometimes called natural monopolies.

Excludability. Table 4–1 shows that two characteristics uniquely define a public good. The first is *excludability.* A good is excludable (e.g., is subject to the *exclusion principle*) if there is some way to "put a fence around it" and exclude somebody from using it. If you don't pay for that shiny new car, the dealer excludes you from using it. Some goods are nonexcludable; once the good has been supplied, its benefits (or damages) accrue to all. Individuals who avoid paying their share cannot be excluded from its use. The reduction in flood damages received by people who live downstream of a dam are nonexcludable. Once the dam is there, it is hard to exclude people from taking advantage of the benefits of the increased flood protection.

Rivalry. The second characteristic of a good is *rivalry.* A good has this characteristic if one person's use of the good prevents others from using the same unit. A good lacks rivalry if the use of one unit of the good by one person *does not* reduce that unit's availability to somebody else. Where rivalry is absent, the additional cost of supplying the unit's services to one more person is zero. As Table 4–1 shows, things like pay television and the services of a lighthouse lack rivalry; that is, the use of those services by one person does not limit use by other people. The bag of groceries you buy at the supermarket has the characteristic of rivalry, because by taking the groceries home, you prevent somebody else from doing the same.

Table 4–1 shows the four kinds of goods according to the presence or absence of excludability and rivalry: public goods, private goods, club goods (i.e., natural monopolies, and common property). The institution of the market is excellent for promoting economic efficiency for *private goods* (i.e., those that have both excludability and rivalry). Markets do not promote efficiency if either or both of those two characteristics are lacking.

Revealed Preferences. In some circumstances, it is within our self-interest to send false signals about our preferences for a public good. Sometimes we have an incentive to pretend we have less support for a given public good than we really have. For example, if I reveal my wish for the use of a public good, I risk being charged my share of the cost. Or I may overstate my support for building of the public good, hoping that my strong vote will help get it built, while also thinking that I can shift the payment responsibilities to somebody else after it is constructed.

A good example of false signaling occurs when I solemnly sign a contract that binds me to pay for my share of a public good, but I withhold hope that I can weasel out of (renegotiate) the contract *after* the public good is supplied. This principle is illustrated by U.S. water reclamation projects, for which irrigation farmers have historically shown a great willingness to support but a reluctance to pay the full contracted costs (Martin, Ingram, and Laney, 1982). The nonexcludable and nonrival characteristics of a public good give rise to a concept known as

Figure 4–3 Impact of a public good

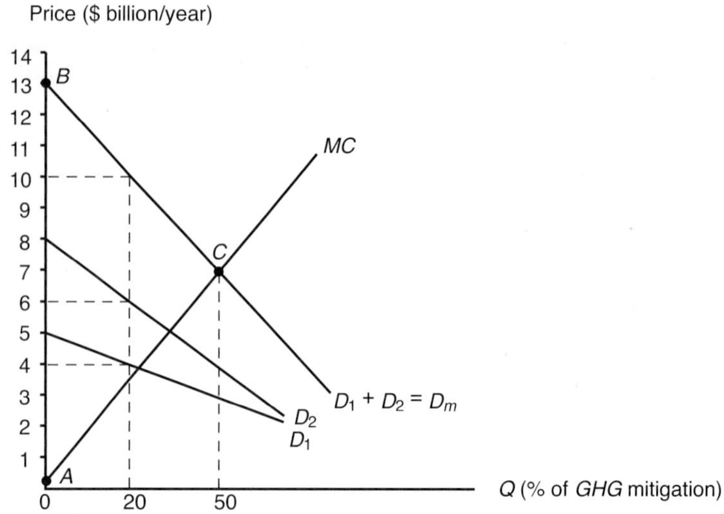

"free riding"; a *free rider* is a person who receives the benefit of a good but successfully avoids paying for it.[7]

The classic problem surrounding public goods is that businesses will supply few public goods, because the free riding that occurs after the good is produced makes it unprofitable to finance the supply. Business owners are not fools; they know about free riding, or they discover it from hard experience. Markets often are reluctant to produce public goods when free riders cannot be excluded from the benefits.

Figure 4–3 shows the impact of a public good, using the example of greenhouse gas (GHG) mitigation in two countries. D_1 is the marginal benefits to country 1 produced by a joint action of all countries that reduces GHGs. D_2 is the same concept for country 2. Reduction of GHG is a public good because it lacks both excludability and rivalry. No one nation (or person) can be excluded from the benefits of reduced GHGs by withholding services once it has been supplied. The cost of supplying services produced by GHG reduction to the additional nation is zero. The figure shows that the first percentage point in controlled GHG emissions produces $5 billion per year for country 1 and $8 billion per year for country 2, for a total of $13 billion annually to both. Note that because of nonexcludability and nonrivalry, the total demand for a public good (D_m) is found by vertically summing individual demands, D_1 and D_2, for any given percentage of GHG emission mitigation. Vertical summation of individual demands is the right method to find the market demand for any public good. Loomis (2000a, b) shows how vertical demand summation is applied to valuing a wide range of natural resource public goods.

Once 20% of GHGs have been reduced, the incremental benefit of one more percentage point reduction is $4 billion for country 1 and $6 billion for country 2, for an aggregate marginal benefit of $10 billion. The figure shows that a treaty (a group activity) that successfully reduces GHGs can produce a net efficiency gain equal to the triangular area *ABC*, the area between the total marginal benefits curve (D_m) and the marginal cost schedule, *MC*. The figure also shows that any institution that blocks the maximum potential efficiency gain of *ABC* (such as no GHG treaty or its weak enforcement) produces a failure to achieve a potentially large benefit. So, it is in society's best interest to limit GHGs by 50% if maximum net efficiency benefits are the goal.

---

[7] Arce and Sandler (2002) illustrate the point with an African example, showing that since any person can benefit from a public good without contributing to its maintenance, incentives to contribute are weak and private markets will not provide it. Efforts by the African nation of Kenya to clean up Lake Victoria's waters can substitute for efforts by Tanzania and Uganda to do the same. If beneficiaries in these three countries cannot cooperate voluntarily, public intervention at the formal multinational level may be the only way to make it happen.

Despite the potential efficiency gains of supplying public goods, it is important to ask this question: What will we observe actually happening? We will see that business will lose money in providing public goods, so government has a potential role in ensuring that economically desirable public goods are built and supplied. Government, through the coercive power of taxation given to it by the people[8] (mutual coercion mutually agreed upon), can finance public goods for which the benefits exceed the costs. A local government can sponsor a fireworks display and finance the display in the form of a tax. If each person's tax burden is less than the value of the fireworks display to that person, each person is better off. However, if a private business, which lacks the power of taxation, produced the same fireworks display at the same total cost, most people would watch from their porches and the business would lose money. This example illustrates the point that governments can potentially improve market outcomes.

Common Property

Common property is owned (possession with the right to transfer possession to others) by a group rather than an individual. *Common property resources* present a special problem for which there is no easy solution. A good example is a group of water users who all pump from a shared, nonrechargeable underground pool in which each user's pumping reduces the water available for the other users.[9] We want to know if unregulated perfect competition in the use of the common resource, in which each owner uses the quantity that maximizes his or her own income, has the effect of producing the greatest possible total benefits for all (i.e., we want to know if unregulated perfect competition produces economic efficiency). See Example 4–2.

Unless great care is exercised in regulating use of the common pool, incentives facing each individual have the effect of producing behaviors that damage both the resource and all its owners. Although the common pool problem has been recognized for centuries, one of the clearest and most compelling explanations was given by Hardin (1968).

Tragedy of the Commons. Hardin illustrates what is called the *tragedy of the commons* by asking us to imagine a commonly owned pasture, shared by n (more than one) ranchers. Each rancher, being a rational self-interested person, attempts to maximize her economic gain. Each rancher's major decision is whether to stock an extra animal on the common pasture. Each asks, "What is the benefit to me versus the cost to me of stocking one more animal on the common pasture?" Each sees both a personal benefit and a personal cost of stocking the extra animal. Since each rancher receives all the revenue from the sale of the animals' additional weight produced by the stocking, the positive benefit is B (added net income). Society also receives the same benefit, B, from the rancher's decision to stock, because the rancher's net income from the sale of the animal equals society's net willingness to pay for the animal products.[10]

The cost component, C, is the additional overgrazing created by one more stocked animal, typically equal to the money value of reduced weight on existing and future animals stocked. When there are few animals on the pasture this cost is zero, but these costs escalate with higher animal density. More important, the cost of any one rancher stocking one more animal is shared by all the ranchers, because they all have less grass for their animals. Therefore, the cost for any

---

[8] The people do not always give that power to the government; sometimes the government takes it. History is full of examples where the people gave no authority to the government whatsoever to supply public goods or do anything else.

[9] Nonrechargeable means that there is no recharge from a renewable source like a river, so each gallon pumped is a gallon lost for the future. Since each person's use imposes costs on others, common property resources are characterized by rivalry. It is not the same as a public good, which lacks rivalry.

[10] The rancher's demand for grass (added net income) is a demand for an input, which is derived from the demand for additional animal products (e.g., weight gain) the grass produces. That is, the change in the rancher's net income is the value of the grass's marginal product in producing animal products, equal to the animal products' price (willingness of society to pay) times the added physical output of animal products. See Chapter 9 for details.

Example 4–2
Wildlife Management in Zambia

Communal land in Zambia is an environmental common property resource. A lack of established private property rights to that land blocks incentives to take care of that land. Slash-and-burn agriculture can lead to deforestation. So why do farmers do it? Economic theory gives us the answer. The losses to farmers produced by logging trees are seen as minimal, since those farmers have no property rights to future uses of the land. The result is the tragedy of the commons.

Negative externalities associated with wildlife hunting pose special problems in Zambia. The production of agricultural goods, horns, ivory, and tourism all result in private costs being incurred as well as private benefits. Typically, the market price of a good or service reflects these private costs and benefits. But production and consumption of goods and services can involve additional external costs, or negative externalities experienced by people other than those who are directly producing or consuming the good.

Killing a rhino and selling its ivory horn produces considerable income and brings economic benefit to the hunter's family. However, the cost of killing the rhino to future generations is ignored. Self-interest, an important guiding principle of the free market, fails to take into account external costs of wildlife hunting imposed on future generations.

What policy choices are available to address these market failures? For the common property problem, private property rights could be assigned to the affected communities. For example, private land ownership could be assigned; this would give citizens a greater incentive to manage the area's wildlife to ensure sustainable populations. For the negative externality problem, the ivory industry could be regulated by banning or otherwise limiting the ivory trade. This has actually occurred, so the future looks better. Zambia signed the Convention on International Trade in Endangered Species (see United Nations, 2004), in which rhinos were classified as receiving the highest possible protection.

Source: Biz Ed (2004). *Zambia: Environmental market failure.* Available from *http://www.bized.ac.uk/virtual/dc/wildlife/theory/th3.htm* [Accessed October 20, 2004].

single rancher from stocking one more animal is only C/n, where n is the number of ranchers. From society's view, additional animals will be stocked as long as $B > C$, if the goal is economic efficiency. However, after stocking has increased to the point where the marginal animal produces $B = C$, the herd size is economically efficient and additional stocking will stop.

Table 4–2 shows the incentives facing any single rancher as well as the impact of his or her behavior on society. Notice that from the view of any single rancher, the benefits of stocking one extra animal (B) are typically much higher than the costs (C/n). Also, the benefits to the rancher of not stocking the extra animal (C/n) are typically much lower than the costs (B). The table's top row shows that each rational rancher finds that the only economically sensible course to pursue is to add another animal to her herd as long as $B > (C/n)$. With a large number of ranchers (large n) with whom to share the costs, C/n is small, so even if B is small, each rancher still has an economic incentive to continue stocking extra animals even if the number stocked is already very large.

The bottom row of the table also shows that any single rancher sees little sense in investing in conservation methods, such as reduced stocking or other pasture improvements. For that one rancher, the costs of reduced stocking (B) are typically much larger than the benefits (C/n).[11] Because each rancher reaches the identical conclusion, each rancher will foster an aggregate stocking behavior that results in the tragedy of the commons. Each rancher is locked into a system that compels her to increase the herd size without limit, even when the grass is limited. Bankruptcy is the end result toward which each rancher rushes, each pursuing her own best interest in a society

[11] More generally, any user of a common property resource sees little reason for paying for, economizing on, investing in, improving, or conserving a resource without assurance of secure and exclusive rights over that improved resource that ensure the recovery of those costs through use, lease, or sale and without assurance that those rights will be enforced.

Table 4–2 Economic incentives facing each owner of a common property resource

Decision	Benefit (*B*)		Cost (*C*)	
	Individual	Society	Individual	Society
Stock extra animal	*B*	*B*	*C*/*n*	*C*
	Extra revenue accrues entirely to individual	Extra revenue accruing to individual = society's willingness to pay	Extra costs to individual of overgrazing are small, shared with all ranchers	Extra costs of individual's overgrazing to society are high
Don't stock extra animal	*C*/*n*	*C*	*B*	*B*
	Extra costs saved by animal not grazed are shared with all ranchers	Extra costs saved by animal not grazed are borne by society	Extra revenues foregone from animal not grazed accrue entirely to individual	Extra revenue foregone from animal not grazed is society's displaced willingness to pay

n = number of ranchers.

that believes in the freedom of the commons. Complete freedom in the use of a commons ruins the resource and bankrupts everybody.

Examples of the Tragedy. A review of the World Bank's experience in promoting sustainable economic development in third-world countries showed some surprising results (Bruce and Mearns, 2001). The bank discovered that forests and pastures in many poor developing countries have common property characteristics because they are difficult to split into individual allotments. The same factors that give rise to assigning individual rights cause these resources to be subject to the free-for-all use pattern that Hardin described. Three situations occur in which the bank's advice is solicited on management of common property resources: (1) when a resource has been subject to open access in the past, but now there is a wish to create resource-user incentives for sustained use and management; (2) when a resource has been managed under common property institutions, but the system faces challenges; and (3) when a resource has failed when managed under direct state management. Compounding the bank's problem is the fact that the poorest people in developing countries typically receive a larger part of their livelihoods than others from common property resources. Typically, women and other vulnerable groups rely on these resources heavily.

Another example of Hardin's tragedy involves certain popular U.S. national parks. The parks are limited in quantity and quality: There is only one Yosemite Valley (U.S. Department of Interior, 2004a, b), but the number of visitors who wish to visit Yosemite grows every year. After the park's carrying capacity is reached, each additional visitor's trip to the park contributes to crowding, degrades the resource, and lowers the value of the experience for other visitors. When the park is congested, even though one more visitor gets a larger personal benefit than personal cost, the gain in total benefits to all visitors is less than the gain in total cost. The problem of managing the commons is particularly acute in a democracy, in which it is a shared value that nobody should be excluded from enjoying a community's (e.g., a nation) treasures.

Still another example comes from an internationally shared supply source for *groundwater*. The Hueco Bolson is a depletable *aquifer* that lies under the border of the United States and Mexico and is shared by the two large cities of El Paso, Texas, and Ciudad Juarez, Chihuahua. Pumping by each city is unregulated by the other. Neither city has exclusive rights to use its own part of the aquifer. Pumping reductions on the part of each city save water for the future. However, each city knows that part of the water it saves will be used by the other city. From each city's view, its own costs of conservation are high and its benefits are low.

Carryover storage at water storage reservoirs provides a compelling example of the tragedy of the commons. Elephant Butte Reservoir in southern New Mexico supplies about 800,000 acre-feet of water per year; 57% of the rights to use the water are owned by New Mexico property owners and 43% by Texas owners. If New Mexico decides to save (carry over) 100,000 acre-feet for use next year, it reduces its reservoir releases and water use by 100,000 acre-feet this year. However, there is little incentive for saving: The saved water is a common pool resource, which means that New Mexico only has rights to use 57% of 100,000 next year and must share the other 43% with Texas. Evaporation also takes 10 feet per year off the top, creating savings at a highly negative interest rate. Not surprisingly, neither user saves much water for future use, even in periods of serious drought.

Oil pools provide another example. Unregulated oil pools are a common resource when many oil producers pump from the same formation. Each producer faces the same benefit and cost incentives faced by Hardin's cattle rancher who grazes the commons. Heard stories of new oil field discoveries that prompted many wells to be sunk by independent producers? Each owner was afraid to conserve for the future, for fear of losing the conserved oil to the co-owners. Without regulation, huge quantities of oil and gasoline were dumped onto the U.S. market in the 1930s at bargain prices, and consumers experienced a great short-run benefit. However, consumers incurred the long-run cost of considerable reduction in total oil supply that can be recovered economically.

Merely understanding the dangers of an unregulated commons does not guarantee reform. Because nobody owns the earth's atmosphere, it is treated as a common disposal ground into which everyone may discharge his or her wastes. Unwanted consequences of this behavior include the greenhouse effect, acid rain, and damaging the earth's protective ozone layer. Each industry, region, and nation that could limit its air emissions views the limit as too expensive. Continuing to defend the freedom to discharge pollution into the air may ultimately lead to ruin for all. In the case of global warming (see Chapter 15) and the ozone layer, nations are just beginning to evolve controls to limit damage.

Crowded toll-free roads are another example of a common property resource. Where roads are privately owned, owners often charge tolls (prices). They may implement *peak-load pricing*; that is, they charge higher prices during times of peak demand and lower prices at other times. Owners charge prices and people take the price into account in deciding whether, when, and how often to use them. But because governments own roads that they finance with tax dollars, they typically charge no price. The next vehicle to enter an already crowded road has no incentive to take an alternative route or enter at an alternative time, for the costs to her of her own added crowding are less than the benefits of her time saved. However, the costs of added crowding and increased travel time imposed on other drivers may greatly exceed the benefits. The government makes roads into a commons; the result is crowding.[12]

Figure 4–4 shows impacts of a common property institution using Hardin's example of numbers of cattle stocked in a commonly shared pasture. Under common ownership, greater numbers of cattle stocked, shown by increased Q, increase cattle weight and cattle sales revenues (TB) up to point D where Q^m cattle numbers are stocked (m = maximum sales revenue). As stocking increases beyond point Q^m, however, total weight over all animals produced by stocking begins to fall as the grass's carrying capacity is reduced. Larger numbers of animals begin to produce less total weight gain in the face of greater competition for scarce grass.

Under sole private ownership, the pasture's owner realizes this problem of diminishing marginal returns and limits stocking to the economically efficient level of Q^*, for which total bene-

[12] In February 2003, London instituted a road-pricing scheme to relieve congestion in the downtown area (City of London, 2004). Its rationale for the fee was based on the need for a more efficient use of a heavily congested common property resource, central London's streets. The intent was to reduce the amount and cost of wasted travel time.

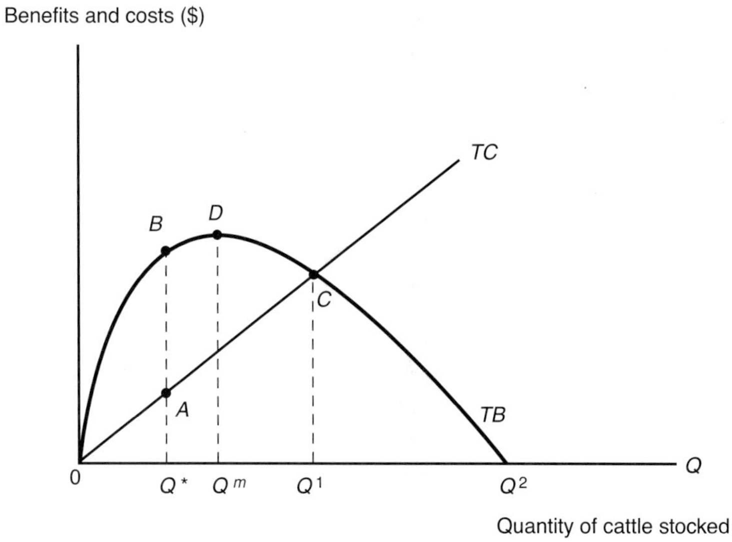

Figure 4–4 Impact of a common property institution

fit (*TB*)[13] at point *B* exceeds total cost (*TC*) at point *A* by the largest amount. The problem produced by common ownership is that it takes away an individual owner's incentive to conserve grass for fear of losing the conserved grass to rivals. Common ownership produces herd expansion to the level where there are zero profits, which occurs where total cost equals total revenue at stocking level Q^1 shown by point *C*. Acheson (2000) shows that privately owned resources can also be overexploited, even when there are secure property rights.

Mixed Messages of Conservation Appeals. Is there any way to motivate people to stop exploiting the commons? We could ask them to stop and explain that by stopping they would earn the respect and admiration of everybody by being a good citizen. But Hardin reminds us that asking someone to be a good citizen sends a mixed message: Practice moral behavior, and act against your own self-interest.

What is our intended message? We intend to ask the individual to act responsibly by not wasting the resource. The billboard encouraging people to be good citizens by limiting water use in a drought is a good example. The actual message received may be this: If you do practice good citizenship, everybody knows you can be shamed into standing aside while the rest of us exploit the commons. People who hear the actual message discover that "she who conserves most is punished the most." For not only does she lose her current use of the resource, but she sees others benefit from their use, and their use takes the resource away from her. Emphasizing the importance of good citizenship as a method to motivate people to conserve common resources is an attempt to influence a free person into acting *against* her own self-interest. Without major legal sanctions, it will have little effect on behavior.

Managing the Commons through Public Action. If we believe the costs imposed on others brought about by someone's unrestricted behavior are too high (e.g., stealing cars), we establish laws with punishments that requires him to face the real cost of his actions. Two ways to carry out this coercion are to create rules and regulations that limit individual behavior or to raise the price of undesired behavior through taxation or other pricing mechanisms.

[13] In this case *TB* refers to the total revenue earned by all users of the pasture. The single owner will maximize the difference between revenue and cost.

Rules and Regulations. People will not conserve common property resources unless we set up laws to limit their waste, and these laws must be enforced or they will be ignored. Hardin refers to this as *mutual coercion mutually agreed upon.* In an efficient society, only those regulations would be passed that limited exploitation of the commons for which the benefits exceeded the costs. In the example of the rancher, regulations would be agreed on to limit animal stocking to the point where society's marginal benefit of the added animal equals its marginal cost, but no more.

The problem presented by rules and regulations is an old one: How do you design institutions to give governments enough power to enforce the laws, but not so much that they abuse the people? Government agencies and their individual workers who are paid to carry out society's sanctioned regulations are liable to corruption.[14] The famous truth is *power corrupts;* a government agency with too much power will use it. For example, many U.S. landowners fear what they see as attempts by federal government to seize control of private lands for broadly mandated public purposes, such as the Endangered Species Act. Elected leaders limit government abuses, but these leaders are busy people: Abuses have to be large to get their attention.

Taxation. The English economist Arthur Cecil Pigou (1877–1959) proposed a tax as an instrument for raising a resource owner's private costs until it equals social costs. The idea is to raise private costs, C/n, in Table 4–2 by enough so that they equal social costs, C. The size of the tax would therefore be C minus C/n. Such taxes, known as *Pigouvian taxes,* are seen as a mechanism for confronting the resource owner with the real cost of his or her use of the common resource. For example, see the analysis of Jaeger (2002) who calculates impacts of carbon taxes designed to compensate for negative externalities of greenhouse gasses.

The premise of imposing a Pigouvian tax on each common resource user is that the cost of using the resource will rise and the quantity used will fall. That reduction in use leads to a reduction in external costs[15] imposed on the other co-owners. It also encourages each user to economize on the use of the common property resource. Finally, it encourages each user to substitute other inputs where the real cost of those other inputs is less than that of the commonly owned resource.

One problem with implementing Pigouvian taxes is that it is difficult to know how high the tax should be, as the social costs of using a common property resource are typically difficult to measure. Another problem is that the co-owners will almost never support Pigouvian taxes politically, especially when the use of the resource is an important input for their livelihoods (see Arnason, 2000 Brett and Keen, 2000). For example, Pigouvian taxes have been proposed as a way to improve (reduce depletion) commonly owned commercial fish stocks (see Chapter 12), but they have never received much support from commercial fishers.

Managing the Commons through Private Property Rights. Efficient private property rights systems (see Alchain, 2002) have four characteristics: (1) universality: all resources are privately owned; (2) exclusivity: benefits and costs of resources accrue only to their owners; (3) transferability: property rights freely transferable between people, and (4) enforceability: property rights protected from seizure or encroachment. When property rights have these four characteristics, the owner internalizes all costs and benefits. The result is economic efficiency. If any or all of these characteristics are absent, the resulting weakened structure of property rights creates a

[14] Some argue persuasively that many of the problems with government abuse of power originate in the private sector, with markets themselves and with individuals who make up those markets. They argue that the corruption in government feeds off corruption by private business. That is, unhindered markets can create an influential force, such as lobbying and financing, that hampers progress toward social welfare goals at the expense of special interests.

[15] Figure 4–1 shows that a Pigouvian tax set at BC raises the private costs of electric power to equal the social cost for the case of a negative externality, which reduces actual output so that it equals the efficient level, Q^* at a price of P^*. For a positive externality, a subsidy of CE reduces private cost of an environmentally friendly technology to make it equal to the comparatively low social cost, thus expanding output of the environmentally friendly output to the economically efficient level, Q^{**} at a price of P^{**}.

serious impediment to private investments made in resource conservation, development, improvement, and protection.

One way to reduce the exploitation of common property is to develop creative ways to make it private property with the four attributes just described, in conjunction with a stipulation that this property may be transferred to other owners through the market system. This puts efficient incentives into place to protect the resource sustainably. After the resource is privatized, there is a single owner, and each n term in Table 4–2 takes on a value of 1. In this case the private costs and benefits of additional stocking are equal to its social costs and benefits, thereby giving the individual owner strong incentives to manage the resource efficiently and sustainably.

Privatization harnesses the power of the market to protect natural resources, but not all resources can be privatized. The costs are simply too high. Cattle ranches can be privatized by fences, but it is expensive or impossible to privatize the air we breathe or privatize our own piece of a congested highway, or identify our share of an underground aquifer. Another problem with privatization is the problem of equity. Once the principle of privatization is agreed upon, the initial distribution of the property may be unfair. For example, the property may end up in the hands of people with the most power or influence. Whoever receives the initial private right gets a better deal than the rest of us who may get nothing.

A good example of the problems of privatization is presented by the U.S. Taylor Grazing Act, which was passed in 1934 to reduce overgrazing and abuse of the open ranges in the western United States. Soon after, what had been common property was effectively privatized through a permit system: Individual ranchers who owned private land were given a permit that entitled them to graze up to a specified number of cattle on a given parcel of federal grassland. Ranchers who received the original federal grazing permits got a nice gift from the U.S. government. If you buy a ranch today with federal grazing permits, you will be asked to pay full market price for the right to graze on these federal lands. Typically, the value of these lands is appraised at \$3 to \$5 an animal unit month (AUM), or the amount of forage needed per month by an adult cow with a calf (800 to 1,000 pounds). This price holds firm even though the federally administered price to graze each animal is slightly more than \$1. The original owner of the permit received a subsidy, and since then received the added benefit of having the fair market value of the permit (difference between income earning value and administered price) capitalized into the market value of the ranch. Although it is true that the initial apportionment of the private property may be unjust, Hardin argued that this injustice is preferable to total ruin that would occur if the resource continued as common property. To give one person the incentive to take care of the resource is more efficient than taking away everybody's incentives to do so.

When Common Property Works. Many economists and other researchers have believed for some time that Hardin's original paper was too pessimistic in its conclusion that common property tends to be degraded and that the only solution is privatizing the resource or setting up government regulations and control. A huge body of research since the 1980s has focused on taking a closer look at the commons problem (e.g., Arrow et al., 2000; Baker, 1998; Beck and Ghosh, 2000; Cavendish, 2000; Holland and Moore, 2003; McCay and Jentoft, 1998; Pandit and Thapa, 2003; Potkanski and Adams, 1998; Sathirathai and Barbier, 2001; Uphoff and Langholz, 1998).

Common property institutions do not always produce tragedy; many such institutions have worked successfully for centuries. These institutions succeed in promoting economic efficiency when population density is low, or harvesting technology is inefficient, and all members of the community share the same goal. But when some or all of these conditions fail to occur, common property risks become subject to overuse and depletion. As population grows and resource harvesting technique gets more efficient, communities, especially in Western cultures, often discover that the people and the resource survive longer by assigning individual private property rights.

Ostrom (1990, 1998, 1999), Agrawal (2001), and Katz (2000) have found that successful common property institutions with several generations of users have certain features. Ostrom describes these features as design principles that predict the success or failure of any proposed common property institution (rule). She notes the importance of incentives. A resource user will only want to change the rules if the benefits of that change exceed the costs. Each user compares the expected net benefits resulting from a new set of rules with the existing situation of no rules or weak rules. Institutional change is most likely to occur when the costs of the change exceed the benefits for most people.

Ostrom identified seven such design principles for defining or changing institutions that she believes produce benefits exceeding costs. Moreover, these principles are versatile enough to be applicable to designing institutions in most cultures:

- There must be clearly defined boundaries of the resource and of people with resource access rights.
- Rules that govern use of the resource and required contributions of labor, material, time, and money adapt to the local culture.
- Arrangements for making collective choice permit most members to participate in designing, implementing, and changing the rules.
- There is cheap and effective monitoring to identify each person's resource use, and the people who monitor are accountable to the monitored.
- Resource users who violate the rules face punishments that increase with the seriousness of the violation.
- Low-cost conflict resolution mechanisms exist.
- External government authorities recognize rights of the resource users to organize.

See Example 4–3.

Challenges. The dilemma facing all attempts to design an institution for coping with common property is that the design needs to possess several characteristics. First, it should protect private property rights where private property is a valued institution. Second, it should protect the public interest from abuses by individual action. Third, it should promote economically efficient and sustainable uses of resources. Finally, it should adapt to changes in society's values.

Either private property or common property can promote efficiency and sustainability in natural resource use. Regardless of which approach is chosen, an effective institution for governing the use of a natural resource will give all resource users maximum freedom of choice, while requiring each user to face the real cost of his or her actions. Applying this principle to the design of institutions that best promote a society's objectives may be the major environmental and natural resource challenge of the future.

Government Failures

When government gets involved in conservation, the instruments for resolving natural resource conflicts enter the political arena. Participants in this arena include elected officials and stakeholders such as interested voters, government agency employees, lobbyists, the press, and affected special interests such as farmers and environmentalists. Doubtless, *some* sort of resolution occurs, but it is not always clear if the results promote more economially efficient decisions.[16] Despite the good intentions underlying attempts to set up government conservation policies and plans, four kinds of *government failures* stand in the way of conservation: rent seeking, taxpayer ignorance, empire building, and abuses of economic analysis.

[16] The hypothesis of government inefficiencies reached its full expression with the works of 1986 Nobel laureate James Buchanan (1988). Historically, economists typically saw the government as the impartial servant of the public good. Buchanan believed it was bad science to seek out *market* failure while never looking for *government* failure (Nobel Foundation, 2004a). Buchanan believed there should be no place in economics for the "bifurcated man," defined as the person who acts in his own interests when in the private sector but who is miraculously transformed into someone with only the public interest at heart when employed by government.

Example 4–3
Can the Tragedy of the Commons Be Averted?

The tragedy of the commons (Hardin, 1968) has influenced debates, thinking, and policy in connection with sustaining environmental and natural resources. *Tavy* is a slash-and-burn cultivation system practiced in or near forested areas. Nearby residents have identified a valuable service provided by the forest, that is, the production of fertilizer through biomass accumulation. After biomass is burned, ash is available for use as fertilizer. Each household chooses a plot of land to be farmed and plants crops immediately to take advantage of the fertilizing ash that will be irrigated into the soil with coming rains. Households unable to meet their annual food needs from their own production have more reason to exploit the remaining *common property resources* by practicing tavy cultivation, harvesting timber, extracting honey from trees, fishing streams for endangered crayfish, and other resource-depleting practices.

In Costa Rica, there are laws against these resource-depleting practices but few enforcement mechanisms. The U.S. Agency for International Development (2004) trained forest guards and locally recruited ecotourism guides in the hope that economic returns from activities related to private forests could reinforce local people's willingness to accept and support establishment of a tourist park. The Costa Rican government realizes that completely prohibiting use of the commonly owned forest resource means that villagers will see no value to themselves in using the park; therefore, they will have little incentive to help preserve it. Villagers with no personal stake in the forest typically ignore timber harvesting by outside businesses rather than report it to authorities. Uphoff and Langholz conclude that local residents need a greater economic stake in taking care of the natural resources. That stake would increase willingness to accept the legal status of the park and contribute to averting the tragedy of the commons. Still, the challenge ahead lies in finding financially feasible methods that will give that greater economic stake to local people. The Costa Rican government has experimented with methods to promote resource-conserving behavior. One policy has attempted to get private landowners to set aside blocks of their land to be protected against degrading uses. An important principle in designing financial incentives is to make them large enough to be taken seriously, yet not so large that they undercut people's ability to act for the benefit of environmental protection.

Source: Uphoff, N. and J. Langholz. 1998. Incentives for avoiding the tragedy of the commons. *Environmental Conservation.* 25 (3): 251–261.

Rent Seeking

Rent seeking occurs when a special interest group uses its private resources in lobbying and other activities aimed at getting government to pass laws that economically benefit the group. Successful rent seeking—getting the group's favorite laws passed—increases net benefits for the special interest group, but typically at a cost to the larger society. One such group sought and received generous financing from the U.S. Congress to build water projects to the western United States. The rent seekers consisted of congressional representatives, agricultural interests in connection with banks and real estate, and federal water construction agencies.

Rational Ignorance

Rational ignorance means that taxpayers often find it more expensive than it is worth to them to stay informed about issues that do not affect them directly. For example, an individual taxpayer in the eastern United States typically has little interest in investing the time and money to stay informed about proposed legislation that would save an endangered species living in the Colorado River. It's simply not worth her time to scrutinize public actions that have little direct bearing on her. Many U.S. federal water projects were built for which total benefits fell far short of total costs, because the local beneficiaries wanted the projects and the rest of the nation paid little attention. Today, many government programs are enacted because a highly vocal and organized minority is anxious to secure the benefits, while the majority of citizens who pay the bill have little incentive to scrutinize the costs.

Empire Building

Empire building, especially when combined with a bureaucratic desire for job security, is another activity that blocks effective government programs aimed at conservation. Few government agencies or employees can resist the temptation to increase their own influence or power.

Why do government agencies get away with building empires? Because there is no profit check to stop them. Private business loses money if people are not willing to pay for the products produced by their expansion. The owner of a bicycle shop will not open new stores unless there is enough customer demand for bikes to support the new locations. But if government builds too many dams, drains too many swamps, or restores too many wetlands, there are no negative profit signals that require them to stop building, draining, or restoring. The only support needed for government projects is a legislative majority, which can be obtained by a minority that provides campaign financing for the needed legislators. Once begun, government programs and agencies can take on a life of their own. After World War II, the U.S. Congress gave the Army Corps of Engineers and Natural Resources Conservation Service large budgets to drain swamps. During the 1990s these agencies received large budgets from Congress to re-create or restore the swamps, more recently called wetlands.

Abuses of Economic Analysis

Another government failure occurs because agencies ignore the economic consequences of their decisions. Sometimes a government agency chooses to ignore important costs or benefits because it is in the business of promoting one type of resource use (Herfindahl, 1961). In fact, stimulating use of the resource may be implied in the agency's mission statement. Beginning in the early twentieth century, hundreds of dams were built in the western United States by the federal Bureau of Reclamation and Army Corps of Engineers, whose missions are, respectively, irrigation development and flood protection.

When confronted with major decisions, decision makers often ignore costs or benefits that cannot be quantified in money terms or can be described only in vague, nonquantitative terms. Several U.S. dams built in the 1950s and 1960s, for example, were justified by flawed benefit-cost analyses that assigned high values to reservoir recreation behind a dam, while presuming that river-based recreation and wildlife habitat displaced by the dam had zero value, simply because those values were hard to measure. (Chapter 7 discusses several methods for quantifying environmental values.)

Only if benefits and costs are properly measured and used to set up, carry out, and evaluate government programs will these programs promote conservation. Only then is the public protected against rent seeking by special interests, taxpayer ignorance, and empire building. Still, economic analysis is expensive and hard to understand; in addition, it is seen by some as unreliable. As we have seen throughout this chapter, objective economic analysis has the potential to promote efficient regulations and resource management. Future chapters examine how economic theories and methods can be used to measure the value of environmental improvements, and how economics can be used to design policies that remedy failures by both markets and governments to protect the environment.

■ Summary

The market institution has the potential to be an impressive mechanism for promoting economic efficiency in the use of environmental and natural resources. If market institutions permit each person to reap the full rewards of his or her own actions, as well as recognize the cost imposed on others produced by those actions, markets produce the most human benefits from society's scarce resources. However, the market system cannot always shoulder the full burden of achieving economic efficiency. When the burden is too heavy, markets may suffer from one of three kinds of market failures: externalities, public goods, and common property.

An externality occurs when a market transaction between a buyer and seller directly affects a third party, and that third party receives no compensation for the impacts. Where negative externalities occur as a result of use of a natural environment or natural resource, the market system produces too much of that resource for economic efficiency. Where positive externalities occur, the market produces too little. The Coase theorem states that if there are well-defined property rights, low bargaining costs, perfect competition, perfect information, and an absence of wealth and income effects, environmental and natural resources will be used efficiently even in the presence of externalities. The search for property rights systems that encourage economic efficiency is an active area of research.

Once a public good has been designed, built, supplied, and made available for use, many people can use it without limiting the opportunity of others to do the same. A public good is a resource that lacks excludability and lacks rivalry. Lacking excludability means there is no way to put a fence around it and exclude people who do not pay for it. A good lacks rivalry if the use of one unit of the good by one person does not reduce that unit's availability to somebody else, so the marginal cost of supplying the unit's services to one more person is zero. Private businesses typically incur losses if they attempt to supply public goods. Therefore, governments have a potential role in ensuring that economically desirable public goods are built and supplied. Much current economic research focuses on methods to identify efficient methods for financing, building, and allocating public goods.

A common property resource is owned by a group rather than an individual. An important question about common property resources centers around whether unregulated competition in the use of a common resource motivated by self-interest produces the greatest possible total benefits for all. Unless great care is exercised in regulating use of a common resource, incentives facing each individual produce behaviors that result in the tragedy of the common, which damages both the resource and all its owners. Nevertheless, many economists continue to search for policy solutions to the tragedy of the commons.

When government gets involved in the environment and natural resources, the instruments for resolving conflicts enter the political arena. Participants in this arena include elected officials and others stakeholders such as voters, government agency employees, lobbyists, the press, and affected special interests. Despite the good intentions of government environmental programs and plans, four kinds of government failures stand in the way of conservation: rent seeking, taxpayer ignorance, empire building, and abuses of economic analysis. Because both government and market failures can neglect to protect environmental and natural resources, there is a need for finding methods by which greater efficiency can be achieved.

■ Review Questions

1. What are externalities and what problems do they present for management of environmental and natural resources?

2. Explain how negative externalities damage economic efficiency that could otherwise occur in a competitive market system. Describe one type of government policy intervention that might correct market failure resulting from negative externalities.

3. Suppose that commercial timberland produces a positive externality by also being a habitat for an endangered species. Explain how market systems answer the question of what to produce. Use your answer to explain why markets fail to supply adequate quantities of habitats for endangered species.

4. If an area's chemical industry has been dumping its toxic waste free of charge into a local stream, explain why government action to ensure a more efficient allocation of resources would raise the industry's price and reduce its output.

5. Explain the Coase theorem. What does its findings mean for environmental and natural resource management?

6. Why does the Coase theorem claim an equal amount of pollution is reduced regardless of who receives property rights to a pollution-free environment?

7. What are the major limits of the Coase theorem?

8. What are public goods and what problems do they present for management of environmental and natural resources?

9. What is common property and what problems does it present for management of environmental and natural resources?

10. Compare the intended message and the actual message received by people who are asked to conserve common property.

11. What are the advantages and disadvantages of taxes and rules for coping with common property?

12. Explain how private property rights can remedy the tragedy of the commons.

13. Explain the limits of privatization as a way to deal with the common property problem.

14. Under what conditions can management of common property be economically efficient?

15. What four government failures can damage environmental and natural resources?

16. Defend or critique the claim that imposing identical emissions standards on all polluters is neither fair nor economically efficient.

17. Defend or critique the claim that environmentalists are elitists: They want to tax the poor and middle class to preserve open spaces and wildlife for their own selfish enjoyment. They care more about animals and plants than they do about people.

18. Suppose a respected think tank just published a study showing that 90% of environmental standards are set too high—beyond the point that maximizes the present value of the net surplus. Using your knowledge of economic theory, show that this result does *not* mean that the environmental standards should be eliminated.

19. Explain why society benefits most when choices are made so that $MB = MC$.

20. Explain why a popular salmon run in a river in the U.S. Pacific Northwest is likely to lead to overfishing. How would awarding monopoly ownership over access to the run solve the problem? What practical issues might accompany such a plan?

■ PROBLEMS

1. Externalities—analysis and policy design: Suppose that in a competitive market, demand is given by the equation $P = 600 - Q$, and supply is given by the equation $P = 160 + Q$, where P is price and Q is quantity of some good or service. Production of each unit of output Q leads to a marginal *external* cost of $50, caused by pollutants emitted by the production of Q. If we add this marginal external cost to the market information, the equation for the social-cost supply curve is given by $P = 210 + Q$.

a. Compute the *unregulated market* level of output and price as well as the *socially efficient* level. By how much does the market output exceed the socially efficient output, and by how much is the market price below the socially efficient price?
b. Compute the monetary value of the deadweight social loss from the market failure that occurs if society lets firms to continue to produce negative externalities without regulation.
c. Suppose a tax per unit output (per-unit tax) is imposed on the production of Q with the intent of making equal the *market* level of output and the *socially efficient* level. How high should that per-unit tax be? What is the gain in net social benefits (consumer plus producer surplus) that results from this per-unit tax?

2. Policy application of the Coase theorem: A chemical company emits pollutants into a fishing stream. Its *marginal benefit* from added emissions (marginal abatement costs avoided), is given in the table below in column *MBc*. A fishing group that likes to fish in the stream suffers a *marginal cost* (called a marginal damage) from the company's pollution, given in column *MCf*.

Tons per year	*MBc*	*MCf*	*MNSB*
	← millions of dollars →		
0	1.00	0.00	
1	0.75	0.50	
2	0.50	1.00	
3	0.00	1.50	

a. How many tons of pollutants per year does the company emit?
b. How many tons per year would the fishing group prefer the company to emit?
c. Calculate the marginal net social benefit from each of the four pollution levels under column *MNSB*.
d. What is the socially optimal level of pollution?
e. Suppose the fishing group has the rights to unpolluted water. Describe briefly the negotiation process predicted by the Coase theorem. How much pollution is emitted after the negotiation?
f. Suppose the chemical company has the right to emit as much pollution as it wants. Describe the negotiation process predicted by the Coase theorem. How much pollution is emitted after the negotiation?
g. Compare the emission levels after negotiation in question e and question f. Are both socially optimal?
h. Suppose there is a transaction cost of $1 million to get the negotiations started. What will be the outcome?

■ REFERENCES

Acheson, J. 2000. Clearcutting Maine: Implications for the theory of common property resources. *Human Ecology.* 28 (2): 145–169.

Agrawal, A. 2001. Common property institutions and sustainable governance of resources. *World Development.* 29 (10): 1649–1672.

Alchain, A. A. 2002. *Property rights.* Available from *http://www.econlib.org/library/Enc/PropertyRights.html* [Accessed October 20, 2004].

Arce, D. G., and T. Sandler. 2002. *Regional public goods: Typologies, provision, financing, and development assistance. Expert Group on Development Issues Studies, Secretariat of the Swedish Ministry of Foreign Affairs.* Stockholm: Almkvist and Wiksell International.

Arnason, R. 2000. Economic instruments for achieving ecosystem objectives in fisheries management. *Ices Journal of Marine Science.* 57 (3): 742–751.

Arrow, K., G. Daily, P. Dasgupta, S. Levin, K. G. Maler, E. Maskin, D. Starrett, T. Sterner, and T. Tietenberg. 2000. Managing ecosystem resources. *Environmental Science and Technology.* 34 (8): 1401–1406.

Baker, J. M. 1998. The effect of community structure on social forestry outcomes: Insights from Chota Nagpur, India. *Mountain Research and Development.* 18 (1): 51–62.

Beck, T., and M. G. Ghosh. 2000. Common property resources and the poor—Findings from West Bengal. *Economic and Political Weekly.* 35 (3): 147–153.

Brett, C., and M. Keen. 2000. Political uncertainty and the earmarking of environmental taxes. *Journal of Public Economics.* 75 (3): 315–340.

Bruce, J. W., and R. Mearns. 2001. *Natural resource management and land policy in developing countries: Lessons learned and new challenges for the World Bank.* Available from *http://www.gm-unccd.org/FIELD/Multi/WB/Lessons.pdf* [Accessed July 22, 2003].

Buchanan, J. M. 1988. The constitution of economic policy. *American Economic Review.* 77 (3): 243–250.

Cavendish, W. 2000. Empirical regularities in the poverty-environment relationship of rural households: Evidence from Zimbabwe. *World Development.* 28 (11): 1979–2003.

City of London. 2004. *Transport for London: Congestion charging.* Available from *https://www.cclondon.com/index.shtml* [Accessed October 21, 2004].

Dasgupta, P. 2000. Population and resources: An exploration of reproductive and environmental externalities. *Population and Development Review.* 26 (4): 643–650.

Diakoulaki, D., A. Zervos, J. Sarafidis, and S. Mirasgedis. 2001. Cost-benefit analysis for solar water heating systems. *Energy Conversion and Management.* 42 (14): 1727–1739.

Hardin, G. 1968. The tragedy of the commons. *Science.* 162: 1243–1248.

Herfindahl, O. C. 1961. *Three Studies in Minerals Economics.* Washington, D.C.: Resources for the Future.

Holland, S. P., and M. R. Moore. 2003. Cadillac Desert revisited: Property rights, public policy, and water-resource depletion. *Journal of Environmental Economics and Management.* 46 (1): 131–155.

International Dark Sky Association. 2004. *Web page.* Available from *http://www.darksky.org/* [Accessed October 20, 2004].

Jaeger, W. K. 2002. Carbon taxation when climate affects productivity. *Land Economics.* 78 (3): 354–367.

Katz, E. G. 2000. Social capital and natural capital: A comparative analysis of land tenure and natural resource management in Guatemala. *Land Economics.* 76 (1): 114–132.

Leggett, C. G., and N. E. Bockstael. 2000. Evidence of the effects of water quality on residential land prices. *Journal of Environmental Economics and Management.* 39 (2): 121–144.

Loomis, J. B. 2000a. Environmental valuation techniques in water resource decision making. *Journal of Water Resources Planning and Management.* 126 (6): 339–344.

———. 2000b. Vertically summing public good demand curves: An empirical comparison of economic versus political jurisdictions. *Land Economics.* 76 (2): 312–321.

Martin, W. E., H. Ingram, and N. K. Laney. 1982. Willingness to play: Analysis of water resources development. *Western Journal of Agricultural Economics.* 7: 133–139.

McCay, B. J., and S. Jentoft. 1998. Market or community failure? Critical perspectives on common property research. *Human Organization.* 57 (1): 21–29.

Nobel Foundation. 2004a. *The Bank of Sweden Prize in Economic Sciences in Memory of Alfred Nobel 1986, James Buchanan.* Available from *http://nobelprize.org/economics/laureates/1986/* [Accessed October 20, 2004].

———. 2004b. *The Bank of Sweden Prize in Economic Sciences in Memory of Alfred Nobel 1991, Ronald H. Coase.* Available from *http://nobelprize.org/economics/laureates/1991/* [Accessed October 20, 2004].

Ostrom, E. 1990. *Governing the Commons: The Evolution of Institutions for Collective Action.* Cambridge, UK: Cambridge University Press.

———. 1998. How communities beat the tragedy of the commons. Paper read at International CBNRM Workshop at Washington, D.C.

Ostrom, E., J. Burger, and C. B. Field. 1999. Revisiting the commons: Local lessons, global challenges. *Science.* 284 (5412): 278–282.

Pandit, B. H., and G. B. Thapa. 2003. A tragedy of non-timber forest resources in the mountain commons of Nepal. *Environmental Conservation.* 30 (3): 283–292.

Potkanski, T., and W. M. Adams. 1998. Water scarcity, property regimes and irrigation management in Sonjo, Tanzania. *Journal of Development Studies.* 34 (4): 86–116.

Samuelson, P. A. 1954. The pure theory of public expenditure. *Review of Economics and Statistics.* 36 (4): 387–389.

———. 1955. Diagrammatic exposition of a theory of public expenditure. *Review of Economics and Statistics.* 37 (4): 350–356.

Sathirathai, S., and E. B. Barbier. 2001. Valuing mangrove conservation in southern Thailand. *Contemporary Economic Policy.* 19 (2): 109–122.

Sky and Telescope. 2004. *Web page.* Available from *http://skyandtelescope.com/* [Accessed October 20, 2004].

U.S. Agency for International Development. 2004. *Web page.* Available from *http://www.usaid.gov/* [Accessed October 20, 2004].

U.S. Department of Interior. 2004a. *National Park Service.* Available from *http://www.nps.gov/* [Accessed October 20, 2004].

———. 2004b. *Yosemite National Park.* Available from *http://www.nps.gov/yose* [Accessed October 20, 2004].

United Nations. 2004. *Convention on International Trade in Endangered Species of Wild Fauna and Flora (CITES).* Available from *http://www.cites.org/* [Accessed October 20, 2004].

University of California Santa Barbara. 2004. *Garrett Hardin oral history project.* Available from *http://www.garretthardinsociety.org/gh/gh_oral_history.html* [Accessed October 20, 2004].

Uphoff, N., and J. Langholz. 1998. Incentives for avoiding the tragedy of the commons. *Environmental Conservation.* 25 (3): 251–261.

Chapter 5

DECISION SUPPORT FOR ENVIRONMENTAL POLICY

Outline

Overview of Benefit-Cost Analysis 93

Economic Principles of Benefit-Cost Analysis 95

Ten Steps to Net Present Value 101

Measuring Benefits 108

Measuring Costs 111

Treatment of Time 112

Issues Affecting Both Benefits and Costs 113

Alternatives to Benefit-Cost Analysis 115

Advantages and Limits of Benefit-Cost Analysis 119

Objectives

After studying this chapter, you should be able to

- Summarize the rationale for doing a benefit-cost analysis and its uses
- Explain economic principles used for conducting a benefit-cost analysis
- Describe the 10 steps for computing net present value
- Identify ways to account for benefits, costs, and time
- Discuss issues that affect both benefits and costs
- Recognize the advantages and limits of benefit-cost analysis

Introduction

On July 24, 2003, near Albuquerque, New Mexico, Sandia Pueblo tribal leaders passed out new computers to all 135 of the community's households. The Pueblo's youthful governor, Stuwart Paisano, said the computers were to provide the community with the resources needed to be productive. In buying the computers, Paisano took action to confront a problem that has challenged governments since ancient times: designing and carrying out *public* programs that improve the welfare of a community of *private* individuals.

It is a daunting task to correctly discriminate among community actions that improve a society's welfare from those that reduce it. The task typically becomes harder with the increased size of the community. This widely recognized difficulty has led economists to think systematically about finding the mix of actions that improve a community's welfare. The problem is especially challenging when dealing with questions about the environment and natural resources.[1]

Most natural resource and environmental policy decisions have effects that cause some people to gain while others lose (Barron, 1998). The term *policy* includes any public action to build, design, implement, or evaluate a project, program, regulation, or legislative action. Policymakers and voters want to know if the proposed decision benefits more than it hurts. This is especially true when outcomes of the analysis fail to suit interests of politically powerful people or groups who stand to lose something. In order to decide whether society as a whole is better or worse off from a proposed government action, we need to somehow compare individuals' gains and losses. We also need some way to sum the preferences of all gainers and losers, so an overall judgment can be made. This chapter deals with decision support, that is, collecting and organizing information on individual preferences so the various branches of government can better design and carry out environmental and natural resource policy decisions based on the public interest.

---

[1] Systematic thinking about evaluation of public policy began in 1844 with a paper by the French engineer and economist Jules Dupuit. His 1844 article can be found at New York University (2004).

■ Overview of Benefit-Cost Analysis

Benefit-cost analysis (BCA) is a technique for comparing the costs and benefits of a government activity, project, or regulation over a relevant time period. It is a practical way to determine the desirability of decisions made by government. In finding out how desirable a government decision is, it is important to take a long view by looking at impacts in both the near and far term, and a wide view by accounting for many kinds of impacts on many people, economic interests, and regions. BCA involves the enumeration (listing) and evaluation (measuring) of all relevant costs and benefits.

Why Measure Benefits and Costs?

Since BCA can be expensive, people often ask why various government organizations are motivated to spend the money and make the effort to conduct it. One answer is that what counts as a benefit or a loss to one part of the economy—to one or more persons or groups—does not necessarily count as a benefit or loss to the whole economy. BCA involves the economy as a whole and not some small part of it. In many respects, BCA is the government's equivalent of the private business's profit-and-loss statement, which measures a business's income. In BCA the question of "net return" is asked about a wider group, namely, society as a whole. Instead of asking whether the owners of a private enterprise will be made better off by a proposal, BCA asks whether the whole society affected by a proposed program will be made better off. Even in cases where a government can make a profit (e.g., mineral leases or polluter fines), that profit may be the wrong yardstick to use for public environmental or natural resource plans.

What Benefit-Cost Analysis Is

Benefit-cost analysis evaluates the merits of public actions by translating positive and negative effects to a common denominator, typically money. In the 1990s, legislation being considered by the U.S. Congress would have expanded the use of BCA in the regulatory process. Various proposals required systematic assessment and balancing of costs and benefits for major regulations affecting human health, safety, and the environment. Because of the extensive number of factors agencies would have to consider in doing BCA, that reform legislation generated considerable debate and controversy. Greater use of BCA would significantly change how government agencies develop and implement certain types of regulations. Debates over the merits of various proposals continue to revolve around the practicality and usefulness of requiring more government agencies to use BCA.

What Benefit-Cost Analysis Does

Benefit-cost analysis compares the desirable and undesirable impacts of proposed policies. It permits analysts to rank the economic performance of environmental and natural resource projects, policies, and programs in which impacts are measured in nontechnical terms and estimated by scientific methods (see McAllister, 1980). It allows comparison of all the gains and losses in a common denominator resulting from some public policy action. A BCA organizes information in a way to promote the conduct of *rational policy analysis*. A complete BCA compares alternative actions to determine which one provides society with the most economically beneficial use of its resources (Loomis, 1993). BCA can be used to provide information needed for three kinds of public environmental and natural resource decisions:

1. Simple ranking of the comparative benefits of several possible actions
2. Optimal size or scale of a project produced by a decision
3. Optimal timing or sequencing of several elements of a decision

BCA got its start in the 1930s for U.S. water projects. In recent years BCA has been applied to a huge range of environmental and natural resource policy questions in many countries and cultures. Here is a short list of recent examples: biological control of invasive plants (Van Wilgen et al., 2004); managed health care (Bloom, 2004); sea lamprey control (Marsden et al., 2003); fuel system technologies (MacLean and Lave, 2003); noise reduction (Becker and Lavee, 2003); power plant emissions (Levy et al., 2002); sulfur restrictions (Lee, 2002); common pool management institutions (Kumar, 2002); clean air emissions control (Krupnick and Morgenstern, 2002); mortality risk reductions (Krupnick et al., 2002); forest conservation (Kniivila, Ovaskainen, and Saastamoinen, 2002); windbreak establishment in Australia (Jones and Sudmeyer, 2002); environmental programs (Freeman, 2002a, b); predator control (Engeman et al., 2002); wild nature conservation (Balmford et al., 2002); ozone pollution control (Yoo and Chae, 2001); climate change policies (Tol, 2001); mesquite control (Teague et al., 2001); ammonia emissions from feedlots (Shi et al., 2001); inland recreational fisheries in Norway (Navrud, 2001); agricultural research programs (Marshall and Brennan, 2001); and drinking water contaminant standards (Gurian et al., 2001).

Benefit-cost analysis uses a simple decision rule. If for some proposed action, the sum of its benefits exceeds the sum of the costs by a larger amount than any other action with the same aim, the proposed action should be adopted if economic efficiency (see Chapter 3) is the objective. Otherwise it should not. This decision rule assumes that a dollar is worth the same to everyone, whether it is a small or large loss or gain (i.e., society's marginal utility of any individual's income is the same for all individuals). An additional dollar of benefit produced by a public program for a rich person is worth the same as a dollar of cost paid by a poor taxpayer who finances that benefit. One advantage of BCA is that the monetary unit is easily understood by everyone and does not require technical specialists to interpret. Properly carried out, BCA provides information that allows government to adopt programs that produce outcomes whose value is greater than the value of resources used in financing the program. This ensures that programs contribute positively to economic efficiency.

For revenue to a private firm, BCA substitutes benefit to society. For private cost incurred by a business, BCA substitutes the *opportunity cost* (i.e., the value of benefits foregone when resources are diverted from other economic activities and brought in to support the resource management decision in question). For profit of a firm, BCA substitutes the concept of benefit minus cost.

In BCA all environmental and natural resource services are valued in a common denominator. Consider just a few of the services produced by such decisions: increased human health and safety, electric power, controlled elevation of global temperature, water quality, habitats for endangered species, timber, forage for cattle and wildlife. Expression of all these outputs in a common denominator, allows government to compare all economic values and all related policy decisions on a level playing field.

How Benefit-Cost Analysis Is Used

Most applications of BCA deal with *ex ante*, or planned, policy questions (i.e., questions that look forward and ask about benefits and costs of actions not yet taken). If we conducted only *ex ante* analysis, we might never learn from past mistakes. By contrast, *ex post* analysis looks backward and asks how well an existing project, program, or regulation performed after it was established. *Ex post* information can be used for three purposes: (1) to analyze the stream of actual benefits and costs produced by actual projects built or policies enacted, and to determine if the previous *ex ante* BCA was accurate or if errors were made; (2) to revise methods, forecasts, and assumptions where mistakes were made; (3) to gain information about the existing economic impact and value relationships on which future BCAs ultimately rest. For example, if *ex post* analysis showed that reducing each ton of particulate in the atmosphere reduced hospital admissions by *X*, that fact tells us something about the future value of controlling particulate emissions by *Y*.

Design, Implementation, and Review. Properly done, a BCA can provide information to help managers design, implement, and review programs and policies. BCA can sharpen the goals of environmental programs by more comprehensively accounting for the economic benefits and costs of various ways of designing policies and setting priorities. In addition, BCA can provide estimates of overall benefits and costs of a proposed policy, as well as identify people who stand to gain or lose. As an example, a BCA was conducted in connection with European air quality targets for sulphur dioxide (SO_2), nitrogen dioxide (NO_2), and suspended particulate matter (PM10) for major European cities. In October 1997, the Commission of the European Communities adopted a proposal that air quality standards for all three pollutants be in force by 2010. One BCA of this proposal showed that these proposed reductions produced benefits at least 50 times greater than costs for all three pollutants, mostly from improved health and longer lives. Thus, the BCA provided important information to environmental policy decision makers (Olsthoorn, 1999).

The usefulness of BCA depends on the stakes involved and the likelihood that the resulting information will influence ultimate decisions. BCA increases in importance when a policy has high stakes. The information provided by a BCA may justify devoting more resources to analyzing policies because the stakes are higher. Increasing the scale of a BCA increases costs, and these added costs are good investments only if they inform and influence policy decisions (Arrow et al., 1996). Two examples of high stakes programs are those involving sizable taxpayer cost and proposed regulations for which the cost of compliance is high.

Informational Contributions. For programs whose anticipated costs far exceed their expected benefits, measured negative net benefits can provide valuable information. Decision makers learn of the costs they incur if they decide to move forward on a program in spite of its weak economic performance. Decision makers often wish to weigh factors other than economic benefits and costs into their policy decisions, such as environmental justice (see Chapter 19) or irreversible environmental damages. Even in these cases, BCA provides valuable information by measuring the economic efficiency benefits lost from implementing an economically weak program, or the economic efficiency benefits that were never realized from failing to carry out an economically strong program.

The fact that BCA focuses on a project's *economic* feasibility helps it contribute to the political process. BCA can be used to measure both the overall relationship between benefits and costs and the distribution of those benefits and costs among major interest groups at various times and in various places. Designing a policy that produces a positive economic efficiency payoff for a wide range of major interest groups may make it more politically attractive than concentrating benefits or costs on a small number of people. In other words, efficiency improvements may get votes. A good example of this is the U.S. Endangered Species Act, which enjoys wide public support. Most Americans believe they pay very little cost while they are receiving large environmental benefits. Most costs are incurred by a relatively small number of landowners and commercial natural resource firms.

■ Economic Principles of Benefit-Cost Analysis

Economic principles can be used to help us meet three important challenges assigned to benefit-cost analysis: (1) defining government policy goals, (2) accounting for people affected by such policies, and (3) identifying incremental impacts resulting from the policy.

Goals

A very old challenge surrounding the design of government policy is the question of what *ends* are served by government activity and what programs can be established to best meet those ends. BCA is one method that provides information for policymakers to better serve the people through government action. Therefore, establishing the goals of a BCA is of special importance.

Economic Efficiency. A BCA is based on the government policy objective of economic efficiency (see Chapter 3). This principle states that economic efficiency is the important standard for *evaluating* government regulations or programs that are proposed for adoption, maintenance, or change. For a government program to meet the efficiency goal, benefits must exceed costs. Efficiency proponents argue that government programs cannot be justified unless their benefits outweigh their costs. In this view, targeting the efficiency objective increases the likelihood that government actions will only place burdens on businesses and consumers that are in proportion to improvements in health, safety, or the environment, which also accrue to businesses and consumers. Opponents of economic efficiency argue that other goals are also important.[2]

Equity. Environmental and natural resource policies typically redistribute economic benefits as well as costs. Critics of the use of BCA state that it ignores the distribution of benefits and costs, and that a simple summing of costs and benefits across all affected individuals leaves out important considerations of *equity* (fairness). When a BCA is based solely on economic efficiency, it assumes that policymakers believe that economic efficiency is the most important aim. In fact, economic efficiency fails to distinguish whose costs and benefits are considered. Frequently the benefits accrue to one sector of society or class of individuals, while the costs are borne by others. (This poses a special problem when society's marginal utility of its individuals' incomes is higher for the poor than for the rich.)

BCA might be more politically acceptable and become more widely used if environmental and natural resource programs are designed to account for equity. One way to accomplish equity is to design programs for which benefits exceed costs to most groups (i.e., people in most tax brackets, time periods, geographic locations, ethnic backgrounds, and economic sectors). For example, a proposed environmental policy that tightens health and safety standards for the manufacture of toxic chemicals may produce *fewer* overall benefits than costs by reducing certain chemicals' effectiveness. However, if the policy redistributes economic opportunity between those who make and use chemical compounds and those who bear the real costs of their use, people will see it as fair, so it may receive political support. What this means is that economically inefficient proposals (those for which the costs exceed the benefits) can still receive considerable political support if people believe that the most deserving people secure the benefits and the least deserving ones pay the costs.

Administrative Ease. Decisions about a particular environmental action depend on several factors: what is known about the public's concern, likely consequences of failing to act, available choices, and the consequences or costs of taking various remedial actions. Public actions to protect public well-being often are implemented on the pragmatic basis of matching public needs and expectations with administrative ease of application. Decisions are also based on prevailing philosophies of government as well as political expediency.

Multiobjective Goals. Conventional BCA is widely criticized on the ground that economic efficiency is not the only criterion by which public programs and regulations should be judged; in some cases decision makers reject efficiency altogether. Decision making involves balancing conflicting objectives of public management. At various times and circumstances these objectives can involve considerations of fairness in how different groups are treated, ease of administration, clear understanding by the public, predictability in implementation, as well as economic efficiency.

---

[2] Many people believe that subjecting decisions about health, safety, and the environment to the cost-benefit tests that influence the outcome of those decisions is wrong because these are "specially valued" things. But beware: When it comes to government policy decisions, the whole idea of avoiding monetary value attachment to "specially valued" things presents the problem that different people specially value different things. Watch what happens in a community of people with different special values. The group will need to recognize and accept as legitimate everyone's special value, without concern for cost. Therefore, everyone inside the group is entitled to as much of the thing she specially values, and it becomes immoral to discuss questions of resource limits. Any group built on that kind of premise can go in either of two directions: It could try to incorporate so many different subgroups and interests that the huge weight of its internal contradictions becomes clear. Or it could limit its membership at some point and decide that the special values of those left outside the group are not legitimate and should be sacrificed to the special values of those left inside the group. See DeLong (2000).

If several legitimate objectives of environmental and natural resource programs can be identified, a method for implementing multiobjective planning is to weight each objective in an overall objective function according to its social importance. However, any weights actually used that are not equal to one for everybody are likely to be subject to considerable scrutiny and controversy.

Scope of Analysis

Monetary values of a proposed program are defined by benefits and costs of individuals who would be affected by the program. For example, if a proposed tightening of air quality standards under the U.S. Clean Air Act produces X dollars of health benefits, that means that there are people out there whose reduced doctor visits, reduced work days lost, and the like are worth X dollars to them. BCA is based on principles of democracy. Cost and benefit values that analysts assign to a program's effects are those of affected people; they are not theoretical values held by politicians, agency analysts, economists, or hosts of radio talk shows. Therefore, a correctly done BCA will define whose benefits and whose costs are being counted. In technical terms, economists say that it will define the *accounting stance.* Major classes of gainers and losers will also be identified. If a national government agency proposes reducing cattle grazing to increase the probability of survival of an endangered species, ranchers and beef consumers are likely losers, while the general public who value survival of the endangered species are gainers.

A private firm is presumed to take a private accounting stance. Individual businesses are motivated to act in accordance with gains and losses as each perceives them, such as income. They are presumed to have less interest in social benefits and costs. When the responsibility for an environmental or natural resource decision rests with a public agency, however, a wider view is required.

For programs that are national in scope, the national accounting stance has a broader role. It should try to measure social opportunity costs and values for all inputs and outputs and for all people in the nation affected by the program, whether those values are correctly or incorrectly priced in the marketplace. BCAs also can be performed for more limited accounting stances (e.g., regional and local; high- and low-income groups; current and future generations). Environmental proposals are more politically attractive if they can be designed to produce a favorable BCA outcome for a wide range of accounting stances.

Incremental Analysis

More efficient policies may be designed by using BCA to identify *added* benefits and *added* costs resulting from a range of various *incremental* changes rather than from a single all-or-nothing proposal. An excellent example is provided in a study by Montgomery and Brown (1992) of the endangered spotted owl in the U.S. Pacific Northwest. These authors found that the typical all-or-nothing proposal presents policymakers with a single plan for species preservation and a single estimate of its economic costs. But an all-or-nothing proposal presents a problem. By choosing only two extreme policies, one that guarantees species survival and one that guarantees species extinction, policymakers are boxed in one of two corners. They have little opportunity to compare various incremental plans by assessing the costs of moving gradually toward more species protection.

Montgomery and Brown (1992) found that the lack of an incremental approach intensifies the adversarial debate between people who advocate species preservation and people who will likely bear the cost of its programs. Their study identified incremental costs from various policies that could incrementally improve the probability of the owls' survival in the wild. Such analysis provides important information to policymakers who may wish to know the added cost of increasing the probability of species survival by a certain amount. The information on added cost can be compared with other possible uses of the same resources and, hence, can be used to make more informed decisions.

Equimarginal Principle. The incremental (marginal) benefit represents the contribution of one more unit to economic efficiency. It is measured by the change in total benefits from one more policy unit. This concept of marginal benefit is important in environmental and natural resource policy analysis for achieving efficiency.

Figure 5–1 Total benefit (*TB*) and marginal benefit (*MB*) of greenhouse gas (GHG) mitigation

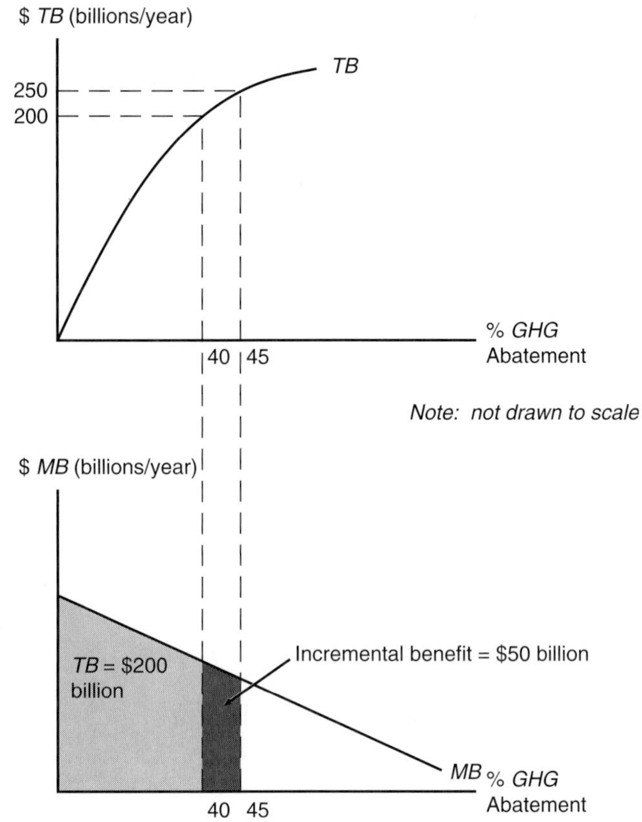

Many government programs are intended to develop a natural resource or expand the scale or scope of an existing regulation. For example, a larger scale occurs if an environmental regulation on air emissions is tightened. A larger scope occurs if the regulation is applied to a wider geographic region. In either case, economic efficiency requires that the development or expansion be undertaken until the marginal benefit of the expansion equals the marginal cost.

Figure 5–1 compares marginal benefits (*MB*) to total benefits (*TB*) using the example of various possible percentage increases in greenhouse gas (GHG) abatement. The graph shows that a program that increases greenhouse abatement from a baseline of 40% of current levels to 45% of current levels increases total benefits from $200 billion per year to $250 billion per year. The top part of the graph shows total benefits increasing from $200 to $250 billion, and the bottom graph shows the same information for marginal benefits. The marginal benefit from increasing abatement from 40% to 41% is slightly above $10 billion (not shown in the graph). If the marginal cost of increased abatement from 40% to 41% is more (less) than $10 billion, than that expansion fails (passes) the test of economic efficiency.

Figure 5–2 shows the same information as Figure 5–1, but incorporates total cost (*TC*) and marginal costs (*MC*), so a complete efficiency analysis can be performed. It shows how the efficient level of a program's output can be found by comparing its marginal benefits to marginal costs, again using the example of greenhouse gas abatement. Note that the decision to expand abatement from 40% to 45% of present emissions produces marginal costs that exceed marginal benefits, so according to the numbers shown in the figure, the proposed program is considered *economically inefficient.* However, the cost and benefit estimates are the subject of considerable uncertainty and debate. For that reason, there is much uncertainty over what level of abatement is the most efficient.

In any case, the efficiency principle just described is known as the *equimarginal principle for program expansion.* If the marginal benefit is greater than the marginal cost from tightening air emission standards from the current regulated level to a more stringent level, then that additional regu-

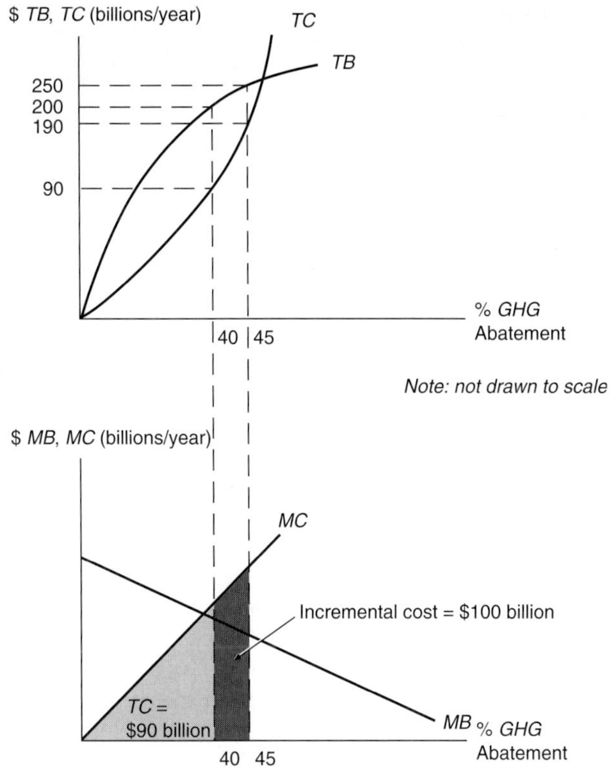

Figure 5-2 Total benefit (*TB*) and marginal benefit (*MB*), and total cost (*TC*) and marginal cost (*MC*), of greenhouse gas (GHG) mitigation

lation improves (but does not reach) economic efficiency. Only when expansion occurs to the point at which marginal benefit equals marginal cost is the program at an economically efficient level.

Some proposals would reallocate a resource, such as scarce water among competing users[3] or scarce taxpayer resources among competing human health improvements. For the reallocation decision, economic efficiency occurs when marginal benefits per unit of the resource are equal for all uses. This principle is called the *equimarginal principle for reallocation.*

With and without Principle. The *with and without principle* states that a program's benefits and costs should be measured *with* the program and compared to benefits and costs *without* the program. Following this rule ensures that measured benefits (or costs) are due solely to the program, rather than to changes that would have occurred anyway, even without the program. So benefits and costs should be measured *with versus without* a policy, not before and after. To use this principle, a clear *baseline* policy must be defined to measure the incremental benefits and costs of a *proposed* policy. Defining a clear baseline helps avoid the double-counting problem. Many environmental and natural resource policy analyses have counted as benefits those changes that would have occurred even without the policy.

For example, an analyst performs an *ex post* BCA of a flood control dam in a desert environment using the before and after principle. The analyst concludes that before building the dam there was a terrible flood in the community, but after its construction the community has been free of flood. However, use of the with and without principle shows that there has been too little precipitation to cause a flood even without the dam, so the with-project (i.e., the dam) results are no different from the without-project results. The analyst who used the before and after principle reached the false conclusion that building the dam was economically efficient. Correct use of the with and without principle shows that building the dam has produced negative net economic benefits to date.

[3] An excellent example of a policy that reallocated scarce water among competing users is the Central Valley Project Improvement Act of 1992, that reallocated 800,000 acre-feet of water from agriculture to improvement of fish and wildlife habitat (see U.S. Fish and Wildlife Service, 2004).

Example 5–1
Benefits and Costs of U.S. Environmental Programs

Economist Myrick Freeman took on the ambitious task of estimating benefits and costs of six major U.S. environmental laws enacted or amended since 1970: Clean Water Act; Clean Air Act; Safe Drinking Water Act; Federal Insecticide, Fungicide, and Rodenticide Act; Comprehensive Environmental Response, Compensation, and Liability Act (Superfund); and the Toxic Substances Control Act. In terms of net economic benefits, Freeman identified the following environmental regulations as winners: removal of lead from gasoline, controlling particulate matter air pollution, reducing lead concentration in drinking water under the Safe Drinking Water Act, maximum allowable concentrations on some volatile organic compounds under the Safe Drinking Water Act, and cleanup of some hazardous wastes under Superfund sites with the cheapest cost-per-cancer-risk-avoided.

What common features did these winners share? Before enactment of the law, all substances had threatened human health and had widespread human exposure. In the case of lead, which is known for its toxic effect on nervous systems, a large part of the monetary benefits came from reduced hypertension and its associated risk of cardiovascular disease.

Which environmental regulations produced negative net benefits? Freeman identified these: mobile source air pollution regulations; regulations of discharges into U.S. waterways (except some especially polluted lakes and rivers); and most regulatory actions taken under the Federal Insecticide, Fungicide, and Rodenticide Act, as well as those taken under the Toxic Substances Control Act, the Safe Drinking Water Act, and Superfund. Freeman says the most important policy implication is that nearly all environmental policies and programs can be improved by making them more cost-effective, that is, by seeking ways to reduce the costs of reaching specified targets.

Can these environmental targets be achieved more cheaply? Freeman's answer is yes. One way is to replace command-and-control policy levers with market-based incentives. Examples include emission taxes, tradeable emissions permits, and deposit-refund systems (discussed in Chapter 16). He found that the overall cost-effectiveness of environmental regulations can be increased by *scaling back* regulations and standards where the marginal costs per unit of measurable performance (e.g., cost-per-cancer-risk-avoided) are high and adopting *stricter* standards where there are low marginal costs per unit of the program's performance.

Sources: Freeman, A. M., III. 2002. Environmental policy since Earth Day 1: What have we gained? *Journal of Economic Perspectives.* 16 (1): 125–146; and 2002. Environmental policy since Earth Day 1: What do we know about the benefits and costs? *Agricultural and Resource Economics Review.* 31 (1): 1–14.

Timing. In the face of growing demands for an environmental regulation (e.g., one that produces better health) or for a project's output (e.g., endangered species habitat), *net present value* (NPV) of benefits may actually be increased by postponing the regulation or project even though it might be found to produce a positive net present value if carried out right away. The benefits of waiting will exceed the costs if net benefits are growing at a percentage rate larger than the *discount rate* (i.e., the interest rate at which future benefits and costs are discounted to express them in present value terms). Suppose the discount rate is 6% and immediate implementation produces a positive NPV equal to X. If waiting one year for implementing the project produces a NPV equal to anything larger than 1.06 X, it is economically efficient to wait. See Chapter 6 for more details about the discount rate.

Sequencing. Program or policy elements are not always independent. Implementing one program may influence the benefits or costs produced by a related program. Sometimes this is expressed by saying that these programs have interdependent elements of value. Introducing environmental regulations that improve human health (e.g., air quality regulations) before introducing a program whose output is mostly aesthetic (e.g., visibility in national parks) may produce greater economic efficiency than introducing the programs in reverse order. If regulatory resources (e.g., EPA's budget) are scarce, then it may be most efficient to consider many different possible time sequences for introducing programs, if one program's outputs influences another program's benefits or costs. See Example 5–1.

■ TEN STEPS TO NET PRESENT VALUE

Every environmental program manager spends a considerable amount of time trying to answer this key question: Should we try out this policy, enact that program, or simply do nothing at all? A benefit-cost analysis is the preferred tool of analysis for econocic efficiency, and it can be broken down into 10 steps.

Step 1: Scoping

A BCA begins by defining the scope of the proposed program or regulation (i.e., the proposed action). Step 1 involves defining the proposed action, identifying different ways it could be carried out, and selecting the range of relevant decisions. The accounting stance should be defined to indicate whose benefits and whose costs will be counted. Major gainers and losers should also be identified. The available budget for the study and the overall importance of the decision should determine how much should be spent on finding the facts. Large sums of money should only be spent on a BCA for policy decisions where the stakes are very high.

Step 2: Benefits per Unit

Step 2 considers how to measure the gains and losses from the proposed action. People could be asked how they would vote for the proposal (yes or no) but their responses would fail to indicate the strength of individual preferences for or against the proposed action.

BCA approaches individual preferences by asking about people's *willingness to pay* (WTP) for the additional services provided by the proposed action.[4] The benefit per unit of a program's output is defined as its "price," denoted as *P*. For any service produced by a natural resource or environmental program whose outputs are traded in actual markets, the analyst should compute *P* as its actual market price, since that is what people show they will pay. Two examples are hydroelectric power produced in a reclamation project, and a family's willingness to pay for health improvement programs from the savings they gain as a result of reduced physician visits.

Many outputs of natural resource or environmental improvement programs are not traded on formal markets. Environmental regulations that save or lengthen human lives by reducing exposure to toxic chemicals, protect a species from going extinct, or moderate global warming from the greenhouse effect are good examples. As an example of a natural resource think of grass on federal land used for livestock grazing. In the United States grazing allotments are rarely bid competitively; only qualified ranchers who own nearby ranches are eligible buyers.

For all these unpriced services, prices must be estimated, because they are not sold in markets. But the correct concept for measuring this price is still willingness to pay.[5] Many economists

[4] In some cases, willingness to accept payment (WTAP) to *compensate the beneficiary for loss of the service* is used instead of WTP by the beneficiary to *compensate others for receiving the service.* WTAP assumes that the beneficiary already owns the right to have the service, whereas WTP assumes the beneficiary does not own that right and must pay to receive it. The difference between the two measures of benefit can be considerable: WTAP can be infinite, for it has no upper bound. The WTP can be no larger than the beneficiary's income. People typically require a larger payment to give up a right than they will pay to acquire the right. So in order to use WTP, the analyst must first decide whether an individual already lacks a right to the good in question. For example, suppose EPA proposes tightening regulations on the use of asbestos, a known carcinogen. The decision of whether to use WTP or WTAP in measuring the program's benefits turns on whether people have a right to a healthy, asbestos-free environment. All this poses a potential circularity problem for the use of BCA to determine property rights. The correct measure of benefits to use depends on who is presumed to have the property rights in the first place. Where property rights are in doubt or are the subject of debate or dispute, the BCA should be conducted and results presented for both measures of benefit.

[5] Some special things like human lives prolonged or protected ecosystem services may have very high values indeed to those who receive them. The use of WTAP instead of WTP gives the beneficiary the opportunity to proclaim that the service is "not for sale," if indeed she would not trade it at any price. Ownership always gives the right to not sell at any price. The statement that something is not for sale may affirm, improve, and protect a service's value. The statement is one way to show that a service is valued for its own sake and that no possible trade in money or other things could compensate for its loss. Trading a service for money (WTAP at a finite price) shows that other things could be traded for it with no loss in satisfaction.

have published research during the past 25 years dealing with how values of nonmarket goods can be estimated (e.g., see Cropper and Oates, 1992). The contingent valuation method, travel cost method, and hedonic method are three widely used techniques to estimate these values. The cost required to estimate willingness to pay for unpriced values can be hundreds of thousands of dollars or more (e.g., several million dollars to measure the economic damages to Alaska's ecosystem resulting from the 1989 *Exxon Valdez* oil spill).

Step 3: Costs per Unit

Step 3 considers the cost per unit of resources used or otherwise displaced to make way for the proposed action. We shall call these costs, *C*, and they include things bought for ordinary government projects (e.g., fire trucks, planted trees, poured concrete, salaries, improved streamside habitats). Costs also include any other unpriced benefits that would be lost to society from a proposed action, such as increased power prices resulting from restrictions on sulfur dioxide emission restrictions or reduced effectiveness of chemicals resulting from safety regulations imposed on their manufacture. Issues regarding how to measure costs are described in greater detail in Chapter 7.

Step 4: Output Quantity

Step 4 lists what the proposed action produces or supplies for the public's benefit; call these quantities *Q*s. For a multiple-purpose water project, for example, these outputs include things like acre-feet of water delivered to farms for crops, kilowatt hours of power, recreation visitor days, and acres of habitat for endangered species. For a proposal to set up an aquifer conservation plan, the quantity is acre-feet of water made available for other current uses or saved for the future. For an environmental regulation that improves workplace safety, outputs could include longer or healthier lives for people affected.

Step 5: Input Quantity

Step 5 lists inputs that cause the proposed action's costs to be incurred; that is the physical resources used in carrying out the action; call these input quantities *I*s. These include things like number of fish stocked, number of boat ramps purchased, hours of workers hired, acres of spotted owl habitat set aside, and time spent by attorneys and others who write and enforce regulations. Steps 1 through 5 assemble the data to be used for the analysis in steps 6 through 10. Table 5–1 shows hypothetical data of the kind you might see after going through the first five steps. The data refer to benefits and costs for a hypothetical reservoir project.

Table 5–1 Data on benefits and costs of a hypothetical reservoir project

Year	Benefits						Costs			
	Power (no. 1)		Agriculture (no. 2)		Recreation (no. 3)		Concrete (no. 1)		Recreation lost (no. 2)	
	Q (kwh)	*P* ($/kwh)	*Q* ac-feet	*P* $/ac-foot	*Q* (visitor days	*P* ($/visitor day)	*C* ($/ton)	*I* (tons)	*C* ($/visitor day)	*I* (visitor days)
1	10,000,000	0.10	100,000	40	100,000	10	80	100,000	$40	50,000
2	10,000,000	0.10	100,000	40	150,000	10	80	0	$40	50,000

Q = output quantity; *P* = price; *C* = cost; *I* = input quantity; kwh = kilowatt hour; ac-foot = acre-foot.

Step 6: Benefits by Period

No data beyond that found in the first five steps are required for steps 6 through 10. In step 6 total gross benefits are computed. That is, once all the prices and quantities are found, defined, and computed, total gross benefits (*GB*) are computed for each time period. They are computed as the sum of prices times quantities for all project purposes and time periods. The formula is

$$GB_t = \sum_{i=1}^{n} P_{it} Q_{it} \tag{5.1}$$

where the *t* subscript denotes the *t*-th period, and *i* denotes the *i*-th project purpose or service. The summation sign means that for any period, that period's total gross benefit is found by summing that period's prices per unit output (*P*) times quantities produced (*Q*), over all services.

For example, suppose we wish to calculate a project's total gross benefits in period 1, GB_1, as well as gross benefits in period 2, GB_2. The hypothetical plan is to build a reservoir that produces three services: power (number 1), agriculture (number 2), and recreation (number 3). The generic formula in Equation 5.1 for gross benefits in any period, GB_p, applied to each of the first two periods is as follows:

$$GB_1 = P_{11} Q_{11} + P_{21} Q_{21} + P_{31} Q_{31} \tag{5.1a}$$

$$GB_2 = P_{12} Q_{12} + P_{22} Q_{22} + P_{32} Q_{32} \tag{5.1b}$$

To see how Equations 5.1a and 5.1b are applied, suppose we know the following information about the operation of a reservoir construction project designed to reduce dependence on coal-fired electricity, thereby reducing the impacts and costs of global warming. Assume there is only a two-year time horizon. Assume the discount rate is 10% (0.10).

The data in Table 5–1 make these assumptions: the price of power is $0.10 per kilowatt hour (kwh), and the project is expected to produce 10 million kwh each year; the income-earning value to farmers per acre-foot of irrigation water is $40 and 100,000 acre-feet will be supplied; the value per visitor day of recreation is $10 and 100,000 visitor days will be supplied in the first year, growing to 150,000 days in the second year. So total gross benefits in year 1 (GB_1) is $0.10 times 10 million, plus $40 times 100,000; plus $10 times 100,000—$6,000,000 Similarly, total gross benefits in year equals $6,500,000.

Step 7: Costs by Period

Step 7 computes total gross costs for each period. After computing the values of all the prices and quantities of each input, total gross costs for each *t*th period, GC_p, is computed as

$$GC_t = \sum_{j=1}^{m} C_{jt} I_{jt} \tag{5.2}$$

where *j* denotes the *j*th input purchased. This equation can be applied to find total gross costs in period 1, GC_1. The data in Table 5–1 say that the reservoir construction consumes two resources of value: concrete (number 1) and recreation lost (number 2). All construction costs are up front, which means there are no construction costs in the second year. Construction costs in the first year are $80 per ton of concrete times 100,000 tons poured in the first year. The other kind of cost is the opportunity cost of $40 per visitor day applied to 50,000 visitor days of recreation displaced in both years. Applying Equation 5.2 to each year's cost produces the following equations:

$$GC_1 = C_{11} I_{11} + C_{21} I_{21} \tag{5.2a}$$

$$GC_2 = C_{12} I_{12} + C_{22} I_{22} \tag{5.2b}$$

Application of these two cost equations shows that GC_1 = $10 million, since $80 times 100,000 plus $40 times 50,000 equals $10 million. There are no construction costs in the second year, so the only gross costs in that year are the lost recreation days, valued at \$2 million (\$40 times 50,000). So GC_2 = \$2 million.

Step 8: Net Benefits by Period

The eighth step requires computing net benefits for each period, that is, the period's total benefits minus total costs. This is computed as a simple algebraic subtraction of each period's costs from the same period's benefits:

$$NB_t = GB_t - GC_t \tag{5.3}$$

where NB_t is gross benefits (GB_t) minus gross costs (GC_t). Application of this formula typically results in several periods that show negative net benefits, usually the earliest periods when costs are being incurred with no benefits yet being produced. Application of the data in Table 5–1 for each of the two periods produces the following equations:

$$NB_1 = GB_1 - GC_1 = -\$4{,}000{,}000 \tag{5.3a}$$

$$NB_2 = GB_2 - GC_2 = +\$4{,}500{,}000 \tag{5.3b}$$

That means net benefits are minus \$4 million for year 1 and plus \$4.5 million for year 2. A negative net benefit for any year means that the year's costs exceed its benefits. The net benefits by year in step 8 are used for the computations required in step 9.

Step 9: Net Present Value

The net present value computed in this step deals with the time factor. In general, there is a time profile of benefits and costs corresponding to each of the time periods under consideration. The following two examples illustrate the need to deal with time in a systematic way. Figure 5–3 and Table 5–2 present the same data for two projects, each with different time streams of net benefits, but each producing identical present values. In the top half of the figure, project 1 produces negative net benefits (−20) during the first three years and positive net benefits (+15) for the next seven years. The outcome produces a *net present value* of +14, which comes through the use of discounting (described in the next section). For project 2, shown in the bottom half of the figure, the time profile is reversed: small positive net benefits (+2) for all 10 years, which also produces a net present value of +14. To

Figure 5–3 Two time streams of net benefits (*NB*) with equal net present values (*NPVs*)

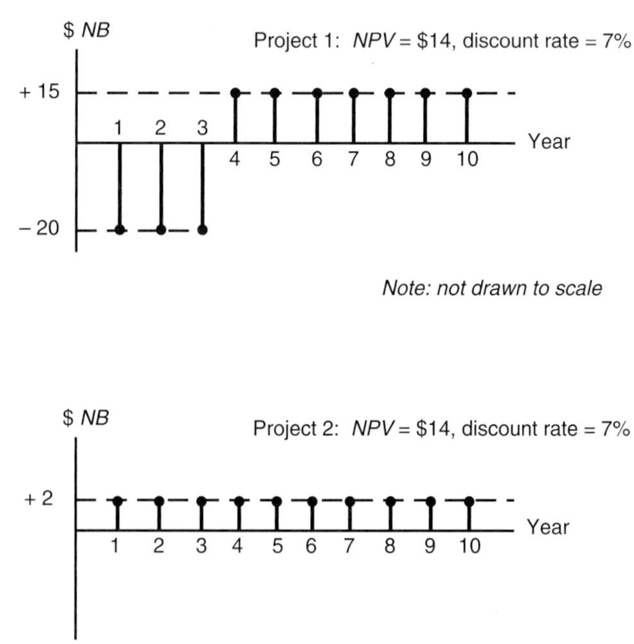

Table 5–2 Two time streams of net benefits with equal net present values

Year	Project 1		Project 2	
	Net benefit		Net benefit	
	(Undiscounted)	(Discounted at 7%)	(Undiscounted)	(Discounted at 7%)
1	−20	−18.69	2	1.87
2	−20	−17.47	2	1.75
3	−20	−16.33	2	1.63
4	15	11.44	2	1.53
5	15	10.69	2	1.43
6	15	10.00	2	1.33
7	15	9.34	2	1.25
8	15	8.73	2	1.16
9	15	8.16	2	1.09
10	15	7.63	2	1.02
Total discounted (rounded)		14.00		14.00

obtain a common denominator comparison of these two very different streams of net benefits, we need to reduce all possible time profiles of many competing programs to a single figure.

Calculation. Having seen the challenge posed by starkly different streams of net benefits, we are ready to return to the hypothetical project and learn more about step 9. Step 9 transforms a sequence of net benefits that occur in several different periods to a single common denominator. The formula for turning that stream of net benefits into a single overall performance indicator is

$$PVNB = \frac{NB_1}{(1 + r)^1} + \frac{NB_2}{(1 + r)^2} + \ldots + \frac{NB_T}{(1 + r)^T} = \sum_{t=1}^{T} \frac{NB_t}{(1 + r)^t} \tag{5.4}$$

where r is a discount rate[6] and T is the last year in the project's time horizon. For some environmental protection programs (e.g., limiting population growth or reducing the risk of global warming), T could be several hundred years.

In step 9 all future benefits and costs are discounted to a single number, called the *present value of net benefits* (PVNB), sometimes called *net present value* (NPV). Its calculation requires an appropriate rate of discount. If that discount rate is 10% per year, *both* a certain net benefit valued at \$1,100 next year, *and* one valued at \$1,210 in two years' time, have an equal discounted value of \$1,000. That is, $[1{,}210/(1.10)^2] = [1{,}100/(1.10)^1] = 1{,}000$. Equation 5.4 says that to find the present value of any amount of net benefit received t years from now, compute it as $NB_t/(1 + r)^t$, where r is the appropriate discount rate. Most of us would agree that it is better to receive a given dollar-valued benefit now rather than waiting for it, so some discount rate greater than zero is usually appropriate.

In the reservoir problem, if you apply Equation 5.4 to the two-period net benefit stream, it produces a single number. That result says that the two-period project produces a discounted net present value of +\$82,645, which means that this reservoir project is a good economic investment when the discount rate is 10%.[7]

Now let us examine some broader applications of the PVNB formula. Figure 5–4 and Table 5–3 compare two different programs, *A* and *B*—one with concentrated costs and diffuse benefits and the other with concentrated benefits and diffuse costs. Both programs have a stream of benefits and costs lasting for seven years.

---

[6] U.S. government environmental programs have seen discount rates ranging from 0 to 20% (i.e., 0.00 to 0.20). See Lind (1990).

[7] However, a quick calculation shows that if you raise discount rate to 15% you get a PVNB of −\$75,614. This result shows the power of a rising discount rate to turn a strong economic performer into a weak one. See Chapter 8.

Figure 5–4 Net present value (NPV) and its distribution

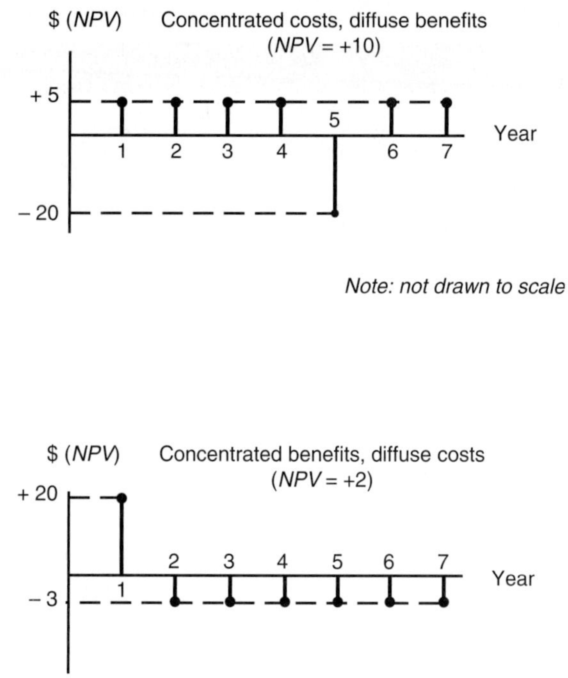

Table 5–3 Net present value (NPV) and its distribution

Year	Discounted net benefits by year	
	Program *A* (concentrated costs, diffuse benefits)	Program *B* (concentrated benefits, diffuse costs)
1	5	20
2	5	−3
3	5	−3
4	5	−3
5	−20	−3
6	5	−3
7	5	−3
NPV	10	2

Both the figure and table show that program *A* produces highly concentrated costs in period 5, whose discounted value is −20, but steady positive net benefits in all other periods, equal to +5. Program *A*'s overall NPV is found by summing the discounted value of net benefits over all seven periods. It is computed as $6\,(+5) + 1(-20) = 10$. Program *B*'s net benefits are highly concentrated in period 1, which has a discounted value of +20. Program *B* produces a steady but slightly negative net benefits spread out over all other periods (discounted value is −3 in each of the periods). Program *B*'s overall NPV is found by summing the discounted value in all seven periods, for an overall NPV of $2 = [1(+20) + 6\,(-3)]$. In terms of economic efficiency, program *A* is preferred to program *B* (compare +10 to +2). However, often programs like *B* are preferred politically because a large number of time periods (or groups of people) each incur low costs, while a small number receive large benefits.

Sensitivity Analysis. The NPV is one, single computed number that indicates the overall economic performance of a project or program. But a huge number of assumptions go into producing that single number. How confident are we of our chosen discount rate? Do we really know how long the project will last? What kind of substitutes will consumers and voters have for our

Example 5–2
Adding Value to Benefit-Cost Analysis: Sensitivity Analysis

Many environmental groups vigorously oppose increased use of benefit-cost analysis (BCA). They fear that BCAs will typically be biased against the environment or that BCA reports will be ignored when the results favor environmental projects. Fears of potential abuses of BCA are so widespread that environmental groups have lobbied for legal requirements that mandate certain environmental programs regardless of cost. The U.S. Clean Water Act and Endangered Species Act are good examples.

Sensitivity analysis (SA) could help promote more widespread use of BCA. This type of analysis investigates the sensitivity of calculated net present value (NPV) to changes in assumptions, price and quantity data, and benefit calculation methods. The purpose of SA is to provide more and better information in an BCA. Suppose a proposed environmental regulation produces a single best estimate NPV of $1 million, based on a large number of assumptions, discount rates, and benefit and cost computation methods. However, SA might show that the same project produces an NPV ranging from negative $0.5 million to positive $1.8 million, depending on the assumptions.

A single "most likely" NPV and BCA finding presented *along with* detailed SA gives everybody a better opportunity to evaluate decisions *comprehensively*. The ability to watch decision makers' respond to BCAs with detailed SAs helps citizens judge the motives of leaders and gives them insight into leaders' beliefs about the future. Including an SA with a BCA makes it harder to ignore politically unpopular results or results that are unfavorable to powerful special interests. It also makes it harder to use a BCA to mislead the public by justifying economically and/or environmentally weak decisions.

Sensitivity analysis can also increase the relevance of a BCA. For example conditions change so an environmental project's supporters ask a rejected project to be reconsidered. Is it possible to avoid doing another complete BCA, thereby saving considerable expense? Yes. The new conditions may lie entirely inside the range included in the original BCA. So a simple rereading of the original report's SA may show the size and direction of how results change as conditions that affect the program's economic feasibility change. Not all assumptions can be varied in a BCA without producing thousands of tables, so assumptions to be varied should be chosen with care. Each additional SA factor varied at least doubles the number of tables of NPVs. Public comments on the original BCA can help identify the most important assumptions under debate, and these assumptions could determine the SA factors.

Source: Merrifield, J. 1997. Sensitivity analysis in benefit-cost analysis: A key to increased use and acceptance. *Contemporary Economic Policy.* 15 (3): 82–92.

---

project 25 years from now? What kinds of environmental control technologies will be available to substitute for the environmental program? Perhaps several possible NPVs that vary according to our assumptions would be helpful.

Sensitivity analysis tests the variability of the NPV results from changes in critical assumptions (see Example 5–2). It identifies those input parameters (assumptions) that have the greatest influence on the outcome. It then repeats the analysis with different parameter values and evaluates the results to see which parameters have large or small influences on NPV.

If a small change in the value of an input parameter changes the NPV enough to change the policy alternative selected, then the results are *sensitive* to that parameter. By contrast, if the value of a parameter going in an assumption has to be tripled before there is a change in the selected alternative, the results are *insensitive* (robust) to that parameter. Sensitivity analysis identifies the most sensitive input parameters, so that a decision can be made on whether it is worth the added cost needed to get better information on that parameter. Sensitivity analysis also adds to the openness and transparency of the exercise, which is usually a good thing. For all these reasons, the conduct of sensitivity analysis increases the acceptability and use of BCA by decision makers, who often do not wish to rely on a single estimate of NPV in judging the overall effectiveness of a program.

Assumptions that are good candidates for varying will depend on the most important benefit and cost elements as well as the areas of greatest uncertainty of the program being analyzed. For example, suppose you are analyzing a program that would change emission regulations for sulfur dioxide. You might want to consider several possible levels of coal-fired electric power, production, future growth in power demand, incidence of acid rain, and the discount rate. Sensitivity analysis, though, suffers from what is called the curse of dimensionality: You cannot have too many values of too many parameters while having a finite number of pages in your report. If you decide to assign four possible values to each of five uncertain parameters, you will have $4^5 = 1{,}024$ possible NPVs to present.

In general, sensitivity analysis should be conducted for estimates of benefits and costs, the discount rate, and assumed distributions of benefits and costs. Parameters used in the analysis should be well documented and available for independent reviews. Properly done, sensitivity analysis has great potential to increase the use and acceptability of BCA by decision makers.

Step 10: Decision Rules

The *maximum* NPV rule is the best way to use information on benefits and costs to maximize overall economic efficiency. The use of NPV to make decisions will help you choose environmental and natural resource actions that maximize the present value of net benefits. So, use of the NPV rule says we should build the reservoir. Application of the maximum NPV rule will produce the decision that maximizes the economic performance of the available resources.[8]

Another rule says that the benefit-cost ratio must exceed 1.0, which means that resources should be committed to a decision if the ratio of its implementation produces a ratio of present value of benefits to present value of costs exceeding 1.0 (this is the same as requiring NPV to equal zero or larger). This rule is a conservative standard, because its use permits agencies to implement programs for which benefits are *no greater* than costs. A private business cannot stay in business if its revenues fail to exceed its costs considerably. Similarly, for government environmental and natural resource decisions, maximum economic efficiency requires that benefits exceed costs by the greatest amount possible.

You may have heard this commonly asked question: Does the benefit-cost ratio exceed one? Although it does not promote maximum economic efficiency, this simple statement is easily understood by politicians and the public because it is expressed as dollar of benefit per dollar of cost. When Congress and other decision makers enact laws or make appropriations for their projects, they often wish to know if the benefit-cost ratio exceeds 1.0. Once the ratio exceeds 1.0, decision makers may consider the economics to be acceptable, and then use political, moral, or other criteria to make the final decision.[9]

Measuring Benefits

When environmental programs produce benefits that are traded in markets, the market price can be a good measure of the program's efficiency benefits, but market transactions are rare for environmental or natural resource programs: We rarely buy an improved environment at the local grocery store. Because of the physical, economic, and institutional characteristics of most natural resources and environmental programs, few market prices are involved. Rarely is it easy to measure the value of an environmental improvement from observed market prices. Even when markets exist, the observed price must be interpreted with care. Consider the case of water policies that would alter supplies of water available for agriculture or for growing cities. Sometimes water is made available through a rental market. The title to a water right typically gives the owner a right to use perpetually an annual quantity of water, but the owner may rent out the use of water to a city for a limited time period. Here, the observed price of water is a private, short-run de-

[8] The internal rate of return is another economic performance measure for a project. It is measured as the discount rate, which if applied to a proposed action would make the NPV equal zero. The internal rate of return for the reservoir project is 12.5%. Higher discount rates make NPV negative; lower rates make NPV positive.

[9] A BCA presents results in a clear-cut manner. Either benefits do or do not exceed costs. But decision makers may value the flexibility of their decisions; a BCA that removes that flexibility might not be welcome. At its worst, the cynic would say that politicians will do whatever they please anyway, and they do not wish to be troubled by a BCA whose benefit-cost ratio is less than one: A BCA report containing a single result makes it hard to be flexible (Pearce, 1998).

mand for water. The price may have limited validity in evaluating decisions about long-term public water development, reallocation, or quality improvement.

Another kind of market transaction occurs when a resource's price is set under an administered price system. Many prices of irrigation, minerals, and grazing permits on some U.S. federal lands are set by political forces, not economic ones. If the resource buyers (e.g., farmers, mine owners, or ranchers) are not free to adjust use to meet all their demand at the specified price, the administratively set price understates the efficiency benefits produced by the resource.[10] Likewise, in countries or cultures where the price of medical care is subsidized, but subject to widespread shortages, the market value of reduced medical expenses is one way to value environmental programs that improve human health and safety (e.g., reduced air pollution). However, this market value is likely to be much less than the willingness to pay for health-improving environmental policies.

Marginal and Nonmarginal Changes*

Where market prices can be observed and measured, they can be used to value both costs and benefits of a program. The economic logic behind this principle is that the willingness to pay for one more unit of a project's output or one more unit of a project's input (the marginal unit) equals the market price as long as buyers are free to buy all they want at that price. There are no institutional constraints that limit consumer purchases. Suppose, though, that institutional constraints set the price at or near zero (e.g., subsidies of basic foods), but consumers are restricted from buying all they wish at the subsidized price. In this case, the marginal benefit from one more unit supplied by the program is not the subsidized price. In fact, it is the higher price consumers could be charged.

Many environmental policy proposals are large enough to affect price. National sulfur dioxide emission regulations on coal-fired power will raise the price of electricity by increasing the cost of producing power from the next cheapest source. Global agreements on greenhouse gas emission abatement will likely change prices for products that require burning of fossil fuels in the short run (increased prices) and agricultural products in the long run (decreased prices). Special difficulties arise when government programs, projects, or regulations are large enough to affect total supplies of services enough to change market prices. The incremental benefits accruing from a public program that changes the supply of market goods cannot be measured by multiplying the additional quantity of output by the old or the new price. The former overestimates and the latter underestimates the incremental benefit.

On the benefits side, if the demand schedule is linear (i.e., price is a linear function of quantity supplied, other things equal), a simple unweighted average of prices with and without the programs can be used.[11] For *intermediate goods* (producer inputs), like water for crop irrigation or commercial fisheries, the demand curve is derived from the final product and from the intermediate input's efficiency in producing that product. It is measured as the product price times the input's marginal physical product. It can only be an accurate measure of the social benefit of a program providing more of the resource if the price of the output (crops or fish) to which the input will be applied reflects what people are actually paying. If this condition is satisfied, the gross benefit arising from a public program that supplies intermediate products (Equation 5.1) is measured by the producer's change in net income due to the added supply. That is, the producer's gain in net income measures the social value of programs that supply that input.

On the costs side, when inputs are used by the project or program or other outputs are displaced by it, there is a double problem. First, it is necessary to adjust prices of inputs to eliminate

---

[10] Here is an example of an administratively set price being less than the value of the resource: Prices of U.S. federal grazing forage are set by the Congress at just over $1 per animal unit month (AUM). However, ranchers or environmental users are unable to buy all they wish at that price, so forage is quantity constrained and not price rationed. Several economic studies conducted in the 1990s have shown ranchers can pay a "fair market value" of $8 to $10 per AUM for federal grass, even when the administratively set price is just over $1.

[11] For a nonlinear demand, the analyst must find the added area beneath the demand schedule brought about by the program (see Appendix B on total benefit, marginal benefit, and consumer surplus).

*The asterisk indicates advanced material that can be skipped if you are not specializing in economics.

any payments to them that exceed their *opportunity costs*. Second, there is a problem analogous to one on the benefits side. As greater amounts of an input are used in any one line of output, the price of the alternative product which it might have been making (its opportunity cost) continues to rise. Again we confront the choice of valuing the input at its without-policy price, its with-policy price, or at some intermediate level. If we assume a linear demand schedule, a price half-way between the old and new levels will meet our needs; as on the demand side, the same adjustment is required if it is nonlinear.

Mathematical Methods*

In public programs that propose to supply an intermediate good (i.e., a service that is used as an input by producers), direct benefits can be measured as the producer's change in net income, if the producer's output is supplied in competitive markets. One widespread approach to estimating values of intermediate goods is to estimate the derived demand function for that input. Water supply is a classic case of such an intermediate good for which the demand function is the value of the marginal product.[12]

Mathematical programming methods can be used to derive the value of publicly supplied program inputs when that input is used by a business. Classic examples come from the many water policy studies in which scarce water constrains the actions of an income-maximizing water user, such as a farm, city, or environmental project. The increment in net producer income resulting from adding the publicly supplied input to a production process is defined as the producer's willingness to pay for the added input supply. See Appendix C for technical details.

Alternative Cost Saved

Another method for estimating benefits of a public environmental program comes from the principle of alternative cost (see Chapter 7). This principle recognizes that the upper bound on willingness to pay for a publicly supplied service is the cost saved by not supplying the service from the most likely economically feasible alternative. The method must be used carefully because there are usually many possible alternatives, including private alternatives to public projects, public alternatives to each component of multiple-purpose programs, and the like. And many, if not all alternatives, might be economically infeasible. The technique could be applied correctly to cases in which a private alternative would have been built without the public program in question.

An example is the cost of added medical facilities that would have been supplied to serve an illness if an environmental health regulation had not been in place. Using the principle of alternative cost saved, the benefits of the public program are measured by the cost avoided of the least expensive alternative. The approach is attractive because in many cases lack of demand data makes measurement of a program's benefits through estimating a demand schedule difficult or impossible. The primary advantage of the alternative cost method is that, for cases in which demand data are hard to find, estimation of maximum willingness to pay can be accomplished without estimating demand functions.

An important disadvantage of the alternative cost method is that it can be used to justify a weak program by pretending that there is always a more expensive method with a higher alternative cost

---

[12] Readers with a calculus background will recognize that the value of the marginal product is the derivative of the production function relating output to the input in question multiplied by the price of the final product. For the case of irrigation water, economists sometimes use econometric crop-water production function estimates to find the marginal physical product. Ideally, this is accomplished by conducting field experiments in which water and other inputs are applied to the crop in varying proportions. Resulting crop-water surfaces are then used to statistically fit production functions in which water is one of the inputs. For environmental resources, one might fit a production (dose-response) function in which the program's publicly supplied intermediate good is one of several inputs. The marginal benefit of the intermediate good is measured as the value of the marginal product of that input. Practically, the scarcity of data needed to estimate production relationships, together with the fact that the publicly supplied input may be swamped by other production costs, has typically required analysts to use other methods to estimate benefits. See Chapter 7.

for meeting the program's objective. For example, one could measure the benefits of a government program that improved water quality by claiming that the resulting water quality improvement saves the private sector the large amount of resources that it would have spent on achieving the same water quality. One way to guard against false justification is to only count the cost of the alternative method if there is strong evidence that it would be built without the program.

User Surveys

Sometimes the demand for a public program must be measured when no market transaction and no use in production occur, so it is hard to estimate benefits. Good examples are cleaner air and recreation in natural surroundings, in which cases the resource in question is a public good. Since the early 1980s, analysts have come to rely on user surveys to derive estimates of the value of programs, especially the value of an environmental resource, such as water or wildlife habitat to the user experience. Two of these approaches have been pursued: the expenditure function approach and the income compensation method. The expenditure function approach relies on data in related markets where the quantity of the publicly supplied resource is a variable in the demand for some private market good. Under some conditions, an empirical estimate of the benefit for the publicly supplied resource can be derived. The well-known travel cost method, which is described in detail in Chapter 7, is one example. The income compensation approach is based on estimating the compensation required to offset, or in some cases to take the place of, the proposed policy change. Measuring that compensation often depends on a direct asking approach (contingent valuation) to estimating changes in economic surplus. As described by Mitchell and Carson (1989), Rubin, Helfand, and Loomis (1991), and in Chapter 7 of this book, consumers are asked directly about their willingness to pay for certain environmental or natural resource improvements. In some cases they are asked about their willingness to vote for ballot initiatives in which they would be charged a percentage of the total program cost.

Benefits Transfer

Decision makers and policy analysts often need information on the benefits of environmental programs or improvements. Internally consistent sources of this needed information are ideally obtained from a market or nonmarket survey of affected users. However, collection of the data needed to estimate environmental demand and benefits usually requires an expensive user survey. For *proposed* sites or programs, there are no resource users from whom to collect data. In such situations, *benefits transfer* involves using data on resource use at sites in one region (a surveyed region) to predict visitation and benefits at sites in a separate region for which resource use or benefits data are unavailable (the target region). Benefits transfer is a practical method for evaluating management and policy impacts of environmental proposals when primary research is not possible or justified because of budget constraints, time limits, or the presumption that impacts will be small. Many economists consider benefits transfer essential to the future development and use of BCA. Methods for performing benefits transfer as well as their limits, validity, and reliability, are described for the case of outdoor recreation by Rosenberger and Loomis (2001).

■ MEASURING COSTS

Costs are usually easier to estimate than benefits. The direct effects of compliance with environmental regulations or natural resource management usually involve actions that are measurable in conventional monetary terms and that are relatively easy to identify. In dealing with costs, the analyst attempts (1) to place monetary estimates on what is given up or (2) to determine the resource cost to the economy as a result of an environmental regulation or action aimed at maintaining or improving some aspect of public well-being. In a number of circumstances, cost estimation presents special difficulties: predicting technical innovations that may reduce long-term compliance costs, which can overstate costs of environmental regulation; indirect or hidden costs caused by uncertainty facing people who cope with the regulatory process; costs of an environmental regulation borne by producers who substitute from the regulated form of pollution to an unregulated form.

Underused Resources

The presence of unemployed and underused resources must be faced in any attempt to estimate the real social cost of a public environmental or natural resource program. In this case, the opportunity cost of an underused resource is less than its market price, since little forgone production occurs if such resources are used in connection with a proposed environmental program.

Early analysts acknowledged the problem, with regard to labor but were skeptical of attempts to forecast unemployment over the long interval after implementation of the proposed program. For correct measurement of the opportunity cost of labor, it is not sufficient to measure the labor force status *before* and *after* the proposed program; rather, condition *with* program and *without* must be compared. The latter may be difficult to achieve empirically because of the potential for labor migration from and to other regions.

Opportunity Cost of Unpriced Resources

Market prices may be poor indicators of opportunity costs or might not even exist in the case of land for project sites. Outdoor recreation, ecosystem functions, or grazing benefits displaced by a new reservoir are good examples. If the site is publicly owned, no outlay from a public agency's budget for its purchase is required, but the unpriced alternatives of recreation, ecosystem function, or grazing may be sacrificed. Even if land for the project is bought, the higher private discount rate and risk aversion of land developers may produce an assessed or market value much different from opportunity costs required for a BCA. Particularly in cases where a proposed reservoir damages ecosystem functions to develop irrigated farmlands to feed people in a poor country, the opportunity costs of the site should be assessed with the same discount rate, commodity prices, and production costs as are benefits of the development.

Some costs of public actions may be lost opportunities for other uses of a resource that may involve hard-to-measure values. Market prices of these resources may poorly reflect how the public values such a loss. One example is the damming of a geologically and aesthetically unique canyon where a public program of natural resource development produces benefits measurable in monetary terms (e.g., water supply, recreation, and hydropower). The costs of building and operating the dam and acquiring land are directly measured in monetary costs. However, the public value of enjoying and preserving a unique geographic feature is more difficult to include in the cost estimate. Valuation of lost recreational, scientific, or ecological values may be attempted, but valuing uniqueness is hard and subject to considerable controversy. In this type of situation economic analysis can indicate what is lost in conventional economic output if the resource development project takes place. This information can be viewed as the minimum value society puts on preserving a unique resource and is a valuable reference point for debating the merits of preservation.

■ Treatment of Time

Estimating the life of a project or program is difficult, subjective, and subject to error. It depends on the assessments of the program's physical life, technological changes, shifts in demand or fashion, emergence of competing products, and the general state of the world many years ahead. If a government program had been established in the year 1900 to subsidize buggy whips and had been assigned a project life of 100 years, the estimation of project life would have been too high by a considerable margin.

The effect of any errors in estimations will depend on the discount rate: The higher the discount rate, the less errors of estimation matter. However, for many environmental programs, such as large investments in mitigation of global warming, estimation errors matter considerably. Estimation of project life is an example of the need for sensitivity analysis, where calculations are repeated many times for different possible values of the project's life. Sensitivity analysis is quite important when analysts' estimates of costs and benefits are uncertain (refer to Example 5-2).

The net present value formula discussed earlier in this chapter made it clear that it is necessary to discount future benefits and costs to determine the present value of net benefits. Discounting reflects the time value of money. Benefits and costs are worth more if they are experienced sooner. All future benefits and costs, including nonmonetized benefits and costs, should be discounted. The higher the discount rate, the lower is the present value of future benefits and costs. For typical public investments in construction projects, with costs concentrated in early periods and benefits following later, raising the discount rate tends to reduce the net present value. This may not be true for environmental regulations, because costs are not always concentrated in the earliest years.

The discount rate to select depends on whether the benefits and costs are measured in real or nominal terms. A real discount rate is adjusted to correct for the effect of anticipated inflation and should be used to discount constant-dollar or real benefits and costs. A real discount rate is approximated by subtracting expected inflation from a nominal interest rate. For example, if expected inflation is 3% and the nominal market interest rate is 7%, the real rate is 4%.

Benefits and costs incurred in future years should be assigned lower weights in decision making than ones experienced currently whenever the discount rate is positive.[13] A higher discount rate means that people would rather receive a $1 benefit today than have to wait to receive it. Recall why this is true. On the benefit side, a 10% discount rate means that people are indifferent between receiving a $1 benefit today and receiving $1.10 a year from now. So the present value of a benefit of $1.10 to be received a year in the future is only $1. On the costs side, $1 in cost incurred today imposes the same burden lending out the $1 at 10% as paying a $1.10 cost a year from now. So the present value of a $1.10 cost to be incurred a year from now is only $1.

The discount rate should be based on how individuals affected by the proposed policy are willing to trade off current consumption for future consumption. Common methods for selecting a discount rate are the market rate of interest, social opportunity cost of capital, and social rate of time preference. For reasons discussed in Chapter 6, there is considerable uncertainty over the correct discount rate. Because the discount rate is so hard to identify, the present value of net benefits of a proposed policy should be computed for several possible discount rates.

Issues Affecting Both Benefits and Costs

Several existing conditions in the overall economy can have a major influence on the conduct and outcome of a benefit-cost analysis: (1) secondary effects, (2) employment impacts, (3) inflation, and (4) risk and uncertainty.

Secondary Effects

It is unlikely that any environmental or natural resource program will affect the overall long-run level of employment in an economy. In a normally functioning economy, a new investment project or environmental regulation yields no net benefits beyond the willingness to pay by beneficiaries minus the opportunity cost of resources displaced. Any expansions in income from related activities, termed *secondary effects,* are usually assumed to be offset by income declines elsewhere. Even though secondary impacts are almost always balanced out elsewhere in a national economy, their economic and political impacts are typically important to affected regions. Much of the political motivation for public natural resource or environmental programs results from an attempt to capture such regional effects, which in many cases are reflected in large gains in real property values.

[13] A dollar of benefit to be received a year from now is worth less than the dollar in one's pocket today, because of opportunities forgone during the year. For this reason, it is preferable to accrue benefits earlier rather than later and to delay costs until later because there are many competing uses for current resources.

The notion that some secondary benefits or costs should be included as benefits or costs in a BCA has a long history in U.S. water resource programs. Controversy about secondary effects began in the 1950s; discussion focused on benefit estimation procedures used by water development agencies in connection with irrigation projects. The essential question is illustrated by reviewing the case of a proposed irrigation development project that results in an increase in grain production. The primary (direct) benefits are measured as the value of the increase in grain output less the associated increase in farmers' costs of production in the project area, (in other words, the gain in farm income).

Local supporters of the project point out correctly the increased grain output results in increased economic activity by grain merchants, transportation suppliers, millers, bakers, retail food stores, and restaurants. All these increases will involve an increase in income for various resource owners. In the irrigation example, if the ratio of total income in all these activities to the value of grain at the farm is 60% then secondary benefits of 60% of the value of the increase in grain output would be credited to the irrigation project as a secondary benefit. These have sometimes been called stemming secondary benefits since they "stem" directly from the project. "Induced" secondary benefits, by contrast, are the additional incomes paid to resource owners from activities that sell to farmers, for example, implement dealers, bankers, and farm labor. If the total income paid to these activities averages 40% of farmers' purchases, then an additional 40% of farmers' input purchases associated with the project would be credited to the project as a secondary induced benefit.

Much analytic effort has gone into forecasting regional income gains, often with the use of input-output models.[14] Since the 1990s in the United States, these models have been developed to be applied at county levels and are available through various commercial vendors. Changes projected by these or similar methods are, nevertheless, *income transfers* from one region to another, and it is an error to label them as benefits or real income gains from a national view.

In national efficiency analyses, secondary benefits or secondary costs should only be counted if market prices of outputs fail to reflect opportunity costs. Increases in output resulting from expansion of sectors connected to the project or program are not a separate benefit attributable to the project from the national view: Those output increases cause resource owners to be paid, but those resources have alternative uses displaced by the project. The only time additional income payments should be counted as benefits is when their surplus (payments exceeding their opportunity cost) is greater than surpluses lost in regions of the country where the taxpayer dollars were diverted from private consumption (or other government programs) to build the project. Thus, if competitive conditions in the economy do not occur, such as presence of long-term unemployment or immobility of resources, then real secondary benefits can be correctly attributed to the program.

Employment Impacts

Economic analysis typically presumes that a policy has no effect on overall labor employment. This may seem counterintuitive, but remember that a project's cost reflects the value of the resources like labor consumed or diverted in its implementation. If the project makes wage payments to labor for construction, operation, or maintenance, that reflect the opportunity cost of labor in its next best use, then there is no net job creation and the opportunity cost equals the wage payment made. This means that the wages earned (paid) are a cost and not a benefit. Every analyst knows, however, that projects financed from outside sources are politically popular whenever the wage payments exceed the opportunity cost of the labor. When that labor is unemployed, the opportunity cost is zero. When wage payment exceeds labor's opportunity cost, that wage

---

[14] Wassily Leontief won the 1973 Nobel Prize in economics for developing and applying input-output models. In the input-output model, final demands of products for consumption, investment, government, and exports are treated as known (exogenous) quantities. The input-output model is used to find out how much production needs to be increased in each producing sector of the economy to satisfy a given known increase in final demand brought about by a proposed policy or program. The increased production in each sector must meet both the change in final demand and the derived changes in demand for intermediate products in the various production sectors. Important applications of the input-output model include a range of environmental and natural resource policy issues, particularly policy measures to limit pollution. See Nobel Foundation (2004).

overstates the labor cost. If the labor has low or zero opportunity cost, and the project is financed from an outside source, any net additions to the region's spending and income stream due to the respending of those labor payments are a benefit from the view of the project region. Employment multipliers that claim to measure the secondary effects of government expenditures on employment should not be included in measured social benefits.

Inflation

An important issue is the correct treatment of inflation when performing BCA. Consistency is required in the treatment of prices and interest rates. In other words, either real or nominal prices and discount rates must be considered in making projections. It is usually easiest to use real values whenever possible, that is, prices expressed in constant dollars. The alternative is forecasting the inflation rate for many years into the future over the long life of an environmental project or program, which is a daunting task.

Risk and Uncertainty

Risky programs, those in which benefit and cost streams are largely uncertain or risky, are less attractive than programs that produce equivalent expected values of benefits and costs but less risk or uncertainty. What this means is that risk and uncertainty are costs, and these costs should be reflected in the evaluation of public programs. Two common approaches to reflect risk and uncertainty are conservative rules of thumb and probability analysis. Rules of thumb include measures to penalize riskier proposals: reducing the project's life, including a risk premium to raise the discount rate, and measuring benefits conservatively.

The probabilistic approach typically assigns probabilities to uncertain outcomes. The expected value of benefits is compared with the expected value of costs, in which expectations are weighted by their probabilities of occurring. The expected values of the distributions of benefits, costs, and net benefits can be obtained by weighting each outcome by its probability of occurrence, then summing across all potential outcomes.

Estimates that differ from expected values, such as worst-case estimates, can also be used. For example, studies of past activities affected by environmental regulations or natural resource programs have documented tendencies for costs to grow beyond initial expectations, which would provide a worst-case scenario that is considerably more pessimistic than forecasts of expected values.

Alternatives to Benefit-Cost Analysis

Two commonly used decision support methods that are alternatives to BCA are cost-effectiveness analysis and environmental impact analysis. Each is an attractive alternative to BCA because each requires less information than BCA. However, the price paid for these alternatives is less useful analytical capability to inform and support environmental decisions.

Cost-Effectiveness Analysis

Benefit-cost analysis is an analytical technique used to guide environmental and natural resource policy by organizing information to help find the most economically efficient set of actions. That information requires an ability to measure benefits and costs. But what if no reliable information is available on benefits? Existing U.S. environmental regulations designed to save lives establish values ranging from $70,000 to $132 million per life saved (Farber and Hemmersbaugh, 1993). This huge range shows the difficulty of assigning a specific monetary value to human lives saved. Similar problems hinder efforts to establish a value for the continued existence of an endangered species. When benefits cannot be measured, or their measurement is undesirable or prohibited by custom or law, it is still possible to draw on the structure, content, and power of economic analysis to efficiently achieve a policy target that has been set by political or other forces.

Global climate change is a good example of an important policy question for which economics can help in implementing efficiently. For example, what methods should be used to carry out negotiated reductions in existing greenhouse gas emissions to guide global energy and economic policies (see Chapter 15)? Since the Rio Earth Summit of 1992 and the Kyoto protocol in 1997, various international negotiations have established percentage reduction targets for various countries. Whether or not economics played a major role in *originally* setting those targets, the targets have been set and we must figure out an efficient way to meet those targets.

The use of emissions trading to reach a preset goal is another example. Emissions trading has harnessed the power of incentives to reduce the costs of achieving predetermined program aims for reducing harmful emissions. In the United States, emissions trading has been used to reduce sulfur dioxide believed to cause acid rain. Because SO_2-permit trading *rewards* behavior that complies with regulations rather than *mandating* compliance, it has achieved program aims at reduced cost. Trading has reduced emissions at a lower cost than command-and-control regulations that require power plants to use a mandated technology such as installing scrubbers (Arrow et al., 1996).

In these two examples, and many other situations, it is not possible to make an economically efficient choice (maximizing net present value), because the policy target has already been set. The question is, can economic analysis still inform and guide policy? The answer is a resounding yes. *Cost-effectiveness (CE) analysis* is a method that looks for the cheapest method of accomplishing some predetermined target. The idea behind CE is to look for programs that increase a targeted output by the most possible for a given expenditure of resources, or to find the least expenditure of resources needed to accomplish a fixed targeted output. For example, CE could be used to find the cheapest methods to reduce SO_2 emissions from coal-fired electric power plants at a given specified percentage below existing levels.

What is useful about CE analysis is that it focuses on least-cost methods to accomplish a given aim *without the need to measure the program's benefits.* CE analysis can be especially attractive when a program is desirable on moral, political, or other grounds that make its benefits undesirable, illegal, or impossible to quantify economically.

Once the policy target is set for an environmental program, economic analysis can make a major contribution in connection with measuring the costs of each of several methods for hitting that target. Knowing costs of any proposed method for reaching an aim is important for several reasons:

- Cost information enables you to compare costs incurred for *each* proposed method of reaching the target.
- By comparing costs to your budget, you know if you can reach your target with your available resources.
- Cost information tells you what you must sacrifice to reach your target when that target is set for political, ecological, health, safety, or other noneconomic reasons. Remember, resources typically can be used to achieve different aims in the same time period or used to achieve the same aim in different time periods, or for different aims in different time periods.

Marginal analysis (also called incremental analysis) can be an important part of CE analysis. When there are many methods available for reaching a single target, people have a natural desire to know which method is the cheapest. Is there any kind of equimarginal principle available for knowing when those costs are minimized? Again the answer is yes. The cheapest means for reaching an environmental target is achieved when the marginal costs for all possible achievement methods are equal. Mathematically, this can be stated as:

$$MC_1 = MC_2 = \ldots = MC_n \qquad (5.5)$$

where MC is the marginal cost from using any given method (e.g., first, second, third) of hitting the target. To illustrate the use of the equation for environmental policy analysis, consider the results of a well-known study that stated marginal costs of U.S. federal health and safety regulations ranged from \$200,000 to \$6.3 trillion per added life saved (Hahn, 1998). Analysis of these results in connection with Equation 5.5 shows that even if we know or want to know nothing of the benefits of a human life saved by a regulation, *more lives could have been saved for the same cost, or the same number of lives could have been saved at a lower cost.* Since this finding is so important for human health, safety, and the environment, let us analyze why the finding is true.

We know that we can do *better* than we are doing now whenever the marginal cost per life saved is *different* for different kinds of regulations. When those marginal costs are different, as found by the Hahn study, we know that for each added life saved by added regulations of the $200,000 type, the added lives saved costs industry and others only $200,000. Other regulations are saving lives at a much higher marginal cost, for example, $6.3 trillion.[15] Total costs can be reduced or total lives saved can be increased by assigning more life-saving capacity to the lower marginal cost sources and less to the high marginal cost sources. Only after marginal costs from all regulations are made equal is it impossible to save more lives with available resources, or to save the same number of lives with fewer total resources. Given that industry has limited economic and political capacity to be regulated, and given that society has limited resources overall, more lives will be saved when the total cost of saving those lives is kept as low as possible. We know when we have found the cheapest *total cost* set of means for hitting the target when the *marginal cost* of accomplishing a given target is equal among all possible means.

Cost-effectiveness analysis can also be used to inform decision makers about the added total costs of establishing more stringent or less stringent environmental regulations. CE helps to determine how much more stringent environmental standards can be made without increasing total costs. This can be done by reallocating responsibilities among the various means for hitting the target so that the marginal costs are made more *nearly* equal when those marginal costs are currently far apart (for details, see Appendix D). The equimarginal principle for CE analysis has major implications for environmental policy: Marginal costs of various measures for hitting a given environmental target are typically far apart for the simple reason that cost-effectiveness analysis is rarely performed.[16] When those costs are far apart, CE offers a powerful method for informing decision makers on ways to improve the environment consistent with a fixed budget by reallocating responsibilities among competing measures for hitting the target.[17]

Table 5–4 presents hypothetical data for two proposed environmental programs that would save lives: Program 1 improves workplace safety and program 2 reduces air emissions. For each program, stricter regulations would incur greater costs but save more lives. Weaker regulations cost less but save fewer lives. The table shows that the workplace safety program could save between 0 and 405 lives per year depending on the scale of budget allocated to it, ranging from 0 to $9 billion. The emissions control program could save between 0 and 600 lives, also for potential budgets ranging from 0 to $9 billion. Suppose you have $6 billion to spend on the two programs together. How much should you allocate to each program to maximize the number of lives saved? The answer comes from applying the equimarginal rule for cost-effectiveness. Marginal cost per life saved should be equal among all programs.

---

[15] Schleiniger (1999) compared the cost-effectiveness of three proposals for reducing nitrogen emissions in Switzerland: (1) taxing fossil fuels at their marginal cost, (2) a CO_2 emission charge, and (3) waste incinerator cleaning. Results showed that the cost (in million Swiss francs per year) for the fuel charge was 11; for the CO_2 charge, 31; and for the waste incinerator cleaning, 91. This means that the fossil fuel charge produced nearly three times as much nitrogen emission reduction as the CO_2 charge and more than eight times as much as waste incineration cleaning.

[16] Some critics of CE analysis attack it as a method of analysis when applied to decisions that are "beyond money," such as programs that would improve human health or save lives. These criticisms may be misplaced. In fact, the gains from adopting consistent regulatory standards for various health improvement programs can be dramatic. If costs and benefits are not balanced systematically in making decisions, the returns per dollar in terms of health and safety will likely be small for some programs and large for others. Such programs present opportunities for saving lives, and cost-effectiveness analysis will reveal those opportunities.

[17] Cost-effectiveness analysis is often approached using formal optimization methods, that is, systematic approaches for finding the least-cost way of accomplishing a specified aim. There are many kinds of software available for implementing optimization methods. Most use linear or nonlinear programming in which an objective is specified in addition to economic, institutional, or physical constraints which limit the capacity of the system to optimize that objective.

Finding the marginal cost per life saved requires the preliminary step of computing the marginal number of lives saved for each budget increment. Beginning with program 1, notice in Table 5–4 that increasing its budget from 0 to $1 billion increases saved lives from 0 to 85 lives, that is, 85 marginal lives saved. Increasing the budget again from $1 to $2 billion increases saved lives from 85 to 160 lives, a marginal gain of 75 lives. Column 3 shows similar marginal values of lives saved for each possible program level. Consistent with economic theory, the marginal number of lives saved decreases with an increased budget. Five additional lives are saved as the budget increases from $8 billion to $9 billion.

Next calculate the marginal cost per life saved program 1, column 4. It is computed as the reciprocal of the marginal number of lives saved per 1 billion added to the budget. So, in the workplace safety program, increasing the budget from 0 to $1 billion saves 85 lives for the first billion, which means that the marginal cost per life saved is $1 billion/85 lives = $0.012 (billion), or $12 million per life saved. The second budget increment from $1 billion to $2 billion saves 75 more lives, for a marginal cost per life saved of $0.013 billion, which is $13 million per life saved. The marginal cost per life saved increases to $0.200 billion, which is $200 million when the program's scale reaches $9 billion. A similar set of calculations is performed in columns 7 and 8 for the emissions control program, for alternative program scales running from $1 billion to $9 billion.

Having computed the marginal cost per life saved for all possible combinations of budget allocations to the two programs, we are ready to apply the equimarginal principle for CE analysis. Among all possible ways of spending the $6 billion between the two programs, the most cost effective is to spend $2 billion for workplace safety and $4 billion for emissions control. That combination of program scales equalizes the marginal cost per life saved at $0.013 billion, (see the highlighted rows in Table 5–4). For any other way of spending the $6 billion, marginal cost per life saved is not equal for the two programs, so total lives saved are fewer.

CE analysis is a technique of considerable analytical power. Despite its power, it is still a two-edged sword: CE's biggest limit is that it fails to analyze whether the benefits of the preset targets are greater than or less than the costs. Through exclusive use of CE, a community could spend huge amounts of scarce resources to implement expensive targets, regulations, and other programs whose economic value is unknown and whose impact on other social values is ignored. Overall, it is best to use CE analysis when it is desirable to achieve a noneconomic goal at minimum cost. CE analysis is the right method to use when it is impossible to measure benefits credibly or when laws, culture, or morality say it is unacceptable to measure those benefits.

Table 5–4 Effectiveness of two environmental programs

Program 1: Better workplace safety				**Program 2: Increased emissions control**			
1 Cost (billions of dollars)	2 Lives saved	3 Marginal lives saved (lives/$ billion)	4 Marginal cost/life saved ($ billion/ life saved)	5 Cost (billions of dollars)	6 Lives saved	7 Marginal lives saved (lives/$ billion)	8 Marginal cost/life saved ($ billion/ life saved)
0	0	–	–	0	0	–	–
1	85	85	0.012	1	105	105	0.010
2	**160**	**75**	**0.013**	2	200	95	0.011
3	225	65	0.015	3	285	85	0.012
4	280	55	0.018	**4**	**360**	**75**	**0.013**
5	325	45	0.022	5	425	65	0.015
6	360	35	0.029	6	480	55	0.018
7	385	25	0.040	7	525	45	0.022
8	400	15	0.067	8	560	35	0.029
9	405	5	0.200	9	585	25	0.040

Environmental Impact Analysis

Sometimes environmental policy analysis must be performed when there is no information on either benefits or costs. Nevertheless, environmental policymakers want to be informed on results of major actions before making choices that could have major consequences. One method of analysis that deals with these questions is *environmental impact analysis* (EIA), which attempts to measure environmental impacts associated with proposed environmental policies. Compared with BCA or CE, EIA makes no effort to convert consequences into common denominators such as dollars of cost or benefit. Because no common denominators are attempted, large amounts of raw information are placed in the lap of decision makers and the public who must make their own common denominator comparisons (e.g., compare various kinds of program costs to various environmental improvements produced).

Environmental impact analysis is widely used in policy circles. In fact, it is mandatory for some kinds of environmental programs in many countries. In the United States for example, the National Environmental Policy Act of 1969 requires all agencies of the federal government to account for environmental impact, unavoidable adverse environmental effects, alternative courses of action, and irreversible resource commitments that would occur if a proposed federal action were implemented. This legislation was the impetus for the environmental impact statement (EIS), designed to account for all environmental impacts, now widely recognized in the United States. Much of the history of the EIS, however, has been to produce reports that are so large and unwieldy that they are difficult to comprehend, analyze, and apply.

Some analysts think EIA is an improvement over BCA and CE: EIA avoids hidden value judgments, in contrast to common denominator comparisons like BCA and CE. These critics believe that the decision to place benefits and costs in common denominators makes major assumptions about relative values of inputs used as well as the comparative value of outcomes produced by environmental programs. EIA supporters believe that a properly done EIA summarizes consequences of policy proposals without assigning values to any outcome. By letting the public and decision makers assign their own values to consequences, EIA results in a more democratic decision process. Nevertheless, for all its strengths, EIA presents limited information about helping society make policy choices, because it ignores all nonenvironmental impacts (Hanley, 2001).

■ Advantages and Limits of Benefit-Cost Analysis

Benefit-cost analysis is an accepted method for organizing information and clarifying trade-offs, Nevertheless, its use presents many difficulties in interpretation, measurement, ethics, fairness, and reliability.

Advantages

Benefit-cost analysis provides a framework for organizing and cataloging commonsense hunches about impacts of environmental policy proposals. Its main strength as an appraisal method is that it tests for economic efficiency in resource allocation and for efficiency improvements in policies that would change that allocation. Many actual or proposed policy changes impose large costs on society, so decision makers want a sense of whether they produce similarly large benefits. For some environmental policies, costs may be found to be too high compared with expected benefits. This property of efficiency testing is a major advantage of BCA over other policy appraisal methods such as cost-effectiveness analysis or environmental impact analysis.

Benefit-cost analysis requires analysts, decisionmakers, or the public to identify all benefits and costs of a proposed policy. By presenting information in a structured way, BCA raises relevant questions and provides possible answers in a consistent way. When a policy is selected for which costs exceed benefits, BCA is a source of information for efficiency losses incurred by that policy.

Benefit-cost analysis is attractive because it accounts for social values systematically in environmental decision making. The BCA process incorporates values of the public and not those of experts or leaders, so BCA is a form of economic democracy.[18] BCA also has the advantage that it predicts impacts of policy proposals based on rational human behavior. In addition, BCA allows for environmental and natural resource impacts to be directly compared with financial impacts using a common denominator. Otherwise the value of the environment might inadvertently be counted as zero.

The greatest strength of BCA may be as an organizing tool to force consistency and rigor into environmental policy debates. The process of quantifying and assigning monetary value forces discussion about assumptions and makes opposing sides confront the substance of each other's arguments. Benefit-cost analysis also organizes data that influence policy decisions. It does so in a way that informs decision makers about the important elements of a problem and permits them to test the sensitivity of their decisions in relation to changes in those elements. What this means is that the BCA framework for organizing information is more important to the policy process and outcome than any single net benefit number it produces.

Limits

Few decision makers understand the economic principles that underlie BCA. Many critics of BCA perceive it as being too technical and requiring too much economic theory. So people who read a BCA report either must accept the numbers on faith or ignore them. One solution to this problem may be to make the main part of the BCA short, intuitive, and clear and relegate the heavy math, economic theory, and tables to appendixes.

Measurement. Critics of BCA applied to environmental policies point to its many measurement problems, such as measurement of the benefits of environmental improvements. This is particularly true of what may be the most common method for valuing the environment, the *contingent valuation method* (see Chapter 7). Many nonspecialists are uncomfortable with the idea of attaching monetary values to environmental improvements. Supporters of BCA say it is better to value the environment out in the open, rather than hide implied values of the environment behind policy agendas.

Another measurement problem with BCA is that measured benefits depend on the existing distribution of income. BCA is democratic because it treats all benefits gained and lost equally, regardless of who receives or pays them. However, each person's "money vote" depends on her income. The willingness to pay for most environmental improvements is greater as people's incomes increase, so an environmental program produces smaller benefits when poorer people are the intended beneficiaries. This shortcoming of BCA becomes more significant when we see that measured benefits exceed costs for most programs that would export pollution or other environmental insults from rich to poor countries. One way around this difficulty is to sum the distribution of benefits and costs for all major income, ethnic, and other relevant groups affected by a proposed program.

Still another measurement problem occurs with the discount rate. Simply put, nobody to date has put forth a convincing argument on what the correct discount rate is, let alone shown how that rate varies with culture, income, and other human conditions important to environmental policy.

---

[18] This is an important point. Special interests often resist efforts to perform a benefit-cost analysis for a proposal in which they hope to secure gains at the public expense, claiming that their program produces "special values" or that it meets a "moral standard" that is beyond valuation. But is there really a common moral standard that every regulator will independently arrive at through deliberate reflection? Not likely. Without a systematic BCA or CE analysis, what is far more likely is a system in which bureaucratic decisions reflect the preferences, not of the citizens, but of those in a special position to influence decisions. Those who resist honest benefit-cost analysis, with the intent of making concessions to special interests, often claim that it is degrading to make explicit the trade-offs reflected in the decision. Many grand programs of limited social worth are rationalized by an appeal to "public values" that ignore values revealed by individual choices.

Ethics. Many critics of BCA believe it is morally wrong to value in monetary terms programs such as environmental programs that save or prolong human lives through increased safety. When a decision maker sees money values placed on such a project, it may confuse or anger rather than enlighten.

The discount rate raises important ethical issues. Critics of BCA argue that any discount rate greater than zero is morally questionable because it assigns a lower value to unborn future generations than to current generations. Also, BCA uses values of the current generation as the center of the exercise, so discounting that leads to lower valuation for future generations is morally objectionable. Despite considerable debate over determining the correct discount rate, its supporters believe it should be larger than zero: As long as the current generation prefers current to future benefits, it is difficult to mount a strong argument that no discounting should occur, especially when future generations are expected to be richer than people living today. A more productive debate might be how to implement discounting.

Equity. Some believe that conventional BCA fails the equity (fairness) test because it ignores the distribution of benefits and costs. The *Hicks-Kaldor compensation principle* states that a policy improves society's overall welfare if the gainers could compensate the losers and still be better off, for this is what occurs when benefits exceed costs. The environment presents special problems to this principle. Some people will accept no finite compensation for environmental losses produced by a policy; so BCA is not valid because there is no mechanism for compensation.[19] Typical applications of BCA only measure total benefits and total costs of a proposed decision. This problem can be confronted by computing a proposal's net present value from as many viewpoints as there are interest groups. However, the additional work greatly increases the data requirements and cost of the analysis, and for that reason it is rarely done.

Another fairness issue deals with use of willingness to pay as a criterion for valuing program outputs. Some consider this criterion unfair because it depends on a program recipients' ability to pay. For example, an environmental regulation that improved safety in the workplace for low-income workers may fail a BCA simply because people who get the benefits cannot afford to pay much for them. So an environmental or natural resource management plan may make the rich richer and the poor poorer and still perform well on a BCA.

Conflicts of Interest. Critics of various government agencies that conduct BCAs of their own programs view the agencies as foxes guarding the hen house. These agencies may have an incentive to mislead by inflating the benefits of their programs and ignoring costs, because inflating their own program's benefits may give them a larger future budget and greater job security. Hahn (2000) has argued that regulatory agencies in general, such as the U.S. Environmental Protection Agency, are likely to overstate benefits and understate costs in their analyses for these reasons.

Resolution

Environmental policymaking that ignores accounting of benefits and costs will not likely be accepted by the public or affected industries when resources are scarce. The importance, identification, and measurement of environmental costs continues to increase around the world. None of the alternatives to BCA can do everything that BCA can do, even with its limits. Its technical problems (e.g., accuracy of environmental valuation) and conceptual problems (the way benefits and costs are defined) might be overcome with continued advances in economic research. The method may gain wider respect and acceptability among decision makers if BCA practitioners are open and honest about what it can and cannot do. Although formal BCA is neither required or sufficient by

[19] The existence of a single person with an infinite compensation requirement for environmental damages means that a BCA of that policy will always produce a negative net present value no matter how high the benefits. The potential that a single person's cost could tyrannize the vast majority's benefits will be rejected in most societies (Hanley, 2001).

itself for designing good public policy, it can provide a powerful framework for consistently organizing diverse information. In this way, it can improve the process, and therefore, the outcome of natural resource policy analysis (Arrow et al., 1996). Although BCA is impressive in concept, analytical power, and formulas, much work remains before a consensus is reached on how to estimate environmental benefits and costs objectively and how to deal with the method's limits. Still, even when the data available for a BCA are poor, the steps needed to carry out BCA require analysts to raise many of the right questions.

SUMMARY

Benefit-cost analysis (BCA) is an analytical procedure economists use to compare the costs and benefits of a government activity, project, or regulation over a relevant time period. It is a practical way to determine the desirability of government decisions. BCA involves the enumeration (listing) and evaluation (measuring) of all relevant costs and benefits. It evaluates the merits of public actions by translating positive and negative effects to a common denominator, typically money. A complete BCA compares alternative actions to determine which one provides society with the most economically beneficial use of its resources. A simple decision rule is used. If for some proposed action, the sum of its benefits exceeds the sum of the costs by a larger amount than any other action with the same aim, the proposed action should be adopted if economic efficiency is the objective. Otherwise it should not.

Economic efficiency is the important standard for *evaluating* government regulations or programs that are proposed for adoption, maintenance, or change. More efficient policies may be designed by using BCA to identify *added* benefits and *added* costs resulting from a range of various *incremental* changes rather than a single all-or-nothing proposal. Frequently benefits accrue to one sector of society or class of individuals, while the costs are borne by others. BCA might be more politically acceptable and become more widely used if environmental and natural resource programs are designed to account for equity. One way to accomplish equity is to design programs for which benefits exceed costs to most groups (i.e., people in most tax brackets, time periods, geographic locations, ethnic backgrounds, and economic sectors).

Ten steps are involved in using BCA. The first five steps are scoping, measuring benefits per unit output, measuring costs per unit output, measuring quantity of each output, and measuring quantities of each input. The next five steps are calculating gross benefits for each period, calculating gross costs for each period, calculating net benefits for each period, calculating net present value for all periods, and finally, selecting projects for which net present value is maximized for the available resources.

Benefits can be measured using market transactions, mathematical models, alternative cost saved, and user surveys. Costs can also be measured using similar methods, but special care must be exercised if there are underused resources. Typically the opportunity costs of those resources are lower than their market price.

Time is an important factor in the conduct of BCA. Estimating the life of a project or program depends on the assessments of the program's physical life, technological changes, shifts in demand or fashion, competing products that emerge, and the general state of the world many years in advance. The net present value formula requires discounting future benefits and costs to find the present value of net benefits. Four important issues can affect both benefits and costs: secondary effects, employment impacts, inflation, and risk and uncertainty.

Two alternatives to BCA are cost-effectiveness analysis and environmental impact analysis. Cost-effectiveness analysis looks for the least expensive method of accomplishing some predetermined target. It is attractive because it does not require the measurement of a program's benefits. Environmental impact analysis attempts to measure environmental impacts associated with proposed environmental policies without convering consequences into common denominators such as dollars of cost or benefit. Supporters of environmental impact analysis believe that properly done, it summarizes consequences of policy proposals without assigning values to any outcome. By letting the public and decision makers assign their own values to those consequences, its supporters believe its use results in a more democratic decision process.

Overall, BCA is attractive because it provides a framework for organizing and cataloging commonsense hunches about impacts of environmental policy proposals. Its main strength is that it tests for economic efficiency in proposed policies. However, BCA has several limitations: comprehension by nonexperts is difficult, measurement of many benefits and costs of environmental policies is unreliable, and nonzero discount rates are considered unethical by some. Other problems include ethical issues that result from ignoring the distribution of benefits and costs and conflicts of interest that occur when government agencies do their own benefit-cost analyses.

REVIEW QUESTIONS

1. Define a *benefit* and a *cost* of an environmental and natural resource action and provide an example of each.

2. Explain how *ex post* evaluation of the economic performance of a past environmental program can contribute to the *ex ante* evaluation of a proposed program.

3. Explain how benefit-cost analysis, an *economic* evaluation method of analysis, can help in the design *politically* attractive programs.

4. What is the accounting stance and why is it important?

5. Why might incremental analysis be preferable to the analysis of a single all-or-nothing proposal?

6. Compare the equimarginal rule for expanding a project versus reallocating a fixed resource among competing demands.

7. What is the with and without principle and how is it different from the before and after principle?

8. Define the term *present value of net benefits* (also called net present value).

9. Explain the difference between the net present value and the benefit-cost ratio for evaluating an environmental and natural resource program.

10. Describe how the use of a discount rate allows comparison of benefits and costs that take place in different time periods.

11. Why do some people believe that benefit and cost information should be used to design policies that maximize present value of net benefits? Do you agree with them?

12. What are the 10 steps for conducting a benefit-cost analysis?

13. What is the alternative cost method for estimating the benefits of environmental and natural resource policies? What are its strengths and limits?

14. Explain the difference between benefit-cost analysis, cost-effectiveness analysis, and impact analysis. What are the strengths and limits of each?

15. What is the equimarginal rule for cost-effectiveness analysis? What does its application accomplish?

16. What limits of benefit-cost analysis make some people mistrust it?

17. Citizens everywhere demand action to deal with emerging environmental and natural resource issues, but they want the actions justified, openly and plainly. Governments of the Organization for Economic Cooperation and Development (OECD) countries (industrialized) supported this principle by agreeing to use regulatory impact analysis. The resulting document included a checklist responding to the need for more effective regulations. The sixth item on the list asked, "Do the benefits of the regulation justify the costs?" (OECD, 2000, 2001). Your government just appointed a panel of experts to make a recommendation on the use of benefit-cost analysis in support of policymaking. You, as one of the experts, are asked to respond to the following questions:

a. Is it desirable to use benefit-cost analysis for new regulatory proposals?

b. Are there any major problems in carrying out a benefit-cost analysis?

c. What, if any, should be the role played by benefit-cost analysis in the process of regulatory decision making?

18. What are some limitations of making environmental policy decisions based solely on the outcome of benefit-cost analysis? Under what conditions and why might cost-effectiveness analysis provide a suitable alternative to benefit-cost analysis?

Class Exercises: **1.** Finding an equitable solution for the famous Klamath Basin conflict presents a challenge. The Klamath Basin, straddling the Oregon-California border has been a national battleground between farmers, Native American tribes, fishermen, and environmentalists over how to allocate scarce water resources among competing interests. The drought in 2001 resulted in the U.S. Bureau of Reclamation shutting off water to most of the Klamath project, which irrigates 200,000 acres of farmland, to protect endangered suckers and threatened coho salmon. That action led to millions of dollars in crop losses and tense confrontations between farmers and the government. After irrigation was restored in 2002, 33,000 chinook salmon which squeezed into the lower Klamath River on their fall spawning migration died from gill rot disease, which the California Department of Fish and Game blamed in part on low flows in the Klamath river as a result of the Klamath project. The fish kill led to a lawsuit brought by the Pacific Coast Federation of Fishermen's Associations, the Yurok and Hoopa tribes, environmentalists, and Rep. Mike Thompson (D-California) against the federal government. Can you think of any solution for which the benefits will exceed the costs to most affected groups? For emerging details, see the Web page of the U.S. Geological Survey (2004) which contains links to most of the major players.

2. Search your newspaper for an important local environmental or natural resource issue or debate. Use the methods of benefit-cost analysis or cost-effectiveness analysis to support either side of the debate or to help design an efficient or equitable policy solution.

PROBLEMS

1. The accompanying table shows benefits and costs produced by two different policies for coping with drought: law of the river (LR) and water banking (WB). LR means that water users with the most recent claim on the river (most junior) receive no water in low flow periods. WB means that people with the oldest claim (most senior) are paid to deposit their water into a water bank in a drought. Juniors can pay to withdraw water from the bank. The table summarizes the benefits and costs of implementing LR and WB over a three-year drought cycle. Benefits refer to drought damages avoided; costs are the costs of administering the program. This table has done the first 5 of the 10 steps in determining NPV. Your task is do the remaining 5 steps. For your analysis, compute the PVNB for each policy using a discount rate of 0.10 (10%). Based on your computations, explain which of the two programs produces a higher PVNB.

Benefit-cost analysis of drought

Year	Law of the river (LR)						Water banking (WB)					
	Benefits				Costs		Benefits				Costs	
	Agriculture		Cities				Agriculture		Cities			
	Q (sections of land irrigated)	*P* ($/section irrigated)	*Q* (acre-feet/year)	*P* ($/acre-foot)	*I* (man hours)	*C* ($/man hour)	*Q* (sections of land irrigated)	*P* ($/section irrigated)	*Q* (acre-feet/year)	*P* ($/acre-foot)	*I* (man hours)	*C* ($/man hour)
1	160	100,000	100,000	300	100,000	100	160	100,000	100,000	300	100,000	40
2	160	100,000	50,000	300	100,000	100	140	100,000	100,000	300	100,000	40
3	160	100,000	0	300	100,000	100	120	100,000	100,000	300	100,000	40

Q = quantity; *P* = price; *I* = input quantity; *C* = cost.

2. Two proposals have been advanced for improving a region's air quality, which has been degraded by motor vehicles: (1) putting added restrictions on existing automobile engines and (2) developing an entirely new kind of engine that burns fuel cleaner. The first program will take 2 years to put in place, and the second program would require 10 years of development. How would high versus low discount rates affect the benefit-cost comparison of the two programs?

3. An environmental pollution control program is expected to have a lifetime of 50 years. The costs of the program are estimated at $50 million per year for the first 10 years and $20 million per year for years 11 to 50. Benefits are expected to be 0 for the first 10 years and $40 million per year from years 11 to 50. At a discount rate of 5%, what is the discounted net present value of the program? What is its discounted net present value at 0%? How do you interpret the difference? Are there any conditions under which you justify the use of a 0% discount rate for this program?

4. Two existing environmental programs save lives: one improves workplace safety and one controls particulate emissions. For each program, stricter regulations would incur greater cost but save more lives. Weaker regulations cost less, but save fewer lives. The accompanying table shows that the workplace safety program currently costs $2 billion per year and saves 1,100 total lives per year, the emissions program also costs $2 billion and saves 1,800 lives annually (see highlighted areas).

Improved workplace safety				Increased particulate emission control			
Lives saved	**Cost ($ billion)**	**Marginal lives saved (lives/$ billion)**	**Marginal cost/life saved ($ billion/life)**	**Lives saved**	**Cost ($ billion)**	**Marginal lives saved (lives/$ billion)**	**Marginal cost/life saved ($ billions/life)**
0	0	–	–	0	0	–	–
600	1			1		1,000	
1,100	**2**			**2**		**1,800**	
1,500	3			3		2,400	
1,800	4			4		2,800	
2,000	5			5		3,000	
2,100	6			6		3,000	
2,200	7			7		3,000	

Fill in the blanks in the table; then answer the following questions:

a. Is the current $4 billion in total costs well spent on the two programs? Or should more be spent on workplace safety regulations (reducing spending on emissions control)? Or should more be spent on emissions control? Hint: Compare the marginal cost per life saved for the two programs.
b. Suppose both programs could be expanded so that $6 billion in costs could be allocated to these two programs. Use the equimarginal rule for cost-effectiveness to find out how much should be spent on each program to maximize total lives saved.
c. Suppose an inquiry finds the economic "value of a statistical life" is $2.0 million. (That finding would generate much controversy.) If the value is credible, how much cost should be incurred in each program to maximize economic efficiency? Explain how you would justify the changed level of total cost.

References

Arrow, K., M. Cropper, G. Eads, R. Hahn, L. Lave, R. Noll, P. Portney, M. Russell, R. Schmalensee, V. Smith, and R. Stavins. 1996. Is there a role for benefit-cost analysis in environmental, health, and safety regulation? *Science.* 272: 221–222.

Balmford, A., A. Bruner, P. Cooper, R. Costanza, S. Farber, R. E. Green, M. Jenkins, P. Jefferiss, V. Jessamy, J. Madden, K. Munro, N. Myers, S. Naeem, J. Paavola, M. Rayment, S. Rosendo, J. Roughgarden, K. Trumper, and R. K. Turner. 2002. Ecology—economic reasons for conserving wild nature. *Science.* 297 (5583): 950–953.

Barron, J. C. 1998. The university and controversial research. *Choices.* (2): 9–12.

Becker, N., and D. Lavee. 2003. The benefits and costs of noise reduction. *Journal of Environmental Planning and Management.* 46 (1): 97–111.

Bloom, B. S. 2004. Use of formal benefit/cost evaluations in health system decision making. *American Journal of Managed Care.* 10 (5): 329–335.

Cropper, M. L., and W. E. Oates. 1992. Environmental economics: A survey. *Journal of Economic Literature.* 30 (2): 675–740.

DeLong, J. 2000. Cost-benefit analysis, an ethical critique: Comment. In *Economics of the Environment: Selected Readings, 4th ed.,* edited by R. N. Stavins. New York and London: Norton.

Engeman, R. M., S. A. Shwiff, B. Constantin, M. Stahl, and H. T. Smith. 2002. An economic analysis of predator removal approaches for protecting marine turtle nests at Hobe Sound National Wildlife Refuge. *Ecological Economics.* 42 (3): 469–478.

Farber, D. A., and P. A. Hemmersbaugh. 1993. The shadow of the future: Discount rates, later generations, and the environment. *Vanderbilt Law Review.* 46: 267–304.

Freeman, A. M., III. 2002a. Environmental policy since Earth Day 1: What have we gained? *Journal of Economic Perspectives.* 16 (1): 125–46.

———. 2002b. Environmental policy since Earth Day 1: What do we know about the benefits and costs? *Agricultural and Resource Economics Review.* 31 (1): 1–14.

Gurian, P. L., M. J. Small, J. R. Lockwood, and M. J. Schervish. 2001. Benefit-cost estimation for alternative drinking water maximum contaminant levels. *Water Resources Research.* 37 (8): 2213–2226.

Hahn, R. 1998. *Risks, Costs and Lives Saved: Getting Better Results from Regulation.* Oxford, UK: Oxford University Press.

Hahn, R. W. 2000. The impact of economics on environmental policy. *Journal of Environmental Economics and Management.* 39 (3): 375–399.

Hanley, N. 2001. Cost-benefit analysis and environmental policymaking. *Environment and Planning C: Government and Policy.* 19 (1): 103–118.

Jones, H. R., and R. A. Sudmeyer. 2002. Economic assessment of windbreaks on the south-eastern coast of Western Australia. *Australian Journal of Experimental Agriculture.* 42 (6): 751–761.

Kniivila, M., V. Ovaskainen, and O. Saastamoinen. 2002. Costs and benefits of forest conservation: Regional and local comparisons in Eastern Finland. *Journal of Forest Economics.* 8 (2): 131–150.

Krupnick, A., A. Alberini, M. Cropper, N. Simon, B. O'Brien, R. Goeree, and M. Heintzelman. 2002. Age, health and the willingness to pay for mortality risk reductions: A contingent valuation survey of Ontario residents. *Journal of Risk and Uncertainty.* 24 (2): 161–186.

Krupnick, A., and R. Morgenstern. 2002. The future of benefit-cost analyses of the Clean Air Act. In: *Annual Review of Public Health.* Palo Alto, CA: Annual Reviews.

Kumar, S. 2002. Does "participation" in common pool resource management help the poor? A social cost-benefit analysis of joint forest management in Jharkhand, India. *World Development.* 30 (5): 763–782.

Lee, M. 2002. The effect of sulfur regulations on the U.S. electric power industry: A generalized cost approach. *Energy Economics.* 24 (5): 491–508.

Levy, J. I., J. D. Spengler, D. Hlinka, D. Sullivan, and D. Moon. 2002. Using CALPUFF to evaluate the impacts of power plant emissions in Illinois: Model sensitivity and implications. *Atmospheric Environment.* 36 (6): 1063–1075.

Lind, R. C. 1990. Reassessing the government's discount rate policy in light of new theory and data in a world economy with a high degree of capital mobility. *Journal of Environmental Economics and Management.* 18 (2 Part 2): S8-S28.

Loomis, J. B. 1993. *Integrated Public Lands Management.* New York: Columbia University Press.

MacLean, H. L., and L. B. Lave. 2003. Evaluating automobile fuel/propulsion system technologies. *Progress in Energy and Combustion Science.* 29 (1): 1–69.

Marsden, J. E., B. D. Chipman, L. J. Nashett, J. K. Anderson, W. Bouffard, L. Durfey, J. E. Gersmehl, W. F. Schoch, N. R.

Staats, and A. Zerrenner. 2003. Sea lamprey control in Lake Champlain. *Journal of Great Lakes Research.* 29: 655–676.

Marshall, G. R., and J. P. Brennan. 2001. Issues in benefit-cost analysis of agricultural research projects. *Australian Journal of Agricultural and Resource Economics.* 45 (2): 195–213.

McAllister, D. M. 1980. *Evaluation in Environmental Planning.* Cambridge, MA: MIT Press.

Mitchell, R. C., and R. T. Carson. 1989. *Using Surveys to Value Public Goods: the Contingent Valuation Method.* Baltimore: Johns Hopkins.

Montgomery, C., and G. M. Brown, Jr. 1992. Economics of species preservation. *Contemporary Economic Policy.* 10: 1–12.

Navrud, S. 2001. Economic valuation of inland recreational fisheries: Empirical studies and their policy use in Norway. *Fisheries Management and Ecology.* 8 (4–5): 369–382.

New York University. 2004. *The history of economic thought.* Available from *http://cepa.newschool.edu/het/* [Accessed October 22, 2004].

Nobel Foundation. 2004. *The Bank of Sweden Prize in Economic Sciences in Memory of Alfred Nobel 1973, Wassily Leontief.* Available from *http://nobelprize.org/economics/laureates/1973/* [Accessed October 22, 2004].

Olsthoorn, X. 1999. Cost-benefit analysis of European air quality targets for sulphur dioxide, nitrogen dioxide, and fine and suspended particulate matter in cities. *Environmental and Resource Economics.* 14 (3): 333–351.

Organization for Economic Cooperation and Development (OECD). 2000. *OECD reference checklist for regulatory decision-making.* Available from *http://www1.oecd.org/puma/regref/pubs/rco95/chcklist.htm* [Accessed October 22, 2004].

———. 2001. *Regulatory quality and public sector reform.* Available from *http://www1.oecd.org/puma/regref/pubs.htm* [Accessed October 22, 2004].

Pearce, D. 1998. Cost-benefit analysis and environmental policy. *Oxford Review of Economic Policy.* 14 (4): 84–105.

Rosenberger, R. S., and J. B. Loomis. 2001. *Benefit Transfer of Outdoor Recreation Use Values.* A technical document supporting the forest service strategic plan. Fort Collins, CO. U.S. Department of Agriculture, Forest Service, Rocky Mountain Research Station. General Technical Report RMRS-GTR-72.

Rubin, J., G. Helfand, and J. Loomis. 1991. A benefit-cost analysis of the northern spotted owl—results from a contingent valuation survey. *Journal of Forestry.* 89: 25–30.

Schleiniger, R. 1999. Comprehensive cost-effectiveness analysis of measures to reduce nitrogen emissions in Switzerland. *Ecological Economics.* 30 (1): 147–59.

Shi, Y., D. B. Parker, N. A. Cole, B. W. Auvermann, and J. E. Mehlhorn. 2001. Surface amendments to minimize ammonia emissions from beef cattle feedlots. *Transactions of the ASAE.* 44 (3): 677–682.

Teague, W. R., R. J. Ansley, U. P. Kreuter, W. E. Pinchak, and J. M. McGrann. 2001. Economics of managing mesquite in north Texas: A sensitivity analysis. *Journal of Range Management.* 54 (5): 553–560.

Tol, R. S. J. 2001. Equitable cost-benefit analysis of climate change policies. *Ecological Economics.* 36 (1): 71–85.

U.S. Ecological Survey. 2004. *Klamath Basin Issues.* Available from *http://or.water.usgs.gov/projs-dir/or180/links.html* [Accessed February 6, 2005].

U.S. Fish and Wildlife Service. 2004. *Digest of federal resource laws of interest to the U.S. Fish and Wildlife Service.* Available from *http://laws.fws.gov/lawsdigest/central.html* [Accessed October 22, 2004].

Van Wilgen, B. W., M. P. de Wit, H. J. Anderson, D. C. Le Maitre, I. M. Kotze, S. Ndala, B. Brown, and M. B. Rapholo. 2004. Costs and benefits of biological control of invasive alien plants: Case studies from South Africa. *South African Journal of Science.* 100 (1): 113–122.

Yoo, S. H., and K. S. Chae. 2001. Measuring the economic benefits of the ozone pollution control policy in Seoul: Results of a contingent valuation survey. *Urban Studies.* 38 (1): 49–60.

Chapter 6

The Discount Rate

Introduction

In January 1964, the U.S. Bureau of Reclamation unveiled a "dam sandwich" proposal in which it would build two hydroelectric dams on opposite edges of the Grand Canyon National Park in Arizona. The proposal cited several benefits to the region if the dams were built. One benefit would be a considerable cost savings to power buyers of hydroelectricity compared to coal, its closest competitor. The dams also would produce important revenue generated by that power that could help pay for further water development in the water-scarce Southwest. However large the benefits, one major cost did not escape the public's notice: the irreversible environmental damages to current and future generations produced by two large dams that would be visible to the park's visitors. Three years after the unveiling, and after a sustained national uproar over the environmental consequences of these dams, plans for the dams were scrapped. They have never been seriously discussed since.

The 1967 decision to cancel the Grand Canyon dams bestowed an expensive gift from the then-current generation to future generations: an undeveloped Grand Canyon. Future generations may continue contributing to the high cost of paying for this gift if they so desire. As long as the dams remain unbuilt, the cost to each future generation of keeping the canyon wild will be incurred as the higher price of producing and buying power from a more expensive source.

Most government environmental policy decisions produce effects whose benefits and costs last for several years (Marglin, 1963). The up-front costs of a program for bringing a species back from near extinction are incurred mostly in the present, but the benefits and environmental impacts will likely extend well into the future. Therefore, economists are faced with the question of how to go about comparing future streams of costs and benefits with those of the present.

This chapter's discussion of the discount rate provides insights into decisions affecting long-term investments that aim to sustain, protect, or otherwise alter an environmental or natural resource. Discounting is directly relevant to long-term environmental issues such as biodiversity, climate change, and endangered species as well as shorter-term environmental issues such as human health and pollution. It also is relevant to natural resource issues such as land, water, forests, minerals, and fisheries.

Outline

Using the Discount Rate to Find Net Present Value 128

Search for a Discount Rate 136

Objectives

After studying this chapter, you should be able to

- Show how the discount rate is used to compute net present value
- Describe the method of monetary cost of government borrowing for finding the discount rate
- Explain how the social opportunity cost of capital is used to find the discount rate
- Evaluate the social rate of time preference in determining the discount rate

USING THE DISCOUNT RATE TO FIND NET PRESENT VALUE

The basic principle underlying *discounting* is simple: A dollar today is worth more than a dollar at some time in the future. Given the choice of receiving with certainty a given sum of money right now or receiving the same sum a year from now, most of us would never make the second choice. A dollar received a year from today is worth considerably less than one received now. How much less?

Suppose Lucy lends Betty $1,000 in year 1, to be repaid in year 2. Lucy willingly foregoes use of the $1,000 for the year only if Betty pays her a premium for that foregone use when Betty repays the loan a year later. That premium is interest. If Betty and Lucy agree that Betty will pay $1,100 in year 2 for the use of Lucy's $1,000 during year 1, the simple interest rate is 10%. If we asked Lucy how much $1,100 in year 2 is worth to her today, she says $1,000 if Lucy discounts the money she expects to receive in the future by 10%. This calculation shows the time value of money principle: Y dollars one year from now is worth less than Y dollars today.

The term *present value* describes the current value to the recipient of a benefit that requires waiting to receive. In the example, the present value to Lucy of $1,100 in year 2 is $1,000. The 10% rate Lucy uses to discount the money she expects to receive in year 2 is called the *discount rate*. This analysis also applies to costs to be incurred in the future. Holding everything else equal, Lucy is indifferent between paying a cost of $1,100 in year 2 or $1,000 today, because Lucy discounts future costs at a simple interest rate of 10% per year.

This comparison can be shown algebraically. Let P_0 be the amount Lucy needs to pay out in cash now to receive $1,100 a year from now, assuming that P_0 grows at a 10% interest rate in the intervening year. If Lucy invested P_0, then a year from now she will receive something more in return, namely, the original cash plus 10%. In symbols, the total amount received a year later is the beginning principal, P_0, added to the interest, which is 10% of P_0. So we can solve for P_0 algebraically. It is the amount Lucy needs to pay out now to receive the $1,100 a year from now:

$$P_0 + 0.10\,P_0 = \$1{,}100 \qquad (6.1)$$
$$P_0 = \$1{,}000$$

With the opportunity to earn a guaranteed 10% annually, a rational lender is equally satisfied to receive $1,000 now or $1,100 a year from now. So the present discounted value, P_0, of $1,100 is $1,000.[1] We have just calculated the equivalent *present value* benefit to you of a guaranteed $1,100 *future value* benefit to be received a year from now. It is $1,000. Next we ask what is the equivalent *present value* cost you face now, P_0, from signing a contract today that commits you to pay out a *future value* cost of $1,100 a year from now? The answer, of course, is the same, namely $1,000. This is because any rational person is equally satisfied with paying out a *future cost* of $1,100 a year from now and paying out a *current cost* of $1,000 now. She can put the $1,000 in the bank now, earn 10% then withdraw $1,100 a year from now and pay off the obligation. The exact same formula is used to compare present values and *future values* whether the money is *received* as a benefit or *paid* as a cost.

As a general principle, given the rate of interest (i) and the future cash inflow (benefit) or outflow (cost) a year from now as P_1, the present value (P_0) of the future cash flow is the amount that would have to be put into the bank in order to obtain P_1 a year from now. That present value can be calculated for any known values of P_1 and i as

$$P_0 + P_0 i = P_1 \qquad (6.2)$$
$$P_0 (1 + i) = P_1$$
$$P_0 = \frac{P_1}{(1 + i)}$$

[1] Because $1 received in the future is worth less than the same dollar received in the present, if you want to know what the future $1 is worth now, you must discount it. Discounting is the opposite of compounding. As you would expect, the rate at which a future value is discounted is closely related to the rate at which present values are compounded, namely, the interest rate. If the interest rate is 10%, then $1 placed in the bank today will be worth $1.10 a year from now. So if the interest rate is 10%, $1.10 to be received next year is worth only $1 today.

The equation says that the present value paid or received, P_0, plus the interest paid or received on P_0 in the intervening year, P_0 times i, equals the future value paid or received, P_1. The second and third lines show the algebra that solves for the present value.

The arithmetic of discounting is more complicated when more than one period is involved. As benefits or costs require waiting further in the future, compound interest decreases their present value geometrically. The formula for determining the present value of a sum to be conferred in some future year is

$$P_0 = \frac{P_t}{(1 + i)^t} \tag{6.3}$$

where P_t is what the beneficiary will receive in future year, i is the discount rate, and t is the number of years from the current period when the beneficiary receives the money. Some people discount the distant future at nonconstant discount rates; see, for example, Azfar (1999), Weitzman (2001), and Gollier (2002).

By substituting the monetary value of the benefit to be received in some future year for P_t, Equation 6.3 can be used to determine the present value of any future benefit that can be expressed in monetary terms. Analysts can also discount future costs expressed in monetary terms to present value. Costs and benefits of an environmental or natural resource program typically extend over many years. A policy typically distributes those costs and benefits unequally over time, so simple comparison of undiscounted costs and benefits errs by ignoring the time value of money. So, benefit-cost analysts discount all costs and benefits to present value before comparing them.

Suppose we know we will receive a cash flow two years from now, defined as P_2. The present value of P_2 can be found with the help of the following equations:

$$\begin{aligned} P_0 + P_0 i + (P_0 + P_0 i)\, i &= P_2 \\ P_0 (1 + 2i + i^2) &= P_2 \\ P_0 (1 + i)^2 &= P_2 \\ P_0 &= \frac{P_2}{(1 + i)^2} \end{aligned} \tag{6.4}$$

The first line has three terms on the left-hand side of the equal sign: first, the initial present value, P_0; second, the interest earned at the end of one year on that initial present value, P_0 times i; and third, the interest earned at the end of two years applied to the sum of terms 1 and 2, which is (P_0 + P_0 i) times i. The remaining three lines show the algebra needed to solve for P_0, which is what we are after. Equation 6.4 is a simple example of Equation 6.3. Equation 6.3 shows that this same algebraic process can be performed to find the discounted present value of a benefit or cost received or paid *any number of years* from now. If \$2,000 is to be received 15 years from now and the interest rate is 10%, then P_0 is \$2,000 divided by $(1.10)^{15}$ = \$479.

Now we are ready to introduce a new idea: the notion of a series of cash flows paid or received *for each of many years into the future.* Suppose the interest rate is 10% and you want to find the present value, P_0, of a series of three cash flows: \$600 received one year from today, \$1,200 received two years from now, and \$1,400 received three years into the future. The present value of this sum is

$$P_0 = \frac{600}{(1.10)^1} + \frac{1{,}200}{(1.10)^2} + \frac{1{,}400}{(1.10)^3} \tag{6.4a}$$

The summed terms on the right of the equal sign illustrate a concept known as the net present value (NPV) of any stream of costs and benefits. As shown in Chapter 5, NPV is measured as

$$NPV = (B_0 - C_0) + \frac{(B_1 - C_1)}{(1 + i)^1} + \frac{(B_2 - C_2)}{(1 + i)^2} + \cdots + \frac{(B_n - C_n)}{(1 + i)^n} \tag{6.5}$$

NPV applies the concept of discounting to the calculation of public environmental policy impacts. The costs (C) and benefits (B) of a given government policy typically extend over much more than a year. A policy generally distributes those costs and benefits unequally over time, so a undiscounted comparison of gross costs and gross benefits ignores the time value of money. Consequently, as shown in Equation 6.5, benefit-cost analysts discount all costs and benefits to present value before comparing them. A positive NPV (benefits exceed costs) suggests that an efficiency-minded government would adopt an environmental regulation or program producing that NPV, whereas a negative NPV suggests that it would not.

This NPV formula can find the net present value of any project or program, even if it is characterized by a complex time path of costs and benefits that extend years into the future. It yields a single common denominator that expresses the overall performance of one well-defined project or program, even if it requires many kinds of inputs and produces a wide range of outputs. Moreover, as stated in Chapter 5, each project or program has a single NPV. The formula for that NPV allows comparison of the economic performance for each of many different programs that compete for the same taxpayer resources. Notice that each year's benefit enters with a plus sign, while cost enters with a minus sign. Equation 6.5 can be expressed using the following more compact summation notation:

$$NPV = \sum_{t=1}^{T} \frac{(B_t - C_t)}{(1 + r)^t} \tag{6.5a}$$

where the subscript t refers to any tth (first, second, third, etc.) year and the Σ symbol means to sum each term in both the numerator and denominator over all years; T refers to the last year in which the program still has some lasting effect (e.g., year 100).

Tables of Discounted Values

You can probably guess that applying Equation 6.5 or 6.5a to the task of calculating NPV by hand, particularly for a project that will last many years, is tiring and messy, even with the help of a computer. Suppose you signed a contract that gives you \$10,000 with certainty five years from now. Table 6–1 shows that the present value of this \$10,000 payment, discounted at 15%, is \$4,970, based on the use of Equation 6.4: $\$4{,}970 = \$10{,}000\left[\frac{1}{(1.15)^5}\right] = \$10{,}000[0.497]$.

Table 6–1 Discounted value of a *single* payment of \$1

Year	Discount rate					
	0.01	0.03	0.05	0.10	0.15	0.20
1	\$0.990	\$0.971	\$0.952	\$0.909	\$0.870	\$0.833
2	0.980	0.943	0.907	0.826	0.756	0.694
3	0.971	0.915	0.864	0.751	0.658	0.579
4	0.961	0.888	0.823	0.683	0.572	0.482
5	0.951	0.863	0.784	0.621	0.497	0.402
10	0.905	0.744	0.614	0.386	0.247	0.162
15	0.861	0.642	0.481	0.239	0.123	0.065
20	0.820	0.554	0.377	0.149	0.061	0.026
25	0.780	0.478	0.295	0.092	0.030	0.010
50	0.608	0.228	0.087	0.009	0.001	0.000
100	0.370	0.052	0.008	0.000	0.000	0.000
200	0.137	0.003	0.000	0.000	0.000	0.000

Table 6–2 Discounted value of a *single* payment of $1 million

Year	Discount rate					
	0.01	0.05	0.07	0.10	0.20	0.30
0	$1,000,000	$1,000,000	$1,000,000	$1,000,000	$1,000,000	$1,000,000
1	990,099	952,381	934,579	909,091	833,333	769,231
3	970,590	863,838	816,298	751,315	578,704	455,166
5	951,466	783,526	712,986	620,921	401,878	269,329
10	905,287	613,913	508,349	385,543	161,506	72,538
20	819,544	376,889	258,419	148,644	26,084	5,262
50	608,039	87,204	33,948	8,519	110	2
100	369,711	7,604	1,152	73	0	0
200	136,686	58	1	0	0	0

The table vividly illustrates the large effect of the discount rate in eroding a future benefit received or cost paid out.[2]

Because of the need to discount streams of future benefits and costs, it is important to understand the impact of the choice of discount rate on the results of benefit-cost analysis. Table 6–2, the discounted value of a one-time payment of $1 million, reaffirms that discounting can dramatically affect the value of a proposed regulation's costs or benefits, depending on the size of the discount rate and the time period before society realizes the costs or benefits. Because society often incurs the costs of environmental programs long before the benefits, compound discounting generally has a greater impact in the calculation of the present value of benefits than of costs. Note in Table 6–2 the differences between the present value of a one-time benefit or cost of $1 million dollars at various discount rates and time periods. It should be no surprise that methods of discounting are critical to benefit-cost analysis and often pivotal in environmental policy debates and decisions.

Figure 6–1 is a different way of looking at the same thing. The *discount factor* is a number between zero and one that tells you how much you need to shrink a future benefit or cost to make it equivalent to its present value. Equation 6.3 shows that at a zero discount rate, there is no erosion at all. The future value of a benefit or cost of *X* produces a present value also equal to *X*, so the discount factor is one. But higher discount rates or longer times into the future cause the discount factor to shrink progressively. Note that for any discount rate larger than zero, the discount factor falls closer to zero as the number of future years increases.

Many environmental programs incur current costs so future health can be improved or so future lives can be saved. Suppose saving a life *today* is worth spending $1 million *today.* But today's new environmental program may require several years' gestation before it starts saving lives. (See Example 6–1.) If that occurs, we face this question: How much should we spend *today* to save a life in 20 years? The answer, according to the U.S. Office of Management and Budget (OMB, 1992), is just under $260,000 (the cell in Table 6–2 where the 7% column, OMB's current

[2] The power of a rising discount rate to dramatically alter the payoff of an investment was observed by Kenneth Boulding in 1964, in Young and Haveman (1985), who penned this verse:

> Around the mysteries of finance
> We must perform a ritual dance
> Because the long-term interest rate
> Determines any project's fate:
> At two percent the case is clear,
> At three, some sneaking doubts appear,
> At four, it draws its final breath
> While five percent is certain death.

Example 6–1
What Discount Rate for Saving Lives?

Environmental programs produce benefits to the public that can last for hundreds of years. Environmental cleanups often require major costs that are incurred at the beginning of the project, but they produce a stream of benefits, often lives saved or lengthened, that are not seen for many years. An important question raised by long-term programs is, Should lives saved in the future be discounted for the purpose of comparing the cost per life saved in different time periods? In attempting to answer this question, Cropper, Aydede, and Portney went to the public and described this scenario: Without new programs, X people will die this year from pollution and Y people will die Z years from now. The government has to choose between two programs that cost the same, but there is only enough money for one. Program 1 will save X lives now, while program 2 will save Y lives Z years from now. Which program would you choose?

The authors' data from three surveys showed a remarkably large percentage of respondents could not be motivated to choose the future-oriented program, even when it saved 50 times the number of lives saved by the present-oriented program. Naturally, the authors wanted to know why people discounted the future so heavily. What they found was that about a third of the respondents who showed a strong bias toward saving current lives believed that society will figure out other ways to save people in the future. Those respondents simply ducked the question. They would not accept or believe the trade-off with which the authors tried to confront them.

If a respondent is indifferent between saving one life today and saving two lives one year from now, he shows a discount rate of 100%: $1 = 1/(1 + 1)^1$. Use of this formula allowed the authors to observe each respondent's current-future trade-off to compute that person's discount rate. The accompanying table shows estimates of the computed discount rates for various time horizons, based on the survey response data.

Social rate of time preference for saving lives

t (years)	Mean observed discount rate	Predicted discount rate[a]
5	0.168	0.163
10	0.112	0.115
25	0.074	0.073
50	0.048	0.051
100	0.038	0.036

[a]Based on this equation, discount rate $= B_0 t^{B_1}$, where $B_0 = 0.37$ and $B_1 = -0.50$. The equation explains more than 99% of the authors' computed discount rates.

Several interesting findings emerge: (1) First, discount rates are significantly greater than zero, even at horizons as long as 100 years. So for the authors' policy question, a zero social rate of time preference does not fit the facts. (2) People do not discount at a constant rate. Discount rates are much higher for short time horizons than for long ones. In fact, the table shows that a simple declining exponential discount rate function fit the data well by explaining more than 99% of the authors' measured discount rates.

Source: Cropper, M., S. K. Aydede, and P. R. Portney. 1992. Rates of time preference for saving lives. *American Economic Review.* (May): 469–472.

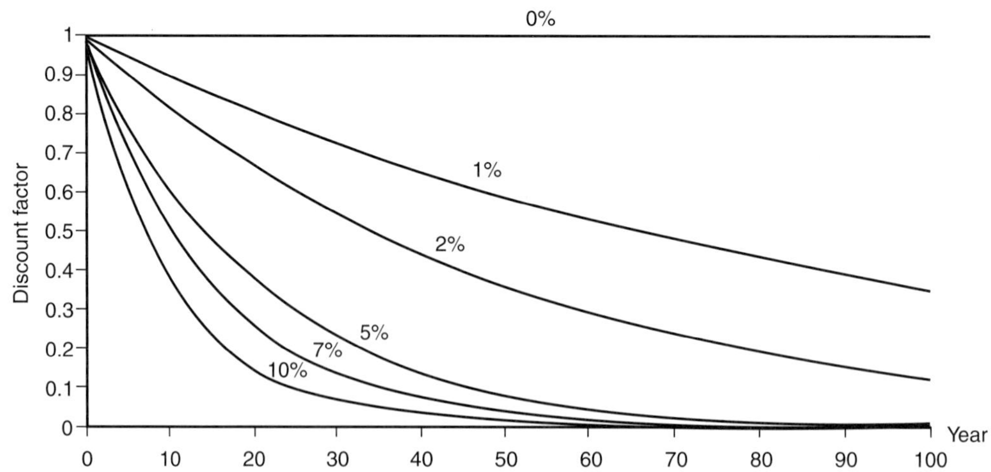

Figure 6–1 Discount factors at selected discount rates

Table 6–3 Discounted value of a *sustained* $1 flow

Year	Discount rate					
	0.01	0.03	0.05	0.10	0.15	0.20
1	$0.990	$0.971	$0.952	$0.909	$0.870	$0.833
2	1.970	1.913	1.859	1.736	1.626	1.528
3	2.941	2.829	2.723	2.487	2.283	2.106
4	3.902	3.717	3.546	3.170	2.855	2.589
5	4.853	4.580	4.329	3.791	3.352	2.991
10	9.471	8.530	7.722	6.145	5.019	4.192
15	13.865	11.938	10.380	7.606	5.847	4.675
20	18.046	14.877	12.462	8.514	6.259	4.870
25	22.023	17.413	14.094	9.077	6.464	4.948
50	39.196	25.730	18.256	9.914	6.661	5.000
100	63.029	31.599	19.849	10.000	6.667	5.000
200	86.333	33.243	19.999	10.000	6.667	5.000

discount rate, intersects the 20-year row). Since 1992, OMB has used a 7% annual discount rate to convert future regulatory costs and benefits into their present value. Because government regulations of cancer-causing agents (carcinogens) are not expected to affect the cancer rate for at least 20 years, OMB's choice of discount rates has dramatic implications for regulatory policy. Its choice of discount rates has even larger impacts on long-term global environmental issues such as ozone depletion and the greenhouse effect. If, for example, increased global temperatures produced by the greenhouse effect will cost society $100 billion 200 years from now, OMB's current discount rate of 7% would indicate that it is not worth spending $200,000 today to avert the harm 200 years from now.[3]

Private business uses the interest rate to make private decisions that maximize the NPV of its income. The same process is used by policy analysts to evaluate public projects, plans, or investments affecting natural resources or the environment. One big difference between private and public decisions is that the discount rate for public projects is less likely to equal a market interest rate.

Consider another example. The same $10,000, to be received 10 years from now, discounted at 5% is $6,140, equal to $\$10{,}000/(1.05)^{10}$; discounted at 20%, the same $10,000 has a present value of a much smaller $1,620 (so its discount factor is 0.1620). Because of its importance in indicating the present value of a sum of money paid or received in the future, the discount rate must be selected with great care.

Table 6–3 deals with the present value of a stream of cash flows over many years. It shows for several discount rates and time periods the present value of a $1 per year constant annual cash flow. The table can be applied for a case where you want to know the present value of $1,000 to be received *once each year for 25 years,* always discounted at 10%. The table itself shows that at a discount rate of 10%, a guaranteed $1 received every year for 25 years has a present value of about $9.077. Therefore, the required calculation is $1,000 times 9.077, which is $9,077. Similarly, the present value of paying or receiving $10,000 per year for 15 years at a discount rate of 20% is $46,750 = $10,000 [4.675].

To confirm that higher discount rates greatly erode future benefits as well as future costs, look at Figures 6–2 and 6–3. The figures compare visually an undiscounted stream of benefits and costs (0% discount) with the same stream discounted at 7%. The hypothetical program has a 100-year time horizon in which benefits are produced in 25 of those years. For costs, the program incurs (1) construction costs of $100 million per year in years 1 through 5, (2) operation and maintenance

[3] The calculation is $\dfrac{\$100{,}000{,}000{,}000}{1.07^{200}} = \$132{,}814.$

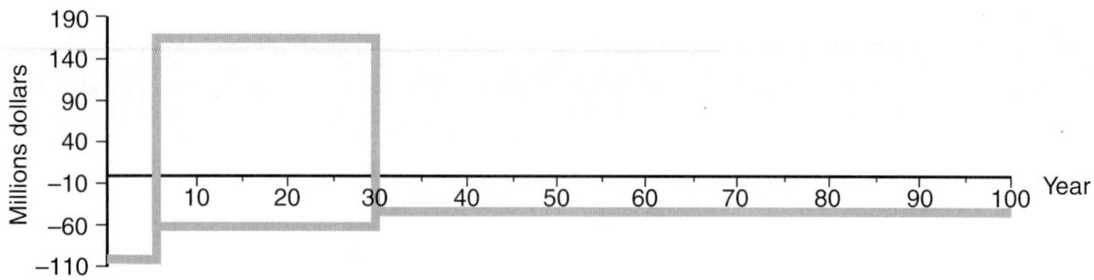

$$NPV = \sum_{t=6}^{30} \frac{B_t}{(1+r)^t} - \sum_{t=1}^{5} \frac{C_{1t}}{(1+r)^t} - \sum_{t=6}^{30} \frac{C_{2t}}{(1+r)^t} - \sum_{t=31}^{100} \frac{C_{3t}}{(1+r)^t} = -\$800 \text{ million}$$

where

B_t = Benefits in year t = \$160 million, years 6–30

C_{1t} = Construction costs in year t = \$100 million, years 1–5

C_{2t} = Operating and maintenance costs in year t = \$60 million, years 6–30

C_{3t} = Environmental costs in year t = \$40 million, years 31–100

r = Discount rate = 0%

Figure 6–2 One hundred year stream of costs and benefits discounted at 0%. Lines above the horizontal axis are benefits; lines below are costs.

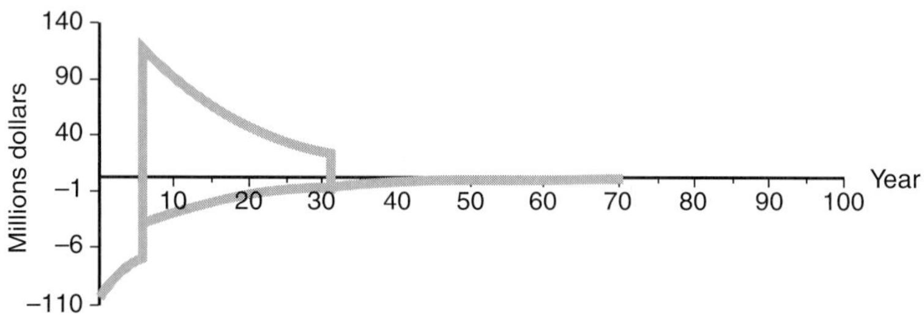

NPV = +\$346 million

Figure 6–3 One hundred year stream of costs and benefits (same as in Figure 6-2) discounted at 7%. Lines above axis are benefits; lines below are costs.

costs of \$60 million per year in years 6 through 30, and (3) remaining environmental costs after the project's life is over equal to \$40 million per year lasting from years 31 through 100. Benefits of \$160 million per year occur in years 6 through 30. Figure 6–2 shows how the NPV is computed at −\$800 million using a discount rate of zero. Figure 6–3 and Table 6–4 (discount rate of 7%) show that the same stream of benefits and costs produces a very different NPV equal to a much higher +\$346 million. That sharp difference in NPV between −\$800 million and +\$346 million shows that it pays to think carefully before selecting the discount rate. The reason that NPV is higher using a 7% discount rate than it is using a 0% discount rate is that the environmental costs that occur late in the program are discounted to near zero values at the 7% rate.

Application to Private Investments

Suppose that the owner of a natural foods store in a hot, humid climate believes she will lose all her customers unless she replaces her dying air conditioner. She wants to know whether she should buy a cheaper but more energy-using model or a more expensive model that saves energy. Both models require the same maintenance and have the same 10-year expected lifetime. The more energy-efficient model costs an extra \$2,000 up front but saves \$500 per year in the electricity bill. Neither model has an effect on the selling price of the business. If the owner must borrow the extra \$2,000 up-front cost at 15% interest is it a good economic decision to purchase the more energy-efficient model?

Table 6–4 Discounted versus undiscounted stream of benefits and costs

Year	Undiscounted ($ millions)			Discounted at 7% ($ millions)		
	Benefits	Costs	Net benefits	Benefits	Costs	Net benefits
1	0	100	−100	0	93	−93
2	0	100	−100	0	87	−87
3	0	100	−100	0	81	−81
4	0	100	−100	0	76	−76
5	0	100	−100	0	71	−71
6	160	60	100	106	40	66
.	.	.	.			
.	.	.	.			
.	.	.	.			
30	160	60	100	21	8	13
31	0	40	−40	0	5	−5
.	.	.	.			
.	.	.	.			
.	.	.	.			
100	0	40	−40	0	0	0
NPV	4,000	4,800	−800	1,329	983	346

Use the present value formula to answer this question by comparing the present value of each decision. Compare the added up-front cost ($2,000) of the energy-efficient air conditioner with the present value of the savings of $500 each year, discounted at 15% over every year of its 10-year life. Refer to Table 6–3 which shows that the present value of this stream of benefits of $500 saved per year is $2,509.50, that is, $500 times the table value of 5.019 (see year 10, 0.15 discount rate). Since $2,509.50 of the discounted electricity bills saved is higher than the additional up-front cost of the more energy-efficient air conditioner ($2,000), the expensive but energy-efficient air conditioner is the better choice.

Suppose now that the store owner expects to move from her current location in 3 years instead of staying for the entire 10 years. She believes that the more energy-efficient, but more expensive, fuel-efficient air conditioner will increase the sale value of her store by $1,000. So, what is the better choice this time? This time, discounted costs and benefits are measured as follows:

$$NPV = \sum_{t=1}^{3} \frac{(B_t - C_t)}{(1+r)^t} \tag{6.6}$$

$$= -\$2{,}000 + \frac{\$500}{1.15} + \frac{\$500}{(1.15)^2} + \frac{\$500}{(1.15)^3} + \frac{\$1000}{(1.15)^3}$$

$$= -\$2{,}000 + \$500\,(2.283) + \$1{,}000\,(0.658)$$

$$= -\$200.50$$

In this case (where the owner expects to move in three years) first use Table 6–1 to measure the present value of $1,000 three years in the future at 15%. Then use the Table 6–3 to find the present value of $500 per year each year for three years, again discounted at 15%. This time, the NPV is negative, so the extra expenditure is more costly but more fuel-efficient air conditioner is the economically *inefficient* decision, so she should not buy the air conditioner.

Application to Environmental Policy

Public policy analysts who use benefit-cost analysis (described in Chapter 5) rely on the discount rate to support several kinds of resource allocation decisions that face governments: Which public programs should be eliminated? Which ones should be started up? Which ones should become larger? Which ones should be smaller? How much money and other resources should be spent on each program and each program element?

Use of the discount rate lets benefit-cost analysts rank the economic performance of environmental resource projects, policies, and programs. By applying the discount rate to calculate net present values of various proposals using the NPV formula, BCA compares benefits and costs in many different time periods[4] to promote the conduct of rational policy analysis. Rational policy analysis attempts to consider all the relevant alternatives, identify and evaluate all the consequences that would follow from each alternative, and select that alternative and its associated consequences that are preferable in terms of society's most valued ends.

The discount rate can be used by analysts to help decide whether any local, regional, or federal program should be initiated. It also can be used to decide whether a number of proposed projects should be introduced and, if funds are limited, which one or which ones should be selected. The discount rate supports benefit-cost analysis in three types of public environmental decisions: simple ranking of the comparative benefits of several possible actions, determining optimal size or scale of a program, and deciding optimal timing or sequencing of several program elements.

SEARCH FOR A DISCOUNT RATE

Benefit-cost analysis requires discounting future benefits and costs to present value. Because BCA has such a practical need for discounting, an understanding of the impact of the choice of discount rate on the results of a BCA takes on special importance. As Table 6–1 shows, discounting dramatically affects the value of a proposed policy's costs or benefits, depending on the size of the discount rate and the number of years before society realizes the costs or benefits. Because society often incurs the costs of environmental programs and regulations long before the benefits, discounting often has an even greater impact in the calculation of the present value of benefits than on costs. Selecting a discount rate must be done carefully because of its central role in deciding whether a given project or program is viewed as economically efficient (economically feasible). The choice of a discount rate is critical in economic analysis of environmental policy.

Discounting reduces the weight of future damages, even catastrophic ones, that could be avoided through current action. Some economists have proposed using very low discount rates in analyzing the feasibility of current actions that could be taken to avert catastrophic future damages. Others have argued that meeting the needs of future generations is best done if usable resources are channeled to programs with the highest NPV, and that setting the discount rate artificially low may reduce the economy's supply of usable resources available for the future. In any case, an economically inefficient decision results from using the wrong discount rate (i.e., resources gravitate from a higher-valued to a lower-valued use). Moreover, using the wrong discount rate impairs the credibility of environmental management plans in a world where resources are scarce. For all these reasons, the discount rate continues to receive much scrutiny by economists, environmentalists, and decision makers. Three approaches have been used to select the appropriate discount rate: monetary cost of government borrowing, social opportunity cost of capital, and social rate of time preference. Feldstein (1964), was the first to propose this three-part classification scheme.

---

[4] In the United States documentation of the use of discount rates for economic analysis of public programs goes back to at least 1958, with the publication of *The Green Book* (U.S. Interagency River Basin Committee, 1958). This book is an early example of development of economic standards for designing, funding, and selecting water projects. It reminds us that the monetary values of benefits and costs that accrue at varying times are comparable only if all are adjusted to a consistent time basis. The use of discount rates provides a way to convert benefits or costs of different times to a common time period. The book also states that the supply of individual savings available for investment is limited mostly by the preference of individual savers for present over future goods. Because of the higher values that individual potential savers place on present goods, interest payments are needed to provide incentives to save that compensation for the lost current consumption. So discount rates are a way of expressing the exchange rate between present and future consumption. Measuring the present worth of goods available only by waiting involves discounting their future values. Updates were published by the U.S. Water Resources Council (1973, 1983).

Monetary Cost of Government Borrowing

The government's long-term real (inflation-adjusted) borrowing interest rate can be an appropriate number to use as the discount rate for public projects with investments for improving environmental quality, if government financing is directly-required to pay for the project (Lind, 1990). Where this is the correct number to use, analysts should compute the present value of net benefits using the government's borrowing rate as the discount rate.

Investors are willing to pay savers to defer consumption because investment is productive: Putting $1 today into good investments in new plants, new training, or renewed environmental quality can yield more than $1 of benefits in the future. In principle, the monetary cost of government borrowing (rate of interest) balances the time preferences of savers with the demand for savings by investors. Many economists agree that this interest rate is generally is on the order of 2% to 3% (in the United States), as reflected in the inflation-adjusted rate of return on short-term Treasury bills since the early 1960s. (For a history of rates for 3-month U.S. Treasury bills, see U.S. Federal Reserve System, 2004.) The Water Resources Planning Act of 1965 and the Water Resources Development Act of 1974 require U.S. federal water agencies to determine annually a discount rate for federal water resources planning. It is based on average yield during the preceding fiscal year on long-term interest rate (more than 15 years) of financing government debt. For 2003 that nominal rate was 5.5%, reflecting a real rate of about 2.5%.

The government's borrowing rate has another important feature: Suppose the government uses it as a discount rate to finance multigenerational environmental improvements. Say the present value of benefits when using that discount rate exceeds the present value of costs. Future generations receiving those benefits could use the fruits of those investments to retire the accumulated debt plus interest, and they would still be better off than if the investment had not been made (Lind, 1990). Thus, people who support using the government's real borrowing rate as a discount rate observe that it is a good candidate for the discount rate in long-run, intergenerational, environmental investments.

Social Opportunity Cost of Capital

The social opportunity cost of capital is based on the productivity of investment capital in its current uses. Typically, investment requires resources taken out of today's consumption stream, but it generates a larger quantity of consumption available for the future. So the return produced by investment taken from today's consumption adds to the future consumption stream. The marginal return earned on that invested capital (measured as an interest rate) reflects the opportunity cost of lost present consumption. The social opportunity cost of capital measures what society gives when consumption is displaced from private consumption through taxation or other means. In the conduct of cost-benefit analysis, social opportunity cost of capital is expressed as an annual rate of return.

Arguments Supporting the Social Opportunity Cost of Capital. Economists who support use of the social opportunity cost of capital as a discount rate support the principle of economic efficiency. By this they mean that the flow of resources between competing projects should be channeled into the private and public sectors according to their relative rates of return. Whichever sector produces the higher rate of return is where the resources should flow. For example, Table 6–5 shows that if the interest rate is 10% in the private sector and analysts set the

Table 6–5 Dangers of setting the discount rate too low

Discount rate	
9.9%	Resources used for programs that earn these rates of return in the government sector crowd out private sector productive uses, where they earn a 10% rate of return.
9.1%	
8.1%	
7.1%	

Table 6–6 Dangers of setting the discount rate too high

Discount rate	
12.9% 12.1% 12.1% 10.1%	Funds for projects earning these rates of return used in government projects crowd out resources in the private sector, where they earn a rate of return equal to 10%.

discount rate at 7% for public projects, society faces the risk of using resources in public projects that yield rates of return somewhere between 7.1% and 9.9%. But society also loses the higher 10% return from resources diverted from the private sector, which produces a net loss to society.

By comparison, Table 6–6 shows what happens if analysts set the public discount rate at a rate higher than the private rate of return of 10%, for example 13%. In that case, public projects will not be selected unless net present value is greater than zero computed at a 13% discount rate, and society runs the risk of giving up public projects that yield rates of return between 10.1% and 12.9%. The potential gain is larger than the 10% rate of return lost by bidding resources away from the private sector. Losing the potential for a return exceeding 10% in the public sector at a cost of only 10% return that is displaced in the private sector, again produces a net loss to society.[5]

At first, these two cases look like opposites; but look closer. For both cases, policy analysts select a public discount rate that differs from the private discount rate. The result of either choice produces an inefficient allocation of resources between the private and public sectors. This analysis leads to an important conclusion: *Economic efficiency for a public environmental or natural resource project requires using a discount rate equal to the social opportunity cost of capital of private sector resources displaced by the public project.*

Government Aims. People who support use of the opportunity cost of capital believe that the government in its role as investor for the public good should attempt to *maximize the well-being of the nation* measured in NPV. They propose that no public investment be undertaken that produces outputs of less value than of the alternative use of the resources it displaces. They conclude society's return on a public program must equal or exceed the private return the program displaced.[6]

Measuring the Social Opportunity Cost of Capital. Because government taxes apply to both consumers and businesses, both consumption spending and investment spending are displaced by taxation. Hence, the private sector interest rates relating to household saving (private time preference) and business investment decisions (pretax rates of return) must be reflected in the social opportunity cost rate of discount. Estimating an actual number for the social opportu-

[5] Grout (2003) argues that public and private projects should be discounted at different rates. Kirby et al. (2002) show that individual discount rates of people who live in Bolivia's tropical rain forests vary widely with age, literacy, and wealth.

[6] Lind (1990) sees not one but four ways to accommodate changes produced by government projects or policies. Often the four occur in some combination: *Changed taxes:* A public program is paid for by a tax increase. An increase in income tax falls partially on consumption and partly on savings-investments. In fact, changes in taxation are made rarely. At the margin, taxes are not adjusted up or down in the short term to compensate for changes in the level of government spending. *Changes in government borrowing:* Changes in the level of the budget are reflected in one-to-one changes in the deficit. In this case we must ask to what extent increases in the deficit crowd out private investment, if at all. *Changes in other government programs:* Government budgets are fixed so that any additional program will come at the expense of some other program. *Changes in private sector regulations:* Programs are incorporated in which the benefits and costs fall entirely on the private sector or some other level of government. All costs and benefits are private and opportunity costs can be measured by market changes in private market benefits displaced.

nity cost of capital requires that the relevant taxes be traced to their source of the household and business sectors. The observed interest rates in each of these sectors must then be weighted (multiplied) by the amounts of each sector's spending that is displaced by federal taxes. The social opportunity cost of capital is the weighted rate of return on private sector spending displaced through the taxes raised to support the public investment expenditure. For example, suppose $1 million spent on an environmental program to save lives is raised by taxing two sectors: $300,000 is taken from sector 1 that earned a return of 10% and $700,000 is taken from sector 2 that earned 20%. The weighted opportunity cost is $[(300{,}000 * 0.10) + (700{,}000 * 0.20)]/1{,}000{,}000 = 0.17$.

In calculating social opportunity, cost assumptions must be made about the sources of the funds that finance public investments: calculate the observed before-tax rates of return[7] prevailing over the economy as a whole, giving appropriate weight to both the private consumption and investment that is foregone.[8] Given these rates and the significance attached to each of them, we have, in effect, an exercise in weighted averaging: calculation of the social opportunity cost of capital rate of interest representing the opportunity cost at the margin of spending displaced by taxes.[9]

Haveman (1969) was one of the first to make this calculation; he assigned weights equal to income tax percentages paid by each income level to finance a new public program. He estimated the social opportunity cost of capital at just over 7%. More recently, U.S. federal agencies have been required to perform benefit-cost analysis in their justifications for proposed public investments and regulation. Since 1992, the U.S. Office of Management and Budget (OMB) has directed all federal agencies to use a 7% real discount rate in their analyses. The logic is that the 7% rate approximates the real (inflation-adjusted) marginal pretax rate of return on an average investment in the private sector in recent years (Bazelon and Smetters, 1999). The OMB rate reflects rates of return on low-yielding forms of capital, such as housing, as well as higher rates of returns yielded by corporate capital. OMB also recognizes the merit of other methods for selecting the discount rate and recommends conducting sensitivity analysis with respect to the discount rate.

Social Rate of Time Preference

The *social rate of time preference* is a discount rate the reflects society's *collective* preferences for trade-offs between present and future generations' consumption.[10] The social rate of time preference is usually lower than either the social opportunity cost of capital or the market rate

---

[7] Why use before-tax rates of return? Without taxes, the savings-consumption interest rate should equal the opportunity cost. But the income tax system drives a wedge between the two and makes them different. For example, if consumers have a 2% time preference discount rate for personal consumption, they will save an additional dollar only if given a 2% return. But to generate a 2% return after taxes to consumers, business firms must invest in projects producing higher returns. If business income is taxed at 50%, it needs to earn a 4% return to produce 2% after-tax return to shareholders. In this example, the savings-consumption interest rate is 2% and the opportunity cost of capital is 4%.

[8] One recently proposed way to implement the social opportunity cost of capital concept is to estimate what's called the shadow price of capital, defined as the present value of the stream of future consumption displaced by a dollar's worth of private investment. The idea entails tracing the future returns including reinvestments that is displaced because a government project or mandated regulation has diverted capital. That is, the method expresses the opportunity cost of capital as a flow of returns to consumers displaced. The shadow price analysis reduces this flow to present value of consumption, so we effectively use consumers' savings rate preferences to determine the discount rate. See Lyon (1990) for details and formulas.

[9] Prest and Turvey (1965) used a metaphor to illustrate how hard it is to track down public fund sources. Whatever method is used to trace these displaced resources to their original source, "one is trying to unscramble an omelette, and no one has yet invented a uniquely superior way of doing this."

[10] A similar definition for the social rate of time preference is the discount rate that expresses the consensus of the electorate concerning the rate of discount they wish to be applied to future costs and benefits in evaluating government projects for future generations (Bain, 1960).

Example 6–2
Are Low Discount Rates Good for the Environment?

Do lower discount rates protect the environment? Environmental policies that depress the discount rate without finding adequate resources to support the investments through increased savings will act like any other price ceiling: They will cause shortages and reduce the overall level of investment. So trying to help future generations through reduced discount rates may be like trying to help the poor through setting food prices artificially low, which reduces the profitability of farmers' food production and ultimately leads to food shortages (see Chapter 11).

For example, Krutilla and Fisher (1975) explore the case of a river valley containing a unique wilderness area for which there is a proposed hydroelectric development. The value of preserving that wilderness area is part of the opportunity cost of the proposed development. However, the authors go to some length to avoid adjusting the discount rate. Instead, they approach the future benefits issue squarely by recognizing that the benefits from preserving the wilderness area are likely to grow over time. Because the supply of substitute wilderness areas stays constant or falls, the demand for their amenities in this location grows with income and population growth. Furthermore, the demand to have such areas preserved from development, even by those who do not intend to use them is growing; that is, passive use values are increasing. Krutilla and Fisher make a similar adjustment for the benefits of developing the wilderness for commercial uses. They recognize that technological change will reduce the benefits from developments such as hydropower because cheaper electricity-generating technologies and/or more environmentally friendly technologies (e.g., solar power generation) will emerge over time.

A conclusion is that no special adjustment to the discount rate is required to show that the net benefits from preservation can exceed the net benefits of development.

Sources: Krutilla, J. V., and A. C. Fisher. 1975. *The Economics of Natural Environments.* Washington, D.C.: Resources for the Future; Markandya, A., and D. Pearce. 1991. Development, the environment, and the social rate of discount. *The World Bank Research Observer.* 6 (6): 137–152; Price, C. 1990. Deforestation and economic criteria. *Project Appraisal.* 5: 159–166.

of interest. Those who support the social rate of time preference as a discount rate say that the interest rate required for people to privately save and invest for the future is too low. The correct rate is based on the value an altruistic society places on gifts that enable future generations to avoid losing rare environmental resources that the present generation's consumption would otherwise displace. (Otherwise, *displace* means consumption by the present generation that would have foreclosed the gift, for example, the Grand Canyon in an earlier example.) See Example 6–2.

Intergenerational Responsibility. Discounting favors programs that confer benefits in the present or near future over programs whose benefits society realizes later. In fact, one purpose of discounting is to favor present benefits over future benefits that could be produced with the same current cost. Discounting typically favors programs that produce short-term benefits and long-term costs. Even a small discount rate will favor smaller benefits produced today over much larger benefits produced in the distant future.

Consider the following simplified example from Farber and Hemmersbaugh (1993): Suppose society is considering two programs for building a nuclear waste storage facility. Assume that both programs have the same monetary costs, that society would incur either program in the same year, and that construction of either facility can be finished in a year. The first option is to build a facility lasting 500 years. But that facility will leak radiation at the end of the 500-year period and produce 1 billion deaths. The second option is to build a safer storage facility that results in no lost lives after 500 years. But the second option, because of added construction hazards, will lead to the death of one worker during the construction period. Finally, assume a 5% discount rate, 2% lower than what OMB uses.

In choosing between the two options, most of us would likely prefer to save 1 billion future lives at the cost of one current life, even though society would not receive the benefit for 500 years. However, if a policymaker applied benefit-cost analysis using discounted benefits, she will choose the first option because 1 billion lives 500 years hence incurs a smaller loss than one life today (the discounted value of a billion lives lost 500 years from now is 0.025 lives). This simple hypothetical example vividly shows how discounting discriminates against the future. In fact, discounting often reduces large benefits in the far future so much that they become trivial when considered in present value terms; this is true regardless of what discounting method is used. Something needs to be done to bring future generations into significance.

Willingness to Bear Collective Burden. The justification for using the discount rate called a social rate of time preference that is lower than private market interest rates is this: Each person's willingness to sacrifice current consumption for future consumption depends on everyone else's willingness to do the same. Economists who adopt this position believe that society's willingness to pay *current* costs for *future* benefits would be significantly higher if governments took the lead in organizing the process through its power of taxation, so that the sacrifice of present benefits is shared by all. Taxpayers are motivated to pay for government projects for two reasons. First is the private benefits argument, in which the project benefits the same taxpayers who financed it. We spend our tax dollars to irrigate the desert so we can eat today. Second is the public benefit argument; namely, government projects are a gift to future generations who paid nothing for them. For example, future generations will be the main beneficiaries of all nations agreeing today to reduce greenhouse gas emissions.

Marglin (1963) introduced the concept of the social rate of time preference as the rate of return necessary to allocate resources efficiently between current and future generations. It is the collective discount rate that the current generation applies to a dollar invested on future generations' benefits compared to a dollar invested on its own benefits. The social rate of time preference goes beyond people's private selfish motives in providing for their own future. Instead, it accounts for collective motives in providing for collective futures. The concept uses these terms:

- *M* is the individual's marginal benefit of saving $1 for her own future, 1.0.
- *A* is the individual's marginal benefit from a program that produces $1 benefit to *future* generations in comparison to her own consumption (e.g., 0.100).
- *B* is the individual's marginal benefit of a program that produces $1 benefit to *current* contemporaries in comparison to her own consumption (e.g., 0.105).

Few of us are enthusiastic about investing our dollars to benefit either current or future generations beyond our direct descendants. So $M>>A$ and $M>>B$. However, each of us would like to see others invest, so *A* and *B* are both much smaller than *M;* but because each has a public goods' dimension, both are greater than zero.

With respect to private goods, intergenerational questions raise no special problems: The decisions of each of us acting alone to save for our own descendants satisfies society's total obligation to future generations. But for public goods like environmental program, the issue of gifts to future generations is more complex. Each member of the current generation may be willing to sacrifice some current consumption in isolation to assure his own descendants' access to public goods. The problem arises because each member cannot do this for his own descendants without providing a free ride to other people's descendants. The economically efficient social decision requires the current generation as a whole to sacrifice the collective consumption needed to provide the desired level of public goods in the future. In other words, with respect to public environmental programs, there is a serious risk of inefficient (inadequate) supply to future generations.

Taxes as Public Gifts. Each of us wishes to see *others* invest for the benefit of future generations, but each of us would also be willing to make that investment *provided others did so too.* Why? If each of us invests $1 in isolation to give to somebody else in the future, our individual net payoff is ($A - M$) a very negative number. So when we invest for the future, in isolation, we each feel like we lose nearly our whole dollar.[11]

If there are 100 people with similar preferences, who *all* invest $1 in a one-way gift to the future, the effect for each of us is ($100A - M$). This provides a net benefit of $(100 \times 0.10) - 1.0 = 9.0$ for *each person.* So each of us feels good by being a small part of a large total contribution to providing an important gift otherwise not available to a future generation. Examples of such collective gifts include institutions like the United Nations, civil society, structures like the Eiffel Tower, environments like the Great Barrier Reef, and endangered species like the California condor, rescued from the edge of extinction. So, in a large society, if we each ante up our share, we all feel good by being a part of something that none of us would have volunteered to finance or could have afforded in isolation.

Despite this great potential for a collective gift to the future, a market consisting of thousands of individuals, each acting in his or her own selfish interest, provides no way to break the deadlock: None of us is willing to invest by ourselves for people we don't know. But each may be prepared to if we all do. In fact, each of us looks on the net loss of personal benefits on her own investment as the price she pays to get others to invest, which is what each of us really wants.

These social motivations to provide for the future sound noble, civilized, and humane. But how do we channel them into public action? Would voluntary associations work for giving gifts of taxes to pay for programs received by future generations? Unfortunately, voluntary associations present a big problem. Like many other collective undertakings, the ease and attraction of avoiding one's own share of the cost of investment forces us to turn to the political process.[12] The coercive power of the state is one way to require that each of us bears our share of the investment burden instead of trying to shift it to others.

Role of Government. Government, according to the English economist Pigou, is the guardian of the interests of future generations as well as of the interests of the present generation. Privately executed preferences of the present generation, taken one person at a time, will be selfish and fail to adequately provide for future generations. So if Pigou's moral principle is accepted, government should guard against taking actions affecting future generations that are simply an aggregation of current, individual, privately motivated preferences.[13]

---

[11] The 1998 Nobel laureate in economics, Amartya Sen (see Nobel Foundation, 1998) refers to this process as the isolation paradox (Sen, 1961). Although the individual acting in isolation is not prepared to make the sacrifice, he is perfectly prepared to do it if others are ready to join in. This apparent paradox is present whenever his relative valuation of others' consumption is such that he would prefer them to sacrifice some consumption for the future generations; and sacrificing something himself might be a way to compensate the loss that he would incur from his own act of sacrifice. That is, each member of the present generation may want to raise the current savings rate of society but may not wish to raise that of himself in isolation.

[12] According to Hardin, the possibility of avoiding one's own share requires public action: "mutual coercion mutually agreed upon" (see Chapter 3). Prest and Turvey (1965) observe that economists have recognized for years that any attempt to get consumers to reveal their preferences regarding public goods, such as gifts to future generations, founders because each consumer rationally understates his demand in the hope that he would thereby be relieved of part of all of his share of the cost without affecting the quantity obtained.

[13] The view that social discount rates should be lower than private rates to justify economically public investments beyond the economic feasibility of private ones is reflected in policy of the United Kingdom. Generally, the U.K. government requires state-owned industries that compete with the private sector, or provide services that could be undertaken by the private sector, to use an 8% discount rate when appraising their investment decisions. However, projects believed to yield high long-term social benefits are allowed to use a 6% discount rate, thereby improving the prospects of investments with delayed benefits showing positive NPV. The U.K. treasury recently announced that they are considering a further reduction in discount rates to 3.5%.

If the number of individuals is large, each individual's personal sacrifice in the community's investment is negligible compared with the gains in satisfaction produced by knowing a large collective gift to the future is made possible. Each member of the future generation can simultaneously enjoy the benefits produced by the gift, benefits that would have been impossible had the gift been withheld.

Calculation and Application. Recall that each person today compares the value of the current tax burden spent for one of two groups:

- *A* is the individual's benefits received from a $1 gift to future generations.
- *B* is the individual's benefits received from a $1 gift to current generations.

Each of us has a social rate of time preference, for which we substitute future for present consumption on the assumption that our sacrifice is accompanied by everybody else's sacrifice. Marglin (1963) shows that the individual's social rate of time preference is as follows:

$$r^* = B/A - 1 \tag{6.7}$$

What does this mean for the social rate of time preference required by the NPV formula? Suppose one of us values equally a gift to *the current* or *any future* generation resulting from $1 of our taxes spent, that is $B = A$. Then

$$r^* = (A/A) - 1 = 0 \tag{6.8}$$

If we value a gift equally to each future generation, then the social rate of time preference equals zero.[14] Equation 6.7 shows that social rate of time preference measures how heavily we collectively discount the current generation compared to future generations. If we value future and current generations' benefits from our tax dollars equally, then each of our social time preferences is zero.[15] This means, as a matter of public policy, we should spend dollars one for one in paying current costs for future benefits.[16] Where this occurs, a $10 million investment by current taxpayers to preserve $10 million of endangered species or to mitigate $10 million of future damages from global climate change is an economically efficient investment.

The U.S. Congressional Budget Office (CBO) uses a 2% real discount rate for policies having long-term social implications, with sensitivity analysis of estimates based on rates of 0 and 4%. The procedure is described by Hartman (1990) and with the rationale is explained by Lind

[14] If there are only two generations, current and future, we may assign a slightly selfish personal marginal benefit for gifts made to current generations, since it's the one with most of our friends. For example, if we each value *A* at 0.100 and *B* at 0.105, then $r^* = (.105/.100) - 1 = 1.05 - 1 = 0.05$.

[15] Weitzman (1994, 1998) argues that even if a nonzero rate of time preference is used for environmental projects, environmental quality will become more scarce in future generations compared to ordinary goods. So he calls for a systematically falling discount rate. Benefits far into the future would be discounted at rates closer to zero than discount rates for today's benefits. In fact, analysts should use discount rates as close to zero as possible for projects with large and/or uncertain impacts felt many years in the future.

[16] Baumol (1968) observes a problem with using the social rate of time preference as a way to increase our one-way gifts to future generations. He asks whether there are so few vulnerable and underprivileged today, so few of today's people who move our sympathy, that we must look to the wealthier future for targets worthy of receiving our gifts. He says that by and large the future can be left to take care of itself, since people are likely to be richer than we are, and that to sacrifice present for future generations is like taking from the poor and giving to the rich. Why should we (the poor) sacrifice more, by lowering the discount rate, for the benefits of the rich future generations for whom the marginal benefit of the added income we give them is likely to be much lower than the marginal benefit of ours lost? Despite this argument, Baumol realizes that even if the future is richer than we are, it may still suffer a poverty of natural environmental assets. If current generations provide the future with a unique environmental asset it otherwise would have lacked, future generations become richer in something of high scarcity value to them. And their high incomes may raise their value of an undeveloped Grand Canyon even further, thus making it more important to them for us to preserve.

(1982). The 2% rate is considered to represent the social rate of time preference. The Intergovernmental Panel on Climate Change (IPCC) also suggests that rates as low as 2% may be appropriate for environmental decision making (Arrow et al., 1996).

Long-Term Decisions. Long-term discounting is controversial in environmental benefit-cost applications. For many government program proposals, especially in the areas of energy, health, and the environment, effects are often felt far into the future—indeed, sometimes across many generations. Global warming issues are a good example. At nearly any commonly accepted or used discount rate, distant future effects of current policy proposals have little influence on the present discounted value of the plan because, as shown by Equation 6.5a, the impact on NPV of large distant future changes is swamped by the force of discounting.

Preventing a Catastrophe. Nordhaus (1999) uses the following example to show the nearly zero NPV of an action that could prevent a costly event far in the future, a cost that could be completely prevented with a small investment today. He asks us to imagine that scientists discover an asteroid on a 200-year orbit that will soon pass by the Earth. Calculations show that on the next orbit, which will occur in 200 years, the asteroid will smash into and destroy Florida. We can knock the asteroid off course today by launching a missile with explosives that render the asteroid harmless. What action would a conventional benefit-cost analysis recommend? The estimated current value of land and capital in Florida is $2 trillion. Use of a 7% discount rate shows that actions taken today that would prevent Florida's destruction 200 years from now is worth less than $3 million today.[17] Therefore, if the missile launch costs more than $3 million, current standard economic analysis advises against saving Florida—a conclusion difficult to defend.[18]

Cooperation from Future Generations. Some programs take a while to finish. Completion of the majestic Notre Dame Cathedral in Paris required more than 180 years of cooperation by artists, builders, financiers, and governments. When many cooperating generations are required to carry out a single program, the distribution of benefits and costs across these generations is important (see Lind, 1999). The unadjusted NPV may fail to capture our concern for fairness to future generations. Furthermore, intergenerational transfers, either forward or backward, can only be made if each of the intermediate generations cooperates with the first by passing to the future the benefits bestowed to it by the previous generation. Designing and implementing such transfers is difficult; the ease by which any one intervening generation can break the chain of transfer makes such schemes hard to implement. In benefit-cost terms, this means it is difficult to design a plan by which the beneficiaries pay compensation to those incurring the costs.

For example, think about an appropriate discount rate for evaluating a program in which current generations invest heavily in climate change mitigation. Future generations that would enjoy the benefits of such a program cannot easily compensate near-term generations that will pay the costs. See Example 6–3. Nor can the present generation choose to avoid mitigation and compensate future generations by (1) investing the savings (from not incurring costs) at a market rate of return and (2) giving the accumulated wealth to future generations as compensation for their added costs incurred from climate change. Therefore, the inability to provide compensation in either direction negates the logical foundation behind the use of NPV for evaluating public programs.[19]

In the case of climate change, near-term generations who pay the costs cannot be compensated by future generations, so any expenditure now for mitigation is a transfer from near-term

---

[17] The calculation is 2 trillion / $(1.07)^{200} = 2{,}656{,}283$. The calculation assumes that the value saved from destruction in 200 years is still $2 trillion.

[18] Like Nordhaus, economists Newell and Pizer (2001) observe that if the plausible discount rates on very-long-term investments range from 2% to 7% the corresponding difference in discounted values range from 200 million to 1; compare $(1/(1 + 0.02)^{400})$ to $(1(/1 + 0.07)^{400})$ for a 400-year time horizon.

[19] That logical foundation is the Potential Pareto Improvement compensation test, in which gainers could compensate losers and still be better off.

Example 6–3
What Discount Rate for Controlling Climate Change?

Four related economic and political questions center on climate change policies: (1) By how much should current greenhouse gas emissions be limited? (2) How many years from now should reductions begin? (3) With what technologies and institutions should emissions be reduced? (4) What countries and people should bear responsibility for reducing emissions? The answers to the first two questions depend on what discount rate is used to design the policies. The accompanying table is an analysis by Conrad (1997) in which a higher discount rate justifies delaying investments in abating greenhouse gases.

Economically efficient timing of investments in greenhouse gas (GHG) reduction

Discount rate	Time to begin GHG reductions (years from now)
0.00	0
0.01	0
0.02	0
0.03	0
0.04	92
0.05	122
0.06	140

Choosing the appropriate discount rate for global warming implies an ethical stand on how one generation values future generations, as distinct from the former's evaluation of its own future. At the same time, the choice expresses positive concern about the degree to which current investments in greenhouse gas abatement displace current consumption and/or other investments in the economy. Higher discount rates assign lower importance to future generations, and for a given cost imposed on current generations to reduce greenhouse gasses, delay the economically efficient number of years to wait before starting abatement.

For global warming, where each of us is concerned about the impacts of public policy spanning several generations, there may be little ethical justification for a positive social rate of time preference. Any future generation is no less important than is the current generation. Rawls (1971) favors a zero social rate of time preference based on intergenerational justice. As he says it, the just society will guard against "the living take advantage of their position in time to favor their own interests" (p. 295). If we agree with Rawls, the only ethically defensible discount rate for long-term environmental protection programs is zero.

Sources: Khanna, N., and D. Chapman. 1996. Time preference, abatement costs, and international climate policy: An appraisal of IPCC 1995. *Contemporary Economic Policy.* 14 (2): 56–66; the table is from Conrad, J. M. 1997. Global warming: When to bite the bullet. *Land Economics.* 73 (2): 164–173. Copyright 1997. Reprinted by permission of the author and the University of Wisconsisn Press; Rawls, J. 1901. *A Theory of Justice.* Cambridge, MA: The Belknap Press of Harvard University.

generations to those in the distant future. Like Marglin's social rate of time preference concept, climate change mitigated is a gift to somebody else, not an investment for one's own future consumption. In this situation, the appropriate question for weighing costs and benefits in a long-term process is how much are people willing to pay today for the benefit produced by the knowledge that we leave options open for mitigating climate change in the future with the information and technology available to that future society? The use of any social discount rate higher than zero, therefore, rests on weak moral principles.

Lind (1995) challenges us to consider two hypothetical cases. For the first case, suppose today we are faced with two investment options that will benefit people living 200 and 300 years from now, respectively. Assume that over each future year, the marginal rate of return on all private and public investments is 10%. The first investment being considered will earn a certain internal rate of return of 0% and will benefit people living 200 years from now who will be on the verge of starvation. So the proposed program will save them. The preferred decision may well be to make that investment and transfer the resources to the future generation even though it earns a zero rate of return. One might think that the taxpayers could have invested that money at 10% and made those people better off. But we are unable to set aside investment funds and commit intervening generations to investing and reinvesting those funds for eventual delivery to the generation two centuries from now. So the usual assumption we make in benefit-cost analysis, that the benefits can be continually reinvested throughout the life of the project, is violated.

The second hypothetical policy considers making an investment that transfers resources to a generation living 300 years from now. This investment has an internal rate of return of 20%.

However, the current generation knows that the generation living 300 years from now will be 100 times wealthier than people are now. So, in what seems to be an easy decision, they reject the investment. Why doesn't the current generation make the investment and keep 10% of the future generation's 20% payoff, leaving the rest for the future, so both generations are better off? After all, future anticipated gains in income that have not occurred yet are sold all the time; a good example is the market for mortgages. However, where several generations are involved this is nearly impossible, in part because of the problem of getting and enforcing the commitment of intervening generations to carry out this intergenerational transfer.

The logic of the benefit-cost criterion requires the possibility to trade income from one period to another at a known discount rate. When some institution, such as a uncooperative generation in the middle, breaks the flow of a gift from the present to the future, the logic behind the NPV criterion and its use of a positive discount rate fails.[20]

Debates over the correct discount and the correct method to find it have raged for more than a hundred years. So economists generally recommend that analysts perform calculations to determine how sensitive their results are to changing the discount rate. For example, if $r = 10\%$, one might choose a lower bound of 5% and an upper bound of 15% to determine the effects of different discount rates on the project's economic desirability. Thus, policymakers make judgments regarding the appropriate measure of the discount rate.

Summary

The U.S. government currently requires discounting of future benefits and costs in benefit-cost analysis of federal investment programs. The choice of a discount rate is crucial because it is a major factor influencing (1) whether the net present value of a single proposed project is positive, (2) the relative attractiveness of competing projects, and (3) the optimal timing of projects. If the discount rate used by government agencies is too low, there will be too much public investment in long-term projects. However, if the government's discount rate is too high, there will be too little long-term project investment.

This chapter described three methods for finding a discount rate for public programs and summarized the limits of each: the money cost of government borrowing, the social opportunity cost of capital, and the social rate of time preference. The government's long-term, real borrowing interest rate can be an appropriate discount rate to use in public projects for short-term investments in improving environmental quality, if government financing is directly required to pay for the project.

The social opportunity cost of capital measures what society gives up by consumption displaced through taxation or other means to achieve an improved environment. People who support use of the opportunity cost of capital believe that no public investment should be undertaken that produces outputs of less value than that of the alternative use of the resources it displaces. They conclude society's return on a public program must equal or exceed the private return the program displaces. Since 1992, the U.S. Office of Management and Budget has directed all federal agencies to use a 7% real discount rate in their analyses. The logic is that the 7% rate approximates the real, marginal, pretax rate of return on an average investment in the private sector.

The social rate of time preference is a discount rate that reflects society's *collective* preferences for trade-offs between present and future generations' consumption. It is usually lower than either the market interest rate of borrowing or the social opportunity cost of capital. It goes beyond people's private, selfish motives in providing for their own future by accounting for collective motives in providing for collective futures. If a society values a gift equally to each future generation, then its social rate of time preference is zero. The U.S. Congressional Budget Office uses a 2% real discount rate for policies having long-term social implications. This 2% rate is considered to represent the social rate of time preference. The Intergovernmental Panel on Climate Change also suggests that rates as low as 2% may be appropriate for proposed policies mitigating climate change.

The need to compare present and future benefits and costs using a common denominator have been recognized and understood by economists and policy analysts for years. But there is no general consensus, either conceptually or empirically, as to what this rate should be.

---

[20] Debates surrounding the discount rate led Resources for the Future to hold a conference (Portney and Weyant, 1999). Most agreed that it is acceptable to discount future benefits and costs at some positive rate. In particular, for time horizons of 40 years or less, many accepted the social opportunity cost of capital. Debates arose over time horizons exceeding 40 years. Many wished to explore some form of nonconstant discounting, for example, Weitzman, 1999; Kopp and Portney, 1999; and Cropper and Laibson, 1999.

REVIEW QUESTIONS

1. Explain, using a numerical example, why the taxpayer suffers an economic loss when policy analysts use the wrong discount rate for evaluating environmental and natural resource programs.

2. Explain what kinds of projects perform best (show the highest net present value) when a low discount rate is used.

3. Why would supporters of local environmental and natural resource programs want analysts to use low discount rates? Explain why this is or is not good for the nation.

4. Why does the choice of the discount rate matter so much?

5. Define the money cost of borrowing and explain the advantages and limits of using it as a discount rate.

6. Explain the meaning of the social opportunity cost of capital as a discount rate, and discuss the advantages and disadvantages of using it as a discount rate.

7. What is the social rate of time preference? Describe one method for finding it.

8. Explain how we give gifts to future generations through taxation.

9. What is the role of government in providing for future generations?

10. Explain why gifts to future generations made through voluntary associations rarely work.

11. Why does the social rate of time preference typically produce a lower discount rate than the opportunity cost of capital?

12. For what kinds of environmental and natural resource programs should we use the social rate of time preference for appraisal?

PROBLEMS

1. The current heating bill for a house is $1,200 per year. Determine the maximum amount that could be afforded for a solar heating system that would pay for itself in 20 years. Assume that the discount rate is 10%. What is the maximum amount you would pay if the discount rate is 5%?

2. Consider an environmental project with the following stream of benefits and costs.

Year	0	1	2	3
Benefit	0	40	40	40
Cost	100	0	0	0

a. Calculate the NPV of this project using a discount rate of 5%.
b. Graph the NPV of the project for varying discount rates between 0% and 10%. What pattern do you see and what does it suggest about the influence of discount rates on the NPV for projects that incur early costs and later payback costs?
c. Calculate the benefit-cost ratio of this project (present value of benefits divided by present value of costs) using a discount rate of 10%.
d. Graph the benefit-cost ratio of this project for varying discount rates between 0% and 10%.
e. Calculate the internal rate of return for this project, that is, the discount rate, which if used, produces a zero NPV.

3. Your city receives $100,000 at the beginning of period 0 which you must completely spend for environmental projects in periods 0 and 1. After period 1, the money is lost. You can save some of the money in a bank between period 0 and 1 and receive an interest rate of 10%. You can spend the unspent principal plus interest in period 1. Your total environmental benefits function shows diminishing marginal benefits from increased spending in *either* period. That function is given by this equation: $B(C_0,C_1) = C_0C_1$, where C_0 is that part of the 100,000 you spend in period 0 and C_1 is the part spent in period 1.

a. Write down your city's budget constraint, assuming the prices of consumption in both periods are 1.
b. Solve for your city's optimal levels of consumption in period 0 (C_0) and period 1 (C_1). Hint: Find the budget allocation of the initial $100,000 between the two periods that makes *marginal* benefits per dollar equal in both time periods.
c. How does your answer change if the interest (discount) rate falls to zero? Why?

4. Discount rates help you allocate resources efficiently across time. A two-period model illustrates the idea. Suppose you wish to allocate a depletable resource between two periods. The price (P) you receive per unit in any period falls with increased quantity supplied (Q) to that period. The equations are as follows: P_i = marginal benefits = $100 - 2Q_i$ (same demand for both periods, i = 1,2) reserves = total quantity available for both periods = 50; discount rate = 10%.

a. What is the optimal allocation of reserves over the two periods? Hint: Find the allocation of the reserves that makes discounted marginal benefits equal in both periods.[21]
b. Now redo this problem for a discount rate of 0%. What happens to your allocation between periods as the discount rate falls? Can you explain the economic logic for this? What does this say about the role of the discount rate in affecting decisions about depletable resource conservation?

[21] An advanced mathematical technique, known as optimal control theory, can also be used to find the answer. It enables you to find a way to allocate a resource over several time periods that optimize an objective function defined over those periods. That optimal time path can be solved as an explicit closed form function of time or as a numerical solution. See Dorfman (1969) and Sethi and Thompson (1981).

References

Arrow, K. J., W. R. Cline, K. G. Maler, M. Munasinghe, R. Squitieri, and J. E. Stiglitz. 1996. *Intertemporal equity, discounting, and economic efficiency. In Climate Change 1995: Economic and Social Dimensions of Climate Change,* Edited by J. P. Bruce, H. Lee, and E. F. Haites, Cambridge, UK: Cambridge University Press, pp. 125–144.

Azfar, O. 1999. Rationalizing hyperbolic discounting. *Journal of Economic Behavior and Organization.* 38 (2): 245–252.

Bain, J. S. 1960. Criteria for undertaking water-resource developments. *The American Economic Review.* 50 (2): 310–320.

Baumol, W. J. 1968. On the social rate of discount. *American Economic Review.* 58 (4): 788–802.

Bazelon, C., and K. Smetters. 1999. Discounting inside the Washington D.C. beltway. *Journal of Economic Perspectives.* 13 (4): 213–228.

Cropper, M., and D. Laibson. 1999. The implications of hyperbolic discounting for project evaluation. In *Discounting and Intergenerational Equity,* edited by P. R. Portney and J. C. Weyant. Washington, D.C.: Resources for the Future.

Dorfman, R. 1969. An economic interpretation of optimal control theory. *American Economic Review.* 59 (5): 817–831.

Farber, D. A., and P. A. Hemmersbaugh. 1993. The shadow of the future: Discount rates, later generations, and the environment. *Vanderbilt Law Review.* 46: 267–304.

Feldstein, M. 1964. The social time preference discount rate in cost-benefit analysis. *Economic Journal.* 74: 360–379.

Gollier, C. 2002. Time horizon and the discount rate. *Journal of Economic Theory.* 107 (2): 463–473.

Grout, P. A. 2003. Public and private sector discount rates in public-private partnerships. *Economic Journal.* 113 (486): C62–C68.

Hartman, R. W. 1990. One thousand points of light seeking a number: A case study of CBO's search for a discount rate policy. *Journal of Environmental Economics and Management.* 18 (2, part 2).

Haveman, R. H. 1969. The opportunity cost of displaced private spending and the social discount rate. *Water Resources Research.* 5 (5): 947–957.

Kirby, K. N., R. Godoy, V. Reyes-Garcia, E. Byron, L. Apaza, W. Leonard, E. Perez, V. Vadez, and D. Wilkie. 2002. Correlates of delay-discount rates: Evidence from Tsimane' Amerindians of the Bolivian rain forest. *Journal of Economic Psychology.* 23 (3): 291–316.

Kopp, R. J., and P. R. Portney. 1999. Mock referenda for intergenerational decisionmaking. In *Discounting and Intergenerational Equity,* edited by P. R. Portney and J. C. Weyant. Washington, D.C.: Resources for the Future.

Lind, R. C., ed. 1982. *Discounting for time and risk in energy policy.* Washington D.C.: Resources for the Future.

———. 1990. Reassessing the government's discount rate policy in light of new theory and data in a world economy with a high degree of capital mobility. *Journal of Environmental Economics and Management.* 18 (March): S8–S28.

———. 1995. Intergenerational equity discounting, and the role of cost-benefit analysis in evaluating global climate policy. *Energy Policy.* 23: 379–389.

———. 1999. Analysis for intergenerational discounting. In *Discounting and Intergenerational Equity,* edited by P. R. Portney and J. P. Weyant. Washington, D.C.: Resources for the Future.

Lyon, R. M. 1990. Federal discount policy, the shadow price of capital, and challenges for reform. *Journal of Environmental Economics and Management.* 18 (2, part 2): S29–S50.

Marglin, S. A. 1963. The social rate of discount and the optimal rate of investment. *Quarterly Journal of Economics.* (February): 95–111.

Newell, R., and W. Pizer. 2001. *Discounting the Benefits of Climate change Mitigation: How Much do Uncertain Rates Increase Valuations?* Washington, D.C.: Pew Center on Global Climate Change.

Nobel Foundation. 1998. *The Bank of Sweden Prize in Economic Sciences in Memory of Alfred Nobel, 1998, Amartya Sen.* Available from *http://nobelprize.org/economics/laureates/1998/index.html* [Accessed October 27, 2004].

Nordhaus, W. D. 1999. Discounting and public policies that affect the distant future. In *Discounting and Intergenerational Equity,* edited by P. R. Portney and J. P. Weyant. Washington, D.C.: Resources for the Future.

Portney, P. R., and J. P. Weyant, eds. 1999. *Discounting and Intergenerational Equity.* Washington, D.C.: Resources for the Future.

Prest, A. R., and R. Turvey. 1965. Cost-benefit analysis: A survey. *Economic Journal.* 75: 685–705.

Sen, A. K. 1961. On optimising the rate of saving. *The Economic Journal.* 71 (283): 479–496.

Sethi, S., and G. L. Thompson. 1981. A tutorial on optimal control theory. *Infor.* 19 (4): 279–291.

U.S. Federal Reserve System. 2004. *Rate of interest in money and capital markets.* Available from *http://www.federalreserve.gov/releases/h15/data/a/tbaa3m.txt* [Accessed October 27, 2004].

U.S. Interagency River Basin Committee Subcommittee on Benefits and Costs Report to the Federal Inter-Agency River Basin Committee. 1958. *Proposed Practices for Economic Analysis of River Basin Projects, "The Green Book."* Washington, D.C.: U.S. Government Printing Office (originally published in 1950).

U.S. Office of Management and Budget. 1992. *Guidelines and Discount Rates for Benefit-Cost Analysis of Federal Programs.* Circular No. A-94 (revised). Washington D.C.: U.S. Government Printing Office.

U.S. Water Resources Council. 1973. *Water and Related Land Resources: Establishment of Principles and Standards for Planning.* Washington D.C.: U.S. Government Printing Office.

———. 1983. *Economic and Environmental Principles and Guidelines for Water and Related Land Resources Implementation Studies.* Washington D.C.: U.S. Government Printing Office.

Weitzman, M. L. 1994. On the "environmental" discount rate. *Journal of Environmental Economics and Management.* 26 (2): 200–209.

———. 1998. Why the far-distant future should be discounted at its lowest possible rate. *Journal of Environmental Economics and Management.* 36: 201–208.

———. 1999. Just keep discounting, but. . . . In *Discounting and Intergenerational Equity,* edited by P. R. Portney and J. C. Weyant. Washington, D.C.: Resources for the Future.

———. 2001. Gamma discounting. *American Economic Review.* 91 (1): 260–271.

Young, R. A., and R. H. Haveman. 1985. Economics of water resources: A survey. In *Handbook of Natural Resource and Energy Economics.,* edited by A. V. Kneese and L. L. Sweeney. New York: North-Holland.

Chapter 7

VALUING THE ENVIRONMENT

Outline

Relevance of Environmental Values 151

Individual Values and Social Welfare 152

Methods for Measuring Environmental Values 166

Objectives

After studying this chapter, you should be able to

- Describe how information on economic values of environmental quality supports environmental policies
- Explain the relation between individual values of environmental improvements and their resulting social values
- Identify several methods for measuring the value of environmental improvements
- Discuss the contingent valuation method and explain why it is controversial

Introduction

In 1870 mystery and speculation surrounded the U.S. Yellowstone area in what is now the northwest corner of Wyoming. People could neither see nor imagine its scope and scale. Landscape artist Thomas Moran asked to be included as part of Ferdinand Hayden's Great Western Survey of 1871. Moran, an admirer of the artist J. M. W. Turner, joined the famous photographer William Henry Jackson in an effort to record for posterity the journey to Yellowstone. Later, after seeing Jackson's black and white photos and Moran's watercolor,[1] the U.S. Congress voted to set aside Yellowstone as America's first national park in 1872. Moran's artwork may have been the first to show Congress and the public the value of preserving the nation's natural treasures, even when setting aside a natural treasure displaced opportunities to harvest commercial natural resource products such as timber and minerals. Today's decision makers require more than great paintings to frame debates on the benefits of preserving, improving, or protecting the natural environment. Unlike the case of Yellowstone, for which there were virtually no established uses displaced by establishment of the park, today there are nearly always costs associated with measures taken to improve the environment.

Now consider an example of a current environmental policy issue that involves questions of economic values and trade-offs: Meeting the air and water pollution control goals established by the U.S. Congress requires huge expenditures by both the private and public sectors. To meet these goals, resources must be taken away from the production of other goods and services. So is meeting the pollution control goals making us better off or worse off?

This chapter presents the economic theory for identifying the value of environmental improvements and describes techniques used for measuring the values of those improvements. We begin by examining the relevance of environmental economic values; then we analyze the relationship between individual values and social welfare. Most of the chapter concentrates on specific methods for measuring environmental values.

[1] Moran's watercolor is on the cover of this book. See U.S. National Park Service (2004). For information about Moran, see National Gallery of Art (2004). For information about Turner, see National Gallery at Trafalgar Square London (2004).

■ Relevance of Environmental Values

Preservation, protection, restoration, and improvement of the natural environment face tough competition whenever resources are scarce. Public and private decision makers must deal with limited budgets—yet growing demands—for environmental action. Their choices about how to protect and restore the natural environment are not easy.

In making difficult decisions, environmental program managers consider many objectives, including environmental quality, threats to ecosystem integrity, and outcomes on people's quality of life. This chapter describes the use of economic principles and methods of analysis in making environmental decisions.

Uses of Information on Economic Values of the Environment

Information about the economic value of the environment has many uses:

- Supporting decisions on environmental improvements or restorations
- Helping resource managers allocate their budgets
- Promoting more informed decision making
- Facilitating common-denominator comparisons
- Providing a baseline for factual analysis and measurement of needs
- Attracting public or private investors, sponsors, and partners
- Promoting legislative support and influence funding decisions
- Promoting environmental conservation and protection by defining carrying capacity
- Helping business, industry, and landowners forecast demand for their environmentally related services
- Playing an educational role by helping people discover cultural similarities, determine community values, and learn about environmental values and multiple uses of natural resources

Economic valuations of the environment suggest how to allocate or prioritize public spending on conservation, preservation, or restoration by comparing the benefits of different projects or programs. Such comparison allows the maximization of total benefits of the environment from the available budget (benefit-cost analysis) or the minimization of costs from accomplishing a given environmental aim (cost-effectiveness analysis).

In the United States, information on economic values of the environment has been used for five major purposes in environmental decision making: (1) benefit-cost analysis (BCA) of projects, (2) BCA of new regulations, (3) assessments of natural resource damage, (4) environmental costing, and (5) environmental accounting. In Europe, there have been considerably fewer environmental valuation analyses, because less importance is attached to economic efficiency in the environmental decisions of most European countries. Thus, few valuation studies from Europe have served as information sources for environmental policy and regulations. However, there will likely be increased use of valuation techniques in Europe because the European Union (EU) has adopted these goals: to establish environmentally adjusted national accounts and to apply BCA to environmental policy and regulations. See Example 7–1.

Monetary Measures of Environmental Values

Economics typically presumes that measures of economic value are based on what people want, and values based on preferences of the individual members of a society are the relevant values for public decision making in a democratic society. According to this principle of social value, the maximum amount that a person is willing to pay (typically in monetary terms) to receive a service is considered an accurate measure of the value of that service. In assessing the economic value of an environmental resource, however, it is important to recognize that measuring the value of the resource in monetary terms does not require that the service be traded in formally functioning markets. It does require estimating how much people are willing to pay (in monetary terms) for access to the resource.

Example 7–1
Environmental Valuation: United States versus Europe

The United States has a long tradition of applying benefit-cost analysis (BCA) at the level of project appraisals. One of the keystones in the development of BCA was *The Green Book* (U.S. Federal Interagency River Basin Committee, 1958–originally published in 1950) which represented a systematic, consistent, and theoretically sound framework for the economic analysis of river basin projects. By 1960 BCA became a standard tool of the federal government's evaluation of water resources programs. The application of BCA was later widened from water resources to other government activities such as transportation, health care, education, and job training. In 1994, President Clinton instructed the Bureau of Economic Analysis to develop a green-adjusted economic accounting for use in BCA. Ludwig, Mangel, and Haddad (2001) describe some of the challenges in securing cooperation among ecologists, economists, and other environmental scientists.

In Europe, the history of research and applied work in BCA is much shorter than in the United States, and the development of environmental valuation methods has been limited. In most European countries environmental impacts are rarely valued in monetary terms. When studies have been conducted in Europe to evaluate an environmental regulation, they have either not been applied at all to affect the final decision or they have been used to justify the environmental goal already set by environmental authorities. Navrud and Pruckner found few cost-benefit studies in Europe that could add information to the design, review, or implementation of environmental regulations or programs. Why not? First, BCA has received much criticism in Europe because of the difficulty of measuring environmental benefits accurately. In particular, contingent valuation models based on surveys are criticized with respect to their accuracy and meaning, for reasons described in this chapter. Second, the recognition that environmental policy measures are associated with economic costs and benefits may be poorly understood among European voters. Third, there is no institution in Europe like the U.S. EPA, that employs large numbers of applied economists who can close the gap between academic research and environmental decisions. Increased cooperation of scientists and economists who could assess risks and weigh costs and benefits is needed to inject greater economic content into Europe's environmental regulation.

Sources: Navrud, S., and G. J. Pruckner. 1997. Environmental valuation–To use or not to use? A comparative study of the United States and Europe. *Environmental and Resource Economics.* 10 (1): 1–26; Hanley, N. 2001. Cost-benefit analysis and environmental policymaking. *Environment and Planning C–Government and Policy.* 19 (1): 103–118; Reuss, M. 2003. *Federal water resources planning.* U.S. Army Corps of Engineers. Available from *http://www.usace.army.mil/inet/functions/cw/hot_topics/24nov_waterresources.htm* [Accessed October 30, 2004].

■ Individual Values and Social Welfare

Benefit-cost analysis is a way to make environmental policy decisions that serve a society's welfare. The purpose of a theory of social welfare is to provide an objective rationale for making social decisions when individual members of the society have different views on or interests in the alternative choices. A good decision about public environmental policy will account for the interaction of individual preferences with the range of alternative decisions actually available. Those alternative decisions may change from time to time because of variation in the community's income, wealth, or technology. The application of BCA to environmental policy is based on the idea that society's values, derived from individual preferences, are expressed for all economically possible choices; then the most preferred alternative is selected from the choices available.

Valuation of Environmental Systems

Decisions affecting the protection of environmental resources often involve trade-offs in allocating resources (i.e., giving up something to gain something else). These are economic decisions that are based, either explicitly or implicitly, on society's values. Economic valuation can provide the information needed to justify and set priorities for programs, policies, or actions that protect or improve environmental resources. Several important definitions and concepts are worth reviewing to better understand how economists approach environmental valuation.

Nature of Environmental Values. Environmental values summarize the worth of environmental services to people. The value of any resource to a person, in the economic sense, measures what that person is prepared to sacrifice or is willing to accept in payment for access to the services that resource provides. An environmental resource that provides functions of high value to a person is one for which that person is prepared to pay a considerable sum of money rather than go without, or is prepared to accept in payment a considerable sum to take the place of the resource.

While some services of the environment, like harvested fish or timber, are traded in formal markets, many environmental services produced by natural systems (e.g., water, wildlife, oxygen, carbon dioxide absorption, the products of photosynthesis) are not traded in markets. Thus, people are not paying money in formal markets for many environmental services. Furthermore, people do not commonly experience the purchase of environmental services; therefore, they may not be able to characterize their willingness to pay simply because it is not clearly defined in their minds.

Environmental services need not be traded in markets in order to measure their value in monetary terms. What is required is a measure of how much money people are willing to give up to receive a service of the environment, or how much people need to be paid to give up a service. Observing trades in formal markets allows us to see those values. The lack of observed market trades makes environmental valuation harder.

Use versus Nonuse Values. Two important kinds of environmental values are *use values* and *nonuse values*, sometimes called passive-use values. Use values come from direct human uses of the environment; nonuse values do not require direct use of an environment or its services. The total economic value of an environment's services is the sum of its relevant use and nonuse values.

Use value is the value derived from the human use of an environment's services such as fishing, bird-watching, or other forms of outdoor recreation.[2] An *option value*, a kind of use value, is the value people assign to keeping open the option for future use of a resources, even though they do not currently use it. For example, a person may be willing to pay something to block off the commercial development of a wilderness area to ensure the option of being able to visit the area sometime in the future. *Bequest value*, another kind of use value, is the value people assign to knowing that future generations will have the option to use an environmental resource. Bequest value is measured by people's willingness to pay to preserve the natural environment for the use and benefit of future generations.

Nonuse values are independent of current or anticipated uses of a service. *Existence value*, a nonuse value, is the value people assign to merely knowing that an environmental resource exists, even if they do not expect to see, feel, touch, or use it. Somebody may be willing to pay into a fund to guarantee the continued existence of a specified wilderness area, even though she never plans to use the resource. Existence value has characteristics of a public good: Many people can simultaneously receive benefits from the same environmental system without any person's enjoyment reducing those benefits enjoyed by others.

Expenditure Function

Imagine you inherit a sum of money so large that you never have to work again. You decide to invest in a low-risk, stable financial instrument, which should produce a steady standard of living even in the face of future economic shocks. That means you want to ensure a steady *real* income (utility) to sustain and withstand any future changes outside your control, that is, changes in your *money* income, changes in the *prices* you face, and changes in your surrounding *environmental quality*. Therefore, your investment program must adjust your money income to

---

[2] Use values can include indirect uses. For example, watching a television show or reading a magazine article about some wilderness area produces an indirect use value. People can also receive indirect use values from a resource that contributes to services they directly use. For example, various organisms in the aquatic food chain supply indirect use values to people who fish.

offset any economic or environmental changes. You want that adjustment to reduce your income worries to mere background noise so that you are free to sample the world's offerings.

Two questions arise in connection with your investment program:

1. What does your real income stream look like?
2. How must your program adjust your money income stream to produce that constant real income stream?

Your real income stream should stay constant. Your investment program should be set up to watch external conditions and your money income should be adjusted to those conditions to meet your aim: a steady real income. Here is an example of how your money income must adjust to an outside shock: Suppose the price per airline ticket to your favorite wildlife reserve increases from \$500 to \$1,000. You are accustomed to taking four annual trips. How much must your *money income* increase to compensate for the higher ticket price? The answer: It should rise by enough to let you continue taking your four annual trips at the higher price without reducing your consumption of anything else. An approximation of the added money income you need is (\$1,000 − \$500) times four, or \$2,000.[3] That added money income is your *willingness to accept payment* to offset your losses produced by the higher ticket price.

Here is another example: Suppose air pollution increases where you live, so that maintaining your real income requires incurring moving costs of \$10,000 to find an equivalent place to live with acceptable air quality. Your money income needs to increase by \$10,000 that year. The added money income is your willingness to accept payment to offset your losses produced by the reduced environmental quality.

The two examples hold real income constant by adding money income to compensate for negative shocks. The next example holds real income constant by *subtracting* money income to compensate for an environmental improvement. Without that subtraction, the improvement would increase real income, which, oddly enough, would fail to meet your aims of a steady real income. Remember, your goal is an investment plan that maintains a constant real income.

Suppose water quality improves at a nearby stream from its old, polluted level. You can now fish that stream instead of bypassing it for a more distant stream of acceptable quality. While the nearby stream was degraded, your price per fishing trip at the nearest acceptable stream was \$100 and you made 20 trips per year. Now, the presence of the acceptable close-by site lets you fish for \$20 per trip, so you can make 20 trips per year for \$400 instead of \$2,000. Reducing your money income by approximately \$1,600 per year maintains a constant real income. The \$1,600 is your *willingness to pay* for your gains produced by the improved fishing.

Economics has invented a name for a process that adjusts *money* income to keep constant *real* income: *expenditure function.* It adjusts expenditures up or down in the face of price changes or environmental quality changes so that real income is maintained at a baseline level. The expenditure function is defined as the minimum monetary expenditure needed to reach a fixed, real income (utility) level to compensate for changes in prices or environmental qualities. This income-adjusting expenditure function is expressed as

$$E = E(P, Q, U) \tag{7.1}$$

where E is monetary expenditure needed to hold real income constant, P is a list of goods' prices, Q is environmental quality, and U is the ordinal utility reached under given baseline prices and environmental quality levels. You can use that expenditure function to shift your money income, E, so that you keep a constant U in the face of changes in P or Q. But it has another use: to measure the benefits of environmental improvements. Appendix E a presents an expenditure function in which prices change but environmental quality is unchanged.

[3] The \$2,000 compensates you if you are constrained to continue taking four trips per year at the higher price. But you are free to adjust those trips, so actual compensation required is slightly less than \$2,000. See Appendix E.

Compensating and Equivalent Variation

Consider the airline ticket example again but with some changes. Instead of rising, the ticket price falls from $1,000 to $500. Compensating variation (CV) is the change in income that compensates for (offsets) the welfare change produced by the changed price. If you were charged after the price decline more than the value of the CV produced by the cheaper tickets, you would be worse off. If you were charged less, you would be better off paying the charge and taking advantage of the price reduction. For the gain from a price reduction to equal the consumer's willingness to pay to take advantage of it, the gain must be exactly equal to the CV.

Equivalent variation (EV) is another measure of the benefits produced by the change. It is the amount of money paid to the consumer that *would* keep her at the same new level of welfare *actually* reached by access to the reduced price. It is the amount of income paid to the consumer that produces an *equivalent* welfare change produced by the price change. Most conclusions in connection with welfare change made for compensating variation can also be made for the equivalent variation. In this chapter the CV will be described in more detail, but references to the EV will be made when needed.

The CV and EV can be better understood in terms of benefits individuals receive when they make choices in the face of changes in opportunities. After measuring individual benefits, we can think about the benefits received by a society.

Figure 7–1a shows an analysis for a consumer who faces choices over two representative goods, X_1 and X_2. The straight lines are budget constraints, reflecting the consumer's possible

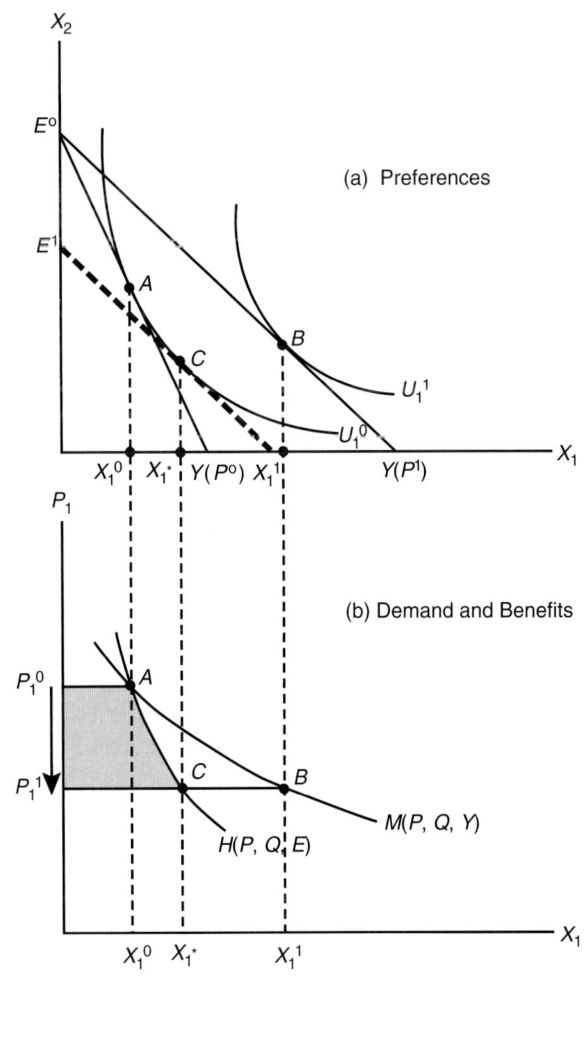

Figure 7–1 Preferences, demand, and benefits from a price change

choices that could be made facing the constraint of a fixed income and known prices for X_1 and X_2. The consumer can afford less of both X_1 and X_2 when her income falls, so reductions in income with unchanged prices shift the budget constraint inward toward the origin and parallel to the original budget constraint line, E^0B. The income level shown by the dashed line through point E^1 is lower than the level shown by its parallel line through point E^0B. The slope of the consumer's budget constraint indicates how many more units of X_2 are available in the marketplace when the decision is made to buy one less unit of X_1. This means that the slope of any single budget constraint is negative and equal to the price of X_1 divided by the price of X_2. As an example, if the price per unit of X_2 is \$50 and the price of X_1 is \$100, the budget constraint has a slope of -2: Two units more of X_2 can be had for each unit less of X_1 bought.

If the price of X_1 rises relative to that of X_2, the budget constraint becomes more steeply sloped, shown by pivoting inward clockwise from the solid line E^0B to the solid line E^0A. When the price of X_1 increases and the budget line pivots inward and clockwise around point E^0, a larger number of units of X_2 can now be gained for each unit of X_1 the consumer is willing to give up. An increase in the absolute price of X_1 with no change in the absolute price of X_2 means that the *relative* price of X_2 falls.

The curved lines in Figure 7–1a are the consumer's *indifference curves.* All points on any one indifference curve show combinations of goods X_1 and X_2 that provide the consumer with equally preferred (indifferent) consumption bundles of X_1 and X_2. Thus, the consumer is equally satisfied (indifferent) between points A and C on indifference curve U_1^0 and on all points on indifference curve U_1^1. A more preferred condition, higher utility or satisfaction is produced by consumption on an indifference curve that is more distant from the origin.

Thus, our consumer prefers *any* point on indifference curve U_1^1, such as point B to *any* point on indifference curve U_1^0, such as point C. This should not be a surprise, because point B shows greater consumption of both goods than point C. Indifference curves are negatively sloped. This occurs because movement between any two points on a single indifference curve requires that increased consumption of one good occurs at the expense of reduced consumption of the other by enough to leave the consumer no better off. If our buyer consumes *more* (less) of both goods, then she moves to a clearly *more* (less) preferred position: An indifference curve cannot slope upward.

Indifference curves are convex in shape, which means they bend inward toward the origin. That convexity reflects diminishing rates of substitution between any two goods. As our consumer consumes more of one good, she becomes increasingly less willing to sacrifice consumption of yet one more unit of the other good. At high levels of consumption of any one good (X_1), added amounts of it become poorer substitutes for reduced consumption of the other good (X_2). For example, the convex shape of indifference curve U_1^0 says that at point A our consumer is willing to sacrifice more units of X_2 to consume an added unit of X_1 than she is at point C. Here is an example: When the quality of the environment is good, an individual might accept a given reduction in quality for a small amount of money. The individual may vote to allow a polluting industry to come to town if it produces a few high-income jobs. But when the environmental quality is already poor, the same individual requires a much higher compensation to accept an equal incremental loss in quality.

Suppose our consumer initially faces the budget constraint going through point E^0 and $Y(P^0)$. She will choose point A on indifference curve U_1^0. Point A is the consumer's equilibrium because she cannot get to a more preferred position by moving to any other point on her budget constraint, so our consumer has no incentive to move.[4] Now suppose an environmental program results in the building of an outdoor recreation site (good X_1) closer to home, so our consumer's price of good X_1 is reduced. This changes her equilibrium to point B on a flatter-sloped budget constraint, going through point E^0 and $Y(P^1)$ and the more preferred indifference curve, U_1^1. Consumption

[4] At that point, the rate of substitution, the slope of the highest indifference curve reached, equals the price ratio, the slope of the budget line.

of good X_1, the recreation site, increases from X_1^0 to X_1^1. The consumer is definitely better off going from *A* to *B*.

Now suppose our consumer is charged a lump sum of money sufficiently high to eliminate her entire gain produced by the reduced price (remember the expenditure function). This loss in money income shifts her budget constraint in a parallel movement from *B* to *C*, pulling her back to the original indifference curve, U_1^0. That was the highest curve she reached under facing the original high price without the new recreation site. However, in the face of *reduced income* combined with a *reduced* price of X_1, our consumer chooses point *C*, consuming X_1^* of good X_1. Point X_1^* lies between X_1^0 and X_1^1 of X_1. The vertical distance between the two parallel budget constraints (difference between points E^0 and E^1) is the maximum income the consumer can be charged to offset the welfare gain produced by the reduced price. This distance measures the CV produced by the price reduction.[5] It is the maximum amount the consumer is willing to pay to take advantage of the reduced price.

Income and Substitution Effects. The information contained in Figure 7–1a shows the total effect on the buyer's consumption of X_1 produced by the decrease in the price of X_1 as the change from X_1^0 to X_1^1. That change can be decomposed into two separate effects: a substitution effect and an income effect.

The substitution effect is represented in Figure 7 1a as the change in quantity demanded from X_1^0 to X_1^*. That change tells us the effect of a change in the price of X_1 on the consumption of X_1 when the consumer's income is *compensated* in such a way that she remains on the original indifference curve U_1^0. For example, if her actual income is $50,000, but that income *could be* reduced to as low as $45,000 *after* the price reduction with no satisfaction loss, then the opportunity to take advantage of the reduced price has produced for her a monetary benefit of $5,000. Her compensating variation of the price change is $5,000. The substitution effect always causes the demand for a good to change in the opposite direction from the price change. The substitution effect of a reduced price of X_1 is an increase in quantity demanded of X_1.

The income effect is shown in Figure 7–1a as the change in quantity demanded from X_1^* to X_1^1. It occurs because the reduced price of X_1 increases the consumer's *real income.* The figure shows that the increase in real income increases the demand for X_1.[6] Therefore, like the substitution effect, the income effect produced by a price reduction causes the consumer to increase her demand for X_1 from X_1^* to X_1^1.

Demand Schedules. A demand schedule (see Chapter 3) tells us how utility-maximizing consumers respond to changes in opportunities. The economic forces underlying those demand schedules reveals much about the value of environmental programs. In fact, both the CV and the EV resulting from an environmental program are closely related to the demand schedule. So learning more about the demand schedule will help measure the benefits of programs that influence individual behavior. In turn, measuring benefits can better contribute to environmental policy choices when they are compared to costs.

The slopes of the budget constraints in Figure 7–1a show old and new prices of X_1 and indicate the points tangent to these budget constraints with indifference curves. Those tangent points indicate how much of X_1 (and X_2) the consumer chooses to buy at each price of X_1. So the figure tells us everything we need to know about two points on the consumer's demand schedule for X_1. In fact, we know the quantity of X_1 our consumer demands after the price change, both if her real income (*welfare*) is held constant and if her *money income* is held constant. We have the information to produce two kinds of demand schedules, which appear in Figure 7–1b between points *AB* (money income constant) and between points *AC* (welfare constant).

---

[5] The vertical distance in the graph is the CV divided by the price of X_2, assumed to equal 1.

[6] Increases in real income increase the demand for a *normal* good. But people buy reduced quantities of *inferior* goods (e.g., used clothing, poor cuts of meat, and low-quality housing) as their real income increases.

The first demand schedule, which comes from the curve in Figure 7–1b between points *A* and *B* is a conventional demand schedule of the kind that appears in most microeconomics texts. This schedule is called the Marshallian[7] demand schedule and includes both the substitution and income effects produced by changes in the price of X_1. Efforts by economists to statistically estimate relations between the price of a good and quantities bought—holding constant money income, prices, and preferences—are attempts to empirically estimate Marshallian demand schedules. The label for the Marshallian demand schedule, *M* (*P, Q, Y*), in Figure 7–1b, means that the Marshallian demand for X_1 depends on goods' prices, *P;* their environmental qualities, *Q;* and money income, *Y.*

The second demand schedule in Figure 7–1b connects points *A* and *C*. It comes from adjusting money income[8] to any *new* price, so that adjusted money income (the budget line) stays tangent to the indifference curve reached under a given set of initial conditions. Initial conditions are the prices, income, and environmental quality that produced tangency at point *A* on indifference curve U_1^0. This special demand schedule strips away the income effect, leaving only the substitution effect produced by changing conditions. This may seem like a strange and unnatural thing to do, but you will see how it can be used for analysis of the benefits of environmental improvements.

This second demand schedule is called the income-compensated or Hicksian[9] demand schedule. Because Hicksian demand schedules have their income effects stripped away, they are typically more steeply sloped than Marshallian demand schedules. That is, the Hicksian demand is less sensitive to a price or quality change, as shown by the demand schedule passing through points *A* and *C* in Figure 7–1b.

The label for the Hicksian demand, *H*(*P, Q, E*), means that the Hicksian demand for X_1 depends on prices, *P*; qualities, *Q*; and expenditure, *E* (not constant money income). That expenditure budget line is varied so that it is always tangent to the original indifference curve, U_1^0. Because expenditure compensations are rarely performed for a policy that hold individuals' welfare constant at fixed levels, it is hard to find the data to support Hicksian demand schedules. Consequently, there is little opportunity to estimate them directly using statistical methods. But Hicksian demands can sometimes be calculated from Marshallian demands, which are directly estimated from observed data (see Appendix E).

Consumer Surplus Approximation

Movements down the Hicksian demand schedule, say, from prices P_1^0 to P_1^1, are a result of the price changing while compensating the consumer with a money income change of just the right size to permit her to remain tangent to her original indifference curve. This lump-sum payment can be measured graphically as the vertical distance between the two parallel budget constraints in Figure 7–1a, or the difference between money income at points E^0 and E^1. It can also be measured as the compensating variation, the shaded area bound by the two prices, P_1^0 and P_1^1, on the Hicksian demand schedule in Figure 7–1b, or the area $P_1^0P_1^1CA$.

The change in expenditure that offsets the welfare change resulting from the price change can be measured with the Hicksian demand schedule as the consumer's CV. It is the maximum amount the consumer is willing to pay to take advantage of the price reduction. Hence, the Hicksian demand schedule permits measurement of the CV associated with price changes. However,

---

[7] Named after the British economist Alfred Marshall (1842–1924). See the University of Cambridge (2004) Alfred Marshall Society.

[8] It shifts the budget line, placing it tangent to the point on the original indifference curve that keeps the line parallel to prices that occur under new conditions.

[9] Named after the British economist John R. Hicks (1904–1989). See Nobel Foundation (2004) John R. Hicks.

it should be no surprise that estimates of the Marshallian schedules are much more easily available for use in the conduct of a benefit-cost analysis for environmental policy.[10]

These two demand schedules usually differ, although in many cases the difference is small. Where they do differ, the use of the area between two price horizontals under a *Marshallian* demand schedule (area $P_1^0 P_1^1 BA$) to measure a consumer's welfare gain, called the consumer surplus (CS), produces an incorrect estimate of CV and therefore misestimates willingness to pay for a price reduction. Figure 7–1b shows that the two areas left of each of the two corresponding demand schedules in fact produce different measures of the benefits of the price change. The difference between the two areas is the triangular-like area *ACB*.

In the case of one good's price decrease, the Marshallian CS is larger than the Hicksian CV whenever the income effect is positive. The error in estimated willingness to pay that results from using Marshallian, rather than Hicksian, demand schedules to measure consumer benefits depends on the size of the income effect associated with a price change. For most services the income effect is small, so the CS is close to the CV. This occurs whenever the price change is small and the service in question absorbs a small amount of the consumer's income. When the income effect is zero, the two are identical.

For this reason, BCAs of policies that affect most homogeneous commodities will be close to correct by the use of Marshallian rather than Hicksian demands. However, the error can be of considerable importance for a BCA of a policy that would produce large price changes (e.g., in housing or in major changes in the quality of important, unique environmental services). Mathematically, CV, the difference in the consumer's actual money income and the expenditure function required to sustain an initial (prepolicy) utility level U^0, can be expressed as

$$CV = Y - E(P, Q, U^0) \tag{7.2}$$

The second welfare measure, equivalent variation (EV), is defined as the difference between the expenditure function required to sustain a terminal (postpolicy) utility level, U^1, and the consumer's money income. Its equation is

$$EV = E(P, Q, U^1) - Y \tag{7.3}$$

The CV can be compared with both the EV and the CS in terms of four criteria: practicality, uniqueness, consistency with the person's actual welfare change, and implied property rights.

1. A practical issue is whether each welfare measure can be calculated from observed market data. The answer is a qualified yes for the ordinary CS. If the Marshallian demand curve can be estimated statistically, calculating the area between any two price lines is straightforward. But measurability is harder for the EV[11] and CV, because they are both areas under Hicksian-compensated demand curves. All three welfare measures meet the standard of being in monetary units, which is preferred to measuring benefits as a unitless index. So the measured value produced by each can facilitate comparison with real monetary resource costs of a proposed environmental program.

---

[10] If a Marshallian demand schedule is estimated consistent with choice theory (maximizing behavior and a budget constraint), its corresponding Hicksian demand can be computed by subtracting income effects. This means that if the Marshallian demand is known to be consistent with choice theory, the original utility function, up to a monotonic transformation of the utility index, can be calculated. From that, the Hicksian demands can be calculated and the compensating variation can be measured for any desired price change. Note that this method does not directly *estimate* the Hicksian demand, but rather *infers* it from the Marshallian demand and its underlying behavioral assumptions (see Appendix E).

[11] The EV is not shown in Figure 7–1b to reduce clutter. Like the CV, it is a similarly shaded area passing through point *B*, parallel to *AC*, and intersecting the initial price P_1^0. The EV measures the change in money income that produces the same welfare gain as the policy change actually produced. It is the amount by which the original budget line needs to be adjusted to be tangent to the new indifference curve.

2. Both the CV and EV are unique. Once a price change or quality change in the environment is known, both the CV and EV assign only a single monetary value to the change. However, depending on the order in which the price or quality changes occur, the CS can produce many values. For this reason, the CS is sometimes said to be path dependent, whereas the CV and EV are path independent.

3. The CS is not consistent with any definition of welfare change. That's the bad news. The good news is that the CS typically lies between the CV and EV measures for most proposals. And the CS is equal to both CV and EV when the income effect is zero.

4. Whatever the direction of the price or quality change, the CV takes the status quo utility as the reference point. If the CV is used to value the improvement, it is presumed that the individual has no right to make purchases at the new set of prices or qualities, but must instead pay for that right. By contrast, the EV is that payment, which if given to or taken from the individual, would make her willing to take money and forego the price or policy change. Use of the EV presumes that the individual has a right to the new price or quality and should be compensated if her access right is taken away. The choice between the two welfare measures is an ethical one; that is, it depends on which distribution of property rights is believed to be ethically correct.

Individual Demand for Environmental Improvement

An individual's demand for and benefits from environmental improvement is based on individual preferences, in which environmental quality affects her welfare directly. Suppose her preference ordering looks like this:

$$U = U(X, Q) \tag{7.4}$$

where X is a list (vector) of private goods quantities $(X_1, \ldots X_n)$, and Q is an objective measure of environmental quality. Including environmental quality in her preference ordering means the individual perceives the effects of changes in environmental quality. For example, if an increased number of parts per million of an air pollutant stings her eyes, she must be aware of the irritation so that she notices the effect of reduced pollution, prefers it, and feels better. She does not necessarily know the underlying cause of the air pollution nor does she need to know its measured level. This discussion about perception is important because if an individual is unaware of the effects of changes in quality, her benefits from the improvement cannot be measured directly using the methods described here. As a counterexample, individuals may not perceive the effects of reduced long-term exposure to air pollutants on death or sickness, but they may still benefit from those reduced rates.

Sometimes the demand function for a private good[12] contains environmental quality as a variable:

$$X_i = X_i\,(P, Q, E) \tag{7.5}$$

Suppose this demand equation for some ith good has been estimated statistically and that it came from the conditions of maximizing X utility subject to a budget constraint. If the private good and environmental quality are *weak complements,* this demand function can be used to determine willingness to pay for an environmental improvement. Weak complementarity between the private good and environmental quality means that when the quantity demanded of the private good is zero, the marginal benefits from additional Q is zero. Weak complementarity occurs in a number of useful situations. For example, people who fish a particular stream have positive marginal benefits from using the resource (use value) when an environmental policy improves the water quality. For people who do not fish the stream, or make any other use of it, the marginal benefits of improved water quality are not weak complements to the resource.

---

[12] The demand should be Hicksian compensated (see Appendix E).

Analytically, weak complementarity requires two conditions. First, there must be a price, P_1^{**}, for X_1 so that at that price, its demand function looks like this:

$$X_1(P_1^{**}, Q, E) = 0 \tag{7.6}$$

This means there must be a price that is suitably high to reduce demand to zero. The second condition requires that at that high price, P_1**, for the expenditure function defined as

$$E = E(P_1^{**}, Q, U^0), \tag{7.7}$$

the change in E is zero when changes in environmental quality, Q, occur at that high price, P_1^{**}.

What happens when both conditions hold true? Suppose that for a given level of Q, the demand curve for X_1, has been estimated statistically. This demand curve is labeled $H(P, Q^0, E)$ in Figure 7–2.

The market price of X_1 is P_1^1 and stays constant throughout our analysis. The CV (total benefits) associated with the use of X_1 is the area ABC under the demand curve $H(P, Q^0, E)$. Now let the environmental quality be improved to Q^1. The effect of that increased quality is seen in the demand for X_1, shifting it outward to $H(P, Q^1, E)$. Measuring the benefits produced by the quality change involves three steps. First, beginning with the inside demand curve at the old quality, $H(P, Q^0, E)$, let the price increase from P_1^1 to P_1^{**}, which eliminates the consumer's use of X_1. In moving from point C to zero comsumption of X_1, in order to be no worse off after the price increase, the consumer must be compensated by an amount of income equal to area ABC. Second, suppose a policy action improves environmental quality and the demand curve shifts out to $H(P, Q^1, E)$. Under the assumption of weak complementarity, the consumer's welfare is unaffected since the high price has reduced her consumption of X_1 to zero. At this point she has no requirement for compensation: She gains nothing from the improved quality because she is consuming none of the private good tied to that quality. Third, after the quality improvement, reduce the price back to P_1^1. The consumer is better off by a monetary amount equal to area ADE. The total effect of these three changes is a net increase to the consumer by the area $BCED$, which equals area ADE minus ABC. That four-sided area, the area between the two demand curves, measures the benefit of the change in Q.

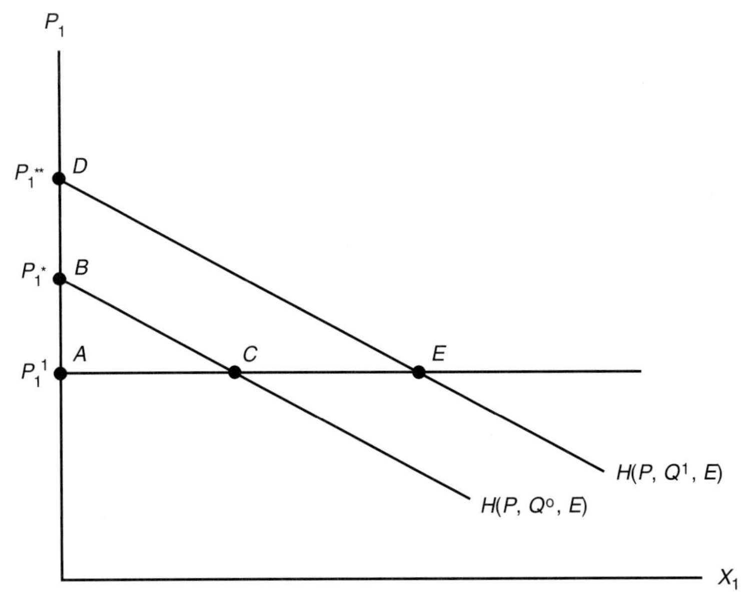

Figure 7–2 Benefits, demand, and quality

Figure 7–3 Expenditure function and benefits

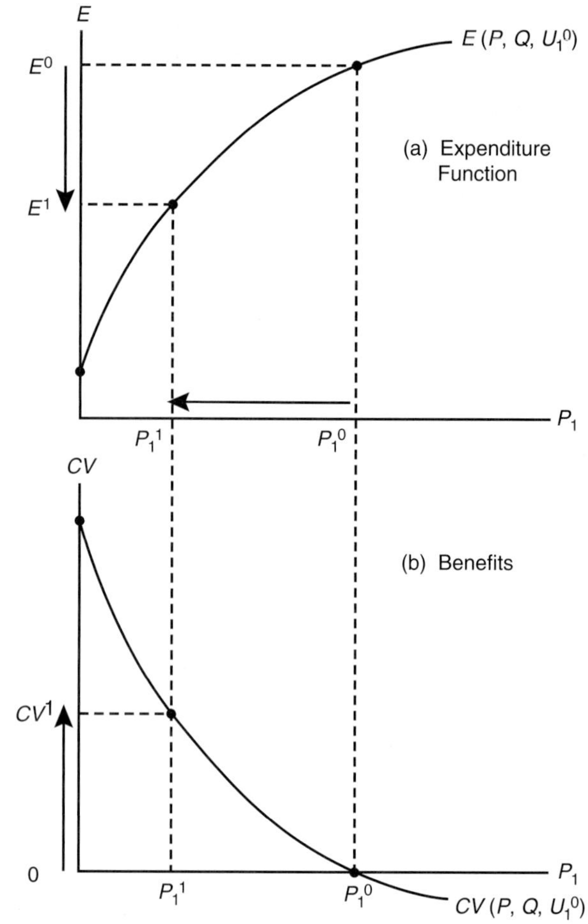

Valuing Price and Quality Changes

The expenditure function adjusts income to compensate for changes in prices or qualities. Figure 7–3a shows a typical expenditure function for a single individual, labeled $E(P, Q, U_1^0)$. The figure shows what you would expect: Beginning at the initial price of good 1, P_1^0, suppose the price falls to P_1^1. This price reduction increases the individual's welfare by increasing the purchasing power of her constant money income. Reducing her expenditure from E^0 to E^1 offsets her welfare change, as shown in the figure. The variables inside the parentheses mean that the expenditure function adjusts money income for changes in prices (P), changes in environmental quality (Q), or changes in anything that affects the individual's initial standard of living U_1^0.

Figure 7–3b illustrates that the CV is a mirror image of the expenditure function. The CV, which is money income minus the expenditure function, translates the expenditure function into measurable benefits, which supports benefit-cost analysis of environmental policy proposals. Starting again from the prepolicy price of P_1^0, the CV is zero until there is some price change. As the price falls to P_1^1, the CV has a value of CV^1. Since the CV is simply the baseline constant money income minus the expenditure function, the CV label $CV(P, Q, U_1^0)$, shows that it is a mirror image of the expenditure function and depends on the same factors.

Aggregation and Social Welfare

The four panels in Figure 7–4 show how the analysis of the benefits of environmental improvements can be used to formulate economically efficient environmental policies. Figure 7–4a shows how the expenditure function varies with environmental quality changes (extending the analysis

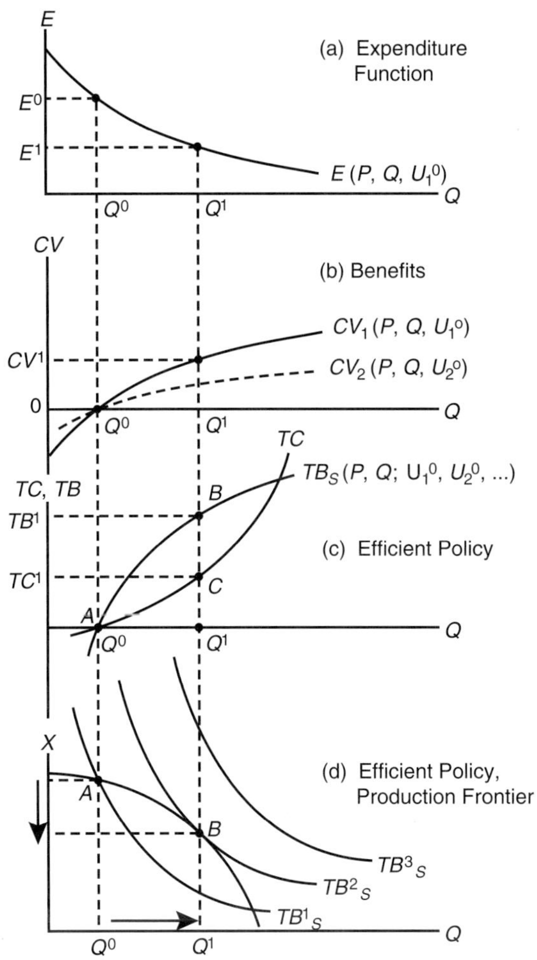

Figure 7–4 Efficient environmental quality improvements accomplished through analysis of benefits

of Figure 7–3a which dealt only with price changes). Raising environmental quality from Q^0 to Q^1 reduces the expenditure needed to sustain the individual's standard of living $U_1{}^0$. That expenditure reduction is shown falling from E^0 to E^1.[13]

Figure 7–4b shows the CV produced by the quality change as the mirror image of its expenditure function in panel a. Beginning at a CV of zero, when baseline quality is Q^0, the CV increases to CV^1 when quality increases to Q^1. The CV is increasing at a decreasing rate with quality increases, reflecting decreased marginal benefit (but increasing total benefit) from added increments of environmental improvements.[14]

The dashed line in Figure 7–4b indicates that not all individuals have identical preferences for and values of environmental improvements. Different individuals exposed to the same environmental quality improvement will receive different benefits from that improvement for many reasons. Individual 2 is shown as having a smaller CV response (CV_2) than individual 1 to the

---

[13] The slope (partial derivative) of the expenditure function with respect to quality measures the amount that expenditure or income must change to hold real income at a constant level as quality changes. That slope traces out a marginal willingness to pay (or willingness to accept payment) for quality improvements, sometimes called a compensated inverse demand function for quality. The CV and EV can be measured as areas under that inverse demand function between the prepolicy and postpolicy quality levels. The CV (EV) is based on prepolicy (postpolicy) real income.

[14] We expect the slope (marginal benefit) from the environmental improvement to fall at high levels of environmental quality. However, when existing quality levels are poor, the first few increments of improvement may increase the slope, reflecting increasing marginal benefits.

same environmental improvement from Q^0 to Q^1. That smaller response is traced back to where analysis began in Figure 7–1a. Individual responses to a single environmental quality improvement are a result of many factors: individual preferences, income levels, demographic factors, locations, substitute prices, existing environmental and economic conditions, and personal sensitivities. As a practical matter, one could imagine the estimation of a family of CV response functions shown in Figure 7–4b, one for each individual or possibly one for each of several groups of people.

Suppose we have measured the CV, either positive or negative, for all individuals, for which CV_1 in Figure 7–4b is an example. How can we use this information to make choices about environmental policy decisions? Or to pose the question more broadly, what is the appropriate method to connect the welfare of *individuals* to produce a number that reflects the *social* welfare? Welfare economists have found four ways to approach this question. All four methods assume that social values are the sum of individual values. Nyborg (2000) raises important questions in connection with individuals who wear two hats: (1) as a consumer pursuing self-interest and (2) as an ethical observer or citizen judging matters from society's point of view.

1. The first way to approach the question is based on the *Pareto criterion,* which states that no policy change shall be enacted unless at least *one* person is made better off (receives a positive welfare change) and *no* person is made worse (suffers a negative welfare change). Acceptance of that standard rules out any attempts to make common-denominator comparisons of gainers and losers, because it says there can be no losers. This is certainly an admirable policy evaluation standard, but it places the policymaker in chains: Few if any policies will ever be enacted, because virtually all policies produce some welfare losses to some people.

2. A second approach is to say that a policy should be enacted if people who gain from it could fully compensate for the welfare losses of those who lose by the policy, and still have something left over. This is called the *Potential Pareto Improvement* criterion, since nobody loses if the compensation is paid. Should that compensation actually be paid? If you answer yes, then this criterion becomes essentially the same as the Pareto criterion and government should enforce taxes or other transfers to be sure nobody suffers a welfare loss. If you answer no, then you believe that all individual welfare changes are morally equal (that is, all gains and losses have equal moral standing). Thus simple algebraic sum of gains and losses for all affected people should be the basis for policy decisions. This is the simple rule of measuring total economic efficiency, which compares total benefits to total costs—the method most often used in a benefit-cost analysis (see Chapter 5).

3. A third approach to the question deals with the question of fairness in the distribution of benefits and costs. First, does the policy pass the Potential Pareto Improvement criterion? Second, is the distribution of those benefits and costs fair? This approach requires definition of the term *fair,* which poses a difficult challenge indeed.

4. The fourth approach is an outgrowth of the third. It attempts to make specific social judgments regarding equity (fairness) and to introduce equity criteria systematically into the evaluation of environmental policy. One way this can be done is to set up weights in the evaluation of benefits and costs according to the relative deservingness of different people. People who are most deserving of receiving benefits (e.g., the deserving poor or the vulnerable) receive the highest weights on benefits gained or costs incurred. A problem with this approach is deciding what the weights will be and who will determine them. Remember, the second approach assigns a weight of one to everybody.

Figure 7–4c shows how the common aggregation method of ordinary benefit-cost analysis can be used. It shows that a simple algebraic summing of individual CV responses over all affected people produces a total social benefits response, labeled TB_s. Begin at the baseline quality level in which there is no change (Q^0). Then for each quality improvement, such as an increase to Q^1, add vertically the total CV over all affected individuals. The resulting total social benefit for

Example 7–2
Does Valuing the Environment Support Environmental Policy?

Economically efficient environmental regulation requires measuring the benefits and costs of environmental standards and policies. Benefits and costs are correctly measured by the willingness to pay or willingness to accept payment. Environmental economist Maureen Cropper examined several economic benefits studies conducted between 1974 and 1999 to see if they were being used to support environmental policy.

Recreation benefits constitute one of the largest categories of benefits from improving the quality of surface water. Improvements in water quality and/or in fish catch may increase participation in water-based recreation and can enhance the value of a recreation experience. By 1974 the travel cost method had been developed and used to estimate the value of recreation sites. However, it had not been used to value quality improvements, which were a far more important environmental policy issue.

In 1974, Karl Goran Maler broke new ground by showing that a shift in the demand for site visits in response to a change in the site's environmental quality could be used to value quality change (Mäler, 1974). He showed that the change in compensating variation corresponding to the shift in the demand curve measures the value of site quality where the site's quality is weakly complementary to the demand for site visits. Weak complementarity means that there is a finite price that reduces visits to zero, and at that price, the marginal benefits from improved site quality is zero to the potential visitor.

Work by Carson and Mitchell (1993) has been used by the U.S. EPA to value national water quality improvements. Their study asked a sample of U.S. households to value improvements in water quality at several U.S. sites from boatable to fishable and from fishable to swimmable. Respondents were then asked to allocate a part of that value to water quality improvements in their home state. Benefit estimation has come a long way but it has far to go. Policymakers have called for a greater number of applied studies to value the benefits as well as the costs of pollution control.

Source: Cropper, M. L. 2000. Has economic research answered the needs of environmental policy? *Journal of Environmental Economics and Management.* 39: 329–350.

quality level Q^1 is shown to be TB_s. The variables inside the brackets of TB_s indicate that existing prices (P), existing quality (Q), and existing standards of living of all affected people ($U_1^0, U_2^0, \ldots$) are important factors influencing total social benefits of a given environmental quality improvement.

The bottom curve in Figure 7–4c, TC, is the total cost of the proposed environmental improvement. If that cost can be measured and if economic efficiency is a morally acceptable standard for decision making, then Q^1 is shown to be the new quality level, improved from Q^0, that maximizes net economic efficiency benefits (benefits minus costs).[15] See Example 7–2.

Figure 7–4d shows the same economically efficient environmental policy, but with a different emphasis. This panel shows what we sometimes forget, namely, that environmental improvements typically come at a cost of something else, typically reduced output of certain private goods (e.g., goods that pollute or public programs that are displaced by public environmental programs). Figure 7–4d shows the highest possible total social benefit that can be reached starting at point A at the current mix of private goods (X) and environmental quality (Q^0). Point B is where benefits exceed costs by the largest amount, in which total (net) benefit level TB_s^2 is reached[16] when quality is improved to Q^1 at the cost of private goods being reduced from X^0 to X^1.

[15] The figure is drawn so that the most efficient policy improves the environment from Q^0 to Q^1. In fact, there is nothing sacred about the configuration of these diagrams. Depending on the nature of the policy under consideration and the resulting shape of the benefit and cost curves, the most efficient policy may produce *greater* quality, *no change*, or *reduced* quality compared with existing levels.

[16] Each TB contour level in Figure 7–4d includes total social benefits defined over Q and X.

■ Methods for Measuring Environmental Values

Valuing the environment is an important task of the discipline of economics. Holl and Howarth (2000) and King and Mazzota (2002) provide excellent overviews of measurement methods. All approaches for estimating monetary values of the environment are based on estimating compensating variation, consumer surplus, or equivalent variation. Three approaches for measuring these values are in widespread use:

- Revealed values
- Imputed values
- Expressed values

The values of some environmental services can be measured from people's revealed behavior, by simply observing market prices. Some ecosystem products, such as fish or timber, are traded in markets. Thus, their values can be estimated by estimating consumer surplus as with any other market good. Other ecosystem services, such as clean water, are used as inputs in production; their values may be measured by their contribution to the profits made from the final good. Some ecosystem or environmental services, like aesthetic views or many recreational experiences, may not be directly bought and sold in markets. However, the prices people are observed paying in markets for related goods can be used to estimate their values. For example, people often pay a higher price for a home that has a view of the ocean. People incur added costs and time to travel to a special spot to fish or watch birds. These kinds of expenditures can be used to place a lower bound on the value of the view or the recreational experience. Economists refer to methods involving revealed values as revealed preference methods, since people's values are based on preferences revealed in the marketplace from their actual behavior. Four such methods are the *market price method, productivity method, hedonic pricing method,* and *travel cost method.*

Imputed values are values of an environmental service found by estimating what people are willing to pay for *costs avoided* by the presence of an environmental service. People might be willing to pay to avoid damages, to replace damaged resources, or substitute other resources for damaged resources, all of which might compensate for lost services if the resource were gone. Spash (2000) and Vatn (2000) examine the difficulties in interpreting imputed values when people are not willing to make substitutions.) For example, wetlands provide important services by protecting people from floodwaters that can damage homes, lives, and property. Can imputed values be used to value costs avoided by existence of a wetland? The amount that people pay to build dams, buy flood insurance, and move homes away from flooded areas are evidence of the value of flood protection offered by a natural wetland. By providing the wetland these costs avoided are a benefit of protecting the wetland. This method for valuation is sometimes known as the *avoided costs method.*

Expressed values are measured through the use of surveys. The economics literature refers to these as stated preference methods because values of the environment are based on what people state when they are surveyed. Most environmental services are not traded in markets and are not closely related to the purchase of any traded services, so people are typically unable to reveal what they are willing to pay through their market activity. Where this inability to reveal values occurs, surveys are sometimes used to ask people directly what they are willing to pay for an environmental service defined in a survey. As a substitute for revealing values, people are sometimes asked to make tradeoffs among different services, from which their willingness to pay can be estimated. This approach is called the *contingent valuation method.* Contingent means true only under certain conditions; contingent valuation refers to a valuation exercise contingent on the presence of an actual market. Another method that transfers results from old to new locations is called the *benefit transfer method.*

Market Prices

Market prices are good measures of economic values of environmental services that are bought and sold in commercial markets. Market prices can be used to value changes in the quantity or quality of an environmental service. Where quality differentials are reflected in price differences, the increase in market price is a measure of the economic value of a higher-quality environment. The price of services traded in commercial markets measures the incremental benefit of one more unit supplied by an environmental program or policy. For example, suppose an oil spill or other chemical discharge into a water body causes the closure of a commercial fishing area. Policy analysts wish to compare the benefits of cleanup to the benefits of having a policy that would prevent future discharges. In this case the market price method can be used, because the main resource affected is commercially harvested fish, and data on market prices of fish are likely to be available.

Implementation. The aim is to measure total economic gain from the additional fish population and fish harvest that would occur if the pollution is cleaned up or if a policy is enacted that blocks future pollution episodes from occurring. The market price method uses prevailing prices for environmental services that are traded in markets, such as commercially harvested fish or timber. The market price represents the value to consumers of an additional unit of that service; the service is sold in a competitive market in which buyers can buy all they wish at the going price.

Assuming that the added fish harvest produced by the policy is too small to affect the price of harvested fish, there is no impact on fish consumers but there is gain in net income for fish harvesters.[17] That gain in net income is the gross benefit of the policy, measured as the increase in revenue from the added fish harvested, minus the increased cost to producers of harvesting the larger number of fish (e.g., more boats, larger boats, larger crews, more fishing gear). Results produced by the analysis can be used to make more economically efficient decisions. Proposed decisions can be designed by comparing the gross benefits of programs that would increase harvestable fish populations to the gross costs of such programs (see Chapter 5).

Advantages. A major advantage of using market prices to measure environmental protection, conservation, preservation, or protection values is that those prices, where relevant, reflect an individual's willingness to pay for benefits of services that are traded in the marketplace. Good examples include timber, power, fish, hunting expeditions, and firewood. Thus, measured benefits of environmental programs whose outputs are commercial products are likely to be easy to measure. Another advantage is that data on price, quantity, and production costs are easy to find where there are well-established markets. A third advantage is that the method uses data of actual consumer preferences based on actual choices revealed in the marketplace. This advantage is important, for choices made in the marketplace are hard for consumers to fake: What the consumer buys and fails to buy provide a signal of what she values. By contrast, when environmental values are based on hypothetical choices, individuals may behave strategically by inflating values for programs they support or by deflating values for programs they do not support (see Appendix F). The fourth advantage is that the method of market prices uses standard methods that have been in widespread use for years.

Limits. For many natural environments, a large percentage of the services provided to people are not available in commercial markets. In fact, conserving or protecting a natural environment often means keeping nature's products from reaching the market. Even if the service in question is commercially marketed, heavier burdens are placed on the market price method when it is used to measure the economic value of larger scale changes likely to affect price.

---

[17] If the environmental improvement affects harvested fish by enough to reduce the market price of fish, then the gain in consumer surplus must be added to the change in producer net income. The gain in consumer surplus can be approximated by the reduction in price multiplied by the average of the existing and new level of fish supplied to the market if the demand is linear. If it is not linear, a demand function must be estimated statistically to measure the gain in consumer surplus. This may require time series data on the quantity demanded at various prices, in addition to data on other factors that might affect demand, such as income and demographic factors.

Productivity

In some cases the environmental service in question is an input and is used in the processing of a commercial product. Use of that service increases the producer's productivity and raises her net income. For that reason, the producer's change in net income is a good measure of the service's economic productivity. So change in net income can be used to estimate the economic value of environmental services that contribute to the productivity of commercially marketed goods. The productivity method can be applied when the products or services of an environmental system are used, jointly with other inputs, to produce a marketed good. A good example is a water quality improvement which (1) increases the productivity of crop-irrigated lands, (2) reduces the costs of treating municipal drinking water, and (3) increases the productivity of industries that require clean water. The economic benefits of a policy that increases water quality can be measured by the increased farm income from increased agricultural productivity, the reduction in costs of supplying drinking water, and the increase in net income of a computer-chip maker who uses higher-quality water as an input.

Suppose a river that provides water for a city's drinking water system is being polluted by toxic chemical releases from an upstream source. Policy analysts wish to determine the economic benefits of a program that would improve the river's quality by regulating those chemical releases. Ultimately, analysts wish to compare those benefits to the added costs associated by restricting the polluter's activities. The change in net income method can be used because in it environmental quality is a productive input (factor of production) that affects the cost of supplying a commercially supplied good, in this case, municipal drinking water.

Implementation. We start by determining how the product (in this example, pure drinking water) depends on various inputs. This could be done by estimating a production function for the output, that is, drinking water purified to an acceptable level. The *production function* is the mathematical relation between the inputs (river water of a known quality, chemicals, filtration) and the output (treated drinking water).

Next, we estimate how the cost of purified water decreases when the quality of river water increases, using the production function estimated in the first step. The idea is to estimate the reduced quantities of purification chemicals and filters required to deliver the known product, acceptable drinking water, as the quality of river water increases. These reduced quantities of complementary inputs are then multiplied by their prices.

Economic benefits of protecting the river from chemical pollutants are calculated as the reduction in costs of delivering acceptable quality drinking water. For example, if all pollutants are eliminated, the river water will need the least treatment and purification costs for drinking water will be the lowest. These minimum costs can be compared to the costs of purifying water where pollutants are progressively less controlled. The savings in purification costs are an estimate of the benefits of reducing the river pollutants. The results of the analysis can be used to compare the benefits of achieving different levels of water quality in the river with the costs of programs to improve water quality by reducing or eliminating the polluting runoff.

Information Requirements. To apply the productivity method, data must be collected about how changes in the quantity or quality of the environmental service affect net income by producing the final good. The method can be applied in two situations:

1. The environmental service in question is a perfect substitute for one or more inputs. Increased water quality in a river means that less chemical treatment and less filtering are required for treating the water to meet drinking standards. In this case, an increase in water quality results in decreased costs for the substitute inputs. So the benefits of increased water quality can be measured by the decreased chemical and filtering costs.

2. The environmental service is an imperfect substitute for other inputs, which means that the proportions of several other inputs used to produce the output are changed. For example, the improvement in water quality may improve the productivity of irrigation water for agriculture, even if there are no simple substitutions. Instead, greater crop yields, shifts in cropping patterns favoring higher-valued, more water-sensitive crops, and entrance of marginal land into crop production may all result from better quality water. If the market price of crops is unaffected by water quality improvement, water quality benefits should be estimated from changes in net income resulting from increased productivity of several of the nonwater inputs.

Advantages and Limits. The productivity method has these advantages: It is widely accepted, well understood, and fairly straightforward to apply. Data requirements are limited and relevant data may be readily available. The method also has shortcomings: It is limited to valuing environmental services that can be used as inputs in further production of marketed goods. When there is a need to value the services of an environmental system, not all those services will be inputs to the production of marketed goods. Whenever this is the case, use of change in net income risks understating the total economic value of the improvement to society. Data may be costly: Information is required on production functions, that is, physical relations between actions that improve quality or quantity of the environmental service and outcomes on production. In many cases these relationships may be poorly understood; even if they are understood in concept, the actual production function needed may be missing or expensive to estimate.

Hedonic Pricing

For purposes of this book, hedonic pricing refers to methods used for attaching value to environmental assets that directly affect market prices. (The term *hedonic* suggests pleasure; increased pleasure is influenced by reduced prices.) Hedonic pricing is most commonly applied to variations in housing and other real estate prices that reflect the value of local environmental attributes. Hedonic pricing can be used to estimate economic benefits produced by (1) environmental quality, such as controlling air, water, or noise pollution; and (2) environmental amenities, such as access to outdoor recreational areas.

The underlying premise of the hedonic pricing method is that the economic value of an environmental improvement can be inferred from the price differential of certain environmentally sensitive characteristics associated with commercially marketed products. For example, the price of a home surrounded by polluted air may sell at a discount compared with the price of an equivalent home surrounded by unpolluted air. Therefore, we can value the individual characteristics of a home (e.g., size, number of rooms, number of garages, air quality) by looking at how the price varies with changes in its characteristics. The hedonic pricing method is most commonly used to value environmental improvements that affect the price of residential property.

For example, suppose that policy analysts wish to find information on the benefits of a program that would preserve an area's open space as it continues to shrink in the face of urban development. The hedonic pricing method is well suited for this kind of application for two reasons: (1) Housing prices in many areas increase in the face of greater access to open space. (2) Factual data on real estate sales and open space parcels are typically available. Other methods to determine economic values are possible. For example, if the open space in question is used mostly for outdoor recreation activities, the travel cost method may be considered. Survey-based methods such as the contingent valuation method are also possibilities. However, both of those alternatives are likely to require greater cost and time to apply.

Implementation. In implementation of hedonic pricing, a good place to start is the collection of data on residential property sales in the area: (1) selling prices and locations of houses; (2) property characteristics relevant to selling prices, such as lot size and number and size of rooms; (3) characteristics of the neighborhood likely to affect a house's selling price (e.g., transportation access, shopping centers, distance to central work areas, crime rates, school quality); and (4) various measurable environmental characteristics thought to affect price.

After the data are collected, assembled, and compiled, the next step is to use the data to statistically estimate a mathematical function that relates property values to the various property characteristics. The usual method for doing this involves *econometrics.*[18] Using "distance to open space" as the criterion, the intent is to separate out all factors that influence housing prices from the effect of access to open space. The mathematical equation produced by this exercise attempts to identify the part of the price attributed to each characteristic. Access to that function enables the analyst to estimate the value of preserving open space by seeing how the value of various kinds of houses change when the amount of open space nearby rises or falls. Results of the analysis can be used to evaluate public programs or other actions that invest in preservation of open space. If several land parcels are being considered for inclusion in the program, the hedonic value function can be used to estimate the benefits of taking each parcel out of commercial use compared to its cost.

Advantages. The hedonic pricing method offers several advantages. It is used to estimate values of environmental improvements based on actual choices made in the marketplace. If home buyers place a small value on a higher-quality environment, then home prices will be only slightly higher in environmentally improved areas. Another advantage of the method is that property markets are often highly responsive to information about characteristics, so these markets can be a good indicator of the importance of those characteristics. Sales data about property are typically detailed, reliable, and cheap. The method is flexible and is adaptable to a wide range of interactions between commercial market goods and environmental quality.

Limits. In the hedonic method, measurable environmental improvement values are limited to resources whose values are closely related to housing prices. The method only captures the willingness to pay for perceived differences in environmental attributes. So if home buyers are unaware of connections between the environmental service's quality and direct benefits to them, the value will not be observed in home prices. The conduct of hedonic pricing analysis presents a number of challenges. First, there is no reason to believe that the mathematical relation between a property's price and its characteristics is linear. For example, consider the following simple estimated hedonic price relation, which is likely to be estimated with econometric methods using multiple regression analysis:

$$P = \beta_0 \, S^{\beta 1} \, Q^{\beta 2} \tag{7.8}$$

where P is selling price of a house; S is square feet; Q is environmental quality; $\beta_0 \ldots \beta_2$, are coefficients (numbers) normally larger than zero; and β_1 and β_2 typically lie between -1 and $+1$ where diminishing marginal impacts are likely to be expected.

Even this simple equation shows that the incremental value of increasing environmental quality is not independent of a house's size. Furthermore, the incremental value of a given im-

---

[18] Econometric analysis using multiple regression offers a method for disentangling the effects of the various factors that produce a single result. In multiple regression, variables are permitted to vary under natural conditions. The values of all relevant factors are recorded. An observation consists of a single value for each variable in the regression model. Each observation provides a small piece of useful information, because what occurs at any one time depends on the values taken by all those factors. Determination of the relative impacts of the separate factors comes only from an analysis of a number of observations. Here is an example of a regression model: $Y_t = B_0 + B_l X_{1t} + B_2 X_{2t} + B_3 X_{3t} + U_t$. In this example Y_t is the dependent or endogenous variable; X_{1t}, X_{2t}, and X_{3t} are independent or exogenous variables; B_0, B_1, B_2, and B_3 are parameters; and U_t is the error or disturbance term. The variable for which you want to predict a value is called the dependent variable. The variables used as the basis for prediction are called the independent variables. The subscript t refers to the time period (or other unit of observation, such as zone of visitor origin for a travel cost model or neighborhood location for a hedonic price model). For each unit there is one observation on each independent variable and one observation on the dependent variable. If there are 10 units observed for which you have data, you have 10 observations. Each observation consists of a single value of Y and a single value of each of the Xs. Parameters are constants, which are presumed to have fixed numerical values in any observation situation. Parameters quantify the systematic and presumably fixed impact of each independent variable on the dependent variable. Multiple regression is a statistical technique that provides estimates of the parameters for a given set of observations on Y and the Xs. Each parameter estimate measures the effect that its own independent variable has on the dependent variable, holding all other independent variables constant.

provement is smaller when the environmental quality is already high than when it is low. In addition, square feet and environmental quality may be highly correlated; for example, larger houses may be located in areas with a higher environmental quality. This correlation makes it hard to separate the effects of environmental quality and house size on price, since both tend to move together. In any case, a fair amount of statistical sophistication is required to piece together the correct value of the environmental improvement. When two or more independent variables are highly correlated, the analyst has the problem of multicollinearity, and it is difficult to isolate the impact of each on the dependent variable. Econometrics deals with a wide range of problems facing regression analysts; Gujarati (2003) and Greene (2003) are two popular texts.

Travel Cost Model

Economists use *travel cost models* (TCMs) to estimate values produced by environmental systems that serve as outdoor recreation sites. Four kinds of environmental policies can be valued with TCMs: (1) changes in the cost of accessing a recreational site, (2) closure of an existing recreational site, (3) opening of a new recreational site, and (4) increases or decreases in environmental quality at a recreational site. The TCM is based on the principles that travel cost is an expense people incur to visit a site and that greater distances from a site effectively raise the price of access. Observation of the impact of increased distances on visit rates results in a demand function for the site. Based on this logic, visitors' benefits produced by the environmental resources at a site can be estimated by observing the reduction in visit rates at various increased travel costs.

Suppose a site used for sport fishing is threatened by activities that damage the site's environment. For example, pollution would damage fish habitat at the site, resulting in reduced fish populations and a loss of fishing activity produced by the reduced capacity of the site to support fishing. Policy analysts wish to estimate the economic value of programs that protect or improve the site's fish habitat. The travel cost model is a good selection for two reasons: (1) The site's value is as a sport fishing area; there are no rare or endangered species at risk that could produce significant nonuse values. (2) Major expenditures are not required to protect the site, so a relatively inexpensive valuation method like the TCM may be suitable. Other candidates are contingent valuation or contingent choice methods. These methods may produce greater precision in estimated values for special-site characteristics, and they can capture nonuse values. However, often they are much more complicated, controversial, and expensive to use.

Travel cost models have been estimated since the 1960s for valuing a variety of environmental programs. Three debates about the method are ongoing: (1) accounting for the opportunity cost of travel time, (2) handling multidestination and multipurpose trips, and (3) viewing travel time as a benefit instead of a cost. Three ways to apply the TCM are currently in widespread use: the zonal model, the individual model, and the random utility model.

Zonal Models. The zonal model can most easily value the entire package of recreational services produced by the site. It is more difficult, however, to use this model to value changes in a recreation site's quality, and it may be unable to consider some important factors that influence a site's value. The zonal model is typically applied by collecting data on visitation patterns to the site from a number of different market areas (distances).[19] Because travel cost per trip increases with increased distance, the natural experiment produced by varying distances permits the analyst to calculate the number of visits at different prices. Information on the response of visitation to increased distance is used for two purposes: to construct a site's demand function and to calculate consumer surplus received by visitors who travel from closer distances compared with consumer surplus of visitors from the edge of the market area.

[19] The TCM only works if site visitors come from more than one distance. Unless price varies, it is not possible to know how price changes are affecting use.

Table 7–1

Zone	Total visits/year	Zone population	Visits/1,000
0	800	1,000	800
1	800	2,000	400
2	800	4,000	200
3	800	8,000	100
Beyond 3	0		
Total	3,200		

Table 7–2

Zone	Round trip travel distance (miles)	Round trip travel time (minutes)	Distance times (cost/mile $0.60)	Travel time times (cost/minute $0.30)	Total travel cost/trip ($)
0	0	0	0	0	0
1	10	15	6	4.50	10.50
2	20	30	12	9.00	21.00
3	40	60	24	18.00	42.00

Data Assembly. The first step is to define a series of zones (market areas) that surround the site. These could be defined as concentric circles or geographic units (e.g., counties, zip code aggregations, or census units) that allow analysis of different visit rates at varying distances from the site. The second step is to find data on visit numbers sampled from each zone, as well as total actual visits made in the relevant time unit (e.g., year or season). Resource managers often keep good records on site use. However, visitor origin is often unknown, so some sampling may be required, such as counting license plates, conducting mail surveys, or stopping visitors at an entrance area. The third step is to estimate visitation rates (e.g., per 1,000 total population) by zone. Visitation rates are commonly measured as total visits per year from each zone divided by the zone's total population (in 1,000s). Table 7–1 shows an example.

The fourth step is to measure average travel distance and travel time to the site by zone, beginning with visitors who live at the site and face zero travel distance at zone zero. Visitors at all other zones face increased distance. Constant average travel costs per mile and per hour of travel time are used to compute travel costs per trip from each zone. In this example, cost per mile is \$0.60. Because visitors' time usually has alternative uses, those costs also need to be reckoned. How time should be valued is controversial. One way to value recreational travel time is to observe the additional expenses people voluntarily incur to reduce their travel time, for example, added fees they are willing to pay to travel on a fast toll road compared to a slower free road. Another approach is to use a figure close to half the average hourly wage, which reflects the fact that leisure travel displaces something less than the full opportunity cost of lost income from work. Suppose this is \$18 per hour, or \$0.30 per minute, for all zones. Table 7–2 shows the calculations.

Analysis. Analysis starts with estimation of an equation that explains factors affecting visitation. The usual approach is regression analysis, where results are used to estimate the typical visitor's demand function. The zonal travel cost model shown in Tables 7–1 and 7–2 presumes that price is the only variable affecting visit rates. Calculating that equation with travel cost as the sole predictor shows

$$Visits / 1{,}000 = 660 - 15.51 * (travel\ cost) \qquad (7.9)$$

Table 7–3

Zone	TC = travel cost plus $10	Visits/1,000	Population	Total visits
0	10.00	505	1,000	505
1	20.50	342	2,000	684
2	31.00	179	4,000	716
3	52.00	0	8,000	0
Total				1,906

Table 7–4

Entry fee	Total visits
10	1906
20	820
30	258
40	40
50	0

This equation can now be used to predict visit rates per 1,000 population at a given site for the travel cost of any zone of visitor origin.[20] For example, when travel cost is zero, the equation tell us that visits per 1,000 are 660 − 15.51 (0) = 660. When travel costs are $10, visits per 1,000 are 660 − 15.51 (10) = 505.

If we need to predict visit rates at a nonzero price (entry fee), we simply add that price to the actual travel cost and see what the equation predicts in new visits per 1,000. The ability to use the equation to predict visits at a nonzero price allows us construction of a demand function for visits to the site. The first point on the demand curve is the total actual visitation to the site at current travel costs with a zero entry fee. Table 7–1 shows those total visits amount to 3,200 per year. All remaining points on the demand schedule are found by computing total predicted use under various nonzero entry fees. The accuracy of this method depends on the assumption that visitors respond to an increased entry fee in the same way they respond to increased travel costs. A first nonzero point in the demand curve can be found by determining the number of visits expected at a low nonzero entrance fee, say $10.[21] Plugging this into the estimated regression Equation (7.9) and substituting TC (i.e., actual travel cost plus the entry fee) for *travel cost* in the equation, gives the results shown in the Table 7–3.

This exercise produces the first nonzero price point on the demand curve, which is 1,906 visits that occur with the $10 entry fee. In the exact same way, the number of visits produced under a range of increased entry fees can be found to produce Table 7–4. Those points make up the site's demand schedule.

Next, total economic benefits are found by measuring their consumer surplus, defined as the total area under the demand curve, of which five points are shown in Table 7–4. That area measures the total economic benefits produced by recreational use of the site. Those calculated benefits are about $46,000 annually, or about $14.38 per current visit ($46,000/3,200). Finally, the proposed policy or program is analyzed by comparing the TCM-measured benefits to its costs. Recall that the policy debate questions whether it is economically worthwhile to incur the costs

[20] Factors influencing visit rates could include demographic variables such as the zonal average of age, family size, income, and education levels. If census data are available, large numbers of variables can be used. Smaller defined zones typically produce greater natural variability in variables.

[21] A more precise method is to increase the entry fee by a series of many very small increments, say $1 each, raising the price until total visits fall to zero.

required to protect this site. If the program's total cost is less than $46,000 per year, the cost is less than the benefits and the resulting action is economically sound—otherwise it is not.

Individual Models. Another way to apply the TCM is through the individual model. This approach uses the same economic principles as the zonal model, but analyzes individual visitor data rather than averaging data from different origin zones. So typically, the individual model analyzes larger data sets and may require more extensive analysis; however, it produces more precise results on the benefits of environmental improvements at outdoor recreation areas. For example, say we are conducting a survey of the anglers at a site used for sport fishing. The survey asks for several kinds of information: an angler's home location, how often he fishes the site, travel time, on-site time, family income, demographic characteristics, other sites visited, fish catch and harvest rates, and characteristics of alternative sites.

After obtaining the survey data, we would likely use regression analysis to estimate the site's demand, just as we did with the zonal model. The individual model explains the relationship between number of (individual) visits per unit time, travel costs, travel time, possibly environmental factors, and other visit predictors. The estimated regression equation produces a demand function for the typical site visitor using the same list of individual and site characteristics. For each visitor, the area below the demand schedule measures the estimated consumer surplus. Assuming a random sample of the visitor population, consumer surplus of the total sample is multiplied by the inverse of the sampling rate to estimate the site's total consumer surplus. For example, if a random 1% of the site's fishers are sampled, total sample consumer surplus is multiplied by 100 to estimate total surplus produced for all the site's fishers.

Random Utility Models. Random utility models (RUMs) are the most extensive and powerful of the travel cost models. They are designed to estimate benefits for site characteristics or changes in their quality, rather than for the site as a whole or for any particular trip. The RUM is especially discriminating when a given site has many substitutes. The RUM model is based on the principle that on one occasion, an individual chooses one preferred site among several. The individual's choice is based on trade-offs between site quality and the cost of travel. A higher-priced (more distant) site is chosen if its quality is sufficiently higher to make up for its greater cost of access. For any given visitor's choice, the dependent variable takes a value of zero or one: The site visited receives a one, and each of the nonvisited sites receives a zero. The price paid for this greater elegance and power is that the estimation of a RUM model requires information on prices at all sites and characteristics of all sites. That estimation could require hundreds of thousands of pieces of information, each of which adds to the cost of the analysis.

The RUM approach is flexible and can predict activities and sites under various conditions. For example, an estimated RUM might estimate *which* of 50 fishing sites is chosen for a visitor who has a certain kind of fishing preference, faces a certain set of economic and demographic conditions, and faces a certain set of site prices and qualities. If all quality characteristics of all sites are included in the RUM, the model can estimate values for changes in one or more characteristics at one or more sites. For this reason, the RUM is an attractive method for analyzing regional or national environmental policy proposals that might affect environmental characteristics in a large geographic region.

Advantages of the Travel Cost Model. The travel cost model is attractive because it can be used to estimate economic values of unpriced environmental improvements. The method is based on people's actual behavior in actual conditions, rather than on what people say they would do under hypothetical conditions. On-site surveys provide opportunities for large sample sizes, and they can be enjoyable to administer.[22] Visitors are usually enthusiastic about being sampled as long as their recreation is respected. Results are fairly easy to calculate, explain, interpret, and apply.

---

[22] On-site samples also present the statistical problem that observations are truncated at positive trip demand rather than at zero, since people who did not visit are not included. They also have the problem of endogenous stratification; that is, the likelihood of a visitor being sampled depends on the frequency of visiting the site. Various statistical methods have been developed to deal with both issues.

The travel cost model has several advantages because it is based on visitors' choices made under a budget constraint. For example, two visitors who travel the same distance to a site might be presumed to have the same consumer surplus. But proper use of the travel cost model shows that if one visitor faces several substitutes, but still travels to the given site, that person's consumer surplus is higher.

Limits. The travel cost method assumes that people perceive and respond to changes in travel costs the same way that they respond to changes in admission price. The simplest models assume that individuals take trips for a single purpose. Thus, if a trip has more than one purpose, the value of the site may be overestimated. Another limit is that the researcher still has to calculate the travel cost, which introduces error into the analysis. Also, travel cost data are often gathered through the use of surveys, and respondent recall may be inaccurate (Randall, 1994).

It can be difficult to apportion travel costs among the various recreation activities. Estimation of a site's demand function requires significant differences between distances traveled, and these different distances must affect trip rates. So the model cannot be applied to sites such as urban parks where visits come from a single zone of origin. The travel cost method requires visitor participation, and it is unable to support visitor decisions that focus on off-site values. In addition, the model is unable to measure nonuse values that accrue to people whose benefits are not based on direct use of the site.

Avoided Costs

Under some special conditions, the value of improving an environment can be measured by the future costs avoided by making that environmental improvement. Three applications of the avoided costs method are (1) damage costs avoided, (2) replacement costs avoided, and (3) substitute costs avoided. These three applications are based, respectively, on the costs of avoiding damages suffered by lost services, costs of avoiding replacing environmental services, and costs of avoiding supplying substitute services.

These methods bypass the need to measure beneficiaries' willingness to pay for an environmental service. Instead, they recognize that the costs of avoiding damages or replacing ecosystems or their services can sometimes be at least as large as the willingness to pay for these services. The validity of the method rests on the assumption that if people voluntarily incur costs to avoid damages from lost environmental services, then those services must be worth at least what people are paying to protect, replace, or substitute for them. Here are a few applications: valuing water quality improvements by costs of controlling (avoiding) pollutants into the water; valuing a forest's erosion protection services by the avoided cost of cleaning up downstream muddy waters; valuing a wetland's water treatment services by the avoided cost of artificially treating water; and valuing fish habitat by the avoided cost of fish hatcheries.

Consider this example: A community is debating whether a degraded wetland should be restored to improve its capacity to protect a nearby area from flooding. Voters want to know the benefits provided by the improved flood protection. The avoided costs method can be used because flood protection is the most important service of the wetland. Begin by performing an ecological assessment of the flood protection services the wetlands supply. This assessment compares the existing level of flood protection to the anticipated level supplied after complete restoration of the wetlands. The next step varies according to which of the three implementation methods is chosen:

1. The method of damage costs avoided could be applied using the following approach: Identify the price similarly located property owners have paid to protect their property from the risk of flood damages (e.g., insurance and flood proofing). Sum these costs for all affected properties protected by the proposed restored wetlands. Those avoided costs are an estimate of the benefits provided by the wetland's increased flood protection.

2. Replacement costs avoided can be measured by estimating the costs of replacing the wetland's flood protection services. Perhaps a dam or levee could be built at a distant location that would supply equivalent flood protection services. Maybe the river channel could be deepened or straightened. Perhaps an existing system of dams could be operated differently. In summary, value the wetland's flood protection services by the least-cost method to replace those services.

3. Substitute costs avoided (a special kind of replacement costs avoided) can be estimated as the costs of providing a substitute for the affected wetland services. Imagine building a retaining wall to protect nearby properties from flooding, which is a substitute for the services of the wetlands. Estimate the cost of constructing, operating, and maintaining the wall. This method assumes that people who live in the flood-prone area perceive the wall as a perfect substitute for the restored wetland. The monetary value of the avoided costs of the wall provides an estimate of the value of flood protection services produced by restoring the wetlands. That value can be compared to the costs of the wetlands restoration to decide whether it is economically efficient to restore the flood protection services of the wetlands.

Advantages. Avoided costs provide a good measure of the economic value of an environmental improvement when the costs avoided would actually be incurred without the proposed program. In fact, if those costs avoided are a *lower* bound on the program's benefits—benefits could be higher. One especially attractive feature of the avoided costs method is that it is usually much easier to measure the costs of protecting the benefits than to measure the benefits themselves. For example, it is easy to measure the avoided costs of the fence not built. By contrast, valuing the willingness to pay for the benefits protected by the fenced area could be expensive, difficult, or impossible.

Limits. If a proposed program avoids the need for a costly substitute method to protect the environment, the cost saving is a real benefit. People are certainly willing to pay up to X to avoid incurring costs of X for receiving the same service. The danger is that it is too easy to assume that society would voluntarily incur those costs (X) without the program. In fact, any environmental program can be falsely justified economically by simply assuming that a more expensive protection method is waiting in the wings to be implemented without the program. And any proposed program can be justified by suggesting a more expensive program that produces the same services less efficiently. So it is not hard to claim that by displacing the more expensive (and implausible) substitute, society is getting a good deal by implementing the proposed program. The avoided costs method should only be used if it can be demonstrated that society will incur the costs of a more expensive substitute without the project. Otherwise, the evidence is slender that the value to society of the service provided by the proposed program exceeds its cost.

Contingent Valuation

The contingent valuation method (CVM) measures the benefits of environmental improvements by eliciting people's willingness to pay (WTP) or willingness to accept payment (WTAP) for changes in quantities of environmental services. The usual way to elicit those benefits is to survey people. Many kinds of environmental assets, such as water quality at recreation sites, improved landscapes, hunting and fishing, wildlife habitat, and air and water pollution reductions have been valued with CVM surveys. The surveys have also been used to value more complex environmental programs such as hazardous waste reduction, endangered species' habitat protection, and human lives saved. Carson (2000) and Carson, Flores, and Meade (2001) provide excellent reviews of the CVM. The European Commission (2003) summarizes recent European environmental valuation studies; see also Navrud and Pruckner (1997).

The use of CVM as a method for estimating both costs and benefits continues to grow. For example, in cases involving assessments of natural resource damages, U.S. federal courts have

Example 7–3
Guidelines for CVM: Valuing Environmental Improvements

Lawsuits over natural resource damages, such as oil spills, have led to heated debates over the use of the contingent valuation method (CMV) for valuing environmental protection. In 1992 the U.S. Department of Commerce's National Oceanic and Atmospheric Administration (NOAA), announced that it was establishing a panel under the chairmanship of Nobel laureates Kenneth Arrow and Robert Solow to advise on the use of CVM in damage assessments. That panel concluded that CVM studies can produce estimates reliable enough to be the starting point for measuring natural resource damages. To be acceptable however, the following guidelines (a partial list of standards) should be followed:

- For a single dichotomous question (yes or no), a total sample of at least 1,000 respondents is required.
- Face-to-face interviewing produces the most reliable results.
- Full reporting of data and questionnaires is needed for others to evaluate results.
- A conservative design likely to underestimate willingness to pay (WTP) is preferred to one likely to overestimate WTP.
- The WTP format is preferred to willingness to accept payment.
- The valuation question should be posed as a vote on a referendum, that is, dichotomous choice question related to the payment of a particular tax level.
- Accurate information on the valuation situation should be presented.
- Respondents should be reminded of the status of any undamaged potential substitute resources.
- A "no answer" option should be explicitly allowed in addition to the yes and no options on the main valuation question.
- Yes and no responses should be followed by this question: Why did you vote yes/no?

Sources: Bateman, I. J., I. H. Langford, R. K. Turner, K. G. Willis, and G. D. Garrod. 1995. Elicitation and truncation effects in contingent valuation studies. *Ecological Economics.* 12: 161–179; National Oceanic and Atmospheric Administration. 1993. *Report of the NOAA Panel on Contingent Valuation.* January 11, 1993. Available from *http://www.darp.noaa.gov/library/pdf/cvblue.pdf* [Accessed May 6, 2005].

elevated results of citizens' environmental valuations to the level of rebuttable presumption.[23] A panel of social scientists convened by the National Oceanic and Atmospheric Administration (NOAA, 1993) supported the use of CVM in cases of natural resource damages by finding that it could be the basis for estimating damages (benefits lost). See Example 7–3 for their suggested guidelines.

A CVM analysis typically follows these steps:

1. An environmental policy question is identified for which there is a need to measure benefits and/or costs.
2. A sample of respondents is identified from the affected population.
3. The sampled individuals are asked questions about their valuations of services affected by the proposal.
4. The information from each response is used to estimate that respondent's benefits provided by the policy.
5. Benefits computed from the sample are extrapolated to estimate total impact to the population of affected people.

[23] A fact assumed to be true under the law is called a *presumption*. A presumption is a valuable resource because it is used to relieve a party from the cost and burden of proving the truth of the presumed fact. However, once a party relies on the truth of the presumption, an opposing party may be allowed to present evidence to falsify (rebut) the presumption; in much the same way, science discovers truth by rejecting a false hypothesis. When a fact reaches the status of a rebuttable presumption, the person supported by that fact has a valuable property right. The person who has the rebuttable presumption puts the obligation of finding rejecting evidence on the shoulders of whomever lacks that presumption.

Four methods of CVM have been used at various times: the open-ended, willingness-to-pay method; the closed-ended, iterative bidding method; and the contingent ranking method. The dichotomous choice, or referendum method, has been advocated by the NOAA panel as the best method for most cases.

Open-Ended Valuation. The earliest application of the CVM relied on open-ended valuation. In that method, survey respondents are asked to state their maximum WTP for the services provided by the proposed environmental policy. For example, "What is the most you would be prepared to pay in additional monthly electric bills to guarantee that the nearby wildlife habitat will remain protected from future development?" Although popular at one time, this kind of question has disadvantages. Because it is open ended, people may have a hard time answering it. Nobody is actually asking for real money, so the respondent might answer with a shrug. If the respondent thinks she might be charged for part of the cost, she may give a low number. Similarly if she thinks she will not be charged, she may give a high number. Generally, this method has fallen from favor to guard against unrealistic responses.

Closed-Ended Valuation. Closed-ended valuation means that respondents are asked whether they would pay some particular amount of money (X_1) for the services the proposed policy would provide. If a respondent says yes, then the amount is incrementally increased (X_2, X_3, etc.) until the respondent finally says no. At that point, the question is over. Likewise, if the respondent's first response is no, the interviewer continues reducing the amount until the respondent changes to yes. The idea is to keep iterating up and down until you reach the respondent's willingness to pay. The result is a precise measure of WTP from questions that are easy to answer. Although closed-ended iterative bidding was once fashionable, it is rarely today, because evidence has shown that the final results have a *starting point bias.* That is, the final WTP result depends on the starting value of X_1. A respondent's WTP at a starting bid of \$5 may be different from her WTP at a starting bid of \$20.

Contingent Choice. Contingent choice (sometimes called contingent behavior) is a method for finding values in which the respondent is asked to rank various combinations of environmental services and payments charged. (See Hanley, Mourato, and Wright, 2001, for an example for a program to protect rock climbing in Scotland.) The result of a respondent's ranking is used to *infer* her WTP for the service. Suppose the goal is to measure a respondent's WTP for improving water quality from level A to an improved level B. Assume the respondent is given two choices: (1) water quality level A with zero annual tax and (2) water quality level B with an annual tax of \$100. If the respondent prefers choice 1 to choice 2, the respondent's WTP for the improved water quality is less than \$100. If the respondent is indifferent between choice 1 and choice 2, her WTP equals \$100. If she prefers choice 2 to choice 1, her WTP exceeds \$100.

Each respondent might be asked to rank a wide range of choices on water quality levels and associated prices to pay for them. From those survey responses, a mathematical equation can be developed to determine each respondent's WTP for *several levels* of water quality improvement. After estimating each respondent's WTP for various increments of quality improvement, the analyst may generalize from the sample to the population. The analyst calculates the WTP of the entire affected population for each increment of quality improvement. If the goal is economic efficiency, policies can be established at whatever level maximizes the relevant population's WTP minus the cost of implementation (net benefits).

Contingent choice is unique because it asks respondents for an *ordinal ranking* (a ranking from most to least preferred) for various environmental improvement levels and the cost of financing each level. In contrast, both the open- and closed-ended valuation methods require cardinal evaluation; that is, respondents are asked to state what the environmental improvement is worth to them in absolute monetary terms (e.g., \$10, \$20). Ordinal rankings do not directly ask people to state their values. Instead, values are *inferred* from the choices people make. As long as one of the characteristics of the environmental program is price, it is possible to derive the WTP for changes in the levels of program characteristics.

Contingent choice focuses on trade-offs among scenarios with different price and quality characteristics. Therefore, it is well suited to economic analysis of policy decisions in which a set of possible actions might result in different impacts on natural resources or environmental services. Contingent choice is particularly useful in valuation of environmental resources such as ecosystem changes where several ecosystem services may be jointly affected. For example, a policy that improves water quality in a lake will improve the quality of several services provided by the lake, such as drinking water supply, fishing, swimming, and biodiversity. Although contingent choice can be used to estimate dollar values, it can also be used to simply rank policy options, with no attempt to translate choices into monetary values. This option is useful if all programs cost about the same or if estimated values are not desired.

Because responses focus on trade-offs, rather than on direct expressions of dollar values, contingent choice may reduce some of the problems associated with contingent valuation (described later in this chapter). Often, relative (ordinal) values are easier and more natural for respondents to express than absolute (cardinal) values. In other words, contingent choice may minimize some of the biases that can arise in open-ended or closed-ended contingent valuation studies where respondents are asked to put prices on environmental improvements. Contingent choice actually measures respondents' preferences in the form of attitudes instead of intended behavior.

Contingent choice does have limitations. For example, if the number of program characteristics is high, the sample size and/or number of comparisons each respondent makes must be increased. When presented with a large number of trade-off questions, respondents may lose interest or become frustrated.

Dichotomous Choice. Dichotomous choice means the respondent has only two choices, yes or no. The dichotomous choice method for valuing environmental improvements was developed by Bishop and Heberlein (1979). When voters cast their ballots for a referendum item, such as whether or not to finance a new public program, they are faced with a dichotomous choice. In the dichotomous choice (referendum) method, respondents are asked if they are willing to pay a particular price to receive an environmental service. Each respondent receives one randomly drawn price. Respondents are then asked to state whether they would be willing to pay that price for the service (e.g., protecting a wilderness area from development). Like a referendum vote, a yes response means they are willing to pay the price and a no means they are not.

Respondents are given a take-it-or-leave-it offer of the kind they face in many private market decisions. In this approach a researcher typically uses a large number of predetermined prices, one for each respondent, that includes the anticipated maximum WTP amounts. Sometimes the researcher guesses badly at the range of predetermined prices. Pretest surveys are usually administered to catch and fix errors before administering large surveys. Respondents are asked to vote for or against a particular level of taxation, just as she would be asked in most real referendums. For example, a referendum on a ballot typically asks a voter to vote yes or no to be taxed X dollars to pay for a public structure or program. In surveys, the dollar amounts of X, sometimes called bid prices, that are presented to respondents vary over a range selected by the researcher. Based on responses, the researcher analyzes observed probabilities of "accept versus reject" for each bid price.

An estimated statistical model is used to predict the probability that an individual with specific characteristics will accept a particular bid price. The estimated model, often a logistic regression with bid price as one of the response predictors, allows the analyst to sketch the relationship between bid price and the probability of a yes response. Using advanced probability theory, the analyst uses that relationship to approximate the average WTP for respondents. To find the aggregate WTP for the entire population, the mean for each group is multiplied by that group's share of the relevant population, then the result is summed over groups.

An advantage of the dichotomous choice approach is that each respondent is asked only one valuation question that is simple to answer. Smith (2000) concluded that dichotomous choice is desirable for valuing the benefits of health care programs because it offers the most realistic market and because it reduces the mental load on survey respondents who take part in the valuation exercise. The major disadvantage is that the survey researcher only learns whether each respondent's WTP is above or below the surveyed dollar amount. A large sample is required for this approach. For each dollar amount included in the survey design, a large sample (more than 30) is necessary to obtain good estimates of the true WTP.

Survey Administration. Three methods for administering CVM surveys are in widespread use: in-person interviews, telephone interviews, and mail questionnaires.[24] A fourth method, the Internet, may become popular in the future. In-person interviews are attractive because they permit the interviewer to convey complex and detailed information to respondents through the use of graphs, drawings, maps, pictures, videos, and other visual aids. (Recall that a painting played a major role in convincing the U.S. Congress to show its willingness to pay for Yellowstone National Park.) Such interviews also permit the interviewer to respond directly to the respondent when more information is needed. One limit of in-person interviews is that they introduce a risk of interviewer bias, as respondents react to the interviewer's voice cues and other personal characteristics. Another downside is that in-person interviews are hard to monitor and are expensive, particularly for studies that require sampling over a large geographic area.

Telephone interviews are the most common method for doing CVM surveys. They are much less costly than in-person interviews, even if several callbacks are necessary. Telephone surveys allow the drawing of nearly random samples through random-digit dialing, thereby eliminating the need for the address lists required for mail surveys. Still, there are two disadvantages of telephone surveys: (1) They require the caller to explain the policy issues, choices, options, and prices with little chance for graphic reinforcement. (2) Telephone communication introduces potential interviewer bias, as the respondent may react to voice cues and perceptions of the interviewer's personality characteristics. Also, not everybody has a telephone and many people have unlisted numbers, so there is no guarantee of securing a representative sample.

Mail surveys have the advantage of being very inexpensive. They also allow the researcher to provide various visual aids to respondents. If a random sample of the relevant affected population can be found, and accurate mailing lists can be found, costs decrease more, and reliability of results increases. Even if response rates are low, several mailings can be done (possibly with incentives) so that return rates reach acceptable levels. The good news is that without interviewers, all respondents experience the same stimulation from the question, so there is no interviewer bias. The bad news is that mail surveys are capable of conveying issues and choices of only limited complexity.

The Internet has seen comparatively few applications of CVM surveys to date (see Berrens et al., 1998, 2004) but it offers great potential. As more adults worldwide become sophisticated Web users, the Internet may provide an opportunity for administering CVM surveys and offer several advantages. Direct interviews are not needed, so the incremental cost per survey is nearly zero. This low marginal cost permits the researcher to invest considerable time in the fixed costs of designing an excellent and detailed survey. As interviewers are not needed, Internet surveys have very low marginal costs, perhaps lower than mail surveys; and like mail surveys, Internet surveys avoid interviewer bias. Furthermore, the interactive nature of the Internet permits the researcher to convey complex and interactive information to respondents, thereby approaching the power of in-person surveys.[25]

---

[24] Actually survey administration is a big issue for both stated preference methods such as the CVM and revealed preference techniques such as the travel cost method. Survey administration has evolved into its own science, dealing with issues of pretests, sample splitting, payment incentives, remails, response rates, data entry, preparing data for use, and the like. It's much more than quickly assembling a survey and mailing it out.

[25] Although Internet access grows daily, a digital divide exists among age and racial groups, income levels, and geographic settings. Ensuring that each potential respondent has an equal chance of being selected to participate poses a major challenge in conducting a scientifically sound economic valuation survey. This is especially true in the environmental area, where electronic access to specific respondents cannot easily be obtained. Rigorous sample selection procedures should be followed if results are to be generalized to a population and sources of coverage and sampling error are to be minimized. Unfortunately, the sampling procedures reported in many electronic surveys reflect unknown samples. When respondents are recruited by targeting news groups or search engines, it is difficult to determine the distribution of the sample population. Another concern unique to conducting electronic surveys is the variation of the level of computer literacy among respondents and the capabilities of their computers.

Willingness to Pay versus Willingness to Accept Payment. In principle, the correct measure of lost value for damage to an environmental asset that has already occurred is the minimum level of compensation that the affected individual would be willing to accept. That is, WTAP is the correct measure of value when a respondent already has an established property right to the asset. In contract, WTP is the correct measure when the respondent has no current property right to the asset and must pay to get that right. Despite the clear distinction between the two measures of value, WTAP may present a special problem for empirical CVM studies. Willingness to accept payment to compensate for an environmental insult that has already occurred *has no upper bound.* So because of concern that respondents inflate their answers to such questions, most existing CVM studies are based on policy and compensation scenarios in which respondents are asked their WTP to prevent future occurrences of *similar* environmental insults.

Predictions of economic theory are based on rational behavior. So it should make little difference whether respondents to a CVM survey are asked their WTP for receiving a service or their WTAP for the loss of a comparable service (see Willig, 1976). Likewise, it should make little difference whether they are asked their WTP to prevent a loss or their WTAP for the same loss. Despite what economic theory predicts, however, many empirical studies have found that most people require much greater monetary compensation to give up services they already have than they are willing to pay to acquire the same service (see O'Brien et al., 2002, for an example for the valuation of environmental programs that improve human health). In experiments that actually require people to trade goods for money, peoples' required WTAP amounts range from 3 to 15 times greater than WTP amounts. Thus, WTP estimates are used more commonly than WTAP estimates, because WTP responses from CVM surveys are closer to estimates derived from methods based on revealed preferences than are WTAP responses. This is true even in situations for which WTAP fits the case better than WTP (e.g., in cases where compensation is sought for a lost asset).

Strategic Response. Even if respondents know their WTP for an environmental improvement, will they answer honestly when asked about it in a survey? Honesty may be the best policy but many people face incentives to misrepresent their true preferences in response to CVM questions. Why would they lie? They may see it as a way to increase the chances of getting something they want, that is, they see the benefit of lying as exceeding the costs. If a respondent sees her response as influencing the likelihood of having access to a program for which no payment will be required, then the lie may have elements of a public good, for which there are well-known incentives for free riding.

Designing CVM surveys to produce honest answers requires knowing something about the nature of the strategic behavior of respondents to questions likely to be encountered in the surveys (see Appendix F). Expect respondents' behavior to be predicted by economic theory only if respondents believe that their response to the survey may influence an outcome about which he or she cares. Otherwise economic theory has little to say about their behavior. Questions are called incentive compatible if they deal with outcomes respondents care about that provide incentives for them to reveal their preferences truthfully. Incentive compatibility requires that questions be limited to binary responses (e.g., yes or no).

Questions that allow more than two responses (e.g., open-ended WTP responses) are *not* incentive compatible. That is, these questions give respondents an incentive to lie about their preferences in order to obtain an outcome more desirable than they would receive by being honest. For example, suppose a respondent values at \$100 per year a policy that would reduce air pollution in her region by a known amount.[26] If she believes (1) she will actually be charged nothing for the policy if enacted and (2) a higher bid on her part increases the probability the policy will be enacted, then she has an incentive to tell a CVM surveyor that she would pay a very high value, say \$10,000 per year. After all, she believes she will not be charged and the high stated value may raise the likelihood that she will be getting cleaner air. This misrepresentation of her

[26] Many psychologists believe most of us know too little of our own preferences to precisely value the benefits we receive from an environmental improvement. Clark, Burgess, and Harrison, (2000) analyze why this may be the case.

preferences produces an outcome more desirable to her than the outcome produced by sending a truthful signal about her preferences. Honesty is the morally acceptable choice, but because lying pays, it is best economically. Multiple response formats, such as those used in the open-ended WTP method, are vulnerable to strategic responses that reward people who misstate their true preferences.

The dichotomous choice method to elicit WTP for a proposed public (nonexcludable) program funded with respondents' coercive payments (required taxes) meets the requirements for incentive compatibility. Respondents have an incentive to reveal their true preferences; that is, respondents maximize their net benefits by telling the truth. In this case, honesty is the best policy, and it pays. Because CVMs are often the only way analysts can estimate WTP for many environmental programs, this result is comforting.

In some cases researchers use inference in CVM surveys. They estimate the average value for an individual or household in the sample and extrapolate this to the relevant population to calculate the total benefits from an environmental program site. For example, if they find that the mean willingness to pay is $0.10 per capita, the total benefits to all citizens would be $26 million for a relevant population of 260 million.

Advantages. Contingent valuation can be used to estimate the economic value of virtually any environmental improvement. Nevertheless, it is best suited for estimating values for environmental services easily identified and recognized by respondents. The CVM is the most widely used method for estimating total economic value, including use values and the various nonuse values. Although CVM requires well-trained survey analysts to produce defensible estimates, the nature, results, and interpretation of CVM studies are easy to understand, analyze, and describe. Monetary values can be presented in terms of a mean value per capita or per household, which can easily be aggregated to total values for the affected population.

Limits. Even though the CVM has been popular for years, controversy remains over whether it really measures people's willingness to pay for environmental improvements. See Example 7–4. For example, many people are not accustomed to placing monetary values on environmental services, therefore, they may not have a realistic basis for stating their value, if they even know it. Responses to a willingness to pay question in a CVM survey may be biased because the respondent answers a different question than the surveyor had intended: Rather than expressing a willingness to pay for the service, the respondent may be expressing feelings about the policy or the valuation exercise itself.

Strategic bias occurs when a respondent supplies a biased answer with the intent of affecting a particular outcome. If a decision to protect a stretch of river for fishing depends on whether the survey produces a sufficiently large measured WTP for fishing, respondents who like to fish may have an incentive to provide an answer that ensures a high value, rather than an honest lower value. For all these reasons, most who have studied the CVM believe it has considerable potential but that it has not yet reached that potential (see Appendix G for a detailed critique of the CVM approach).

Benefit Transfer

The benefit transfer method estimates economic values of environmental services by transferring information. Brouwer (2000) describes several strict standards that a benefits transfer exercise needs to meet to be useful and credible for environmental policy. For example, sport fishing values needed for one river basin in one country may be estimated by adapting existing sport fishing values taken from another basin. The aim of a benefit transfer exercise is to adapt estimated benefits from an *original* site to the valuation question of a similar *target* site. The need for this method is greatest when it is impractical to conduct an original valuation study at the target site, yet some measure of benefits is needed to make a decision there. In the case of outdoor recreation, the cheapest type of benefit transfer is to transfer unit day values

Example 7–4
CVM's Acid Test: Hypothetical versus Actual Cash Payment

In a celebrated study, economists Richard Bishop and Thomas Heberlein subjected the contingent valuation method (CVM) to the acid test: They wanted to compare hunters' *stated* willingness to accept payment (WTAP) for a resource to actual cash payments offered to hunters for the identical resource. Hunting permits were issued free via a lottery. For the 1984 lottery, a large number of applications were received, but only 150 permits were issued. Since not all hunters willing to pay the zero price received the permits, the zero price was obviously too low to avoid demand outstripping supply. But what was the (unpriced) permit worth?

To measure WTAP, half the hunters received a letter explaining the intent of the study along with a check. To cash the checks, the hunters had to return their permits. The denominations of the checks, which ranged randomly from \$18 to \$518, corresponded to the dichotomous choice method described in this chapter. The other half of the hunters received a similar letter with a hypothetical payment offer drawn from the same range as the first half. They were asked if they would be willing to turn in their permits for the hypothetical amounts stated in their letters, but they were not actually paid the amount when they said yes.

To measure willingness to pay (WTP), 150 unsuccessful lottery participants were selected at random. Again, 75 hunters received a letter explaining that a hunting permit was available if they paid the amount stated in the letter. The amounts covered the same range as described previously. The other 75 hunters were asked the same question hypothetically; that is, they were asked the CVM question instead of revealing their preferences through purchases.

Take a guess at what the researchers found. For the WTP analysis, the hypothetical valuations to buy the permit (\$35) were only slightly higher than values measured by selling a real permit to a hunter for cash (\$31), a statistically insignificant difference. But as theory predicts, WTAP valuations were much higher. The WTAP hypothetical average valuation (\$420) was significantly higher than values revealed by trading in the permit in hand for the cash (\$153).

Source: Bishop, R. C., and T. A. Heberlein. 1979. Measuring values of extra-market goods: Are indirect measures biased? *American Journal of Agricultural Economics.* 61: 926–930.

from the original site to the target site. Unit day values, such as *X* dollars per visitor day of coldwater fishing, are used to value the same activity at the target site. A more defensible but expensive benefits transfer would transfer a *complete benefit function* from another study. That function might be estimated statistically using regression methods and might look like this: WTP = *f*(site characteristics, activity, demographics). Willingness to pay is related to the site's characteristics, the kind of activities supported, and the visitor demographics. The advantage of transferring the complete function is that adjustments are made for differences in all three classes of variables between the original site(s) and the target site. The larger the number of original sites and the wider range of characteristics included in the estimated equation, the greater is the likelihood that the target area will have environmental and human characteristics inside those of the previous study areas.

The benefit transfer method is most reliable when the original sites have a range of conditions that include the study site in terms of quality, location, and population attributes. It is also more reliable when the old and new levels of environmental quality are similar for original and target sites, and when the original study was conducted using reliable valuation methods.

Suppose a new beach is proposed for a lake in a city park. Policy analysts wish to know the benefits of the upgrade so they can compare them to costs. The benefit transfer method could be used for this case for two reasons: First, often there is little money for a site-specific valuation study. Second, many recreation valuation studies have been conducted and published since the 1960s; even fairly large databases of these studies that might be required for the transfer are fairly easy and cheap to build. Begin by searching for existing studies that include values that could be transferred. If possible, find previous studies that valued beach recreation. Suppose that a search turned up several primary travel cost analyses. Next, decide whether the existing values can be transferred. One important question is likely to be, Is the service being valued equivalent to the service valued in the already completed study?[27]

Advantages. The benefit transfer method is typically easier to apply than an original valuation study. Economic benefits can be computed more quickly than when beginning from scratch in an original valuation study. The benefit transfer method can be used as a screening measure to decide if a more detailed, original valuation study is needed.

Limits. Transferring benefit estimates from other studies to the target site may be inaccurate. Accuracy increases as the target site shares more site and user characteristics with the study sites. People at the local target site may say its people and resources are unique. This means that equivalent existing studies for the policy or issue in question may not be available and might not be believed locally even if they are available. Another problem is that it may be hard to find appropriate transferable studies. Original studies are rarely conducted for the purpose of being transfered to other places, so the reporting of these studies is often inadequate to make the needed transfers. Extrapolation outside the range of site and human characteristics of the original study is dangerous and unreliable. Benefit transfers can be no more accurate than initial values estimated.

Summary

Information on economic values of the environment have been used for five major purposes in environmental decision making to date in the United States: (1) cost-benefit analysis (BCA) of projects, (2) BCA of new regulations, (3) assessments of natural resource damage, (4) environmental costing, and (5) environmental accounting. While there have been much fewer environmental valuation analyses in Europe, this could change in the future with the new European Union goals of establishing environmental adjusted national accounts. Economics typically presumes that measures of economic value are based on what people want, and values based on preferences of the individual members of a society are the relevant values for public decision making in a democratic society.

Benefit-cost analysis is a way to make environmental policy decisions that serve a society's welfare. The application of benefit-cost analysis to environmental policy is based on the idea that society's values, derived from individual preferences, are expressed for all economically possible choices; then the most preferred alternative is selected from the choices available. Compensating variation, equivalent variation, and consumer surplus are three measures of the benefits of an environmental improvement. All three welfare measures are measured in monetary units and can be summed over affected individuals.

Several methods are in widespread use for measuring the benefits of environmental improvement market prices, productivity, hedonic pricing, the travel cost model, avoided costs, contingent valuation models, and benefits transfer. Each of these methods has advantages and limits. Contingent valuation has the greatest flexibility and sensitivity for measuring benefits of environmental improvements. However, because of concerns about possible errors in measuring environmental values, its use to support environmental policy decisions is controversial. For these reasons, contingent valuation is one of the most active research areas in environmental economics.

---

[27] Other factors could be used to decide on comparability: similar kind of sites, similar site facilities, and similar substitute prices and access. The affected populations' characteristics are also important. How could all that information be used systematically? Here is an example. Find several existing travel cost studies on water-based beach recreation. From information on those studies, assemble a multiregional database, from which something like the following regression equation could be estimated:

$$WTP_{ij} = 10\, W_j^{0.15}\, Q_j^{0.25}\, B_j^{0.40}\, TC_{ij}^{-0.5}\, Y_i^{0.2}$$

where WTP_{ij} is willingness to pay per capita population in the ith zone for any jth water-based recreation site; W_j is an objective indicator of the jth site's water quantity; Q_j is an indicator of water quality; B_j is number of beaches; TC_{ij} is travel cost from the ith zone to the jth site; and Y_i is average per capita income in the ith zone. If the WTP from the original studies was based on sound economic theory, and if the right-hand side variables included values for the proposed site, then this equation can be applied to estimate WTP for the proposed site improvement to visitors from each zone in the market area. Predicted WTP is multiplied by each zone's total population for both the old and proposed level of beaches. The predicted change in WTP, summed over the zones, gives an estimate of the value of the added beach.

Review Questions

1. Why do preservation, protection, restoration, and improvement of the natural environment face tough competition whenever resources are scarce?

2. Describe at least five uses that can be made of information on the economic value of the environment.

3. Why do economists measure environmental values of public programs based on preferences of affected citizens rather than on views of opinion leaders?

4. Distinguish use values and nonuse values of natural environments and illustrate with an example.

5. What is the expenditure function? How can it be used to measure the economic value of an environmental policy?

6. Define compensating variation and equivalent variation in income and explain how they can be used to measure the economic value of an environmental policy.

7. What is the relationship between the expenditure function and compensating and equivalent variation?

8. What are the income and substitution effects of a price change? How do they work together to lead a utility-maximizing hiking enthusiast to take more hikes as the price of each hike falls? What kind of policy might reduce the price of a hiking trip?

9. Explain the difference between a Marshallian and a Hicksian compensated demand curve. What are the uses of each in valuing environmental programs?

10. Distinguish the compensating variation, the consumer surplus, and the equivalent variation resulting from a price decline. Under what conditions are all three equal?

11. What errors could result from using the consumer surplus for valuing an environmental program that reduces the price of an environmental asset?

12. Explain the concept of weak complementarity. How can it be used to measure the benefits of a program that improves the quality of the environment?

13. Describe the four methods welfare economists have used to connect the welfare of individuals to produce a number that reflects the social welfare effects of an environmental policy.

14. What are the strengths and limits of measuring social welfare as a simple addition of individual welfare gains and losses?

15. Distinguish the following three methods for measuring an individual's value of an environmental quality improvement: revealed values, imputed values, expressed values.

16. Describe and explain the strengths and limits of each of the following methods for measuring the value of an environmental improvement: (1) market prices, (2) productivity, (3) hedonic pricing, (4) travel cost models, (5) avoided costs, (6) contingent valuation, and (7) benefit transfer.

17. How do the willingness-to-pay and the willingness-to-accept-payment measures of valuing an environmental improvement differ? Under what conditions are each the preferred measure?

18. Describe the dichotomous choice approach to contingent valuation and explain why it is a preferred method for conducting a contingent valuation analysis.

Problems

1. Valuing environmental improvements—the expenditure function: An expenditure function for a program that would reduce air pollution measures the minimum expenditure required to hold the citizen's welfare constant at the level reached under initial quality (Q^0) and income (Y) when the consumer faces improved higher quality (Q^1). Suppose that function is

$$E(Q^1, Q^0, Y) = Y\left(\frac{(Q^0)^B}{(Q^1)^B}\right)$$

where Y is income equal to \$50,000 and the coefficient B has been estimated by an environmental economist to equal 0.5. The compensating variation (CV) produced by a quality change from the old Q^0 to the new Q^1 is

$$CV = Y - E(Q^1, Q^0, Y)$$

a. What is the value of the expenditure function when both old and new quality are 4.0?
b. What is the value of the expenditure function produced by a policy that increases quality from 4.0 to 5.0? What are the benefits (CV) of the same environmental policy?
c. What are the benefits of an environmental policy that increases quality from a 3.0 to 4.0?
d. Both policies (improving quality from 3 to 4 as well as 4 to 5) increase quality by one unit. Can you think of a good reason why increasing quality from 3 to 4 (an increase of 1) produced a greater CV than increasing quality from 4 to 5 (also an increase of 1)?
e. Suppose the efficiency goal is to increase environmental quality as long as the marginal CV gain exceeds the marginal cost. The current quality level is 4.0. If each marginal unit of quality improvement costs \$3,000, to what level should quality be improved?

2. Value of open space protection—weak complements: Suppose a hiker's annual (compensated) demand for trips to an outdoor recreation site is given by the following equation:

$$\text{Demand for trips} = 6.0 - 0.5\,P + 0.0001\,Y + 1.0\,Q$$

Trips demanded are zero whenever the equation predicts negative trips; P is the hiker's travel cost (price), Y is expenditure, and Q is environmental quality, measured in acres of open space. Trips and environmental quality are weak complements, so when trips are zero the marginal benefit of a site quality improvement is zero. Travel cost is $10, expenditure is $40,000, and the site has 10 acres of open space.

a. At what price will demand for trips fall to zero?
b. The hiking area is currently under threat from urban expansion. At what price will demand for trips fall to zero if proposed urban expansion reduces acreage to 8?
c. If the price per trip is $10, how many trips are demanded annually and what is the hiker's compensating variation (CV) produced by current acreage?
d. If the price remains at $10, but urban expansion reduces the area to 8 acres, what is the CV?
e. What is the impact on the hiker's benefits (lost CV) from reducing acreage from 10 to 8?
f. What is the total loss (CV) from the land loss if 10,000 hikers have this demand function?
g. If the urban expansion produces $1 million per year in benefits, how large does the number of hikers have to be to justify economically protecting the open space from urban encroachment?

3. Using market prices to value oil spill prevention: The benefits of an oil spill prevention program include avoided costs, fewer disruptions to normal shipping operations, better community relations, and a cleaner environment. Every environmental manager dreads an oil spill. Cleanup is time consuming and expensive. A typical 1,000 gallon oil spill might cost $600,000 to clean up or $300,000 to prevent in the first place. Suppose a 1,000 gallon oil spill closes a commercial fishing area for a time. Your aim is to measure the total net economic gain from a program that prevents a future spill. The gain in net income to commercial fish harvesters from cleaning up the pollution is additional revenue minus additional cost:

$$\text{Added revenue} = P_f * H$$
$$\text{Added cost} = C_f * H$$

where P_f = price of fish at $5 per pound; C_f = cost of harvesting fish at $4 per pound; and H = added fish harvested per avoided oil spill = 400,000 pounds of fish. Calculate the annual gross benefits and annual net benefits per oil spill prevented. How low can the price of fish go for future prevention programs and still be economically feasible?

Extra credit: Suppose an oil spill prevention program increases the supply of fish by enough to reduce the price of fish from $5 to $4.50 per pound, with no impacts on costs of production. Is the program economically feasible at the lower price? Why or why not? (Hint: Think of the impact on consumer surplus.)

4. Valuing river restoration—hedonic pricing: Rivers can help develop economically depressed regions by providing water, transportation, and various environmental improvements. Environmental economists Hitzhusen, Ayalasomayajula, and Lowder (1999)[28] conducted a hedonic analysis of the benefits of upgrading the residential corridor near Ohio's Muskingum River. Among other things, they wanted to determine what establishing zoning ordinances would do to the value of river-based residential properties. Remember, the hedonic technique is a method for estimating implicit prices of characteristics that differentiate closely related products. The authors estimated a hedonic pricing function in which property value was the dependent variable and various property characteristics were independent variables; a simplified version is

$$P = 2{,}000 * \text{size}^{0.50}\ \text{zoning}^{0.25}$$

where P is selling price of a typical house in dollars; size is square footage; and zoning takes on a value of 1 or 2 (1 means the house is in a nonzoned area, 2 means it is zoned).

Suppose there are two typical house sizes, 1,500 square feet and 3,000 square feet; they are all near the river and there are 20 of the smaller homes and 30 of the larger ones. The entire area is currently nonzoned (zoning = 1). What are the total benefits resulting from establishing and enforcing zoning ordinances for the typical owner of the smaller home? What about for the larger home? Can you think of an explanation of why zoning ordinances increase the property value of a large house by more than a small one? What are the total benefits to owners of all 50 homes produced by the zoning ordinance? Zoning can be expensive and controversial. If the zoning ordinances costs $500,000, should it be done based on the efficiency criterion? Explain.

5. Valuing wetlands—dichotomous choice contingent valuation: Your city council is debating whether it should vote for a proposed environmental program to buy farmland and turn it into a wetlands capable of supporting migratory birds. You have promised to measure the benefits of the wetlands and have designed a survey to implement the dichotomous choice CVM method. Your results are shown in the accompanying table.

Would you vote for the program at the following stated (bid) price?	Proportion of respondents saying yes
0	1.0
5	0.9
10	0.8
15	0.7
20	0.6
25	0.5
30	0.4
35	0.3
40	0.2
45	0.1
50	0

[28] Hitzhusen, F. J., R. Ayalasomayajula, and S. Lowder. 1999. Economic valuation of a River Corridor: Integration of Natural Resource and Development Economics. Paper read at American Agricultural Economic Association, Nashville, TN, August 8–11, 1999.

These results can be expressed as the following equation that predicts the probability of a yes response based on the various bids:

$$\text{Prob (yes)} = 1.0 - 0.02 * \text{bid}$$

a. What is the mean willingness to pay for these respondents?
b. These bid responses came from a random sample of 1,000 (one tenth of 1% of the affected population). What are the total estimated benefits of the program to the population?
c. If the total cost is $30 million, what are the net benefits to the population?
d. How large must the population be to justify the proposed policy based on efficiency?

References

Berrens, R. P., A. K. Bohara, H. C. Jenkins-Smith, C. L. Silva, and D. L. Weimer. 2004. Information and effort in contingent valuation surveys: Application to global climate change using national Internet samples. *Journal of Environmental Economics and Management.* 47: 331–363.

Berrens, R. P., D. Brookshire, P. Ganderton, and M. McKee. 1998. Exploring nonmarket values for the social impacts of environmental policy change. *Resource and Energy Economics.* 20 (2): 117–137.

Bishop, R. C., and T. A. Heberlein. 1979. Measuring values of extra-market goods: Are indirect measures biased? *American Journal of Agricultural Economics.* 61: 926–930.

Brouwer, R. 2000. Environmental value transfer: State of the art and future prospects. *Ecological Economics.* 32 (1): 137–152.

Carson, R., T. Groves, and M. J. Machina. 2000. *Incentive and informational properties of preference questions.* Department of Economics, University of California, San Diego. Available from *http://econ.ucsd.edu/~rcarson/cgm.pdf* [Accessed October 28, 2004].

Carson, R. T. 2000. Contingent valuation: A user's guide. *Environmental Science and Technology.* 34 (8): 1413–1418.

Carson, R. T., N. E. Flores, and N. F. Meade. 2001. Contingent valuation: Controversies and evidence. *Environmental and Resource Economics.* 19 (2): 173–210.

Carson, R. T., and R. C. Mitchell. 1993. The value of clean water: The public's WTP for boatable, fishable and swimmable quality water. *Water Resources Research.* 29: 2445–2454.

Clark, J., J. Burgess, and C. M. Harrison. 2000. "I struggled with this money business": Respondents' perspectives on contingent valuation. *Ecological Economics.* 33 (1): 45–62.

European Commission. 2003. *Environment.* Available from *http://europa.eu.int/comm/environment/enveco/studies2.htm* [Accessed October 28, 2004].

Greene, W. H. 2003. *Econometric Analysis, 5th ed.* New York: Prentice Hall.

Gujarati, D. N. 2003. *Basic Econometrics, 4th ed.* New York: McGraw-Hill.

Hanley, N., S. Mourato, and R. E. Wright. 2001. Choice modelling approaches: A superior alternative for environmental valuation? *Journal of Economic Surveys.* 15 (3): 435–462.

Hitzhusen, F. J., R. Ayalasomayajula, and S. Lowder. 1999. Economic Valuation of a River Corridor: Integration of Natural Resource and Development Economics. Paper read at American Agricultural Economics Association, Nashville, TN, August 8–11, 1999.

Holl, K. D., and R. B. Howarth. 2000. Paying for restoration. *Restoration Ecology.* 8 (3): 260–267.

Kemp, M. A., and C. Maxwell. 1993. Exploring a budget context for contingent valuation estimates. In *Contingent Valuation: A Critical Assessment.,* edited by J. A. Hausman. Amsterdam: North Holland.

King, D., and M. Mazzotta. 2002. *Valuation of ecosystems.* Available from *http://cbl.umces.edu/~dkingweb* [Accessed October 28, 2004].

Ludwig, D., M. Mangel, and B. Haddad. 2001. Ecology, conservation, and public policy. In *Annual Review of Ecology and Systematics.* Palo Alto, CA: Annual Reviews.

Mäler, K. G. 1974. *Environmental Economics: A Theoretical Inquiry.* Baltimore and London: Johns Hopkins University Press.

National Gallery at Trafalgar Square London. 2004. *Ulysses deriding Polyphemus—Homer's Odyssey.* Available from *http://www.nationalgallery.org.uk/cgi-bin/WebObjects.dll/CollectionPublisher.woa/wa/work?workNumber=ng508* [Accessed October 28, 2004].

National Gallery of Art. 2004. *Moran.* Available from *www.nga.gov/feature/moran/early4.html* [Accessed October 28, 2004].

National Oceanic and Atmospheric Administration. 1993. *Report of the NOAA Panel on Contingent Valuation.* January 11, 1993. Available from *http://www.darp.noaa.gov/library/pdf/cvblue.pdf* [Accessed May 6, 2005].

Navrud, S., and G. J. Pruckner. 1997. Environmental valuation—To use or not to use? A comparative study of the United States and Europe. *Environmental and Resource Economics.* 10 (1): 1–26.

Nobel Foundation. 2004. *John R. Hicks 1972.* Available from *http://nobelprize.org/economics/laureates/1972/* [Accessed October 28, 2004].

Nyborg, K. 2000. *Homo economicus* and *Homo politicus:* Interpretation and aggregation of environmental values. *Journal of Economic Behavior and Organization.* 42 (3): 305–322.

O'Brien, B. J., K. Gertsen, A. R. Willan, and L. A. Faulkner. 2002. Is there a kink in consumers' threshold value for cost-effectiveness in health care? *Health Economics.* 11 (2): 175–180.

Randall, A. 1994. A difficulty with the travel cost method. *Land Economics.* 70: 88–96.

Smith, R. D. 2000. The discrete-choice willingness-to-pay question format in health economics: Should we adopt environmental guidelines? *Medical Decision Making.* 20 (2): 194–206.

Spash, C. L. 2000. Multiple value expression in contingent valuation: Economics and ethics. *Environmental Science and Technology.* 34 (8): 1433–1438.

U.S. Interagency River Basin Committee Subcommittee on Benefits and Costs Report to the Federal Inter-Agency River Basin Committee. 1958. *Proposed Practices for Economic Analysis of River Basin Projects, "The Green Book."* Washington, D.C.: U.S. Government Printing Office (originally published in 1950).

U.S. National Park Service. 2004. *Grand Canyon of the Yellowstone 1872. Thomas Moran.* Available from *http://www.cr.nps.gov/museum/exhibits/moran/grand.htm* [Accessed February 14, 2005].

University of Cambridge. 2004. *Alfred Marshall Society.* Available from *http://www.cam.ac.uk/societies/marshal/marshall.htm* [Accessed October 28, 2004].

Vatn, A. 2000. The environment as a commodity. *Environmental Values.* 9 (4): 493–509.

Willig, R. 1976. Consumer surplus without apology. *American Economic Review.* 65: 587–597.

Chapter 8

FORESTS

Introduction

Forests have provided people with shelter, food, fuel, medicines, building materials, and clean water throughout recorded history. About half the Earth's land area was covered by forests at the inception of agriculture 8,000 years ago, compared to about 30% in the year 2000 (Food and Agricultural Organization, FAO, 2001). Historically, increases in human population have been linked with reduced areas of forests. Agriculture expands and trees are cut down. Harvesting trees for timber goes back at least to the days of the ancient Greeks and others who roamed the Mediterranean region 2,500 years ago. A primary use of the area's forests was to build warships. Greece was one of the early cultures to pay the price of unsustainable forestry; they cut down too many trees too fast. Their rocky soils slowed tree growth, and by the fifth century B.C. there were few trees left.

Six major uses of forests are especially important today, and all ultimately rest on the ability of forests to support functioning ecosystems (see FAO, 2001):

1. Timber contributes to the economy and culture of forested regions of the world. It also serves as firewood in developing countries and places where substitute fuels are scarce.
2. Forests produce grazing forage for livestock.
3. Forests produce *carbon sinks,* absorb other air pollutants, and produce oxygen.
4. Forests provide shelter for wildlife, which in turn provides benefits to people who value watching or hunting wildlife.
5. Forests contribute to water production.
6. Forests act as a resource for outdoor recreation.

Even when timber is the only economical product of a forest, the forest manager faces the huge challenge of deciding how to optimize yields on a given amount of land and when to harvest and replant trees. But timber is not the only forest resource that people value. A delicate balance must be struck among the various possible uses of forest resources as well as on the joint management of forests' environmental resources. Since harvesting trees now often displaces other valuable resources to people—future timber, water, recreation,

Outline

Forest Management 190

U.S. Forest Legislation 211

Designs for Efficient, Equitable, and Sustainable Forestry 212

Objectives

After studying this chapter, you should be able to

- Describe the biological characteristics of forests
- Compare different economic principles for supporting economically efficient and sustainable forest management
- Describe selected U.S. forest policy legislation
- Identify a range of institutional approaches that have been applied internationally to improve efficiency, equity, and sustainability of forests

and wilderness–establishing a balance requires some means of comparing the economic value of competing uses. This chapter describes how the use of economic concepts in connection with forest ecology can promote improved management of forest resources.

Forest Management

Like many other renewable resources, *timber* is both a final output that people directly use and a capital good that is used to make future output. Trees, when harvested, provide a marketable good; but left standing, trees are a capital resource, providing for additional growth the next year. Every year the forest manager must decide whether to harvest a particular stand of trees or wait for additional growth. In contrast to many other living resources like crops, the time period between initial investment (planting) and recovery of that investment (harvesting) is long.

Forestry is typically subject to a large variety of unpriced benefits and costs, which are associated with either the standing timber or harvested timber. Unpriced benefits and costs make it difficult for the private sector to implement economically efficient forest management. Unpriced effects place special challenges in designing institutions to promote the forestry objectives of society.

Single-Use Forest Management: Timber

Timber production occurs over long time periods, whether the timber is used for recreation, commercial purposes, or endangered species habitat. For example, growing a tree to maturity can take from under 20 years in humid climates[1] to 300 or more years in some dry mountain areas. Time is not important in most production cycles because the production time between planting and harvesting is sufficiently short for time to make little difference. For biological processes like crop production, planting typically occurs in the spring and harvest takes place in the fall, so time does not produce an added cost. But for timber production, time is costly (i.e., the cost of financial capital tied up in waiting for an economic return): Planting costs, other uses of the land, and the opportunity to sell the land for other purposes are tied up for many years (Rideout and Hesseln, 2001). A central *economic* question of timber production, one for which economic theory offers important insights for public and private forestry, is that of determining which trees should be cut and taken to the sawmill and which ones should be left standing in the woods (Gregory, 1955, 1987).

Forest Ecology. Tree growth is typically measured in volume (e.g., cubic feet per acre of the tree), excluding the bark and limbs, between the bottom and a four-inch top diameter. Only standing trees are measured (Tietenberg, 2003).

Trees typically go through several growth periods. Young trees may grow fast in height, but their volume growth is slow. Later, rapid and sustained volume growth occurs, with volume growing at an increasing rate. In the last phase, as a stand of trees matures, volume increases but by smaller amounts each year. Any given stand of trees has a unique growth pattern that depends on several factors: the tree species, weather, climate, soil, tree spacing as influenced by the competition for sunlight, air pollution, vulnerability to fire, previous harvesting, and management intensity.

[1] Shorter rotations are possible. For example, rotations as short as seven years are common in the U.S. Pacific Northwest for trees of the genus *Populus*, which are grown for paper pulp. An experimental tree-planting project in Ojinaga, Mexico, has shown that rotations as short as five years are possible with adequate water supply.

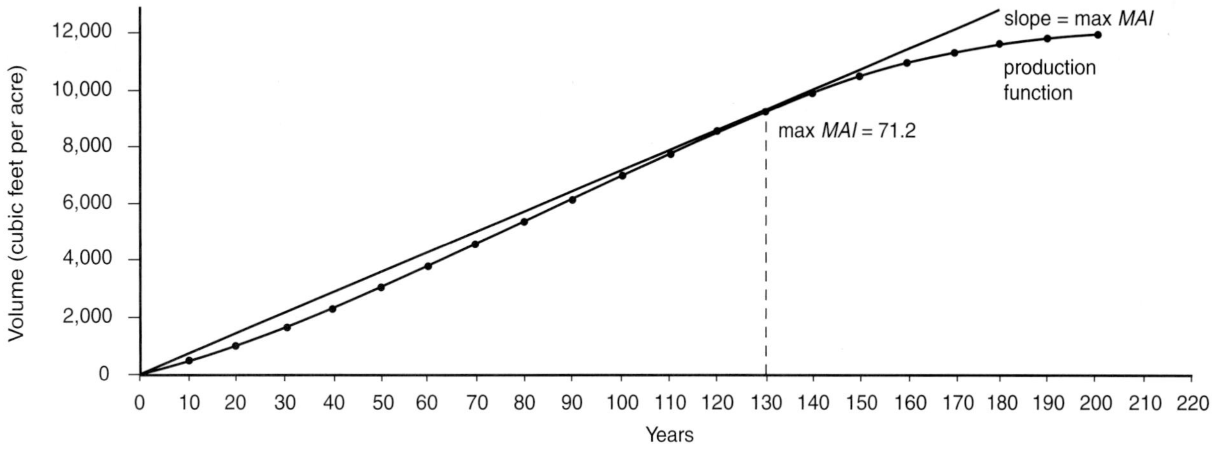

Figure 8–1 Tree stand growth pattern, Douglas fir. *MAI* = mean annual increment.

Despite the considerable differences that exist from one species and stand to the next, all have similar biological growth properties. Figure 8–1 shows the pattern of a typical stand of Douglas fir trees in British Columbia in the North American Pacific Northwest. The figure shows all the growth phases described, from years 0 to 200. In the early years, growth is slow, much faster growth occurs in the middle years, and growth nearly stops after 200 years.

The question facing the forester is that of the *year* in which the stand be harvested. If replanting occurs after the harvest, that year is called the *optimal rotation* period. Answering this question requires comparing the benefit to the cost of delaying a harvest. By delaying the harvest for one more year, the forester derives a future benefit in terms of a larger volume of salable timber. However, the wait imposes several costs: (1) value of interest foregone on timber that could have been sold now and banked, (2) lost opportunity to use the land cleared of timber for other purposes such as urban development, and (3) lost income from being unable to sell the cleared land.

The biological answer to the forester's question is based on a concept known as the *mean annual increment* (MAI). The MAI is an important concept, both as a stand-alone biological principle and for the contrast it provides with the economic way of thinking. The MAI is computed by dividing the stand's cumulative volume at the end of each time period (e.g., 20 years) by the total number of years the stand has been growing since being planted.[2] The MAI is measured in volume per unit land per unit time (e.g., cubic feet per acre per year). It is the average productivity for the variable input time, and it is computed like the average productivity of any other scarce resource (e.g., labor, capital, or water). The MAI, which measures the average growth of wood per year, changes in different growth phases in a stand's life: It is lowest in the early years, highest in the middle years, and slowly decreases with age.

The point at which the MAI is highest is commonly used by biologists to identify the biological maturity of the stand and its readiness for harvesting. By maximizing the MAI, the forester harvests at the period in which the volume of timber the stand produces per acre per year is maximized. In Figure 8–1, the maximum MAI is found by drawing a series of straight lines from the origin to the production function, then picking the single line whose slope is as high as possible. The line's slope for any given year (total volume divided by total years) is that year's MAI. By finding the year in which the line's slope is maximized, the forester identifies the rotation period for which volume divided by age is as large as possible.

[2] Analyzing the volume of trees over time presents its own challenges. Bar-Hen (2002) illustrated some of difficulties in estimating tree growth of *Eucalyptus* in the Congo, Africa.

Table 8–1 Biological rotation analysis, Douglas fir

(1) Year	(2) Volume[a] (cubic feet/acre)	(3) MAI[b] (cubic feet/acre/year)	(4) Annual Increment[c] (cubic feet/acre/year)
10	448	44.8	–
20	984	49.2	53.6
30	1,596	53.2	61.2
40	2,272	56.8	67.6
50	3,000	60.0	72.8
60	3,768	62.8	76.8
70	4,564	65.2	79.6
80	5,376	67.2	81.2
90	6,192	68.8	81.6
100	7,000	70.0	80.8
110	7,788	70.8	78.8
120	8,544	71.1	75.6
130	9,256	**71.2**	71.2
140	9,912	70.8	65.6
150	10,500	70.0	58.8
160	11,008	68.8	50.8
170	11,424	67.2	41.6
180	11,736	65.2	31.2
190	11,932	62.8	19.6
200	12,000	60.0	6.8

[a]Data are from the Province of British Columbia (2001). The equation expressing the relation of wood volume (V) to age in years was obtained by fitting a polynomial regression equation to the original data using $V = 40 \text{ years} + 0.5 \text{ year}^2 - 0.02 \text{ year}^3$.

[b]MAI = mean annual increment, equal to the average product of time, defined as volume in column 2, divided by years in column 1.

[c]Annual increment is equal to the marginal product of time, defined as added volume in column 2, divided by the added number of years in column 1 (i.e., 10 years).

A related concept known as annual increment (AI), is the *added growth per year.* The AI is measured as the gain in volume divided by the added number of years that produced the added volume, shown in Table 8–1, column 4. The AI is the year's *marginal* growth while the MAI refers to the *average* growth from the planting year to the current year.

Table 8–1 shows that rotation based on maximum MAI occurs when the stand is 130 years old, which is also shown in Figure 8–1. Columns 3 and 4 of the table indicate why this age is so important. As long as AI is larger than MAI, the marginal growth exceeds the average growth. AI is larger than MAI for the first 130 years, which means that the MAI must be increasing. (Think of your grade point average: As long as your semester's (marginal) GPA exceeds your cumulative (average) GPA, your cumulative GPA increases.) However, when the AI is less than the MAI, which happens in all years after 130, the MAI must be falling. The MAI reaches a maximum when the AI equals the MAI, and this occurs when the stand is 130 years old.

Forest Economics. The forester who cuts the stand at the year of maximum MAI uses only the physical aspects of the forest to arrive at an optimal rotation. He ignores the economic factors affected by increasing the waiting time to harvest. In fact, the price of timber, the costs of planting trees and harvesting timber, and the discount rate all should have a crucial influence on the income produced by harvesting timber.

Example 8–1
Forest Policy Challenges in Eastern Regions of Russia

Timber harvesting is a major income producer in far eastern regions of Russia. Forest resources there are huge, making up nearly 40% of Russia's total. Backman and Zausaev analyzed the balance between supply and demand for forest products there. Official data report that forest harvests have continued to decline since 1989. However, Backman and Zausaev believe the official data may understate much activity because of lack of reporting to avoid taxes. Little of Russia's far eastern forest sector has developed strong manufacturing capabilities. Instead, the region exports most of its wood for logs and poles.

Eastern Russian forest resources are harvested under a system in which prices *monitor* forest products but do not *allocate* them; that is, forest product prices do not reflect actual scarcities. This forest sector's capacity to develop in the future will be successful only if prices are high enough to pay for the costs of production, including the cost of supplying capital to meet expansions of the forest manufacturing base. Prices based on demand and supply of forest products will have an important impact only if the region's forests are economially accessible to support a domestic manufacturing industry. Letting forest prices adjust to match supply and demand will encourage market forces to access the forests and develop manufacturing industries.

Sources: Backman, C. A., and V. K. Zausaev. 1998. The Forest Sector of the Russian Far East. *Post-Soviet Geography and Economics.* 39 (1): 45–62. Krasnoyarsk Center for Forest Protection. 2004. *All about Russian Forests.* Available from *http://www.forest.ru/* [Accessed September 30, 2004]; World Wildlife Fund. 2004. *Pskov model forest project.* Available from *http://www.wwf.ru/pskov/eng/* [Accessed September 30, 2004].

The economically rational forester maximizes income by choosing a rotation that *maximizes the present value of net income produced by the wood.*[3] See Example 8–1. In attempting to maximize net income, the forester could plan for either the short run or the long run. The short-term view consists of a single rotation in which the stand is harvested once and the forester walks away with the income and leaves the land to regenerate on its own. The long-term view consists of an infinite planning horizon of repeated cycles marked by harvesting, planting, and waiting for future growth.[4]

Single Rotation. In the short-term view, a forester considers only a single rotation, harvests the stand once, takes the income, and ignores later land uses. Although this is a simplified explanation, it brings out important insights and raises important questions about the difference between unsustained and sustained forestry. Single-rotation analysis incorporates essential economic information into the optimal harvest decision. Table 8–2 shows how economic information is combined with biological factors from Table 8–1 to make economic harvesting decisions.

Recall that income is maximized by choosing a rotation that maximizes the present value of net income produced by the wood. This analysis requires knowing the present value of revenue and the present value of cost (see Chapter 5). Let us begin by analyzing revenues. Revenue analysis

[3] The principle of economically rational forest management has been around for a long time. For example Appuhn (2000) documented the emergence of economic rationality in Italian forest management as early as the 1400s. Maximization assumes that forestry is the only use of the land. If other land uses, such as farming, compete with forestry, their potential income is an opportunity cost of producing wood. For example, Angelsen and Shitindi (1999) found that even subsistence farmer-foresters in Tanzania, Africa, will expand their farms onto forested lands if crop prices increase by enough to make commercial farming profitable. Price (1997) presents an insightful analysis of the history of forest management objectives. One such objective that was widespread in England in the late nineteenth and early twentieth centuries was the undiscounted mean net annual income, sometimes called forest rental. Forest rental is equal to the net present value evaluated at a zero discount rate. As Price explains, the use of forest rental as a management objective ignores alternative uses of capital tied up in the forest.

[4] Taking the long view requires security of property rights to future forest revenues. For example Boscolo and Vincent (2000) illustrate the importance of property rights in their findings from Malaysia, where harvesters adopted more sustainable logging practices when they had renewable harvest rights to state forests.

Table 8–2 Economic analysis of single rotation, Douglas fir

		Revenue		Cost			Income	
(1) Year	*(2)* Volume (cubic feet/ acre)	*(3)* Revenue ($)	*(4)* PV revenue[a] ($)	*(5)* Planting cost ($)	*(6)* Harvesting cost ($)	*(7)* PV total cost[b] ($)	*(8)* Income ($)	*(9)* NPV income[c] ($)
10	448	896	811	500	179	662	217	149
20	984	1,968	1,613	500	394	823	1,074	790
30	1,596	3,192	2,368	500	638	974	2,054	1,395
40	2,272	4,544	3,052	500	909	1,110	3,135	1,942
50	3,000	6,000	3,648	500	1,200	1,230	4,300	2,419
60	3,768	7,536	4,148	500	1,507	1,330	5,529	2,819
70	4,564	9,128	4,549	500	1,826	1,410	6,802	3,139
80	5,376	10,752	4,850	500	2,150	1,470	8,102	3,380
90	6,192	12,384	5,058	500	2,477	1,512	9,407	3,546
100	7,000	14,000	5,176	500	2,800	1,535	10,700	3,641
110	7,788	15,576	5,213	500	3,115	1,543	11,961	**3,671**
120	8,544	17,088	5,178	500	3,418	1,536	13,170	3,642
130	9,256	18,512	5,078	500	3,702	1,516	14,310	3,562
140	9,912	19,824	4,923	500	3,965	1,485	15,359	3,438
150	10,500	21,000	4,721	500	4,200	1,444	16,300	3,277
160	11,008	22,016	4,480	500	4,403	1,396	17,113	3,084
170	11,424	22,848	4,209	500	4,570	1,342	17,778	2,867
180	11,736	23,472	3,915	500	4,694	1,283	18,278	2,632
190	11,932	23,864	3,603	500	4,773	1,221	18,591	2,383
200	12,000	24,000	3,280	500	4,800	1,156	**18,700**	2,124

Discount rate (r) = 0.01. Price = $2 per cubic foot.

[a]PV (price times volume)/$(1+r)^t$

[b]Planting cost + (harvesting cost times volume)/$(1+r)^t$

[c]NPV = present value of revenue − present value of total cost

attempts to understand the benefits of the timber harvested at different possible times. *Stumpage* (i.e., standing timber) revenue, or simply revenue for short, is defined as price, assumed for now to equal to $2 per cubic foot, multiplied by the volume of output. This revenue is what the forester or stand owner receives by harvesting at different possible rotation ages. Revenue can be increased by increasing output, which comes from delaying the harvest and letting the stand grow.[5]

Table 8–2, column 3, shows undiscounted revenues at different possible rotation ages, increasing from a low of $896 if the stand is harvested in year 10 to a high of $24,000 if harvested in year 200. For any given rotation, only one of the rows applies. For example, once the owner recoups $4,544 in revenue by harvesting in year 40, the $6,000 in revenues that could have been earned by delaying the harvest until year 50 no longer applies, for the trees are now cut and gone. Discounted revenue (column 4) is computed by discounting nominal revenue (column 3) using a discount rate, which is assumed for now to be 1% (0.01).

[5] Berck and Bentley (1997) found marginal net revenue (marginal revenue minus marginal cost) that could have been earned by keeping California's Redwood National Park in private hands grew at about the rate of compound interest. This spectacular redwood (*Sequoia sempervirens*) is the world's tallest known plant species. In California's north coast, the perfect combination of latitude, climate, and elevation limits the redwoods' range to a few hundred coastal miles.

Next consider two important costs: planting and harvesting. Planting costs are a fixed cost per acre incurred in year 0 at the time of planting; for that reason, the costs are not discounted. Table 8–2, column 5 shows the example of $500 planting costs, equal in all possible harvest periods.

Harvesting costs assumed to be $0.40 per cubic foot harvested, are variable costs and are not incurred until the trees are harvested. Column 6 shows undiscounted harvesting costs, which increase from a low of $179 if the stand is harvested in year 10 to a high of $4,800 if harvested in year 200. Because these costs are not incurred until the stand is harvested, they are discounted back to the present to tell the forester the current equivalent cost of incurring those costs so many years in the future. The logic of discounting is based on this realization: If harvesting costs will be paid out as X dollars to harvest the stand 100 years in the future, the forester only needs to find $X/(1 + r)^{100}$ dollars now, put the money in the bank for 100 years earning r per year in interest, then withdraw the money in year 100 to pay the costs that need to be incurred then. Column 7 shows the present value of total cost, equal to $662 in year 10, which is equal to $500 (planting costs) plus $\$170/(1 + 0.01)^{10} = \$162$. It reaches a maximum of $1,543 in year 110, then falls back to $1,156 by year 200.

Now we are ready to compute net present value. We have all the information on revenues and costs to compute the NPV of income that would be produced from harvesting the stand at each possible age. Table 8–2, column 9 shows this NPV, obtained by subtracting the present value of total costs (column 7) from the present value of revenue produced by harvesting and selling the timber (column 4). Net present value of income is maximized at a rotation period of 110 years.

Now that we have computed NPV, we can perform an analysis using it. Table 8–2 illustrates three important economic principles of forest management.

1. The MAI which is so important to the biology of forest harvesting, bears little relation to economic decision making. Remember, by harvesting at the period that maximizes MAI, the forester maximizes the average annual volume of harvested wood. Economics tells us to ask, What is the addition to *discounted net* revenue produced by waiting one more period? By asking that question, economics lets us account for all the biological facts plus the all-important information on costs, revenues, and the discount rate.

2. The following results illustrate a second important economic principle: If the discount rate is zero, then whenever the added revenue per cubic foot (price) exceeds the added harvesting cost (variable cost per cubic foot), it pays to delay the harvest even if the added growth is small. Table 8–2 shows that if the discount rate is zero, income is maximized by delaying harvest until year 200. This rather remarkable result comes from comparing the *added* cost and *added* revenue produced by a wait. The addition to revenue of waiting from year 190 to year 200 is 68 extra cubic feet of wood from the added stand growth (12,000 − 11,932 in column 2) multiplied by its $2 (undiscounted) price per cubic foot: $68 \times \$2 = \$136$. The added (undiscounted) cost is the variable harvest cost per cubic foot, $0.40, times the added volume that the wait produces in harvested timber. That added volume, 68 cubic feet, requires us to incur $27 more in harvest cost ($4,800 − $4,773 in column 6). Again, the wait has no influence on fixed planting costs. Where the discount rate is zero, the biological optimum will always produce an earlier harvest than the economic optimum.

3. The third economic principle centers on the remarkable power of the discount rate itself to influence long-term decisions where growth occurs. The discount rate shortens the optimal rotation, and a higher discount rate shortens it even more. Even the very low discount rate of 1% shortens the optimal rotation by *nearly half* from 200 to 110 years. Table 8–2 shows that nominal (undiscounted) net income is maximized by harvesting in year 200, but the very slight change of discounting net income at only 1% shortens the income-maximizing rotation by 90 years. At this discount rate, the NPV of income is maximized by harvesting in year 110, which produces a value of $3,671.

For the case of a single rotation, the following formula shows why higher discount rates depress the NPV of income:

$$NPV = \frac{SV - HC}{(1 + r)^t} - PC \tag{8.1}$$

where

NPV = net present value ($3,671)

SV = undiscounted stumpage value ($15,576)

HC = undiscounted harvest cost ($3,115)

PC = planting cost ($500)

r = discount rate (0.01)

t = rotation period (year 110)

Numbers in parentheses are the values for year 110. Equation 8.1 is used to produce net present value, the last column in Table 8–2.

The secret to understanding the power of the discount rate to erode future revenues and costs is learned from watching how it affects the marginal cost versus marginal benefit of delaying the harvest. Even if the harvest cost is zero, watch what happens by delaying harvest from year 110 to year 120. Nominal (undiscounted) revenues increase by $1,512 from $15,576 to $17,088; the 10-year wait increases volume by 756, from 7,788 to 8,544, which is applied to the price of $2 per cubic foot. But watch out for the power of the discount rate.

The *marginal benefit* of delaying the harvest from year 110 to year 120 is $458. The marginal benefit is the discounted added revenue brought in by letting the timber stand grow for 10 more years. It is computed as $\$2\,[8{,}544 - 7{,}788]\left[\frac{1}{1.01^{120}}\right] = \$458$. It equals the price ($2) times the gain in volume (8,544 − 7,788) multiplied by the discount factor applied to the period in which the added revenues are finally received $\left[\frac{1}{1.01^{120}}\right]$.

The *marginal cost* incurred by this same delay from year 110 to 120 is $494. Marginal cost from waiting occurs for reasons that are less obvious at first but are important. It is incurred because of increasing the waiting time to receive nominal receipts of $15,576. Without the wait, these nominal revenues would have been received by cutting and selling the 7,788 cubic feet in year 110. However because of the 10-year added wait, even the first 7,788 cubic feet cannot be received until year 120.

Waiting an extra 10 years to receive the same nominal revenue imposes important costs: Year 110 revenue is tied up in the ground waiting for trees to grow; this is revenue that could be working to earn a return in those same 10 years. That marginal cost is measured as $\$494 = \$2\,[7{,}788]\left[\frac{1}{1.01^{110}} - \frac{1}{1.01^{120}}\right]$. It is the price ($2) times volume in period 110 (7,788) multiplied by the added discounting erosion caused by waiting[6] that delays the harvest from year

[6] Waiting until year 110 erodes nominal receipts of $15,576 to a value of 0.3347 times $15,576, or $5,213. But delaying the harvest even more to year 120 increases the erosion on unchanged nominal receipts of $15,576 to the more highly eroded 0.3030 times $15,576, which is $4,719. Therefore, the harvest delay from year 110 to year 120 erodes the already heavily discounted revenue by an added $494, although nominal revenues are unchanged. The $494 is the marginal cost of the delay.

Table 8–3 Net present value analysis, sustained harvesting, Douglas fir

	Discount rate = 0.01				Discount rate = 0.03			
	Single rotation		Infinite rotations		Single rotation		Infinite rotations	
(1) Year	(2) Constant P[a] ($)	(3) Growing P[b] ($)	(4) Constant P ($)	(5) Growing P ($)	(6) Constant P ($)	(7) Growing P ($)	(8) Constant P ($)	(9) Growing P ($)
10	149	179	1,572	1,892	33	58	130	228
20	790	923	4,379	5,112	372	461	833	1,033
30	1,395	1,710	5,404	6,624	552	727	**939**	1,236
40	1,942	2,519	5,913	7,673	**614**	878	886	**1,266**
50	2,419	3,331	6,170	8,497	595	**937**	771	1,214
60	2,819	4,121	**6,270**	9,167	523	925	630	1,114
70	3,139	4,869	6,257	9,705	422	861	483	985
80	3,380	5,553	6,159	10,118	308	761	340	840
90	3,546	6,156	5,994	10,405	193	640	207	688
100	3,641	6,660	5,776	10,567	83	508	87	536
110	**3,671**	7,054	5,517	**10,603**	−18	374	−18	389
120	3,642	7,328	5,225	10,514	106	241	−109	252
130	3,562	7,479	4,909	10,306	−183	124	−187	126
140	3,438	**7,504**	4,574	9,983	−247	14	−251	14
150	3,277	7,407	4,227	9,555	−301	−83	−304	−83
160	3,084	7,194	3,872	9,033	−344	−166	−348	−168
170	2,867	6,875	3,515	8,427	−380	−237	−382	−239
180	2,632	6,460	3,159	7,754	−408	−296	−410	−297
190	2,383	5,965	2,806	7,026	−431	−344	−432	−345
200	2,124	5,405	2,461	6,261	−448	−383	−449	−384

Entries are net present values from year zero to the given year.

[a]Timber price is assumed to be a constant $2 per cubic foot for all years and all rotations.

[b]Timber price is assumed to increase linearly with volume per acre: price = $2 + (1/6,000) volume.

110 to year 120, equal to the term $\left[\frac{1}{1.01^{110}} - \frac{1}{1.01^{120}}\right]$. In summary, the wait from year 110 to year 120 incurs a larger marginal cost ($494) than a marginal benefit ($458), which makes it an economically poor timber management decision. The NPV of revenues (as well as revenues minus cost) is maximized by harvesting at year 110.

Now let us see how a single, optimal, timber harvest rotation changes when either a new discount rate or a new timber price pattern occurs (Table 8–3). Say the discount rate is increased from 1% to 3%. Comparing column 2 with column 6 shows the dramatic change in both harvesting cost and stumpage revenue when the discount rate is raised. The power of that increase in the discount rate to erode revenue and cost streams is shown by comparing those two columns. For example, at the 10-year rotation, you can see a large decline in NPV of income from $149 to $33.[7]

[7] Table 8–3 shows only net present value, but its underlying calculations can be replicated. For harvest year 10, discounted revenues fall from $811 at a 1% discount rate to $666 under a 3% rate, reflecting the greater impatience of waiting the same amount of time (10 years) for the same amount of revenue. Similarly discounted harvested costs fall from $162 to $133. Planting costs are not discounted since they occur in the first year. Net present value of income falls from $149 (811−162−500) to $33 (666 −133−500).

Column 6 of Table 8–3 illustrates the important economic principle that raising the discount rate has two effects: First, it reduces net income for any given rotation length; second, it reduces the number of years for which the optimal rotation occurs.[8]

The first effect of raising the discount rate is shown by comparing the net present income in column 6 with its much larger value in column 2 under a 1% rate. Not only is net present income much smaller with the higher discount rate, but it turns negative after year 110, reflecting the fact that a wait that long at a 3% rate simply cannot pay back the fixed planting costs that occurred 110 years earlier.

The second effect of the higher rate is shown by the optimal rotation period itself, which falls dramatically from a 110-year rotation at 1% to a much reduced 40-year rotation at 3%. These two effects of a rising discount rate suggest its impact on decision making. Higher discount rates will cause foresters to harvest their trees sooner. Moreover, because absolute discounted returns are lower, investable capital will be removed from long-term renewable resource operations like forestry and will be put in activities that produce a faster return and show more tolerance of higher discount rates. The history of private forestry practices bear out these theoretical insights: Periods of higher interest rates are typically associated with greater timber harvests and associated lower timber prices produced by shortened optimal rotations.

So far we have assumed a constant price of timber for all volume levels. In fact, the price of timber may rise with the stand's age. Aging can boost the *stumpage value* (e.g., price per cubic foot) because a higher percentage of the tree can be used for highly valued products such as sawtimber and veneer logs. (Sawtimber is highly-valued timber suitable for sawing into lumber; veneer logs are also highly valued because wood is peeled off the log into thin slices that are used to cover inferior wood.) Where timber price does rise with the stand's age, economic theory tells us that this price increase should make it pay to wait a few more years to the optimal harvest. Indeed, Table 8–3 bears out this prediction. Column 3 shows that a growing timber price[9] under a discount rate of 1% both increases the discounted net revenues in all periods and increases the optimal harvest date itself. Column 7 shows the same principle applied to a discount rate of 3%. In column 6 the optimal harvest date is 40 years; harvest is increased to 50 years if the forester expects the added volume to produce a higher price per cubic foot.

Even with a rising timber price, the eroding effect of the discount rate is pronounced, as shown by comparing Table 8–3's columns 3 and 7. At a 3% discount rate, the forester waits 50 years to the optimal harvest, but the much longer and more patient 140-year rotation period occurs at a discount rate of 1%. Higher discount rates shorten the waiting time before the stand is harvested. They also shorten the wait-lengthening time produced by expectations of rising prices.[10] Why? Because high discount rates raise the cost of waiting to harvest. Therefore, the added benefit of waiting for the higher price is squeezed when the penalty is higher.

Sustained Management. Despite the simplicity of the single-rotation model just described, it is unrealistic in several ways. Its biggest shortcoming is that it considers the decision to harvest the stand as a one-time event, rather than as part of more realistic *sustained* management, which consists of an infinite sequence of harvesting, replanting, and waiting for growth. This sequence of sustained management is sometimes called continuous management. Even under an infinite sequence of repeating actions, the economically rational decision remains to choose the plan that

---

[8] In some cases, the optimal rotation can fall to zero years; that is, the forester should not plant in the first place, allowing the land to be converted to other uses or left fallow. A zero optimal rotation can be realistic whenever there are large risks associated with timber plantations. Establishing a plantation requires assumptions about future market conditions years into the future, which introduces considerable uncertainty.

[9] For illustration purposes, stumpage price increase with the stand's age is assumed to follow the pattern of starting at $2 per cubic feet for the smallest trees, and increasing linearly to $4 per cubic feet when the stand reaches 12,000 cubic feet per acre. This pattern assumes that price varies only with volume and not with time itself.

[10] The 30-year wait-lengthening time, shown by comparing columns 2 and 3, is reduced to 10 years when comparing columns 6 and 7.

maximizes the present value of net income. The first person to analyze continuous forest management was the German forester Martin Faustman (1849); Samuelson (1976), Hartman (1976), and Bowes and Krutilla (1985) expanded the analysis.

You might ask, "Isn't sustained management really the same problem as the single-rotation case, but simply repeated an infinite number of times?" The answer is no. Sustained management considers the long-term future, where as single-rotation management ends the time horizon at the first harvest. The same principle tells you to expect that the elected official who announces she is not running again will not worry much about consequences of her actions that occur after she is out of office. So she may make political decisions that are very different from those of a person who is seeking reelection. Similarly, the forester who bears responsibility for results long after a stand is first harvested will consider the impacts of a current decision on the *infinitely recurring sequence* of costs and revenues.

Under sustained management, the decision to delay a harvest incurs an added cost that does not exist in the single-rotation model, namely, the cost of the future stream of income lost by delaying the start of the next cycle of planting and harvesting. Remember, under single-rotation management, the forester maximizes NPV by harvesting in year 110 when there are constant prices and a 1% discount rate. This forester who wishes to *maximize* NPV knows that any wait for future growth must be paid for by added revenue produced by the added growth. In any period earlier than year 110 (e.g., harvesting in year 100), the marginal benefit of added tree growth more than pays for the marginal cost of delaying the net revenues. In any period after year 110 (e.g., harvesting in year 120), the marginal benefit fails to pay for the added growth.

Under sustained management, there is one important marginal cost of waiting that does not exist under single-rotation management. It is the marginal cost of delaying *all future incomes* by one period because *all future income-earning cycles* are now delayed by one period. Under sustained management, the efficient forester will harvest the stand earlier in order to start earning future incomes sooner.

The formula for computing NPV under sustained management (derived in Appendix H) is given by

$$NPV = \frac{SV - HC}{(1 + r)^t - 1} - \frac{PC}{1 - (1 + r)^{-t}} \tag{8.2}$$

where all variables are defined as in Equation 8.1. This equation is used to derive the four infinite rotation columns in Table 8–3. The table compares sustained management to single-rotation management for both a discount rate of 1% and 3% for the case of constant timber prices and growing prices. Results illustrate two important economic principles.

The first principle states that NPV under single-rotation management is always lower than the NPV of an identical decision made an infinite number of times under sustained management. This makes sense. Whatever NPV of income is earned by harvesting under single-rotation management will be earned an infinite number of times under sustained management. (This result also is true if management produces a negative NPV: If a forest owner loses a small amount of money under single-rotation management, losses are magnified with an infinite number of rotations.) However, future NPVs are heavily discounted as the rotation periods lengthen, so any rotation after the first will have a much smaller discounted value than its value at the time of first harvest. Table 8–3 illustrates this principle: Compare column 2, NPV of income under single-rotation management, to column 4, the equivalent outcome produced by sustained management. Even for the very short rotation of 30 years under constant timber prices, application of the NPV formula from Equation 8.1 for a 1% discount rate is $1,395 for the first rotation. (Table 8–3, column 2 equals Table 8–2, column 9.) However, because $1,395 will be earned in year 30, year 60, year 90, and so on, the total discounted NPV for all those future years is $5,404, which is shown in column 4.

The second important principle is that the optimal rotation period is *reduced* by shifting the time horizon from single-rotation to sustained management, because sustained management raises the marginal cost of waiting. That marginal cost of waiting is the marginal cost of delaying *all future incomes* by one period because *all future income-earning cycles* are now put off by one period. Table 8–3 illustrates this principle by showing the higher marginal cost of waiting under sustained management. At a 1% discount rate with no price increases (column 2), the optimal single-rotation management is 110 years, producing a maximum NPV of $3,671. Introduction of sustained management with the same discount rate and price pattern reduces the optimal rotation from 110 by nearly half to 60 years. If the manager simply repeated the 110-year optimal single-rotation plan forever, discounted net income would rise from $3,671 (column 2) to $5,517 (column 4). But the manager can do considerably better by taking advantage of an infinite horizon of earlier income earning opportunities by reducing the optimal rotation from 110 years to 60 years. This action raises the NPV of the infinite cycle of future incomes from $5,517 to $6,270.

Table 8–3 shows that these two principles apply regardless of the discount rate and timber price growth pattern. For a 1% discount rate, column 3 shows that growing timber prices produce a 140-year optimal rotation, earning a $7,504 in net present value. Under sustained management, the optimal rotation falls to 110 years, earning a much higher $10,603. A similar pattern is shown for the higher discount rate of 3%, although the higher rate reduces the difference between single-rotation and sustained management. The optimal rotation under constant timber prices falls from 40 years (single-rotation management) to 30 years (sustained management); under growing prices, the optimal rotation falls from 50 years (single-rotation management) to 40 years (sustained management).

The discussion so far has assumed the forester has only a single stand of trees marked by long periods of waiting combined with short periods of harvesting and replanting. Every efficient forester knows, however, that sustainable harvests can occur every period, as long as they occur on different stands. A forester who builds a sawmill, lumberyards, other wood processors, and a wood-dependent society needs a steady stream of timber harvested year after year. One would expect the efficient forester to divide the forest into a number of stands of different aged timber and to practice what is called *fully regulated forestry.* For example, if the optimal rotation on one stand is 100 years, the forester would divide the forest into 100 similarly sized stands, each of a slightly different age. A different stand reaches maturity each year. In each year, that year's total forest growth is harvested, although the entire forest's annual growth is concentrated into the output of the single harvested stand. A single stand is completely harvested at the end of its rotation, but as long as replanting continues, the stand's complete harvest need never reduce the forest's principal. Sustained yield harvesting can be practiced successfully if the forest's annual growth (interest) is harvested each year without eroding the principle.[11]

Multiple-Use Forest Management

Most forests produce more than timber, and the multiple-use concept is the foundation of modern forestry. In the United States, since multiple use became required by various acts of Congress in the 1960s, the public and most foresters have focused on outputs other than timber. As a result of legislation, court actions, journalism, television, and special interest groups, society today expresses considerably more interest in public and private forestry decisions than in years before 1970. Then, the harvested timber industry exerted significant economic and political influence.

Multiple Goals. Federal forest managers who are required to carry out the U.S. Congress's mandate of multiple-use management are challenged by conflicting goals that make it difficult to carry out Pinchot's (1905) concept of conservation: "the greatest good for the greatest number for the longest time." It is not possible to establish plans that simultaneously maximize the benefits of each use, the number of people receiving benefits, and the number of years over which the benefits are received. Furthermore, maximizing any single physical output such as timber is usually

---

[11] Fully regulated forestry may be economically inefficient when managing small plots, like small woodlots. For large plots, the theory and applications of the fully regulated forest were stated by Pinchot, as requiring a regular supply of trees of the right size at the right time ready for harvesting (see Pinchot, 1905).

Example 8–2
Why Are Tropical Forests Undervalued?

Many of the uses of forests involve unpriced public goods that produce little local benefit. Tropical forests contain much of the world's biodiversity and sequester huge amounts of carbon, which absorb greenhouse gas. Rising concerns as a result of species loss and climate change have focused attention on protection of tropical forests for global benefits. This protection would provide a form of global insurance. Financing this global benefit, however, requires that a large part of the insurance premiums be incurred by local people who live and farm near the protected forests. These people would stand to lose access to one of the forests's most important local, risk-mitigation properties: insuring against a bad year.

Pattanayak and Sills and Takasaki, Barham, and Coomes wanted to know more about the extent to which local people relied on nontimber forest products (NTPs) to reduce the risk associated with subsistence agriculture. They examined households who live on the margin of the Tapajós National Forest in the Brazilian Amazon region. Tropical forests provide NTPs of considerable value (e.g., vines, honey, sap, nuts, fruits, bark, fuelwood, thatch, and rubber). The poorest people have the fewest resources and opportunities and often rely most heavily on forest collection as insurance to compensate for lost income in bad times. In the Brazilian Amazon, 1.5 million people live on farms that engage in forest collection activities. These households typically have little access to formal insurance. Also, they usually do not have access to government-subsidized credit because they lack clear title to their land so they cannot use it as collateral for credit applications.

What does this mean for *global* forest policy? It means that excluding local people from public forests in order to protect global public goods like biodiversity and global climate may impose greater costs than simply the local income lost from forest products. By *providing access* to public forests for local people, policymakers may improve local welfare and recruit local people as supporters for the goals of global forest conservation. Public forest conservation complemented by *local NTP collection rights* may be productive for commercial agricultural policies that improve access to credit, crop insurance, and savings. Local access to local forests can provide local insurance to offset the local opportunity costs often imposed by policies that protect global biodiversity, global climate, and other global forest services.

Sources: Pattanayak, S. K., and E. O. Sills. 2001. Do tropical forests provide natural insurance? The microeconomics of non-timber forest product collection in the Brazilian Amazon. *Land Economics* 77 (4): 595–612; Takasaki, Y., B. L. Barham, and O. T. Coomes. 2003. Risk-coping strategies in tropical forests: Floods, illness, and resource extraction. *Environmental and Development Economics.* 9 (part 2): 203–224.

an economic mistake, because managers invariably reach the point where producing one more unit (e.g., timber) will cost more than it is worth; that is, the marginal cost will exceed marginal benefit. To promote balanced use of forests, managers typically consider several forest uses in their mix of outputs.

Nontimber revenues can have important impacts on forest management decisions, both for public and private foresters. For example, in some private forested areas, managers generate revenues by leasing easements to fishing or hunting clubs or other organizations that value an unharvested landscape. These revenues provide important economic incentives to change harvest patterns to preserve or improve the beauty of the land. Where nonharvesting revenues are important, harvests may be delayed or even eliminated so as to avoid marring the beauty or ecological functions of an area. In any case, where nontimber values are important, forest managers look for the pattern of land uses that maximizes NPV at some discount rate (see Example 8–2).

That principle is the essence of forestwide, or nationwide optimization models like the U.S. Forest Service's FORPLAN (forest model for planning), which is a large-scale linear programming model to help foresters maximize NPV, subject to natural resource, legal, institutional, and budget constraints (Barber, 1986).

Challenges of Multiple-Use Management Although the idea of multiple-use forest management has been around since the early 1900s, the first writer to apply economic principles systematically to this type of management was G. Robinson Gregory in 1955 (Gregory, 1955). Yet

Example 8–3
Science versus Culture in Finland's Forest Policy

Forests have been Finland's most important natural resource for centuries. Even today, Finland lives off its forests, despite considerable economic and political changes. Through the 1990s, views held by both industry and the state emphasized that if the timber is not harvested, all the years of waiting for the trees would go to waste by letting timber rot. Unharvested timber was viewed as having a zero opportunity cost. Today, debates focus on the questions of how much forest, and what kinds, to protect from industrial use, and how to manage forests that are destined for use in paper and pulp manufacture—a major income producer in Finland.

Finland's culture places heavy emphasis on objective knowledge to support major decisions. Finnish people believe that major decisions about forests should be based on truths that can be found in the natural sciences and that acceptable policy should reflect these truths. However, science provides few answers to questions about peoples' values and beliefs. This issue arises in debates over what constitutes old-growth trees, whether or not enough of Finland's forests are in protected areas, and the value of ecosystem services and aesthetics.

Finland's culture produces widespread support for policy through reliance on experts and technical data. Still, reliance on technical experts creates a gulf between the industry, the state, scientists, and ordinary people. The experts have access to the data and their interpretation, but ordinary people without the data contribute to the policy debate and ultimately to important forestry decisions. Much remains to be worked out to give nonexpert citizens access to better data, so that they can support informed decisions through good science, in accord with Finland's diverse culture and values.

Sources: Berglund, E. 2001. Facts, beliefs and biases: Perspectives on forest conservation in Finland. *Journal of Environmental Planning and Management.* 44 (6): 833–849; Runesson, U. T. 2002. *Finland's forest resources.* Available from *http://www.borealforest.org/world/world_finland.htm* [Accessed August 29, 2003]; Ehrnrooth, C. 2002. *Finland lives from forests.* Available from *http://www.publiscan.fi/bl14e-7.htm* [Accessed August 29, 2003].

even now there are many practical difficulties of managing multiple uses to obtain maximum NPV. One such difficulty is widely differing mixes of outputs that yield similar NPVs. Managing a piece of land for 100,000 board feet of timber and 20,000 recreation visitor days may produce essentially the same total benefit as managing for 50,000 board feet and 35,000 visitor days. Given the considerable uncertainty in benefit and cost calculations, several proposed plans may produce approximately equal NPVs. Each plan may produce very different levels of benefits to various parts of a country, as well as different distributions of benefits over time and among user groups.[12] So even successful NPV maximization does not answer the ethical and politically important question of who should get what, where, and when; that is, it sidesteps important distributional questions. See Example 8–3.

Asking forest managers to maximize the NPV may solve an important efficiency problem. However, there are several reasons why maximization of NPV by itself does not solve many multiple-use forest management problems:

- Different ways of managing a forest can produce similar NPVs.
- Good data on physical trade-offs are expensive to gather or simply are not there. As a forest output (e.g., increased timber harvest is changed, little is known about how other outputs change (e.g., reduced probability of survival of an endangered species).
- There is little agreement on the correct discount rate (see Chapter 6).
- Many benefits and costs produced by forests are hard to measure.

[12] Barreto et al. (1998) illustrate from analysis in Brazil that economically poor forest management can result when information on costs or benefits of any of the forest uses is scarce. Similarly, Barrett (1999) showed how links between food price policy, farmer behavior, and deforestation provide insights into the vicious cycle of peasant poverty and deforestation in low-income tropical regions. Barbier and Burgess (1997) analyzed some of the special challenges for multiple-use management for tropical forests.

Example 8–4
How Can Multiple-Use Forestry Serve Nepal?

For centuries, intense property-right conflicts have occurred over the use of Nepal's forests. Historically, the Nepalese government found greatest benefits from forests used for timber or converted to farmland. Villagers, in contrast, relied on the forests for fuelwood and livestock feed. Nepal's famous Himalayas, the world's highest mountains, have attracted mountain climbers and tourists for years. So it is no surprise that forests provide a major source of Nepal's income, producing nearly half of its gross domestic product and most of its jobs.

Houghton and Mendelsohn examined the potential for multiple-use forestry in Nepal by measuring the economic value of animal feed and fuelwood in villages in the middle hills of Nepal. Their results show that fodder and fuelwood are the most important products in this forest. Even if the trees were dedicated entirely to timber, instead of a combination of fuelwood and timber, the value of nontimber products would still account for about a half of the value of the forest. Revenues from the various forest products show that multiple-use forestry is the most economically efficient way to manage Nepal's forest land. By taking advantage of fodder, fuelwood, and timber, public and private landowners can put their scarce land to best use.

Even though the authors estimate considerable benefits for multiple-use forestry, their figures may understate the social value of conserving forests. In addition to the products the authors measured, forests provide other benefits the authors could not measure. For example, there are additional nontimber forest products such as bamboo, charcoal, fruits, and medicines; the authors found evidence these products are traded and used in the area. In addition, there are nonmarket forest services, which society values even if the services are unpriced; for example, forests provide habitat for game and nongame species as well as a reliable supply for drinking water. The forests serve as essential soil-erosion control in an area with serious potential problems because of its steep lands with high mountains. Finally, the forests serve as carbon sinks, thereby providing a worldwide service for controlling greenhouse gases.

Overall, the authors found that multiple-use forestry could have an important place in sustainable land management in Nepal. For the local landowner, multiple-use forestry has the potential to provide fuelwood, timber, and fodder. For the village as a whole, multiple-use forestry can provide wildlife habitat, nontimber forest products, soil-erosion control, and fresh water. For the world, forests can provide biodiversity and carbon sinks.

Sources: Houghton, K., and R. Mendelsohn. 1996. An economic analysis of multiple-use forestry in Nepal. *AMBIO.* 25 (3): 156–159; Lichtenberg, Z., and R. Upadhya. 1998. *Saving Nepal's forests: The promise of community management.* Available from *http://lnweb18.worldbank.org/sar/sa.nsf/v* [Accessed August 28, 2003].

- Even if benefits and costs of a proposed action are measured correctly, NPV alone[13] fails to answer questions about how they should be distributed. And politicians do not vote for actions that produce an unfair distribution of benefits and costs. Forest management trade-offs between absolute NPV and the distribution of impacts, especially environmental impacts, have been debated heatedly in Finland (Berglund, 2001). See Example 8–3.
- Building very large computer models that maximize NPV might cost more than the added benefits of the maximized NPV.

So forest managers must somehow produce an acceptable quantity of total benefits while trying to maintain positive NPVs of projects and their separable parts. If NPV is negative, managers need to ask if the noneconomic results, such as an increased probability of an endangered species' survival, are worth the loss in economic output. Despite all these qualifications, the NPV or quantitative models built for its maximization provide important information to forest managers. All other things being the same, high NPVs are preferrable to low NPVs, and its good economics to avoid policy actions that produce a negative NPV. Furthermore, if actions can be taken that increase the NPV for all major user groups and/or regions in a country, those actions are worth considering seriously. See Example 8–4.

[13] Many question whether maximizing NPV for multiple uses is ethically defensible. For example, Booth (1997) takes the nature-centered view (described in Chapter 3) in which old-growth forests may be valuable in their own right, independent of their value to people. Where this is true, benefit-cost analysis implied for maximization of NPV is the wrong analytical technique. Booth examines what the nature-centered view means for the policy debate over the preservation of spotted owl habitat in the U.S. Pacific Northwest.

Moreover, finding the NPV-maximizing policy is an extremely powerful method for organizing data and formulating mental experiments for testing results of different plans. So measurement of NPV is an efficient way to weed out bad projects from good ones. Determining the NPV of parts of multiple-use plans is also useful.

Trade-Off Analysis

Trade-off analysis shows how much of one or more outputs must be given up to gain more of another. It lets managers compare gains and losses from proposals that would change the mix of a forest's outputs. Through trade-off analysis, managers pose two closely related questions:

1. What is the physical change in each output X_1, X_2,. . . X_n resulting from a proposed policy that would supply one more unit of output *Y?*
2. What is the economic value of increasing or decreasing all those *X*s, and how do those values compare to the money value of the added unit of *Y?*

So the manager uses trade-off analysis to compute the physical costs and gains (trade-offs) from a proposed plan and translates those physical impacts into money impacts.

Endangered Species versus Timber. Figure 8–2 shows the typical shape of a production curve illustrating a trade-off between maximum possible net present value of timber harvests and numbers of spotted owls (an endangered species) on a forested area. With initial increases in timber harvests, very few owls are lost, especially if there is no clear-cutting. But the number of owls drops quickly as the unconstrained maximum NPV of timber harvest is approached (Montgomery and Brown, 1992). At any combination of owls and timber harvest, the slope of the trade-off curve tells you the marginal rate at which owls are traded for NPV: dollars of timber lost per owl gained. What is important is not owls gained from timber dollars lost, but the underlying economic principle. You can apply this way of thinking to any two outputs from a forest: timber versus livestock, recreation versus endangered species, and so forth.

Important policy choices dealing with timber and spotted owls were examined by Loomis and Ekstrand (1997). They looked at the benefits and costs of setting aside forested areas as critical habitat for the endangered Mexican spotted owl, which is related to its more famous relative the northern spotted owl by biology and political controversy. Adding the Mexican spotted owl to the endangered species list in 1993 severely restricted logging of old-growth forests in the four-corners states of Arizona, Colorado, New Mexico, and Utah. In 1995 the U.S. Fish and Wildlife Service designated 4.6 million acres as critical habitat for the Mexican spotted owl. The lands included 3.6 million acres of national forests and 871,000 acres of Native American lands.

Figure 8–2 shows impacts of setting aside the owl habitat, by moving from the current point *A* to the proposed point *B* (less timber and more owls). Loomis and Eckstrand wanted to test two competing theories, each of which is significant to environmental policy. One theory proposes that the decision to increase owls and reduce timber produces more benefits than costs to the nation.

Figure 8–2 Benefits and costs of more owls and less timber

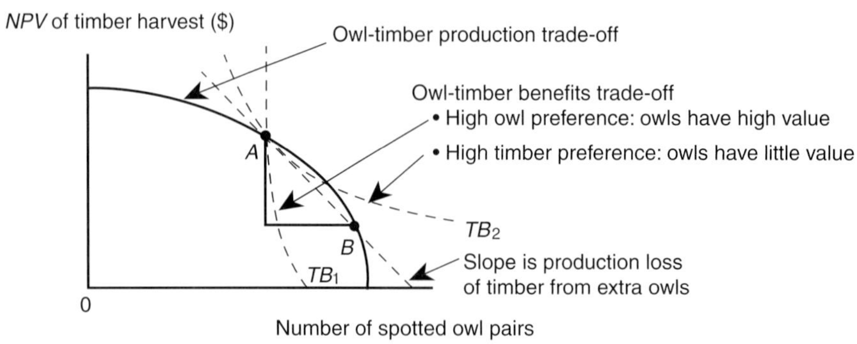

The value of owls gained is greater than the value of timber lost and the public's preferences are something like the dashed line TB_1. If the public's preferences are summarized by line TB_1, reducing timber and increasing owls by moving from point *A* to point *B* reduces total benefits, because a total benefits curve higher than TB_1 would pass through point *B* (higher benefits curve not shown). The competing theory says that the decision to increase owls and reduce timber produces smaller benefits than costs. That theory says the public's preferences are summarized by the dashed line TB_2. If those preferences are summarized by TB_2, then the policy moving from *A* to *B* reduces total benefits, because a total benefits curve lower than TB_2 passes through point *B* (lower benefits curve not shown).

Under U.S. federal laws concerning endangered species, economic analysis may *not* be used to make the decision to add a species to the endangered species list (the listing decision). However, economic analysis *can* be used to evaluate decisions about proposed critical habitats. One cost associated with setting aside the owl habitat is the net benefit of timber harvesting foregone. According to the U.S. Fish and Wildlife Service, $4 million in total costs is incurred each year for 10 years to support the owl's recovery.

A natural question that arose in the debates over designating critical habitat for the Mexican spotted owl related to the benefits of protecting the species and its habitat. Given the irreversibility of extinction by failing to protect, some segments of the general population may receive nonuse values such as existence and bequest values from owl protection. Existence value refers to the benefits people receive from knowing that a species and its natural habitat exist, even if no visitation or viewing is planned. Bequest value refers to the benefits gained from knowing that preservation provides this species and its habitat to future generations.

The passive use value—a component of total economic value—that is provided by preservation is a public good. As such it can be simultaneously consumed by all citizens, whether living in the four-corners region or anywhere else. So small values per U.S. household produced by owl habitat protection potentially translate into large total national values or worldwide values. At present, the contingent valuation method (CVM) is the only method available to measure passive use values. CVM relies on respondents' *stated* willingness to pay (WTP; see Chapter 7) rather than actual cash paid. Considerable controversy exists over the validity of peoples' stated WTP, and this controversy continues to generate heated debate among economists, policymakers, and affected businesses.

Loomis and Ekstrand wanted to measure the total economic value of the Mexican spotted owl per household. In the process of this inquiry, they wanted to test the competing theories that raised the important policy question of whether the benefits of protecting the critical habitat for the Mexican spotted owl are worth the costs (Figure 8–2). The authors found some surprising results: Total public benefits produced by owl habitat protection were in the range of $1.8 billion to $2.6 billion in present value terms. These are very large values compared to the cost the species listing added to the value of timber displaced by the owls. This means that public preferences for owl habitat protection look more like Figure 8–2's TB_1 and less like TB_2. According to the authors, their estimated benefits greatly outweigh the estimated costs of displaced timber.

Timber versus Recreation. Figure 8–3 shows several hypothetical plans for managing timber and recreation on a forest with limited capacity for both. Note that the timber-recreation production trade-off curve is much like the timber-spotted owls trade-off curve. Each point on this curve shows a different mix of timber and recreation, so in theory there is an infinite number of mixes for which the forester could manage.[14] In practice, analysts would consider only a small number of mixes for policy comparisons.

The upper-left point of the timber-recreation production frontier in Figure 8–3 (the curve concave to the origin) shows all timber stocking/harvesting and no recreation. Movement down

[14] Production trade-off curves are used for environmental policy analysis. For example, Aylward and Echeverria (2001) estimate the cost of displaced timber as well as the environmental benefits produced by expanded beef cattle output for Costa Rica.

Figure 8–3
Optimizing timber and recreation

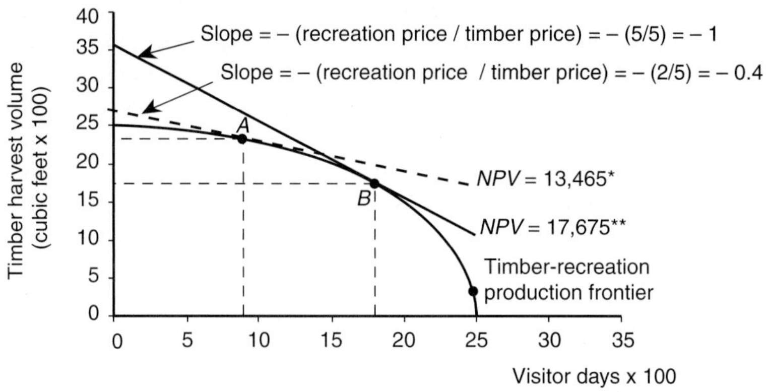

* Timber price = \$5 per cubic foot; recreation price = \$2 per visitor day; *NPV* (\$13,465) maximized at 900 days, 2,333 cubic feet

** Timber price = \$5 per cubic foot; recreation price = \$5 per visitor day; *NPV* (\$17,675) maximized at 1,800 days, 1,735 cubic feet

and to the right on the production frontier shows a gradual reallocation of the forest's scarce productive capacity to larger and larger numbers of recreation visitor days and less and less timber harvest. This description assumes that reduced timber density makes more recreation access and use possible, which is not always true. Some kinds of recreation may increase with greater tree density, especially recreation in connection with hunting and fishing in which catch and/or harvest rates increase with habitat density.

Policymakers want to know what mix of timber and recreation maximizes total economic benefits from available resources. That optimum mix occurs when the following equimarginal condition is satisfied:

$$\frac{P_{\text{recreation}}}{P_{\text{timber}}} = \textit{Slope of production trade-off} \tag{8.3}$$

where $P_{\text{recreation}}/P_{\text{timber}} = \$2/\$5$. It is the willingness to pay in timber per visitor day gained. This ratio equals the willingness to pay per recreation visitor day divided by the value per cubic foot of timber harvested: *slope of production trade-off* = marginal volume of timber lost per recreation visitor day gained.

The price ratio of recreation to timber, the left-hand term in Equation 8.3, is the willingness to pay in timber per visitor day gained, that is, the marginal benefit of added recreation measured in timber units. Consider a simple example. Suppose each visitor day of recreation is worth very little to people, perhaps because the forest is isolated, is expensive to reach, is difficult to access, or has low-quality recreation. Assume the price of recreation (\$2 per visitor day) divided by the price of timber (\$5 per cubic foot) is \$2/\$5. This low relative value of recreation might well occur in an economic depression when incomes are low and few people have time for luxuries like recreation. People are not willing to pay much in lost timber for recreation. They are willing to give up (pay) only a little in timber per visitor day gained, namely \$2/\$5.

The equimarginal equation (8.3) says that the total value of a multiple-use forest is maximized when the forest manager chooses a mix of activities so that the *marginal benefit* of recreation (or any other forest activity) is equal to its *marginal cost.* That tangency point between the relative price line and the production trade-off is shown by point *A* in Figure 8–3. If you set up a forest management plan that produces any other mix of recreation and timber on the production trade-off curve, when the relative price of recreation is \$2/\$5, you will discover that the total benefit falls below \$13,465 (total benefits = price of recreation × recreation visitor days + price of timber × cubic feet harvested). The maximum of \$13,465 is produced at point *A*. At that point, the forest manager is satisfying the equimarginal condition, which assures us that the greatest total benefit is produced by the managed forest.

What happens to the optimum mix of timber and recreation if timber prices are unchanged while recreation benefits per visitor day *increase?* The value per visitor day of recreation could increase if incomes increased, if people became more health conscious and wanted more outdoor experiences, or if site access is improved. Intuitively, if recreation becomes worth more, other things being the same, you would expect the forest to be managed more intensively for recreation, giving up some timber production.

To illustrate, Figure 8–3 shows a new, steeper relative price line pivoting clockwise from the original price line when the price of recreation increases from \$2 to \$5 per day. Given unchanged timber prices, the ratio of the price of recreation to the price of timber increases. This increase is shown by the now steeper price line in Figure 8–3. The new optimal forest management plan is based on a new higher price of recreation, which occurs at point *B*. What this means to the forest policymaker is that *a higher price of recreation produces a new optimal management plan with more recreation and less timber stocking.* It also means that the total benefits over both forest uses are increased. For our example, they increase from the old \$13,465 on the first price line to the new \$17,675 on the second, steeper price line.

In summary, these factors determine the optimum multiple-use forest management plan:

- Output and input prices
- Physical trade-offs among outputs
- Economics, physical, or institutional factors that influence those prices or trade-offs

Because of these factors, optimal multiple-use management plans are rarely constant. Forest managers continually do new analyses as they confront changes in the economic environment, weather or climate, or institutions or publicly held values that would alter timber or recreation price. We have considered only a small percentage of all possibilities facing forest managers. More comprehensive analyses might include many other forest activities, for example, water, recreation, hunting and endangered species as well as grazing and timber. Managers who aim to maximize the total economic efficiency produced by a forest want to find the mix of land uses and policy actions that maximizes the total sustained economic benefits from the land.

For most multiple-use management or policy questions, physical trade-off data are typically poor. For example, if you decrease annual timber harvest by 20%, 40%, or 60%, how do you know precisely how much recreation use will increase? And at the same time, there are other outputs: By how much will water yields or deer populations change? Nevertheless, rough estimates can be made, and examining the problem using simple examples has an important benefit: It forces managers to ask the right questions even if the answers are hard to find or cannot be answered with much precision.

The Equimarginal Principle

The equimarginal principle is a powerful tool for forest managers who wish to allocate resources among competing forest outputs. A manager who seeks cost-effectiveness can maximize economic output for a fixed budget by allocating that budget so that the *added* return on the *last* dollar spent brings the *same added return in each activity.* Application of the equimarginal rule for cost-effectiveness is accomplished by experimenting with various allocations of the scarce resource until that condition holds. (A problem at the end of this chapter illustrates the idea.)

Figure 8–4 shows how to apply the equimarginal principle to the optimization of a single forest's \$140,000 budget. This budget is efficiently allocated by spending \$20,000 on planting trees, \$70,000 on developed recreation, and \$50,000 on wildlife management. With that mix of spending, the last dollar spent in each activity brings the same marginal return of 12,000 added visitor days.

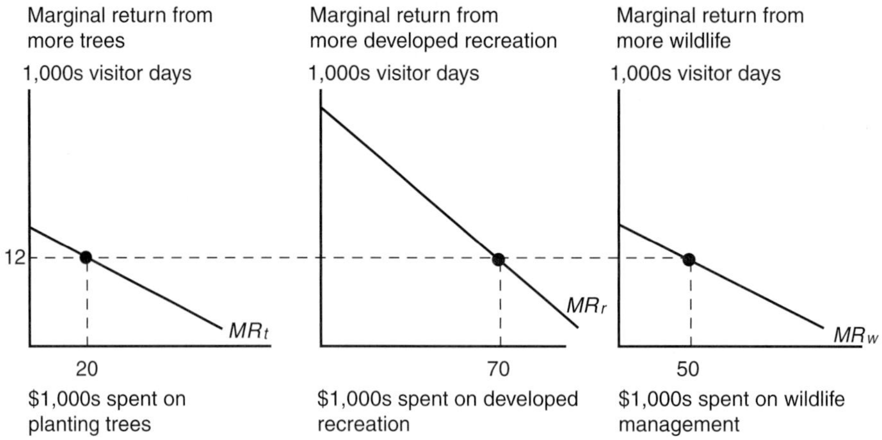

Figure 8–4 Cost-effective budget allocation by the equimarginal principle. MR = marginal return; t = trees; r = recreation; w = wildlife

Table 8–4 Cost-effective budget allocation using the equimarginal principle[a]

(1) Budget ($1,000s)	*(2)* Marginal return from tree planting (1,000s visitor days)	*(3)* Marginal return from developed recreation (1,000s visitor days)	*(4)* Marginal return from wildlife management (1,000s visitor days)
0	16	40	25
5	15	38	24
10	14	36	22
15	13	34	21
20	**12**	32	20
25	11	30	19
30	10	28	17
35	9	26	16
40	8	24	15
45	7	22	13
50	6	20	**12**
55	5	18	11
60	4	16	9
65	3	14	8
70	2	**12**	7
75	1	10	6
80	0	8	4
85	−1	6	3
90	−2	4	2

[a]Any reallocation of a $140,000 budget increases total losses by more than it increases total gains.

Table 8–4 shows the same data. To see that this budget allocation is the best (most cost-effective), look at the table and try to reallocate the budget. For example, try putting $5,000 more in one activity and $5,000 less in another. Begin with column 2 in Table 8–4 (left panel in Figure 8–4): Reducing tree-planting expenditure from the optimum $20,000 to the new $15,000, will produce a change of −13,000 visitor days. Move to column 3, developed recreation; begin at the optimum, which is $70,000 being spent currently, then add $5,000—the added budget freed up from reduced tree planting. The result is less than 12,000 visitor days gained, in this case +10,000. The net result of this management experiment was to reduce visitor days by 13,000 and increase them by 10,000, for a net change of −3,000. This is not a cost-effective way to allocate the budget. You will see that any budget reallocation produces an incremental loss in one area that exceeds the incremental gain in any other area. Although only three activities are shown here, many others are possible.

The equimarginal principle can be applied to real-life problems if several conditions hold true:

1. Physical production responses are known (e.g., how many more hunter days or harvested deer from a given wildlife management expenditure).
2. The economic value of all forest outputs, from endangered species habitat to recreation visitor days, can be expressed in a common denominator such as money.
3. *Marginal* (not total) returns from added size of an activity gets smaller with higher levels of investment.
4. The distribution of benefits and costs produced by using the equimarginal principle is politically acceptable.

If all four of these conditions occur, simple spreadsheet analysis is an easy method for applying the equimarginal principle, and it can handle a wide range of alternatives that actually face forest managers.[15]

The equimarginal principle is quite powerful in allocating budgets or other resources even when outputs from each activity are not measured in money. You can still maximize total benefits from a scarce resource by spending in a way that the last dollar spent in each activity brings the same added benefit even if the benefit is not measured in money (this is an example of cost-effectiveness analysis, described in Chapter 5). For example, imagine a budget management plan that allocates money to three different programs for increasing visitor days of forest recreation. The equimarginal method works even if the economic value of a visitor day is not measurable in dollars, but only in visitor days (for an example, see Cowley et al., 2003).

The general format for applying the equimarginal principle is always the same: The horizontal axis has the same input (e.g., time, money, trees planted, fish stocked, or endangered species habitat) spent in different pursuits (as many as are needed), and the vertical axis is some common added benefit unit (e.g., income, lives saved, fires extinguished, or timber produced).

Linking Benefits and Costs to Decisions

A common mistake made by analysts is to improperly link benefits to forest management actions being considered. Two common examples of this mistake are (1) ignoring important linkages between management and outcomes and (2) inventing linkages between management and outcomes that do not exist.

---

[15] Mathematically implementing the equimarginal rule for a system of n uses of a single scarce resource for several competing activities that use that resource requires solving a system of simultaneous equations: For the n resource uses, the first $n-1$ equations are written to require that the value marginal productivity of each activity's use of the resource must be equal. The last equation summarizes the total supply of the scarce resource (e.g., money, water, land) available for allocation among the n resource uses. Suppose there are three activities, as in Figure 8–4, each with a linear schedule reflecting the value of its marginal product (price × marginal physical product). For the linear marginal products with constant prices, the three equations to solve are

$$P_1 (B_{01} - B_{11} L_1) = P_2 (B_{02} - B_{12} L_2)$$
$$P_2 (B_{02} - B_{12} L_2) = P_3 (B_{03} - B_{13} L_3)$$
$$L_1 + L_2 + L_3 = 140$$

where P_1, P_2, and P_3 are the known prices of each activity, equal to 1 when the output is expressed in common nonmonetary terms for cost-effectiveness analysis. The terms B_{01}, B_{02}, and B_{03} are known intercepts, and B_{11}, B_{12}, and B_{13} are known slopes. The three equations are solved to find the efficient quantities of the scarce resource allocated to each of three uses of the resource: L_1, L_2, and L_3. The first two equations are the equimarginal rule for the three activities. The third equation says that the total supply of L is fixed at, say, 140. Table 8–4 illustrates the calculation.

Sometimes important linkages are ignored. For example, planting new trees sometimes bring benefits of recreation, better fishing, and soil and water protection. But in some regions, little or no investment on the same sites will yield other trees or other vegetation that will bring the same nontimber benefits. In such cases, it is a mistake to include the nontimber outputs in the tree-planting benefit-cost analysis to bump up the computed rate of return or net present value. The tree-planting program has to stand on its own, since the nontimber benefits would occur even *without* the planting.[16] It is valid, however, to attribute to forestry investments any nonmarket benefits that would otherwise *fail* to occur *without* the investment.

At other times, unimportant or illusory linkages are made to look significant. The allowable cut effect is a good example. In past years the U.S. Forest Service has mistakenly attributed immediate increases in allowable timber cuts to new trees just planted. This mistake occurred when current harvest rates for large timber stands were set to equal the long-run annual timber growth of a national forest. Suppose more trees are planted today to serve as a sink for greenhouse gasses, and today's annual harvest is therefore increased by the amount of that planting and boosts long-run annual growth. If this immediate, continuing, higher annual harvest income is falsely attributed to a single tree-planting investment, the rate of return appears extremely high, even for a weak reforestation investment with high costs on a poor site. This is a mistake because the mature timber harvest could have been increased anyway without the reforestation. In summary, the rate of harvesting today's tree inventory has nothing whatsoever to do physically with planting on other places.

The two cases just described illustrate the pivotal importance of the *with and without principle* in benefit-cost analysis in forest management planning. As stated in Chapter 5, the net benefit from any added investment is the NPV of the system *with* the investment minus the NPV *without* the investment (not before and after the investment). Applying this principle is vital, although sometimes quite difficult, when evaluating separate pieces of large, multiple-use plans.

Distribution of Benefits and Costs

Many proposed multiple-use plans emphasizing nontimber values produce benefits far exceeding costs; for that reason, they merit consideration on economic efficiency grounds. Nevertheless, a common political problem occurs when the costs of the proposal are concentrated in certain politically important sectors like timber harvesting or ranching. These industries often have access to the money and/or time to launch public relations and lobbying efforts against a policy. On the other side of the debate, the impressive large gains to the general public are often scattered over so many millions of people at low benefits per person that few people outside the most-affected local area are motivated to organize and raise money to support the policy. The problem is compounded when gains, (e.g., from genetic diversity or control of greenhouse gasses) extend across international boundaries. An example is the spotted owl versus old-growth preservation controversy in the U.S. Pacific Northwest, where losses are concentrated in the forest products industry, its employees, and dependent communities, while gains are diffuse and hard to measure. This is the problem of concentrated costs and scattered benefits. When it occurs, as it often does in forestry, policies with large, positive NPVs to the nation may be politically hard to implement.

The opposite problem can occur when a proposed policy yields concentrated small benefits and scattered high costs. This is an issue if total losses exceed total gains. For example, at one time U.S. Forest Service policy allowed and even encouraged large clear-cuts, in which large chunks of land had all their trees harvested. The benefits in terms of low logging costs were concentrated in

---

[16] However, all external costs associated with deforestation are counted in a properly done benefit-cost analysis. For example, Barbier and Strand (2002) illustrate the importance of correctly linking mangrove deforestation policies in Thailand with their impacts on damaged breeding grounds for small-scale fisheries.

small groups, so gainers could easily mobilize to support their position. These gains were often offset by larger but more diffuse losses to millions of recreationists who were too poorly organized to bring their political influence against the policy of allowing the clear-cuts. A continuing problem is that little work is done to quantify diffuse losses, precisely because they are diffuse.

Small numbers of vocal businesses that incur high gains or losses sometimes have a tremendous influence in the legislative process. In recent decades, however, people involved with diffuse benefits and costs have increased their political clout. Examples are the recent successes of groups supporting endangered species and opposing clear-cutting for unique ecosystems in the United States.

U.S. Forest Legislation

Several pieces of legislation have been enacted in the United States since the 1960s, all of which have had a major impact on past forest management decisions and will affect future management. Some of the more important legislation is discussed next. Space limits prohibit a comprehensive treatment.

Multiple Use Sustained Yield Act

The Multiple Use Sustained Yield Act of 1960 provided a legislative mandate for multiple uses of forest lands. It has helped the U.S. Forest Service protect the general public interest from special interests who seek congressional support to manage forests for their special benefit. Unfortunately, the U.S. Forest Service has never quite figured out how to manage the nation's forests using multiple-use principles to maximize NPV. There are many reasons why they have not fully succeeded:

- A lack of market incentives is one problem. In the private sector, companies that do not attempt to maximize profits—or something pretty close to that—sooner or later go out of business. Failure to maximize NPV does not make government planners go out of business. The U.S. Forest Service gets a budget every year, and that budget has been getting larger not smaller.
- A scientific problem is that both economics and biology are immature sciences. Economics has much to learn about measuring values, and biology has a long way to go in measuring biological effects of proposed actions.
- A political problem is that many special interests go to great lengths to make sure the U.S. Forest Service does not maximize NPV to the nation; rather, the interest groups want the Forest Service to maximize the NPV on the interest group's income statements.

Wilderness Act

By the early 1960s, the issue of permanent legislative protection of forested wilderness and wild lands became a priority of many conservation groups. Conservation groups also wanted the wilderness concept applied to national parks and wildlife refuges and Bureau of Land Management (BLM) lands administered in the Department of Interior. After several years of congressional debates, the Wilderness Act passed in 1964 over the opposition of the U.S. Forest Service and the timber industry. The Forest Service was against the act because of fear that wilderness would be managed as single use, thereby limiting the multiple-use philosophy. The Wilderness Act provided for several allowed and disallowed uses of wilderness:

- Lands must retain their primeval character, with no permanent developments or human habitation.
- Primitive recreation, hunting, fishing, backpacking, and rafting can take place.
- Livestock grazing can continue in the same manner and degree as before designation.
- No new roads or permanent structures can be built.

National Environmental Policy Act

A decade of industrial and economic growth in the 1950s also produced increased pollution in America's waterways and air. After the publication in 1962 of Rachel Carson's famous environmental warning *Silent Spring,* environmental issues began to receive widespread attention (see Chapter 16). The existence of pollution problems and possible solutions received widespread attention in the 1960s, and after considerable debate, the National Environmental Policy Act (NEPA) was enacted in 1969. This act ensured that environmental issues and values would receive consideration in all major decisions carried out by federal agencies that had an effect on the environment. However, NEPA does not require selection of the most environmentally beneficial management plan. NEPA requires that every federal action be evaluated in terms of its potential impact on the natural and human environment. Federal action can mean a direct action by a federal agency (e.g., building a dam) an indirect action such as issuing a federal permit to a private or public entity for a commercial activity (e.g., allowing a utility to construct a power line across federal lands).

The NEPA process begins with a simple environmental assessment (EA). An EA is a cursory evaluation of the likely magnitude and significance of environmental impacts of any proposed federal action. If an EA finds of no significant impact, the process stops there. But if results show a significant impact, then a full environmental impact statement (EIS) must be prepared. The EIS typically takes several months or years to prepare, and it is expensive in time, money, and public involvement. First, a federal agency's proposed action must be written down in sufficient detail to allow its staff to predict environmental effects. Next, a wide range of alternatives to the proposed action must be evaluated and compared to the proposed action. Finally, written justification for selecting the proposed action over alternatives must be made. Agency actions affecting forests, such as timber sales or range or wildlife improvement projects, require at least an EA. A major effect of NEPA has been to considerably raise the cost, requirements, and size of the bureaucracy required to manage public lands. It has also contributed to a high demand for college graduates in all environmental and natural resource fields, such as forestry, hydrology, various environmental sciences, and economics.

NEPA has greatly expanded the opportunities for public participation in making decisions about public lands managed by the Forest Service and BLM. Public meetings and formal public comment are a standard part of the NEPA process. Federal agencies must respond to public comments in preparing a final environmental assessment or EIS. This public comment process has had a major influence on how public lands are managed.

National Forest Management Act

The U.S. Forest Service was originally established to focus primarily on timber and water production and protection. However, by the early 1970s, the public put increased value on recreation, wilderness, wildlife habitat, and other land uses.

After the National Forest Management Act was passed in 1976, the Forest Service became legally obligated to manage federal forests according to changing public demands and values. As would be expected, this law has been considerably more popular with environmental interests than with private landowners and traditional commercial users of public lands. The Forest Service by law must manage to maximize net public benefits summed over all uses and all time periods, rather than limiting management aims for commercial benefits of timber. This additional legislative mandate requires obtaining information on and managing for noncommercial values (e.g., recreation, clean water, and endangered species habitat).

■ Designs for Efficient, Equitable, and Sustainable Forestry

Several institutional and policy innovations have been developed internationally to promote efficiency, equity, and sustainability for forests. Each has potentials and drawbacks.

Green Certification

Continued global deforestation[17] has motivated the idea of green certification for commercial logging activities. Green certification is based on the concept that some people may be willing to pay a premium for sustainable forestry through their purchases of sustainably produced wood products. (Aravanopoulos, 1998, has shown that environmental pollution has important implications for forests that go beyond marketing of wood products; it can produce effects that alter a forest's genetic evolution.) The idea of certification is to let buyers express their preference for environmentally sustainable forest management through the marketplace. Green certification on a product label tells a buyer that her purchase comes from a forest that will not favor unsustainable practices.[18] Becoming certified could qualify a landowner for access to a market driven by certification and may allow the landowner a higher market price for timber when sold.

Green certification is gradually being introduced to the world market but it is not yet known if it will gain in popularity. Some trends to watch include participation of major lumber retailers, such as Home Depot, and large-scale certification of publically owned forest lands. If certification catches on, it may alter standards on how forests are managed worldwide. Many details about certification standards need to be worked out. For example, a practice may be sustainable from a nation's view but unsustainable viewed from a region. The appropriate sustainability time frame could be a single year to centuries. Even sustainability itself has many definitions (see Chapter 19).

Another challenge centers around how rigorously standards for certification should be defined, which influences how much certification will cost. Initial certification for a parcel of land can cost tens of thousands of dollars plus smaller amounts for annual renewals. A strength of green certification is that it has the potential to harness market forces to promote sustainable forestry. Because some consumers will pay a premium for certified wood products, its market demand shifts out and to the right, while the demand for conventional wood products shifts to the left (see Chapter 3). Market signals reward timber producers who are able to shift from conventional forestry to sustainable forestry in the way illustrated in Table 8–2 and 8–3. Freedom of expression in the marketplace allows consumers of wood products the choice of paying a premium for green-certified wood products, as well as encouraging environmentally cost-efficient producers to make an income based on sustainable management. So like any other extra cost option added to a base product, green certification is a characteristic designed to appeal to buyers who are willing to pay for the premium and to reward producers financially who can meet that demand cheaply.

By placing added restrictions on landowners on how the forest can be managed, green certification normally increases the cost of producing certified wood products compared with the

---

[17] Comparisons across countries show a surprising positive relation between deforestation and national income. Bhattarai and Hammig (2001) examined this relation, known as the environmental Kuznets curve (EKC), in a 66-country, 3-continent comparison across Latin America, Africa, and Asia. Results show a strong connection between growing income and increased deforestation for all three continents. However, improvement of the environment with income growth depends on institutions to accompany that growth. Income growth can create the conditions for environmental improvement by raising the demand for improved environmental quality and make the resources available for supplying it. Whether environmental quality improvements occur or not, depends on government policies, social institutions, and functioning of markets. Thus, although agricultural and population expansion displaces forests in many developing countries, that does not always occur in the industrialized countries. In the United States and elsewhere in the twentieth century population growth did not cause forest loss, especially if the rate of improvement in agricultural productivity exceeded the rate of population growth (Food and Agricultural Organization, 2001).

[18] More comprehensive certification standards can be imagined; for example, products are certified harvested under the following conditions: forest workers receive a fair wage and/or an acceptable percentage of indigenous people controls forest decisions.

equivalent uncertified product.[19] If certification becomes required for forest products through government regulations, it will lead to more expensive forest and wood products, which will, in turn, encourage substitution in the marketplace by materials such as steel, concrete, masonry, and plastics (encouraging substitution is an objective of green certification). However, if certification remains an option to be chosen by the buyer of wood products, the price of certified wood will increase and the price of uncertified wood will remain unaffected. Each wood product buyer then conducts a private benefit-cost analysis of the decision to buy certified or uncertified wood. Certified wood products will only be purchased if the benefits received by the buyer from the knowledge of sustainable forestry practices outweigh the added cost.

Green certification has received some support in policy circles. In January 2000, New York became the first U.S. state to receive National Wildlife Federation green certification for all of its multiple-use public forest lands managed for timber, wildlife, water quality, and recreation. More than 700,000 acres of the state's forest land were green certified. To become certified, New York's decision makers invested considerable resources to permit forests to meet rigorous standards that shows they are managed for long-term ecological, social, and economic health. For example, forestry operations had to show that they protect streams and wetlands and conserve fragile or rare wildlife, including endangered species. After the required evaluation process was completed and certification was awarded, forest products from these lands could be labeled as green certified, thereby providing that information to wood products buyers who value that premium.

Various gradations of sustainability could be offered in the marketplace rather than restricting choice to a simple presence or absence of a green label. Wood products labeled with the highest (most intensive) level of sustainable management would charge the highest price. The market would adapt to a wide range of levels of environmental commitment by both consumers and producers.

Conservation Easements

A *conservation easement* is a contract that offers financial inducement to landowners to limit certain kinds of environmentally damaging development on their private land, while allowing the landowner to continue to own, occupy, and use the land for environmentally benign purposes. By offering the landowner a financial reward to reassign some of property rights to a qualified organization, it permits landowners to achieve a financial goal and allows the qualifying organization to achieve environmental *stewardship* goals. The landowner sells the easement to a qualified organization such as a government agency, which pays for the resources necessary to monitor and enforce the terms of the easement.

Recent U.S. government actions, such as the Farm and Forest Conservation Act in 2000, have increased the financial attractiveness of conservation easements for many forest landowners. Increased funding for federal programs that provide cash for purchase of conservation easements include the Forest Legacy Program and the Conservation and Reinvestment Act. Since these actions, landowners have been able to realize significant tax benefits as well as direct cash payments for entering into conservation easements (Fledderman, 2002). Each conservation easement describes specific activities that are incompatible with environmentally sustainable actions and limits them. The loss in market value of the land without these restrictions, compared to value with them, is what determines the amount of the tax benefits or payment.

---

[19] A difficult problem occurs when added costs of producing green-certified wood products are less than added revenues. One Oregon landowner (Erickson, 2003) observed that certification in the local area makes landowners feel good but is an economically weak investment under current conditions. The mills pay the timber grower a little more for a certified log than for a noncertified log. Few buyers in the area are willing to pay a premium for the certified lumber. For landowners who are good land stewards, certification provides weak economic incentive for sustainable activities. For poor stewards, certification offers little incentive to become better stewards. Under current economic conditions, if the added cost of certification is more than 50 cents per acre, the added revenues fail to justify it economically.

Conservation easements are a legal measure to permanently protect the important natural values of a forest property and provide management consistency, while keeping the land in private ownership and use. By preventing forests from developments such as apartments, protecting significant environmental values, and supporting sustainable economic uses, conservation easements can benefit both landowners and the public (Pacific Forest Trust, 2002). There are several specific benefits of such easements:

- By offering landowners a financial incentive to refrain from environmentally damaging actions, easements reduce the price of unsustainable management actions by enough to eliminate them. The multiple-use forestry concepts described earlier in the chapter can be used to explain the incentives produced by conservation easements. The easements reduce the price of unsustainable activities by enough to reduce their production to zero in the optimal mix of forest-managed activities.
- Conservation easements promote long-term sustainable management of private working forests. When landowners commit to conservation, they are paid for protecting public benefits such as open space and habitat quality on their private property. The agreements can prohibit some activites entirely, modify existing uses, and/or introduce new ones (Farm Bureau, 2000).
- If the easement buyer is willing to pay more than the landowner's loss in commerical value (opportunity cost) produced by the restrictions, there is a mutual gain from trade. Suppose that a conservation organization will pay up to $X on behalf of the public for the added public goods (e.g., wildlife habitat) produced by limiting the owner's commerical activites. Also suppose the landowner's loss in land value in present value terms from the restriction is no more than $Y. If X is larger than Y, then the conservation organization (society), landowner, and buyers of forest products all benefit from the easement.

Extractive Reserves

The creation of *extractive reserves* is a recent approach to conservation, initiated in Brazil in the late 1980s, in which economic incentives are designed and applied to promote sustainable resource use of forest resources. Extractive reserves are forest reserves held in trust and managed by people who live in the nearby forest communities, for example, the Brazilian rubber tappers who make a living collecting wild rubber latex in the forest. These reserves are large areas set aside in Brazil for the harvesting of latex, nuts, fruits, oils, and other products with the intent of creating sustainable economic benefits from the rainforest without cutting it unsustainably.

Brazil's program is based on the principle that traditional communities that gather forest products are positioned to have incentives to provide sound stewardship of natural resources. Because these people have lived in the forest for centuries, they have learned to coexist with the forest. A culture of coexistence encourages avoidance of the environmental impact of intensive types of unsustainable forestry such as cattle ranching and timber harvesting. Beginning in the 1970s, a popular movement was organized to protect the rights of forest people who depend on nontimber forest products for their livelihood (e.g., as rubber and nuts). In 1989, the Brazilian government recognized extractive reserves in which traditional communities receive long-term property rights to use the natural resources in exchange for carried out promises of sound stewardship. More than 20 reserves have been created in the past decade by the Brazilian government, covering some 9 million acres (0.8% of Brazil's land), and some local communities are calling for more.

Despite the great potential of extractive reserves, it is important to guard against overselling that institution as the perfect path to sustainable regional development. Extractive reserves will support few people on a sustainable basis, at least for the near future. The 9 million acres currently established in Brazil contain only 20,000 families. If rubber is to be maintained as the principal product, then at least 1,500 acres are needed per family to be economically feasible according to the principles described earlier in this chapter.[20] If the families of this area were to live by extractive activities alone, only about 1.5 million people could be supported in the region, compared with the approximately 10 million people who currently live there.[21] Sustainable forest management through creation of extractive reserves presents a considerable number of challenges.

Community Management

Community-based management is an institutional innovation that could encourage both forest owners and users of forest products to accept a higher percentage of the costs associated with their production and consumption actions. This innovation could provide incentives for sustainable economic efficiency as well as greater equity. The idea is to assign ownership of public forests and to vest forest management decisions in people who have the greatest stake in seeing that those decisions are carried out sustainably, thus mitigating the tragedy of the commons (see Chapter 4). Production of nontimber forest products such as nuts, wildlife, fuel wood, and construction materials may have considerable economic value in use, especially to the poorest people who live and work small land plots (Arnold, Michael, and Ruiz Perez, 2001). In other words, production of these forest services may produce higher net incomes defined as usable products harvested than could be earned from commercial activities. Under some conditions, application of optimal multiple-use forest management principles may show that this kind of activity produces a greater sustainable income than commercial activities like timber harvesting. See Example 8–5.

Community management has the power to tap the knowledge of indigenous people in applying sustainable forestry. For example, in the Philippines, property rights to more than 300,000 hectares (740,000 acres) of national forests have been assigned to indigenous communities. By giving forest communities greater say in forest activities, enforcement costs for guarding against and punishing poachers are reduced, and in community support to reduce these illegal activities increases. Similar decentralized community management systems are being successfully adopted in Nepal and Papua New Guinea. Reforestation programs have had considerable success in reducing the loss of forest cover in Thailand and elsewhere.

Governments that experiment with community management need to guard against making mistakes such as these:

- Assigning forest property rights to local people without adequate searching for the legitimate holders of property rights.
- Ignoring to check if the new owners embrace a sufficiently democratic form of government to let all individual preferences shape group decisions, rather than just the rich and powerful.
- Allowing "pressure shift," that is, damages produced by shifting unsustainable timber harvesting from current forests to substitute forests with high potential for commercial timber harvest. Because these substitute forests are typically excluded from the ownership transfer, new depletion pressures are initiated that could have been avoided had the new property rights scheme been avoided.

Baral (2002) observed that these mistakes have been made by several donor agencies that worked with the national government of Nepal to transfer several public forests over to indigenous tribes. Still, despite the potential for these errors, community-based forestry seems to have considerable potential for improved efficiency, equity, and sustainability in forest management. The economic concept involved is multiple-use forestry in which the assign-

[20] Work by Brown and Rosendo (2000) confirms this caveat. They found that alliances between rubber tappers and environmentalists have been more successful in making people feel good and in enabling political alliances rather than in producing economic improvement. They conclude that advances in political empowerment have not yet resulted in improvements in livelihood conditions of the poor forest dwellers.

[21] Similarly, Baumann (1998) analyzed emerging forest policy trends in India. Since 1988, the Indian government has changed its policy from the old emphasis on industrial forestry to forestry that meets local subsistence needs.

Example 8–5
Reducing Poverty through Sustainable Forestry in India

About one fifth of India's 1 billion people depend on forest resources for their livelihoods. Strong links exist between the locations of these people, their poverty, and forests. A classic policy debate facing forest policymakers in India is the question of the national interest versus interests of commercial and subsistence forest harvesters. The Indian government is charged with managing forests in the national interest, but defining the national interest has been hard and definitions have changed in recent years. One important factor has been the issue of how forests contribute to poverty and inequality.

Reddy and Chakravarty examined data from rural India to learn more about the role of forest access in contributing to or reducing poverty and inequality. What they found was surprising. Policies for increasing forestry conservation could actually increase poverty, unless other measures were taken such as developing community-based projects for forest conservation.

Their survey focused on valuing the contribution of forestry products to household incomes and wealth as well major forest uses by households. The authors found that many households depended on fuelwood and fodder; these findings point to danger from conservation policies directed toward restricting common property rights in the forest. Some of the poorest people depend heavily on the forest for subsistence. These people own the least land and depend on forests for the largest part of their income. If common property rights were restricted to combat the well-known tragedy of the commons, the resulting increase in poverty could not be reduced by increasing pay in occupations in which some of the poor are engaged.

Reducing rural poverty and conserving forest resources in India pose two major economic and policy challenges for India's government and for external agencies that design and fund forest conservation initiatives. The poor depend on common property resources in the forest, so it is not in their interest to allow the forests to degrade further. Historically, as long as population pressures were modest, traditional arrangements for managing common property resources worked well. Sustainable forestry can make a major contribution to reducing poverty. Village institutions could be established to identify existing forest users, to lay ground rules ensuring everybody's participation, and to start dialogue with the poorest and most vulnerable, especially poor women.

Sources: Reddy, S. R. C., and S. P. Chakravarty. 1999. Forest dependence and income distribution in a subsistence economy: Evidence from India. *World Development.* 27 (7): 1141–1149; Khare, A., M. Sarin, N. C. Saxena, S. Palit, S. Bathla, F. Vanya, and M. Stayanarayana. 2000. *Joint Forest Management: Policy, Practice, and Prospects. India Country Study, Policy That Works for Forests and People.* International Institute for Environment and Development Series No. 3. Available from *http://www.iied.org/ptw/india.html* [Accessed August 23, 2003].

ment of property rights gives incentives for forest managers to adopt the view of long-term, multiple-use, income maximization.

Ecotourism

Ecotourism, travel to natural areas that conserves the environment and sustains the well-being of its consumers and producers, offers considerable potential to increase producer income while preserving sensitive areas that might otherwise suffer deforestation. In many tropical forests of developing countries, commercial logging and mining present major deforestation threats. Economic weaknesses, political instability, and insecure property rights in these countries often increase pressure for exploitation of forest resources for short-term commercial benefit. Weak local institutions lack the capacity to adequately manage natural resources. In some areas, heavy poaching and commercial exploitation of wildlife pose serious threats. Ecotourism is one economic alternative to the exploitation of these forests. It offers the potential for sustainable forestry and the development of nontimber forest-based industries that are based on the unique biodiversity of the region and can be protected over the long term. If managed correctly, ecotourism offers the potential for some indigenous land issues to be dealt with comprehensively, sustainably, and equitably.

The 24,000 acre Monteverde Cloud Forest Preserve is an example of a successful ecotourist destination. Through a series of private nature preserves, the community has preserved part of the cloud forest that would have otherwise been cleared for cattle, ranching, and coffee farms. The preserve sits on a low mountain on the Continental Divide in central Costa Rica. Its Atlantic and Pacific slopes provide habitat for remarkable biological diversity. The Audubon Society counted a record of 369 different bird species in one 24-hour period. Until recently, nobody would have expected that birds not hunted, not exported, not cooked, but only watched, would become an important economic industry for Costa Rica. Ecotourism has approached banana and coffee exports to grow into a major source of Costa Rica's foreign exchange, earning as much as $700 million in one year. The Monteverde Cloud Forest Preserve supports itself through donations and an entrance fee. Biologists say the heavy volume of visitors has had a low impact on the fragile environment. People are willing to stay on trails, don't throw their garbage, and stay quiet when requested. The tourism has also generated a considerable number of related businesses that line the main road leading to the preserve: hotels, restaurants, snack shops, gift shops, horse stables, and art galleries. An artisans cooperative employs 150 women who sew specialty clothing to sell to tourists.

Ecotourism can be included in the mix of multiple-use forest management activities. But like other activities, ecotourism needs to be managed carefully. Unmanaged ecotourism sometimes gains uncontrollable momentum and leads to foreign buy-ups of land, rising land and food prices, destructive levels of tourist visits, and failure of builders and tourists to comply with codes. Although ecotourism can generate considerable income, it can also produce costs of garbage, waste disposal, and firewood shortages. Careful preparation and organization are required to guard against disrupting local lifestyles and values. Still, under the right conditions and management, ecotourism may offer economic potential for sustainable forest management.

Mitigation of Global Warming

Although the verdict is still out, the preponderance of scientific evidence points to deforestation as a contributor to global warming. Since trees absorb carbon dioxide, a major greenhouse gas, deforestation eliminates a potentially significant resource for limiting the rise in carbon dioxide emissions. Furthermore, burning trees—an activity commonly associated with agricultural land clearing—adds carbon dioxide to the air. Why is deforestation occurring so rapidly when the benefits conferred by a standing forest are so significant? See Example 8–6.

The concepts of externalities and public goods provide an insight to this question. Global warming benefits are largely external to the nation that contributes to those benefits through their forests. In contrast, the costs of preventing deforestation or planting new trees are mostly paid for by the nation with the forest. The willingness to pay for reduced biodiversity caused by deforestation is highest by residents of the wealthy, high-income nations, which are not usually the nations that have the forests. And those benefits are a public good that accrue to everybody in current and future generations, whether or not their country or generation pays for them. The economic challenge comes from the fact that reducing deforestation typically comes at the cost of lost jobs and incomes derived from commercial tree harvest or farms developed after the forest is cleared. Because of large differences in the distribution of costs and benefits brought about by these economic incentives, we would expect the most vocal opposition to biodiversity loss to come from wealthy nations, not the nations that have the forests. The existence of these global externalities points to an excellent reason for international cooperation in designing innovative institutions, such as a global warming treaty.

Gender, Property, and Health

The gender issue is central to forest management in environments where large numbers of poor women and men directly depend on access to common property resources for meeting their survival needs. Gender roles define how women and men perform tasks for the household, earn income, control their incomes, allocate time, exercise rights, and obtain and use knowledge. Trees are especially important in the lives of women in the developing world. Far more women than men cut and use firewood, collect and sell nontimber forest products, and tend livestock. When these products cannot be collected or grown, scarce cash must be used to buy them. Shortages caused by disappearing forest resources have a severe impact on the lives of women and their children (Canadian International Development Agency, 2001).

Example 8–6
Can Economic Incentives Sustain Tropical Forestry?

Tropical deforestation occurs because people find it profitable. A major factor blocking attempts to encourage sustainable management of tropical forests is that it is rarely a financial success. By contrast, forest exploitation and deforestation continue to be highly profitable. Weak incentives pose increasing challenges for governments and international organizations to find methods to *produce financial value to standing forests,* particularly forests with important worldwide "public goods" values like carbon sequestering for mitigating global warming and biodiversity conservation, which includes genetic benefits and values of those willing to pay for knowing that biodiversity exists, whether or not they plan to use it themselves. The major challenge is how to design workable financial incentives for users of tropical forests while strengthening and developing the regulatory framework for private sector forestry. To date, the tropical forest sector has been the victim of both market failures and policy failures. Deforestation continues.

The concept of *opportunity cost* presents a powerful framework for thinking about policies for reducing deforestation. The idea is to look for ways to raise the opportunity cost (private income lost) to people who engage in deforestation activities. Raising the opportunity cost would reduce the deforestation activities entering the mix of multiple-use forestry management activities. Much deforestation actually is produced because of policies outside the forestry sector. For example, many farmers in tropical forest areas with high income from farm employment opportunities find it expensive to put their time and resources into forest-degrading activities. If farmers make a good living at farming, then they pay a higher cost from forest-depleting activities and are less economically motivated engage in them. Policies could focus on increasing productivity in labor-intensive agriculture, thereby making it economically attractive for people to reduce depletions of vulnerable forest areas.

Roads facilitate access to forest areas and raise land values, both of which could encourage increased deforestation. Reducing subsidies to roads that raise land values near forest areas has a potential to limit deforestation. The same concept could be applied to policies that would invest in rural education and other forms of social capital such as light manufacturing, all of which increase the opportunity cost of deforestation. Another possible policy based on raising opportunity cost is to set higher forest concession fees charged to timber harvesters with the aim of driving up the relative profitability of standing tropical timber and slowing the harvest. This kind of policy could benefit producer countries through higher prices for their exports and improved trade opportunities for forest products, which would offset revenue losses from reduced sales of tropical timber to industrialized countries.

Sources: Richards, M. 2000. Can sustainable tropical forestry be made profitable? The potential and limitations of innovative incentive mechanisms. *World Development.* 28 (6): 1001–1016; Gray, J. 2002. *Forest Concession Policies and Revenue Systems: Country Experience and Policy Changes for Sustainable Tropical Forestry.* Washington, D.C.: World Bank.

Poor rural women have little property ownership, so they have a much greater dependence on common property for meeting their survival needs. In many rural forested areas of the world, the use of land, water, and trees is governed by social practices much more than by an individual's legal ownership. Property rights and access to resources can be greatly affected by gender differences. For example, culture and history differentiate the ways to use a forest: women gather fruit and men harvest logs. Gender differences in resource access significantly affect the use of natural resources in three main areas: sustainability, efficiency, and empowerment.

1. Chapter 4 emphasized that secure property rights in natural resources encourages investment in sustainable resource management. Villages and national governments that promote gender equality have found that women who are provided long-term access to resources have more economic incentive to sustain those resources, and are motivated to develop better ways to preserve and sustain them.

2. Secure property rights also encourage increased efficiency; however, efficiency of use is in part defined by gender. For example, a study by Meinzen-Dick et al. (1997) compared ownership and productive activities by gender for a forest that produced both fruit and logs. Logs, harvested and sold by men, had a higher market value and were preferred by men in comparison to fruit. However, where medical facilities and markets are lacking, child health is an unpriced service that has considerable value. If a price could be assigned to improved child health, fruit can have a greater value than logs despite the lower market value of fruit. When women are assigned property rights in forest activities, the total economic value of the family enterprise, defined by applying the equimarginal rule for multiple-use forest management, can increase if women use the fruits to improve their children's nutrition.

3. Acquiring property rights can significantly increase a woman's empowerment by giving her a greater voice in the household and increases her associated power to bargain in the household as well as in the community. Meinzen-Dick et al. found that collective action by women's support groups is the most effective method of strengthening women's access to natural resources. They also found that use rights are probably more important goals for policy interventions than formal ownership rights.[22]

Stewardship Incentives

Historically, U.S. forest landowners have been reluctant to replant or maintain private forests, especially on smaller sized holdings, because the returns on investments require many years to be realized. For example, suppose a landowner has perfect certainty in the stream of future benefits and the market interest is a relatively low 5%. This chapter's earlier analysis of single-use timber management shows that a small $1,000 investment made today by the landowner in improving a forested land area must return more than $7,000 in 40 years to break even economically. Because trees take a such long time to mature, tree planting, maintenance, preservation, and improvements can appear to produce weak economic returns.

Forestry programs that provide stewardship incentives can help overcome the economic disincentives facing landowners that discourage them from replanting or maintaining their private forests. For example, the Forestry Incentives Program supported by the U.S. Department of Agriculture (USDA) offers financial incentives that pay 65% of the total costs of tree planting and improving the timber stands. For practices implemented on some private lands, financial incentives of up to $10,000 are offered for landowners to plant and maintain forests on acreage up to 1,000 acres for a minimum of 10 years. Since the USDA program inception in 1973, it has had some success in encouraging landowners to undertake various sustainable forest improvements that would have not been undertaken without the program. The USDA program supports three principle forest management activities: tree planting, forest stand improvement, and site preparation for natural regeneration. In all three instances, the principal goal is to build or restore the productive capacity of nonindustrial private forest lands to produce public goods (e.g., watershed protection, wildlife habitat, aesthetics, recreational activities, reduced wind and soil erosion, enhanced water quality, and sequestering of carbon to help reduce greenhouse gases).

Food Security

Sustainable food production and earning adequate incomes are closely linked to the sustainable use of forests and other renewable resources in many poor countries. For example, in Ethiopia more than 85% of the people live in rural areas and depend on subsistence agriculture (Food and Agricultural Organization, 2003). Sustainable food production can reduce the economic pressures on deforestation that sometimes are carried out to compensate for poor harvests. Intense economic activities and unsustainable forest and farming practices pose serious threats to the sustainability of natural resources and maintenance of ecological balance. Many poor, forested countries face problems related to intensive cultivation, overgrazing, and deforestation, and the resulting soil erosion and decline in soil fertility, water scarcity, and depletion of livestock feed

[22] Brown and Lapuyade (2001) reported similar findings in Cameroon. Sources of livelihood for men diversify in the face of economic crisis, whereas women have reduced room to maneuver. Women were increasingly dependent on using nontimber forest products for cash to meet their needs.

and for wood. These factors often interact and produce a viscous cycle of poverty, food insecurity, and forest depletion.

Potential policy solutions rest with community-based integrated agricultural and forest management. Policies would attempt improvements in these areas: soil and land productivity, agroforestry development and other rural energy sources, low-cost rainwater harvesting, livestock care, and the livelihood base. The community-based approach would emphasize stabilizing yields and reducing vulnerability by broadening the livelihood base among small-scale farmers who live in marginal, degraded, and fragile ecosystems. Community-based organizations would help empower local people and provide greater economic incentives to manage and use resources sustainably. Approaches for improving food security would involve developing markets for locally produced commodities and legalizing land tenure for local people.

Sustainable Management and Property Rights

Chapter 4 emphasized that well-defined property rights exert a major influence on decisions about sustainable resource management. If a piece of forest land is believed to belong to an individual for a considerable amount of time, it has the potential to be managed sustainably because the individual receives the benefits from and pays the costs of decisions affecting current and future productivity. The opposite also holds true. For example, forests surrounding villages in parts of India have been controlled by the government without regard for current or future needs of nearby villagers. Forests were harvested for national commercial benefit (i.e., timber production) cleared for grazing and farming, and used for fuel wood. This form of government control damaged incentives for local villages to use economically efficient or sustainable practices.[23] The forests became a common property resource and a race occurred to see who could get the largest benefit from this disappearing resource. Each "owner" felt he must get his share of the resource before another took it away.

The solution to the problem was to involve local villagers in managing the lands so that they would benefit. Land management rights were given back to people, who received a benefit from a sustainable forest. This method of management has been used with considerable success in Andhra Pradesh, a large state on the east coast of India. Local villagers are given the resources and incentives for protecting the forests. In return, they receive the benefits resulting from sustaining the forest ecosystem.[24]

Summary

For a forest that produces only timber, the economically efficient time to harvest a timber stand is when net present value is maximized. And NPV is maximized when the marginal gain from delaying the harvest one more year is equal to the marginal cost of the delay. For longer-than-efficient harvests, the additional costs outweigh the increased benefits, for earlier-than-efficient harvests, more timber benefits are given up than costs saved.

A manager cannot manage everything at once, but he or she can optimize output levels over multiple uses so as to maximize NPV. Nevertheless, NPV maximization often presents problems when complex, multiple-use plans are compared: Very different plans often have the same NPV, unvalued benefits and costs might be ignored, and NPV cannot conclude who should get what, when, and where. If there are two outputs, trade-off analysis tells indicates much of one output must be given up to gain more of another. Sometimes managers can produce more of both outputs, if they are willing to give up some of a third output.

Marginal analysis can be used to show the combination of forest outputs that maximizes NPV. A limited budget is efficiently allocated to multiple-use investments when the last dollar in each activity brings the same rate of return. Efficient forest managers make sure that the benefit they attribute to an expenditure like planting trees would not have occurred anyway without the expenditure. The correct measure of the net benefit from a

[23] The analysis of Aviles Benitez (2001) provides another example from rural Spain in which economic and environmental objectives for forest management conflict.

[24] Bach (1999) asked similar questions; policy options were examined for promoting low-impact logging through the use of economic incentives in Ghana, Africa. Results showed that direct subsidies to low-impact logging activities were far more efficient than subsidizing prices of tropical timber.

policy action in a multiple-use forest is the NPV of the forest *with* the policy action minus the NPV *without* the action.

Economically efficient policies, those with positive NPVs, are politically hard to implement if small costs are concentrated among a few people while large benefits are scattered among many. Economically inefficient policies, ones with a negative NPV, are sometimes carried out anyway, when small benefits are concentrated among the few or large costs are scattered among many or are hard to measure.

In the United States, federal legislation such as the Multiple Use Sustained Yield Act of 1960 and the National Environmental Policy Act of 1969 will influence the course of forest management will into the twenty-first century. Internationally, several institutional approaches are in fairly wide use to promote sustainable, efficient, and equitable forstery: green certification, conservation easements, extractive reserves, and community-based management. Other approaches are also in evidence: ecotourism, reversing global warming, gender issues, food security initiatives, and property rights.

Review Questions

1. What are three important characteristics of forest ecology?

2. How does economics help answer central questions of forest management?

3. Describe the strengths and limits of NPV as a forest management objective.

4. Explain multiple-use forest management. Why it is important and hard to accomplish?

5. What is trade-off analysis and why do forest managers use it?

6. Describe a production trade-off curve and how forest managers can use it.

7. Why do forest managers maximize economic benefits by producing at the point where a forest's production trade-off equals its price ratio?

8. Explain why society wants efficient forest managers to manage for less timber harvest and more endangered species when the price of timber falls.

9. What is meant by incorrectly linking benefits and costs with forest programs? Provide an example.

10. Why do concentrated but small costs or benefits make it politically hard to manage forests for economic efficiency?

11. Which piece of U.S. forestry legislation do you believe has produced the most economic benefit to date? Explain why.

12. Identify an important forest management issue for which you have an interest. Of the 10 institutional approaches to forest policy reform mentioned in this chapter, which one(s) do you believe would best encourage economically efficient forest management and why?

Problems

1. Cost-effectiveness analysis: You are a forest manager who wants to allocate a $10 million budget between forest recreation and timber production for a multiple-use forest. You have estimated the total benefits for each program associated with various levels of possible expenditure based on the accompanying table.

Spending on timber output	Total timber benefit	Added timber benefit	Spending on recreation	Total recreation benefit	Added recreation benefit
millions of dollars					
0	0	–	0	0	–
1	22		1	18	
2	42		2	34	
3	60		3	48	
4	76		4	60	
5	90		5	70	
6	102		6	78	
7	112		7	84	
8	120		8	88	
9	126		9	90	

a. Your aim is to maximize total benefits for the $10 million budget. Describe how you would allocate the budget. Hint: Use the equimarginal principle so that the marginal benefits are equal in both uses of the budget.

b. After you have allocated that $10 million efficiently, what would be the ratio of *marginal benefits* to *marginal costs* if you could find another $1 million to add to your $10 million budget?

c. Based on that ratio, can you justify asking for a larger budget? Explain.

d. How large of a budget should you ask for to maximize economic efficiency? Hint: Keep trying a larger budget until the last million dollars produces a marginal benefit of $1 million in both uses. How would you defend that budget request to a room of skeptical taxpayers? Hint: At that optimum point, the marginal benefit per dollar of expense equals one for all programs.

2. Trade-off analysis: A multiple-use forest manager wants to know the economic impact from a proposal that would increase recreational use by 400 visitor days at a cost of 200 acres of lost timber stocking. The lost timber stocking would result from thinning trees to provide for additional camp grounds. Suppose that the price per recreation visitor day is $40 and the price (value) per acre of timber stocked is $50. Remember that the optimum mix of any two forest products, such as recreation and timber, occurs when the equimarginal condition occurs, that is,

$$\frac{P_{\text{recreation}}}{P_{\text{timber}}} = \text{Slope of production trade-off}$$

Calculate the value of each of the two terms in this equation. Would this proposed addition to recreational use move the manager closer or farther away from an economically efficient mix of recreation and timber? Based on your answer, what are the benefits and costs of this proposed plan? Can you guess the size of net benefits associated with a proposal that would change the mix of products on a forest when the equimarginal condition is *already fulfilled?*

References

Angelsen, A., and E. F. K. Shitindi. 1999. Why do farmers expand their land into forests? Theories and evidence from Tanzania. *Environment and Development Economics.* 4 (3): 313–331.

Appuhn, K. 2000. Inventing nature: Forests, forestry, and state power in Renaissance Venice. *Journal of Modern History.* 72 (4): 861–889.

Aravanopoulos, F. A. 1998. Environmental Pollution Effects on the Evolutionary Genetics of Forest Trees. *East-West Series in Economics, Business, and the Environment.* 2: 29–52.

Arnold, J. E. Michael, and M. Ruiz Perez. 2001. Can non-timber forest products match tropical forest conservation and development objectives? *Ecological Economics.* 39 (3): 437–447.

Aviles Benitez, A. 2001. Gestion des ressources naturelles et viabilite des exploitations agricoles, Le cas de l'agriculture de dehesa en Andalousie (Management of natural resources and viability: The example of "Dehesa" farms in Andalusia). *Economie Rurale.* 1 (263): 48–62.

Aylward, B., and J. Echeverria. 2001. Synergies between livestock production and hydrological function in Arenal, Costa Rica. *Environment and Development Economics.* 6 (3): 359–381.

Bach, C. F. 1999. Economic incentives for sustainable management: A small optimal control model for tropical forestry. *Ecological Economics.* 30 (2): 251–265.

Barber, K. H. 1986. Large FORPLAN Models: An Exercise in Folly. Paper read at Workshop on Lessons from Using FORPLAN, April 29–May 1, 1986, at Denver, CO.

Barbier, E. B., and I. Strand. 2002. Do open access conditions affect the valuation of an externality? Estimating the welfare effects of mangrove-fishery linkages in Thailand. *Environmental and Resource Economics.* 21 (4): 343–367.

Barbier, E. B., and J. C. Burgess. 1997. The economics of tropical forest land use options. *Land Economics.* 73 (2): 174–195.

Bar-Hen, A. 2002. Influence of missing data on compact designs for spacing experiments. *Journal of Applied Statistics.* 29 (8): 1229–1240.

Barreto, P., P. Amaral et al. 1998. Costs and benefits of forest management for timber production in eastern Amazonia. *Forest Ecology and Management.* 108 (1–2): 9–26.

Barrett, C. B. 1999. Stochastic food prices and slash-and-burn agriculture. *Environment and Development Economics.* 4 (2): 161–176.

Baumann, P. 1998. The persistence of populism in Indian forest policy. *Journal of Peasant Studies.* 25 (4): 96–123.

Berck, P., and W. R. Bentley. 1997. Hotelling's theory, enhancement, and the taking of the redwood National Park. *American Journal of Agricultural Economics.* 79 (2): 287–298.

Bhattarai, M., and M. Hammig. 2001. Institutions and the environmental Kuznets curve for deforestation: A crosscountry analysis for Latin America, Africa, and Asia. *World Development.* 29 (6): 995–1010.

Booth, D. E. 1997. Preserving old-growth forest ecosystems: Valuation and policy. *Environmental Values.* 6 (1): 31–48.

Boscolo, M., and J. R. Vincent. 2000. Promoting better logging practices in tropical forests: A simulation analysis of alternative regulations. *Land Economics.* 76 (1): 1–14.

Bowes, M. D., and J. V. Krutilla. 1985. Multiple use management of public forestlands. In *Handbook of Natural Resource and Energy Economics,* edited by A. V. Kneese and J. L. Sweeney. Amsterdam: North-Holland.

Brown, K., and S. Rosendo. 2000. Environmentalists, rubber tappers and empowerment: The politics and economics of extractive reserves. *Development and Change.* 31 (1): 201–227.

Brown, K. and S. Lapuyade. 2001. A livelihood from the forest: Gendered visions of social, economic, and environmental change in Southern Cameroon. *Journal of International Development.* 13 (8): 1131–1149.

Canadian International Development Agency. 2001. *Community participation in forest conservation.* Forestry Advisers Network. Available from *http://www.rcfa-cfan.org/english/issues.7.html* [Accessed August 20, 2003].

Carson, R. 1962. *Silent Spring.* Boston: Houghton Mifflin.

Cowley, D. E., F. A. Ward, R. Deitner, and M. D. Hatch. 2003. Optimizing the allocation of hatchery-produced fish among multiple stocking sites. *North American Journal of Fisheries Management.* 23: 216–229.

Erickson, R. K. 2003. Tree farm certification, personal communication with author, September 2, 2003.

Farm Bureau. 2000. *Guide to conservation easements.* Available from *http://www.fb.com/news/nr/nr2000/nr1208.html* [Accessed August 20, 2003].

Faustman, M. 1849. On the determination of the value which forest land and immature stands possess for forestry. In *Martin Faustmann and the Evolution of Discounted Cash Flow.* Institute Paper 42, edited by M. Gane (1968). Oxford, UK: Commonwealth Forestry Institute, Oxford University.

Fledderman, R. J. 2002. *Conservation easements, State of South Carolina.* Available from *http://www.scforestry.org/issues/easements.htm* [Accessed August 20, 2003].

Food and Agricultural Organization. 2001. *Global forest resources assessment 2000.* Available from *http://www.fao.org/forestry/fo/fra/main/index.jsp* [Accessed August 19, 2003].

———. 2003. *Integrated natural resources management to enhance food security: The case for community-based approaches in Ethiopia.* Available from *http://www.fao.org/sd/2003/EN11013_en.htm* [Accessed September 30, 2004].

Gregory, G. R. 1955. An economic approach to multiple use. *Forest Science* 1 (1): 6–13.

———. 1987. *Resource Economics for Foresters.* New York: John Wiley.

Hartman, R. 1976. The harvesting decision when a standing forest has value. *Economic Inquiry.* 14: 52–58.

Loomis, J., and E. Ekstrand. 1997. Economic benefits of critical habitat for the Mexican spotted owl. *Journal of Agricultural and Resource Economics.* 23: 356–366.

Meinzen-Dick, R. L., R. Brown, H. S. Feldstein, and A. R. Quisumbing. 1997. *Gender, property rights, and natural resources.* Available from *http://www.ifpri.org/divs/fcnd/dp/dp29.htm* [Accessed August 20, 2003].

Montgomery, C., and G. Brown. 1992. Economics of species preservation: The spotted owl case. *Contemporary Policy Issues.* 10 (2): 1–12.

Pacific Forest Trust. 2002. *What are conservation easements.* Available from *http://www.pacificforest.org/services/easements.html* [Accessed August 20, 2003].

Pinchot, G. 1905. *A Primer of Forestry, Part II: Practical Forestry.* Washington, D.C.: U.S. Department of Agriculture. Bulletin no. 24.

Price, C. 1997. A critical note on a long-running debate in forest economics. *Forestry.* 70 (4): 389–397.

Province of British Columbia. 2004. *TIPSY model development.* Available from *http://www.for.gov.bc.ca/hre/gymodels/TIPSY/model.htm* [Accessed September 30, 2004].

Rideout, D. B., and H. Hesseln. 2001. *Principles for Forest and Environmental Economics.* 2nd ed. Fort Collins, CO: Resource and Environmental Management.

Samuelson, P. A. 1976. Economics of forestry in an evolving society. *Economic Inquiry.* 14: 466–492.

Tietenberg, T. 2003. *Environmental and Natural Resource Economics.* 6th ed. New York: Harper Collins.

Chapter 9

RANGELANDS

Introduction

The world's rangelands are places where it is too dry, too steep, too infertile, too cold, too rocky, or too high to be anything else. Rangelands, which include most deserts and natural grasslands, consist of unsettled and uncultivated lands that can provide habitats for browsing and grazing animals—free-ranging native animals as well as domestic livestock (Batabyal, Biswas, and Godfrey, 2003). Other kinds of land areas can outproduce rangelands in livestock, but rangeland areas are at less of a comparative disadvantage with livestock than with other products. These generally unproductive lands are grazed by livestock when there is no productive alternative use, that is, where the opportunity cost of the land is at or near zero (Stoddart, Smith, and Box, 1975). Rangelands comprise about 70% of the world's land area and about 42% of the land area of the United States. (DiTomaso, 2000). The world's growing population will put added pressure on these lands and on the value of the services the lands provide. Rangelands are likely to play an increased role in the future by providing people with valued goods and services.

Range economics aims to optimize returns from rangelands for uses that are desired by society through sustained management of range ecosystems. Five important principles govern the management of rangelands:

1. A rangeland is a renewable resource.
2. The sun's energy captured by the green plants of a rangeland can be harvested by browsing and grazing animals.
3. A rangeland's productivity is determined by climate, soil, and topographic factors as well as by use.
4. In comparison with cultivated lands, rangelands provide people with food and fiber at low energy costs.
5. A variety of goods and services such as livestock, wildlife, food, minerals, recreation, and endangered species habitats are produced on rangelands.

This chapter shows how principles of economics and range management are applied to determine the economic consequences of decisions involving the use, development, and/or protection of rangelands.

Outline

The Rangeland Resource 226

Scarcity and Competition 226

Desirable Outcomes 227

Unregulated Market Outcomes 228

Designs for Efficient, Equitable, and Sustainable Rangelands 241

Emerging Issues 252

Objectives

After studying this chapter, you should be able to

- Characterize the rangeland resource
- Describe the competition for the world's rangelands
- Define three desirable outcomes for guiding rangeland resource policies
- Describe rangeland outcomes produced by unregulated markets
- Analyze institutions and policies that influence rangeland outcomes
- Identify emerging international rangeland issues

THE RANGELAND RESOURCE

If managed carefully, rangelands can be a renewable resource. As long as rangelands are not misused, grazing during one period does not adversely affect the level of forage produced and available for use during succeeding seasons. Most rangelands of the world can be managed sustainably in perpetuity (Workman, 1986). Rangelands, whether publicly or privately owned, are important because they serve many uses and users. Beneficiaries include livestock producers, hunters, guides and outfitters, consumptive and nonconsumptive wildlife enthusiasts, recreationists of various kinds, and mineral and timber producers. The wide diversity and interests of these groups has led to conflicts and competition for scarce rangeland resources. Designing institutions and management plans that resolve these rangelands conflicts efficiently, equitably, and sustainably are the main objectives of range managers.

Forage is often measured by the animal unit month (AUM). One AUM is the amount of forage required by an animal unit (AU) for one month. An animal unit is defined as a mature (1,000 pound) cow or the equivalent, based on an average consumption rate of 26 pounds of dry forage per day. That makes an AUM equal to about 800 pounds of dry forage. Conversion factors are used to measure the forage requirements of other animals, such as horses, elk, deer, or antelope. A rough rule of thumb often used to determine the amount of safe grazing, or *carrying capacity*,[1] is to estimate total forage production on a parcel of land per unit time, save half of that production for the future, and treat the other half as available for the current period's grazing.

SCARCITY AND COMPETITION

Like other natural resources, rangeland forage has the potential to be scarce in comparison to the demand for it. When there is no price mechanism to allocate it and/or when there are no clear private property rights for it, some other allocation mechanism is required. In many parts of the world, livestock producers allow overgrazing, which damages rangelands.[2] One reason for overgrazing occurs where the range is an open-access resource, so livestock producers have little stake in conserving or improving the range. They do not own rights to an improved resource and they are not the sole beneficiary of resources invested in improvements. For these reasons, rangeland degradation currently occurs in parts of the world, such as sub-Saharan Africa, where the system of property rights permits individual ranchers to reap benefits from commonly owned forage but discourages costly investments that could improve the land.

The history of grazing in the western United States since the late nineteenth century, as well as policy responses, provides a case study for understanding the motivations of individual actors, identifying forces that deplete resources, and examining policy actions available to the state. In the early nineteenth century, the United States had few users of its rangelands, ample grazing areas, and few competing uses for lands. Open access to grazing for people who were brave enough to settle an unsettled West presented few problems. However, the growing population in the late nineteenth century brought increased settlement and increased numbers of livestock. From the 1880s to the 1930s, the common property nature of the open range gave ranchers an incentive to overstock the range. Recall from Chapter 4 that overstocking a common property can be an economically rational way to keep somebody else's livestock off the commonly owned property. The only way to prevent another livestock producer from using the open range was to strip the land. Overstocking combined with forage shortages due to severe winters and prolonged drought damaged range plants and exposed the thin, fragile layer of organic topsoil, which eventually was washed or blown away by rain and wind.

---

[1] Carrying capacity is a complex issue; some of the important research questions it poses for range managers in Mexico are summarized by Manzano et al. (2000).

[2] Vesk and Westoby (2001) describe some of the difficulties of identifying the response of plant species to grazing pressure in Australia.

Another factor that contributed to overgrazing was the Homestead Act. The Act provided settlers 160 to 640 acres of land for ranching, but this amount was too small to raise enough cattle to make a living. The Act produced an incentive to stock more cattle than the land could support. Water was then, and remains today, a limiting resource. There is not enough water in most of the West to produce supplemental feed to make up for scarcity in native grasses.

Common property factors and inadequate land acreage assigned to each ranch led to loss of what little good soil there was, thereby preventing plants from getting established. By the 1930s, livestock producers began feeding supplemental feed to their cattle to compensate for shortages of range forage. Nevertheless, the open range remained in poor condition. (DiTomaso, 2000, summarizes the mechanisms by which overgrazing over long periods of time changes the plant composition of rangelands.)

Increased demand for meat brought on by World War I raised meat prices in the United States and elsewhere, which also contributed to overgrazing on rangelands. (As you will see later in this chapter, a higher price for livestock raises the profitability of increased stocking.) Finally, a prolonged drought from 1930 to 1935 dealt a serious blow to the open rangelands.

In 1934, the U.S. Congress passed the Taylor Grazing Act, which gave authority to the U.S. federal government to administer 80 million acres of land. The act defined property rights to grazing lands, which many ranchers still have. By providing the increased security of tenure, the Act gave ranchers more incentive to take care of the land, and that helped improve range conditions over the long term. Since 1934 there has been increased competition for rangelands. The forces of settlement, tourism, demands for endangered species protection, and a national environmental ethic continue to grow. These forces have increased the values from and number of uses of rangelands.

■ Desirable Outcomes

Many of the world's rangelands have the potential for multiple uses such as livestock grazing, wildlife, watershed protection, and recreation. There is growing conflict and controversy over how rangelands should be managed for efficient, equitable, and sustainable uses. Fowler (2000) provides a rich analysis of the history and land ownership patterns of ranches, factors affecting ranch profitability, rancher characteristics, and current and potential issues and conflicts.

In many parts of the world lack of infrastructure and lack of investment in resource management have made rangelands more sensitive to the impacts of stresses brought on by drought and climate change. Some options, such as multiple-use land management, could be carried out regardless of infrastructure and technological factors. In some cases options could be carried out most effectively through active involvement of communities in rangeland management.

Efficiency

Economic efficiency means managing rangelands to maximize their total economic benefits summed for all uses and all time periods. It means allocating rangelands across uses and periods to produce the greatest possible long-term economic benefit (i.e., the discounted net present value of benefits described in Chapters 5 and 6). When the required data are available, rangeland management can be measured objectively in regard to economic efficiency.

Three components contribute to the total economic benefits from a rangeland management plan. The first component consists of the incomes of livestock producers and other industries linked to the producers' incomes. These incomes are measured as total revenue minus total cost of livestock, wildlife, and related enterprises. The second component of rangeland benefits is consumer surplus from marketable products. Consumer surplus is accrued by buyers of beef, wool, wildlife hunting, and related products. It is equal to the cost savings of rangeland products purchased by members of the larger society compared to what they are willing to pay. For example, reduced beef prices produce an increased consumer surplus for beef. Consumer surplus benefits typically are spread out in a nation, usually at a considerable distance from where the rangeland use occurs. The third component of rangeland benefits is unpriced public goods produced by a rangeland; these include willingness to pay for recreation, wildlife, water, or endangered species. Unpriced values are hard and/or expensive to estimate; thus they are rarely measured.

Equity

Managing rangelands for equity means allocating rangeland resources fairly across users, uses, and time periods. Fairness means different things to different people, so it is hard to measure objectively. But since many rangeland management decisions affect who gets what, fairness is important to politicians and policymakers. In fact, equity is often more important than efficiency in the political arena.

Here are a few examples of rangeland actions that are considered by some people to be fair:

- Ranchers who make financial investments in range improvements on public lands have a property right to those improvements.
- Taxpayers who pay for the upkeep of public rangelands have a right to access the land for hunting, wildlife watching, and recreation—and for the financial return produced by uses of those lands.
- Governments that own rangelands have a right to maximize the financial return of those lands to provide for their citizens.

Sustainability

The 1987 World Commission on Environment and Development defined sustainability as development that meets the needs of the present without compromising the ability of future generations to meet their own need (Brundtland, 1987). For rangelands, sustainable actions are those that avoid overgrazing range forage in order that the soil, water, and plant resources serve current and future generations. See Example 9–1.

Sustainable actions have a strong link with conservation and maintenance of soil and water resources. Soil influences hydrologic processes by providing the medium for the capture, storage, and release of water. The flow of water through rangelands influences the physical and biological properties of the soil. In many of the world's rangelands, water is the limiting resource, and overgrazing during periods of drought damages forage.

Sustainable actions on rangelands also have an important connection to the conservation and maintenance of rangeland plant and animal resources. These resources reflect the degree to which the integrity of the soil and the ecological processes of rangelands are sustained. Herrick et al. (2002) describe the role of rangeland monitoring systems for measuring sustainability. Sustaining biodiversity can also be a part of this process. *Biodiversity* is defined as the variety of life and its processes that encompass the variety of living organisms, the genetic differences among them, and the communities and ecosystems in which they occur.

Sustainable rangeland activities maintain productive capacity on rangeland ecosystems. Productive capacity refers to the opportunity of those lands to provide the current generation of livestock producers, ranching-dependent communities, and society at large with a wide variety of goods and services, depending on the mix desired by society at any particular time. It also refers to assuring that future generations will be able to obtain their desired mix of goods and services. Bartlett (2003) describes ongoing cooperation between the U.S. Forest Service and Colorado State University to develop indicators of sustainable rangeland activities. See Example 9–2.

■ UNREGULATED MARKET OUTCOMES

Four important economic questions face private livestock producers: First, what is the optimum grazing intensity: How many animals, livestock, or wildlife should be grazed per unit of land now and in the future? Second, what combination of inputs should be used? For example, hay and win-

Example 9–1
Challenges in Sustaining Rangelands in the Himalayas

The Himalayas are the world's highest and most populated mountains. Extending 3,500 kilometers (2,170 miles) from east to west and home to 140 million people, the mountains cover an area of which about one third is rangeland. Population pressures in this part of the world cause considerable emphasis to be placed on the ability of regional institutions to manage rangelands sustainably.

Since the early 1900s, Western range science has developed principles and methods to describe the state of rangelands. One basic principle is *range condition class,* which describes a rangeland's health. Describing range health involves measuring the composition of vegetation at any given site in comparison to an ideal climax (stable, mature) plant community on the same site. A second principle is *carrying capacity,* defined as the number of animals that can safely be stocked without damaging the range. Carrying capacity estimates are normally based on assumptions about the impact of livestock on plants and plant succession. Heavy livestock grazing leads to a decline in range condition, and reducing or removing grazing pressure restores the range to an improved condition. The challenge facing range managers in the Himalayas is how they should regulate livestock numbers in migratory areas where neither kind of data described in the two principles can be determined.

Over hundreds of years, migratory livestock producers (pastoralists) in the Himalayas have learned, taught, and applied ecological knowledge of their ecosystems. Being able to get and use this knowledge has been important in sustaining their livestock economies. (Fernandez-Gimenez, 2000, examines the relation between historical knowledge of range ecosystems and social institutions for Mongolian nomadic pastoralists.) One new view that might be applied to regions like the Himalayas, where precipitation is highly erratic, is the concept of *shifting vegetation states.* According to this view, there is no such thing as a stable (climax) vegetation state, and vegetation on the ground follows the highly variable precipitation patterns.

The concept of "opportunism" may also prove to be economically efficient and sustainable in the Himalayas. The producer could *estimate* the annual grazing pressure that can be sustained on that year's forage production, with no special reference to its potential climax state. This philosophy suggests doing the best you can this year with the resources you have, rather than adjusting stocking pressure to produce an ideal vegetation state that may never occur.

Source: Bhatia, A., S. R. Chalise, and C. Richard. 1998. *Common property resources of the Hindu Kush-Himalayas.* Available from *http://www.icimod.org/focus/cpr/cpr.htm* [Accessed June 12, 2004].

ter range (i.e., grasses grazed in winter) are substitute inputs for many private ranches. The rancher who optimizes those two inputs asks what is the optimum mix of hay and winter range that minimizes the costs of producing a given output, such as animal weight gain. Third, what mix of outputs should be produced? For example, a private livestock producer might want to compare how many cattle should be stocked on a piece of land versus how many big-game licenses should be sold to hunters who would harvest wildlife on the land. Larger numbers of livestock compete with wildlife for the same forage. How does the economic loss of reducing the output level of one compare to the economic gain made possible by increasing the other output?[3] Fourth, what mix of inputs and outputs over time maximizes the discounted value of current and future net income? For instance, how much money should be invested in various range improvements and stocked livestock at various future time periods to maximize the present value of net income?

The remainder of this section deals mostly with the first question by analyzing optimum grazing intensity from the view of a private working ranch. This is a most significant question today in considering rangeland policy. In simplified terms, the first question is seeking to determine the economic optimum *stocking* rate. The example that follows asks how steers should be stocked per 100 acres of land to maximize profits.

[3] Little et al. (2001) describe the process of choosing the mix of outputs for ranchers in northern Kenya and southern Ethiopia.

Example 9–2
Coping with Drought on Africa's Rangelands

The village of Banizoumbou in Niger Africa is a typical community of the Sahel with low rainfall (450 millimeters, or 18 inches) and sandy soils. (The Sahel is a region of semiarid lands bordering the southern edge of the Sahara Desert in Africa; it is characterized by a long dry season and a short wet season from July to September.) Selling range-stocked animals is the main source of income for the village's livestock producers. There are competing theories about what makes livestock production factors contribute to sustainable rangeland. One theory says that overgrazing has no major long-term effect on forage production. This theory is supported in areas where there are sandy soils, such as in Banizoumbou, where short-term grazing does not hurt future range conditions. A competing theory says that overgrazing can damage future forage production. Supporters of this theory observe that intensive grazing during the rainy season has lead to a change in the grass species composition toward lower quality and less productive forage.

Livestock producers in Niger try to anticipate and react optimally to inevitable droughts. Still, two exceptional droughts of 1973 and 1984 reduced the national livestock by half. This led many migratory livestock producers to fatten their livestock as much as possible, anticipating the next drought, so that cattle can survive the drought with minimum weight loss. This risk-management approach reflects producers' preferences for building up larger herds as a major source of wealth and for protecting their food security in drought years. Selling cattle can mean a significant loss of future income. Another reason to avoid selling livestock is that under the widespread practice of open access grazing, a larger herd has a better claim on community grass than a smaller herd.

The issue of settled versus nomadic livestock production is widely debated. Not everybody believes that more settled production systems are the solution. Some analysts believe that more settled systems are environmentally unsustainable in low-rainfall and drought-prone areas like Banizoumbou. There are simply too many livestock producers in one place with too little rain. Some analysts prescribe policies that promote and reward continued migration in search of better forage, policies that promote coevolutionary adaptions to natural processes. However, it is not clear that returning to the traditional patterns of resource management is an economically sustainable choice given the livelihood needs of a larger and rapidly growing rural population. Resolution of these conflicting views will require a serious analysis of the choices open to communities such as Banizoumbou, including a design and assessment of the long-term consequences for sustainable rangeland resource management.

Sources: Burbier, B., and P. Hazell. 2002. *Implications of population growth and declining access to transhumant grazing areas for the sustainability of agropastoral systems in the semiarid areas of Niger.* International Food Policy Research Institute. 2004. Available from *http://www.ifpri.org/* [Accessed June 12 2004]; Batterbury, S. and A. Warren. 1999. *Land use and land degradation in southwestern Niger: Change and continuity.* London School of Economics. 1999. Available from *http://www.regard.ac.uk/research_findings/L320253247/report.pdf* [Accessed July 6, 2004].

The economics of private ranching aims to understand how to maximize profits on working ranches. It also identifies impacts on ranches of economic changes that are beyond their control. A common assumption made when attempting to understand and predict a private livestock producer's decisions is that the rancher's goal is to do whatever maximizes profit.[4] Recall from Chapter 3 that profit is defined as total revenue minus total cost.

Economically Optimum Stocking Rate

The most important problem facing a private livestock producer who wishes to profit from the use of rangelands is determination of the economic optimum stocking rate. The stocking rate, defined as the number of animals grazed on a given land area for a fixed period of time, is the primary decision over which a private rancher has control. Table 9–1 shows a production function in which

[4] Some studies have found that profit is becoming less of a motivation for many ranchers. Martin (1966) and Pope (1988) found that quality of life values influence the ranch real estate market at least as much as the land's income-earning power from livestock production. Harp, Rimbey, and Darden (2001) found that social factors, such as being attached to the community where a rancher grows up, is also an important motivation.

Table 9–1 Profit-maximizing stocking level of steer[a]

(1) Variable input (steers/ 100 acres)	*(2)* Total physical product (TPP) (weight gain/ 100 acres)	*(3)* Marginal physical product (MPP)	*(4)* Price per unit output (P_O)	*(5)* Total value product (TVP)	*(6)* Value of marginal product (VMP)	*(7)* Price per unit variable input (P_I)	*(8)* Variable input cost (VCI)	*(9)* Fixed input cost (FCI)	*(10)* Total net revenue (TNR)
	← pounds →		← dollars →						
3	1,215	405	1.00	1,215	405	75	225	1,200	−210
6	2,260	349	1.00	2,260	349	75	450	1,200	610
9	3,141	294	1.00	3,141	294	75	675	1,200	1,266
12	3,850	236	1.00	3,850	236	75	900	1,200	1,750
15	4,398	182	1.00	4,398	182	75	1,125	1,200	2,073
18	**4,779**	**127**	**1.00**	**4,779**	**127**	**75**	**1,350**	**1,200**	**2,229**
19.2	4,884	75	1.00	4,884	75	75	1,444	1,200	2,241
21	4,984	68	1.00	4,984	68	75	1,575	1,200	2,209
24	5,031	16	1.00	5,031	16	75	1,800	1,200	2,031
27	4,912	−40	1.00	4,912	−40	75	2,025	1,200	1,687

[a]Input is grazing intensity, measured in steers stocked per 100 acres nonirrigated rangeland. Output is total seasonal weight gain in pounds per 100 acres summed over all animals stocked.

Source: *Data are adapted from Torell, L. A. 2004. Unpublished data about stocking trials in Wyoming and Colorado; and Hart, R., and M. Ashby. 1998. Grazing intensities, vegetation, and heifer gains: 55 years of shortgrass.* Journal of Range Management. *51: 392–398.*

numbers of steers stocked is the *variable* input. A known quantity of grazing forage on a fixed parcel of rangeland is the *fixed* input, assumed equal to 800 pounds of dry vegetation (1 AUM) per acre. The livestock producer's *output* is total weight gain for the season summed over all animals stocked.

Optimum Input Use. Table 9–1 shows several possible stocking rates for a five-month spring-summer season in which varying numbers of steers can be stocked on a fixed, 100-acre parcel of land. The data indicate the law of diminishing marginal returns.[5] An increase in stocking rate from 0 to 3 steers (column 1) increases total output from 0 to 1,215 pounds. Column 2 shows total output as *total physical product* (TPP). Further increases in steer numbers result in output increases, but at a decreasing rate. Increasing steer numbers from 3 to 6 allows an additional weight gain of 1,045 pounds (1,215 to 2,260), but the increase from 6 to 9 produces only 881 pounds of gain (2,260 to 3,141). Continued increases in numbers of animals stocked on a fixed acreage of rangeland brings smaller additions to total weight gain, due to increasing competition for scarce forage. In the extreme, there will be some number of animals stocked on the land parcel (27 in this example) that causes competition for available forage to be so severe that total weight gain for the available forage declines.

The *marginal physical product* (MPP) of additional animals stocked is defined as the *added* weight gain (output) produced by *one more* unit of the variable input applied (steers stocked). Sometimes shortened to simply *marginal product,* the marginal physical product is expressed by the following equation:

$$MPP = \frac{\Delta TPP}{\Delta \text{ Variable input}} \tag{9.1}$$

where ΔTPP means change in total physical product and ΔVariable input means change in variable input, steers stocked. The change in the variable input is often assumed to equal one. When

[5] Sources: Ideas similar to those in this section are described in Workman, J. P. 1986. *Range Economics.* New York: Macmillan; Torell, L. A., and R. H. Hart. 1997. Economic Considerations for Efficient Stocking Rates on Rangeland. Paper read at Fort Keogh Research Symposium, Miles City, MT, September 1997.

the stocking rate is increased from 6 to 9 steers, MPP is (3,141 − 2,260) / (9−6) = 294, shown in column 3. That column shows that MPP continues to decrease throughout the range of stocking shown in the table. Increasing the number of stocked steers from 24 to 27 produces an MPP of minus 40: larger numbers of animals compete for a fixed amount of forage.[6]

Output price, P_O, is defined as average revenue per unit sold. In a competitive market for beef, it is assumed the producer can buy or sell all animals desired at the going market price of $1.00 per pound.[7] No individual producer can influence the price per pound of animals, because the individual's purchases or sales are too small a part of the world's total supply to alter the market price. So the purely competitive market assumption is reasonable. Column 4 shows the results of this assumption: Output price per pound is $1.00—constant over the range of potential stocking rates, which means that the sales price per pound is not affected by the individual rancher's stocking rate.

The *total value product* (TVP) is the monetary value of total weight gain, equal to the market price times total physical product. It also equals total revenue. Because TVP multiplies TPP by a constant, the numbers in column 5 are directly proportional to column 2; and in this case they are equal because the price is $1.00 per pound.

The value of the marginal product (VMP) in column 6 is the *addition* to TVP produced by an increment of the variable input, animals stocked. It is the monetary value of the marginal physical product. At a stocking rate of 9 animals, VMP is $294: ΔTVP / ΔVariable input is (3,141 − 2,260) / (9−6) = $294. The VMP is important because it measures *additional* revenue produced by applying one more unit of the variable input.

Costs are just as important as revenues to the livestock producer, because when costs are higher than revenue the producer will not survive economically. The input price, P_I, is the cost paid by the producer for using one more unit of the variable input, animals stocked. The price of the input includes costs for things such as purchased feed, grazing fees paid for forage, interest paid on purchased animals, fences, medicine, and hired labor. Variable costs increase with larger amounts of the variable input used. Think of variable costs as costs the rancher can avoid paying by reducing the number of animals stocked. Assume the price of stocking each additional animal for the season is $75. The stocking rate of 9 animals, at a variable input cost (VCI), is 9 times $75 per animal = $675 (column 8). VCI is calculated by multiplying column 1 by column 7.

Fixed costs, which pay for the fixed inputs, include cash costs such as land rent per acre, property tax, insurance, and mortgage payments. They also include opportunity costs such as family labor and interest that could have been earned on the money that is tied up in the ranch operation's fixed capital. Fixed costs do not change with a change in use of the variable input or from changes in output. Fixed costs do not increase with increased production, and the rancher cannot avoid paying these costs in the short run by reducing use of the variable input (number of animals stocked) to zero. In the example shown in Table 9–1, fixed costs are labeled fixed cost of the inputs (FCI) and consist of an annual charge of $1,200 for the 100-acre parcel of land, column 9.

Total net revenue (i.e., profit) is defined as total revenue minus total cost. In this example, *it is maximized by stocking whatever number of animals equates the input price* (P_I) *to the value of the marginal product of the variable input (VMP).* The point of profit maximization is the economically optimum grazing intensity. That optimum grazing intensity, where P_I = VMP, is found by compar-

---

[6] Increasing the number of animals stocked on a fixed amount of forage increases grazing pressure. Grazing pressure is the relationship between the number of animal units and the weight of forage dry matter per unit land area at a point in time. Increased grazing pressure reduces the average weight gained per day of grazing per animal, called the average daily gain. At low stocking levels, grazing pressure is low enough so that average daily gain does not fall with increased stocking levels, so marginal productivity stays constant. Declining marginal product results from declining average daily gains.

[7] In fact, the rancher's buying price per pound of lighter animals stocked on the range is typically higher than the selling price per pound of heavier animals removed and sold. Assuming an equal buying and selling price per pound lets us measure revenue as price times weight gain. It also lets us ignore the need to compare the cost of buying the light animal with the revenue from selling the heavier one.

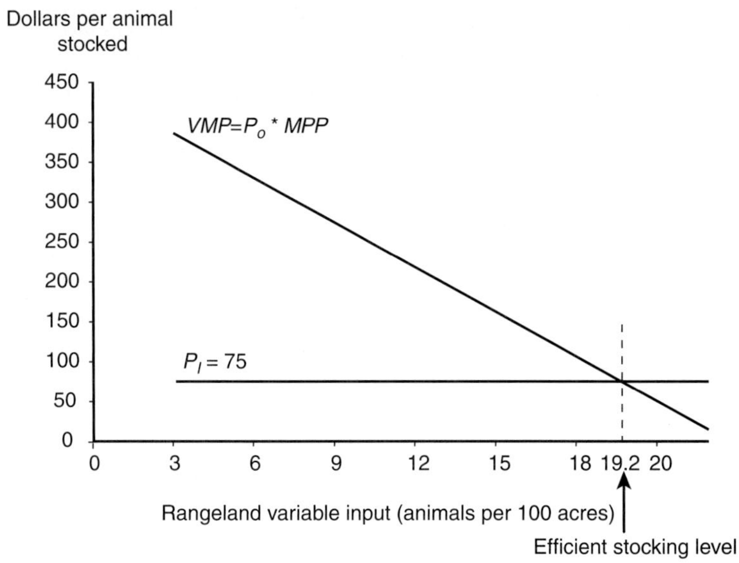

Figure 9–1 Profit-maximizing stocking level

ing columns 6 and 7 in Table 9–1. The optimum stocking rate is slightly more than 18 steers per 100 acres, since more net revenue is earned at this stocking rate than at any other.[8] This 18+ steer stocking rate occurs where the price of the input equals the value of its marginal product.

The logic underlying the equating of P_I and VMP is easy. Between 6 and 9 animals, additional herd weight gain from higher stocking intensity adds more to revenue than to costs (P_I = \$75 while VMP = \$294), so it pays the livestock producer to stock more animals. By increasing stocking from 21 to 24 animals, the additional weight gain adds more to costs than to revenue (P_I = \$75 while VMP = \$16), so profit could be increased by stocking *fewer* animals. Only at the optimum stocking rate of 18 animals is there no economic incentive to *either* increase *or* decrease animal numbers. At the stocking rate of 18 animals per section, the VMP (\$127) approximately equals P_I (\$75), so we know that 18 animals is the profit-maximizing or optimum stocking rate.[9] It is also called the economically efficient stocking rate, sometimes shortened to the efficient stocking rate. Profit at this stocking rate is maximized at \$2,229.[10] Figure 9–1 shows that the downward sloping VMP must equal the constant sloping input price, which cross at the efficient stocking level.

Notice that the *economic* optimum stocking rate (18 animals) is lower than the level at which total output is maximized. The stocking rate at which total output is maximized is sometimes called the *biological* optimum. In this example, the biological optimum is 24 animals, which produces 5,031 pounds of weight gain. The only time that the economic optimum stocking rate equals the biological optimum is when the variable input is priced at zero. In that unusual and special case, it pays to use the variable input until it has a VMP of zero. Otherwise, the economic optimum will be at a *lower* level of variable input use, a *lower* level of output, and a *lower* level of forage consumption than the biological optimum. See Example 9–3.

---

[8] The *exact* optimal stocking rate is 19.2, shown in Table 9–1 as a shaded row between 18 and 21. (Computation is based on the mathematical equation for the total physical product and is available from the author on request.)

[9] This analysis shows that the profit-maximizing stocking rate varies with output (cattle) price, variable costs, and the productivity of the variable input. Profit-maximizing stocking rates also vary accordingly with many other factors, including precipitation and temperature, the operator's property rights associated with the land and forage, ranch debt structure, length of the planning horizon, and terms of grazing leases. Winder et al. (2000) show how severe drought can reduce the economically optimal stocking rate in the case of New Mexico.

[10] Table 9–1 shows alternative stocking rates that vary in large increments of three animals, so the value of the marginal product never exactly coincides with the input price. The theoretical optimum variable input use is 19.2 animals stocked, at which point the VMP = P_I = \$75, and profits are \$2,240.99.

Example 9–3
What Do Iran's Ranchers Want to Know?

The economy and culture of Luristan province in western Iran are based on dry farming and on irrigated agriculture. Illiteracy plagues the province, and lack of schools are a major barrier to people in villages who want to pursue an education. Livestock producers face special obstacles in their search for sustainable grazing systems. Such grazing systems often require the ability to think counterfactually, namely, the ability to imagine a future never seen and to take steps to achieve that future when it seems worth pursuing.

Chizari and Noorabadi wanted to learn more about the education needs of Iran's rural livestock producers. Aims of their study were (1) to find out what ranchers wanted to know about rangeland management systems, (2) identify their educational needs, and (3) determine ranchers' preferred educational delivery method. They found that nearly 90% of the region's ranchers were illiterate. Most had small landholdings; only 3% grazed more than 300 hectares (approximately 740 acres) of rangeland at any one time. On the other hand, more than one third grazed 151 to 200 sheep and goats. Most were migrants who move in a continued search for forage for their livestock.

Most ranchers expressed a need for training in a range of grazing systems, learning more about grazing intensity, and acquiring the ability to compare with and without project analysis (see Chapter 5). Nearly half expressed a desire to read and write. More than half of the ranchers wanted to know more about factors leading to rangeland degradation (84%). The ranchers had the least need for information on seedbed preparation (40%), fencing (38%), seeding techniques (36%), and fertilization (30%). Educational methods believed to be most useful in delivering information were practical, hands-on learning approaches.

What conclusions can be drawn from the study of Chizari and Noorabadi? Ranchers in rural Iran wish to learn a variety of rangeland conservation and management principles. A majority of the study respondents indicated a need to learn specific rangeland conservation principles as well as techniques for saving water and for preventing soil erosion and general deterioration of rangeland. The low level of literacy, migratory patterns, and significant learning needs of ranchers in the study pose a special challenge for government programs.

Sources: Chizari, M., and R. Noorabadi. 1999. Perceived learning needs and program delivery preferences of ranchers in Noorabad township of Luristan province, Iran. *Journal of International Agricultural and Extension Education.* (3): 39–45; Aagesen, D. 2000. Crisis and conservation at the end of the world: Sheep ranching in Argentine Patagonia. *Environmental Conservation.* 27 (2): 208–215.

Optimum Supply Response. The previous section described the livestock producer's optimal use of the variable input. The same decisions made by the producer can be shown by looking at *outputs* instead of inputs. Table 9–2 shows the choices facing the livestock producer who seeks the economically optimum level of *output.* In the table the optimal stocking rate is determined by searching for a stocking rate that produces an output level at which the last unit adds the same to cost as it does to revenue. The table illustrates the concept of marginal cost, that is, the additional cost resulting from producing one more unit of output.

Suppose the livestock producer currently produces 3,141 pounds of weight gain, produced by stocking 9 animals. Raising the stocking level to 12 increases total weight gain from 3,141 pounds to 3,850 pounds. In other words, output is raised from 3,141 pounds to 3,850 pounds. The resulting additional herd weight gain adds more to revenue than to costs, since the price per pound is \$1.00 and the marginal cost of the added output is only \$0.32. The marginal cost of the added output is the additional variable cost of the inputs, ΔVCI, divided by the additional output produced, ΔQ (column 9). It is calculated as (900 − 675) / (3,850 −3,141) = \$0.32. That marginal cost is the change in column 4 divided by the change in column 1. Since the price exceeds the marginal cost, it pays the ranch manager to increase the herd's weight by stocking more animals.

Increasing output from 4,984 to 5,031 pounds (produced by raising the stocking rate from 21 to 24), adds more to costs than to revenue (P_O = \$1.00, while MC = \$4.74), so profit can be increased by producing *less* output, which means stocking fewer animals. Only at the economically optimum output level of 4,779 pounds of gain is there no economic incentive to either increase or decrease output. At the output level of 4,779 pounds, P_O = \$1.00 and approximately equals

Table 9–2 Profit-maximizing output level[a]

(1) Output (*Q*) (total physical product)	(2) Price (P_O)	(3) Total revenue (TR)[b]	(4) Total variable cost (TVC)	(5) Total fixed cost (TFC)	(6) Total cost (*TC*)[c]	(7) Average variable cost (AVC)[d]	(8) Average total cost (ATC)[e]	(9) Marginal cost (*MC*)[f]	(10) Total net revenue (TNR)
pounds	←				dollars				→
1,215	1.00	1,215	225	1,200	1,425	0.19	1.17	0.19	−210
2,260	1.00	2,260	450	1,200	1,650	0.20	0.73	0.22	610
3,141	1.00	3,141	675	1,200	1,875	0.21	0.60	0.26	1,266
3,850	1.00	3,850	900	1,200	2,100	0.23	0.55	0.32	1,750
4,398	1.00	4,398	1,125	1,200	2,325	0.26	0.53	0.41	2,073
4,779	**1.00**	**4,779**	**1,350**	**1,200**	**2,550**	**0.28**	**0.53**	**0.59**	**2,229**
4,884	1.00	4,884	1,444	1,200	2,644	0.30	0.54	1.00	2,241
4,984	1.00	4,984	1,575	1,200	2,775	0.32	0.56	1.10	2,209
5,031	1.00	5,031	1,800	1,200	3,000	0.36	0.60	4.74	2,031
4,912	1.00	4,912	2,025	1,200	3,225	0.41	0.66	infinite	1,687

[a]Output is weight gain per hundred acres over all animals stocked for the season.

[b]Column 2 × column 1.

[c]Column 4 + column 5.

[d]Column 4 ÷ column 1.

[e]Column 6 ÷ column 1.

[f]Change in column 4 ÷ change in column 1.

Source: *Date are adapted from Hart, R., and M. Ashby. 1998. Grazing intensities, vegetation, and heifer gains: 55 years on shortgrass.* Journal of Range Management. *51: 392–398.*

MC = \$0.59, so 4,779 pounds of gain is the economic optimum level of output.[11] Profit at this output level is maximized at \$2,229,[12] shown in Figure 9–2 as occurring at point *A*, the output level where beef price (\$1.00) intersects a rising marginal cost.

Shifts in Supply. The profit-maximizing rancher stocks about 18 animals and produces about 4,779 pounds of weight gain when the output price is \$1.00, when the input price is \$75, and when the forage supply is 800 pounds per acre. Will the rancher produce 4,779 pounds of output year after year? Probably not, because output price, input price, and forage supplies change from time to time. The alert rancher watches all three like a hawk and adjusts stocking rates so that profits continue to be maximized. Any of these three items can change because of changes in consumer tastes and preferences, consumer incomes, events in related markets, and impacts of nature itself.

---

[11] For the technically inclined reader, here are the details. The producer seeks the stocking level, S^*, that maximizes profit, where profit = $[P_O * Q(S, F_0)] - P_I * S - FC$. One production function commonly used by range economists is $Q(S, F_0) = D_0 * (B_0 + B_1 * (GP) * S)$, where GP (grazing pressure) = $D_0 * (S/F_0)$. Q = weight gain, the production function; D_0 = days in season (153); P_O = output price (\$1 per pound); P_1 = input price (\$75 per additional animal stocked); F_0 = forage supply (800 pounds per acre); FC = fixed cost (\$1,200); B_0 = empirical grazing pressure intercept = 2.83; B_1 = empirical grazing pressure slope = −0.32. If all remaining resource constraints are included in fixed costs (e.g., access to water, fences), then the fixed input forage is the most important limiting resource. That is, over a wide range of stocking rates, doubling of F_0 doubles the optimal stocking rate, doubles maximum returns over variable costs (revenues minus variable costs), and does not affect the optimal grazing pressure. So the forage supply effectively sets the scale of the ranching operation as well as the income it can produce. Profitable private-ranch investments and efficient public-range policies will increase the supply of forage whenever total benefits exceed total costs. See the subsequent section on the economic value of forage improvements.

[12] Price is not exactly equal to marginal cost because of the large decision increments. The above answer is correct if stocking rates can only be varied in increments of three. For infinitely small stocking increments, optimal output is 4,884 pounds, price and marginal cost both equal 1.00, and profits are \$2,241, as shown in Figure 9–2.

Figure 9–2
Profit-maximizing output level

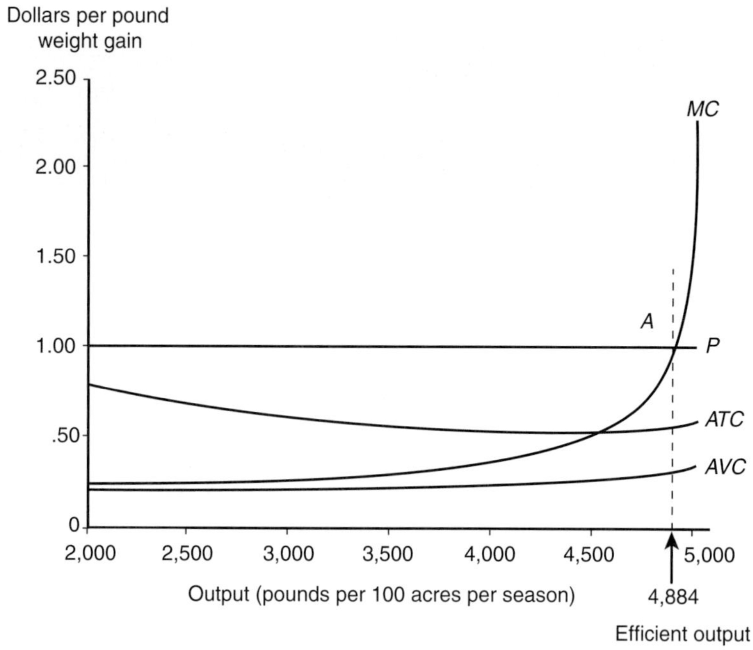

Recall from Chapter 3 the things that shift demand and supply. All of these demand and supply shifters influence ranchers' stocking decisions. Here are some examples:

- If consumers decide they prefer more beef, the price of beef will increase, which increases output price, input price, and the economically optimal stocking rate. Increases in consumer income can have similar effects.
- Forage supplies are strongly influenced by nature, especially events that affect the timing and amount of precipitation. Reduced precipitation reduces forage supplies and reduces the optimal stocking rate by shifting to the left the marginal productivity of animals stocked. Severe droughts can reduce forage supplies to a fraction of normal levels, thereby causing ranchers to reduce their optimal stocking rates, often by selling large parts of their herds.
- Government policies have a major effect on the ranchers' stocking decision. Any policy that influences output price, input price, or forage supply will influence the optimal stocking rate, as well as ranch income and output supply. Policies that subsidize the variable input reduce the price of the input and increase the optimal stocking rate, because livestock producers find that an unchanged VMP schedule combined with a reduced variable input price requires a larger stocking rate to maximize profits. If the public sector charges a grazing fee to use public forage, any policy that raises that fee increases the variable input price and reduces the optimal stocking level. In the United States many environmental organizations have supported policies that raise the grazing fee on western public lands in hopes that rancher stocking rates on public lands will fall, thereby reducing overgrazing, and increasing other uses of the land, such as wildlife production and protection. Governments in livestock-dominated economies often look for policies to help ranchers cope with forage losses produced by severe droughts. Government policies that subsidize nonanimal inputs can partly compensate for severe drought by mitigating the loss of ranchers' income otherwise suffered by lost forage.

Regardless of actions by the economy, nature, or the government, profit-maximizing ranchers attempt to stock the *variable input* where the value of the marginal product equals the input price: equivalently, they attempt to *produce the output* at which price equals marginal cost. Table 9–3 shows the profit-maximizing economic behavior for all combinations of two input prices, four output prices, and two forage supplies for a total of 16 possible economic conditions

Table 9–3 Rangeland supply shifts

(1) Variable input (steers/100 acres)	(2) Output, supply (pounds/100 acres)	(3) Net revenue ($)	(4) Marginal value of forage ($/100 pounds)[a]	(5) Price per unit of variable input ($/steer)	(6) Price per unit of output ($/pound)	(7) Forage supply (100 pounds/acre)
19.2	4,884	2,241	4.30	75	1.00	8
20.6	4,969	4,708	7.39	75	1.50	8
21.3	4,998	7,201	10.50	75	2.00	8
21.7	5,012	9,704	13.63	75	2.50	8
17.9	4,767	1,777	3.72	100	1.00	8
19.7	4,916	4,205	6.76	100	1.50	8
20.6	4,969	6,678	9.85	100	2.00	8
21.1	4,993	9,169	12.96	100	2.50	8
16.8	4,274	1,811	4.30	75	1.00	7
18.0	4,347	3,970	7.39	75	1.50	7
18.6	4,373	6,151	10.50	75	2.00	7
19.0	4,385	8,341	13.63	75	2.50	7
15.7	4,171	1,405	3.72	100	1.00	7
17.2	4,302	3,529	6.76	100	1.50	7
18.0	4,347	5,693	9.85	100	2.00	7
18.5	4,369	7,873	12.96	100	2.50	7

[a]The marginal value of forage is defined as the additional net income produced by increasing forage supply from its current level by 100 pounds per acre. For the interested reader, grazing pressure can also be calculated in this table. Grazing pressure is the number of animals stocked (column 1) times days in season (153 days for this example) divided by hundreds of pounds of forage (column 7). Information on grazing pressure is important to track for dynamic analysis, in which current grazing intensity affects future forage production and future optimal grazing intensity.

Source: *Data are adapted from Hart, R., and M. Ashby. 1998. Grazing intensities, vegetation, and heifer gains; 55 years on short grass.* Journal of Range Management. *51: 392–398.*

facing the livestock producer. The table shows responses to all 16 of those changes in which four important producer behaviors are targeted for analysis:

- Optimal use of the variable input (column 1)
- Optimal output supply (column 2)
- Optimal profit (column 3)
- Marginal value of forage (column 4)

Results of the table show that whenever the price of the variable input increases from $75 per animal stocked to $100 per animal stocked, the number of animals stocked falls, for any constant output price and for any constant forage supply. This result simply reflects the law of input demand: When the price of a variable input increases, other things equal, the use of that input decreases (in some cases, the input's use is unchanged). The livestock producer moves up or down the unchanged VMP schedule whenever the input price increases or decreases.

Table 9–3 also shows that whenever the price of the output increases from $1 to higher levels, the livestock producer expands output. Therefore, the producer moves up the rising marginal cost schedule. This action results in the producer adjusting output to make the new higher price equal to the higher marginal cost. The table also shows that when output price increases, other things equal, the optimal supply response produces a higher profit-maximizing level of income. The direction of the livestock producer's response to an output price change is the same, no matter what the input price is and no matter what the forage supply is. This behavior reflects the law of supply: Recall from Chapter 3 that when output price increases, other things unchanged, the profit-maximizing quantity supplied increases.

Figure 9–3 uses the results of Table 9–3 to show four changes in supply as well as two movements along each supply response curve. Beginning with the supply farthest to the right (S_1), notice that it shifts left to S_2 when the input price increases from $75 to $100 per animal stocked.

Figure 9–3
Rangeland supply shifts

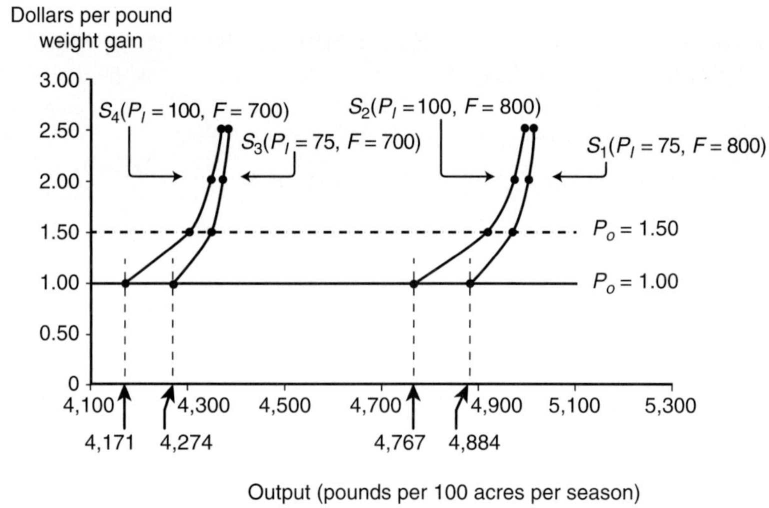

Notice that when the input price returns to \$75, S_1 shifts much more to the left, all the way to S_3 when the fixed input (forage, F) is reduced from 800 to 700 pounds per acre. The livestock producer's supply response shifts even farther to the left, all the way to S_4, when forage is reduced to 700 pounds *and* the variable input price increases from \$75 to \$100. Finally, for each of the four supply schedules, Figure 9–3 shows that the livestock producer moves up and to the right along an unchanged marginal cost schedule as output price increases from \$1.00 to \$1.50 whenever input price and forage supply are constant.

Economic Value of Forage

In the short run, rangeland forage is often assumed to be the fixed input to which varying amounts of stocked animals can be applied. But in the long run, forage is no longer fixed; actions can be taken by the livestock producer or the state to increase its productive potential. Water improvements and range seeding, for example, cause the livestock producer to incur costs, but they also increase forage production typically by a certain amount per acre. If the possibility of making this long-run improvement occurs, the livestock producer and the policymaker who may share in paying for that improvement must decide how to compute the economic value of that additional forage.

The economic value of the additional forage produced by the range improvement is defined as the additional profit it enables the rancher to earn. Recall that when there was 800 pounds of forage per acre the producer optimally stocked 19.2 animals for the 100-acre land parcel,[13] which produced 4,884 pounds of weight gain, incurred \$2,644 in total costs, produced \$4,884 in total revenue, and produced \$2,241 in profit. Of the \$2,644 in total costs, variable costs were \$1,444 and fixed costs were \$1,200. When forage is the limiting resource, and all other inputs are paid for, increasing (decreasing) forage supplies simply scales the entire operation up (down).

For example, if drought reduces forage supplies from 800 to 700 pounds per acre, and no other prices or costs change, the profit-maximizing stocking rate becomes seven eighths (7/8) of 19.2 animals (16.8), variable costs are 7/8 of \$1,444 (\$1,263), and total revenue is 7/8 of \$4,884 (\$4,273). Do profits fall to 7/8 of their original \$2,241 (to \$1,961)? The answer is no, because the livestock producer must still pay fixed costs of \$1,200, even when forage supplies fall. But even though fixed costs do not change, all variable costs and revenues fall to 7/8 of their original level because of the scaling effects caused by the reduced forage on production and optimal stocking.

---

[13] Recall the 19.2 figure was the exact optimal stocking rate; 18 was optimal when only 18 or 21 animals can be stocked. In fact, 19.2 animals cannot be stocked on a 100-acre parcel, but 192 animals can be stocked on 1,000 acres.

Therefore, total revenue minus variable costs fall to 7/8 of the original value, from $4,884 − $1,444 ($3,440) to $4,273 − $1,263 ($3,010). So what does this tell us for calculating the economic value of forage? Measure the *change in net income divided by the change in forage supply.* In this case, that forage value is $4.30 per hundred pounds lost ($3,440 − $3,010) / 100. That $4.30 is the marginal value of forage and it is shown in column 4 of Table 9–3's first row.

Information on the economic value of additional forage is important to the livestock producer, the public lands manager, and to the natural resource policy analyst because it allows the value of additional forage to be compared to its cost. For example, if economic conditions of Table 9–3's first row occur (conditions summarized in the last three columns), and if the range improvement costs anything less than $4.30, the improvement is economically efficient (profitable). So any range improvement that prevents a forage loss from 800 to 700 pounds is profitable, and hence economically desirable to the livestock producer, if it can be found at a cost less that $4.30.

Is the economic value of forage always set at a value of $4.30 per hundred pounds? Absolutely not. Economic conditions change and those changes influence the value of forage. Any increase (decrease) in the price of the variable input reduces (increases) the economic value of forage, because that price increase (reduction) reduces (increases) the power of forage to contribute to the rancher's income. Table 9–3 shows that whenever the variable input price increases, the economic value of forage falls. Similarly, increases (decreases) in the price of cattle increase (decrease) the economic power of the forage to produce income. Table 9–3 shows that increased cattle prices increase income and raise the economic worth of forage to the ranching operation.

Economic Value of Range Improvements

According to Murphy and Bartlett (2004), range improvements are structures, practices, or developments that increase the sustained yield of benefits produced by rangelands.[14] Improvements can vary from complex grazing systems to range seeding, to clearing undesirable plant species, to simple water developments. Such improvements maintain or improve range conditions for a particular rangeland use or combination of uses. For either a private rancher or the public land manager, range improvements typically must produce tangible benefits before the investment will be made. One important standard for judging the effectiveness of a range improvement is to produce a high enough addition to profit to pay for the investment.

The previous section described the economic value of forage at a point in time, such as a season, but most range improvements are long-term investments. The costs for improvements are paid out immediately, but additions to income (returns) come back to the livestock producer or to society at large in increments. *Present value determination* is the method most commonly used to determine the profitability of range improvements. Present value determination compares the present value of the improvement's costs with the present value of the additional returns it produces. Comparing these two elements tells the livestock producer whether the net present value of the range improvement exceeds zero (is economically efficient). The analysis of a range improvement using present value analysis can be applied through the following steps:

1. Describe the range improvement and its intent.
2. List the sources of additional costs and revenues from the improvement. Typically the largest costs are additional fixed costs in the early years. Returns typically occur as shifts in the value of the marginal product function described earlier (i.e., increased weight gain for a given amount of stocking).
3. Calculate the present values of additional costs over the life span of the improvement.
4. Calculate the present values of additional revenues over the life span of the improvement.
5. Subtract the present value of the total additional costs from the present value of the total additional added revenues to determine the net present value for the improvement.

---

[14] Source: Murphy, J. S., and E. T. Bartlett. 1994. *Economic Analysis of Range Improvements.* Colorado State University Cooperative Extension Service Report 3756.

Grazing and Forage Dynamics

So far this chapter has viewed animal weight gain as an outcome resulting from the application of a fixed input (forage) and a variable input (number of animals stocked). Forage input has been considered something controlled by a manager that is separate from the number of animals stocked. For a short time period, that is true. Forage exists and the livestock producer must decide how many animals to stock. If the time horizon is one season and the future is ignored, no one worries much about the impact on the next period's forage availability when the number of animals stocked in this period is increased. In most real rangeland situations, the horizon is wider and the number of animals stocked in this initial period can have a major impact on how much forage is available in the next period. Once a certain level of stocking is reached, heavier stocking in the same period *reduces* forage supplies available in the future.

Stocking rates found to be economically optimal over a single period may not be optimal when the current period's stocking reduces future forage production. When future forage production depends on current stocking rates, a simple static analysis is inadequate. Instead an explicitly dynamic analysis should be conducted in which the relationship between stocking and forage over several periods is considered to find optimal stocking rates.[15] In a dynamic analysis, time matters very much. The significant change in dynamic analysis is that stocking decisions during the current period are recognized as directly affecting forage production, potential animal weight gain, and revenues and costs in all future periods. In a dynamic setting, a rancher must choose the *time path* (often used interchangeably with time stream) of rangeland use rates that maximize the *discounted net present value.* If the rangeland is irreparably damaged, no time path of actions can maximize discounted NPVs. The rangeland cannot be improved (Torell, Lyon, and Godfrey, 1991).

Forage production changes through time, depending on the grazing history of the pasture. If heavy stocking occurred last year, this year's forage production may be reduced. Forage production over time depends on the previous period's grazing pressure, the previous period's forage production, and productivity changes not related to grazing rates, such as rainfall and invasive plant growth (see Masters and Sheley, 2001, for a summary of principles for managing invasive plants).

In both static and dynamic analyses, grazing pressure directly affects beef production in the current period through effects of grazing pressure on average daily gain. However, in dynamic analysis, the effect of stocking intensity in previous years is captured through its impact on forage production during the current period. Profit at any point in time is influenced by the amount of forage at that point, regardless of what previous actions produced that amount of forage. Forage production in any period, *t,* evolves from the level in the starting period, t_0, depending on how heavily the pasture is grazed between the starting and current period. Three classic works on dynamically optimal rangeland management are those of Burt (1971), Pope and McBryde (1984), and Torell, Lyon, and Godfrey (1991).

Profit Motive and Overgrazing

Many range scientists, environmentalists, and others fear that if the private market system is allowed to function without regulation the search for profits will lead to irreversible degradation of private rangelands.[16] Although ignorance (and optimism) concerning rangeland carrying capacity

---

[15] Batabyal and Godfrey (2002) describe why sustainable range policies require dynamic approaches to range management. Bestelmeyer et al. (2003) describe the application of dynamic rangeland management models to a wide range of climate and ecological conditions. Scarnecchia (2003) presents a dynamic and optimistic vision of how the science and art of range management could evolve.

[16] Carpenter et al. (2001), Ludwig et al. (2001), and Gunderson (2000) review the literature on the resilience of rangelands in recovering from grazing impacts. See the United Nations (2004) Web page for a history of its programs to combat rangeland desertification around the world. Desertification is the degradation of land in arid or semiarid areas into desert, resulting from climatic variability and human activities. Often it occurs because of the demands of increased populations that settle on the land in order to grow crops and graze animals. See the World Bank (2004) Web page for a summary of its rangelands program.

has damaged much private rangeland, this damage has occurred in spite of the profit motive, not because of it. Producers who overgraze are typically not behaving in their economic self-interest. Based on the production functions described earlier, biologists recommend stocking rates that maximize sustained yield, a point that occurs under conditions in Table 9–1 at about 24 steers per 100 acres. To stock fewer than 24 steers would waste forage that could be converted to a useful product; stocking more than 24 steers reduces total weight gain and causes long-term range damage.

The stocking rate recommended by economic analysis is based on the profit motive of the private landowner. This rate is only 18 animals. Thus, the profit motive leads to a lower stocking rate than maximum sustained yield and one that poses less risk of range degradation than the biologist's recommendation. The only condition under which the economic optimum (maximum profit) would correspond to maximum biological sustained yield is if additional steers could be stocked at zero extra cost. As long as there is any cost associated with variable input (numbers of animals), the economic optimum of maximum profit produces a stocking rate that is less than maximum sustained yield. Note that both the economic and biological optimum could produce unsustainable range practices.

Multiple Uses for Rangelands

Alternative uses for rangelands are under consideration by land managers around the world. In the western United States, for example, two significant uses of rangelands are nonpriced outdoor recreation and commercial livestock. Wildlife and livestock are two important competitors for grasses produced on rangelands. This competition for forage occurs on federally administered lands and on private ranches where forage supports publicly owned wildlife. For example, policy decisions on the pricing and quantity of state-supplied licenses for elk hunting are especially important where elk compete with cattle on private ranch lands. Important commercial economic choices face commercial ranchers who must decide how much forage to allocate to cattle versus fee-hunted elk on private ranches.

An analysis in Arizona raised an important management question regarding the competition for grass by elk and cattle (Cory and Martin, 1985). The authors wanted to know about policies that changed the size of a wildlife elk herd and how that change should be compared to the value of altering the size of a cattle herd. Typically, a reduced wildlife herd allows for a larger cattle herd; likewise, an increased wildlife herd allows for a smaller cattle herd. The authors recognized that knowing the total benefits generated by all hunting activity, or the average benefits per hunter trip, gives no useful information for estimating the optimum mix of livestock and wildlife to be produced on a given rangeland parcel. Where livestock and wildlife compete for the same grass, the economically efficient combination of cattle and livestock is an important objective.

To make the economically efficient decision, the range manager must know the value of the marginal wildlife animal, the value of the marginal domestic animal, and the production possibilities trade-off curve. One very important issue facing decision makers is the ability to relate the size of wildlife herds and hunting success. A related issue is the relationship between hunting success and total recreational benefits of that added success. Neither of these two issues have received much attention from economists, but studies that estimate recreation values associated with increased wildlife population sizes (e.g., sizes of elk herds) and compare them to ranching income values of cattle herd reductions could have an important influence on decisions by public and private range managers everywhere.

Designs for Efficient, Equitable, and Sustainable Rangelands

In designing rangeland policies, the challenge for range managers and policymakers is to fully capture the contributions of livestock and other rangeland uses that satisfy current and future human needs while sustaining the natural resource base. Economically efficient, equitable, and sustainable rangeland policies will correct for environmental effects on livestock production and other rangeland uses not reflected in product and input prices. Such policies also will reward resource conservation efforts that are excluded from market compensation by encouraging actions that promote sustainable grazing and maintenance of environmental resources. As described, for example, by Steinfeld, DeHaan, and Blackburn (1997), policies are most effective when they are flexible, site specific, and targeted.

Because livestock production is almost always one competitor for use of the world's rangelands, any sustainable rangelands policy will understand the objectives that govern the behavior of livestock producers. The maximization of present value net income is a major target of livestock producers in developed countries. However, among poor producers—particularly in poor countries—less tangible future sustainability of the rangeland is often sacrificed for immediate needs. See Example 9–4.

This section discusses five policy instruments available to promote sustainable rangelands that are compatible with objectives of livestock producer: (1) market incentives, (2) pricing, (3) regulations, (4) institutional development, and (5) development of infrastructure. Each of these instruments can be used to establish, guide, and reward the behavior of livestock producers who ensure that the use of rangelands is consistent with overall social objectives.

Market Incentives

Market incentive policies rely on market forces. The more a rangeland system is exposed to the market, the more it will be influenced by changes in input or output prices, such as variable costs of animals stocked or meat prices. These prices can be adjusted by increasing, reducing, or eliminating subsidies or, in some situations, introducing taxes that raise the price of inputs whose use is unsustainable.

Livestock producers around the world need to sell their products, so their behavior is directly influenced by market forces. By introducing policies that influence those prices, livestock production systems can be channeled into more sustainable patterns of resource use. For example, grazing pressure is reduced by increasing the price of variable inputs, which increase economic incentives to stock fewer animals, as described in detail earlier in this chapter. Here are a few examples of market incentive policies that reduce grazing pressures:

1. Full-cost recovery for inputs to livestock enterprises, especially for water, provide an incentive for reduced stocking of rangelands when the supply of forage resources are scarce or falling. In many places, water is a free good, supplied publicly. Full-cost recovery, including construction costs, reduces the number of large boreholes, thereby reducing local degradation around these water points.[17]

2. Charging grazing fees for common property rangelands[18] is another step that can reduce grazing pressure. The introduction of such fees requires institutional capabilities as well as the political will to implement fees consistently and the political machinery to do it openly and transparently.

3. Efficient benefit-sharing systems may be designed to protect biodiversity where use of rangelands contributes to unpriced social benefits. For example, in the European Alps of Austria and Switzerland, farmers receive income support for grazing actions that support social benefits created through landscape maintenance. Benefit-sharing systems also can be

[17] Boreholes are wells drilled in rangeland areas that lack surface water to look for groundwater development potential, which can promote use of a previously unused forage resource. Not all boreholes are inefficient or unsustainable. Often the forage near these wells is heavily damaged by the tendency of cattle to congregate near water. If borehole development permits a mostly sustainable use of a previously unused forage resource, it can be economically efficient. See Pallas (1986) and Food and Agricultural Organization of the United Nations (2004).

[18] Ho (2000) describes the challenges of establishing common property grazing systems in China. Mishra, Prins, and Van Wieren (2001) describe similar challenges for high mountain areas of India.

Example 9–4
Rangeland Policy Challenges in Ethiopia

Rangelands cover well over half of Ethiopia. High economic dependence of the Ethiopian rural population on natural resources (particularly land) and weak private property rights are major reasons for degradation of land and other natural resources. Because of limited off-farm job opportunities, pressure on natural resources continues to increase as the population grows. Policies that can support programs that take into account the creation of alternative livelihoods need to be designed and implemented to avoid destruction of the land's productive potential as well as to regenerate its capacity. Historically, land reallocations have created strong tenure insecurity among land users. Insecure land rights have promoted a culture of heavy grazing today for fear of lost grazing rights tomorrow.

Ethiopia's population is growing at an annual rate of 3%. If this growth rate continues, pressure on the already stressed natural environment will increase, thereby adversely affecting natural resource sustainability. Growing demand for domestic fuel has resulted in extensive use of crop residues and cow dung for fuel. Forest resources also are being depleted because of high demand for fuelwood and timber. More land is being deprived of plant cover due to the ever-increasing demand for crop production. Depletion of top soil has reduced water retention capacity of land and caused erosion of plant genetic resources and loss of habitat.

Ethiopia has several options: It could develop policies designed to alleviate poverty, promote growth, and encourage the sustainable use of natural resources. The government could ensure that communities are provided with land ownership and use rights in which those who care for the land are given a property right for its future use. People who take care of the land could be allowed and encouraged to determine what land uses, to decide on the rules of access, and to determine the distribution of benefits from common property resources. The state also could profitably ensure that further development of common natural resource management is supported by changes in state regulations regarding tenure on common property. The authors suggest the establishment of marketing cooperatives that support producers' capacity to sell and buy products, so that livestock producers can more readily reap the rewards of their labors.

Source: Federal Democratic Republic of Ethiopia. 1998. *National action programme to combat desertification environmental protection authority.* United Nations Convention to Combat Desertification. Available from *www.unccd.int/actionprogrammes/africa/national/2000/ethiopia-eng.pdf* [Accessed July 5, 2004].

established for protection of global common property resources. Brown et al. (1993) estimated damages produced by a doubling of atmospheric carbon dioxide[19] in terms of global warming at 1% of gross world product. Results of this analysis suggest that carbon dioxide sequestration can be valued at $10 per ton of CO_2 trapped. This means that South American savannas can be valued at $500 per hectare if the preserved or improved savanna produces 50 tons of sequestered carbon per hectare.

4. Taxation can be implemented for pasture and crop land in rainforest areas to reveal the correct economic cost of converting forest to crop and pasture land. However, implementing this kind of taxation may require well-developed land registry services, which are in fact often scarce and/or unreliable.

The Right Prices

Getting the prices right means adjusting input prices so that they reflect the real opportunity cost of their use, as well as adjusting output prices to reflect the value of those outputs to society. Economically efficient prices of animals grazed on rangelands include all economic and environmental costs. Those efficient prices, if they can be established, offer economic incentives by providing market signals that reward the practice of sustainable grazing. Economically efficient

[19] Campbell et al. (2000) review the literature on impacts of global climate change on the world's pastures and rangelands. Dube and Pickup (2001) describe the challenges of separating impacts on rangelands in southern Africa produced by grazing from those produced by short-term and long-term climate change.

prices encourage efficient use of resources and guide technologies to anticipate and overcome future scarcities. Efficient prices promote technologies that recycle wastes and increase the supply of resources, especially future forage supplies.

Price is a powerful economic instrument, and it may be the preferred instrument if regulatory institutions are weak and if their constraints on individual behavior make them politically unpopular.[20] Here are a few examples of actions that can influence input or output prices:

- Reduce subsidies for inputs such as water, concentrate feed, fossil fuel, and fertilizer.
- Look for ways to support livestock production that do not promote unsustainable grazing practices.
- Require cost recovery for common property water and grazing resources.
- Charge for waste disposal.
- Subsidize reuse of waste products like methane and resource conversion facilities that use those products as inputs.
- Eliminate tax advantages for large livestock enterprises that fail to promote sustainable use of rangelands.
- Subsidize the production of socially desirable or environmentally valuable services.

Regulations

Regulatory policies can be used to promote sustainable resource use by prescribing (1) limits for animal numbers, or input use, (2) technical methods, and (3) access to resources. For grazing systems in wet areas, such as tropical rainforests, the establishment of protected areas for the preservation of biodiversity can be one workable measure. Flather and Sieg (2000) describe data requirements for measuring biodiversity in the world's forests and rangelands. Regulations work best where the behavior in question can be clearly identified, where governments have the financial resources to establish the relevant machinery, and where there are reliable institutions to enforce regulations.

Institutional Development

The livestock sector is under pressure from growing demands to feed a growing human population (see Chapter 14), so there is a growing need for legal systems that promote well-defined and enforceable rules and institutions for governing the use of natural resources. A major cause of negative externalities on rangelands is poorly defined access to resources such as common property grazing land. Janssen et al. (2000) describe the interactions of the individual behaviors of pastoralists, biological processes, and social institutions in a rangeland system.

Authority and responsibility for resource management should be put at the lowest level at which it can be exercised effectively. Direct participation of the community in decision making enables local knowledge to be harnessed and promotes the identification of problems and discovery of solutions at local levels.

Insecure tenure can be a prime cause of rangeland degradation. In some parts of the Brazilian Amazon, for example, security of tenure has been granted only after rainforests are cleared. The result has been that significant forested areas are cleared solely as a means to receive free land from the government. Although this government action has the desirable effect of promoting an inexpensive method to encourage settlement and reduce the political and economic costs of population pressures, it also establishes a zero price for local people who could invest in measures to sustain tropical forests. Land tenure arrangements are an essential policy instrument for rewarding long-term investment in sustainable rangeland grazing. Livestock producers will initiate sustainable grazing activities only if they can expect to reap the financial rewards from their investments.

---

[20] Brown and Shrestha (2000) describe the effects of market prices of livestock products on sustainable grazing practices in Nepal.

In developing countries, droughts reduce the supply of forage for livestock, thereby increasing the competition for decreasing supplies. The reduced supplies of forage make it more difficult to sustain animals. Livestock producers may have to reduce the size of their herds, even eliminate them. Livestock markets become oversupplied, forcing the prices down to levels of perhaps only one tenth of the levels before the drought. In regions of the world where livestock are an important part of the agricultural economy and a major part of household assets, programs are needed to sustain the livestock or, at least, support the price of the livestock. One option described by Borton and Nicholds (1994) is setting up cattle camps into which cattle can be held, fed, and watered on behalf of their owners until the drought has passed. In this way a crucial farm household asset can be kept alive and dramatically improve the speed of household recovery after the drought.

Infrastructure Development

Government investments in infrastructure are economically efficient if the benefits exceed the costs. When that occurs, the building of infrastructure is an economically efficient method for protecting and enhancing economic development. Infrastructure development is essential for achieving a better regional balance between livestock and the rangelands on which they depend. Investments in roads,[21] markets, slaughterhouses, and cold storage can help livestock producers quickly remove their livestock in times of drought or in times of financial crisis. Infrastructure also is a powerful instrument to support policies that reduce animal densities in environmentally sensitive areas. It provides market access, an important complement that supports market incentive policies.

U.S. Grazing Fees

In the western United States, the grazing fee is an important and highly publicized government intervention into the rangelands economy; Table 9–4 shows its history. A federal grazing fee is charged to livestock producers who graze their livestock on federal lands. The two federal land agencies that charge a fee are the U.S. Bureau of Land Management (BLM) and U.S. Forest Service (Forest Service). The fee is set by the U.S. Congress and implemented by federal land agencies. Strictly speaking, the U.S. grazing fee is an example of "getting the prices right." The fee has a long and colorful history and has been subjected to widespread controversy and analysis for nearly 100 years.

Fees have been charged for domestic livestock grazing on U.S. federal rangelands and forest lands since 1906. The fees were originally set at 5 to 8 cents per head for the summer season for sheep and 20 to 35 cents per head for cattle and horses (Nielsen and Wennergren, 1970). The authority provided in U.S. federal law for grazing fees is the Taylor Grazing Act.

The setting of grazing fees for almost 100 years has been influenced by changing philosophies about the aims of public land management as well as by changing economic forces. The philosophical concepts underlying the purposes or goals of public land use have changed as U.S. national needs have changed. At one time, the settlement and development of western U.S. land resources was in the national interest. Since the 1970s, the historical need for promoting settlement through economically viable livestock production has been reduced, and the range has seen intense competition for multiple uses of what has become a scarce resource.

Major areas of contention over the grazing fee include how much should be charged, how grazing fees should be adjusted through time, and whether fees should vary in different parts of the country. All three of these issues were part of the intense political debate at the time the Public Rangelands Improvement Act (PRIA) formula was adopted in 1978 (Torell et al., 2003). The debate will likely continue for some time to come.

---

[21] Road construction may be the single most important contributing factor to deforestation. Discouraging road construction in forested areas may be the most economically efficient deterrent to establishing ranches where forests currently exist. Such a policy could, however, be politically sensitive because proper targeting is difficult and the preservation effect of blocking infrastructure development needs to be balanced against the affected population's development needs.

Table 9–4 History of U.S. federal grazing fees

Study or legislation	Description
1906 Forest service fees	Charged fees on grazing of U.S. Forest Service lands, averaging \$0.05 per animal unit month (AUM)
1924 Ratchford Appraisal	Set fees based on the price of beef and lamb; prevailed in the Forest Service from 1928 to the mid-1960s
1936 Fees on BLM lands	Set fee of \$0.05 per AUM on Bureau of Land Management (BLM) land; had no stated economic rationale except to cover administrative costs of the federal land agencies
1966 Grazing Cost Study	Made a total grazing cost comparison of 10,000 public and private interagency task force land forage lessors (owners) and lessees (tenants). Results of the study showed that a weighted average grazing fee of \$1.23 per AUM equalized total grazing costs on public and private lands. That value became the base value used in the 1978 Public Rangeland Improvement Act (PRIA). Study also proposed adjusting the base fee annually by an index of private grazing land lease rates, known as the forage value index (FVI).
1976 Federal Land Policy and Management Act	Established precedent-setting public lands management objectives and mandated that a grazing fee study be submitted to Congress within a year. A grazing fee committee proposed using the FVI to set fees, since it measures income earning capacity of forage.
1978 Public Rangeland Improvement Act	Established the fee formula shown in Equation 9.2.
1986 Executive Order 12548	Indefinitely extended the PRIA fee formula and set a minimum fee of \$1.35 per AUM.

Source: *Adapted from Torell, L. A., N. R. Rimbey, L. W. Van Tassell, J. A. Tanaka, and E. T. Bartlett. 2003. An evaluation of the federal grazing fee formula.* Journal of Range Management. *56: 577–584.*

Clearer direction about what the grazing fee is meant to accomplish is contained in the Federal Land Policy and Management Act (FLPMA) of 1976. The Taylor Grazing Act said only that the fee was to be "reasonable." FLPMA defined reasonable as a fee that was fair to both the public land livestock forage permittee (the producer grazing the livestock) and the owner (the taxpayer). What actually is fair to both the permittee and the federal government is a matter of long-time debate. One element of the debate centers on the meaning of a fair market value grazing fee.

PRIA Formula. The PRIA legislation uses a formula for setting the grazing fee. The U.S. Congress established this formula to represent both the economic value of the forage to the permittee and the fair market value for public land grazing. The formula bases grazing fees in the current year on a cost equalization base fee of \$1.23 per AUM. That base fee is updated annually by indexes reflecting the relative values of private grazing land rental rates, beef cattle prices, and costs of production in the immediately preceding year. Next year's fee is calculated from this year's forage rental rates, costs, and returns together with the \$1.23 base fee. The formula is

$$\text{Fee}_{t+1} = 1.23\,\frac{[FVI_t + BPI_t - PPI_t]}{100} \tag{9.2}$$

where

1.23 = base fee

FVI = forage value index

BPI = beef prices index

PPI = prices paid index

t = current time period

$t+1$ = following time period

The \$1.23 base fee is intended to represent the amount that BLM and Forest Service permittees would have been charged in a base period (1966) to equalize the total costs per AUM of livestock forage on federal and private lands in 11 western states. That base fee was derived from the results of a 1966 survey of some 10,000 ranch operations and 500 financial institutions in the western United States (see Table 9–4).

Economic changes over time affect livestock producers' input and output prices; the \$1.23 base fee is updated to account for those changes. In the PRIA formula, the base fee is multiplied by three indexes to account for emerging economic changes facing the livestock industry. The forage value index (FVI) is an index of private grazing land prices in the 11 western states. It is derived from the results of an annual survey conducted by the National Agricultural Statistical Service (NASS). The other two indexes are intended to reflect the rancher's changing ability to pay for federal rangeland forage based on changing economic or environmental conditions. The beef prices index (BPI), which had a value of 100 in 1966, reflects average weighted selling price of cattle in the 11 western states. The prices paid index (PPI), which also had a value of 100 in 1966, is computed from national beef production input costs, modified to reflect costs of production for ranching enterprises.

Permit Value. Historically, federal grazing fees have been set at a price lower than the equivalent price ranchers charge each other for leased forage. Remember, the fee is an administered price; it is not a fee set by market forces. Economic theory tells us that because it is administered, the fee will have a special effect. And the theory is correct: The fee has had the predicted effect for more than 100 years. Use of federal forage typically confers an exclusive property right for the livestock producer to graze on a piece of federal land—a right that no other producer shares. Even in the early 1900s observers noticed that grazing permits on national forest lands had a *commercial* value to the rancher over and above the grazing fee charged by the government. Private ranches are sold in the private real estate market where the forces of supply and demand have their full power of expression. In the case of federal grazing fees, the purchasing rancher is willing to pay the fee to graze on the acquired lease, plus an additional amount to gain sole access to the grazing permit if the permit was transferred with the ranch. For many years the privilege of grazing on federal lands became so valuable that it became an important factor influencing land prices for private lands adjacent to the federal lands. For example, if a rancher owns private land near the ranch's forest grazing allotment, that private land sells for a premium compared with similar private land located far from an allotment. So grazing privileges on federal forest lands have been valued by ranchers as a capital asset for many years.

Permit value (PV) is the capitalized difference between the total cost of access to grazing forage on public lands and the total cost equivalent of grazing forage on private lands. The formula for PV is based on the present value of an infinite series of payments to be received in future years (see Chapters 5 and 6).

$$PV = \frac{[(P_2 + E_2) - (P_1 + E_1)]}{i} \tag{9.3}$$

where

PV = permit value

P_2 = private lease rate

E_2 = other private user, nonfee costs

P_1 = public grazing fee

E_1 = public user, nonfee costs

i = discount rate

The relationship in Equation 9.3 between the level of grazing fees (P_1) and the permit value (PV) has important implications for rangelands policy. The equation shows that any conditions that alter the *difference* between the production costs of public and private grazing will be reflected in private real estate markets in changing permit values. *Raising* the federal grazing fee *decreases PV*. In fact at a discount (interest) rate of 5%, an increase in the grazing fee by $1 per AUM reduces the market value of the permit by a 20-fold factor, $20. That $20 loss is calculated by using the Equation 9.3 in which the public grazing fee, P_1, is increased by $1 and other factors are unchanged. Several U.S. environmental groups have taken an active interest in this equation. According to these groups, only when PV is zero does the taxpayer receive full market value of the forage. And the environmentalists are right. As long as ranchers are willing to buy and sell grazing permits that are issued at below-market prices, the taxpayer is not capturing all of the forage value through the fee.

For many years, livestock producers who held grazing permits realized capital gains when permit values appreciated. The permit holder received this capitalized value because he or she gained control of public grazing based on prior use and location of the base property. Any current-day rancher who buys a ranch with the grazing privilege will be asked to pay current market prices for the permit. The current-day rancher who buys access to federal forage receives no subsidy whatsoever, since he or she must pay for any difference between its administered price and its economic value. Two important questions in the fee debate are debated widely: What is the potential effect of increased fees on the wealth position of present permit holders? Is this impact a relevant or legal concern to the government agencies or to the taxpayer?

Permit value is important to the public lands livestock producer for two reasons. First, the permit's recognized commercial value is evidenced by its acceptance as collateral by banks and other financial institutions. It is accepted as collateral because the person owning the private ranch typically holds sole access to the permitted grazing land. Second, even as long ago as 1970, Forest Service and BLM permits held by ranchers in the western United States had an estimated capitalized value of $343 million (Nielsen and Wennergren, 1970). Setting the grazing fee at full market value, a value which would reduce permit value to zero, would impose losses of this magnitude adjusted to current dollars. Capitalized incomes of public lands ranchers would also decrease by an amount equal to the increase in grazing fees. In addition, secondary community losses would likely occur due to the multiplier effects associated with reduced rancher incomes and reduced rancher spending in local communities.

Fee Design Standards. Recall the three standards for public resource policy analysis: efficiency, equity, and sustainability. Each of these standards can be applied to the search for the fair market value of federal rangeland livestock forage.

One way to define an economically efficient grazing fee is a market price determined under conditions of free competition. (This statement is true only if all opportunity costs of the land are included in that competitive market; it requires that no environmental or other costs are displaced by grazing.) In fact, the federal grazing fee is an *administered* price set in a market that is neither open nor competitive. The market is characterized by a single seller, the federal government, and a few eligible buyers, usually livestock producers who own base property or water in the immediate vicinity. An efficient fee cannot be discovered in an inefficient market unless all of the restrictions on competition are relaxed, which is not permitted under existing U.S. public land law. For the sake of efficiency in a restricted market (public land grazing), a price discovered in an efficient market (private land grazing) cannot be imposed in the restricted (i.e., inefficient) market with the expectation that aggregate or social efficiency will be increased.

Equity means fairness or justice and is an important economic standard for public resource policy analysis. Two federal laws, FLPMA and PRIA, say that the federal grazing fee should be equitable to two parties: the federal government as the owner and the permittee as the user. This suggests that the federal grazing fee cannot be higher than the economic use value of the federal rangeland forage to the user, nor less than the cost to the federal government as the proprietor that produces the forage and administers the grazing program.

Sustainability, a term with both ecological and economic meaning (see Chapter 19), is an important standard for public resource policy analysis. It is especially relevant to management and policy for renewable natural resources like rangeland forage, both to sustained long-term productivity of the (federal) rangeland resource and to sustained stability in the (federal) rangeland-dependent livestock sector of the economy. What is now called sustainability was defined in the preamble to the Taylor Grazing Act as managing federal rangelands in a way that guarantees long-term future use and supports a stable livestock industry that depends on the public range. The same two dimensions of sustainability apply to all renewable resource policy analyses. A sustainable grazing fee would, on the one hand, discourage overgrazing and, on the other, be a fee that the rancher can afford to pay where grazing is sustainable.

Forage Pricing Schemes. The policy options discussed next fall into two general categories. First, some have argued that there should be no grazing fees at all, to be fair in relation to other unpriced uses of federal land resources. Second, at least four methods for determining positive grazing fees have been proposed: (1) fees based on the agencies' costs of producing federal land livestock forage and administering their domestic livestock grazing programs, (2) fees determined by market forces, (3) fees based on cost equalization formulas, and (4) fees based only on economic value to the permittee.

No Fee. Although federal land livestock grazing at no fee may be politically impractical, it may have economic merit in some conditions. As a matter of national policy, FLPMA states that charges should be levied on all uses of federal lands and their resources. However, the Act also allows Congress the option of writing laws that make exception to the policy. Grazing was free on U.S. forest reserves prior to 1906 and remained free on the rest of public lands prior to passage of the Taylor Grazing Act in 1934. Free livestock grazing would be equitable relative to other unpriced, but valued, uses of federal lands such as the many kinds of free outdoor recreation.

A zero grazing fee does not necessarily lead to overgrazing, because stocking rates and grazing seasons can be regulated by federal land managers who are charged with maintaining or improving range conditions. Furthermore, if federal rangeland grazing is free but a permit is required to qualify for the free use of the federal forage, property rights are secure from theft by stocking by another producer. So there is no economic incentive to damage one's own property right by overgrazing. No rational producer who has control over forage will want to damage his or her own future income-producing opportunities by overgrazing in the short run. Despite all these appeals to profit-maximizing logic, no one has taken the free grazing alternative seriously.

Administration Costs. Recovering the taxpayers' cost is a common objective of federal resource pricing programs. Although cost recovery is an excellent concept, it has been proven difficult for land management agencies to carry it out because it is hard to separate the cost of administering federal grazing programs from other programs. In the western United States, both the BLM and the Forest Service administer many programs on the public lands (e.g., timber, recreation, wildlife habitat, wilderness, water, and minerals). There is no clear-cut, unambiguous method for determining the incremental cost of federal administration on grazing programs. A large part of the costs are fixed (joint) costs that cannot be attributed to any single program.

Market Fees. There are two widely debated versions of the market-determined federal grazing fee: direct negotiation and competitive bids. Direct negotiations about fees could occur between the permittee and the local federal rangeland manager. In 1951, BLM director Marion Clawson argued that this was the only way to accurately discover the true, full value of public rangeland livestock forage. Clawson noted that the permittee and the local federal manager know the allotment or grazing area, its conditions, and the characteristics of the permitted livestock operation (Clawson, 1951). These two parties—the local manager and the permitted user—should be the ones to negotiate in good faith what the forage on that allotment or permit actually is worth to the livestock producer as permittee and to the federal government as proprietor. In principle, good faith negotiation between two fully informed parties would result in an efficient, equitable, and sustainable federal grazing fee. The fee would be unique to the allotment or permit under negotiation. Even though this argument makes some economic sense, it has never been considered seriously in policy circles. Competitive bidding, the second market-based approach to fee determination, works this way: A federal rangeland forage market is created and the interaction of buyers and sellers in that market leads to the discovery of a fair market value grazing fee. This method has been tried experimentally on a small scale with some success. One example is an experimental auction held at the McGregor range in southern New Mexico (Rimbey and Nelson, 1994).

Cost Equalization. Many methods discussed for setting the grazing fee are based on the belief that the fee should be set so that total cost of livestock forage is equal for public and private forage. This belief underlies the actual PRIA formula fee. The principle is applied by measuring the average forage price per AUM for the user of private forage. For the user of public forage, the total forage cost *excluding* the fee is measured. The difference between the first and second is the amount that private forage exceeds public forage and forms the base grazing fee for the year in which the measurements were taken. The base grazing fee, shown in Equation 9.2, averaged $1.23 per AUM in 1966 for the 11 western states. Implementation could occur by modifying the base or by adjusting the indexes inside Equation 9.2.

The indexes inside Equation 9.2 could also be modified. Many economists have suggested that a more economically efficient PRIA formula would delete both the BPI and the PPI. Their logic is that use of the BPI and PPI leads to double counting. The FVI, which measures annual private grazing lease prices, already accounts for both cattle prices (the BPI) and costs of production (the PPI). The base fee times the FVI alone equalizes total per AUM costs for forage use on private and public rangelands. Deletion of the BPI and PPI from the PRIA fee would leave the following equation to set the fee:

$$\text{Efficient fee}_{t+1} = 1.23\frac{FVI_t}{100} \tag{9.4}$$

where all terms are as defined in Equation 9.2.

Full-Cost Pricing. Economic efficiency occurs when the use of a natural resource is priced equal to the marginal costs of uses displaced by the use (i.e., at a price equal to the value of other uses displaced). This means that if every permittee paid a grazing fee equal to the cost of current and future land uses displaced by that grazing, that fee would promote efficient and sustainable use of rangeland. In principle, this kind of fee could be implemented by setting the grazing fee equal to the opportunity cost of the next highest-valued use of the land. The result of implementing a full-cost pricing fee is greater economic efficiency, which means greater total economic benefits to the nation. Two possible outcomes could occur: First, where cattle production is the highest-valued use of a particular piece of rangeland, the fee would be set low enough to make cattle grazing profitable. If the grazing fee is set at the value of other uses displaced and the value of those other uses displaces is $\$X$ per stocked steer, the profit-maximizing rancher will stock a certain number of animals per unit of land (as described earlier in this chapter). Second, where the value of other uses displaced per stocked animals is a higher $\$Y$, and the fee is priced at the marginal cost of other uses displaced by grazing, the rancher will stock fewer animals per unit of land. When some other use outside grazing (e.g., endangered species habitat, minerals, recreation, or new subdivisions) produces greater economic benefit to the nation than livestock grazing, then economically efficient policy would eliminate cattle grazing on those land parcels. In that case, the economically rational livestock operator grazes no cattle on these public lands and instead identifies a substitute: grazing cattle on another piece of public land; leasing private forage; buying baled hay; or, if no other choices exist, leaving ranching for another line of work (which means the livestock producer goes broke). An equivalent policy to pricing grazing on federal lands is to remove the grazing allotment for those periods and those pieces of land for which a higher-valued use occurs.

There are several practical problems of implementing a fee program in which grazing is priced at marginal cost. The first problem occurs with measuring opportunity costs. For most nonmarket uses of the land (e.g., recreation, wildlife habitat, or endangered species critical habitat), there is little agreement on how to measure the economic value of nongrazing uses of public rangelands. Another problem is data. Even if we know how to measure all nongrazing values, those data are typically expensive to find and, when they can be found, they will unlikely transfer to other areas or policies without some kind of adjustment to reflect conditions in the new location.

U.S. Legislation on Rangelands

Several legislative acts of the U.S. government have made significant contributions to the discussion of sustainable rangelands.

Taylor Grazing Act. The aim of the U.S. Taylor Grazing Act, passed in 1934, was "to stop injury to the public grazing lands by preventing overgrazing and soil deterioration, to provide for their orderly use, improvement, and development, to stabilize the livestock industry dependent upon the public range." This Act and its 1939 amendments made three important changes to western U.S. rangelands. First, grazing districts were set up with grazing advisory boards to define which ranchers had grazing privileges to specific areas of federal land, typically near the rancher's privately owned ranch (base property). Grazing privileges defined by the number of cattle allowed to graze federal lands were based on (1) the range's carrying capacity and (2) historical use levels (which often were not the same) by established ranch units. Second, the Department of Interior was authorized to establish the Grazing Service (later named the Bureau of Land Management) to administer the use of the grazing districts. Third, the U.S. government started charging a grazing fee for grazing livestock on federal lands.

The Act has produced important long-term results. Because effective property rights were established in range forage, damages caused by common ownership and management of a depletable resource (tragedy of the commons) were reduced. Range conditions throughout the western United States have recovered considerably, and this recovery continues to date. The Taylor Grazing Act is considered by most to be a significant contribution to sustainable rangelands.

Endangered Species Act. The Endangered Species Act (ESA), passed in 1973, blocks federal agencies from authorizing, funding, or implementing any action that would destroy the significant habitat of an endangered species. There are four major provisions of ESA (Loomis, 1993):

1. ESA shifted responsibility for endangered animals from state control to the federal level and included threatened or endangered plants.
2. The ESA set up a process whereby species would be studied for recommendation as either threatened or endangered status.
3. The ESA required that all federal agencies seek to conserve endangered species and take all measures necessary to restore threatened and endangered species and their habitats to population levels that would place them out of danger of extinction.
4. The ESA emphasized protecting habitat, including designation of critical habitats in which the management of endangered species would be the dominant concern.

Congress amended the ESA in 1978 by establishing the Endangered Species Committee (the "God Squad") to consider applications for exemptions from the requirements of the Act. The committee was empowered to grant an exemption for any project that met the following criteria: (1) There are no reasonable and prudent alternatives to the proposed actions. (2) The benefits of the proposed action clearly outweigh the benefits of alternative courses of action consistent with preserving the endangered species or its critical habitat. If these criteria are met, the committee is authorized to grant an exemption to the ESA, provided that it also requires actions that minimize adverse effects of the proposed action on the species and its critical habitat.

Many controversies surround the ESA. The U.S. Fish and Wildlife Service has interpreted the prohibition on harming endangered species as meaning that an agency could not engage in any action on public (or private lands) that would jeopardize the continued existence of an endangered species. The Fish and Wildlife Service hires specialists who render biological opinions on these actions. For example, in June 1990 the service listed the northern spotted owl as a threatened species throughout its range, from northern California to Oregon and Washington. The extent of critical habitat required to protect the spotted owl as part of the recovery effort has still not been determined. The initial estimate of more than 10 million acres of critical habitat have been scaled back to about 7 million. Biologists determined that continued clear-cutting of the remaining old-growth forest was a major threat to the survival of the spotted owl. A panel of scientists determined that preventing extinction of the owl would require preservation of several million acres of old-growth habitat as habitat protected from logging.

Federal Lands Policy and Management Act. An important provision of the FLPMA, passed in 1976, is that land and resource management must follow principles of multiple use and sustained yield. The BLM is required to inventory specific roadless areas as suitable or unsuitable for wilderness designation. The FLPMA stated that the BLM would became a multiple-use land manager rather than a custodian of federal lands in preparation for future disposal, which had been its role before 1976. Multiple-use advisory councils were established to reflect all resource users. Recreational, environmental, and wildlife uses were given more consideration in rangeland management decisions; in contrast, the influence of livestock producers on rangeland uses was reduced. This reallocation of mission and influence is still controversial and debate is likely to continue in the future.

■ Emerging Issues

A common criticism of rangeland policies in most parts of the world is that policies are often enacted with weak scientific knowledge of potential impacts. Obtaining objective data to develop policies with a stronger scientific base, however, is not easy. Relationships among wildlife, livestock, forage, and people often are not known and, in most cases, are difficult and/or expensive to measure. Better-timed and more objective research has the potential to help considerably in sustaining rangelands, but such research is expensive and often unavailable at the right time to help policymakers. Several emerging issues will likely be confronted by rangeland economists in the near future:

1. Much more needs to be known about what policies would increase security of tenure in grazing rights to give livestock producers effective economic incentives to conserve private or public rangelands sustainably. Livestock producers' time horizons on common property rangelands are short because of their insecure tenure, that is, policies that can change from one government to the next.

2. Linkages between livestock and other industries are an important question, about which there is little factual data. How strong is the economic connection between livestock grazing and the industries with which they are connected? If livestock producers provide few benefits to suppliers and to people who buy livestock products, economic efficiency suffers from policies that subsidize livestock production. Where strong linkages are discovered, such as between livestock production and rural communities, there is a convincing economic justification, nationally for subsidizing livestock production.

3. Another issue involves the economic merits of privatizing wildlife. It is widely recognized that people have incentives to take better care of a resource when it is privately owned. According to those who favor privatization, if wildlife is privately owned, the owners take better care of the resource, so the quality and quantity of the resource increases over time. Those who disagree say that the public has a right to access wildlife and that the agents of the public are government wildlife departments. According to this argument, private ownership would limit access to hunting and fishing for the common person, which violates many nations' principles of democracy.

4. Generally, better information is needed on the impact of wildlife and livestock on the health and sustainability of riparian areas. (DeSoyza et al., 2000, summarize early warning indicators of problems in rangeland health.) What is the impact of endangered species designation on riparian areas? Few systematic studies have been done on these relationships. Research that identifies impacts of grazing near riparian areas (near rivers and streams) could provide important facts to help make more informed decisions. A significant policy question centers on how these areas should be managed for maximum benefit to wildlife, livestock, recreation, and the environment. In many dry places, riparian areas consist of a small percentage of the total land area, but produce most of the carrying capacity for livestock and wildlife. The extent of riparian areas may be a barometer of an ecosystem's health and economic sustainability. These areas are economically valuable to livestock and wildlife. The quantity of riparian habitat is fixed by supplies of rivers and streams, so there is little opportunity to produce larger supplies of them. An important issue involves the mix of livestock and wildlife that should be allowed on these areas.

One economic efficiency question is this: Is the long-term benefit to fish, wildlife, and the environment worth the added cost of the fencing where fencing could reduce livestock damages to riparian areas? If the answer is yes, who will pay the cost of the fences and displaced forage? Without financial incentives, livestock producers will not normally invest in fences to exclude their cattle from a cheap source of water. Many believe taxpayers should not have to pay the cost, because, according to their argument, the "polluter pays" principle should be used: The livestock producer's cattle stocking damaged the environment, so the producer should be held accountable.

5. A highly debated issue is the privatization of public rangelands (i.e., the process by which some government agency transfers the land to the highest private bidder). A major question here involves the economic benefits and costs of privatizing public rangelands. Proponents of privatization believe private landowners have more economic incentive to take care of their lands than federal agency managers. They point to the evidence of privately held family ranches that have lasted many generations. Opponents of privatization say that private landowners manage only for profit. So although while these landowners take good care of the land, they may not have the incentives to take equally good care of fish, wildlife, and the environment. Opponents of privatization believe government ownership and management is the only way to ensure that the land gets used for maximum public benefit. Unfortunately, there are few objective studies that help us conclude which view is right under which conditions.

Overall, research is needed that pursues balanced and objective policy analysis: Government regulations, laws, and agency decisions on rangelands have historically relied on the political process; much less attention has been paid to factual data. Also, there has been little analysis on the ecological and economic effects of past policies, so few inferences can be made about future policies based on past experience. Discovering, understanding, and interpreting *facts* will let us formulate regulations, laws, and decisions based, at least in part, on their economic performance, thereby contributing to a more informed political process.

■ SUMMARY

Many rangelands of the world are livestock grazing areas by default; that is, they cannot be used economically for anything else. The production of livestock is an important element of rangeland areas in developed economies. Rangelands in poor countries often face the problems of low and unreliable rainfall, low agricultural productivity, and politically weak livestock producers.

Aiming for economic efficiency means managing rangelands to maximize the total economic benefits of rangelands summed over all uses and all time periods. Managing rangelands for equity means allocating rangeland resources fairly across uses, users, and time periods. Sustainable actions applied to rangelands are those that avoid overgrazing range forage, so that the soil, water, and plant resources serve current and future generations.

Four economic questions confront private livestock producers: (1) What is the optimum grazing intensity? (2) What combination of inputs should be used? (3) What mix of outputs should be produced? (4) What mix of inputs and outputs over time maximize the discounted value of current and future net income? The most important problem facing a private livestock producer who wishes to profit from the use of rangelands is the determination of the economic optimum stocking rate. Profit, which is defined as total revenue minus total cost, *is maximized by stocking whatever number of animals equates the input price to the value of the marginal cost of the variable input.* The point at which profit maximization occurs is defined as the economically optimum grazing intensity. This point also occurs where the output price equals a rising marginal cost of output. The economically optimum stocking rate always stocks an equal or fewer number of animals than the biologically optimal stocking rate. The only time that the economically optimum stocking rate equals the biological optimum is when the variable input is priced at zero.

Any government policy that affects output price, input price, or forage supply will influence the optimal stocking rate, as well as ranch income and output supply. Policies that subsidize variable input reduce the price of the input and increase the optimal stocking rate. If a government charges a grazing fee to use public forage, any policy that raises that fee increases the variable input price and reduces the optimal stocking level. In the United States, many environmental organizations have supported policies that raise the grazing fee on western public lands. Their hope is that rancher stocking rates on public lands will fall, thereby reducing overgrazing and increasing other uses of the land. Governments in livestock-dominated economies often look for policies to help ranchers cope with forage losses produced by severe droughts. Policies that subsidize nonanimal inputs (e.g., low loan interest rates or programs that support livestock prices) can partly compensate for severe drought by mitigating the loss of ranchers' income suffered by lost forage.

In designing rangeland policies, the challenge for range managers and policymakers is to fully capture the contribution of livestock and other rangeland uses that satisfy current and future human needs while sustaining the natural resource base. Economically efficient and sustainable rangeland policies will correct for environmental effects of livestock production and other rangeland uses not reflected in product and input prices. Such policies will encourage and reward actions that promote sustainable grazing and the environmental resources it supports. Policies are most effective when they are flexible, site specific, and targeted.

Five policy instruments are available to promote rangelands: market incentives, getting the prices right, regulations, institutional development, and infrastructure development. The history of U.S. grazing fees is an example of a major government intervention into rangeland economy. Several U.S. legislative acts have contributed to the discussion of sustainable rangelands. Internationally, research is needed to find data that will lead to more informed decisions on emerging rangeland issues.

■ Review Questions

1. What are rangelands?

2. What are the aims of rangeland management?

3. Explain the principles that govern the economics of rangeland management.

4. What are the special problems of people who occupy rangelands in dry, poor parts of the world?

5. Describe three desirable outcomes that guide rangeland economic decisions.

6. What four important economic questions face private working ranches?

7. Explain the economically optimal stocking level and how it is found.

8. How is the economically optimum output level determined?

9. What factors shift the output supply curve?

10. How is the economic value of forage computed and why is it important?

11. Why is the optimal dynamic stocking level different from the optimal static level?

12. Explain each of the following policy instruments and how each can contribute to more efficient, equitable, and sustainable rangeland management: (1) market incentives, (2) getting the prices right, (3) regulations, (4) institutional development, and (5) infrastructure development.

13. Explain the U.S. federal grazing fee and briefly state its history.

14. What is the U.S. Public Rangelands Improvement Act (PRIA) formula that is used to set the grazing fee on federal lands?

15. What alternatives to the PRIA formula have been proposed? Which one do you think would do the best job of promoting economic efficiency and why?

16. Explain a permit value and describe why it is important for rangeland policy.

17. How does the profit motive encourage or discourage rangeland degradation?

18. Name three important pieces of U.S. federal rangelands legislation and explain how each performs using the standards of efficiency, equity, and sustainability.

19. Discuss at least five emerging issues concerning rangelands.

■ Problems

1. Profit-maximizing use of variable input (part 1—static analysis): Use the data in the accompanying table to find MPP, TVP, VMP, VCI, and TNR. From the data you have filled in, compute the optimal use of the variable input, *S*. How do you know when you have found that optimal stocking level?

(1) Variable input (steers/ 100 acres)	(2) Total physical product (TPP) (weight gain/ 100 acres)	(3) Marginal physical product (MPP)	(4) Price per unit output (P_O)	(5) Total value product (TVP)	(6) Value of marginal product (VMP)	(7) Price per unit variable input (P_I)	(8) Variable input cost (VCI)	(9) Fixed input cost (FCI)	(10) Total net revenue (TNR)
	← pounds	→	←			dollars			→
3	1,186		1.00			75		1,200	
6	2,150		1.00			75		1,200	
9	2,890		1.00			75		1,200	
12	3,407		1.00			75		1,200	
15	3,702		1.00			75		1,200	
18	3,773		1.00			75		1,200	
21	3,622		1.00			75		1,200	
24	3,247		1.00			75		1,200	
27	2,650		1.00			75		1,200	

2. Profit-maximizing use of variable input (part 2—dynamic analysis, *advanced problem*): In the table in problem 1, you used the equimarginal rule:

$$VMP = P_I \tag{1}$$

to find the economically optimal stocking rate for a single period. At that optimal rate, the value of the marginal product of the additional stocked animal (VMP) equals the price of stocking the additional animal (P_I). Another way to solve for the optimal input is by using algebraic equations. Remember the definition of VMP: output price times the marginal physical product of the input, which is P_0 MPP. In the previous table, VMP turns out to equal the following simple linear function of the number of animals stocked, S:

$$P_0\,MPP = P_0\,(B_0 + B_1\,S) \tag{2}$$

where the equation's right-hand side terms are $B_0 = 431$ and $B_1 = -24.67$. Check this yourself. Solving the optimal stocking problem algebraically using the $VMP = P_I$ rule requires solving this single linear equation:

$$P_0\,MPP = P_0\,(B_0 + B_1\,S) = P_I, \tag{3}$$

using the known values of P_0, B_0, B_1, and P_I, and solving for S. That solution produces an optimal value of $S = 14.43$, a more precise answer than possible from the table.

Things get more complicated when analyzing optimal stocking for *more* than one period. Multiple periods must account for *two* adjustments: (1) impact of current animal stocking on future weight gain through reduced future forage production, and (2) effect of previous periods' stocking on current weight gain through impacts on current forage. Here we limit to a two-period analysis with the two adjustments, restating equation 3's simple optimal static (one period) stocking rule. The first adjustment tells what the optimal stocking rule must be in the first period:

$$\text{period 1: } P_0\,MPP_I = P_I + MDFC_I, \tag{4}$$

which says that output price times the marginal physical product of animals stocked in period 1, $P_0\,(MPP_1)$, must equal the input price, P_I, *plus* the marginal dynamic factor cost from additional stocking in period 1, $MDFC_1$. This means in period 1 we wish to stock additional animals until the value marginal product of the last animal stocked in period 1, $P_0\,(MPP_1)$, equals its input price (P_1) *plus* the future cost of that stocked animal.

The second adjustment gives an adjusted optimal stocking rule for the second period:

$$\text{period 2: } P_0\,(MPP_2 - \Delta MPP_{21}) = P_I \tag{5}$$

where MPP_2 is the marginal physical product of stocking in period 2, and ΔMPP_{21} is the change in marginal physical product in period 2 from an extra animal stocked in period 1. This equation says that the first period's stocking (normally) reduces forage available in the second period, and shifts MPP_2 to the left by an amount ΔMPP_{21}. With that reduction in the second period's MPP, period 2 gives us a smaller overall VMP. We set $P_0\,(MPP_2 - \Delta MPP_{21})$ equal to that period's variable input price. Here is an example of two algebraic equations for the two-period model:

$$\text{period 1: } P_0\,(B_0 + B_1\,S_1) = P_I + B_2\,S_1 \tag{6}$$

$$\text{period 2: } P_0\,(B_0 + B_1\,S_2 + B_3\,S_1) = P_I \tag{7}$$

a. Suppose $B_2 = 109$ and $B_3 = -15.12$. This says that $MDFC_1 = B_2 S_1$, and $\Delta MPP_{21} = B_3 S_1$. Solve equations 6 and 7 simultaneously to find the optimal stocking rate in *both* periods, S_1 and S_2. What economic logic causes the optimal *dynamic* stocking rate to be less than the optimal static grazing rate?

b. Suppose that stocking in period 1 has no impact on period 2's forage production. If that is true, what are the values of B_2 and B_3? With those special values, how does an optimal *static* stocking program compare with the optimal *dynamic* stocking program?

References

Bartlett, E. T. 2003. *Sustainable rangelands roundtable.* Available from *http://sustainablerangelands.cnr.colostate.edu/2003Report/2003Report.htm* [Accessed October 24, 2004].

Batabyal, A. A., B. Biswas, and E. B. Godfrey. 2003. *On the choice between the stocking rate and time in range management.* Available from *www.dklevine.com/archive/refs42121.pdf* [Accessed June 12, 2004].

Batabyal, A. A., and E. B. Godfrey. 2002. Rangeland management under uncertainty: A conceptual approach. *Journal of Range Management.* 55 (1): 12–15.

Bestelmeyer, B. T., J. R. Brown, K. M. Havstad, R. Alexander, G. Chavez, and J. Herrick. 2003. Development and use of state-and-transition models for rangelands. *Journal of Range Management.* 56 (2): 114–126.

Borton, J., and N. Nicholds. 1994. *Drought and Famine,* 2nd ed. United Nations, Disaster Management Training Programme. Available from *http://www.undmtp.org/english/droughtandfamine_guide/drought_guide.pdf* [Accessed July 5, 2004].

Brown, K., D. Pearce, V. H. Chewings, and T. Swanson. 1993. Economics and the Conservation of Global Biological Diversity. Washington D.C.: The World Bank UNEP, Working Paper no. 2 GEF.

Brown, S., and B. Shrestha. 2000. Market-driven land-use dynamics in the middle mountains of Nepal. *Journal of Environmental Management.* 59 (3): 217–225.

Brundtland, G., ed. 1987. *Our Common Future: The World Commission on Environment and Development.* Oxford, UK: Oxford University Press.

Burt, O. R. 1971. A dynamic economic model of pasture and range improvements. *American Journal of Agricultural Economics.* 53: 197–205.

Campbell, B. D., D. M. S. Smith, and GCTE Pastures Rangelands Network M. 2000. A synthesis of recent global change research on pasture and rangeland production: Reduced uncertainties and their management implications. *Agriculture Ecosystems and Environment.* 82 (1–3): 39–55.

Carpenter, S., B. Walker, J. M. Anderies, and N. Abel. 2001. From metaphor to measurement: Resilience of what to what? *Ecosystems.* 4 (8): 765–781.

Clawson, M. 1951. Memorandum from Bureau of Land Management Director to Regional Administrators and Chief, Division of Range Management, Concerning Study of Grazing Fees in Grazing Districts. Washington, D.C.: Office of the Director, Bureau of Land Management.

Cory, D. C., and W. E. Martin. 1985. Valuing wildlife for efficient multiple use: Elk versus cattle. *Western Journal of Agricultural Economics.* 10 (2): 282–293.

De Soyza, A. G., J. W. Van Zee, W. G. Whitford, A. Neale, N. Tallent-Hallsel, J. E. Herrick, and K. M. Havstad. 2000. Indicators of Great Basin rangeland health. *Journal of Arid Environments.* 45 (4): 289–304.

DiTomaso, J. M. 2000. Invasive weeds in rangelands: Species, impacts, and management. *Weed Science.* 48 (2): 255–265.

Dube, O. P., and G. Pickup. 2001. Effects of rainfall variability and communal and semi-commercial grazing on land cover in southern African rangelands. *Climate Research.* 17 (2): 195–208.

Endangered Species Act of 1973. 16 U.S.C. 1531–1534.

Federal Land Policy and Management Act of 1976. 43 U.S.C. 1701.

Fernandez-Gimenez, M. E. 2000. The role of Mongolian nomadic pastoralists' ecological knowledge in rangeland management. *Ecological Applications.* 10 (5): 1318–1326.

Flather, C. H., and C. H. Sieg. 2000. Applicability of Montreal Process Criterion 1—conservation of biological diversity—to rangeland sustainability. *International Journal of Sustainable Development and World Ecology.* 7 (2): 81–96.

Food and Agricultural Organization of the United Nations. 2004. *Boreholes and wells.* Available from *http://lead.virtualcenter.org/en/dec/toolbox/Tech/1undergw.htm* [Accessed October 24, 2004].

Fowler, J. M. 2000. Historic range livestock industry in New Mexico. In *Livestock Management in the American Southwest: Ecology, Society, and Economics.* Amsterdam: Elsevier Science.

Gunderson, L. H. 2000. Ecological resilience—in theory and application. In *Annual Review of Ecology and Systematics.* Palo Alto, CA: Annual Reviews.

Harp, A. J., N. Rimbey, and T. D. Darden. 2001. *Cohesion, Integration, and Attachment in Owyhee County Communities.* New Mexico State University Agricultural Experiment Station Research Report 737. Available from *http://agecon.lib.umn.edu/* [Accessed February 21, 2005].

Hart, R., and M. Ashby. 1998. Grazing intensities, vegetation, and heifer gains: 55 Years on Shortgrass. *Journal of Range Management.* 51: 392–398.

Herrick, J. E., J. R. Brown, A. J. Tugel, P. L. Shaver, and K. M. Havstad. 2002. Application of soil quality to monitoring and management: Paradigms from rangeland ecology. *Agronomy Journal.* 94 (1): 3–11.

Ho, P. 2000. China's rangelands under stress: A comparative study of pasture commons in the Ningxia Hui autonomous region. *Development and Change.* 31 (2): 385–412.

Janssen, M. A., B. H. Walker, J. Langridge, and N. Abel. 2000. An adaptive agent model for analysing co-evolution of management and policies in a complex rangeland system. *Ecological Modelling.* 131 (2–3): 249–268.

Little, P. D., K. Smith, B. A. Cellarius, D. L. Coppock, and C. B. Barrett. 2001. Avoiding disaster: Diversification and risk management among east African herders. *Development and Change.* 32 (3): 401–433.

Loomis, J. B. 1993. *Integrated Public Lands Management: Principles and Applications to National Forests, Parks, Wildlife, Refuges, and BLM Lands.* New York: Columbia University Press.

Ludwig, J. A., M. B. Coughenour, A. C. Liedloff, and R. Dyer. 2001. Modelling the resilience of Australian savanna systems to grazing impacts. *Environment International.* 27 (2–3): 167–172.

Manzano, M. G., J. Navar, M. Pando-Moreno, and A. Martinez. 2000. Overgrazing and desertification in northern Mexico: Highlights on northeastern region. *Annals of Arid Zone.* 39 (3): 285–304.

Martin, W. E. 1966. Relating ranch prices and grazing permit values to ranch productivity. *Journal of Range Management.* 19: 248–252.

Masters, R. A., and R. L. Sheley. 2001. Principles and practices for managing rangeland invasive plants. *Journal of Range Management.* 54 (5): 502–517.

Mishra, C., H. H. T. Prins, and S. E. Van Wieren. 2001. Overstocking in the trans-Himalayan rangelands of India. *Environmental Conservation.* 28 (3): 279–283.

Murphy, J. S., and E. T. Bartlett. 1994. *Economic Analysis of Range Improvements.* Colorado State University Cooperative Extension Service Report 3756. Available from *www.colostate.edu/depts/coopext/pubs/farmmgt/03756.pdf* [Accessed October 24, 2004].

Nielsen, D., and E. B. Wennergren. 1970. Public policy and grazing fees on federal lands: Some unresolved issues. *Land and Water Law Review.* 5: 293–320.

Pallas, P. 1986. *Water for animals.* Food and Agricultural Organization of the United Nations. Available from *http://www.fao.org/docrep/R7488E/r7488e07.htm* [Accessed October 24, 2004].

Pope, C. A. 1988. Romance of rangeland resource management. *Rangelands.* 10: 168–170.

Pope, C. A., and G. I. McBryde. 1984. Optimal stocking of rangeland for livestock production within a dynamic framework. *Western Journal of Agricultural Economics.* 9: 160–169.

Rimbey, N. R., and J. R. Nelson. 1994. *Current issues in rangeland economics.* Available from *http://climate.usu.edu/publications/Current%20Issues%20in%20Rangeland%20Economics%20—%201994.pdf* [Accessed July 5, 2004].

Scarnecchia, D. L. 2003. Viewpoint: Empowering diversity: Envisioning, designing, and developing range management science. *Journal of Range Management.* 56 (6): 558–569.

Steinfeld, H., C. DeHaan, and H. Blackburn. 1997. *Livestock-environment interactions: Issues and options.* Available from *http://www.fao.org/docrep/x5305e/x5305e07.htm* [Accessed June 18, 2004].

Stoddart, L. A., A. D. Smith, and T. W. Box. 1975. *Range Management.* New York: McGraw-Hill.

Taylor Grazing Act of 1934. 43 U.S.C. 315.

Torell, L. A., and R. H. Hart. 1987. Economic Considerations for Efficient Stocking Rates on Rangeland. Paper read at Fort Keogh Research Symposium, at Miles City, MT.

Torell, L. A., and J. P. Doll. 1991. Public land policy and the value of grazing permits. *Western Journal of Agricultural Economics.* 16 (1): 174–84.

Torell, L. A., K. S. Lyon, and E. B. Godfrey. 1991. Long-run versus short-run planning horizons and the rangeland stocking rate decision. *American Journal of Agricultural Economics.* 73 (3): 795–807.

Torell, L. A., N. R. Rimbey, L. W. Van Tassell, J. A. Tanaka, E. T. Bartlett. 2003. An evaluation of the federal grazing fee formula. *Journal of Range Management.* 56: 577–584.

United Nations. 2004. *United Nations Convention to Combat Desertification.* Available from *http://www.unccd.int/* [Accessed October 26, 2004].

Vesk, P. A., and M. Westoby. 2001. Predicting plant species' responses to grazing. *Journal of Applied Ecology.* 38 (5): 897–909.

Winder, J. A., C. C. Bailey, M. Thomas, and J. Holechek. 2000. Breed and stocking rate effects on Chihuahuan desert cattle production. *Journal of Range Management.* 53 (1): 32–38.

Workman, J. P. 1986. *Range Economics.* New York: Macmillan.

World Bank. 2004. *Land resources management.* Available from *http://lnweb18.worldbank.org/ESSD/ardext.nsf/11ByDocName/TopicsLandResourcesManagement* [Accessed October 26, 2004].

Chapter 10

WATER

Outline

The Water Resource 259
Desirable Outcomes 264
Actual Outcomes 270
Institutional Designs for Economically Efficient Water Use 281

Objectives

After studying this chapter, you should be able to

- Describe the physical nature of water
- Identify water management objectives
- Identify actual outcomes produced by modern water institutions
- Analyze economically effective changes that can be applied to current water institutions

Introduction

In 1880, Los Angeles, California, was a sleepy village of 11,000 people. By 1900, Los Angeles had 100,000 people and began to face water shortages. The city responded to its water shortages in 1907 by borrowing $23 million to build a 233-mile long aqueduct to bring water from the Owens River in the Sierra Nevada Mountains. The project supplied water for a million people and provided power for lighting, manufacturing, and transportation. All water in excess of the city's domestic and manufacturing needs was used for agricultural irrigation. The project was completed in 1913. By 1920, when the city's population neared 600,000, the aqueduct was found inadequate. Los Angeles looked east to the Colorado River for relief. Other states in the Colorado River Basin noticed the city's growing thirst and watched nervously.

Water laws in the western United States raised a difficult question in connection with interstate water allocation. For years the law of the West had been based on the doctrine of prior appropriation, which gives top legal rights to the first water user. That doctrine governed water allocation disputes within each basin state, but nobody knew if that doctrine applied to competing users in two or more states on a common stream. In June 1922 the U.S. Supreme Court, in *Wyoming v. Colorado,* announced that the rule of priority applied regardless of state lines. Now the law appeared to favor faster-growing states like California. Something needed to be done to divide the waters.

In November 1922, the Colorado River Compact was signed. That compact set the stage for many more interstate compacts dividing the waters in the western United States. These agreements provide a framework to resolve transboundary water conflicts in allocating scarce and variable supplies in an orderly and predictable way. As described in more detail by McCormick (1994), 22 such compacts presently divide the waters of western U.S. rivers.

Water is essential to life, yet water may be the most severe constraint on world population growth in the near term. It is already a major problem in most dry places around the world, including the western United States. The United Nations projects that 40% of the world's population will likely experience water shortages (United Nations, 2002). This chapter examines how institutions have allocated water in the past, how they currently allocate water, and how they could improve on water allocation in future years.

■ The Water Resource

Water, a renewable resource, covers nearly three fourths of the surface of the Earth and occurs both above and below the surface. It is the only natural resource that exists naturally as a liquid, solid, and gas. Worldwide, the total supply of water depends on the *hydrologic cycle*, the process by which water continues to circulate (Figure 10–1). Huge quantities of water circulate through this system each year, even though people use only a part of it. As a cycle, it has no beginning or end. Liquid water from the Earth's surface, especially the oceans, evaporates into gas and enters the atmosphere as water vapor that becomes clouds. Atmospheric moisture is eventually returned to the Earth's surface as rain or snow. About 20% of the world's total annual precipitation falls onto land. This freshwater moves over the land's surface on its journey toward the ocean simultaneously *recharging* lakes, wetlands, and/or groundwater aquifers. See Example 10–1. An estimated 42,000 cubic kilometers of precipitation flows back to the oceans through the world's rivers annually.

Some of the precipitation seeps downward to become groundwater. Other water is used by plants and later released into the atmosphere (*transpiration*). The most important source of water in the atmosphere is ocean evaporation, which produces about 90% of the world's total annual supply.

The oceans contain an estimated 1.3 billion cubic kilometers of water—97% of the Earth's water supply. Of Earth's total supply of water, *freshwater* is estimated at 35 million cubic kilometers. About 75% of the world's freshwater is locked in frozen form in polar ice caps and glaciers; 14% is located deep underground, beyond economical reach; 11% is in economically accessible groundwater; 0.35% is in lakes; 0.03% is in rivers; 0.06% exists as soil moisture; and 0.035% is in the atmosphere.

On average, there is plenty of freshwater to supply all present and foreseeable human water needs. But averages are deceiving. Water has a highly skewed distribution around the world, both in location and in timing. Some areas receive large quantities of precipitation each year. In contrast, arid regions face considerable water scarcity. Figure 10–2 and Tables 10–1 and 10–2 show that in the United States there are large differences in the ratio of water consumed to total renewable supplies in various regions. Globally, water resources are also unequally distributed. Arid and semiarid regions account for about 2% of the world's total water runoff, but they form about 40% of the Earth's land surface.

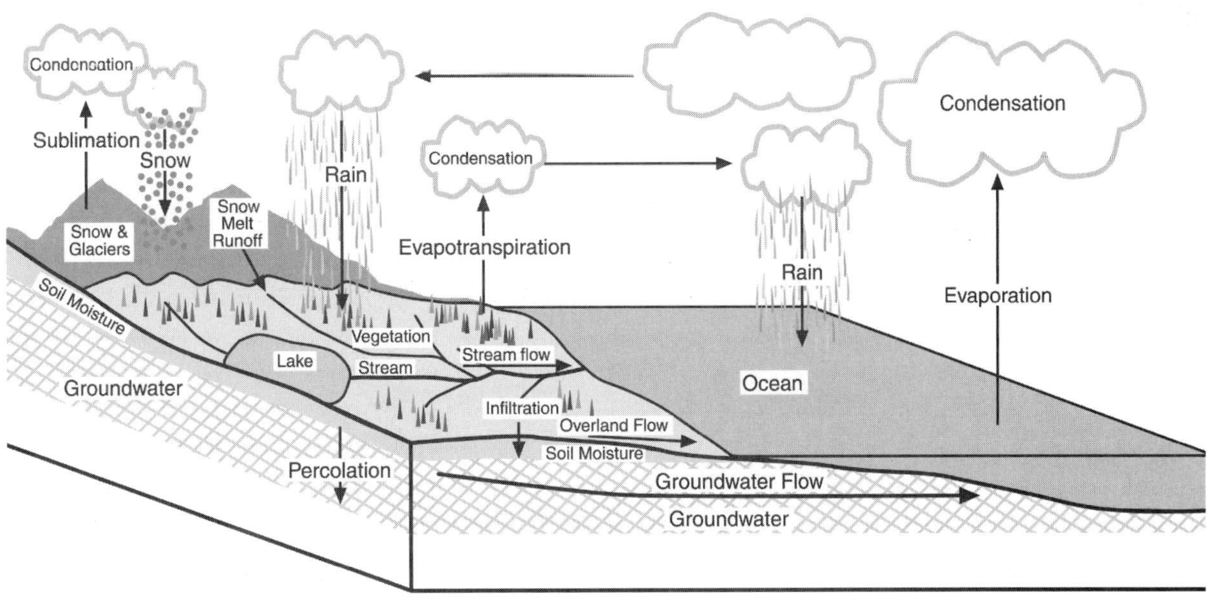

Figure 10–1 Hydrologic cycle

Example 10–1
Water Policy Challenges in the Middle East

Two major uses of water in the Middle East are urban/industrial purposes (which include drinking and waste transport) and irrigation. Irrigation is the region's largest user of water, often accounting for more than 90% of all water usage. Despite the modest contribution of irrigated agriculture to national incomes, the Middle East has seen a huge growth of irrigation since 1960. At that time, total irrigated area was 16.3 million hectares (40 million acres), but by 1998 it had grown to 29.7 million hectares—an 82% increase in 40 years. More than 60% of this growth occurred through expanded irrigation in Turkey, Iran, Iraq, and Saudi Arabia.

To deal with growing agricultural use of water, Beaumont (2002) found that many Middle Eastern governments have a short list of policy choices: (1) They can develop and use a greater part of the renewable water resources of their country. This will be possible for some countries with well-watered highlands, but many countries already use virtually all their water. (2) They can import water. (3) New fresh water could be produced by desalination. Although most countries could make use of this method, the cost of desalination is high. In addition, it can only be used in major cities, and even there only for domestic or high-valued industrial uses. (4) Water could be transferred from one use to another. An obvious choice would be to transfer water from irrigation to urban/industrial use. For example, 1,000 hectares of irrigated land (about 2,500 acres) can supply the domestic water needs of 100,000 people at a Western living standard.

The industrial and service sectors in most Middle East countries are capable of generating more than 100 times more wealth for each cubic meter of water used than the agricultural sector is. To maintain the prosperity of urban systems as engines for economic development, growing quantities of water will be required, and the most economically efficient method for meeting requirements typically is to transfer water from irrigation use to supply urban needs. The way in which water can be transferred from agriculture while avoiding social disruption is uncertain. One promising method for encouraging farm-to-city transfers is using the price mechanism, for example, through water markets. In a market approach, owners of a water source would be able to sell their water or the right to use it to the highest bidder for a desired period of time.

Still, any policy that promotes transfers of water from agricultural to urban users needs to ensure that a vulnerable section of the population (farmers) is protected from worse poverty. Beaumont proposes that use of water in irrigation for the production of low-valued crops is no longer a viable proposition in the twenty-first century as pressure on water resources continues to grow.

Source: Beaumont, P. 2002. Water policies for the Middle East in the 21st century: The new economic realities. *International Journal of Water Resources Development.* 18 (2): 315–334.

Figure 10–2 U.S. water resource regions.
Source: Solley, W. B., R. R. Pierce, and H. A. Perlman. 1998. *Estimated Use of Water in the United States in 1995.* U.S. Geological Survey Circular 1200. Washington, D.C.: U.S. Government Printing Office.

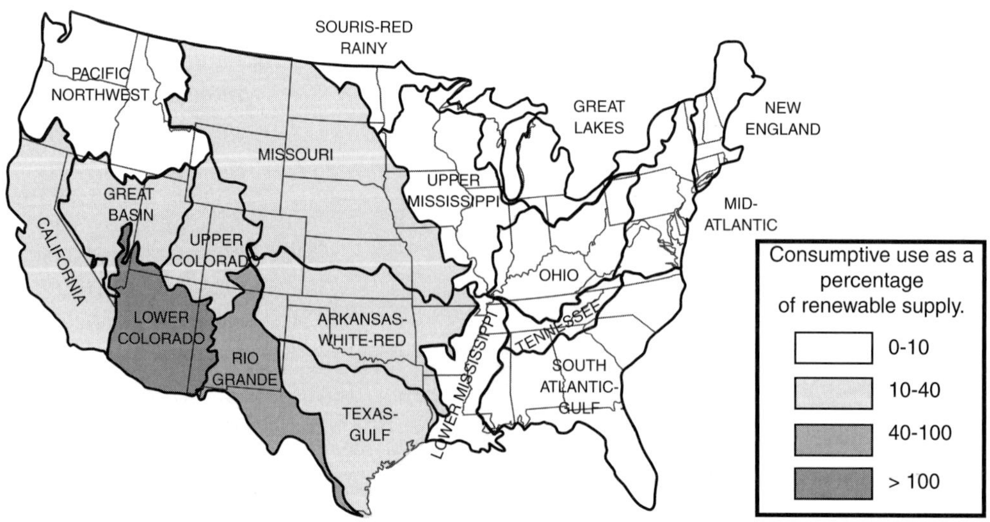

Table 10–1 U.S. offstream water use by river basin in 1995

Basin name	Population (1,000s)	Per capita use of freshwater (gallons per day)	Withdrawals (million gallons per day)									Consumptive use of freshwater (million gallons per day)
			Groundwater			Surface Water			Total			
			Fresh	Saline	Total	Fresh	Saline	Total	Fresh	Saline	Total	
New England	12,849	289	725	0	725	2,980	8,800	11,800	3,710	8,800	12,500	388
Mid-Atlantic	42,412	509	2,690	1	2,690	18,900	20,300	39,200	21,600	20,300	41,900	1,170
South Atlantic, Gulf	37,845	848	7,110	16	7,120	25,000	12,700	37,700	32,100	12,700	44,800	5,570
Great Lakes	21,836	1,500	1,510	5	1,520	31,100	7	31,100	32,700	11	32,700	1,580
Ohio	22,631	1,330	1,980	22	2,000	28,100	1	28,100	30,100	23	30,100	1,870
Tennessee	4,198	2,140	258	0	258	8,730	0	8,730	8,980	0	8,980	289
Upper Mississippi	22,268	1,050	2,570	4	2,570	20,700	0	20,700	23,300	4	23,300	1,660
Lower Mississippi	7,324	2,720	9,180	0	9,180	10,800	0	10,800	20,000	0	20,000	7,740
Souris-Red-Rainy	693	364	115	0	115	138	0	138	253	0	253	122
Missouri Basin	10,664	3,380	9,320	38	9,360	26,700	0	26,700	36,000	38	36,100	14,200
Arkansas-White-Red	8,931	1,800	7,490	284	7,780	8,590	0	8,590	16,100	284	16,400	8,190
Texas Gulf	16,755	1,050	5,960	324	6,280	11,700	4,860	16,600	17,700	5,190	22,900	7,340
Rio Grande	2,566	2,600	1,930	61	1,990	4,740	0	4,740	6,670	61	6,730	2,960
Upper Colorado	714	10,400	116	14	130	7,310	0	7,310	7,420	14	7,440	2,520
Lower Colorado	5,318	1,500	3,000	12	3,010	4,970	2	4,970	7,960	14	7,980	4,520
Great-Basin	2,405	2,510	1,610	56	1,660	4,420	143	4,560	6,030	199	6,230	3,260
Pacific Northwest	9,948	3,220	5,500	0	5,500	26,500	38	26,500	32,000	38	32,000	10,600
California	32,060	1,140	14,600	185	14,800	21,900	9,450	31,300	36,500	9,640	46,100	25,300
Alaska	604	350	58	75	132	154	43	196	211	117	329	25
Hawaii	1,187	853	515	16	531	497	906	1,400	1,010	922	1,930	542
Caribbean	3,858	152	156	0	156	433	2,450	2,880	588	2,450	3,040	189
Total	267,066	1,280	76,393	1,113	77,507	264,362	59,700	324,014	340,932	60,805	401,712	100,035

Source: *Adapted from Solley, W. B., R. R. Pierce, and H. A. Perlman. 1998.* Estimated Use of Water in the United States in 1995. *U.S. Geological Survey Circular 1200. Washington, D.C.: U.S. Government Printing Office. Available from http://water.usgs.gov/watuse/pdf1995/html/ [Accessed November 6, 2003].*

Table 10–2 U.S. fresh surface water withdrawals by use and river basin in 1995

Basin name	Public supply	Domestic	Commercial	Irrigation	Livestock	Industry	Mining	Cooling	Total
	← million gallons per day →								
New England	1,100	1	26	99	13	100	21	1,620	2,980
Mid-Atlantic	4,730	1	65	165	55	1,090	163	12,600	18,869
South Atlantic, Gulf	2,710	0	16	2,320	217	2,010	162	17,500	24,935
Great Lakes	3,830	1	108	145	20	3,900	356	22,800	31,160
Ohio	1,800	5	80	43	81	3,310	212	22,600	28,131
Tennesse	449	0	18	39	187	1,030	7	6,990	8,720
Upper Mississippi	731	0	114	54	39	660	112	19,000	20,710
Lower Mississippi	330	0	21	1,200	272	2,280	2	6,670	10,775
Souris-Red-Rainy	32	0	0	43	3	20	1	38	137
Missouri Basin	926	1	15	16,600	173	50	201	8,770	26,736
Arkansas-White-Red	1,170	0	99	2,590	205	360	26	4,140	8,590
Texas Gulf	1860	0	8	1,170	126	846	79	7,630	11,719
Rio Grande	131	0	2	4,600	9	0	2	2	4,746
Upper Colorado	106	0	1	6,990	50	4	4	146	7,301
Lower Colorado	698	0	8	4,200	7	6	26	17	4,962
Great-Basin	254	2	15	4,020	77	31	3	21	4,423
Pacific Northwest	993	7	1,030	21,700	1,470	866	29	384	26,479
California	2,880	12	319	18,200	222	19	62	202	21,916
Alaska	50	0	0	1	0	51	24	26	152
Hawaii	14	1	0	479	3	0	0	0	497
Caribbean	342	7	2	75	2	4	1	0	433
Total	25,136	38	1,947	84,733	3,231	16,637	1,493	131,156	264,371

Source: *Adapted from Solley, W. B., R. R. Pierce, and H. A. Perlman. 1998.* Estimated Use of Water in the United States in 1995. *U.S. Geological Survey Circular 1200. Washington, D.C.: U.S. Government Printing Office. Available from http://water.usgs.gov/watuse/pdf1995/html/ [Accessed November 6, 2003].*

Some precipitation penetrates the Earth's surface and becomes groundwater. Groundwater is not limited to surface depressions like rivers and lakes; rather, it collects in aquifers, between particles of soil and rock. Groundwater circulates as part of the hydrologic cycle. Precipitation and other water sources can recharge groundwater aquifers as they drain gradually toward a discharge point, often an ocean. The residence time of groundwater (i.e., the length of time it spends underground) varies enormously, from a few days to thousands of years. Thus, the hydrologic cycle is important for the recharge and discharge of groundwater. Groundwater is an important natural resource, especially in those parts of the world that have scarce surface supplies. In the United States, groundwater provides about 38% of the water for use in homes, businesses, and industries. In rural areas, it is the source of drinking water for 99% of the population, many of whom supply water from their own wells.

The amount and timing of precipitation are highly skewed. Precipitation on land averages 26 inches a year, but has a variable distribution. Some places in the United States get less than 1 inch a year and others get more than 400 inches. Some locations receive the bulk of their annual precipitation at one time of the year, but human needs (e.g., for crop irrigation) may occur at a different time. This skewed timing of natural supplies versus human demands is a major motivation for the building of reservoirs and developing other alternatives that allow people to store water for use at a time when they need it.

The world's major rivers have their headwaters in mountains. More than half the world's population relies on the freshwater that accumulates in the mountains for drinking, domestic uses, irrigation, hydropower, industry, and transportation. Although mountain land areas constitute a small part of river basins, they provide the greater part of the river flows.

Use Patterns

Regardless of distribution and timing, adequate freshwater supplies are essential for economic development and human welfare. Water is used for a variety of purposes, all of which have different quantity, quality, and timing characteristics. Demands for water also fluctuate by day, season, month, and year, depending on weather patterns. People can exist on a gallon or so of water a day for drinking, cooking, and washing though we rarely do. In medieval times people probably used no more than 5 gallons a day. By the eighteenth century in Europe, people used about 100 gallons a day. In the United States today the average person uses about 1,500 gallons a day for their needs and wants, including drinking, washing, cooking, cooling, landscape watering, and consuming the products of industry.

Figure 10–3 shows that water withdrawn from streams and rivers has several uses, many of which also return water to the stream; these uses may be domestic, commercial, industrial, and agricultural. The largest use of water in North America is for agriculture; that is agriculture withdraws about 50% of all water used. In Africa and Asia, 85% to 90% of water is used in agriculture. In Asia, continued growth of industry is increasing industrial water demands. Worldwide, domestic and municipal water use accounts for about 8% of total use, but in industrialized countries it accounts for about 15%. Rivers are a major source of the world's water withdrawals and use. Rivers, as an important part of the hydrologic cycle, also influence human settlement, economic development, and patterns of land use.

People intervene in the hydrologic cycle for human benefit. The movement of water is controlled to increase water's value to people through measures such as pumping it upstream, building reservoirs, and even moving it hundreds of miles from where it occurs naturally. Throughout the world, the need for water continues to increase as the population grows. Per capita use of water, especially in industrialized countries, is increasing rapidly. It is management of the precipitation available that determines whether the world will have both the quantity and the quality of water its needs.

Worldwide, the percentage of people served with some kind of improved water supply rose from 79% in 1990 to 82% in 2000 (World Health Organization, 2000). During this period the percentage of the world's population with access to improved sanitation facilities increased from 55% to 60%. At the beginning of 2000, 1.1 billion people of the world lacked an improved water

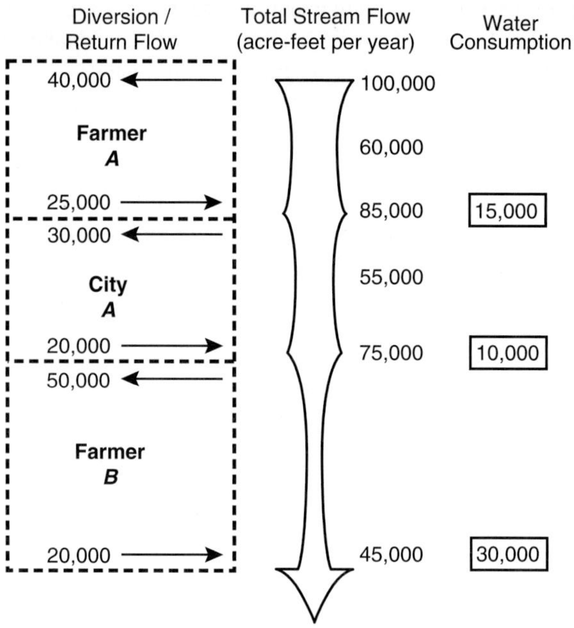

Figure 10-3 Uses of water withdrawn from streams and rivers. Stream flow, diversion, and consumption are shown in acre-feet per year.

supply and 2.4 billion lacked access to improved sanitation. Most of these people live in Asia and Africa, where less than 50% of Asians have access to improved sanitation and 40% of Africans lack improved water supply.

Management

The allocation of water needs to be managed.[1] Standards for good management depend critically on a region's location. For example, people who live where there is much rainfall, adequate surface water, and little human crowding often find it is unnecessary (poor economics) to develop groundwater, recycle, or invest in other water-saving technologies. In contrast, people who live in water-scarce, crowded, dry places often find that it is good economics to invest in various conservation measures such as recycling, groundwater development, and purification. Overall, human actions are an important factor determining the quantity and quality of water available for water uses, as well as for maintaining natural ecosystems. So several questions are important in connection with the hydrologic cycle: (1) how much water to use from various sources, (2) how much to invest in maintaining water quality, and (3) what time periods of use are best for human benefit. Deciding which actions produce efficient, fair, and sustainable water use patterns is the major challenge facing water resource economists and policymakers.

Desirable Outcomes

The objectives of water resource policies serve as standards for judging the effectiveness of the institutions that allocate water. Examples of those institutions range from complete government allocation, to a mix of market and government allocation, to mostly market allocation with background support from government. Institutions that influence the allocation of water within any

---

[1] For early analyses of water management objectives, see Eckstein (1958), McKean (1958), Marglin (1962), U.S. Congress (1962), Hartman and Seastone (1970), Young and Gray (1972), Young and Haveman (1985), U.S. Interagency Committee on Water Resources (1958), and U.S. Water Resources Council (1973, 1979, and 1983).

political unit (e.g., country or state) form a unique system with rules for sharing available water among known sources of demand. Three important standards for judging a water allocation—efficiency, equity, and sustainability—were addressed in an important human rights declaration by the United Nations in 2002.

Human Rights

On November 27, 2002, the United Nations Committee on Economic, Cultural, and Social Rights took the unprecedented step of formally stating that access to safe water is a human right (United Nations, 2002). It approved a declaration stating that the human right to water is indispensable for leading a healthy life in human dignity and that this right is a prerequisite to the realization of all other human rights. The declaration stated the human right to water entitles everyone to sufficient, safe, acceptable, physically accessible, and affordable water for personal and domestic uses. An adequate amount of safe water is necessary to prevent death from dehydration; to reduce the risk of water-related disease; and to provide for consumption, cooking, and personal and domestic hygienic requirements. According to the declaration, several dimensions of water characterize that right:

- Availability: Each person's water quantity should be sufficient for personal and domestic uses. These uses ordinarily include drinking, personal sanitation, washing of clothes, food preparation, and personal and household hygiene.
- Quality: Each person's water quality should be free from hazards that are a threat to their health. Water should be of an acceptable color, odor, and taste.
- Economic accessibility: Water and water facilities and services should be economically accessible and affordable to everyone without discrimination. Charges associated with securing water must be affordable and must not compromise or threaten the realization of other human rights.

The 145 countries that ratified the agreement agreed to ensure that all their citizens have access to safe and secure drinking water and sanitation facilities. Despite these good intentions, progress since 1990 has been slow in providing safe drinking water to the estimated 1.1 billion people who lack access to it.[2] Sanitation progress has been worse: About 2.4 billion people still lack access to a safe latrine. Unsafe water and poor sanitation are major causes of diseases such as malaria, cholera, dysentery, schistosomiasis, infectious hepatitis and diarrhea, which are associated with 3.4 million deaths each year. Inadequate water and poor sanitation are major contributors to poverty.

Efficiency

Allocation of water among users can be judged from the view of economic efficiency, where water allocations are part of a series of investment projects. Water is the limited resource and economic sectors use the resource and produce economic returns. In an economically efficient water allocation, the marginal benefit from the use of the water is equal across uses in order to maximize total benefits produced by the water. The benefit from using one additional unit of the water by one user should be the same for any other. If not, society benefits by allocating more water to the sector where the incremental benefits are highest. For a classic, early attempt to measure the benefits of water quality improvements, see Freeman (1979).

The economic challenge posed by water scarcity is to allocate a known supply of water among all competing users. Where water is a renewable resource, produced mostly by current rain or snow, a major challenge is to allocate the current period's flow among current competing users.

---

[2] Former U.S. Ambassador John McDonald saw the damages produced by unsafe water when he served as U.S. coordinator for the United Nations Decade on Drinking Water and Sanitation: "Nothing I had ever seen in my life prepared me for the day I landed in Africa. I saw villages where people trekked miles in the hot sun just to get clean water for the day. But even more tragic were the children I saw suffering from the lack of clean water. Many seemed to be just hours from death, and others had lost their vision to trachoma, an easily preventable disease caused by contaminated water." Available from *http://www.globalwater.org/whoweare.htm* [Accessed May 15, 2004].

However, water is sometimes a depletable resource that is used by drawing down on a fixed stock, such as pumping from a nonrechargeable aquifer. In this case, the challenge is to allocate water among competing users in each of many time periods. Let us begin with the analysis of allocating a known supply in a single period.

Water Flows. Efficient allocation of a current period's water flows (often surface water) needs to (1) account for many competing users and (2) deal with highly variable flows from one period to the next. The first issue is important because many users typically claim a legal right to use the water, users like farmers, cities, factories, and the like that consume water by taking it from rivers and streams. Other users are recreation and environmental interests that take advantage of water's benefits while it resides in streams or reservoirs. The second issue is important because stream flows vary widely over days, months, and years, so the total supply changes from one time to the next, sometimes unpredictably.

Consider the case of allocating water efficiently among many competing users. Economic efficiency is achieved if the equimarginal rule is followed:

$$MNB_1 = MNB_2 = \ldots = MNB_n \tag{10.1}$$

That is, marginal net benefits (MNB) are equal for all competing uses.[3] The best way to see why this equimarginal rule defines economic efficiency is to think about what happens when the equation is violated. For example, suppose that MNB for municipal and industrial (M&I) users is $200 per acre-foot, whereas the MNB for irrigation is only $50.[4] Figure 10–4 shows that some of the existing water supply can be reallocated from agriculture to cities and total net benefits will increase. The first acre-foot taken from agriculture and put into M&I use produces a marginal gain of $200 for M&I (quantity Q_1^C) and a marginal loss to agriculture of $60, (quantity Q_1^A) for an overall net gain of $140. Because total net benefits can always be increased by reallocating whenever the MNBs are not equal, *total* benefits can always be increased when the *equimarginal* rule is violated. Therefore, the equimarginal rule must hold true for economic efficiency to occur.[5]

When *marginal* net benefits are *not* equal under an existing water use pattern, *total* net benefits can be increased by transferring water from users with low marginal net benefits to users who produce higher marginal net benefits. Any policy or institution that effectively transfers water to users who value the marginal water more also increases total net benefits produced by the water's use. People who lose use of the water (valued at $B+A$) are losing a smaller economic benefit than the water gainers are gaining in benefit (valued at $C+D$). When enough water is transferred so that marginal net benefits are equalized among all users, it is not possible to increase overall total net benefits by more transfers. The equimarginal rule holds at point Q_2^A for agriculture and at point Q_2^C for M&I, when water is valued at $80 per acre-foot for both uses.

Figure 10–5 shows the concept of water allocation efficiency. Two marginal net benefit schedules are shown, one for M&I users (D_C) and one for agriculture (D_A).[6] Total marginal net benefit is shown as the horizontal summation of the two demands (D_T). At high prices that exceed what farmers can afford to pay, aggregate marginal net benefit is the same as M&I's demand, since M&I is the only user under those conditions. If total water supply is S_0, total available quantity is Q_0^T. An efficient allocation of Q_0^T will equate marginal net benefits between the two uses,

---

[3] The marginal net benefit for the last acre-foot of water use is its marginal benefit minus the marginal cost of delivering the water to its point of use.

[4] If there is enough water for both users to buy all they want at the going price, then each user's marginal net benefit is its price minus marginal cost.

[5] Considerable research has searched for policies that increase economic efficiency of water use. One way to conduct this search is through the use of optimization models and associated software packages (e.g., GAMS, 2004).

[6] Crop–water production functions are a common method for measuring the benefits of water in agriculture (e.g., Hexem and Heady, 1978). Johansson et al. (2002) summarized techniques for pricing water in agriculture. Ward and Michelsen (2002) summarized methods for valuing water in agriculture.

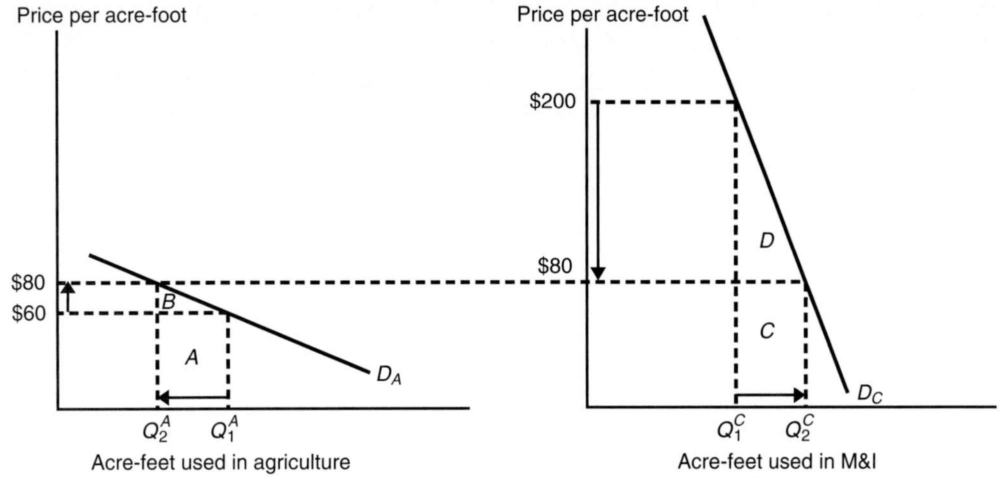

Figure 10–4 Gains from water trades between agricultural uses and municipal and industrial (M&I) uses

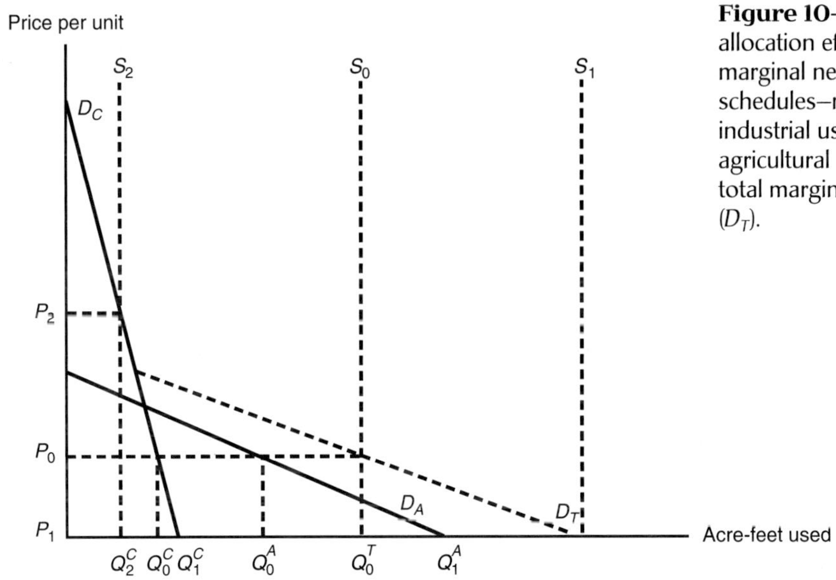

Figure 10–5 Water allocation efficiency for two marginal net benefit schedules–municipal and industrial users (D_C) and agricultural users (D_A)–and total marginal net benefit (D_T).

giving Q_0^C to M&I and Q_0^A to agriculture. Any institution or policy that blocks that allocation of water is said to be economically inefficient.

Next consider how efficient water allocation is defined under conditions of highly variable stream flows. Figure 10–5 also shows the economic impact of increased water supply abundance, where total water supply available for allocation is increased to S_1. In that case, water is so abundant that both users get all the water they need at a zero marginal net benefit (zero price). If the water supply exceeds the quantity where D_T intersects the horizontal axis, then water is not economically scarce and the two demands do not compete. The equimarginal condition holds at a very special case of being zero for both users. See Example 10–2.

Now, think about the effects of falling supplies, as in a drought. The equimarginal principle still works and is still required to produce economic efficiency. In fact, the principle defines a special rule for sharing shortages produced by supply shortfalls. The rule says that the loss in total net benefits produced by a drought is minimized if the more price-elastic user, in this case agriculture, gives

Example 10–2
Peace and the Price of Water

Israel's dry climate, high irrigation demands for water, and military and political tensions all contribute to its growing water scarcity. A better outlook for peace and security in the region would result in greater economic benefit available from the region's scarce water. Currently, Israel's lack of food security requires that it develop much of its food and water from its own domestic sources. Water prices in Israel are cheapest for agriculture, more for industry, and highest for home use. Israel's policymakers set a low price for agriculture to encourage food self-sufficiency in case food imports are cut off. Wide differences in water prices among Israel's major water users suggest that the *total* economic value of water could be increased considerably by equating water prices across sectors. (Raising the price of water to agriculture and reducing it for other uses makes the economic cost of water transferred from agriculture smaller than the value gained from expansions of other water-using sectors.) Equalizing water's prices would shift water's demand from users with low marginal values to users with higher marginal values.

International politics will have major impacts on the water situation. Success in international peace negotiation complements success in international water cooperation. Water pricing reforms raise the total economic benefits produced by scarce water as water gravitates from low to higher valued uses. Improved prospects for peace in the region will encourage more domestic investment and more international lending to finance dams and other water-related investments. These will increase supplies, quality, and reliability. Increases in water supply, in turn, will improve long-run prospects for peace and stability, because existing water shortages in Israel and in neighboring countries are an important cause of political and military tension.

An improved peace outlook would reduce Israel's need for food security generated by food production, since the country could rely on more predictable food imports. Greater food imports would reduce the national need to subsidize agriculture through cheap water prices. Increases in agricultural water prices would free up water that can move to households and industry, both of which have a higher marginal economic value.

A number of international water projects have been proposed as a method for dealing with water scarcity problems shared by Israel, Jordan, and the surrounding region. One project would involve building a Jordanian–Syrian dam on the Yarmouk River to store currently unused winter flood waters for summer use and building canals between the Yarmouk River and Lake Kinneret (the Sea of Galilee) to store winter flood waters in the lake (with partial return water shipments to Jordan). Relaxed hostilities could advance this kind of international water cooperation. Peace would provide three economic benefits: (1) added water supplies; (2) water price equalization across sectors without decreasing Israel's food supply; and (3) the ability to relax food self-sufficiency concerns. The total net economic benefits of the proposed Yarmouk–Kinneret canals is estimated to be about $22.3 million, for which each of the three benefit categories would contribute 80%, 15%, and 5%.

Source: Just, R. E., S. Netanyahu, and J. K. Horowitz. 1999. Water pricing and water allocation in Israel. *Journal of Policy Reform.* 3 (2): 97–119.

up a higher proportion of its use. Figure 10–5 shows water supply when stream flow falls to S_2. Under these dry conditions, efficiency requires marginal net benefits to be equal, but an allocation is required that is different from that under normal conditions (S_0). When supply is S_2, agriculture receives no water at all, while M&I receives everything. (M&I does not get more water with the drought than without, but by getting Q_2^C, it gets the entire supply, whereas agriculture gets nothing.) Could this strange reallocation actually lead to economic efficiency? The answer is yes. The marginal net benefit schedule for M&I use (D_C) lies above marginal net benefits for agriculture at low levels of M&I use. This means that when supplies are very low, cities will pay a much higher price for water than irrigators will. As total supplies fall, the marginal net benefit displaced (opportunity cost) from losses to M&I is higher than equal losses to agriculture. To minimize total economic damages produced by the drought, a larger part of the shortfall is placed on agriculture than on M&I.[7]

[7] Because economically efficient methods for coping with supply shortages require unequal shortfall-sharing patterns, any rule of thumb for sharing shortages, such as proportional sharing of shortfalls, is unlikely to produce economic efficiency. However, well-designed market institutions like water banks or water rental markets may be able to compensate for inefficient shortfall-sharing conventions.

Water Stocks. Sometimes water supply is stock and not a flow: Supply is not continually renewed; rather, it is a nonrenewed, fixed supply whose current use takes away from future use. Two good examples are nonrecharging aquifers and nonrecharging natural lakes. Extending the analysis of economic efficiency to account for depletable water stocks requires that the depletable nature of the resource be considered. When demand exceeds recharge for a particular supply source, the resource is "mined" just like a seam of coal or vein of copper. The mining continues until the supply is depleted or the the marginal cost of supply becomes economically prohibitive. An equimarginal rule guarantees economic efficiency for a depletable water stock. That rule characterizes *dynamic economic efficiency* and is as follows:

$$PVMNB_{11} = \ldots PVMNB_{n1} = PVMNB_{12} = \ldots PVMNB_{n2} = PVMNB_{1m} = \ldots PVMNB_{nm} \quad (10.2)$$

where the first subscript in each term refers to the use of water (e.g., agriculture, cities, recreation) and the second refers to the time period in which it is used (e.g., 1, 2, 3). The equation says that the *present value of (discounted) marginal net benefits (PVMNB) should be equal for all competing uses and in all time periods.* Time itself has a special role in this rule. Since each acre-foot of water used in period 1 can take an acre-foot away from a use in different period, the present value of gains produced by current use should be compared with the present value of losses from future uses displaced by current use. If demand patterns over time change little, then any positive discount rate (see Chapters 5 and 6) will produce a declining water use path over time for which the dynamic equimarginal equation holds true. Over time, the marginal cost of supplying any use tends to rise as depletable supplies fall. Water use for any particular purpose (e.g., agriculture) tends to fall to zero over time when (1) the total supply is exhausted or (2) the marginal cost of supply is either larger than the marginal benefit or larger than the marginal cost of finding water from a substitute source.[8]

Think about what happens when the actual water use path over time is violated by this equation's rule. Suppose that the last acre-foot of water applied to agriculture in period 1 produces a discounted marginal net benefit of \$50. Also suppose that leaving the same acre-foot unused in period 1 makes it available for use in period 5 for M&I uses and produces a discounted marginal net benefit of \$120. Then, some of the existing water supply can be reallocated from agriculture in period 1 to cities in period 5, and discounted total net benefits will increase. The first acre-foot taken from period 1's agriculture and put into period 5's M&I use produces a discounted marginal gain of \$120 for M&I and a marginal loss to agriculture of \$50, for an overall discounted net gain of \$70 (\$120 − \$50). Because total discounted net benefits can always be increased by reallocating water whenever the PVMNBs in the equation are not equal, *total* discounted net benefits cannot possibly be maximized when this equimarginal rule is violated. Therefore, the equimarginal rule must hold true for economic efficiency to occur over time for a depletable-stock resource. Any depletable-resource policy or institution that effectively transfers the water to users or time periods for which discounted marginal values are higher, also increases discounted total net benefits produced by the water's use.

Equity and Sustainability

Equity is an important objective of water resource management. Equity concerns fairness of allocation across economic and political groups. Often it conflicts with the efficiency objective. In

---

[8] It is common to think of water stocks as applying to nonrechargable aquifers and water flows as applying to renewable stream flows. In fact, some aquifers are renewable much like streams, and if managed carefully, can support sustained levels of pumping. Still, where there is a cheap but depletable-stock aquifer coexisting with a higher cost of supply from a renewable stream, a common pattern is to see the groundwater mined from the aquifer in the early years. After the aquifer is economically depleted, economic efficiency shows that it pays to switch to the higher cost renewable stream flow source, from which undiscounted marginal net benefits are equal among all competing uses in each future year. Solow (1974) presents a classic treatment of an economically efficient joint management of a depletable resource that is backed up with a renewable substitute (a backstop technology).

the case of household water, equitable allocation suggests that all households, regardless of their ability to pay, have a human right to safe and healthy water, such as the right defined by the 2002 United Nations Declaration. Policies that meet this objective may require government subsidies, free service, or adjustments of pricing structure to income or family size. As another example, the equity objective may suggest providing enough water to irrigated agriculture at a sufficiently low price so that farmers have enough income to keep their farms. Under the equity objective, the poor, women, and other vulnerable groups in society must be considered for water resources development and management to be socially acceptable. Sustainability, a third important objective of water policy, is discussed in Chapter 19.

Actual Outcomes

Water allocation principles are judged by the important standards of efficiency, equity, and sustainability. Actual water allocation is influenced by laws, supply, demand, and regulations.

Water Laws

Water is supplied in certain quantities, locations, and time periods according to the hydrologic cycle, but typically these deliveries do not match people's demands. How water is ultimately developed and used for human benefit depends critically on the status of the law regarding who has legal rights to own, develop, or use water. Water law, properly designed and enforced, is a powerful institution to compensate for the mismatch between water's natural supply and its human demand. Good laws must evolve in response to changes in human values, technologies, political systems, and settlement patterns. One of the earliest economists to examine economic impacts of water laws was Ciriacy-Wantrup (1952). Hutchins, Ellis, and DeBraal (1971) summarized the water laws of the western United States and Trelease (1965) summarized criteria for economically efficient and equitable water laws.

Surface Water. Much water law in Europe, the eastern United States, and other wet places is based on the rights to use the water that belongs to people who own land on the banks of rivers, streams, reservoirs, or lakes. This law is known as *riparian law,* based on the Latin word *riparius,* which means related to river banks or lakes. Under riparian law, each landowner has the right to a reasonable use of the water in the river; nobody, however, may use the water so that it significantly damages other riparians. People who own no riparian lands have no legal rights to use the water or to influence how it is used. Because the water right is part of land ownership and the water is not a separate piece of property, the right to water use is kept intact even if the owner fails to exercise it. All riparian landowners have equal use rights, and the law accommodates new users, even if riparian lands are subdivided. The right to use water is a relative right; it is not a right to a fixed quantity. It increases in wet periods and decreases as existing riparians find new uses for water. Riparian rights to use a specific quantity of water are insecure in situations in which there is too little water to accommodate all desired uses (e.g., in a drought).

In the United States, when people began leaving the wet East to settle the dry West they found very different climate and stream flow patterns. Water was more scarce and stream flows were more random. After rainstorms, streams flowed in torrents, then dried up completely for months at a time. Some of the western states and territories tried to establish riparian law, but it soon became clear that a water law suitable for wet areas worked badly. Newcomers settled streamside areas first and irrigation was typically required to support a family. A small number of settlers would soon deplete the entire flow of a stream, streams that typically had little natural predictable flow. If the riparian doctrine were practiced, each new settler/landowner would have a right to reasonable use of the stream's water, but soon most of the water would be used. So new settlers would require existing users to reduce their demands. Over time, all settlers would have to adjust their use so that everybody shared existing supplies in a reasonable proportion. If the riparian doctrine had been applied to western settlement, the outcome would have produced inadequate and unreliable supplies for everybody. Because all landowners' water uses could not avoid depleting other riparians' use, it was clear that riparian doctrine would fail to work in the West. Beginning with the needs of gold miners who settled California in the late 1840s, a new kind of water law emerged—the *prior appropriation doctrine*.

Prior appropriation assigns *water rights* to the first person who takes it from the stream and puts it to beneficial use,[9] which is typically interpreted as any use that avoids flagrant waste. Whereas riparian law views all holders of water rights as having equal use rights, prior appropriation creates a system of priorities in which the first user has superior rights over those who come later. So the first person to divert water for irrigation or mining had a *senior* water right, an absolute right to continue all beneficial use. Later users had *junior* rights, meaning they could claim rights over only that part of the water not already used by the senior user. Even today, appropriative rights require continued beneficial use, hence the phrase "use it or lose it." Once beneficial use ceases, the right may be lost.

Groundwater. Groundwater law is based partly on the unique features of groundwater itself. Groundwater resides in and is extracted (abstracted) from aquifers, underground soil and rock formations in and through which water flows. Aquifers range widely in size: Some are small and others consist of large formations that lie beneath land areas of several thousand square miles. Some aquifers have slow or even zero recharge rates; others are recharged must faster. In many cases aquifers are hydrologically connected to surface water sources; that is, they recharge from and to streams, rivers, lakes, and reservoirs.

In the eastern United States, the concept of *reasonable use* is important. The landowner above an aquifer can make reasonable use of the aquifer's water. Reasonable use rules out uses that can be viewed as wasteful and blocks transfer of groundwater extraction rights to landowners who are not above the aquifer. In recent years some eastern states have moved toward systems of withdrawal permits in which public agencies issue permits to extract groundwater that place upper limits on withdrawn quantities.

In the western United States, much groundwater law has developed parallel to the prior appropriation doctrine. Although groundwater pumping is much more recent than surface water use, many of its legal principles are the same as those governing use of surface water. Senior rights to specified quantities can be claimed by the first to extract the resource, and subject to limits, they can be defended against junior claimants. If an aquifer is hydrologically connected to surface water, a new pumping permit may be denied if it reduces the stream flow of the holder of senior rights. Commonly in the West, a public agency grants a pumping permit to any groundwater appropriator who can show there is unappropriated water, that is, the proposed pumping does not impair an existing surface or groundwater right.

Supply Management

Water supply management refers to the implementation of policies or measures that reduce shortage through means that expand water supply, increase the safety or quality of existing supplies, increase supply reliability, or reduce the risk of supply shortages.

Interbasin Transfer. Some of the fastest-growing places in the world are in arid regions where there is little natural water available to serve the increasing population. U.S. western states are growing at nearly double the national average. Colorado, for example, anticipates a 40% increase

[9] Beneficial use is the foundation of a water right. A common expression is that beneficial use is the basis, the measure, and the limit of a water right. Establishing a water right requires a person to make an appropriation. The essential elements of an appropriation are the diversion of water from a stream and its application to a beneficial use. A diversion is typically made by removing water from its natural course. The standard of beneficial use is satisfied by irrigation, mining, industrial application, recreation, stock watering, domestic, municipal, and other nonwasteful economic activities. The opposite of beneficial use is waste. In most of the western United States, an appropriator may remove the water from its source and put it to beneficial use at any location. In contrast to riparian water rights, there is no geographical limit to place of use. Also in contrast to riparian law, the ownership of land bordering a watercourse carries with it *no* right to the use of the water *without* an appropriation.

from 2001 to 2020. California expects an *additional* 15 million people in the same time span. The U.S. western states also face unmet demands to protect and restore instream flows for fish and riparian ecological needs, as well as large numbers of Native American water claims.

An interbasin transfer involves removing water from its river basin of origin and exporting it to a destination basin, typically to places where water is scarcer and/or political power is greater. For example, one interbasin transfer proposal that has been around since the 1960s involves taking water out of the Mississippi River to recharge the Ogalalla aquifer to serve irrigated agriculture in high plains areas. More recently, there have been proposals to load water in tankers to be shipped to dry regions in Asia or in the Middle East. In 2002, a company filed an application with the California Department of Water Resources to remove 30,000 acre-feet of water each winter from two northern California rivers, pump the water into battleship-size bags, and tow the bags to San Diego (Sax, 2002). Although such a shipping project may be technically feasible, it is unlikely to be economically efficient.[10] In any event, history has shown that most interbasin transfer proposals fall short on the economic efficiency standard long before they are blocked by legal, political, or environmental constraints.

Storage. The natural variability in precipitation within a single year and among years produces a highly variable natural supply of water flowing in streams and rivers. In wet periods, high precipitation has the effect of increasing stream flows when farms, factories, and homes least need the water. In droughts, low precipitation reduces stream flows at just the time when water users need more water. Water use also increases with growing population and more economic activity. More water is used in peak periods, such as in summer and in early morning and early evening in homes. Without some mechanism to store unused water for use in dry periods, mismatches between supply and demand create water shortages and floods, sometimes producing disastrous effects. Storage tanks, reservoirs, and recharging underground aquifers moderate these potential troubles through availability of a steady stream of water to match human needs. Storage typically occurs at higher elevations than where the water is used, which raises the water pressure for the water user at the point of use.

Treatment. Water treatment removes undesirable natural materials as well as pollutants caused by people. The importance of treatment illustrates the significance of water quality in comparison with water quantity. One example of treatment is *desalination* of seawater or mineralized water to produce freshwater. Desalination consumes large amounts of energy and presents a waste disposal problem; therefore it is economically feasible only in places where desalination is cheaper than freshwater, such as dry coastal areas and dry inland regions well-endowed with saline groundwater.

Building and operating freshwater treatment plants raises the costs of delivered water. Costs increase according to the kind and range of pollutants in the water as well as the degree of purification required for each pollutant. Because the quality of water required by different kinds of users varies, an important question, known as the dual supply debate, asks whether all customers of a utility should receive water of equally high standards. The debate focuses on whether users with more limited quality needs, such as agriculture, should receive water treated to lower standards, while customers such as households should receive water treated to higher standards. Benefit-cost analysis has considerable potential to contribute to this debate: If the cost added by a building and operating a dual supply system is smaller than the incremental net benefits of limiting treatment for some kinds of customers, dual supply is economically efficient. The incremental net benefit produced by adding a dual system is the cost savings of treating water to a

---

[10] Still, one can imagine a growing population in a dry place with a high enough willingness to pay for imported water so that the exporting region could be paid with cash and both regions would be better off. Howe (1987) provides an insightful and entertaining summary of the trades among various groups required to successfully implement the Colorado–Big Thompson project. The trades had all major regional players benefit at the federal taxpayer's expense.

lower standard minus the reduction in benefits accruing to users of the water treated at a lower standard. If, for example, agriculture suffers no loss from water treated to a lower standard, its reduction in benefits from reduced treatment standards is zero.

Distribution. Water distribution systems are built to deliver water from water sources to final users. A distribution network can consist of pipes, pipe junctions, pumps, valves, and storage tanks or reservoirs. By pooling risks, water supply networks reduce the overall need for maintaining a safety margin capacity, in which normal supply is sized intentionally larger than normal demand. So if there is a shortage of demand in one region and a shortage of supply in another, interregional transfers can be set up. Such supply networks require expensive investments in reservoirs, pipes, pumps, and the like. In urban areas, water for households is commonly delivered by main pipes; separate connecting pipes deliver to each house or to each block of apartments.

The design, maintenance, and repair of water distribution systems is both important and difficult. For new systems, sizing is especially important, as it requires planning for future expansion. The designer must determine current demands and forecast demands as well as know water supplies available. Materials must be chosen depending on local codes, soil types, cost of the water, and consumers' willingness to pay. The advent of fast computers in the 1990s, simplified processes through development of modeling packages for water supply and water distribution systems, risk analysis systems, and water quality analysis (e.g., transport and control of pollutants).[11]

Water distribution systems must deal with the problem of leaks, which can have major economic significance. Leaks occur between the point of removing the water from its source and the point of final use on the water buyer's property. For a given level of final use (demand), the capacity of a water system in terms of removing the water from its source, storage, freshwater treatment, and distribution must be higher to compensate for the economic and environmental costs of leaks. Leakage losses depend partly on the total distance over which water is distributed. But the rate of leakage from pipes also increases directly with water pressure, which is higher in hilly areas and in densely populated areas. Losses are also higher in places where pipes are old and ground conditions are unstable. Infrastructure improvements aimed at limiting losses take several forms, for example, leakage control targets, pressure reduction measures, and discovery, repair, and replacement of broken lines and corroded main pipes.

Limiting water losses through leakage control produces the greatest net benefits where water is scarce and/or where the cost of new capacity is high. For a given demand requirement, a lower leakage rate permits considerable savings by building lower capacity systems. In situations where demand is at or near capacity, loss allows for fewer additions to capacity to meet growing demands that would otherwise occur. That is, as the price or environmental cost of water supply expansion increases, it becomes economically feasible to invest more heavily in leakage control. For any given price or environmental cost of water, there is an economically efficient leakage control level.

Recycling. Some water consumers, especially in the industrial sector, consume only part of the water supplied to them. So they are able to reuse that supply for themselves repeatedly, though water treatment may be required to maintain the process. Environmental benefit occurs in two ways: (1) Use of recycled water requires removing less water from its natural source. (2) Pollution in wastewater is reduced.

Most of the benefits of water recycling accrue to people other than the direct buyer of water. Typically, financial incentives are inadequate to encourage individuals to recycle. The individual usually discovers that it is cheaper to meet her water needs by purchasing utility-supplied, unrecycled freshwater than to make expensive recycling investments. Therefore, economic principles

---

[11] Fast computers and efficient algorithms make even large-system modeling feasible. For example, water distribution models have been implemented for Vancouver in British Columbia with more than 10,000 pipes. Larger cities such as Atlanta, Chicago, and Los Angeles can require more than 20,000 pipe segments in a model of a water distribution system.

have considerable potential in designing efficient recycling programs. For example, economics can help determine how recycled water should be priced compared with its unrecycled substitute. To date, though, little research has been done to establish a consistent framework for analyzing when recycling is economically efficient (see California State Water Resources Control Board, 1997).

Water recycling becomes more economically efficient for any of the following reasons: new water capacity becomes more expensive, the price of recycling falls, recycling efficiency rises, and environmental costs of new freshwater sources are higher. To more reliably identify these thresholds, further work is needed to identify and measure nonmarket benefits and costs, especially environmental damages reduced by avoiding new sources for freshwater. Incorrect measurement of nonmarket impacts of recycled water projects is a serious impediment to analyzing the feasibility of projects and may ultimately block implementation of economically efficient projects.

Demand Management

Demand management for water refers to methods that avoid shortages by limiting demand. To avoid the need for new supplies, demand management focuses on using existing supplies to a greater extent, reducing demand, and altering the timing of demands. It aims to squeeze more beneficial use from existing supplies through better management, leakage reduction, improved revenue collection, use regulations, recycling, technological change, metering, pricing, and markets.

Transfers. Urban populations grow in many dry areas despite hydrologic, environmental, or political constraints that may block development of additional water sources or escalate their costs. For these reasons, there is growing interest in transferring existing water supplies among users in the search for economically efficient ways of meeting increasing demands. For a classic economic analysis of water transfers, see Kelso, Martin, and Mack (1973).

One way to transfer existing water supplies is through legislative or administrative fiat. For example, growing cities could simply condemn water rights of farmers and take it for their needs, stating that the public welfare is served. In May 2003 the Texas State Senate passed a bill (Texas, 2003) that establishes requirements before a city can declare eminent domain over a private landowner's water rights. The bill recognized the potential damages to farmers' water rights posed by condemnation and required a city to prove the need for condemnation by showing that (1) the city really needs the water, (2) cheaper water supply methods are not possible, and (3) voluntary market transfers are not possible.

Administrative (nonmarket) water rights transfers damage existing users of water rights as well as discourage search for the least expensive sources and demand reductions. One less expensive method to find new water supplies involves use of water markets, which can function like other markets: Willing buyers and sellers conclude private agreements to transfer given amounts of water at agreed-on prices. In many U.S. western states, where much water is locked into agriculture, there has been considerable interest in expanding the role for water markets. For example, the 1992 Central Valley Improvement Act allows farmers who receive water from California's Central Valley Project to sell or rent some of their contracted water to beneficial uses either inside or outside the project area.

Water markets also can contribute to fairness by reducing political favoritism. In countries and cultures where water is allocated by government decree, bureaucrats may be pressured to allocate supplies. Powerful people who are friends with the government get the most water. A market system can remove water allocation activity from the political arena and place it with owners of the water rights.

Kinds of Market Transfers. The short-term water market is a powerful institution that encourages efficient response to drought. As supplies of water shrink, cities buy or rent larger amounts from agriculture. Because of the lower supply, farmers benefit by trading water for money that exceeds the value of losses in agriculture. Cities benefit by receiving water for a price that is lower than the price from finding supplies from other sources. The water market has great potential as a way to find supplies for growing populations in dry places by reducing demands for existing uses without developing expensive new supplies. Several types of transfer have the potential to provide economic opportunity for both water sellers and buyers. All of those discussed next use the power of incentives to encourage buyers to look for the cheapest source and sellers to seek out the highest-valued market.

1. Under *contingent transfers,* prospective water buyers may be less interested in acquiring permanent water rights than in increasing the predictability of their water supply system. For both buyers and sellers, temporary transfers during periods of shortage may produce a significant economic efficiency benefit. For example, a city in need of water could pay an irrigation farmer a sum of money for the privilege of exercising the right to use the farmer's water should the need arise. The city might arrange to pay an extra sum if the right is actually exercised. Advantages for the seller are the immediate acquisition of cash when the contract is established and added revenues if the contingent transfer option is carried out. The advantage to the buyer is access to wet water[12] when it is most needed.

2. *Spot market transfers* are short-term transfers, typically agreed to and carried out within a short period such as a single year. These transfers typically set up a bidding process, often with some conditions for transfer (e.g., the water's price or quantity being fixed). The seller has the advantage of immediate infusion of cash when the transfer takes place. The buyer gets the water without having to show future beneficial use or take on other responsibilities required by being the owner of the water right.

3. *Water banks* are a special kind of a spot market organized and operated by a central banker, such as a state or federal government agency or a group of water utilities. The bank is a mechanism through which willing owners of a water right lease water to a bank for release to renters on a short-term basis, such as a year. The bank is responsible for organizing the lease and for tracking the supply and demand for money and water. A water bank is characterized by flexible, temporary transfers of water without changes of ownership and bank participants may differ each year. The California Drought Emergency Water Banks of 1991 and 1992 are classic examples of institutions that harnessed the power of the market to cope successfully with serious drought. Operated by the state of California, the banks acquired water in three ways: by paying farmers for water they would have used to irrigate their fields; by purchasing surplus water from local irrigation districts; and by paying farmers or irrigation districts to use groundwater instead of surface water. These banks taught water managers important lessons: (1) water markets, even when severely constrained, still work; (2) water has a high value for many buyers, and many owners of senior rights are willing sellers; (3) large amounts of water can be found if enough money is put on the table; and (4) third-party interests in market transactions can be protected.

4. *Wheeling* is another kind of transfer. Electric power is often sent through the transmission system between power companies and generation plants to reduce the cost of power and to get it to the time and place it is most needed. Water can be similarly wheeled through water conveyance and storage facilities to reduce economic damages from drought or to supply rapidly growing cities. Seasonal wheeling of water is common in agricultural regions in which different areas have complementary demands for water over time. For example, the city of El Paso, Texas, needs water from the Rio Grande in the winter, while the Elephant Butte Irrigation District (EBID) in southern New Mexico has limited need for its channel capacity in winter. So seasonal wheeling provides opportunities for El Paso to exchange water with EBID

[12] Wet water, sometimes called "real water," is water transferred from a user who reduces use; it is not derived at the expense of another lawful user. Here are two examples: (1) net water savings resulting from not planting and irrigating a crop that would otherwise be irrigated; and (2) stored water released that would not otherwise be released. The opposite of wet water is paper water, that is, water transferred that creates no new supply increase. An example is a proposal to sell water that the seller has not historically used.

during low-flow irrigation demand periods. Repayment can come in the form of added water and/or cash during the high-demand irrigation season. If laws are passed that remove barriers from farmers renting water to cities the cities may benefit: Water freed up from agriculture might be available at a lower cost than the cities' next cheapest source of water. Wheeling for environmental purposes could involve storage facilities that release water for instream flows. For example, agriculture could release water from upstream locations (paid for by environmental interests) to meet recreation or endangered species habitat requirements.

5. Transfers of *reclaimed, conserved, and surplus water* are also possible. Water purchases made available by reclamation or reduced water demands are a form of a water transfer. Under a 1989 agreement, southern California's Metropolitan Water District (MWD) invested more than $100 million in lining irrigation canals and other water conservation projects in the state's Imperial Irrigation District. In return, the MWD received the right to use the conserved water—106,000 acre-feet per year, for 35 years.

Conditions for Efficiency. The ability of water markets to increase economic efficiency requires a property right to the use of a certain amount of water that can be *defined, secure, and transferable.* These three conditions may be hard for some countries to meet for political, economic, or cultural reasons. In many nations, the ownership of water resources is held in trust by the state, but the right of use is granted in several forms (e.g., include perpetual concessions, permits, licenses, contracting rights, and outright ownership). In any case, the use right must be measurable in well-defined terms and by using clear, practical methods.

A *defined* water right means there is no ambiguity about the nature of the property that is owned. For example, when stream flows vary highly from one period to the next, a clearly defined right tells the owner with certainty how much water he has the right to withdraw from the stream under what conditions. Here is an example of a clearly defined right: The owner of the right knows that she is third in line behind two other diverters and that she can divert up to 50 acre-feet if the water is available. One way for the state to establish clearly defined rights is to define titles to water through court adjudications. When the right is clearly defined, the market works better and the market price is higher, because the buyer knows exactly what she is buying from the seller. The buyer is willing to pay a higher price for a water right with a clear title than for a right clouded with uncertainty. Nevertheless, most streams of the world are still not adjudicated because the adjudication process is time-consuming and expensive.

In much of the western United States, water rights often take the form of shares of stock in user-financed companies that developed water supplies with private capital during the late nineteenth century. Constitutionally, all water belongs to the states, but the right to use it is a property right vested in the user through an allocation process. This right of use can be traded within the marketplace, under varying degrees of regulatory control.[13]

Water rights need to be *secure* if water markets are to promote economic efficiency; they need to be enforced and defended from illegal use. Insecure water rights have little private or social value as the following example shows: A buyer who wants to transfer water to a higher-valued use is willing to pay little or nothing for the right to divert the water from a stream if she would be unable to stop illegal use of that diverted water. Without a system of laws accompanied by monitoring and enforcement, there is no way to stop illegal water diversions, and scarce water has little opportunity to gravitate to higher-valued uses and support growing populations or changing demands.

---

[13] Water in most western U.S. states is treated as a private property right that can be marketed, and it is often separated from the land on which it is used. The Colorado system for defining a water right is typical: Water rights are defined by the point of diversion, diversion rate, priority, source, amount of use, and place of use. Diversions are subject to the priority system: first in time, first in right. The earliest diversion from a stream or aquifer put to beneficial use is senior to subsequent (junior) diversions, and more junior water rights are shut off as the stream flow falls. When there is a call on the river, decreed but more junior water rights are shut off to deliver water to the more senior rights.

Controlled supply is needed to establish security. People need water continually, but the source of supply is the hydrologic cycle, which is outside human control. As a result, water must be controlled through storage. In some places, nature has provided storage through snowpack or groundwater; in most instances, however, storage must be developed and managed at considerable cost. Once a water supply is controlled, it can be traded in a market system.

Security also includes a system of laws, government, and administration that has been established to assure the buyer and the seller that their respective rights will be honored and enforced. This involves keeping accurate records of owners and their rights as well as maintaining an accounting and enforcement system to guarantee that owners receive their legally entitled share. Until this kind of system is established and until people have general confidence in its fairness, a market-based transfer program cannot function efficiently. Security also requires adequate infrastructure to transport water supplies to the buyer. In many locations around the world, water rights have little value to places outside the immediate basin until a transportation system is developed to carry the water to the place where it is needed. The value of the right is heavily discounted by the cost to build the canals, pumping plants, and reservoirs.

Markets cannot produce economic efficiency if water or water rights are not *transferable.* Transferability helps create a market for water and efficient allocation of water according to demand and willingness to pay. It promotes flexibility and avoids locking water into current use patterns. Transferability permits scarce supplies to serve growing populations, emerging demands, and changing values. The Colorado–Big Thompson Project, which allows transfer of rights for temporary rentals and for permanent transfers, is a good example. Water users can adjust yearly to meet annually changing needs, and rights can be permanently transferred through the market to meet the changing needs of growing cities, industries, agriculture, and the environment. The only regulatory controls required are that the transfer title be registered and that beneficial use be shown.

Emerging Challenges. For water markets to seriously contribute to meeting growing demands where water supplies are fixed or unstable, a number of policy questions must be resolved:

- Short-term water transfers, such as water banks, may provide an incentive for new water use that would not otherwise take place.
- If cities pay holders of water rights to reduce their water use, some owners may start using as much water as possible to establish a higher baseline level of use.
- Problems may arise from failing to account for the relationship of surface water and groundwater.
- Two-party transfers of water between agriculture and some other user affect third parties, such as local communities and environmental interests. Some institutional mechanism is needed to ensure that all interests are protected.
- Market-based water transfers are likely to work better in places that have extensive conveyance systems and storage facilities and well-coordinated operations. For many other locations, considerable experimentation with better organization and/or development of conveyance and storage facilities may be required to make short-term trading work.

Pricing. Most municipal water in the United States is supplied by public water utilities. These utilities have monopoly status in their service area, subject to regulation by utility commissions. Cost-recovery principles typically determine the pricing structure of these utilities. A widespread belief among utility regulators and the public is that prices should be set so that revenues exactly equal costs. If this belief is enforced by policy, the pricing guideline that emerges is called a cost-based pricing rule. If revenues are too low to cover costs, resulting losses must be subsidized from other sources, for example, general tax revenues or revenues from elsewhere that exceed costs. If revenues are larger than costs, there are profits, which are considered unacceptable for a public utility with monopoly status. The belief that revenues should just cover

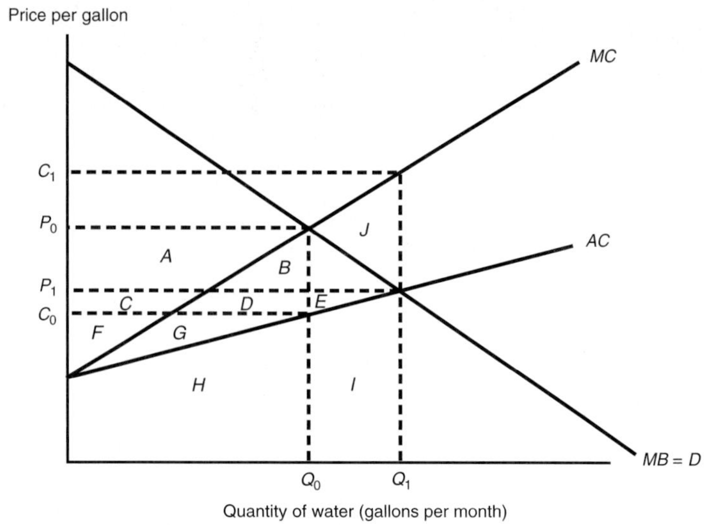

Figure 10–6 Average cost pricing and marginal cost pricing of utility-supplied water.

costs has led to the concept of *average cost pricing.* This means setting a price per unit of water equal to its average cost, which is total costs divided by the total quantity of water delivered. Average cost pricing is desirable for producing total revenues that just cover total costs.

Even though average cost pricing ensures that revenues equal costs, it has undesirable efficiency properties, as shown in Figure 10–6. For utility-supplied water, the figure shows the marginal benefits (*MB*) of delivered water, marginal cost (*MC*), and average cost per gallon (*AC*). Demand (*D*) is downward-sloping: For the first few gallons per month, water for drinking is a necessity and people will pay a high price if required. They will not use water for things like washing unless the price is lower, and use for luxuries (e.g., outdoor landscaping) would require considerably lower prices.

Based on the relationships shown in the figure, pricing water at marginal cost occurs at Q_0 gallons per month. At that quantity, $MB = MC$, and the quantity is supplied at a price of P_0 per gallon. Pricing water where $MB = MC$ is called *marginal cost pricing*; it is economically efficient (as discussed later in the chapter), but problematic: Think of what happens if the utility charges the efficient price, P_0. Total costs of supplying the water is a number equal to area $F + G + H$ that is, average cost times output. But total revenue equals price times quantity, equal to the rectangle $A + B + C + D + F + G + H$, which is larger than total costs. So in marginal cost pricing the utility earns a profit. To avoid profits, utility commissions can require the utility to set price where average cost equals demand, P_1, which is lower than marginal cost. See Example 10–3.

Average cost pricing is economically inefficient because it rewards people who use *high cost water for low-valued uses.*[14] The quantity of water demanded at P_1 is Q_1, which the utility must supply to avoid shortages. When the utility supplies Q_1 units, marginal cost on the last unit exceeds marginal benefit by $C_1 - P_1$. The utility will supply some water ($Q_1 - Q_0$) for which the marginal benefit to buyers is less than the marginal cost. Pricing water at P_1, average cost, encourages economic inefficiency: People use water with a marginal cost of C_1 for a use with a marginal value of P_1. Total loss in economic efficiency produced by average cost pricing is the triangular area *J*, known as a deadweight loss. This triangular area is the economic value of gains that could be had by pricing water at P_0, at which demand and supply are equal at a quantity of Q_0.

---

[14] Imagine a policy that expands supply by taking water from important ecosystem functions, then charges a price low enough to encourage people to use the water to hose down their driveways. The cost is the lost ecosystem function; the benefit is the cleaner driveway.

Example 10–3
Pricing Water for Social Aims in Indonesia

Only a small minority of Indonesia's population has access to piped drinking water. Even for those who have access, water quality is unreliable and often must be boiled before it is safe to drink. Forty percent is lost to leakage, and supply interruptions are common nationwide. Piped water is supplied by the government monopoly PDAM (Perusahaan Daerah Air Minum); typically costs are much higher than revenues.

No matter who runs the water supply system, substantial investments will be needed to bring about major increases in piped drinking water delivered to the masses. One of the bottlenecks to providing more access may be the price of water. Government leaders want to keep prices low enough to be affordable to as many households as possible. Although low prices sound politically and economically attactive, inexpensive water produces low revenues for water suppliers, so the investment opportunity for the private sector is poor.

PDAM has experimented with one innovation: increasing block pricing, a form of marginal cost pricing in which the marginal price of water varies according to the level of demand. The first *X* units of demand (lowest block) are priced at the lowest level, then the next *Y* units are priced at the second lowest level, and so on. (Increasing block pricing only works when the individual consumer can be identified and when resale from one consumer to another is expensive. Both conditions occur for piped water.) This two-part pricing has considerable potential to encourage economically efficient private development and to provide adequate revenue to keep it financially sustainable. The two parts of the price could consist of (1) one part equal to the marginal cost of production combined with (2) a fixed access charge independent of use level. Rietveld, Rouwendal, and Zwart estimated demand functions for water in Indonesia. Their results point to a large price elasticity and a small income elasticity. The average price elasticity is about −1.17 and falls with higher income levels, so as income levels increase people are less sensitive to water price increases. Their work shows that the existing block rate structure does little to help the poor but does favor small households, many of which are rich. The authors see two main paths for improving the water system: (1) increasing access to water by reducing installation costs and (2) reinvestment of PDAM profits into enlarging the water supply network to reduce the number of households without connections.

Source: Rietveld, P., J. Rouwendal, and B. Zwart. 2000. Block rate pricing of water in Indonesia: An analysis of welfare effects. *Bulletin of Indonesian Economic Studies*. 36 (3): 73–92.

Regulations

Government can play a role in influencing water allocations and affecting economic efficiency by regulating various dimensions of water use. Where a market failure blocks efficient water use patterns, regulations can be a means to implement policies that control water use levels, quality or pollution levels, timing of use, place of use, and transfers. Supported by an underlying legal framework, government regulations require, permit, or restrict particular activities or prescribe specific results in connection with water use. Such actions are economically efficient, only if the social benefits of taking the action outweigh the costs.

Because regulatory instruments, sometimes called command-and-control regulations, do not rely on voluntary cooperation, they require public acceptance to ensure compliance. They also require enforcement ability and a general belief that the regulations improve the social welfare.[15] Compliance with regulations is determined by a range of factors, all of which affect economic incentives to comply; these factors include level of enforcement, expected penalty if caught, and likelihood of being caught and convicted of a violation. Where standards are strict, the costs imposed on water users to achieve them may considerably exceed the benefits and noncompliance may become common.

[15] Compliance also requires the belief that the regulations will improve the welfare of the individual who is required to comply. For example, many agricultural users believe that groundwater pumping regulations will improve society's welfare but will damage their own welfare.

One controversial regulatory instrument is water transfer restriction. Recall that efficient water allocation requires that the marginal net benefits are equal over all uses, including uses that produce environmental benefits. If water rights are well defined, transferable, and enforced, this efficiency will occur when water rights are transfered from low-valued uses to higher-valued uses. Water right owners who have a low marginal net benefit from current uses will transfer their rights to people who can put the water to a higher-valued use. The buyer is better off because his payment for the water is less than the net benefits produced by his new use of the water. And the payment the seller receives for the water exceeds the income currently produced by the water. The transfer makes both people better off.

If water or water rights are legally transferable to meet the needs of a growing propulation and changing demands, marginal benefits can be made equal if there are no restrictions on transfers. However, when restrictive regulations limit transfers, the power of market forces to equalize marginal benefits is constrained.[16] One such restriction is the "use it or lose it" element of the prior appropriation doctrine, which requires continued beneficial use. Some owners believe their water rights will be lost if they invest in conservation and offer to sell or rent the conserved surplus. This barrier to a secure property right impedes water conservation acts, such as water transfers from agriculture, and reduces efficiency by considerably raising the cost of new water supplies to growing cities. Legislative bodies could remove this barrier to transfers by promoting greater security of tenure of water rights. One possibility would be to enact laws that assign a right to transfer any water saved through conservation. Another would be to enact laws that declare short-term water transfers from agriculture to cities to be a beneficial use of water. Either of these legislative acts could provide a profit incentive for farmers to conserve water in exchange for cash. During a statewide drought in the early 1990s, California's Water Banking Law included legal guarantees that depositing water into a water bank did not risk damaging farmers' permanent water rights. The legislation was cheap, easy, and economically efficient because it increased farmers' security of tenure.

Blocking transfers from agriculture to growing cities and to environmental demands is inefficient. For example, much water in the western United States is locked into irrigated agriculture; so producers must irrigate low-valued crops when they would prefer to transfer some of that water to a thirsty urban area or to an environmental buyer with cash. Locking large amounts of water into agriculture effectively suppresses its economic value to low levels, often less than \$20 per acre-foot (Ward and Michelsen, 2002, summarize methods for valuing water in agriculture), when there are growing cities that are willing to pay prices exceeding \$300 per acre-foot for the same water.

Transfering water from agriculture to other uses potentially can raise net benefits, but many institutional restrictions limit transfers. The inefficiencies that result from blocking water transfers show up as increased costs:

- Farmers, many of whom would like to increase their income by saving water and selling unused water to a buyer, lose a market.
- Cities and environmental groups, two of the largest potential buyers, are unable to get inexpensive water locally and must seek more expensive sources, such as piping in from great distances or building new projects.
- People who would have benefited indirectly from the buyers' cheaper water (e.g., businesses that pay a water rate) must pass on the higher costs to their customers.

---

[16] Regulations on transfers block the equalization of private marginal net benefits, but they can facilitate efficiency by requiring water traders to account for unpriced effects to third parties. For example, if a water right transfer between a willing buyer and seller is blocked because the transfer would reduce instream flows needed for endangered species habitat, the restriction may promote efficiency.

Institutional Designs for Economically Efficient Water Use

One way to promote economically efficient water use is to establish institutions that confront all water users with the *real* costs of their actions. From an individual's view, each water user implements only those actions for which her own benefits exceed her costs. When a water user faces the real costs of her actions, she considers benefits displaced from other uses. Then, acting in her own self-interest, she implements decisions for which society's total benefits exceed its total social costs. Efficient water supply requires clear price signals that provide incentives for efficient use of water by individual consumers. One method for promoting that efficiency is marginal cost pricing in which consumers pay a price that reflects the incremental cost of their use on the system. Several other methods are also available to promote economically efficient water use.

Marginal Cost Pricing

Marginal cost pricing is an important principle for understanding the efficient management of water supply systems. Setting the price of water equal to marginal cost means that each customer pays a price equal to the incremental costs they impose on the system for providing that service. This price provides valuable information to water buyers and provides a basis for making informed decisions regarding water use. Several widely practiced ways of implementing marginal cost pricing are discussed next. Each of these methods results in consumers paying the marginal costs they impose on the system. Consumers can control their payment of these costs by changing their behavior. For example, they can decide to reduce or increase water use, emit more or less effluent, and connect or disconnect from the system. See Example 10–4.

Connection Charges. Connection charges are imposed for installing water distribution systems in new residential or industrial developments. They usually take the form of lot charges, frontage fees, or development charges. These costs that developers pay, combined with water prices facing the ultimate water user, provide information on costs that new developments impose on water distribution systems. Pricing the new water use at its full marginal cost results in efficient growth in system size. Prices based on full marginal costs may limit urban sprawl and limit expensive self-supplied connections in rural areas.

Long-Run Pricing. Long-run marginal cost pricing uses estimates of future capital costs to calculate water rates, rather than historical or embedded costs.[17] Historical costs are sunk costs; that is costs cannot be altered by changing current behavior. Future capital costs related to system expansion can be changed by increasing or decreasing water demands, especially advancing or delaying capacity expansion.

---

[17] In embedded cost-rate design, prices are set so that water buyers pay for a regulated water utility's costs so that total revenues just equal total costs. One way to implement embedded cost rates is for the utility to use a two-part tariff: a fixed charge that when collected from all customers equals total fixed costs of operation, and a commodity charge set to the utility's average variable cost. Then total revenue equals total cost for any output as long as the utility is operating below the system capacity. If a fixed charge is levied such that this charge times the number of customers equals the total fixed costs of the utility, then the commodity charge can be set equal to the average variable cost, and total revenue will equal total cost as long as the quantity demanded at the price equal to average variable cost is less than the deliverable supply of water. Revenue is stable: For any output below the system maximum, net revenue equals zero. If demand grows beyond system capacity or if a drought reduces available capacity below the quantity demanded at the regulated price, there is a water shortage. In the short run, *embedded cost-rate design has no solution to this problem,* and water use must be curtailed. In the long run, the solution based on embedded cost rate design is to find new water sources and/or expand system capacity to deliver the existing water.

Example 10–4
Efficient Water Pricing in Zimbabwe

Since Zimbabwe's independence in 1980, agriculture has provided many jobs and is a major contributor of raw materials to domestic food processing and textiles. Yet, water is scarce and poorly distributed. Even where water is more abundant, it has many competing uses and is allocated to its final use point at considerable cost. Efficient allocation of any scarce resource occurs when each consumer's marginal willingness to pay equals the marginal cost of supplying to that consumer. Efficiently coping with Zimbabwe's water scarcity requires knowing something about water's opportunity cost when setting price. And efficient pricing is crucial to the nation's future development; water cannot be too cheap or too expensive. Efficient prices guarantee that people who invest in water supply (public or private) will realize adequate revenues with which to operate, maintain, improve, and extend the system.

Correct calculation of the marginal cost of delivered water supply in Zimbabwe requires accounting two factors: the impact of water supply development on the environment and elements of the nonrenewable use of Zimbabwe's depletable aquifers. Efficient water pricing considers the future costs of water supply in current prices charged to water users as well as efficient sequencing investments in future supply expansions.

Economic theory shows that the demand for irrigation water will be influenced by the value of crops it produces. The profit-maximizing farm continues to expand its water use whenever the water's marginal value to the farm is larger than its marginal cost. The theory of diminishing marginal productivity of water indicates that the profit-maximizing level of water used by the farm will be less than the level that maximizes crop yield, unless water is free. (Only when added use can be had at a price of zero will the farm use it until its marginal productivity is zero—total yield is maximized.) The farm's demand for irrigation water is derived from the value of its use in crop production. Although policymakers in Zimbabwe are well aware of efficiency gains from marginal cost pricing of irrigation water, they know that application of this pricing risks jeopardizing many farmers' livelihoods. So policymakers face hard choices between the need for equity and efficiency in deciding on a price for water.

Sources: Nyoni, M. J. M. 1999. Water pricing issues in a developing country: Case of Zimbabwe. *International Journal of Social Economics.* 26 (1-2-3): 441–454; Hall, D. C. 2000. Public choice and water rate design. In *The Political Economy of Water Pricing Reforms,* edited by A. Dinar. Oxford, UK: Oxford University Press.

Externality Charges. Waste treatment systems are sized and designed to treat organic wastes. Some individual users, often industries, discharge wastes that are high in volumes and/or strengths; these users impose added costs on the utility. Emissions of extra strength waste impose higher costs on treatment systems than waste of typical residential users. Externality charges could be levied on industrial wastes that exceed existing purification capacity. This method is often used by large cities that treat large volumes of industrial wastewater.

Geographically Differentiated Rates. Geography can play a role in the cost of water delivery. Customers located at higher elevations, for example, impose significant added pumping costs compared with those on level ground. Also, service customers at the edges of a system produce higher costs than other customers. In principle, each of these factors influences water rates based on marginal costs.

Water Markets. Water markets were described earlier in this chapter as important mechanisms for promoting voluntary water transfers. But water markets also are a way to implement marginal cost pricing: The market price a buyer is willing to pay to a potential seller sends an important signal: That signal is the marginal cost of continuing the existing use of water. The price asked for by the seller reflects the marginal cost of existing use displaced. In either case, the market price incentive encourages the resource to move to its highest-valued use, thereby producing economic efficiency as long as all costs are incorporated in the market price.

Where water markets function efficiently, the market price of water can inform the marginal cost of the last units in current use. Therefore, efficient water markets can serve as a form of marginal cost pricing.[18] Nevertheless, there are disadvantages to water markets such as the high cost of measuring and defining property rights with variable and random stream flows and the cost of enforcing rules on owners. One of the biggest problems related to development of water markets is the high cost of initiating the transition from nonmarket water allocation mechanisms to a market environment.

Peak-Load Pricing. Peak-load pricing refers to the practice of charging higher prices for water at times of peak demand. Its rationale is that much of the capital expenditure on water utilities stems from meeting peak demands, particularly summer demands. Peak-load pricing can promote economic efficiency when customers know the added cost of their use in peak periods. That price information can help reduce demands in peak periods, and consequently lower capacity expansion costs over time.

For most water utilities, peak water use occurs in summer when landscapes are watered, so capacity expansion costs are determined by that summer peaking. That means if peaking requirements are lower, smaller water systems can be built. One efficiency-based allocation of peak costs is to treat all summer users as contributing to the summer peak and to charge them the difference between summer and winter use, typically a higher price that reflects *added* system cost. Peak-load pricing can also be an efficient way to promote water conservation, since it provides incentives for reduced water demands during the period in which water supplies are most likely to be constrained. Because smaller users (e.g., residential users) are responsible for much of the peaking capacity, they contribute to higher costs; if marginal cost pricing is practiced, they will pay higher rates per unit of use than larger users who are less responsible for the peaks.

With growing peak demands, many systems face the prospect of a finite water supply or limited treatment capacity, both of which require significant capital investment. High costs of capacity expansion programs can be magnified by the need to develop new storage, production, or supply. An efficient decision to expand capacity requires a comparison between expected benefits and costs of the new facilities. Marginal cost pricing can significantly contribute to water conservation programs by informing managers of a conservation program's costs and benefits. It may be efficient to delay new facilities through a price signal to consumers in peak use periods. The benefits produced by delayed construction can be compared with the cost of consumer surplus lost from reduced use instituted by the surcharge. When implemented efficiently, a conservation surcharge sends the correct signals to consumers regarding the cost consequences of expanding existing water use.

Here is a way to calculate the *net benefits of delaying a system's capacity expansion.* Determine dates for construction of a capacity expansion program based on peak demand projections *with* and *without* the need to expand capacity to meet growing peak demand. Then calculate the net present value of the cost of the capacity expansion program (NPV_c) for each of the two demand projections ($NPV_{c\,with}$ and $NPV_{c\,without}$). The difference for the two cases, ΔNPV_c, is the benefit of deferring expansion. The cost of deferred expansion is the surplus lost by the conservation surcharge required to limit demand. The following formula summarizes the overall NPV of the program:

$$DNB = \text{Benefit} - \text{Cost} = \Delta NPV_c - \Delta NPV_{cs}$$

where DNB is the discounted net benefit from delaying expansion through a conservation surcharge; ΔNPV_c is benefit, equal to the savings in net present value of capital costs by delaying capacity expansion; and ΔNPV_{cs} is cost, equal to the present value of consumer surplus displaced by the higher price. Economic efficiency is improved by using peak-load pricing to delay capacity expansion if the equation shows positive net benefits. This conclusion assumes that efficiency

[18] This assumes that the market price reflects the opportunity cost of continuing to use water in its current use. If the price is an administered price (e.g., set by a rule or by a bureaucracy, rather than a price set by market forces of supply and demand) the price provides little information about costs of current use. For example, a common practice in U.S. irrigated agriculture is to pay a price for federal reclamation water of $10 to $50 per acre-foot, but producers are limited in their quantities demanded. This price says nothing about the opportunity cost of continuing to use water in agriculture, which could be as high as $200 to $300 per acre-foot if a nearby city needs the water.

Figure 10–7 Water conservation pricing

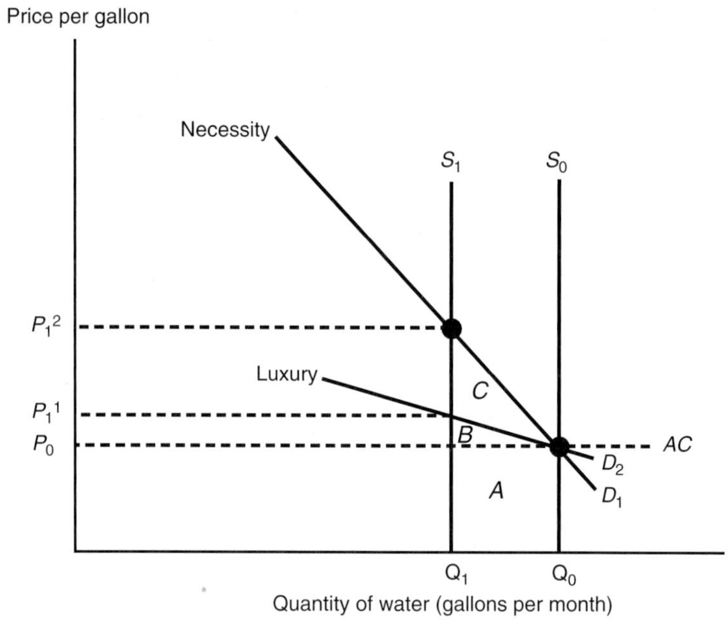

is an acceptable evaluation standard which is a reasonable assumption if only discretionary water use is affected. When the surcharge affects required uses (e.g., for drinking or sanitation), equity is often a more politically acceptable standard than efficiency.

Figure 10–7 shows that if the peak demand for water is highly price inelastic (D_1), the conservation surcharges imposes a high cost in consumer surplus lost (area $B + C$). (Area A is also lost to consumer but it is gained back by the utility from reduced operation costs.) The high cost of lost consumer surplus reduces the *net* benefits of deferring capacity expansion through the conservation surcharge. If water's demand is price elastic (D_2), the surcharge imposes a smaller cost in consumer surplus displaced (area B), which increases *net* benefits of the deferral defined by the equation. Lower price elasticities of demand for water are typically associated with discretionary uses, that is, uses for which nonwater resources (e.g., land, labor, and capital) can be subsituted for water.

When the benefits of delaying capacity expansion (capacity costs saved) exceed the costs (consumer surplus lost), the conservation surcharge provides an efficient price signal.[19] That is, the discounted consumer surplus lost is smaller than the discounted costs saved by delaying expansion. The conservation surcharge can be incorporated into a variety of rate designs based on either embedded or marginal cost. Revenues from a conservation surcharge may be placed in an account to offset future costs incurred by the utility either in adding capacity or in implementing conservation programs. To apply a conservation surcharge successfully, utilities must be able to estimate the marginal (discretionary) use for a variety of customer classes. In addition, implementation may require changes in the customer billing system and in metering of water use.

Drought Pricing. Drought pricing is a form of marginal cost pricing in which water is priced to reflect its real scarcity while producing revenues that meet a desired revenue target. Figure 10–8 shows that under drought conditions water supply falls from normal levels (S_N), where average cost (AC) equals demand (D), to the lower quantity Q_D shown by supply S_D. A de-

[19] When the benefits of the capacity delay are less than the costs, the surcharge also provides a price signal, but it's an inefficient signal. That is, it signals the need to reduce consumption when it is more efficient to expand capacity. Also, remember peak demands grow over time, so consumer surplus losses from delaying capacity expansions continue to grow. At some point, consumer surplus losses from higher prices are too high and it is more efficient to expand. Higher (lower) population growth rates reduce (increase) the time for which capacity expansion is the more efficient choice.

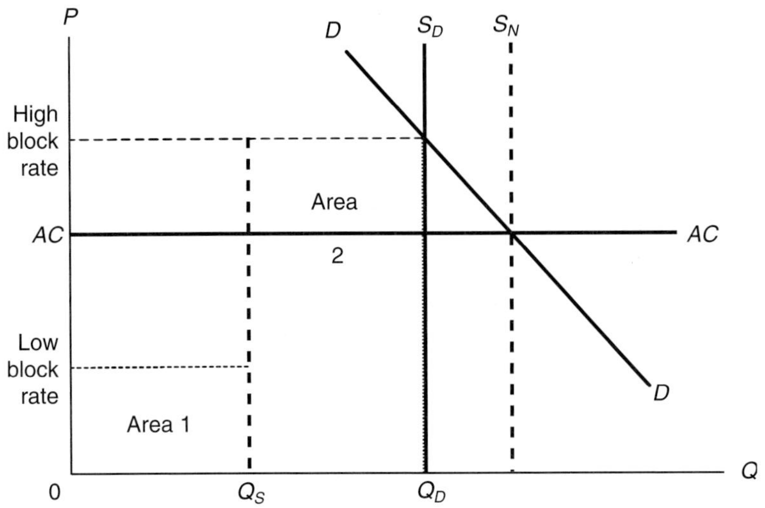

Figure 10–8 Marginal cost pricing in a drought

sign based on embedded cost rate cannot cope with this unexpected water supply shortage: The fact that water is priced at average cost produces a politically dangerous revenue shortfall. Hall (2000) describes the political furor that occurred in Los Angeles in 1991 when city water customers had their rates increased because they voluntarily saved 20% and revenues fell by 20% during a year when there was a shortage of system capacity.

The intractable problem produced by embedded cost rate design supports the argument that marginal cost rates can be used to produce the target revenues in years of normal water supply as well as adjust to drought in periods of water shortages. The idea, as described by Hall (2000), is to set up a system that automatically increases prices on marginal uses so that revenues are constant even though the system's water capacity falls. Figure 10–8 shows that the *high block rate* (the rate charged from the block of Q_S to Q_D) is set to the price at which the shortage supply, S_D, intersects demand; a *low block rate* (the rate charged from 0 to Q_S) is determined by political action to be fair. The high block rate is set above average cost to the market clearing price, is applied to Q_D minus Q_S units, and produces revenue equal to area 2 in the figure. The politically determined, low block rate produces total revenue equal to area 1.

As the drought worsens, total system capacity falls to a vertical supply schedule smaller than S_D (not shown); the high block price must be increased to reflect the new supply–demand intersection to signal increasing scarcity. The increased high block price encourages more conservation. If the price elasticity of demand is larger (smaller) in absolute value than 1.0, total revenue for the high-priced water in area 2 falls (rises) with a higher price. The switch point, Q_S, needs to be reduced (increased) so that area 1 revenue moves in opposite direction but in equal magnitude to compensate for the change in area 2. The change in area 2 is jointly determined by the extent of the drought and by price elasticity. So as the drought is intensified, the high block price increases and the switch point Q_S is adjusted to keep total revenue summed over the two user classes constant. The rate charged for the high block is a true marginal cost because it can reflect the marginal benefit from existing demands that is displaced by the last increment sold to customers (see problem 3 on drought pricing).

Two-Tiered Pricing

A special kind of marginal cost pricing, two-tiered pricing, combines elements of efficiency, equity, and sustainability. Begin by recalling the classic tension between equity and efficiency in water pricing debates. Economic efficiency in water use requires that the price charged on the last unit used equals the marginal cost of supply. Despite the fact that marginal cost of supply may be higher than many poor people can afford to pay, there are both efficiency and financial sustainability gains that results from pricing water at its marginal cost of supply.

Figure 10–9 Pricing water for efficiency, equity, and sustainability

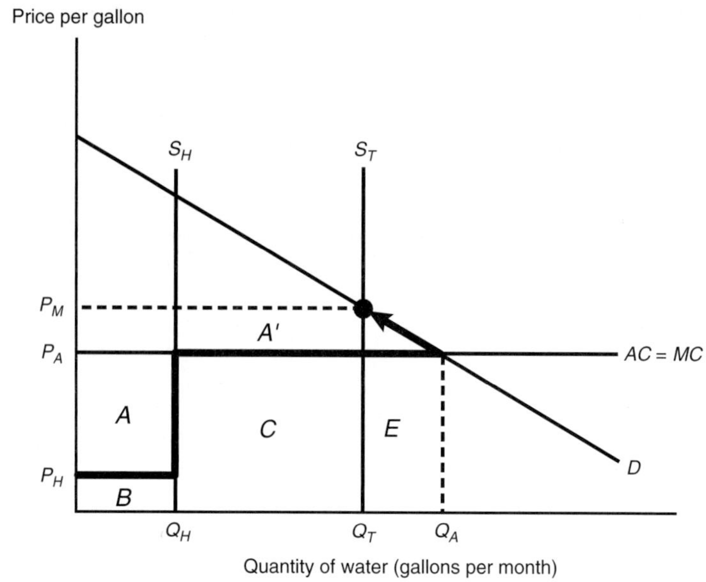

Pricing water at a low or zero price, which may be attractive on equity grounds, makes it difficult for the utility to pay for its costs of operation, and for that reason such pricing puts the long-run sustainable water supply at risk. Still, many debate the hypothesis that total cost recovery adds more to the costs of hardship than it promotes in benefits of sustainable development. In many poor areas of the world where privatization of water has been implemented, millions of people have been cut off because they cannot afford to pay water bills that often are more than 30% of their incomes. Some water cutoffs, for example in parts of South Africa in the mid-1990s, forced poor people to seek water from polluted rivers and lakes. That lead to outbreaks of diseases like cholera in which thousands of people were sickened and hundreds died. South Africa's government spent millions of dollars to control the spread of disease and to truck clean water to stricken areas.

Figure 10–9 shows that if long-run supply expansion operates under constant returns to scale, then average cost (*AC*) equals marginal cost (*MC*). Economic efficiency suggests pricing water at $MC = P_A$. Q_A is supplied and used, and total revenue just equals total costs. The World Health Organization (2000) determined that 50 to 60 liters of water per person per day (about 1,500 gallons per month for a household of four) be considered an acceptable quantity based on human rights, shown as Q_H.

Unless policymakers are careful, a policy that guarantees a minimum quantity, Q_H, be supplied at or near a zero price, P_H, produces a financial difficulty. The low price P_H charged on the human rights requirement (quantity Q_H) combined with a price of P_A charged to the remainder $(Q_T - Q_H)$ causes total revenues to fall short of total cost by an amount equal to area *A*. Total costs for the quantity Q_T are areas $A + B + C + E$, while total revenues are only areas $B + C + E$, since the price P_A is charged only on water use exceeding A_H. The financial loss, area *A*, threatens the financial viability of the water utility and therefore risks undermining its sustainability. Governments are understandably cautious about implementing policies of free water.

One way to deal with the conflict between equity and efficiency–sustainability is to set up a two-tiered pricing system in which the price on all water use in excess of Q_H (the minimum acceptable quantity based on human rights) is raised to a level higher than average cost. How high should that price be? It should be high enough to make up for the losses of area *A*, if long-run financial viability is important. The price that ensures the financial viability of the water utility is P_M, which produces revenues in excess of average costs equal to area A′. At price P_M, revenues lost on subsidized water, *A*, exactly equal excess revenues *A*′ on discretionary water use $(Q_T - Q_H)$. At prices higher than P_M, there are profits; at prices lower than P_M, there are losses.

A two-tiered water pricing system has three advantages:

1. It meets the standard of equity by supporting human rights: Nobody's health suffers from bad water or water they cannot afford.
2. It meets the standard of economic efficiency: By charging a price approximately equal to the marginal cost for all use levels exceeding basic needs, price sends a signal of the real cost of expanding system capacity.[20]
3. It meets the standard of financial sustainability: By producing revenues adequate to cover costs, the utility and its water supply can last for future years.

Integrated Demand/Supply Management

A method called integrated demand/supply water planning and management has considerable potential to improve efficiency of water use. In this method, water supplies and water demands are managed jointly and over long time periods in the continued search for least-cost methods to avoid shortages. Basinwide optimization models have been as a way to accomplish integrated management (e.g., Booker, 1995; Lee and Howitt, 1996). Practically integrated management means linking gradually phased-in development of new water supply to pricing incentives to reduce shortages caused by growing demands.

Through conservation and water recycling programs, utilities can stabilize and sometimes reduce demands to earlier levels. In some cases demands can be reduced despite increased populations, thereby reducing or eliminating needs to implement expensive new supply plans such as developing reservoirs, purchasing farmlands, or building new pipelines. By tying new supply to demands that have been reduced to lower levels, innovative utilities rely on proven conservation capabilities of its customers. Utilities can also avoid incurring the finanical and environmental costs of new supply until such development can no longer be avoided efficiently.

Transboundary Management of River Basins

One way of reducing the cost of water management, as well as increasing the size and scope of its benefits, is for two or more nations to jointly develop, finance, manage, and use common rivers. El ements of responsibility are assigned to each nation based on its comparative advantage, and all nations gain from trade. In the Columbia River Treaty between the United States and Canada, signed in 1961, U.S. money paid for three Canadian dams (Mica, Keeleyside, and Duncan) as well as Montana's Libby Dam, whose reservoir extends into Canada. The United States gained flood control and power benefits from these dams. The U.S. Senate ratified the treaty immediately but the Canadian Parliament refused to ratify the treaty until the United States agreed to purchase excess power generated in British Columbia, power for which the province had no use. See Example 10–5.

This treaty is an excellent example of a coordinated plan that manages the Columbia River Basin as a transational system for mutual benefit. The storage dams built in Canada meant that downstream users were no longer dependent on seasonal river flows. The storage dams ensured the necessary amount of water would be in the river to meet hydroelectric demands regardless of season within the basin and beyond its borders. The United States had a comparative advantage in money and engineering expertise, but suffered from floods. Canada had the comparative advantage in endowments of water but shortages of money. Both nations got what they needed most at a lower cost than either nation could have financed on its own.

---

[20] To see why marginal cost pricing sends an efficient signal, consider the alternatives. Suppose price is less than marginal cost. People will not conserve even when conservation is a cheaper way to cover shortages than system expansion. Now suppose that price is greater than marginal cost. People invest in conservation when supply expansion is a cheaper way to cover shortages. Consider the effects of price being equal to marginal cost. People invest in conservation only when conservation is cheaper than system expansion, and they avoid investing in conservation only when conservation is more expensive than system expansion.

Example 10–5
Dividing the Waters: The Interstate Compact

Economics deals with allocating scarce resources among competing demands. A classic economic problem is faced by people who live in arid regions in which the reliable stream flow is often fully used or even overused (i.e., total claims for the right to use the water exceed total reliable supplies). Transboundary rivers pose special economic and political challenges in resource allocation. Designing an efficient, fair, and sustainable method to allocate scarce and random supplies that meets the needs of all parties is a major challenge, particularly if two or more countries share the same water.

One way to allocate supplies that has been tried with some success is the water sharing agreement. In the United States, for waters extending beyond borders of one single state, interstate compacts have been a common sharing method. An interstate compact is a negotiated agreement among the states which, once ratified by Congress, becomes both a federal law and a contract among the signing states. Twenty-two such compacts currently divide the water of rivers in the western United States.

Under the *storage allocation* method of allocating water, limits are placed on the amount of water that can be stored by the upstream region (state). The downstream region bears the risk of drought, because it receives only what is left over after the upstream region reaches its upper storage allowance. Unless the storage limit is small, the existing quantity of stored water in a dry year is likely to be small, and the upstream region will capture all the flow above the reservoir to raise the volume to the storage limit. No flow is left for the downstream region beyond what is produced below the reservoir. Nevertheless, storage allocation can be attractive to both the upstream and downstream regions. If the upstream region has a storage reservoir, it benefits by being allowed to store flood waters that would have otherwise gone unused. The downstream region benefits by knowing that at least a part of the flow will come downstream for its use, which is especially beneficial in sequential wet years.

A river's annual flow is the total quantity of water per year from all sources reaching the channel. The *flow allocation* method involves dividing up that flow among the regions through which the river passes. Flow allocation deals only with the total flows produced in a period. It is independent of the number or capacity of reservoirs in the system. Several methods of dividing flow have been used; typically they reflect a basin's historical use patterns as well as each state's varying political or economic power. Selection of a method usually depends on the allocation of shortage risks. In basins with highly variable interyear supplies (e.g., the Colorado, Rio Grande, and Nile rivers), each region is willing to take a reduced average flow percentage in exchange for protection against receiving nearly nothing in dry years. Flow allocations can be based on variable proportions of total flow, fixed proportions, fixed allocations, and guaranteed minimum flows. For example, a region with considerable reservoir storage capacity may be willing and able to bear the risk of nearly zero flow in a dry year in exchange for the right to capture high flows in a wet year. A region with little or no storage capacity may wish to require a guaranteed minimum flow, even if it means accepting a low percentage of average total flows and/or giving away all wet year flows to the region with more storage capacity.

Sources: Bennett, L. L., C. W. Howe, and J. Shope. 2000. The interstate river compact as a water allocation mechanism: Efficiency aspects. *American Journal of Agricultural Economics.* 82 (4): 1006–1015; McCormick, Z. L. 1994. Interstate water allocation compacts in the Western United States—some suggestions. *Water Resources Bulletin.* 30 (3): 385–395.

Virtual Water

Virtual water is based on the idea that reducing the demand for water saves just as much water as developing added supplies, and may do so at a lower cost (Allan, 2003; Ioris, Merrett and Allan, 2004). Virtual water development polices look for ways to reduce the water used by commodities that are needed, but can no longer be produced because of a drought or other shortage. An example is increasing the imports of foods or other products because domestic production consumes domestic water.

Virtual water is controversial, as policymakers and hydrologists often balk at the notion that governments should import water-using food crops instead of growing them. Food imports are politically sensitive, even more so than energy imports. People can live without oil but not without food. A danger of virtual water is that importing nations would be vulnerable if food-exporting countries cut subsidies to their farmers, potentially leading to significantly higher prices. Still, many dry regions face water crises and actions need to be taken. So producing virtual water by importing cereal grains that would otherwise be irrigated, could be an economically efficient measure for dealing with shortages.

Managing Links between Poverty and Water

Poor water services have doubly adverse effects on the poor, because poor-quality water promotes the spread of disease, which is magnified in crowded and/or low-income areas. Water scarcity and degraded water quality affect health, food supply, and human living conditions. The poor are particularly vulnerable when water is unclean, scarce, or unreliable. For example, Asia is home to nearly 900 million of the world's poorest people; getting access to adequate clean water takes a large percentage of their energy and income. For many, finding water for their crops is a life-threatening process. Difficulties encountered in accessing water frequently worsen the level of poverty (Asian Development Bank, 2001). Water pollution in many third-world countries increases the incidence of disease. The urban poor spend a large part of their scarce income to buy water from private vendors. Clean running water is unavailable to a high percentage of poor people.

Fetching water for drinking and cooking remains an exhausting and time-consuming task in may rural Asian areas. For example in rural Nepal and parts of the mountainous regions of the Mekong region of southeast Asia, the Asian Development Bank reports that many village people, mainly women and girls, still walk several miles over dangerous ground to fetch water. Employment and schooling are often foregone in the effort to find and fetch water. A World Bank Report (2003) states that the urban poor in densely populated towns and cities in Bangladesh, India, and Nepal line up at public places to access limited quantities of poor-quality water. In India, the poor are often confined to consumption levels below 15 quarts per capita per day for all uses. Because running water is rarely available to them, the poor use community drains in place of toilets. The environmental hazards of poor sanitary practices affect the urban poor more severely than others. The inequity is harsh: The poor have less time to spend on productive work, become sick more often, and spend more on getting well. See Example 10–6.

Where water-services funding from the central government is scarce, more resources may be found in local affected communities. Efforts by researchers could be productively invested to determine the kind and extent of water services the poor want. Research has found the poor, both on farms and in cities, are willing to pay for reliable, high-quality water service. In fact, the poor often pay more for less and lower-quality water, which in many urban third-world countries they receive from street vendors.

Water suppliers that have a financial stake in serving the poor are more likely to provide them with better quality and more sustainable services. Financing services to the poor can be achieved by various measures through which those who are better off subsidize the poor, and through budget transfers to subsidize connections. However, subsidies should be used with caution. Attaching nonmarket objectives to a public enterprise may damage the quality of its services, thereby risking the initiation or expansion of a cycle of poor service and low revenues. Also, policies that alter the ownership of water rights should be approached carefully to guard against harming the poor, since historical and informal water rights may be a major source of household income.

Decentralized Management

Many governments face scarce financial and administrative resources, so they need to choose carefully the responsibilities they take on for water resources. One operating principle is that nothing should be done at a higher level of government that can be done effectively at a lower level. Thus, if local or private capabilities exist and an appropriate regulatory system can be established, central government should take this action: Decentralize responsibilities to local governments and transfer delivery of water services to the private sector and to community organizations such as water user associations. Governments can privatize public water service agencies or encourage their transformation into financially independent organizations through

Example 10–6
Will the Poor Pay a Premium for Safer Drinking Water?

About half the people who live in the developing world suffer from one or more of six diseases associated with water supply and sanitation: (1) diarrhea, caused by pathogens in food and water, (2) ascaris, (3) dracunculiasis, (4) hookworm, (5) schistosomiasis, and (6) Trachoma. An estimated 400 children per hour younger than the age of 5 die in the developing world from water-related diseases. Adding to these grim facts is the economic losses and health costs suffered by an estimated 10 million person–years of time and energy. These costs are incurred by women and girls who carry water from distant, often polluted sources. Longer lives, reduced infant mortality, better health and productivity, and increased standards of living result from improved economic development. People who live in the developing world generally fare badly on all these indicators compared with those who live in industrialized countries. Availability of adequate and safe water for in-home use and adequate sanitation to dispose of wastes are essential to the development process.

Ahmad (2000) found that water prices charged to the poor are enormous in Yemen, Pakistan, Jordan, and several other Middle Eastern countries. In Yemen the burden of inadequate water supplies on the poor occurs in both rural and urban areas. In rural areas less than half of households have access to safe water. The supply provided by National Water and Sanitation Agency (NWASA) is priced low, but it is unreliable and often denied to the poor; new piped water connections often go first to the rich and political well connected. The price charged by public utilities is only $0.04 per cubic meter ($0.16 per 1,000 gallons), but because of inadequate public supply, the poor are forced to buy most of their water from private street vendors, who charge as much as $0.39 per cubic meter ($1.56 per 1,000 gallons). This high price for privately supplied water produces very low per capita consumption.

Ahmad (2000) also found that poor sections of the population have few choices other than to buy from vendors; typically they spend between 5% and 15% of their income on water. Ahmad reached several conclusions that have policy implications:

- Safe water is expensive.
- The urban poor are willing to pay a high price for safe water, even when that price represents a sizable part of their budgets.
- Safe water produces high benefits in health, longer lives, and greater productivity.
- Even the poorest of the poor consumers are often willing to pay a price that reflects the real cost of water as long as they receive a more reliable supply than that delivered by the public sector.
- Private markets often provide reliable supplies of water to the poor, but at a high price.
- The fact that the private sector charges a high price is an indication that the government could supply more reliable services than they currently provide and could do so at minimal financial loss.

Water policy experiments, such as two-tiered pricing described in this chapter, have the potential to promote efficiency, equity, and sustainability, and in so doing, greatly improve health and human welfare in the developing world.

Sources: Gadgil, A. 1998. Drinking water in developing countries. *Annual Review of Energy and Environment.* 23: 253–86; Ahmad, M. 2000. Water pricing and markets in the Near East: Policy issues and options. *Water Policy.* 2: 229–242.

measures such as use of management contracts for service delivery. Decentralization should include arrangements for ensuring performance accountability and for establishing effective regulatory frameworks to encourage environmental protection standards.

These steps toward decentralization should improve incentives for cost recovery and service provision and give users a sense of ownership and participation. In regions or nations where municipal capabilities are unable to manage a complex water resource system, efficient and sustainable government policies could support training and capacity building to improve local management, so that decentralization can ultimately be achieved.

Politically Workable Marginal Cost Pricing

We have seen that pricing water at marginal cost encourages economic efficiency by sending the right conservation signals to the water user. But marginal cost pricing for water can be politically dangerous unless managed carefully. Careless implementation of marginal cost pricing risks pricing a basic human need at so expensive a price that the poor cannot afford their water requirements without shouldering a politically unacceptable burden. In fact, before 1990 only one major American city, Tucson, Arizona, adopted marginal cost rates for water. This occurred after the two-year drought of 1976–1977. One year after adoption of those rates, voters held a recall election over the water rates and voted the entire city council out of office. Similarly, after the six-year California drought of 1986–1991, the mayor of Los Angeles appointed a committee to study water rates (Hall, 2000). Committee members were given Tucson as an example of the political danger of marginal cost rates for water. After much analysis and experimentation supported by constant public participation, the committee designed "politically workable" marginal cost water rates that have stayed intact for the most part since their inception. Politically workable marginal cost pricing is based on this idea: A low price is charged for basic needs such as washing, using toilets, cooking, and drinking. But for "luxury uses" like washing cars, watering lawns, and ignoring leaky pipes, water is priced at marginal cost.

Pricing water at its marginal cost sends a signal of the real (opportunity) cost imposed on other water users resulting from nonessential uses. This pricing scheme has three advantages:

1. It is politically attractive because it moderates the burden of shortages (e.g., droughts) that fall on low-income households and on necessary uses of water.
2. It is economically efficient because price signals the real cost of luxury uses.
3. During shortages, price can be structured to produce enough revenue to pay for total costs of operating the utility; in contrast, average cost pricing produces revenues shortages.

Despite the potential advantages of marginal cost pricing, in many parts of the world there is considerable political pressure to supply irrigation water to farmers as a free good. Many irrigation systems use water heavily because of water's nearly zero price; there is little economic incentive for farmers to invest in technology that would improve efficiency and reduce water use. Despite price incentives for water conservation, governments in poor countries resist what they see as ejecting farmers from their lands and livelihoods by charging them for the water they use. In these countries, what seems to be more politically workable are plans to subsidize more water-efficient irrigation systems. For example, in drip irrigation water use, seepage, and evaporation are reduced by installing perforated pipes just below the soil surface so water can be applied directly to roots of crops. A major challenge lies in finding capital to finance such technological improvements.

Religion sometimes produces an institutional barrier to marginal cost pricing schemes. Analysis by Otchet (1999) describes strong religious objections in the Middle East to charging for water. Since the people are too poor to pay the government for the price of delivering that water, there is no charge.

Stakeholder Participation

Through stakeholder participation, those who have an interest influence policy formulation, alternative designs, investment choices, and management decisions that affect their communities. Participation promotes a sense of ownership of the water delivery services as well as political support for decisions that improve those services. As communities increase their participation in managing water resources, improvements will likely occur in selecting projects, delivering services, and recovering costs.

Governments can promote efficient and sustainable water policies by encouraging participation of beneficiaries and affected parties in planning, designing, implementing, and managing water programs. See Example 10–7. In both environmental and economic assessments, governments may find it in the general interest to consult with affected people and to promote participation of other stakeholders, including the poor, indigenous people and disadvantaged groups. Governments may find it is good policy to encourage special participation of women because they are usually the managers of household water.

Example 10–7
Integrating Political Theory with Water Programs in China

In the early 1980s China mounted a vigorous campaign to improve its water development as part of market reforms and a reassessment of the legal and technical foundations of the water sector. China's efforts lead to large-scale plans to change ancient, government-defined responsibilities for water development, protection, and use. Given the size of China's water problems and the country's long experience in dealing with them, these new ways of thinking and acting are being closely watched to see if the concepts of water conservation and sustainability can be effectively translated into workable programs. China's water policy initiatives test assumptions about the workability of modern policy reforms for water conservation, hazard mitigation, and environmental protection.

Boxer (2001) discusses four challenges to China: (1) applying principles established in a 1988 water law to accommodate new water resource management concepts; (2) facing the policy shift that now treats water as a resource that should respond to changing market conditions; (3) confronting theoretical issues involved in reconciling ancient cultural views on the role of water in mediating relations between the individual and the state; and (4) integrating Chinese and foreign ideas on water science and engineering. The official Chinese view is that water law reforms can help to translate and integrate Western management experiences and policy models to facilitate reform of their water economy. The intent is to better accommodate market-related ideas such as demand management, economic optimization, and rational pricing.

Official policy statements are beginning to formally recognize the need for greater local independence in choosing appropriate instruments that can enhance the efficiency of water use and water projects. Future success in China's water policy may call for the reshaping of public attitudes and responsibilities toward water as a resource and promoting new values of protection, conservation, and improved scientific management as part of reforms to the market economy. Because of China's size and its variable and unknown physical endowments, progress is slow. Another issue is how Marxist concepts of fairness in the distribution of water should be rethought and adjusted to accommodate Western, market-driven ideas. Considerable research and policy debate in China focus on attempts to forge new relationships between private and public water policies.

Source: Boxer, B. 2001. Contradictions and challenges in China's water policy development. *Water International.* 26 (3): 335–341.

■ SUMMARY

Water, a renewable resource, covers nearly three fourths of the surface of the Earth and occurs above and below the surface. It is the only natural resource that exists naturally as a liquid, solid, and gas. Worldwide, the total supply of water depends on the hydrologic cycle, the process by which water continues to circulate. Huge quantities of water cycle through the system each year, even though the world uses only a part of it.

In most other parts of the world there are large differences in the ratio of water consumed to total renewable supplies so water needs to be managed. Good management decisions depend critically on a region's location. Human actions are an important factor in determining the quantity and quality of water available for water uses as well as for maintaining natural ecosystems. Management issues involve deciding how much water to use from various sources, how much to invest in maintaining water quality, and in what time periods to use water for human benefit. Deciding which specific actions produce fair, efficient, and sustainable water use patterns are the major challenge facing water economists, engineers, and policymakers.

Equity is an important standard for judging a water policy. The United Nations stated that the human right to water entitles everyone to sufficient, safe, acceptable, physically accessible, and affordable water for personal and domestic uses. An adequate amount of safe water is necessary to prevent death from dehydration, to reduce the risk of water-related disease, and to provide for consumption, cooking, and personal and domestic hygienic requirements.

In economically efficient water allocation, the marginal benefit from the use of the water is equal across uses in order to maximize total benefits produced by the water. The benefit from using one additional unit of the water by one user is the same for all users when water is allocated efficiently. Where efficiency does not occur, society benefits by allocating more water to the sector where the incremental benefits are highest. Formulas help to identify economically efficient policies for a single period in time (static efficiency) and over several time periods (dynamic efficiency). A sustainable water policy will ensure that human rights and economic efficiency are promoted for the indefinite future.

Actual water allocations are influenced by water laws such as riparian law and the doctrine of prior appropriation. Water allocations are also influenced by supply methods such as interbasin transfers, storage, treatment, distribution, and recycling. In addition, demand management influences allocation by adjusting use through pricing and related incentives. Regulations play a role in allocation by directing various dimensions of water use.

Marginal cost pricing can be used to promote improved economic efficiency through implementations such as connection charges, geographically differentiated prices, water markets, peak-load pricing, and drought pricing. Two-tiered pricing is a special kind of marginal cost pricing that combines elements of equity efficiency, and sustainability to improve water use. Several other institutional innovations could also foster such improvement: integrated demand/supply management, transnational basin management, virtual water, decentralized management, politically workable marginal pricing, and stakeholder participation.

Review Questions

1. Explain why it may be good economics for people who live in water-scarce, crowded, dry places to invest in various conservation measures, while the same investments may be poor economics in more abundant or less crowded places.

2. Distinguish between static and dynamic economic efficiency. Could a particular water allocation or policy pass the test of static efficiency but fail the test of dynamic efficiency? Why or why not?

3. How do we know that a water policy is not economically efficient if marginal benefits are not equal in all uses?

4. Explain how water laws can influence the ownership, allocation, and use of water.

5. Why does the prior appropriation doctrine typically produce more efficient use patterns of water than riparian law in dry places?

6. Wet places like the eastern United States typically have a lower water consumption per capita than similar dry places in the western United States. Why is this true?

7. Explain why marginal cost pricing of water is more economically efficient than average cost pricing.

8. If water is required for human life, how can it be economically efficient to price water the same way we price any other economic good, namely where its marginal benefit equals its marginal cost?

9. What kinds of challenges would a government agency confront in trying to set up and supervise a water bank?

10. What are some of the potential third-party effects in the large-scale transferring of water rights from agricultural uses to uses for endangered species or cities?

11. Suppose that average per capita water use levels in summer are four times as high as those in winter. How would you design an efficient water pricing system if the same rising marginal cost schedule applied to both periods?

12. Show that if the marginal cost of water supply falls with increased output, then pricing water so that marginal cost equals marginal benefit promotes economic efficiency but causes the water utility to experience losses. How might these losses be made up while maintaining a near-efficient water price?

13. Find out what water pricing system is used for your hometown and why that system was chosen. Does the water pricing system affect your water use behavior (e.g., length of showers or lawn watering)? Can you recommend a pricing system for promoting greater efficiency in water use where you live?

14. Suppose you come from a humid climate where the ratio of use (demand) to supply is low. Describe the advantages and disadvantages of public policies that increase demands to make full use of available water.

15. What conditions would encourage governments to make more widespread use of benefit-cost analysis for water resource programs?

Problems

1. Water allocation: A reservoir's water supply of 25,000 acre-feet of water can be used for municipal (m) and/or agricultural (a) purposes. The marginal benefit (MB) and total benefit (TB) relations for the two uses are as follows

$$MBa = 1{,}000 - 0.05\ Wa$$
$$TBa = 1{,}000\ Wa - 0.025\ Wa^2$$
$$MBm = 1{,}500 - 0.2\ Wm$$
$$TBm = 1{,}500\ Wm - 0.1\ Wm^2$$

Where Wa is acre-feet of water used for agriculture and Wm is acre-feet used by municipal purposes.

a. Find the allocation of water between the two uses that maximizes total economic benefits produced by the water. (Hint: Divide the water to make marginal benefits equal for both uses.) Compute the marginal benefits and total benefits to each use produced by that efficient allocation. Suppose 1,000 additional acre-feet of water could be had by constructing new reservoir capacity. Recalculate the efficient allocation of water between the two uses when the new water is available.

b. Recalculate the total added benefits accruing to each user from the reservoir expansion if water continues to be allocated efficiently between the uses. Is the added reservoir capacity an economically good investment if the new reservoir capacity costs \$100,000? Explain. (Hint: Compare the added economic benefit to the cost of the expansion). How high can the cost of expanding the reservoir go before it is economically infeasible?

2. Food security (continuation of previous problem): Suppose a government in an arid region wants to promote national food security by encouraging increased water use in agriculture. The government's food security program allocates 19,000 acre-feet to agriculture of the original 25,000 acre-feet of capacity, leaving only 6,000 acre-feet for municipal use.

a. What is the price (marginal benefit) to each of the two uses after the food security program is initiated? Does the government succeed in reducing the price of water for agriculture compared to efficient levels? What happens to the price of water paid by municipal users?
b. What are the total benefits to each use under the food security program?
c. What is the cost to the nation of the inefficient water allocation produced by the need for food security?
d. What is the economic value of peace in the region in terms of added potential productivity of water? (See Example 10–2, which deals with water policies in Israel.)

3. Drought pricing[21]: A city's water customers normally use 100,000 acre-feet of water per year (S_N in Figure 10–8) when water is priced at as average cost of $300 per acre-foot (about $0.85 per 1,000 gallons). A drought reduces capacity to 80,000 acre-feet (Q_D in Figure 10–8), so use must fall by 20,000 acre-feet. The price elasticity (e) of demand (0.10) is defined by the following formula: $e = (\Delta Q/\Delta P)\ (P/Q)$, where ΔQ is required conservation in acre-feet; ΔP is price *increase* needed to eliminate deficit; Q is normal water use and P is current price.

a. Use the elasticity formula to find what high block price eliminates the deficit.
b. Explain how you can interpret this price as the marginal cost of the 80,000th acre-foot.
c. Total annual cost of operating the utility is $30 million, which must equal total revenue. The city council wants to charge the low block rate of $300 for as much water as possible before raising the price at the break point. At what break point (Q_S) should the price be raised from $300 to your just-calculated higher block rate? (Hint: In Figure 10–8 the area of rectangle 1 added to rectangle 2 must equal total costs.)
d. Show that your two-part pricing scheme produces total revenue that equals total costs.

References

Allan, J. A. 2003. Virtual water—the water, food, and trade nexus useful concept or misleading metaphor? *Water International.* 28 (1): 106–113.

Asian Development Bank. 2001. *Special evaluation study on gender and development.* Available from *http://www.adb.org/Documents/PERs/IN231_01.pdf* [Accessed January 13, 2004].

Booker, J. F. 1995. Hydrologic and economic impacts of drought under alternative policy responses. *Water Resources Bulletin.* 31 (5): 889–906.

California State Water Resources Control Board. 1997. *Water recycling funding guidelines.* Available from *http://www.swrcb.ca.gov/recycling/recyfund/guide.pdf* [Accessed November 3, 2003].

Ciriacy-Wantrup, S. V. 1952. *Resource Conservation: Economics and Policy.* Berkeley, CA: University of California Press.

Eckstein, O. 1958. *Water Resource Development: The Economics of Project Evaluation.* Cambridge, MA: Harvard University Press.

Freeman, A. M. 1979. *The Benefits of Environmental Improvement.* Baltimore: Johns Hopkins Press.

GAMS Development Corporation. 2004. *General algebraic modelling system.* Available from *http://www.gams.com/* [Accessed August 21, 2004].

Hall, D. C. 2000. Public choice and water rate design. In *The Political Economy of Water Pricing Reforms,* edited by A. Dinar. Oxford, UK: Oxford University Press.

Hartman, L. M., and D. A. Seastone. 1970. *Water Transfers: Economic Efficiency and Alternative Institutions.* Baltimore, MD: Johns Hopkins Press.

Hexem, R. W., and E. O. Heady. 1978. *Water Production Functions for Irrigated Agriculture.* Ames, IA: Iowa State University Press.

Howe, C. W. 1987. Project benefits and costs from national and regional viewpoints: Methodological issues and case study of the Colorado-Big Thompson Project. *Natural Resources Journal.* 27: 5–20.

Hutchins, W. A., H. H. Ellis, and J. P. DeBraal. 1971. *Water Rights Laws in the Nineteen Western States.* 3 vol. Washington, D.C.: United States Department of Agriculture. Reprinted in 2004 by The Lawbook Exchange.

Ioris, A. A. R., S. Merrett, and T. Allan. 2004. Virtual water in an empty glass: The geographical complexities behind water scarcity. *Water International.* 29 (1): 119–121.

Johansson, R. C., Y. Tsur, T. L. Roe, R. Doukkali, and A. Dinar. 2002. Pricing irrigation water: A review of theory and practice. *Water Policy.* 4: 173–199.

Kelso, M. M., W. E. Martin, and L. E. Mack. 1973. *Water Supplies and Economic Growth in an Arid Environment: An Arizona Case Study.* Tucson, AZ: University of Arizona Press.

Lee, D., and R. Howitt. 1996. Modeling regional agricultural production and salinity control alternatives for water quality policy analysis. *American Journal of Agricultural Economics.* 78 (1): 41–53.

Marglin, S. A. 1962. Objectives of water resources development: A general statement. In *Design of Water Resource Systems,* edited by A. Maass. Cambridge, MA: Harvard University Press.

McCormick, Z. L. 1994. Interstate water allocation compacts in the western United States—some suggestions. *Water Resources Bulletin.* 30 (3): 385–395.

McKean, R. N. 1958. *Efficiency in Government Through Systems Analysis.* New York: John Wiley and Sons.

Otchet, A. 1999. *An economic mirage.* Available from *http://www.unesco.org/courier/1999_02/uk/dossier/txt33.htm* [Accessed January 13, 2004].

[21] Source: This problem is based on Hall, D. C. 2000. Public choice and water rate design. In *The Political Economy of Water Pricing Reforms,* edited by A. Diner. Oxford, UK: Oxford University Press.

Sax, J. 2002. *Reflections on the Great Lakes.* University of Michigan. Available from *http://css.snre.umich.edu/events/sax_wege_lecture2002.pdf* [Accessed November 3, 2003].

Solow, R. M. 1974. The economics of resources or the resources of economics. *American Economic Review.* 64 (2): 1–14.

Texas State Senate. 2003. *Senate passes critical water condemnation rights legislation.* Available from *http://www.senate.state.tx.us/75r/senate/members/dist28/pr03/p050103a.htm* [Accessed November 3, 2003].

Trelease, F. J. 1965. Policies for water law: Property rights, economic forces, and public regulation. *Natural Resources Journal.* 5: 1–48.

United Nations. 2002. *International Rivers and Lakes Newsletter.* Available from *http://www.un.org/esa/sustdev/sdissues/water/rivers-lakes-news38.pdf* [Accessed May 12, 2005].

U.S. Congress. 1962. *Policies, Standards and Procedures in the Formulation, Evaluation and Review of Plans for Use and Development of Water and Related Land Resources.* 87th Cong., 2nd sess., S. Doc. 97.

U.S. Interagency River Basin Committee Subcommittee on Benefits and Costs Report to the Federal Inter-Agency River Basin Committee. 1958. *Proposed Practices for Economic Analysis of River Basin Projects, The green book.* Washington, D.C.: U.S. Government Printing Office (originally published in 1950).

U.S. Water Resources Council. 1973. Procedures for evaluation of water and related land resources projects. *Federal Register.* 38 (174): 24776–24869.

———. 1979. Procedures for evaluation of national economic development benefits and costs in water resources planning, level C; final rule. *Federal Register.* 44 (242): 72892–72976.

———. 1983. *Economic and Environmental Principles and Guidelines for Water and Related Land Resources Implementation Studies.* Washington, D.C.: U.S. Government Printing Office.

Ward, F. A., and A. Michelsen. 2002. The economic value of water in agriculture: Concepts and policy applications. *Water Policy.* 4: 423–446.

World Bank. 2003. *Urban poverty in East Asia: A review of Indonesia, the Philippines, and Vietnam.* Working Paper No. 11, Urban Sector Development Unit. Available from *http://www.worldbank.org/urban/publicat/urbanpovertyeastasiawout.pdf* [Accessed January 13, 2004].

World Health Organization. 2000. *The global water supply and sanitation assessment 2000.* Available from *http://www.who.int/en/* [Accessed November 6, 2003].

Young, R. A., and S. L. Gray. 1972. *Economic Value of Water: Concepts and Empirical Estimates.* Final Report to the National Water Commission. Fort Collins, CO: Department of Economics, Colorado State University.

Young, R. A., and R. H. Haveman. 1985. Economics of water resources: A survey. In *Handbook of Natural Resource and Energy Economics,* edited by A. V. Kneese and L. L. Sweeney. New York: North-Holland.

Chapter 11

Food

Outline

Food Production and Distribution 297

Desirable Outcomes 300

Market Outcomes 301

Institutional Design for Improving Food Policy Outcomes 315

Objectives

After studying this chapter, you should be able to

- Identify factors that influence the production and distribution of food
- Identify desirable outcomes for food production and distribution programs
- Describe actual outcomes produced by unregulated food markets
- Identify institutions that can improve food production and distribution

Introduction

In 1970, Norman Borlaug won the Nobel Peace Prize for his efforts at promoting peace through increased crop yields to reduce hunger. (See Bailey, 2000; Herdt, 1998.) His Nobel lecture warned that humanity had only a short time to tame the population monster and the environmental costs of trying to cope with it before food shortages would threaten the political stability of many poor countries (Nobel Foundation, 2004). However, because of continuing progress in developing efficient, high-yield crops—known as the Green Revolution—we may have more time. Borlaug, considered by many as the father of that revolution, is responsible for much of the progress made against hunger. Despite occasional local famines caused by war and poor government, food is more abundant and cheaper today than ever before in history. Easterbrook (1997) estimates that Borlaug's work has saved a billion lives.

Borlaug observed that we cannot only increase our food supplies, we must also insure them against biological and physical catastrophes by international efforts to develop food reserves. He recognized that governments can prevent the tragedy of famine in the future instead of attempting to salvage the human wreckage of famine.

Food security has been defined as "all people, at all times, having the physical, social and economic access to sufficient, safe and nutritious food in order to meet their dietary needs and food preferences for an active and healthy life" (Rome Declaration, 1996). Despite the enormous gains made by modern science in food production, an estimated 840 million people, or about 15% of the world's population, lived with food insecurity in 2003, which means they lacked access to the calories necessary for active and healthy lives (Cohen, 2003). Furthermore, micronutrient malnutrition, especially shortages of iodine, iron, and vitamin A, afflicts nearly 2 billion people (Charlton and Rose, 2001). Hunger continues to rob the world of countless human potential. This chapter explores how the concepts and methods of economic analysis can be used to improve the production and distribution of food.

■ Food Production and Distribution

The quest for food security—the avoidance of hunger and famine—is as old as civil society itself. Yet despite continued increases in global food availability and substantial progress in poverty reduction since the 1950s, hunger, food insecurity, and undernourishment remain at high levels, and progress in addressing this dimension of poverty has been weak. Almost 850 million people around the world have too little to eat and in many parts of the world the number is on the increase. Understanding how agriculture can most effectively contribute to food security is an important question, particularly for policymakers as they examine their approach to agricultural development to support increased food security. Inadequate *food production* and inadequate *food distribution* are two competing hypotheses on the causes of continuing hunger. Because food production and food distribution are affected by different factors, we now examine what influences each.

Food Production Mechanisms

Most authorities agree that an adequate amount of food is currently being produced (Johnson, 1997, 1998, and Rosegrant and Ringler, 1997 and the ongoing work of various international development agencies like the Food and Agricultural Organization, FAO, of the United Nations and the International Food Policy Research Insitute, IFPRI). Particularly when a nation's institutions permit food to be allocated by a market system, there appears to be little danger of overall shortage. Recall from Chapter 3 the mechanics of a market response to shortages: Temporary shortages lead to higher prices and a resulting supply response by income-maximizing producers, as well as to a search for substitutes by consumers. For example, in U.S. supermarkets during the 1970s meat and coffee were scarce for a time, and their prices rose considerably to reflect that scarcity. Soon hamburger substitute made from soybean meal appeared in supermarkets. Overall, a dramatic reduction in meat and coffee consumption occurred and consumption of meat and coffee substitutes increased.

What are the mechanics of farmers' supply response to food shortages in a market economy? Shortages produce rising prices and the higher prices signal farmers to take two actions to increase supply: They increase the amount of land under cultivation or they increase crop yields on land currently being cultivated. Usually, both actions take place.

When food prices are low, only the most productive (fertile) land can produce the food at a profit, so lower-quality lands are not used. As food prices rise, the increased profitability of food makes it economically attractive to farm existing lands more intensively while bringing less productive lands into production. In other words, the supply curve for food is upward-sloping—higher food prices lead to greater production. Technological advances in agriculture have played a major role in increasing food production, as has the enormous impact of the Green Revolution. Food production continues to increase faster than population in both the developed and developing countries.[1] Even per capita food production has increased, although the increase has been small overall and even smaller in developing countries.[2] Many studies since the early 1990s have concluded that the evidence indicates that per capita global food production continues to grow. So where does hunger come from?

Food Distribution Problems

Nobel laureate Amartya Sen noticed that even in the severe Bengal famine of 1943 food was available (Nobel Foundation, 2004, Amartya Sen biography). However, many people lacked access to food because (1) they lacked income, (2) the price was too high, (3) or they lacked access

[1] The potential for future global food scarcity is analyzed by Harris and Kennedy (1999).

[2] Kent (1997) describes some of the challenges of maintaining meat production and especially fish production with growing world demand. Smith and Haddad (2001) describe the role of domestic food production in controlling child malnutrition.

to welfare. Today, much of the malnourishment of the world's 850 million hungry people appears to be due more to poor food distribution than global scarcity. This means that the real problem is poverty. If poverty is the problem, we would expect to see that the poorest groups of society are the most malnourished. And, indeed, the poorest countries have the largest percentage of undernourished people.

If poor food distribution is the cause of hunger, the policy solution is quite different than it would be if hunger resulted from inadequate production. Where the problem is poor distribution, the policy challenge is how to deliver food to those who need it most.[3] Eliminating malnourishment by increasing access to food for the hungry is a method that can work. If the problem were simply too little food, this method would not work (Alexandratos, 1999, and Guo et al., 1999).

The evidence in Table 11–1 shows why most food policy experts believe that the problem is one of poor food distribution. Although the data are subject to considerable loss of detail because of averaging across continents, several conclusions can still be drawn. The first two rows show that the average household in both the developed and developing world has an adequate intake of calories, which affirms the experts' conclusions that the food problem is not one of inadequate global production. But the table also indicates that not all countries are equal when it comes to food distribution. For the least developed countries, especially sub-Saharan Africa, many household diets have too few calories to avoid malnutrition.

Even more disturbing is a trend observed between the mid-1980s and mid-1990s. During this period much progress occurred in the developing world in increasing per capita calorie intake; impressive gains were made in Asia and South America. But the situation did not improve in sub-Saharan Africa over the same period. Fourteen of 40 countries in sub-Saharan Africa actually experienced deteriorating per capita calorie intake in the 10-year period.

Strategies are needed to increase the food intake available to the poorest of the poor. Greater domestic food production could help. The Green Revolution began in 1944 when the Rockefeller Foundation founded an institute to improve the agricultural output of Mexican farms. Remarkable results were achieved and Mexico went from needing to import half of its wheat to self-sufficiency by 1956. By 1964 Mexico exported half a million tons of wheat. The technologies were exported from Mexico to regions all over the world, especially eastern Asia. The success in increasing yields was undisputable. The growth of crop yields was such that agriculture has been able to outstrip population growth, and per capita calorie production has increased every year since the mid-1950s. Although the Green Revolution has faced strong criticisms in recent years (see Example 11–1), its ability to increase calorie intake and other living standards of the poorest of the poor is unquestionable. Here are a few examples of the effects of the Green Revolution:

- Short-stalk, fertilizer-responsive rice varieties have spread throughout eastern Asia and varieties of wheat have spread throughout Mexico, India, and Pakistan (see Example 11–2).
- In many areas with access to hybrid crops, productivity has doubled or tripled. Short-duration varieties of crops have permitted farmers to harvest two crops a year instead of one.
- Genetic engineering is viewed by some as a natural continuation of the Green Revolution. Genetically modified foods could be a major contributor to helping poor countries escape malnutrition.

In 2002, the United Nations World Food Program estimated that farmers produced enough food to provide every person with 2,800 calories a day. That is enough energy to meet the daily requirement of teenaged boys and active men according to U.S. government dietary guidelines. But inadequate infrastructure, gender inequality, wars, and corrupt governments have prevented the chronically hungry (i.e., people who eat too little to meet basic standards) from gaining access to this rich harvest. After falling for decades, the estimated number of undernourished in the world increased by 18 million to 798 million between 1997 and 2001 according to 2002 data (Food and

[3] Paarlberg (1999) presents an argument in which most of the world's malnourishment problems could be solved by focusing on Africa.

Example 11–1
Sustainable Food Policy Challenges: The Green Revolution

The Green Revolution improved world food production by relying on two kinds of technologies: (1) breeding of new plant varieties and (2) development of new agricultural techniques to exploit the advantages of those varieties. Despite improved yields and growing food production, the Green Revolution has been criticized on environmental and on social grounds. Here are a few of the reasons:

Loss of biodiversity and food quality: The spread of Green Revolution hybrids and the associated growing techniques has resulted in far fewer crop varieties. Some crops have seen up to a 90% reduction in their varieties. Dependence on such a small number of crop varieties increases the risk to existing plant populations and may impair the ability to improve crops in the future. In addition, Green Revolution crops are bred for growth efficiency, longevity, and appearance, not for health value. Therefore, many hybrid crops have lower nutritional value than their ancestors.

Fossil fuel dependence: As agricultural output increased as a result of the Green Revolution, energy input into the process increased at a greater rate. Therefore, the ratio of crops produced to energy input has decreased over time. Also, Green Revolution techniques also rely heavily on chemical fertilizers, pesticides and herbicides, some of which must be developed from fossil fuels, thereby making agriculture increasingly reliant on petroleum products. As long as fossil fuel is priced at its real opportunity cost, farmers will use more fossil fuels to increase their yields only when society's benefits exceed its costs. But fossil fuels are often priced at less than their real opportunity cost.

Pollution: Fertilizer, pesticide, and herbicide runoff continues to be a significant source of pollution, especially water pollution. Although the most dangerous pesticides of the 1950s have mostly been phased out of agriculture in developed countries, DDT continues to be used in developing countries.

Land degradation: Some critics charge that Green Revolution methods damage soil quality over the long term. This is a result of a variety of factors: increased soil salinity as a result of heavy irrigation, burnt soil caused by heavy use of chemical fertilizers, loss of beneficial soil microbes and other organisms, and soil erosion. These damages can lead to increased reliance on chemical inputs to compensate for deteriorating soil quality, a process that is difficult to sustain.

Income distribution issue: The Green Revolution introduced major changes into a world where the majority of people still depend on farming for their livelihood. Many of the changes financially rewarded large-scale industrial agriculture at the expense of small farmers, who were unable to adapt sufficiently to compete with the high-efficiency of Green Revolution crops. The result has been massive displacement and increased urbanization. Farmers lost their land to large, efficient corporate farms that are able to manage the complex technology and enterprise involved in effectively exploiting Green Revolution techniques.

Table 11–1 World food security

	Cereal production (metric tons per 1,000 population)		Calorie Supply: Average daily per capita		Calorie Supply: Percentage of 2,350 minimum	
Location	**1986–88**	**1996–1998**	**1987**	**1997**	**1987**	**1997**
All developed countries	675	677	3,324	3,240	141	138
All developing countries	252	265	2,453	2,650	104	113
Asia	243	223	2,353	2,463	100	105
Sub-Saharan Africa	127	115	2,145	2,157	91	92
Middle East	145	138	2,935	2,982	125	127
Central America	114	106	2,519	2,493	107	106
South America	287	344	2,505	2,626	107	112
North America	1,447	1,504	3,268	3,409	139	145
Europe	552	537	3,258	3,153	139	134
Oceania	567	682	2,995	3,161	127	135

Source: *World Resources Institute (WRI) in collaboration with United Nations Environment Programme (UNEP), United Nations Development Programme (UNDP), and World Bank. 2003. World Resources 2002–2004: Decision for the Earth: Balance, Voice, and Power. Washington, DC: WRI*

Example 11–2
Economics of the Green Revolution: One Farmer's View

In 2001, Mr. Rajarathinavelu, a farmer from the Tamil Nadu village in India, presented an optimistic view of the Green Revolution and how it had helped him, his family, and his village. Before the 1960s, agriculture was done along traditional lines in his village, with no chemical nutrients or pesticides. Other than human labor, animal power was the sole source of farm energy. Back then, labor was plentiful and working hours were long. However, despite sustained long periods of labor, crops produced low yields, and this had a serious impact on the village economy. Low-yield harvests of rice, the staple food of the farm family, curtailed the funds of family life available to the family for other developmental purposes. Living standards of most farmers and farm laborers were precarious.

Then came the Green Revolution. In the 1960s plant breeders introduced a short-stature rice variety called IR-8. This new, high-yielding rice was a boon to Mr. Rajarathinavelu's family, to his village, and to the country. For the first time, villagers experienced a huge output of grain. Mechanization slowly crept into the village. Installation of an engine to draw water from the well began in 1965. Farmers took advantage of reliable irrigation and the land area under rice cultivation increased. Beginning in 1969, pump electrification reduced much of the drudgery of family life: by 2000, farmers in the district pumped water from 94,000 wells.

By 1974, a large landowner in a nearby village bought the area's first tractor. Soon, tractors occupied a prominent position in performing various agricultural activities as well as reducing the toil of human labor. By 2000, the number of tractors in the district rose to 1,165. The new technology consisting of seeds, fertilizer, plant production, machinery, and irrigation raised rice productivity threefold, and some farmers now produce 8,500 kilograms per hectare (7,580 pounds per acre). The villagers believed the IR-8 variety was a godsend: in gratitude, some villagers baptized their children with the name IR-8.

Increased sustained yields improved village life in many ways. The benefits of the Green Revolution have trickled down to the small and marginal farmers whose standard of living increased. By 2000, some of the village farmers were sending their children to school, even to expensive private schools. Many farmers now own radios, televisions, and small motorcycles.

Many villages can afford improved medical facilities. The basic feature underlying all these improvements was increased rice production in a sustained manner. Had it not been for the Green Revolution, these improvements in living standards would never have occurred in the village.

Source: Rajarathinavelu, K. 2001. In Chapter 16 of Complementary Technologies, *One Goal: Approaches to Sustainable Food Production.* International Food Policy Research Institute. Available from *http://www.ifpri.org/pubs/books/2020conpro.htm* [Accessed October 27, 2004].

Agricultural Organization, 2004). Most of these 798 million live on isolated, unproductive land, and have neither the money to buy food nor the ability to grow it. They are beyond the reach of international feeding programs and also fall through national safety nets. Most of the world's hungry are the very poor, the landless, vulnerable groups of young children, women, and the elderly, and they live mostly in Asia and Africa. The striking contrast between global food adequacy and widespread hunger raises a perplexing question. Because food is so essential to human physiological and emotional needs, why don't societies design mechanisms to distribute food more equally?

■ Desirable Outcomes

Most communities have four basic food policy objectives:

1. Efficient and sustained growth in the agricultural sector
2. Acceptable income distribution, primarily through efficient job creation
3. Adequate nutritional status for everybody
4. Food security that insures against poor harvests, natural disasters, and uncertain world food supplies and prices

Something close to these broad agricultural goals are held by food policymakers in all countries. Achieving these goals requires knowledge of the behavior of food consumers and producers. Understanding their behavior helps identify policy instruments that can alter household and farm decisions to meet society's food policy objectives. Although parts of this chapter are applicable to all countries around the world, the chapter's focus is on the food problems of developing (poor) countries. Food problems are more pressing in the poorest countries where scarcity of resources limits flexibility in reconciling the interests of food producers and consumers.

■ Market Outcomes

Food policymakers face the important question of whether the unregulated market system can accomplish the four desirable food policy outcomes just described. They may also wish to know what kinds of regulations can be implemented to help the market better achieve these desirable outcomes. Several factors that influence those market outcomes are described next.

Complexity of Farm Decisions

Growing food is a complex business that challenges the limits of human decision making. The daily routine of most farmers involved these questions: what crops to plant; what inputs to use in what combinations; when to seed, plow, cultivate, and irrigate; when to harvest; how much to keep for home consumption; how much of what crop to sell now; and how much to store for later sale. Each day's decisions are the result of a string of decisions made in many previous time periods. Agriculture is unique in that millions of individuals and households make these decisions themselves or in cooperation with small numbers of neighbors, friends, or partners.

Farmers are motivated by the potential for economic return. Timmer, Falcon, and Pearson (1983) and Kahan (2004) summarize economic principles of production agriculture. Policymakers who design programs that influence national food policy objectives want to know what influences farmers' decisions. Heated and often fruitless debates have been waged over what incentives are needed to induce change in farmers' behavior. Suggestions have ranged from billboard advertisements to appeals to patriotism, from improved literacy to the availability of household goods for purchase in local markets. But the debate has largely disappeared, for the answer is in. Most farmers respond to opportunities to improve their economic welfare. Evidence from around the world shows that farmers make economic calculations for their agricultural decisions.

When the economic environment changes, farmers change their behavior in directions predicted by economic models of producer behavior. Policies that attempt to adjust a farmer's income-earning opportunities must alter behavior that take advantage of those opportunities. As an example, consider a policy proposal that would limit private decision making in favor of collective action. Collective ownership of land and other resources, and collective decision making about crop patterns and on-farm investments, will shift the farmer's perceived risks. Where farm income is shared collectively, the farmer's perception of reward for individual work and initiative is reduced. To encourage farmers to produce large amounts of output at low cost, socialist agricultural planners have found it necessary to maintain incentives that connect farming effort to rewards.

The considerable evidence showing that farm households are governed by economically rational behavior justifies the use of economic models to explain impacts of food policy actions on rural household decisions. The simplicity of microeconomic models of producer behavior makes them attractive ways to organize and simplify the complex environment of farm decision making. The ability of the models to understand and predict farm behavior enables policymakers to use them in designing responses to the full range of food issues that challenge poor countries.

Farmers' Goals

Understanding the decisions farmers make requires understanding the farm household (Kahan, 2004). The family farm household consists of the farmer and the family and their goals and preferences. The household is both a family and a business. Like most households, it both produces and consumes. Most farm households have common goals:

1. Food security: Many small farmers aim for security rather than earning income by selling to a market. They may set their goal at guaranteeing enough food for family needs. They achieve their goal by ensuring that no matter how bad the weather or how high their costs of production, they will produce enough for their family to consume.

2. Profit maximization: As farmers sell in the marketplace, earning money becomes important to them because of what the money can buy. Most farmers want money to buy products other than food; they want to be able to pay for health care, education costs, and other consumer goods. Even if they do not wish to maximize profits, the competitive nature of farming means that farmers need to make profits to survive economically. Farmers who fail to generate profits eventually go broke.

3. Risk reduction: Even though profit maximization is an important objective, farmers may be more concerned about reducing risks and reducing variability in income from one year to the next. The essential goal of poorer farmers is survival rather than maximizing profit. Still, the goal of reducing risk and increasing profit can be compatible. Larger profits this year give the farm more income to survive bad future years.

4. Social factors: Some farmers may want a comfortable life and may be less interested in harder work to produce more profit. They may want to be sure that farming gives them leisure time. Other farmers might have a goal of social status that brings respect in the community. In some cultures, status is especially important and can be achieved by owning large numbers of livestock, an important symbol of wealth.

Even if profit is not the immediate goal, the use of profit in farming serves a common purpose. It allows the nonfinancial goals of the farm household to be expressed in financial terms. Money is simply a common denominator in making decisions. If a farmer wishes to carry out an enterprise that is not financially justified, there is a cost involved and he or she needs to know this. The difference between the profits a farmer would have earned with and without the enterprise, reflects the cost of the decision. The profit lost by preferring a nonfinancial goal can be calculated and expressed in monetary terms. Such information helps the farmer make balanced decisions.

To survive, market-oriented farming requires a business approach. Profit allows a good living for the farm family. It generates the capital needed for the household and for reinvestment in the farm. The challenge facing farmers in competitive times is how to increase profitability. For all of these reasons, the remainder of this chapter is based on the ideas that farmers either wish to maximize profits, or are required to maximize profits to survive economically as they encounter competition.

Product-Product Relations

A farmer usually can choose from a huge range of possible crops that can be produced, especially where there is adequate time for planning. One important step for making these choices in a rational manner is for the farmer to evaluate the opportunity cost (lost income) of growing more of one crop at the expense of another. Figure 11–1 shows the production possibilities available, namely, *C, J, K,* and *H*. The curve is drawn to show all combinations for two crops that are technically possible for a farm household to grow in a single season using the resources available to it on a plot of land. For this example, the farmer could choose to grow only rice (0*H* kilograms per hectare; 100 pounds per acre = 114 kilograms per hectare) and no beans. (A policy of self-sufficiency in any single crop comes at a very high price; see Coxhead, 2000.) At the other extreme, the farmer could choose to grow only beans (0*C* kilograms per hectare) with no rice. Point *K* is an example of a point where the farmer decides to grow some of each crop: 0*G* kilograms of rice and 0*E* kilograms of beans.

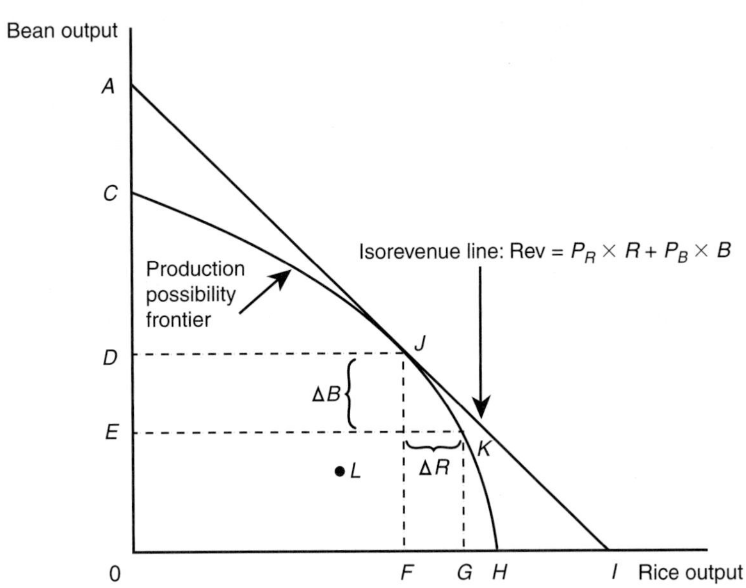

Figure 11–1
Product-product relation for growing beans and rice

A rational farmer who has access to relevant technical information affecting production will consider only points on the production possibility curve. A point like *L,* inside the curve, is an output level less than the available farm resources could produce.[4]

The production possibilities curve is shown as a curve, rather than a straight line, because the farm household's resources do not produce beans and rice equally well. If the two crops are perfect substitutes in their use of the household's resources, the production possibility curve is a straight line. For example, if one unit of beans required twice the resources as one unit of rice, the production possibilities is linear and has a slope of two units of rice displaced by every additional unit of beans. Greater curvature means increasing difficulty in substituting one crop for another.

Nearly all crops can be substituted for others if enough resources can be found to support the substitution. Finding these resources usually comes at a cost, so whether investments in finding the added resources should be undertaken is an important issue, both for the individual farm producer and for national agricultural policy. If policymakers determine that self-sufficiency in rice is an important national aim, it is possible to eliminate bean production altogether and plant only rice. Agronomic constraints can usually be overcome with enough time and resources. Policy aims, economic incentives, and costs and benefits of achieving those aims are typically the determining factors.

A farmer represented in Figure 11–1 who wants to increase the production of beans (*B*), from point *K* to *J,* that is, by ΔB, must forego ΔR, units of rice (*R*). This opportunity cost is shown by the slope of the production possibility curve between *K* and *J.* As drawn in the figure, this slope is measured in physical units (e.g., kilograms). But farmers want to know the relative values they gain and give up, not just physical quantities. To make this comparison, they need to know unit values for each output they consider changing. Whether the decision to grow more beans and less rice produces increased revenue can be determined only by comparing the *revenue lost* from the reduced rice output ($\Delta R \times PR$) with the *revenue gained* by growing more beans ($\Delta B \times PB$), where *PR* and *PB* are the selling prices of rice and beans, respectively. If the gain exceeds the loss (if $\Delta B \times PB > \Delta R \times PR$), the farmer will find point *J* more profitable than point *K.*

The combination of bean and rice output that maximizes revenue for the farm's available resources is tangent to the highest possible isorevenue line. This line is shown as *AI* in Figure 11–1.

[4] People who work in agriculture commonly observe a farmer operating at an interior point like *L.* Reasons can include a lack of knowledge about appropriate production techniques or an experimental technique that failed because the farmer did not know how to apply it correctly. Good agricultural policy requires understanding why some farmers are not on the production frontier to better identify and overcome constraints to expanding output.

It represents the economic value of the output of beans and rice together ($PR \times R + PB \times B$). All points on this isorevenue line produce constant total revenues. The farmer prefers higher isorevenue lines (ones farther away from the origin) to lower ones but is equally well off with any position along any one line, such as point *A* or point *J* or point *I*.

The slope of this isorevenue line is $-PR/PB$, the price of rice divided by the price of beans; it measures the rate at which beans can be traded for rice in the market.[5] When the isorevenue line and the production possibility frontier are tangent (e.g., at point *J*) both lines have the same slope. The production possibility frontier shows the physical trade-off between beans and rice, or $\Delta B/\Delta R$, which is the maximum amount of beans that can be expanded for a given loss in rice. The isorevenue line measures the monetary trade-off, or $-PR/PB$, the rate at which the two outputs trade for each other in the marketplace. Both ratios are equal when the two lines are tangent. At the tangency point, *J*, the revenue lost by reducing production of one crop is just equal to the revenue gained from the producing more of the other crop, because $PR \times \Delta R = PB \times \Delta B$. The marginal cost (i.e., the lost revenue resulting from reducing the other crop's output) equals the marginal revenue, which is the economic criterion for maximizing profit.

The equality of the two slopes reveals important information for consumer household welfare. If the farm household chose a welfare-maximizing menu of beans and rice for its home consumption, based on the budget constraint given by line *AI*, the highest indifference curve it could reach would be tangent to *AI*.[6] This means that relative prices are a major factor determining economic decisions of both producers and consumers. Since government food policy often uses import or export controls to change commodity prices, that policy has a major influence on farm-level decisions on how much of which crops to grow.[7] Policymakers may encourage farmers to grow more rice, but if government price policy favors beans, farmers can be expected to pay little attention to the advice.

Factor-Factor Relations

A farmer decides which crops to grow, then decides the best way to grow them. Factor-factor relations refer to substitutions farmers make in choosing inputs for growing a crop. Farmers have the choice of using many combinations of productive inputs (factors) to produce a given crop. Land, labor, and capital are three important crop-producing inputs. The farmer's decisions about how to use the three inputs have important effects on employment, productivity of agriculture, and the level and distribution of income for rural areas.

Figure 11–2 shows the ease with which labor and capital substitute for each other in production agriculture. The curved line *ABCD* represents all the various combinations of labor and capital that could be used to produce, for example, 1,000 kilograms of rice per hectare for a season. Point *D* is a capital-intensive method. Point *A* uses more labor and less capital to produce the same amount of output. In theory, an infinite number of input combinations can be used to produce the same total output, as is shown on the 1,000 kilogram (kg) per hectare (ha) isoquant (Greek for equal quantity). In most real-life situations, however, only a small number of combinations are likely to be available for the farmer. Figure 11–2 shows four methods of producing the constant amount of rice per hectare: hand labor (point *A*), farm animals (point *B*), a small tractor (point *C*), and large-scale equipment (point *D*). The four points show the four possible technical methods available for growing 1,000 kilograms of rice per hectare of land.

---

[5] This property is the same as the price relationship facing consumers with a given income constraint, as described in Chapter 7. For the simple economic environment illustrated here, isorevenue line *AI* is the farm household's budget constraint.

[6] If the household can buy rice and beans in the marketplace at the relative prices indicated, it is unlikely to buy the same combination it produces. The indifference curve is not shown in Figure 11–1.

[7] Challenges facing policymakers who wish to increase food security by adjusting food prices are described by Jayne and Jones (1997) for Africa and by Jha and Srinivasan (1999) for India.

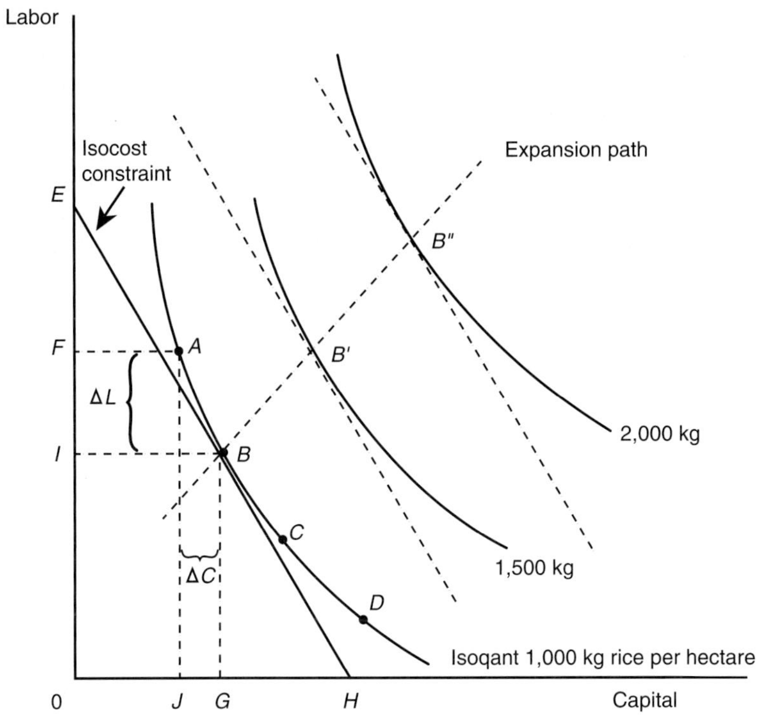

Figure 11–2 Factor-factor relations showing labor and capital used as substitutes in producing rice

All input combinations produce the same output. Which combination should the farmer choose? Does it make any difference? The farmer will choose carefully because, indeed, some combinations are better than others. The economically efficient combination of labor and capital is not fixed in stone, but it is determined by the prices of the inputs. Suppose the farmer is currently cultivating with human labor at point *A* and contemplates using farm animals to reduce drudgery. The farmer wants to know how much labor is saved and how much additional animal time is required. Making this choice moves the farmer along the isoquant from point *A* to *B,* which shows the use of ΔL (less labor) and ΔC (more capital). If the farmer pays a market price for labor and capital in which the expenditure on labor released is larger than the expenditure on additional capital used, then $\Delta L \times PL > \Delta C \times PC$ where *PL* and *PC* represent the prices of labor and capital. In that case, the farmer will find profitable the change to the more capital-intensive rice production.[8]

For known prices of the two inputs, an isocost line can be constructed to define points of equal input costs. This line represents all possible combinations of labor and capital that have equal costs. Like the slope of the isorevenue line the slope of the isocost line is the negative of the two inputs' price ratio. Figure 11–2 shows that the isocost line *EH* is tangent to the 1,000 kilogram isoquant at point *B*. At that point, the farmer has discovered the least-cost combination of labor and capital to produce 1,000 kilograms of rice per hectare. At any other point on the same isoquant (e.g., points *A, C,* or *D*), it costs more to produce the same level of output. When the ratio of input prices is tangent to the isoquant ($\Delta L \times PL = \Delta C \times PC$), a farmer is using the least-cost combination of inputs to produce a given level of output.

What if the farmer wants to produce more than 1,000 kilograms of rice per hectare? Increasing output requires the farmer to use more capital, more labor, or both. Figure 11–2 shows that each level of output has its own isoquant, labeled 1,500 kilograms and 2,000 kilograms. A dashed isocost line is shown parallel to the original isocost *EH* (i.e., tangent to the 1,500 kilograms

[8] Reduced production costs for a given level of output is exactly what many farmers in India discovered in connection with improved rice varieties made available by the Green Revolution. Their isoquants also shifted to favor more capital-intensive production methods.

isoquant) for which the relative price of labor and capital are unchanged. Tangency point B' represents the least-cost combination of labor and capital to produce 1,500 kilograms of rice. Even though it is the least-cost combination of capital and labor, point B' still uses more capital and more labor to increase output from 1,000 to 1,500 kilograms.

Figure 11–2 shows that farmers make decisions about their techniques of production *by comparing the price of each factor to its productivity.* Can government policies affect farmers behavior? The answer is a resounding yes. If government subsidizes an input like capital (reduces its price), for example, farmers will adjust their behavior to choose production techniques that use more capital and less labor than without the subsidy. If a government policy raises the wages of labor, fewer workers will be hired, as farmers substitute away from labor, which is now a higher-priced input. The economic incentives determined by government food and nonfood policies influence hundreds of decisions farmers make about how to plant, cultivate, and harvest crops. These decisions in turn influence how many workers can find productive jobs.

Factor-Product Relations

Policymakers have a natural interest in the outcome of farmers' decisions, for these outcomes determine the quantities of various food supplies, amount of foreign exchange earnings from sales of agricultural output to other countries, and rural area incomes. Policymakers exert an influence by adjusting the economic incentives facing farmers. The *production function* relating the use of inputs to agricultural output is a powerful conceptual tool to understand how farm-level decisions are made and how their outcomes affect variables that are important to a nation. Various economic and agronomic factors are weighted together in the farmer's decision of how much of each factor should be used to produce output, which we label the *factor-product decision.*

Figure 11–3 shows a production function as curve *DJHI*. The curve shows the yield of wheat per hectare that could be expected from applying different quantities of the variable input, fertilizer.[9] The production function shown assumes that other factors of production (like land and labor) are fixed and that any increase in output comes exclusively from increases in fertilizer applied to these fixed inputs. Fertilizer, the variable input, is shown on the horizontal axis. The figure also shows diminishing marginal returns, which means that each extra unit of fertilizer results in a smaller increment in output. (Higher levels of fertilizer do not normally reduce output, but after diminishing returns set in, higher levels of it produce lower increments, gains, in output than lower levels do.) Where no fertilizer whatsoever is applied, a yield of 0*D* results. The physical maximum yield of 0*G* is produced with 0*C* fertilizer applied. Often, agricultural workers in the field can determine what these values are by talking with local farm households.

Curve *DJHI* shows the rate at which fertilizer can be turned into wheat by varying the level of fertilizer applied. This conversion is the on-farm production relationship between wheat and fertilizer. It reflects how the farmer can exchange fertilizer, with nature's help, for wheat. Where fertilizer can be bought on the market for wheat, the trade can be made in the opposite direction. Farmers who sell their produced wheat at the market and buy fertilizer with those receipts are using the medium of money to carry out a market trade of wheat for fertilizer.

Figure 11–3 shows the rate at which farmers can trade wheat for fertilizer as line 0*M*. The line measures the ratio of the price of fertilizer to the price of wheat (*PF*/*PW*). When the price of fertilizer rises, the line 0*M* becomes steeper, indicating that less wheat can be bought for a given amount of fertilizer. Likewise, if *PW* rises, the line 0*M* from the origin rotates clockwise and becomes flatter as each kilogram of wheat buys more fertilizer.

Notice that the price line 0*M* is straight; that is, the quantity of fertilizer applied and the resulting quantity of wheat produced by the individual farmer affects neither the price of wheat nor

---

[9] Cakmak (2002) and Hossain and Singh (2000) analyze the role of fertilizer as a method for increasing food production. Larson and Frisvold (1996) analyze some of the environmental costs and challenges posed by increased use of chemical fertilizers to raise crop yields.

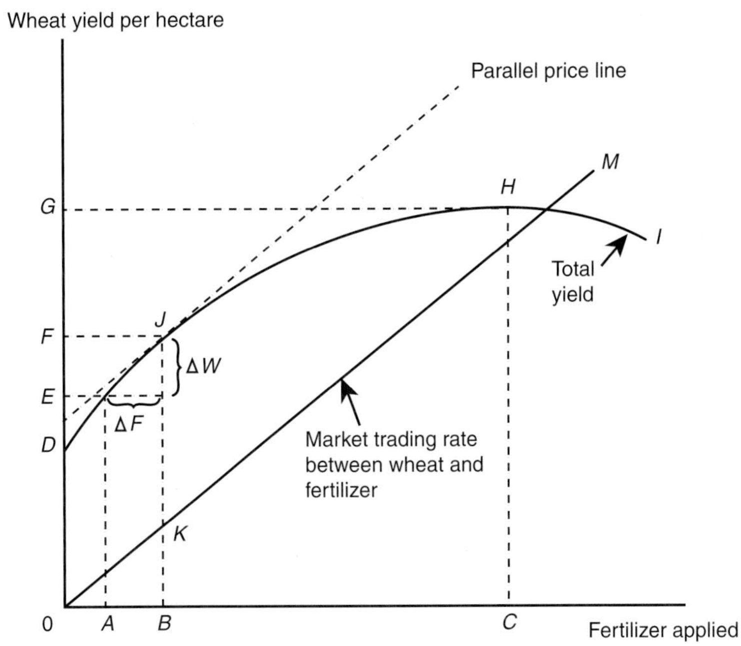

Figure 11–3 Factor-product relations for using fertilizer to increase production of wheat

the price of fertilizer. Because the price of fertilizer is nearly the same whatever the level of use, the line 0*M* represents total variable costs (fertilizer is the only variable input). Costs are measured in the same units as wheat output, so whenever the production function *DJHI* exceeds the total costs 0*M*, the farmer is earning a profit from applied fertilizer. Losses occur because net revenues are negative when 0*M* lies above *DJHI*.

Figure 11–3 can be used to identify how much fertilizer a farmer will use to maximize profit and how much output results from the profit-maximizing decision. Profits are maximized when the vertical distance between output revenues and input costs are the largest. The profit-maximizing use of the variable input is found by shifting the price line 0*M* up and parallel to its original position until it is tangent to the production function at point *J*. The dashed parallel price line shows that this tangency point occurs where total wheat production is 0*F* and fertilizer use is 0*B*. At this profit-maximizing point, an amount of wheat is effectively exchanged in the marketplace (equal to *BK*) to purchase the optimal amount of fertilizer used (0*B*). Think of the farmer as paying *BK* units of wheat to buy 0*B* units of fertilizer at the going market price ratio of wheat for fertilizer. This action by the farmer leaves an amount of wheat (equal to *KJ*) left over to repay the farmer's labor and use of land and other fixed inputs from the total amount of wheat produced (*BJ*).

It is difficult to imagine any individual farmer having enough information, computational power, management skill, and will to find the global profit-maximizing point *J* shown in Figure 11–3. It is hard to find this profit-maximizing behavior in the complex world in which farm households make production decisions.[10] In fact, there is no need to assume the kind of complex global profit-maximizing behavior described. A simpler explanation is that farmers adjust their behavior incrementally (i.e., nonglobally). Farmers only need to adjust their fertilizer use and output decisions in response to mistaken starting points or to changes in input or output prices or in technology. Regardless of where farmers start, they attempt to move in a direction that increases net revenue. When farmers compare additional costs with expected additional benefits—a natural and rational way to make choices—their behavior approximates that predicted by the more complex economic model just described. So policymakers who use the production

[10] Enyong, Debrah, and Bationo (1999) analyze the difficulties facing farmers and policymakers in west Africa who attempt to find the mix of activities and resources that globally maximizes farm profits.

function model to analyze how farmers will respond to changed economic or technical conditions with the profit-maximizing assumption will gain considerable insight into what will actually happen.

By comparing the extra revenue any extra yield brings in with the incremental cost of the added fertilizer required to produce that extra yield, the farmer can decide if the extra fertilizer adds to profits. If the marginal revenue produced by applying the extra fertilizer exceeds the marginal cost (if $\Delta W \times PW > \Delta F \times PF$), the extra fertilizer adds to profits. Otherwise it does not. Thinking and behaving incrementally, the farmer need only continue to expand fertilizer use up to the point where the slope of $0M$ is equal to the slope of *DJHI*. Where the slopes are equal, the following equation holds true:

$$\Delta W/\Delta F = PF/PW \quad (11.1)$$

which is reached at point *J*.

This simple equation is very important, for it defines the farmer's behavior in regard to production inputs. It is also important because the same equation defines the farmer's response to any government food policy decision that affects farm output prices, farm input prices, or on-farm technical production relations. To the extent that this equation approximates the use of on-farm inputs it can be used to design, implement, and evaluate government food policy proposals. When these food policy proposals could reduce hunger and malnutrition, the equation takes on considerable importance in the search for humane food policy programs.

Multiplying both sides of the equation above by *PW* produces the following equation:

$$PW \times MPP = PF \quad (11.2)$$

Remember that $\Delta W/\Delta F$ is the marginal physical product (*MPP*) of the variable input. This new equation says that the economically rational farmer expands the use of the variable input until the value of its marginal product ($PW \times MPP$) equals the price of the input (*PF*). Any government food policy decision that influences any of these three terms will affect the farmer's behavior in the continued search for new opportunities to maximize profits. For example, Equation 11.2 enables us to predict the effects of a government policy that reduces the domestic price of wheat to improve access to basic cereal grains for poor people who cannot afford high prices. By reducing the price of wheat, the policy will require farmers to reduce their applications of fertilizer as they raise their *MPP* of fertilizer by using less of it. The right-hand side of the equation, *PF*, is unchanged; so when the *PW* falls, farmers will increase their *MPP* by applying less fertilizer and producing less wheat so that they can continue to maximize profits.[11] The effect of the policy is to make it *harder* to feed the poor, not *easier*.

Technical Change

Figure 11–4 shows how the production function framework can help in understanding reactions by farmers who face opportunities to use new technology for producing wheat. The development of fertilizer-responsive seed varieties shifts the entire production function up, which allows more wheat to be produced with no change in fertilizer input. But something else happens. Even at a constant fertilizer-to-wheat price ratio a *greater* use of fertilizer is most profitable. After the technical change, the new optimal point occurs at *F* where an increased fertilizer (0*I*) is used to produce the much-increased wheat (0*C*).

The greater wheat output consists of two reinforcing effects resulting from the farmer's access to technical change. The first response is an increase in wheat yields from *A* to *B*, even when the farmer maintains a constant fertilizer use at the old optimal level (0*H*). This increase in yield

[11] Think of a reduced wheat price as rotating the line $0M$ counterclockwise, which provides incentives for the profit-maximizing farmer to reduce wheat production. Chapter 3 explains why producers set output at the level where its marginal cost equals price.

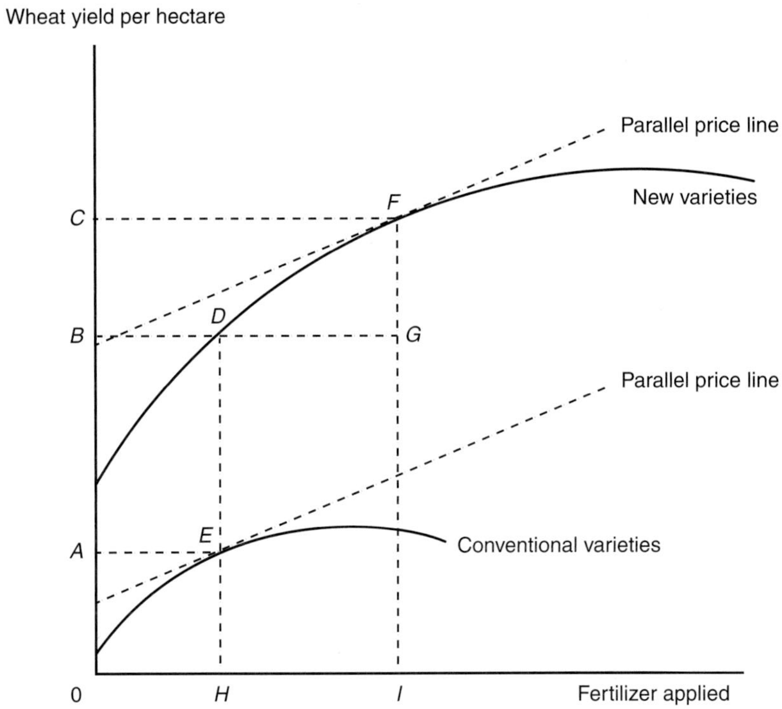

Figure 11–4
Effect of technical change on fertilizer use and crop yields

occurs because the higher yielding varieties shift the production function upward from *E* to *D*. The second response is a behavioral adjustment to the fertilizer-intensive nature of the technical change. Profit-maximizing fertilizer use increases from *D* to *F*, even though the price relation between wheat and fertilizer is constant. Fertilizer use increases from 0*H* to 0*H*,[12] and output reaches its new optimal level of 0*C*. The social dynamics of the shift in the production function could be a matter of simply watching and learning. As farmers see other farmers' outcomes with fertilizer applied to high-yield seeds or experiment through trial and error themselves, they gradually shift upward their production function and increase their demands for fertilizer.

A major factor affecting the farmer's behavior shown in Figure 11–4 is the relative price of fertilizer to wheat. This relationship, shown by the slope of the dashed parallel price line, determines the economic incentives to increase or decrease fertilizer use. In many developing economics, these prices are influenced considerably by government policy. A government policy that addressed hunger by subsidizing fertilizer would rotate the dashed parallel price line clockwise, thereby further increasing fertilizer use and increasing profit-maximizing wheat output. A fertilizer subsidy policy could also increase water pollution. An important policy question centers on the costs to society of additional water pollution versus society's benefit of reducing hunger through greater food production.

Output-Price Relations

Figure 11–5 shows what happens when the price of wheat increases or, equivalently, when the price of fertilizer falls. As fewer units of wheat are needed to trade for a single unit of fertilizer, the farmer receives an important price signal to use more fertilizer to grow additional wheat. The farmer can now convert one unit of fertilizer into a larger amount of wheat than before. As long

[12] How will poor farmers pay for the increased use of fertilizer from 0*H* to 0*I*? With the added revenues from the added wheat yields. The incremental revenues are larger than the cost of the increased fertilizer. Pretty, Morison, and Hine (2003) describe economically efficient input adjustments made by small farms in Africa, Asia, and Latin America.

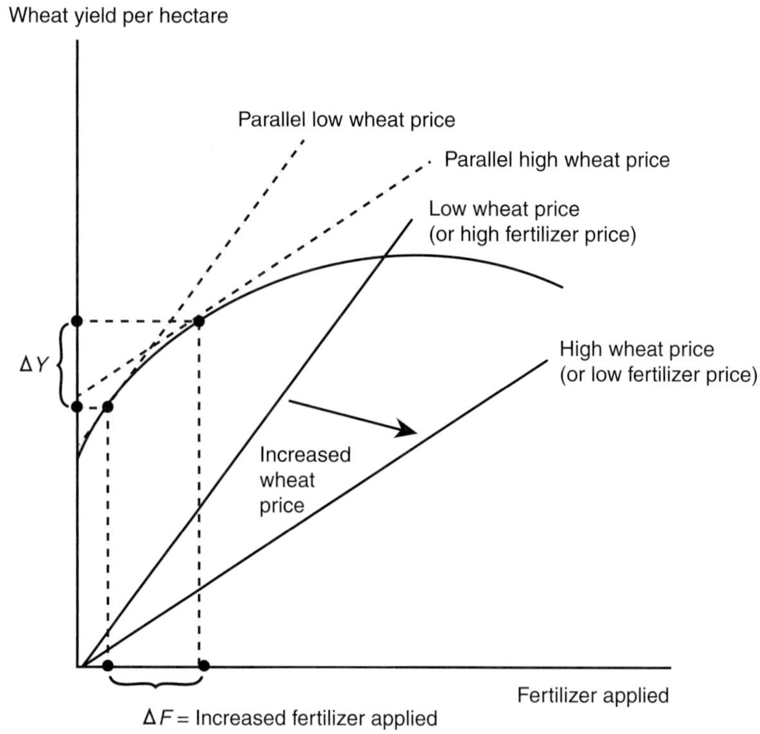

Figure 11–5
Impact of relative price change on fertilizer use and crop yields

as the farmer can exchange a unit of fertilizer for enough wheat to pay for more than the unit of fertilizer, the price system says it is economically attractive to increase fertilizer use.

The farmer will stop expanding only when an extra unit of fertilizer fails to produce enough additional wheat to pay for the fertilizer in extra wheat output. Note on Figure 11–5 that the old level of output occurs where the low wheat price is tangent to the production function. Expansion will continue until the parallel high wheat price line is tangent to the same production function, which indicates higher levels of fertilizer use and more wheat output.

This discussion of output and price relations leads to a general principle: Profit-maximizing farmers can be expected to apply inputs more intensively to increase their commodity output when the commodity's price increases or if the input price falls, if other things are constant. This positive supply response is shown in Figure 11–6, where attention shifts to outputs. The top panel of the figure is built from the technical relationship between input and outputs already shown in the factor-product relation of Figure 11–3. Because the price of wheat is independent of quantities sold in competitive markets, the farmer's total revenue from various possible levels of wheat output is a straight line. If the total revenue line pivots counterclockwise, a given amount of wheat produces a larger revenue, which can only occur when wheat prices increase.

The total cost curve is built from the production function and from a known fertilizer price. Each level of output requires some known level of fertilizer to produce that output. The price per unit of fertilizer combined with the total amount of fertilizer applied determines the cost of producing that output. The top panel of Figure 11–6 shows that the profit-maximizing output, *D,* is at total cost level *E,* where the difference between total revenue and total cost is maximized, which occurs where profits are highest. Figure 11–3 showed that the profit-maximizing wheat output is *F* units, which is identical to *D* units in the top panel of Figure 11–6. At this point, the slopes of the total cost and total revenue are equal, which means that marginal revenue equals marginal cost.

The bottom panel of Figure 11–6 shows the *marginal* conditions as they relate to the farmer's output decisions. The vertical axis now measures cost and price per unit output. Since the farmer's output price is constant for any level of output supplied to the market, price is shown as the horizontal line P_0. Both average costs and marginal costs rise with higher output because inputs are

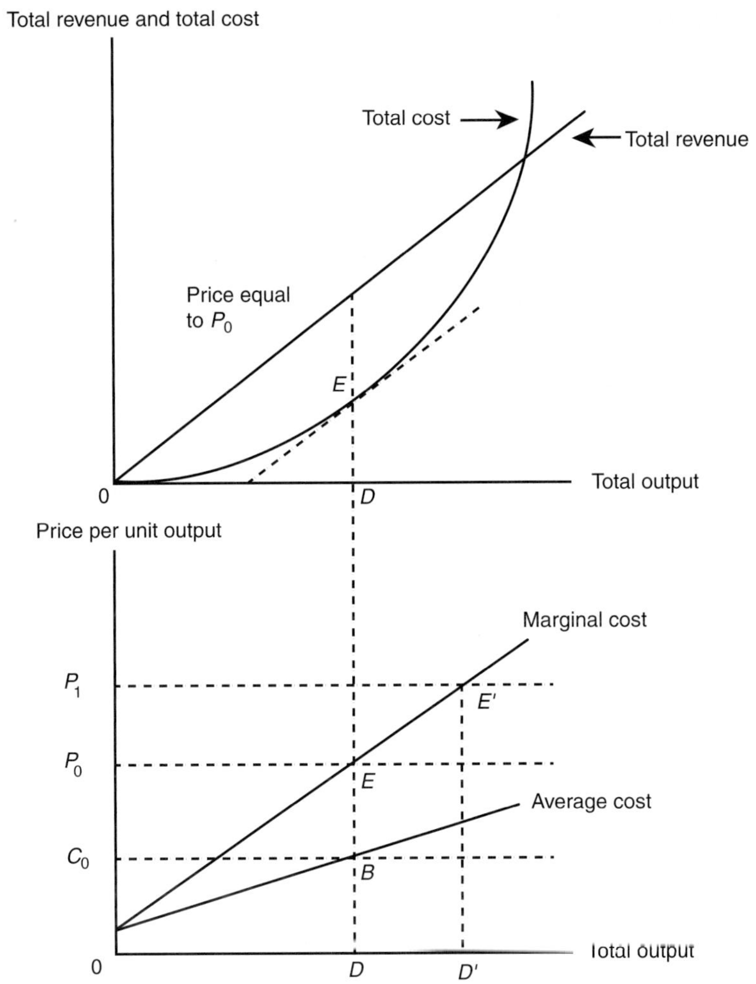

Figure 11–6
Relation between wheat output and the price of wheat

being used more intensively to produce output, which leads to lower marginal productivity. Whenever marginal cost lies above average cost, marginal cost raises average cost.

The farmer's profit-maximizing output level in the bottom panel of Figure 11–6, at output level *D,* shows the same choice as the top panel point *D,* It is also the same choice as Figure 11–3 at output level *J.* At this optimal point, the marginal cost of increasing output equals the marginal revenue gained. Marginal revenue is the output price, P_0; each added unit of output produced and sold by the farmer brings in a market revenue equal to the market price of the output. So the farmer's profit-maximizing output choice is where the marginal cost curve intersects the market price of output (at this profit-maximizing point, marginal cost exceeds average cost). This result is important, both for the farmer and for the design of policy that influences farmer's behavior. If a government policy increases the output price from P_0 to P_1, the farmer's profit-maximizing decision is to increase marginal cost from *E* to *E*', so the farm's output increases from 0*D* to 0*D*'. The farmer's supply response to higher crop price incentives (i.e., the relation between quantity of output supplied and output price) is nothing more than the farmer's marginal cost curve shown in the bottom panel of Figure 11–6.

So, as you might recall from Chapter 3, the farmer's supply curve is equal to the marginal cost curve. Anything that shifts the marginal cost curve will also shift the farmer's supply curve. Developments or imports of new technology, access to irrigation facilities, changes in weather or climate, or building of new reservoirs that permits better management of droughts will also shift the supply curve. Many of these shifts in the marginal cost curve result from government policy or investments, so policy analysts have real interest in farmers' responses.

Impact of Agriculture on the Environment

Despite the importance of increased crop yields to provide adequate food for the world's growing population, modern commercial agriculture can be environmentally destructive, especially in developing countries where environmental regulations and control measures are just beginning. Moxey, Whitby, and Lowe (1998) and Van Huis and Meerman (1997) describe some challenges in the measurement of linkages between agriculture and environmental quality. Negative environmental impacts of agricultural policies include (1) loss of ecosystems, (2) reduced biodiversity, (3) damage to soils, (4) pollution of water, wetlands, and marine environments caused by agricultural chemicals, (6) global warming through deforestation, and (7) loss of land for poor peasants. General linkages between agriculture and the environment are discussed by Hodge (2000), Collins (2001), De Janvry and Sadoulet (2002), and Evrendilek and Ertekin (2002). Linkages in the United State are analyzed by Crosson and Brubaker (1982) and more recently by Plantinga (1996). Now let us consider how several elements of agriculture influence the environment.

Soil Erosion. Most agriculture requires soil which can be either a renewable or a depletable resource. Sustainable agricultural practices replenish soil productivity over time through nutrient recycling from crop residues and from natural fertilizers. Where this occurs, agricultural output can continue at current levels indefinitely. In fact, the situation in many of the world's major agricultural areas is rather different: Soil erosion and degradation are widespread (e.g., see the analysis by Boardman, Poesen, and Evans, 2003). Soil erosion damages crop productivity by reducing the availability of water, nutrients, and organic matter; water resources are also degraded by sediments and pollutants associated with erosion. Soil loss rates are typically highest in developing countries where the costs and dangers of hunger and malnourishment are already highest. The problem of soil erosion grows as more marginal land is brought into production to support growing human populations and less crop residue is returned to the soil. Soil loss rates in Europe range between 10 and 20 tons per hectare per year. In the United States, soil erosion on cropland averages 16 tons. In Asia, Africa, and South America, soil erosion rates on cropland range between 20 and 40 tons (Pimentel 1993).

Often, farmers can greatly reduce erosion and soil degradation in a number of ways such as rotating crops, fallowing the land, alternating crops (e.g., grain and legumes), or allowing periodic resting between production seasons. Recall that individual farmers respond to economic incentives that affect the potential for profits. What are the farmer's costs for taking the actions described? The farmer's costs include lost revenues in any year when the land is out of production and reduced revenues in years when the land does not produce its highest-valued crops. A farmer who considers measures to control erosion makes an economic calculation to see if the immediate costs of controlling erosion are worth the long-term benefits.

Here is a simple example. Suppose a farmer can produce \$100,000 in annual revenues by continually growing the highest-value crops with no provision for rebuilding soils or erosion control. Under these conditions, erosion will cause an annual decline of about 1% in yield. An effective program to control erosion will require the farmer to incur a one-time cost of \$15,000. Is the program worth it to the farmer? The answer depends on the discount rate the farmer uses to balance present versus future costs, as described in Chapter 6. A 1% yield loss means a monetary loss of \$1,000 annually. But this is not a one-shot loss; it continues into the future. How does the farmer evaluate this stream of losses resulting from one year's erosion? In economic terms, the farmer applies a discount rate. Suppose the farmer selects a 10% discount rate. The present value (*PV*) of the stream of losses extending indefinitely into the future equals

$$PV = [-\$1{,}000]\,[1 / 0.10] = -\$10{,}000 \tag{11.3}$$

which is the formula to use for the constant stream of benefits or costs (−\$1,000) accruing into the indefinite future.[13] The present value of the benefits of erosion control are thus \$10,000, which

[13] The formula is $PV = B/r$, where B = annual benefits and r = discount rate. The formula can be derived by methods described in Appendix H.

is not enough, in this example, to justify the one-shot cost of $15,000. Under these conditions, it is economically optimal to continue the erosive practices, but it is hardly ecologically sustainable. According to this economic logic for this particular economic environment, the economically rational farmer will leave severely degraded land for the next generation.

Many farmers face precisely this kind of economic pressure to maximize short-term profits, and high discount rates are part of the problem. Observe what happens if a lower discount rate, say 5%, is used. The benefits of erosion control, now calculated at $20,000, exceed the costs and in principle make erosion control economically attractive. Even so, bearing the short-term losses might burden the farmer. An ecologically sustainable soil management policy depends on the farmer's foresight, low interest rates, and the financial ability and flexibility to invest in erosion control today for long-term benefits.

Off-farm effects of erosion are another issue. In many areas, for example in much of the western United States, large dams have silted up with eroded soil, which ultimately limits their capacity to produce economic benefits for power production, irrigation, and control of floods. Heavy siltation also causes extensive damage to the ecology of rivers. These costs are externalities from the individual farmer's point of view; that is, on-farm investments to avoid these off-farm effects produce little or no benefit to the farmer, so the farmer does not have incentives to invest in soil erosion control.

Irrigation. In arid regions, irrigation is an important factor in increasing crop yields, in some cases even more important than fertilizer. Irrigation has the potential to greatly improve yields, and sometimes it allows for multiple crops in areas with alternating dry and wet seasons. The most optimistic projections of future yield increases in developing nations rely on expanded irrigation.[14] However, short-term benefits of irrigation often bring about long-run environmental damage. Poor drainage causes irrigation water to accumulate underground, eventually waterlogging fields. In some areas, water that reaches the surface evaporates and leaves behind a buildup of dissolved salts that contribute to salinization of soils. Irrigation also transports fertilizer and pesticide runoff which damages area surface water and aquifers.

Irrigation typically occurs where surface water is scarce. It can lead to overdraft of aquifers, if these underground reservoirs are pumped faster than the natural water cycle recharges them. Many farmers pump from a shared aquifer, so no individual has the incentive to refrain from pumping to conserve the resource—the classic problem of a common property resource. The Ogallala aquifer that supports much of the high-plains agriculture from Texas to Nebraska in the U.S. is as much as half depleted in some areas, and its level continues to fall. Continued groundwater declines are taking place in India, northern China, and central Asia.

Water supply limits may be the most significant constraint on future agricultural expansion in dry parts of the world. Irrigation accounts for 65% of total water use worldwide and over 80% in developing countries. Much of China and the Indian subcontinent are near to the limits of their water supply, but urban and industrial water demand continues to rise to meet the needs of growing populations. Powerful economic incentives attract the individual farmer to expanded irrigation. Nevertheless, problems of externality and common property resources mean that this expansion will often produce environmental costs larger than economic benefits.

Fertilizer. The continued increases in crop yields observed in modern agriculture around the world have depended strongly on increased use of fertilizer. Modern agriculture relies on a mix of inputs that includes fertilizer, irrigation, pesticides, mechanization, and high-yielding crop varieties. Although fertilizer is part of the picture, it really serves as a proxy for the complete package. Higher use of fertilizer correlates so strongly with greater use of the other inputs that simply measuring use of fertilizer reveals much about the degree of agricultural modernization.

---

[14] A few of the challenges facing on-farm water users and water policymakers are described by Falkenmark (1999). Heileg, Fischer, and Van Velthuizen (2000) describe the role of irrigation in contributing to China's food security, and Postel et al. (2001) describe its role in reducing production risks in several developing countries.

Fertilizer supplies nutrients to the soil, and from soil to crops. Fertilizers typically supply the three major nutrients of potassium, nitrate, and phosphate. Often, though, many of the nutrients do not reach the crops but leach into surface water and groundwater where they become pollutants. Too much nitrate in water damages human health. Nitrates and phosphates promote algae growth that limits other life in various bodies of water.[15] Another damaging effect of fertilizer is harder to detect but important. As large quantities of nitrate, phosphate, and potassium accumulate in the soil, other important nutrients required in small quantities (micronutrients) are depleted. Over time this depletion reduces crop yields and their nutritional value. Because these impacts are gradual and hard to detect from one year to the next, there may be little incentive for the individual farmer to invest in control measures.

Pesticides. The use of pesticides has also grown extensively with the worldwide spread of modern agriculture. Several health and environmental problems have been caused by this increase:

- Pesticides may damage the health of farm workers. Pesticide poisoning is a serious problem throughout much of the developing world, especially where application of pesticides requires literacy and specialized training.
- Pesticide residues in food may damage the health of consumers. The cancer-causing effects of some pesticides are well publicized, and research has examined their impacts on human reproduction.
- Pesticides damage ecosystems. Polluted groundwater from pesticides occurs commonly in many agricultural areas. Unintended killing of species can result in pest outbreaks that are much worse than the original pest problem (e.g., Makin and Mendelssohn, 1997, describe bat killings by pesticides in Israel).

Despite the environmental damages produced by pesticide use, as described in Chapter 7, the environmental costs are hard to measure in monetary terms. And environmental costs are hard to incorporate into farm-level decision making or into national environmental policy. Furthermore, manufacturers of agricultural chemicals are motivated to promote the expansion of pesticide use as well as to avoid publicizing data on environmental costs. What economists call *information asymmetry* poses a special problem. The producers of a pesticide typically are in a position to know the most about its chemical composition and its potential impacts on human health and the environment. It is likely that the pesticide's external costs will not be accounted for in the pesticide's price, and important environmental consequences of its use will be ignored by the user.

Habitat. As population grows, new agricultural lands are often developed by converting land from other uses, often by clearing forests, wetlands, prairies, and other undeveloped lands. A significant environmental issue occurs when the land converted to agriculture had important ecological values, because the undeveloped lands contributed to wildlife habitat, biodiversity, recreation, watershed protection, and existence values.

The farmer who considers converting land compares the private costs and benefits of conversion (change in net income), but the farmer has no incentive to compare social costs (change in net benefits). Few if any of the environmental costs of wildlands converted to agriculture face the farmer, so even if society's environmental benefits displaced are high, the farmer pays only a small part of those costs. The public goods nature of environmental values is responsible for the loss of significant habitat, especially in developing countries where tropical deforestation from agricultural expansion is a major global ecological issue.

---

[15] Many agricultural areas in the U.S. Midwest suffer from these problems. Large parts of the Mediterranean Sea have incurred ecological damage from agricultural fertilizer pollution, with large areas of algae covering coastlines in the Aegean Sea. Fertilizer use has created environmental problems in the economies of Russia and eastern Europe, including the extinction of several local species.

Agriculture's Multifunctionality. Despite the negative environmental impacts of agriculture just described, agriculture also has the potential to produce positive externalities to the natural and human environment. In addition to producing food, agriculture contributes to several important functions:

- Conservation of land
- Developing of water resources
- Preservation of the natural environment
- Attractive landscapes
- Maintenance of cultural heritage
- Recreation and relaxation
- Viability of rural communities
- Food security

For these reasons, many policy analysts have claimed that agriculture deserves more price support and import protection than other economic sectors. Agriculture is a special sector because it produces nonmarketed externalities and public goods jointly with marketable food and fiber. An important question with significant policy implications is whether the unrewarded positive externalities from agriculture exceed the uncompensated negative externalities by more than the net positive externalities produced by other sectors. If the answer to this question is yes, economic efficiency suggests subsidizing agriculture to boost its production to socially efficient levels. Anderson (2000) and Bonnieux, Rainelli, and Vermersch (1998) present a positive view, Bradshaw and Smit (1997) have a competing view.

Institutional Design for Improving Food Policy Outcomes[16]

The right to food security is one of the most widely proclaimed rights in international human rights law.[17] One definition of food security is good quality, nutritious food available in sufficient quantities all year round and located at the right places at affordable prices for everybody. Maxwell (1996) describes the challenges in attempting to measure food security. In poor countries, several problems characterize the food market:

- Poor consumers have little income, so they are often excluded from the market for food. McCalla (1999) emphasizes the importance of income in producing food security and Minot and Goletti (1998) describe the role of food staples exports in supporting income.
- If food supplies are scarce, their prices rise; the rich and middle class can afford to buy food, but the poor cannot. In addition, the most vulnerable to malnourishment are often hard to identify so that they can be targeted for food assistance. This problem is described by Clay, Molla, and Habtewold (1999) and by Jaspars and Shoham (1999).
- Low prices damage domestic incentives to improve agricultural productivity, which is essential for poor countries to escape food poverty. Pinstrup Andersen and Pandya Lorch (1996, 1998), Pinstrup Andersen (2000), and Tweeten (1999) summarize challenges of eliminating global hunger. A food glut caused by high harvests or cheap imported food causes food prices to drop: In the short run, the small farmer may be ruined because he or she cannot sell her produce for enough to pay her costs. In the long run, the prices are too low to pay for future production costs, so the farmer might not be able to survive.

---

[16] Source: Part of the material in this section comes from the proceedings of a 2001 food security conference held in Bonn, Germany. See International Food Policy Research Institute (2001a, b). Available from *www.ifpri.org/2002conference*.

[17] The following declarations and conventions make that proclamation: the Universal Declaration on Human Rights, Article 25; the International Covenant on Economic, Social, and Cultural Rights, Article 11; and the Convention on the Rights of the Child. For a right to food security to be carried out, it must be accompanied by the power to sanction those who violate that right.

Table 11–2 Measures for improving food security

Policy	Increases demand	Increases supply	Time horizon[a]
Emergency food relief		X	S
Increased production		X	M
Natural resource protection		X	M
Green Revolution		X	M
Increased water use efficiency		X	M
Addressing global warming		X	M
Small-farm friendly		X	M
Biotechnology		X	M
Computer, information literacy	X	X	M
Food stamp programs	X		M
Girls' education	X		M
Health care	X	X	M
Property rights		X	M
Promotion of civil society		X	L
Elimination of corruption		X	L
Open transparent government		X	L
Reduction of violent conflict		X	L
Macroeconomic growth	X	X	L
Reduced population growth		X	L

[a]S-short term; M-intermediate term; L-long term.

Hunger is caused by reduced choices of essential foods available to individual consumers. One way to reduce hunger is to expand available choices, which are a function of incomes, food prices, food supplies, and consumer knowledge.[18] The hunger of any particular individual can be solved directly by providing more income or more food directly. Addressing hunger for a whole society is more complicated because the direct approach is either too expensive or it takes too many productive resources away from other valued uses. Still, the food system in most countries offers mechanisms for policy interventions to reach poor people and provide sustainable improvements in their access to food. Table 11–2 summarizes some of the institutions that could promote food security.

Short-Term Measure: Emergency Relief

On average, about 10% of the total aid given to developing countries is used for emergency food assistance. Although short-term, relief interventions to feed starving people are needed to save lives, they do not offer a long-term solution for the problem of hunger. Figure 11–7a shows how short-term food security measures are designed only to shift the vertical supply of food from a lower to a higher level. The result is a decrease in effective price which makes more food available to buyers in the marketplace.

Intermediate-Term Measures

Figure 11–7b shows how intermediate-term measures increase the supply from S_0 to S_1 calories per day or increase the demand for food from D_0 to D_1 within the existing political and institutional structures. These measures are designed to provide economic incentives to food producers or food consumers.

[18] Chavas (2000) developed a microeconomic model that links food security and a number of policy instruments for reducing malnutrition and starvation. Nelson (2000) describes the strong relation between low income and malnutrition.

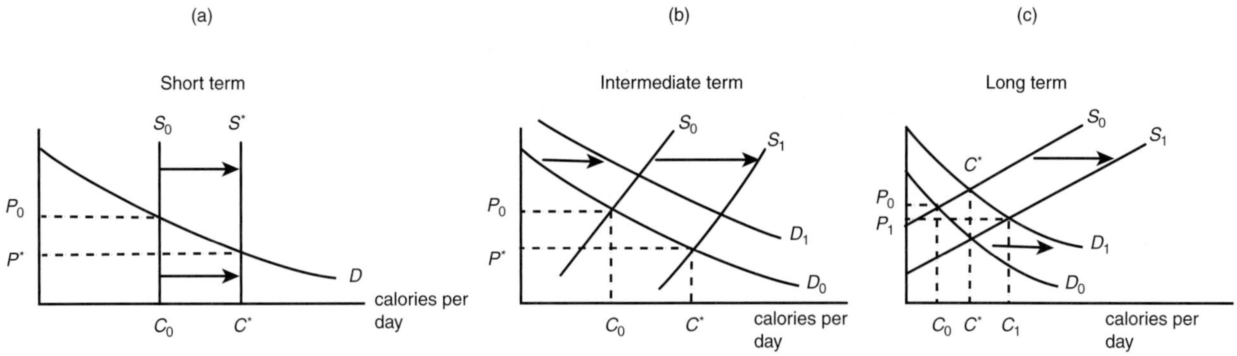

Figure 11–7 Sustainable food security through short-term, intermediate-term, and long-term measures

Increasing Production. A long-debated food policy question centers on whether to increase domestic food production in the less developed countries or to import what is often cheaper food from abroad. There are several reasons for believing that many developing countries best cope with hunger by increasing the percentage of their food consumption from domestic sources. Barrett, Bezuneh, and Aboud (2001) however, describe some limits of expanding domestic food production.

An important reason in favor of domestic sources is that food imports use up scarce foreign exchange. Most developing countries are unable to pay for food imports with their own currencies. Traders require them to pay in an internationally accepted currency, such as the American dollar, that is earned by exporting other domestic products to the world market. Increased foreign exchange used for food imports takes it away from other productive uses such as buildings, machines, engineering expertise, and other capital goods that could raise the productivity and incomes of domestic labor.

Although the scarcity of foreign exchange points to expansion of domestically produced food, it would be a mistake to push the argument too far. Food self-sufficiency is rarely a good policy, because of the law of comparative advantage. Peterson and Valluru (2000) present a method for measuring a nation's comparative advantage in food. A nation provides best for its citizens when it specializes in those products for which it has a comparative advantage. If its comparative advantage is not in food but in cloth, for example, the country earns a larger income for its citizens by manufacturing and exporting cloth and using the earnings to purchase food from nations that can produce food more cheaply. Total self-sufficiency in food for all nations normally produces a loss in each nation's income. Nations that possess a comparative advantage in agriculture because of soil type, climate, quality and quantity of farmable land, and the like (e.g., United States), will be net food exporters. Other nations with a comparative advantage in manufacturing, trading, and services—such as Japan, Taiwan, and the United Kingdom—will be able to consume more and better food by specializing in nonfood items and using the revenues to import food.

Natural Resource Protection. Many developing countries find that poverty, weak agricultural productivity and natural resource and environmental degradation work together in a vicious circle. In contrast, poverty reduction, agricultural development, and environmental sustainability are likely to occur when agricultural development is broad based and inclusive of small- and medium-sized farms, market driven, participatory and decentralized, and accompanied appropriate technological change that improves productivity. Therefore, food security policy should pay more attention to sustainability, to defining and guaranteeing secure property rights, and to collective action to encourage farmers to adopt productive technologies and resource management practices, especially in resource-poor areas. Ruben, Moll, and Kuyvenhoven (1998) and Scherr (2000) analyze the challenges of designing sustainable policies to promote food production while protecting natural resources. Research, technology, and policy actions are needed to guide land management to end soil depletion as well as other kinds of soil degradation, deforestation, water salinity, and chemical pollution.

The FAO is working on the following four-part comprehensive strategy to reduce conflicts between food production and the environment for the purpose of long-term food security and human development (Fresco and Steinfeld, 2004).

- Shifting from input-intensive technologies to knowledge-intensive technologies with the aim of maximizing efficiency in the use of scarce natural resources
- Establishing well-defined property rights to all natural resources. Without such property rights, users of open-access resources such as drylands or rainforests face little incentive to develop and use them in a sustainable fashion.
- Promoting small farmers' participation in decisions that affect the environment and natural resources
- Ensuring that market signals include accurate values of environmental goods, such as land, water, air, and biodiversity

Green Revolution. Another approach to food security is to ensure that changes in the income distribution resulting from agricultural policies are favorable to the poor. One great hope of the Green Revolution was that new varieties of seeds produced by its scientific research would increase food supplies, reduce food prices, and make a better diet accessible to the poor, all while providing income opportunities for poor farmers. These were certainly noble motives but what were the results?

The results were impressive, as reported earlier in this chapter. In most areas,[19] even small farmers benefited from the new varieties of seeds. Labor use and productivity increased as did income received by poor farmers and wage income received by poor farm workers. The Green Revolution increased consumption and nutrition of poor people.

Still, not all news produced by the Green Revolution has been good. Khush (1999) and Lal (2000a, b) describe some of the environmental costs and challenges produced by the revolution. History has shown that relying on a small number of cereal grains raises the risk of being wiped out by plant diseases and pests that attack any single one. One way to manage this risk is to hold a large collection of species in store in case one is attacked by disease or pests. In some farming areas, the Green Revolution may have reduced farmers' ability to cope with risks, because the new higher yielding varieties of seed eliminated competition from some of the lower, yielding more traditional seeds. As discussed earlier in this chapter, the Green Revolution may be promoting environmental consequences that are not yet measurable.

Increased Water Use Efficiency. Worldwide, there are no new water supplies. Increased demands for water resulting from population growth are reducing surface water supplies and depleting some underground aquifers. According to the World Resources Institute (2004), 69% of the world's water use is for irrigated agriculture. Globally, the renewable fresh water supply is about 7,000 cubic meters (1.8 million gallons) per capita per year, but many dry places see much more water supply constraints. In Egypt, for example, the supply is 50 cubic meters per capita per year. So water supply constraints have major implications for future food security. None of the water needed to meet increased food demands through 2020 will come from new supplies, and very little will come from existing uses. Most will come from more efficient water use (i.e., producing larger quantities of irrigated crops with the same or less water).

Comprehensive water policy reform is needed to save water, improve use efficiency, and increase crop yields, while limiting the risk of violent armed conflict and reducing the risk of armed conflict between countries sharing scarce supplies (e.g., Israel and Palestine). Typically, irrigation water is priced at or near zero, since policymakers in many cultures (including the western United States) believe that poor irrigators cannot afford to pay a price for water equal to its real marginal opportunity cost. So irrigation water pricing will be hard to implement in cultures where

---

[19] Singh (2000) analyzes some of the food security benefits and environmental costs of the Green Revolution in India. Mwangi (1997) describes the potential for increased crop yields with greater fertilizer use in sub-Saharan Africa.

inexpensive or free water is an old practice. The best way to implement reforms will vary widely among regions and countries, depending on the existing scarcity of food and water, acceptance of a market culture, level of economic development, nature of production agriculture, and extent of existing water developments.

What would economically efficient, equitable, and sustainable reforms look like? Well-defined water rights are a good place to start. Making these rights transferable to new users increases the economic incentives for economically efficient water conservation measures. Examples of efficient conservation include concrete-lined ditches, recycling, desalination, modern water control systems, leveled fields, and drip and sprinkler irrigation. Another reform would be to price water equal to the environmental costs of its use as shown in Chapter 10. The results would be to increase the lives of nonrechargeable aquifers and reduce other environmental damages produced by irrigation.

Addressing Global Warming. Large numbers of unresolved challenges are posed by increasing carbon dioxide CO_2 concentration, higher temperatures, and other factors associated with global warming. Sanchez (2000) describes linkages between climate change and food security. Farmers can contribute to solving the global warming problem by sequestering carbon in their agricultural systems (i.e., through their pasture, crop, and forest management strategies). Marland, McCarl, and Schneider (2001) analyze the economics of soil carbon sequestration.

Low and reduced tillage practices in wheat–rice rotations can conserve soil and water and mitigate global warming. Planting more trees and reductions in clearing of existing trees will help to absorb CO_2. Where economically feasible, it is estimated that the transfer of croplands to agroforestry and tree enterprises can absorb 2 to 3 tons of carbon per hectare (2.47 acres) per year. In all these cases, the policy challenge is how to make it economically attractive for farmers to invest in enterprises that absorb CO_2, when these enterprises might normally be less profitable than planting and harvesting crops for sale to markets. Other policy challenges are to decide who will finance the environmental improvements and to determine the cost of reduced global warming in terms of food security for the world's poorest consumers.

Small-Farm Friendly. Owners of small farms will benefit from improved human resources and access to productive resources only if they have access to functioning markets for outputs, inputs, consumption goods, capital, and employment. It is especially important that governments ensure that markets are not biased against small farmers, less favored areas, or poor consumers. Capital subsidies as well as infrastructure investments in developing countries must be made available to small farming communities. Rural parts of developing countries that consist mostly of the poor, the landless, and small farmers are often the last regions to get investments in infrastructure; partly for that reason, their food markets are poorly developed. There is a common and often mistaken view that small farms are not competitive (i.e., that public investments in larger farms produce more food per dollar invested). To ensure food security, rural poor people must obtain their fair share of the benefits from globalization of markets. Rural poor people, whether farmers or not, will suffer if they are excluded from participation in the mainstream market economy.

Biotechnology. Biotechnology refers to the use of biotechnical methods to modify the genetic materials of living cells[20] to produce new substances or perform new functions. Serageldin

[20] Biotechnology is based on the discovery of the double-helix structure of DNA in 1953 by Francis Crick (1916–2004) and James Watson (1928–). Crick was working at the University of Cambridge in 1953 when he and Watson seized on the idea that the DNA molecule resembled a twisted ladder. After making the discovery, Crick marched into a Cambridge pub and announced that he and Watson had found the secret of life. Crick and Watson shared the 1962 Nobel Prize for physiology or medicine. The discovery laid the foundation for the biotechnology industry, enabling scientists to engineer bigger tomatoes and doctors to pursue gene therapy to treat disease. In 2003, 7 million farmers in 18 countries grew genetically engineered crops which allowed them to produce food with fewer pesticides and a less reliable water supply.

(1999) describes some of the challenges of designing policies to promote the benefits of biotechnology in agriculture. Some products of genetic engineering include crops that are disease- and pest-resistant, delayed-ripening fruits with a longer shelf-life, and cow hormones that increase the production of milk. Currently, work is in progress to develop foods containing vaccines and antibodies that fight diseases like cholera, and the search is on to modify foods to remove common allergens from them. Worldwide, the following food groups account for 90% of all food allergies: peanuts, tree nuts like walnuts and almonds, legumes, milk, eggs, fish, soybeans, and wheat.

Much of the opposition to genetically modified (GM) foods is based on the concern that natural plants could be forever changed, maybe for the worse, when exposed to genetic material from altered crops. The debate has been especially intense in Europe, where GM crops were banned from 1998 to 2003. The EU also is debating stricter controls over the import of GM foods. In the United States, the debate is escalating. Opposition to biotechnology prevails in other parts of the world: A state in Brazil has declared itself a biotech-free zone, and Japan is taking a hard look at GM foods. These hard choices have elevated the importance of the *precautionary principle* long advocated by environmental groups. That principle calls for governments to restrict any activity that poses risks to the health of people or the environment, even if some cause-and-effect relationships are not fully established.

Adapting GM crop technology to benefit poor farmers and consumers in developing countries will require establishing institutional mechanisms, such as the following to guard against public health and environmental risks:

- Legislation to balance intellectual property rights of the GM developer with farmers' rights to save, reuse, trade, and sell GM seeds.
- Food safety and biosafety regulations to ensure an acceptable risk level for society, including risks of increased hunger associated with failure to use biotechnology.
- Safety measures for the strategies involved in evaluating and managing risks and in informing the public about risks.
- People who are most susceptible to hunger if GM foods are not allowed should be included in the debates and decisions about GM foods.

Despite the controversy surrounding the use of biotechnology in food and agriculture (Phillips 2002), it has great potential to improve food security for the poorest food producers and buyers. Whether it succeeds depends on the economic and social policy environment and on the nature of the intellectual property rights governing development of GM technology. Biotechnology is one measure to be used in comprehensive efforts to address food insecurity, and it should be used where it offers the best available option for addressing a problem.

Computer Literacy. Opportunities abound for using modern information and communications technology in agricultural production and distribution. Currently, most information and communications technology projects in developing countries are small scale because of costs, competing approaches, and lack of interest from both the public and private sector. The Internet, though, has considerable potential to promote rural development and increase food security in the following ways:

- Reducing isolation and marginalization of rural communities
- Facilitating dialogue among communities and people who influence them
- Encouraging participation of communities in decisions that affect their lives
- Coordinating local, regional, and national development efforts for increased food production efficiency
- Delivering information and skills quickly and flexibly
- Overcoming barriers that block farmers from sharing information

If farmers have access to the Internet, they could obtain information from agricultural extension services through electronic mail and agricultural price data through government Web sites.

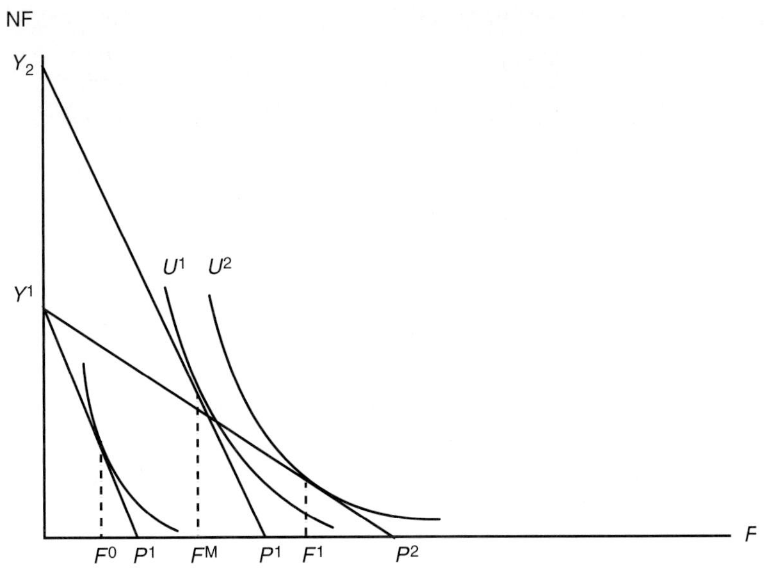

Figure 11–8 Welfare effect of income support programs versus food stamp programs

Food Stamp Programs. Is there a way to reduce the nutritional problems of the poor while maintaining adequate supplies of food? Some countries such as Sri Lanka and the United States have used food stamp programs to subsidize food purchases by the poor in urban areas.[21] Food coupons allow people with low incomes to buy certain foods at a fraction of the market (retail) price. Figure 11–8 demonstrates the impact of raising the purchasing power of people with the greatest need (lowest income) by reducing their effective price of food. The budget line is pivoted counterclockwise for food (*F*) from line Y^1P^1 to line Y^1P^2 while leaving prices unchanged for nonfood (*NF*) items. Such programs provide access to food for those who need it most while protecting the price incentives for domestic farmers to maintain high food production. The result is to increase the demand for food from F^0 to F^1.

An alternative way to make food affordable is for governments to reduce prices of the specified foods to everybody—rich, middle income, and poor. This method, however, incurs a much higher cost to the nation's taxpayers. For example, a food stamp program designed to make cereal grains affordable to the poorest people might have the government pay farmers the market price (*PM*) for *X* tons of grains, then allow the lowest income recipients to buy those *X* tons at a subsidized price (*PS*). The cost of the program to the government is $(PM-PS)$ times *X* tons. However, if the government paid for that subsidy for all tons (*Z*) of grain consumed $(Z >> X)$, then the program cost is $(PM-PS)$ times *Z* tons, a financial burden considerably higher than the food stamp program.

Figure 11–8 also shows that an income support program for the poor guarantees much less food consumption than a food stamp program. Notice that the income support program shifts the budget line from Y^1 to Y^2 without affecting food prices, while producing a lower utility-maxmizing food consumption (utility is u^0, u^1, and u^2) than does the food stamp program with similar program costs. The food stamp program pivots the budget line counterclockwise from point Y^1, thereby assuring greater food purchases. In fact, some food stamp programs operate on a sliding scale in which the poorest people pay the smallest percentage of the market price for program foods.

[21] Gundersen and Oliveria (2001) use principles of insurance to show why the generally successful U.S. stamp program does not guarantee a well-nourished household. A proposed food price cash subsidy program for Bangladesh is analyzed by Dorosh and Haggblade (1997). Ruel, Haddad, and Garrett (1999) describe some of the challenges for food security in urban areas of developing countries. Seeth et al. (1998) describe how family garden plots helped poor and middle-income urban dwellers in Russia improve food security in the early 1990s when Russia was going through a difficult transition period.

Example 11–3
Food Security, Education, and Empowerment of Women

Women with low incomes become empowered as their food security increases. Data from the International Food Policy Research Institute (IFPRI) show that agricultural productivity increases as much as 20% when women are given the same inputs (e.g., land, capital, seed, fertilizer) as men. The IFPRI also has found that household agricultural output can be increased as much as 20% by reallocating inputs from men's agricultural plots to women's.

Increases in women's education and status in the household have had powerful effects: These increases contributed to more than half of the reduction in children's underweight rates in 90% of the population of developing countries between 1970 and 1995. In Brazil, income in the hands of mothers has four times the impact on children's height for age as income in the hands of fathers. In Africa, if men and women had equal education, it is estimated that child mortality would have been 25% lower in the 1990s.

When only half of the potential labor force is able to read and write, to obtain credit, to develop a work skill and obtain work, reduced output can be expected. Systematic exclusion of part of the potential labor force from access to factors of production will affect the level of growth in economic production. If, worldwide, women and men had more equal educational opportunities, incomes and economies would likely grow faster.

The demand for girls' education is more sensitive to changes in prices than demand for boys' education. If governments can establish pricing policies that can increase educational opportunities, especially for the poor, these policies will have a larger influence on girls' enrollment in schools. Price elasticities of demand for girls' education are 12% to 21% higher than for boys', so interventions that reduce costs will have a more pronounced effect on girls' education than on boys'. Investments in rural infrastructure will also make a difference, because girls and women spend a large amount of time in collection activities related to water and fuel. In sub-Saharan Africa, as much as half of a standard work year can be saved by bringing water sources and fuel sources a half hour closer to households. Reduced work time hauling water and fuel frees up the time of girls and women to attend school.

Sources: Quisumbing, A. 2001. *Empowering Low-Income Women.* International Food Policy Research Institute Conference Proceedings; King, E. 2001. *Empowering Low-Income Women: Education.* International Food Policy Research Institute Conference Proceedings. Available from *http://www.ifpri.org/pubs/books/2020conpro/* [Accessed March 7, 2005].

Girls' Education. Today's global economy increasingly places a premium on education. Poor people have greater opportunities to earn income and improve their productivity when they are literate. Providing educational opportunities to girls has had beneficial impacts on family size, healthy spacing of births, child-care practices, child nutrition, and household income. Worldwide improvements in female education during the past 20 years has accounted for a greater share of gains in child nutrition in developing countries than any other factor. See Example 11–3.

Reducing fertility reduces family size and reduces the probability of large numbers of poorly nourished children. A dramatic example is an education project in rural Zimbabwe run by the Cambridge Female Education Trust (CamFed, 2005). Its executive director said that only 5% of the 387 girls who graduated from secondary school with the help of CamFed became mothers between the ages of 18 and 24. This is a dramatic reduction from the national average of 47% of girls in the same age group who gave birth by the time they turned 20. CamFed has found that the value of education is so strong that girls want to continue their education and prefer to secure their economic situation before starting a family.

Mothers use their knowledge, and the additional income earned by applying it, to improve the diets and care of their children. In fact, female education may be the strongest instrument available for reducing infant mortality and child malnutrition. A 1993 World Development Report estimated that a 10% increase in female literacy reduced child mortality by 10% in 13 African countries between 1975 and 1985 (World Bank, 1993). Mounting evidence shows that girls' education also leads to more education for their children, who do better in school because they are healthier. Educated mothers can serve as important role models for girls.

Health Care. Health risks compromise food and nutrition security. For example, in Africa HIV/AIDS has had a devastating impact on agriculture, nutrition, and food security. Efforts to contain the pandemic will improve food security. As agricultural labor becomes depleted, new cultivation technologies and seed varieties need to be developed that do not rely so much on labor, yet permit crops to stay drought resistant and healthy, particularly with respect to micronutrients that support the immune system.

The care of adolescent girls and pregnant women is important for their own health and that of their children. Good maternal nutrition contributes to infants with normal birth weights. Nutrition interventions in young women could cause them to adopt a lifetime approach to nutrition for themselves and their children.

Property Rights. Natural resource management technologies and practices may take several years to give full returns. Without secure rights to resources, farmers have little incentive to adopt conservation approaches. When farmers do not have assured long-term access to land, do not bear the full cost of resource degradation, or are not sure they will receive the benefits of their investments, they are more likely to pursue unsustainable farming practices instead of investing in natural resource conservation.[22] Governments have the ability to ensure that farmers have secure property rights and to remove restrictions that inhibit the most efficient and environmentally sustainable use of private, public, and common property. For example, some community land tenure systems provide farmers with adequate security. Governments could strengthen these indigenous systems and facilitate their adaptation to emerging conditions instead of simply eliminating them. This is especially true of local organizations that manage common properties such as wetlands, rangelands, and woodlands. When adequately empowered, indigenous farmers are often in the best position to take advantage of resource interdependencies and to establish regulations that recognize the rights of existing users of the land. As discussed in Chapter 9, such regulations are especially important in watersheds or drought-prone areas that require the capacity to move animals over wide areas to avoid the risks of inadequate localized rainfall. See Examples 11–4 and 11–5.

Long-Term Measures

Figure 11–7c shows that long-term measures can also be used to increase food supply and food security. These are measures that change the institutional structure in which food producers and consumers operate. Since long-term measures also give producers a longer period in which to adjust to economic changes, the supply curve is flatter, which indicates a larger elasticity of supply response than for intermediate-term adjustments. Several long-term measures are described next.

Governance. Democracy, protection of human rights, and the full and equal participation of men and women are essential for achieving sustainable food security for all. Molnar (1999) examines the importance of good governance that is adapted to meet the needs of local culture. Nobel laureate Amartya Sen emphasized the interdependence of civil and political rights with economic rights, which includes the right to adequate food. Sen found that civil and political rights are not only ethically desirable, but they are a valuable instrument for accomplishing food security and other economic advances (Nobel Foundation, 2004). Democratic governments that uphold freedom of speech, freedom of the press, and the rule of law rarely experience famines because their political structure requires them to respond to popular demands. The rule of law is

---

[22] Chang and Wen (1998) analyze the impacts of insecure property rights in the Chinese famine of 1958–1961. Maxwell and Wiebe (1999) analyze the linkage between land tenure and food security. Aggarwal and Narayan (2004) describe common property groundwater problems in India.

Example 11–4
Did Free Food Contribute to the Great Chinese Famine?

For more than 40 years scholars have debated the cause of the Great Chinese Famine of 1958–1961 that killed between 18 and 30 million people. Chang and Wen hypothesized that the real problem was "consumption inefficiency" (i.e., a combination of reduced food supplies and price incentives that encouraged food consumers to waste what food was available).

To test this unusual theory, the authors looked for empirical evidence of consumption inefficiency during the famine period. They found that a large number of documents, official records, and anecdotes reported wasteful consumption in the free communal dining system during the famine period. Part of this wasteful consumption was due to years of pent-up demand. Before 1958 most peasants in China never had a chance to eat as they wished, so it is easy to imagine how consumption, including wasteful consumption, would surge if a new program provided food free of charge. Some contemporary reports estimate that the waste of food in the communal dining system accounted for 11% of the grain output in 1959. In some places, the pent-up demand was so huge, and property rights to food consumption were so poorly defined, that a three-month supply of food was consumed in only two weeks. Overconsumption drained food stocks quickly. As the communal dining halls ran out of food, some halls began rationing, but by then it was too late. There were too few food reserves left. Malnutrition and low-calorie intake soon resulted in famine. Chang and Wen think the evidence points to free communal dining as the primary cause of China's great famine. Under that system, people lost their most elemental property right: the right and responsibility to consume their own food. When property rights are poorly defined, people do not take care of their property. Inefficiency in food consumption made China's already low food supply inadequate and, according to the authors, produced the famine. What do you think?

Source: Chang, G. H., and G. J. Wen. 1998. Food availability versus consumption efficiency: Causes of the Chinese famine. *China Economic Review.* 9 (2): 157–165.

the principle that every member of a society must follow the law. Governments that fail to operate within the rule of law risk producing a food-insecure future for its citizens. For example, the prosperity of Africa has suffered because weak civil and criminal justice systems or arbitrary and corrupt government administrations have discouraged private investment in food production and transport. Chattopadhyay (2000) describes the effects of poor governance on hunger despite food surplus in Zimbabwe during the 1990s.

The link between the various types of human rights works in the opposite direction as well. U.S. President Franklin D. Roosevelt recognized that people who are hungry and out of a job sow the seeds of dictatorships. In 1980, the U.S. Presidential Commission on World Hunger concluded that where people lose their right to adequate food, protection of other human rights is unlikely to occur because people are putting all their energy into simply maintaining their lives.

Promotion of Civil Society. The right to form civil organizations (i.e., organizations other than markets or governments) that are accountable and advance poor people's interests may contribute to sustainable food security. Organizations of the poor can help ensure that political actions are responsive to their needs.

Organizations of farmers aim to ensure that small farmers have access to inputs, credit, output markets, and opportunities to engage in higher-value crop production. These organizations help reduce the costs of accessing markets and improve the bargaining power of small farmers compared with larger traders (Zeller and Sharma, 2000). When cooperatives are organized by the farmers themselves, they better serve the interests of their members, strengthen the allocation and definition of property rights to individual members, and discourage individual members from engaging in illegal or unethical behavior.

Example 11–5
Can Garden Plots Promote Russia's Food Security?

A transformed political and economic system in Russia during the 1980s was driven by the desire for political freedom and improved living conditions. However, by the mid-1990s many Russians faced poverty. Seeth and coauthors wanted to know more about food security in Russian households during the economic stresses of the transition economy.

By the early 1990s, widespread prediction of food scarcities under an untested market system worried many Russian households. Privatization in the agricultural sector, new food market risks, and the deterioration of the large-scale crop and livestock sectors were accompanied by rapid expansion of home food production. What happened? The average Russian household started producing some of its own food supply to reduce the risk of going hungry.

Three strategies have long governed the economic behavior of the typical Russian household: (1) waiting in lines, (2) cultivating informal relationships to improve access to goods, and (3) producing food for the household. As limited success with the market system reduced the need for the first two coping strategies, home food production grew. A well-managed, efficient garden plot was seen as an advantage in dealing with unknown risks from food market disruptions and an unknowable future. Garden plots were more widespread in rural households than in urban households. Crops grown on the plots included potatoes, vegetables, berries, and sometimes small livestock such as chickens, pigs, sheep, and goats.

Were the gardens successful? You be the judge. Home food production grew from of 24% of Russia's 1990 agricultural output to 38% in 1994. By 1995, Russian households produced 42% of Russian's meat, 44% of its milk, 82% of its potatoes, and 90% of its fruits and vegetables.

Garden plots are more expensive than you might think. Although the plots are a major way to cope with risks of food shortage they also have large opportunity costs. Huge amounts of labor of often highly educated people are allocated to farming these plots. On average, Seeth et al. found that households spent more than 700 hours per year with their garden plots, nearly two hours per day.

Home production of food is a mechanism by which many of Russia's rural and urban households managed transformation risks. But this activity has largely been ignored by public agricultural policy. Few major initiatives exist to facilitate access to inputs and improve marketing conditions for people who maintain garden plots. For many high-valued crops, this sector of agriculture is the largest supply source. Will garden plots disappear as economic growth in Russia increases? Not likely, according to Seeth et al. Under the most likely future conditions of the Russian economy, higher wages would reduce time allocation to household plots only slightly. Historically, household garden plots have brought Russian households through many hard times.

Source: Seeth, H., S. Chachnov, A. Surinov, and J. VonBraun. 1998. Russian poverty: Muddling through economic transition with garden plots. *World Development.* 26 (9): 1611–1624.

Community associations play critical roles in natural resource protection, especially activities such as watershed management, integrated pest management, and irrigation, which require large-scale cooperation and collective action.

Elimination of Corruption. Not all governments are honest. Corruption, a lack of integrity or honesty, is common in some countries; for example, some countries have a history of government officials who are susceptible to bribery. Much of the corruption takes place at the national level (Aziz, 2001). Good governments do not tolerate corruption and persuade the private sector, civil society, and citizens to work toward its elimination. An independent judiciary and a free press are essential to achieving honest public administration.

Competitive civil service salaries—paid on schedule—deter corruption as do good quality training programs for civil servants. Competent public administration is essential for ensuring that contracts and agreements are enforced and that grading and quality control standards are enacted and carried out.

Corrupt governments are blamed for widening the gap between the rich and the poor in many countries. Such governments may block economic development and undermine attempts to fight hunger effectively. In some respects, corruption substitutes for open competition. It retards private sector development and discourages investment. Government-financed roads that could be used to ship food to where it is needed never get built because the money gets siphoned off by dishonest officials. The destructive power of corruption inhibits the implementation of sustainable food programs.[23] In some cases, decisions are made not for public benefit but for show. Expensive, flashy, and prestigious projects are sometimes favored over cost-efficient, community-based food security programs that use technology adapted to local culture.

What causes corruption? Although there are few simple answers to this question, poor and badly governed countries often share a lack of stability in leadership. This often leads to a failure to invest in resources to secure lasting institutions. It is difficult, for example, to develop and implement workable, long-term food security policies against a background of institutional instability. Food policies that serve the people must operate in an existing culture of good governance.

Open, Transparent Government. Open and transparent government is crucial to successful implementation of the rule of law and to promotion of sound public administration and respect and protection for human rights. Democratic governments are more likely to be responsive to the needs of all their citizens, to make food security a high priority, and to welcome community participation. A lack of transparency provides the environment in which poor administration and corruption can flourish. For example, public projects involving deals for friends may take priority over the public interest.

In 1992, a small group of people went to a community in Rajasthan, India, and started asking questions: How much money comes into village? How is it being spent? After they found the information, they set up public hearings in which shared openly with a thousand villagers how much money had reached the people. The revelation generated so much anger and disgust that, for the first time in the history of independent India—without formal recourse to law, bureaucracy, or political support—village officials actually returned the money. This example illustrates the importance of an open and transparent community that did not depend on the formal instruments of law. The community's pressure on corrupt officials was so humiliating that they returned the people's money.

Reduction of Violent Conflict. Between 1970 and 1990, violent conflicts led to hunger and falling food production in 43 developing countries. The opposite is also true. Hunger and lack of basic necessities often are instruments for promoting violent conflicts (Messer, Cohen, and D'Costa, 1998). Creating a hunger-free world requires prevention of violent conflict as well as rebuilding of war-torn countries.

Conflict leads to the destruction of crops and livestock; denial of access to land and water; and damaged infrastructure, markets, and the human resources needed to produce and distribute food. Armed conflict also displaces populations on a large scale, further adding to food insecurity. For example, Sudan could easily feed itself if it were not for its ongoing civil war. Land is plentiful and fertile and the Nile River could be used for irrigated agriculture. But Sudan is chronically food insecure. Combatants use hunger as a weapon of war, starving the opposition into submission. In some cases food aid has been stolen by combatants on both sides, and gifts of humanitarian aid have been used to buy the support of hungry people in support of one fighting group or another. Historically, war blocks access to food markets. Opposing

[23] Aziz (2001) illustrates the damages of corruption: Only systematic corruption can explain the complete failure of the poverty alleviation and food security initiatives in south Asia funded from 1977 to 1988 with 9 million British pounds. According to a report in *The Independent,* a British newspaper, a large number of contractors and suppliers made easy money without doing any work on the project. No significant results in terms of increase in income or sustainable agricultural productivity were achieved. There was no evidence of even a single sapling, although trees were supposed to have been planted under the program. Payments were made for nonexistent works while roads were built that went nowhere. Meanwhile, the number of destitute families increased dramatically during the 10-year period covered by the funding initiative.

armies live off the people, taking available food supplies. In doing so, the armies discourage farmers from producing excess food for market. Food security for all remains elusive while conflict exists.

Macroeconomic Policies. Economic growth provides the resources to fund the science and technology essential for improving agricultural economic productivity. Stable and predictable macroeconomic policies encourage savings and investment, discourage capital flight,[24] and focus private-sector energies and resources on efficient and productive investments. Governments that are not committed to long-term macroeconomic stability are unlikely to sustain food security. Tweeten (1999) and Sanderatne (2001) emphasize the importance of macroeconomic policies in promoting growth in per capita income to promote food security.

Macroeconomic policies can have both a direct impact on a country's food supply and an effect on the price incentives facing domestic producers. Direct effects are most likely to take place through changes in fiscal policy (i.e., government spending or government policies that affect private investment). Attempts to cut public expenditure may affect various agricultural support services such as the provision of educational extension services or the financing of public sector research initiatives, which could lead to improved crop varieties or more effective production techniques. Changes in the foreign exchange rate, typically a devaluation, also can have an effect on the provision of government services to the extent that these services use imported goods, (e.g., fuel or imported equipment). Reliable access to foreign exchange may be particularly important for the maintenance of government-owned infrastructure, such as roads and marketplaces, all of which influence the decision of farmers to provide food products to the market rather than keep them for household consumption.

The foreign exchange rate determines the relative prices of traded goods versus nontraded goods. An overvalued exchange rate tends to cause higher prices for nontraded goods. Most agricultural commodities are internationally traded goods. The foreign exchange rate will influence the price of farm inputs like fertilizer, pesticides, and fuel for tractors. In addition, changes in the domestic price of internationally traded agricultural inputs and outputs will affect the incentives facing producers and traders in the food sector.

Monetary policy can affect the availability of credit to finance the production of food and the purchase and storage of the annual harvest. A tight monetary policy may restrict access to agricultural credit. However, if monetary policy is so expansionary that it encourages high levels of inflation, the long-run impact on domestic food supply is damaging, as investment in productivity-enhancing capital is discouraged. Macroeconomic policy also influences the supply of imported food. Tight balance of payments constraints limit a country's capacity to import food to relieve domestic food shortages.

Developing and sustaining food security requires macroeconomic conditions that favor and reward entrepreneurship. For example, in much of sub-Saharan Africa the food production sector is generally small or has not flourished. It is hampered by limited access to credit, shortages and high costs of inputs, lack of quality inputs, and a low flow of domestic and foreign capital to promote investments in the sector. Macroeconomic growth policies such as the following are key ingredients in attracting investors to commercial agriculture and influencing long-term food security: emphasis on private investment as a source of wealth, guaranteed land security, loan guarantees, promotion of productive farms, producer associations, strategic planning, and existence of transparent regulatory procedures.

Reduced Population Growth. The most important long-term contribution of economic growth to food security may be to reduce population growth—eventually to zero (see Chapter 14). Income from economic growth is essential to fund health clinics that provide immunization,

---

[24] Capital flight occurs when a collapse of confidence in a country's economic policy leads investors and ordinary people to pull their investments out of a country and invest them somewhere else. Capital flight typically causes a sharp depreciation in the value of the currency and makes it difficult to conduct domestic policy.

vitamin and mineral supplements (especially for young children and pregnant women), and family planning. The outlook for reduced population growth is highly debated. The evidence to date suggests that the *Malthusian theory* will turn out wrong. High-yield agriculture has put a brake on population growth rather than accelerated it. High-yield agricultural practices have led the progression from high-birth-rate, high-death-rate societies of feudal cultures toward the low-birth-rate, low-death-rate societies of modern Western nations. Many demographers believe there are several reasons why "development is the best contraceptive."

- It increases female education.
- It expands women's opportunities and promotes their status.
- It changes the cost and benefits of children and the demand for them.
- It decreases infant mortality.
- It reduces the motivation to have large families.
- It multiplies the channels of social interactions so that all countries and most individuals exchange information about the advantages and disadvantages of fewer children.

The evidence so far indicates that all of these factors promote fertility decline. In other words, when the effect of economic development on material aspirations is considered, there seems to be a negative effect on fertility. By contrast, subsistence agriculture promotes large families where children are viewed as sources of manual labor.

SUMMARY

Despite the enormous gains made by modern science in food production, in 2003 an estimated 840 million people, or about 15% of the world's population, lived with food insecurity, which means they lacked access to the calories necessary for active and healthy lives. Most of these people lived in developing countries, primarily in rural areas.

Thanks to the Green Revolution, on average there is enough food supply on hand to meet global demand. Since the mid-1950s, the world's food production has expanded faster than its population. Distribution is the larger problem. Inadequate infrastructure, gender inequality, wars, and corrupt governments have prevented the chronically hungry from gaining access to the rich harvest. After falling for decades, the estimated number of undernourished in the world increased by 18 million to 798 million between 1997 and 2001. Most of these 798 million live on isolated, unproductive lands, and they have neither the money to buy food nor the ability to grow it.

Most communities adhere to the following four basic food policy objectives: (1) efficient and sustained growth in agricultural sectors; (2) acceptable distribution of income, mostly through efficient job creation; (3) adequate nutritional status for all citizens; and (4) community food security that insures against poor harvests, natural disasters, and uncertain international food supplies and prices.

Analysis of unregulated markets focuses on farmers, who are motivated by the potential for economic return. Usually their decisions can be altered only if they perceive incentives favorable to the proposed change. The challenge facing farmers in competitive times is how to increase profitability. *Product-product relations* help farmers make choices in an economically rational manner. They help the farmer evaluate the opportunity cost (lost income) of growing more of one crop at the expense of another. Relative prices are a major factor determining economic decisions of producers and consumers. Government food policies often implement import or export controls to change commodity prices, so such policies have a major influence on farm-level decisions about how much of which crops to grow. *Factor-factor relations* refer to substitutions farmers make in choosing inputs for growing a crop. They can choose many combinations of productive inputs to produce a given crop. Farmers' input decisions have important effects on employment, productivity of agriculture, and the level and distribution of income for rural areas. The *factor-product relation* (production function) relates the use of inputs to agricultural output. It is a powerful conceptual tool for understanding how farm-level decisions are made and how their outcome affects the variables that are important to a nation. The *output-price relation* indicates that profit-maximizing farmers can be expected to apply inputs more intensively to increase their commodity output when the commodity's price increases or if the input price falls, if other things are constant.

Several institutional designs are available for improving food policy outcomes. The primary short-term measure for coping with dangerous food shortfalls is emergency food relief. Intermediate-term measures include increasing production, natural resource protection, further improvements in the Green Revolution, increased water use efficiency, addressing global warming, advances in biotechnology, improved computer literacy, food stamp programs for food buyers, girls' education, health care, and property rights. Long-term measures include various measures to improve governance—such as the promotion of civil society, elimination of corruption, open and transparent government, and reduction of violent conflict—as well as various macroeconomic policies to improve the stability of income and jobs.

■ Review Questions

1. What are product-product decisions and why are they important for individual farmers and for a nation's food policy?

2. Why will a farmer produce a combination of rice and beans for which the slope of the isorevenue line equals the slope of the production possibilities? Can you provide an economic interpretation of what occurs at that point? Explain what happens to the production by farmers of rice and beans when the government imposes a price ceiling on rice so consumers can buy rice more cheaply.

3. What is the cost to the farmer of failing to produce where the slope of the isorevenue line equals the slope of the production possibilities?

4. Why will a farmer choose a mix of inputs so that the slope of the isoquant equals the slope of the isocost line? Provide an economic interpretation for what occurs at that point.

5. When a government policy restricts imports of fertilizer to protect water supplies in environmentally sensitive areas, what will happen to the domestic price of fertilizer, and what will happen to fertilizer use by farmers? Explain, using isocost and isoquant relations, what is likely to happen to the use of substitute inputs and to the output of products that use fertilizer as an input.

6. Explain a factor-product relation. Why does a farm's profit-maximizing use of a variable input require that its use occur where its price equals the value of its marginal product?

7. Suppose a farmer is using fertilizer to produce wheat. Explain why an increase in the price of wheat and a reduction in the price of fertilizer have similar effects on fertilizer use and on wheat output.

8. Why does the profit-maximizing farmer care only about the ratio of the price of wheat to the price of fertilizer and not on the absolute value of either one?

9. Explain why lowering the amount of wheat needed to buy fertilizer (raising the ratio of wheat price to fertilizer price) normally leads to increased fertilizer use and higher yields.

10. Explain why a profit-maximizing farmer produces an output level of wheat for which its price equals its marginal cost of production. What does this mean for government wheat policy that attempts to make bread more affordable by reducing the official price of wheat?

11. Suppose a farmer can obtain \$200,000 in annual revenues from the highest-valued crops with no provision for controlling eroded soil. Under these conditions, erosion will cause an annual loss of revenues of \$2,000 for the indefinite future. An effective erosion control program will have a one-time, up-front cost of \$30,000. Using a 10% discount rate, show the calculations needed to determine if the soil erosion control program is worth it to the farmer. How low must the cost of the erosion control program be to make erosion control desirable to the farmer?

12. Explain how government programs that emphasize natural resource protection on farms can increase current and future food security for developing countries.

13. Explain why a comprehensive system of water rights with clear titles to the right to use water for irrigation can increase a country's food security.

14. Explain why education of girls can be a good investment for promoting future food security worldwide.

15. Why are democracy, protection of all human rights and fundamental freedoms, and full and equal participation of men and women essential for achieving sustainable food security?

16. Describe several of the institutions that promote food security.

■ Problems

1. Agricultural pollution and subsidies: Suppose that the price (marginal benefit) for a crop has the following equation:

$$P = MB = 20 - Q$$

Suppose that the supply function is $P = 3Q$ and the marginal social cost produced by an environmental externality is $MSC = 4Q$.

a. Find the market clearing price and quantity (equilibrium value of P and Q).
b. Find the social losses from the externality that occurs when there is no attempt to regulate the market clearing price and quantity. (Hint: Social loss equals the net surplus that occurs when the marginal benefits intersect the supply function minus the net surplus that occurs when marginal benefits intersects the marginal social cost.)
c. Now assume that a food price support program is enacted in which food's price is supported at \$20. The government buys enough of the crop to raise the price to \$20. Find the additional social losses produced with the price support.
d. What is the reduction in social losses under the food price support policy if the environmental externality can be eliminated at zero cost?

2. Price impacts of food stamp programs: Some people criticize food stamp programs proposals because they believe the programs simply raise food prices with no increase in the quantity of food available to the poor. If you view a food stamp program as increasing effective demand for food, what value does the supply elasticity have to be for the critics to be right? (Hint: Supply elasticity is the percentage change in quantity supplied divided by the percentage change in price). What must the supply elasticity be for a program to increase food availability to the poor with no price increase at all? Use any combination of words, diagrams, or math to answer these questions.

■ REFERENCES

Aggarwal, R. M., and T. A. Narayan. 2004. Does inequality lead to greater efficiency in the use of local commons? The role of strategic investments in capacity. *Journal of Environmental Economics and Management.* 47: 163–182.

Alexandratos, N. 1999. World food and agriculture: Outlook for the medium and longer term. *Proceedings of the National Academy of Sciences of the United States of America.* 96 (11): 5908–5914.

Anderson, K. 2000. Agriculture's "multifunctionality" and the WTO. *Australian Journal of Agricultural and Resource Economics.* 44 (3): 475–494.

Aziz, T. A. 2001. *Governance and food security: Acting in the public interest.* International Food Policy Research Institute. Available from *http://www.ifpri.org/2020conference/* [Accessed March 7, 2005.]

Bailey, R. 2000. *Billions served.* Reason Online. Available from *http://reason.com/0004/fe.rb.billions.shtml* [Accessed October 26, 2004].

Barrett, C. B., M. Bezuneh, and A. Aboud. 2001. Income diversification, poverty traps and policy shocks in Cote d'Ivoire and Kenya. *Food Policy.* 26 (4): 367–384.

Boardman, J., J. Poesen, and R. Evans. 2003. Socio-economic factors in soil erosion and conservation. *Environmental Science & Policy.* 6 (1): 1–6.

Bonnieux, F., P. Rainelli, and D. Vermersch. 1998. Estimating the supply of environmental benefits by agriculture: A French case study. *Environmental & Resource Economics.* 11 (2): 135–153.

Bradshaw, B., and B. Smit. 1997. Subsidy removal and agroecosystem health. *Agriculture Ecosystems & Environment.* 64 (3): 245–260.

Cakmak, I. 2002. Plant nutrition research: Priorities to meet human needs for food in sustainable ways. *Plant and Soil.* 247 (1): 3–24.

CamFed home page. 2005. Available from *http://www.camfed.org/html/about_camfed.html* [Accessed March 7, 2005].

Charlton, K. E., and D. Rose. 2001. Nutrition among older adults in Africa: The situation at the beginning of the millenium. *Journal of Nutrition.* 131 (9): 2424S–2428S.

Chattopadhyay, R. 2000. Zimbabwe: Structural adjustment, destitution and food insecurity. *Review of African Political Economy.* 27 (84): 307–316.

Chavas, J.-P. 2000. The microeconomics of food security. *Australian Journal of Agricultural and Resource Economics.* 44 (1): 1–29.

Clay, D. C., D. Molla, and D. Habtewold. 1999. Food aid targeting in Ethiopia: A study of who needs it and who gets it. *Food Policy.* 24 (4): 391–409.

Cohen, M. 2003. *The right to adequate food and food security.* Available from *http://www.ifpri.org/pubs/testimony/cohen20030521.pdf* [Accessed October 26, 2004].

Collins, A. E. 2001. Health ecology, land degradation and development. *Land Degradation & Development.* 12 (3): 237–250.

Coxhead, I. 2000. Consequences of a food security strategy for economic welfare, income distribution and land degradation: The Philippine case. *World Development.* 28 (1): 111–128.

Crosson, P. R., and S. Brubaker. 1982. *Resource and Environmental Effects of U.S. Agriculture.* Baltimore: Johns Hopkins University Press.

De Janvry, A., and E. Sadoulet. 2002. World poverty and the role of agricultural technology: Direct and indirect effects. *Journal of Development Studies.* 38 (4): 1–26.

Dorosh, P. A., and S. Haggblade. 1997. Shifting sands: The changing case for monetizing project food aid in Bangladesh. *World Development.* 25 (12): 2093–2104.

Easterbrook, G. 1997. *Norman Borlaug. The Atlantic Monthly.* Available from *http://www.theatlantic.com/issues/97jan/borlaug/borlaug.htm* [Accessed July 14, 2004].

Enyong, L. A., S. K. Debrah, and A. Bationo. 1999. Farmers' perceptions and attitudes towards introduced soil-fertility enhancing technologies in western Africa. *Nutrient Cycling in Agroecosystems.* 53 (2): 177–187.

Evrendilek, F., and C. Ertekin. 2002. Agricultural sustainability in Turkey: Integrating food, environmental and energy securities. *Land Degradation & Development.* 13 (1): 61–67.

Falkenmark, M. 1999. Forward to the future: A conceptual framework for water dependence. *Ambio.* 28 (4): 356–361.

Food and Agricultural Organization. 2004. *Eradicating hunger: Moving from pilot projects to national programs to meet the world food summit goal.* Available from *ftp://ftp.fao.org/docrep/fao/007/ae523e/ae523e00.pdf* [Accessed March 6, 2005].

Fresco, L. O., and H. Steinfeld. 2004. *A food security perspective to livestock and the environment.* Food and Agricultural Organization, United Nations. Available from *http://www.fao.org/wairdocs/lead/x6131E/6131e00.htm* [Accessed October 26, 2004].

Gundersen, C., and V. Oliveira. 2001. The food stamp program and food insufficiency. *American Journal of Agricultural Economics.* 83 (4): 875–887.

Guo, X., B. Popkin, T. Mroz, and F. Zhai. 1999. Food price policy can favorably alter macronutrient intake in China. *Journal of Nutrition.* 129 (5): 994–1001.

Harris, J. M., and S. Kennedy. 1999. Carrying capacity in agriculture: Global and regional issues. *Ecological Economics.* 29 (3): 443–461.

Heilig, G. K., G. Fischer, and H. Van Velthuizen. 2000. Can China feed itself? An analysis of China's food prospects with special reference to water resources: *International Journal of Sustainable Development and World Ecology.* 7 (3): 153–172.

Herdt, R. W. 1998. *The life and work of Norman Borlaug, Nobel laureate.* The Rockefeller Foundation. Available from *http://www.rockfound.org/display.asp?context=3&SectionTypeID=17&DocID=87* [Accessed October 26, 2004].

Hodge, I. 2000. Agri-environmental relationships and the choice of policy mechanism. *World Economy.* 23 (2): 257–273.

Hossain, M., and V. P. Singh. 2000. Fertilizer use in Asian agriculture: Implications for sustaining food security and the environment. *Nutrient Cycling in Agroecosystems.* 57 (2): 155–169.

International Food Policy Research Institute. 2001a. *Sustainable food security for all by 2020, vision statement.* Available from *www.ifpri.org/2020conference/PDF/2020conference_visionbonndraft.pdf* [Accessed July 12, 2004].

———. 2001b. *Sustainable food security for all by 2020: Proceedings of an international conference.* Available from *www.ifpri.org/2020conference* [Accessed July 12, 2004].

Jaspars, S., and J. Shoham. 1999. Targeting the vulnerable: A review of the necessity and feasibility of targeting vulnerable households. *Disasters.* 23 (4): 359–372.

Jayne, T., and S. Jones. 1997. Food marketing and pricing policy in Eastern and Southern Africa: A survey. *World Development.* 25 (9): 1505–1527.

Jha, S., and P. V. Srinivasan. 1999. Grain price stabilization in India: Evaluation of policy alternatives. *Agricultural Economics.* 21 (1): 93–108.

Johnson, D. G. 1997. On the resurgent population and food debate. *Australian Journal of Agricultural and Resource Economics.* 41 (1): 1–17.

———. 1998. Food security and world trade prospects. *American Journal of Agricultural Economics.* 80 (5): 941–947.

Kahan, D. 2004. *Farm Economic Handbook: Working with Market-Oriented Farmers.* Food and Agriculture Organization of the United Nations. Available from *http://www.fao.org/ag/ags/subjects/en/farmMgmt/pdf/workingpaper_kahan.pdf* [Accessed July 21, 2004].

Kent, G. 1997. Fisheries, food security, and the poor. *Food Policy.* 22 (5): 393–404.

Khush, G. S. 1999. Green revolution: Preparing for the 21st century. *Genome.* 42 (4): 646–655.

Lal, R. 2000a. Soil management in the developing countries. *Soil Science.* 165 (1): 57–72.

———. 2000b. Physical management of soils of the tropics: Priorities for the 21st century. *Soil Science.* 165 (3): 191–207.

Larson, B. A., and G. B. Frisvold. 1996. Fertilizers to support agricultural development in sub-Saharan Africa: What is needed and why. *Food Policy.* 21 (6): 509–525.

Makin, D., and H. Mendelssohn. 2004. *Insectivorous bats victims of Israeli campaign.* Available from *http://www.batcon.org/batsmag/v2n4–1.html* [Accessed October 26, 2004].

Marland, G., B. A. McCarl, and U. Schneider. 2001. Soil carbon: Policy and economics. *Climatic Change.* 51 (1): 101–117.

Maxwell, D. G. 1996. Measuring food insecurity: The frequency and severity of "coping strategies." *Food Policy.* 21 (3): 291–303.

Maxwell, D., and K. Wiebe. 1999. Land tenure and food security: Exploring dynamic linkages. *Development and Change.* 30 (4): 825–849.

McCalla, A. F. 1999. Prospects for food security in the 21st century: With special emphasis on Africa. *Agricultural Economics.* 20 (2): 95–103.

Messer, E., M. J. Cohen, and J. D'Costa. 1998. *Food from Peace: Breaking the Links Between Conflict and Hunger.* International Food Policy Research Institute, Brief 50. Available from *http://www.ifpri.org/2020/briefs/number50.htm* [Accessed July 22, 2004].

Minot, N., and F. Goletti. 1998. Export liberalization and household welfare: The case of rice in Vietnam. *American Journal of Agricultural Economics.* 80 (4): 738–749.

Molnar, J. J. 1999. Sound policies for food security: The role of culture and social organization. *Review of Agricultural Economics.* 21 (2): 489–498.

Moxey, A., M. Whitby, and P. Lowe. 1998. Agri-environmental indicators: Issues and choices. *Land Use Policy.* 15 (4): 265–269.

Mwangi, W. M. 1997. Low use of fertilizers and low productivity in sub-Saharan Africa. *Nutrient Cycling in Agroecosystems.* 47 (2): 135–147.

Nelson, M. 2000. Childhood nutrition and poverty. *Proceedings of the Nutrition Society.* 59 (2): 307–315.

Nobel Foundation. 2004. *Amartya Sen biography.* Available from *http://nobelprize.org/economics/laureates/1998/sen-autobio.html* [Accessed October 26, 2004].

———. 2004. *Norman Borlaug, Nobel lecture.* Available from *http://www.nobel.se/peace/laureates/1970/borlaug-lecture.html* [Accessed October 26, 2004]. *Biography.* Available from *http://www.nobel.se/peace/laureates/1970/borlaug-bio.html* [Accessed October 26, 2004].

Paarlberg, R. L. 1999. Politics and food insecurity in Africa. *Review of Agricultural Economics.* 21 (2): 499–511.

Peterson, E. W. F., and S. R. K. Valluru. 2000. Agricultural comparative advantage and government policy interventions. *Journal of Agricultural Economics.* 51 (3): 371–387.

Phillips, P. W. B. 2002. Biotechnology in the global agri-food system. *Trends in Biotechnology.* 20 (9): 376–381.

Pimentel, D., ed. 1993. *World Soil Erosion and Conservation.* Cambridge, UK: Cambridge University Press.

Pinstrup Andersen, P., and R. Pandya Lorch. 1996. Food for all in 2020: Can the world be fed without damaging the environment? *Environmental Conservation.* 23 (3): 226–234.

Pinstrup Andersen, P., and R. Pandya Lorch. 1998. Food security and sustainable use of natural resources: A 2020 vision. *Ecological Economics.* 26 (1): 1–10.

Pinstrup Andersen, P. 2000. Food policy research for developing countries: Emerging issues and unfinished business. *Food Policy.* 25 (2): 125–141.

Plantinga, A. J. 1996. The effect of agricultural policies on land use and environmental quality. *American Journal of Agricultural Economics.* 78 (4): 1082–1091.

Postel, S., P. Polak, F. Gonzales, and J. Keller. 2001. Drip irrigation for small farmers: A new initiative to alleviate hunger and poverty. *Water International.* 26 (1): 3–13.

Pretty, J. N., J. I. L. Morison, and R. E. Hine. 2003. Reducing food poverty by increasing agricultural sustainability in developing countries. *Agriculture Ecosystems & Environment.* 95 (1): 217–234.

Rome Declaration on World Food Security. 1996. *Rome Declaration on World Food Security, World Food Summit, November 13–17, Rome Italy.* Available from *http://www.fao.org/docrep/003/w3613e/w3613e00.htm* [Accessed July 22, 2004].

Rosegrant, M., and C. Ringler. 1997. World food markets into the 21st century: Environmental and resource constraints and policies. *Australian Journal of Agricultural and Resource Economics.* 41 (3): 401–428.

Ruben, R., H. Moll, and A. Kuyvenhoven. 1998. Integrating agricultural research and policy analysis: Analytical framework and policy applications for bio-economic modelling. *Agricultural Systems.* 58 (3): 331–349.

Ruel, M. T., L. Haddad, and J. L. Garrett. 1999. Some urban facts of life: Implications for research and policy. *World Development.* 27 (11): 1917–1938.

Sanchez, P. A. 2000. Linking climate change research with food security and poverty reduction in the tropics. *Agriculture Ecosystems & Environment.* 82 (1–3): 371–383.

Sanderatne, N. 2001. Food security: Concept, situation, and policy perspectives. *Sri Lanka Economic Journal.* 2 (2): 22–37.

Scherr, S. J. 2000. A downward spiral? Research evidence on the relationship between poverty and natural resource degradation. *Food Policy.* 25 (4): 479–498.

Seeth, H., S. Chachnov, A. Surinov, and J. VonBraun. 1998. Russian poverty: Muddling through economic transition with garden plots. *World Development.* 26 (9): 1611–1624.

Serageldin, I. 1999. Biotechnology and food security in the 21st century. *Science.* 285 (5426): 387–389.

Singh, R. B. 2000. Environmental consequences of agricultural development: A case study from the Green Revolution state of Haryana, India. *Agriculture Ecosystems & Environment.* 82 (1–3): 97–103.

Smith, L. C., and L. Haddad. 2001. How important is improving food availability for reducing child malnutrition in developing countries? *Agricultural Economics.* 26 (3): 191–204.

Timmer, C. P., W. P. Falcon, and S. R. Pearson. 1983. *Food Policy Analysis.* Baltimore and London: The Johns Hopkins University Press for the World Bank.

Tweeten, L. 1999. The Economics of Global Food Security. *Review of Agricultural Economics.* 21 (2): 473–88.

United Nations World Food Program. 2002. *Reducing poverty and hunger: The critical role of financing for food, agriculture, and rural development.* Available from *ftp://ftp.fao.org/docrep/fao/003/Y6265E/Y6265E.pdf* [Accessed March 7, 2005].

Van Huis, A., and F. Meerman. 1997. Can we make IPM work for resource-poor farmers in sub-Saharan Africa? *International Journal of Pest Management.* 43 (4): 313–320.

World Bank. 1993. *World Development Report: Investing in Health.*

World Resources Institute. 2004. *World resources 2002–2004.* Available from *http://www.wri.org/.* [Accessed October 26, 2004].

Zeller, M., and M. Sharma. 2000. Many borrow, more save, and all insure: implications for food and micro-finance policy. *Food Policy.* 25 (2): 143–167.

Chapter 12

FISHERIES

Introduction

In April 1948, Edward "Doc" Ricketts wrote an article in a Monterey, California, newspaper revealing the declining state of Monterey's sardine industry (Ricketts, 1948). Ricketts was the real-life marine biologist who had inspired several famous characters in John Steinbeck's novels, including Doc in *Cannery Row.* Ricketts observed that the local sardine canning industry had continued to grow even though sardine catches had experienced steady declines for 12 years. Nobody knew how much of the decline was caused by natural migration patterns of fish, changes in cycles of fish population, or overfishing. Ricketts had friends who relied on fishing and canning for their livelihoods, and he worried for their futures. He believed that although fishing might not drive the sardine into *biological* extinction, a continued lack of fishing regulations could drive it into *economic* extinction. See Example 12–1.

Doc's article was written more than 50 years ago, but it raises the question of how to regulate potential overfishing, a question that still challenges fishers, the fishing industry, fishery managers and regulators, and sport fishing organizations. Fisheries have provided a way of living for coastal communities for thousands of years. Fish provide the world's only major food source harvested from the wild. Currently, fish provide the fifth-largest food resource and the world's major source of protein (Figure 12–1).

Total tonnage of fish caught around the world has continually increased from 1950 through 2000. However, the growing total absolute fish catch may hide the problem of falling populations of valuable commercial fish populations such as halibut, cod, and haddock. Fish have been a major part of human diets for thousands of years; people who live in seafaring nations typically get most of their animal protein from fish. Thus, it is important to understand more about the causes of declines in fish populations as well as impacts of policies for coping with these declines. Analysis of fisheries provides the added benefit of teaching lessons about how to manage a wide range of renewable common property resources. Analyzing fisheries teaches us about fish, fish management, and management of other renewable biological resources.

Outline

Biological Model of a Fishery 335

Economically Efficient Model of a Fishery 337

Market Allocations 339

Institutional Designs for Real-World Fisheries 342

Objectives

After studying this chapter, you should be able to

- Identify a fishery's economically efficient catch
- Describe the unregulated market as a mechanism for allocating fish catch
- Analyze institutional designs for promoting efficient and sustainable fisheries management

Example 12–1
Doc's Theory of Overfishing

The following quotations from Doc Ricketts' original 1948 article in the *Monterey Peninsula Herald* implicate overfishing in the decline of the sardine industry in Monterey, California. See also the accompanying table.

> A large waterfront element continues to advance explanations for the increasing scarcity of sardines. The two most fantastic involve the dumping of munitions and the effects of the atom bomb, although obviously neither of these were in operation twelve years ago when the total landings started their downward slide. But the most likely explanation (and still by far the most unpopular!) is that the sardines we are searching for have already been canned and reduced. In other words, that the total number HAS decreased.
>
> At the same time a plausible theory is being entertained that the sardine population center has moved south, and I understand that at present Dr. Clark is investigating that phase. Whatever her investigation discloses, the unpleasant fact stares us in the face that the industry is over-expanded. For many years we have been increasing the numbers of our canneries and reduction plants, while at the same time the sardine population, if not actually decreasing, certainly can't have been increasing. Even if the fish were to come back every year from now on in their greatest recorded numbers, as in the peak season twelve years ago, still there wouldn't be enough to keep all our plants running at capacity.
>
> By continuing to take undersized specimens and by concentrating our fishing activities in the area, where the animals gather preparatory to spawning in Southern California, we can come pretty close to accomplishing this objective. How wonderful it would be, in this connection, if only Southern California and Baja California could be scared as thoroughly as we have been in Monterey, and as the operators have been in San Francisco, Oregon, Washington, and Vancouver Island! Perhaps then we could count on the sardine making a quick come-back.
>
> In the meantime, I have been listening to the waterfront gossip. It is being complained along the row that the Division of Fish and Game should have awakened to the danger long ago, that at least they should have prevented the taking of undersized specimens. But I reason this way: Suppose they should try (as I believe actually they did). A suave lobbyist says to the individual legislators: "Don't pay any attention to these scientists—Good fellows of course; but their heads are in the clouds. Let's listen to the hard-headed down to earth businessmen. There's no chance of depletion. There's just as good fish in the sea as any ever's been caught. Both the fishermen and the canners are agreed on that. And they know. They've had years of actual experience. Let's not have any professional new-dealing in our fisheries." So the legislators refuse to listen to the Fish and Game Division scientists. Sometimes even the Division itself refuses to support its own specialists.
>
> And the sardine populations DO decrease (although not necessarily only through overfishing). Until finally the canners themselves (of northern California for instance)—these down-to-earth hard-headed businessmen—get concerned over the scarcity. Their profits fade, their business decreases. In desperation they appeal to the Division for protection, for a statewide regulation to prevent the taking of small sardines. What happens! There's a meeting. The canners of Southern California scream their heads off. They say: "In this region we get only mixed sardines. The proportion of small ones is often high. Don't just take OUR business too, just because the unfortunate operators in the north are suffering." Again the Division fails to act, confronted with disagreement. And in the meantime still more of the vital and decreasing breeding stock is being wiped out.
>
> When warnings were first sounded fifteen or more years ago, if such a program had been put into effect, the industry perhaps could have been standardized at 400,000 tons per season. Instead, once we hit nearly 800,000. The 1947–48 landings were less than 100,000. If conservation had been adopted early enough, a smaller but streamlined cannery row in all likelihood this month would be winding up a fairly successful season, instead of dipping as they must be now, deeply into the red ink of failure.

Ex poste analysis of the California sardine industry

Year	Conditions	Regulations	Harvest (tons)	Profits
1936	Actual	Without	800,000	(++)
1936	Hypothetical	With	400,000	(+)
1948	Actual	Without	100,000	(−−)
1948	Hypothetical	With	400,000	(+)

Source: Ricketts, E. 1948. *Scientists report on sardine supply: Investigator blames industry, nature for shortages.* Available from *http://www.geocities.com/yosemite/gorge/5604/ed_ricketts1948sardine.htm* [Accessed October 2, 2004].

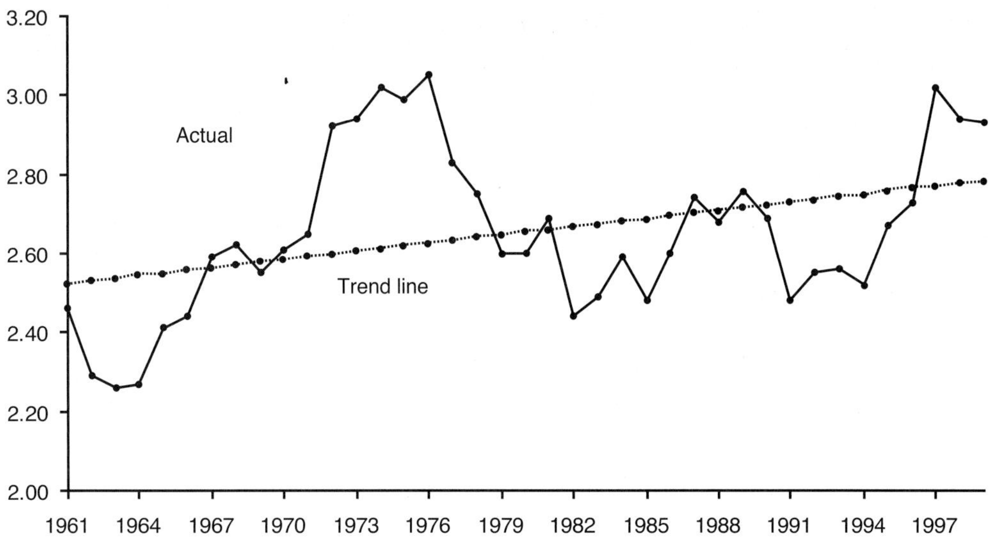

Figure 12–1 Worldwide per capita marine fish consumption: metric tons per 1,000 people. Source: Adapted from Food and Agriculture Organization of the United Nations 2000. *The State of world fisheries and aquaculture.* Available from *http://www.fao.org/docrep/003/x8002E/x8002E00.htm* [Accessed October 2, 2004].

Biological Model of a Fishery

Let us begin with a simple biological model of a fishery that shows a relation between the *growth* and *size* of a fish population. The growth-size function indicates a general pattern that applies to many fisheries, but it ignores important details unique to each fishery such as water temperature, biological productivity, predation, competition for food, and the population's age structure. Figure 12–2, called an effort-yield relation, is a graphic representation of the model. It avoids characterizing any particular fishery on any particular day, but it reveals a pattern common to many fisheries in a wide range of conditions in many parts of the world.

The horizontal axis shows the population's size (S) measured in tons. The vertical axis shows the population's growth (g) in tons per unit of time. The population size, which exists at a point in time, is a *stock.* For example, the size of your personal wealth at a point in time is a stock. Growth is the amount by which the population increases between two time periods. Growth is a *flow,* which cannot exist at a point in time. Your income flow is measured as an amount earned per unit time, such as X per month or per year. Generally, the flow is the change in the stock over time. Figure 12–2 shows a wide range of population sizes (S^e to S^m) where growth increases with increasing population size. However, at population sizes larger than S^m, population increases lead to a reduced (but still positive) growth rate.

Examine the two points (S^e and S^n) where the growth-size function crosses the horizontal axis, showing zero population growth. The point S^n is called the *natural equilibrium,* because it is the stable population at which the fishery settles without human intervention. Falling populations of fish resulting from mortality or other losses are offset by population growth resulting from births, growth of the remaining population, and other gains. The natural equilibrium is stable and tends to sustain itself at that level. At a stable equilibrium, shocks that increase or decrease the fish population produce counterbalancing forces that restore that equilibrium. If the population temporarily shrinks below S^n, it falls below the capacity of its habitat. This temporary excess in the habitat's capacity reduces mortality for the very young and very old. This mortality reduction continues until the population rises to its habitat capacity at point S^n.

Forces that return the fish stock to its equilibrium at point S^n work equally in the opposite direction. If population grows above S^n, the stock is now larger and there is greater mortality pressure

Figure 12–2
Fishery population size (*S*) and growth (*g*)

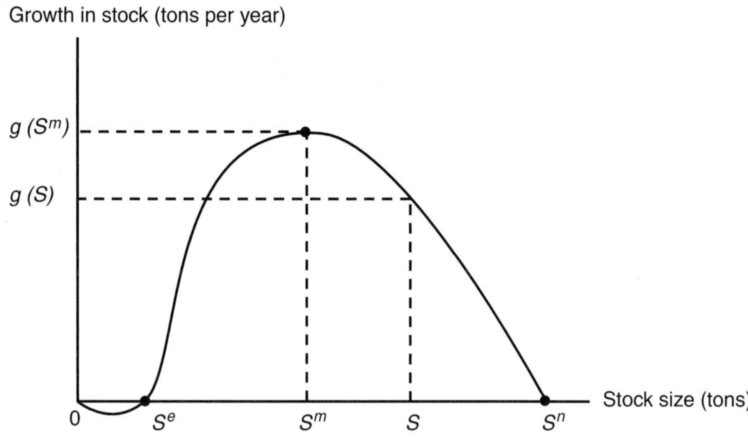

on the very young and old brought about by scarce food resources. Over time, after the temporary population growth, the fish population shrinks to the left until point S^n is again achieved.

Point S^e is called the *minimum sustainable population,* the population level beneath which population growth is negative. Negative growth means that deaths and other losses exceed births and other gains (also called the critical level; Ciriacy-Wantrup, 1952). Compared to S^n, this equilibrium is unstable. Population levels larger than S^e set forces in motion to produce positive growth, moving along the growth-size function towards S^n. However, when the fish population falls below the critical level S^e, it continues to fall until it becomes extinct. There are no biological forces set into motion to bring the population to a sustainable level.

This simple biological background is influenced by humans, who harvest the fish population for consumption. The total number of fish caught is called the *catch.* A fish catch is sustainable when it equals the number of new fish that result from population growth. If fishers in a particular year harvest only that year's population *growth,* without reducing the population's level, that population can be forever sustained. Living off the population growth is the same idea as living off your bank account's interest. As long as you spend only the annual interest on your bank account, you never need to dip into your principal. As long as the absolute size of the fish population does not shrink and fishers harvest only that year's growth, the catch is sustainable. Many catch levels are sustainable. Figure 12–2 shows that when population is S, fishers can forever harvest the annual population growth equal to $g(S)$. The vertical height of the growth-size function for any population level between S^e and S^n is that year's growth as well as its sustainable harvest. At any given population level S, the annual population growth, equal to $g(S)$, is a sustained harvest. Since the year's catch equals its growth, population gains equal losses; therefore, population size and its future growth are constant.

Based on Figure 12–2 by imagining many possible populations, biologists introduce the concept of *maximim sustained yield.* Figure 12–2 shows that S^m is the population that permits the maximum sustained yield, equal to the largest possible annual harvest, $g(S^m)$, that can continue forever. In thinking about all possible populations in Figure 12–2, it is clear why $g(S^m)$ produces the maximum sustained harvest. Fishers could take a larger catch by harvesting a vertical distance larger than the year's growth,[1] but that practice reduces next year's popula-

[1] An important policy question centers on the amount by which annual harvest can *temporarily* exceed annual growth. A community may wish to temporarily promote cheap food or affordable homes by encouraging resource harvesters to "borrow" from the resource's population while it develops a food sector. After the sector is established, the community reestablishes depleted populations. History shows that frontier nations (e.g., United States in the nineteenth century) practiced this borrowing. Before establishing efficient food systems, nations often practice unsustainable renewable resource practices. Later when the nation is settled and food is cheaper and more sustainable, it may make good economic sense to reestablish depleted populations.

tion. If harvests continue to exceed annual growth, the population continues to fall over time, possibly to the critical level and even extinction.

■ Economically Efficient Model of a Fishery

Each yield produced on the curve in Figure 12–2 is biologically sustainable. By biologically sustainable, we mean, since only the annual growth of fish stock is being harvested with a given level of effort, that level of effort can be sustained year after year.

Biologists often view maximum sustained yield as the ideal policy target for which fishery management should be based. Economists are interested in economic efficiency: Does the effort level that produces the *biological* maximum sustained yield equal the same effort level that produces *economic* efficiency?

Classic analyses of the connection between economics and fisheries biology include those of Gordon, 1954; Scott, 1955; Beverton and Holt, 1957; Schaefer, 1957; Clark, 1973; Anderson, 1977; and Munro, 1982. Economic efficiency is examined by bringing people into the analysis. Where people count, the economic value of decisions take on great importance. Finding the economically efficient level of effort requires knowing the value of the harvested fish, as well as the cost of the effort needed to produced that harvest. The economic analysis makes three assumptions:

1. The harvested fish are sold in competitive markets, so the price of fish is unaffected by any individual fisher's actions[2] and is not depressed when that fisher sells greater quantities.
2. The cost per unit of effort is known and constant (i.e., cost per unit-effort is the same at all levels of effort).
3. The quantity of fish caught per unit-effort is proportional to the size of the fish stock (i.e., a smaller fish stock produces a lower catch per unit-effort).

In Figure 12–3 these three assumptions are used in building total cost and total revenue functions. The total benefit (i.e., total revenue) function is the effort-yield curve of Figure 12–2 multiplied by the price per unit of fish harvested. Since the total revenue curve is yield, multiplied by a constant price, Figure 12–2 and Figure 12–3 have similar shapes. Increasing *fishing effort,* going from right to left in Figure 12–2, results in larger harvests and smaller population sizes. In Figure 12–3, increases in harvests are shown by going from left to right, up to harvest size $g(S^m)$ from Figure 12–2, which produces revenue $B(E^b)$ that is, benefits (B) at biologically efficient level, E^b. Further increases in effort, continuing from right to left in Figure 12–2, reduce harvest, which is shown by reduced total revenues in Figure 12–3, moving from left to right. The total cost curve is a straight line going up and to the right from the origin; its slope equals the cost of one more unit-effort.[3] Higher costs per unit-effort produce a steeper total cost curve. The slope of the cost curve has major implications. It means that as a fishery is depleted over many years the cost curve becomes steeper and rotates to the left. It also means that as harvesting technology advances, the cost curve flattens out and rotates to the right.

---

[2] That price is assumed to represent the social value of the fish, that is, the added social benefit produced by each harvested fish equals its market price. If harvested fish produce unpriced values or unpriced costs, that assumption is violated.

[3] It is assumed that the total monetary cost of effort is the same as its opportunity cost, which means that $X units of extra money cost to the fishers equals $X of extra resources displaced by that effort. This requires assuming that all resources used for fishing effort are priced at their real cost to society.

Figure 12–3
Sustainable fishery benefits (B) and costs (C)

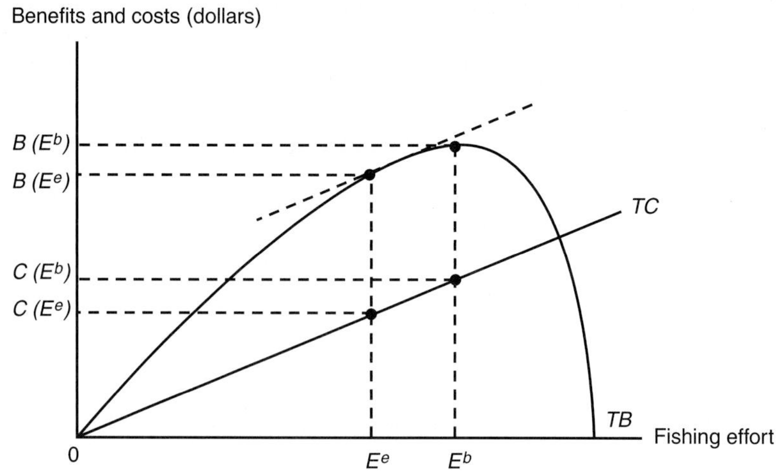

The net income to fishers produced by any given effort equals society's net benefits[4] and is measured by the vertical distance in Figure 12–3 between total revenue (TB) and total cost (TC). The effort level that maximizes net income is E^e; that effort level produces net income equal to $B(E^e) - C(E^e)$. Mechanically, this economically efficient optimum can be found by drawing tangent lines to the total benefits (TB) function at each of several effort levels, then choosing the effort level whose tangent lines (marginal benefit) has the same slope as the total cost curve (marginal cost). Once this effort level is found, you can see that any other effort level produces smaller net income, defined by a smaller vertical distance between the curves.

Now compare the fishery's biology and economics. The net benefits produced by the sustained yield level of effort of E^b (biological efficiency), equal to $B(E^b) - C(E^b)$, are lower than at the effort level E^e at which economic efficiency occurs. Why does this unexpected result occur? The answer is that even though *yield* is greater at E^b than at E^e, the *added* cost of the added fish, $C(E^b) - C(E^e)$, is larger than the added benefit, $B(E^b) - B(E^e)$. Therefore, *net* economic benefits are lower under the maximum sustained yield (MSY) level of effort than under the economically efficient level. This means that the point of MSY incurs *greater effort* and produces *more harvested fish* than the maximum economic yield. The economically efficient effort level is *lower* than the MSY and, therefore, produces a *higher* permanent stock of fish. What all this means is that economically efficient policies will produce a permanent stock level of fish, such as S in Figure 12–2, that is larger than the level S^m resulting from policies favoring harvest levels of maximum sustained yields.

We have defined economic efficiency as the maximization of net income for a single period only, which is termed *static economic efficiency.* A broader definition of economic efficiency is *dynamic economic efficiency,* which is the level of effort that maximizes the present value of net income produced by the fishery over all future periods. We showed that the static efficient sustained yield incurs less effort level and produces a larger fish stock than the maximum sustained yield. More advanced treatments of fisheries economics (e.g., Clark, 1990) show that once discounting is introduced, the dynamic efficient sustained yield incurs more effort and smaller fish stock than the static efficient sustained yield, and typically (but not always) less effort and a larger fish stock than the maximum sustained yield.

---

[4] This is an important point to remember when reading the rest of the chapter. In a competitive market for fish, the fisher's net income from a given effort level equals the fisher's revenues minus her costs. Revenues are price times total catch. Where the fisher's input prices reflect the value in their next best use, private costs to the individual fisher measure the value society loses as a result of the fisher using those resources. So in a competitive market, the fisher's net income is the net benefit received by society from that effort level, equal to society's benefits minus its costs.

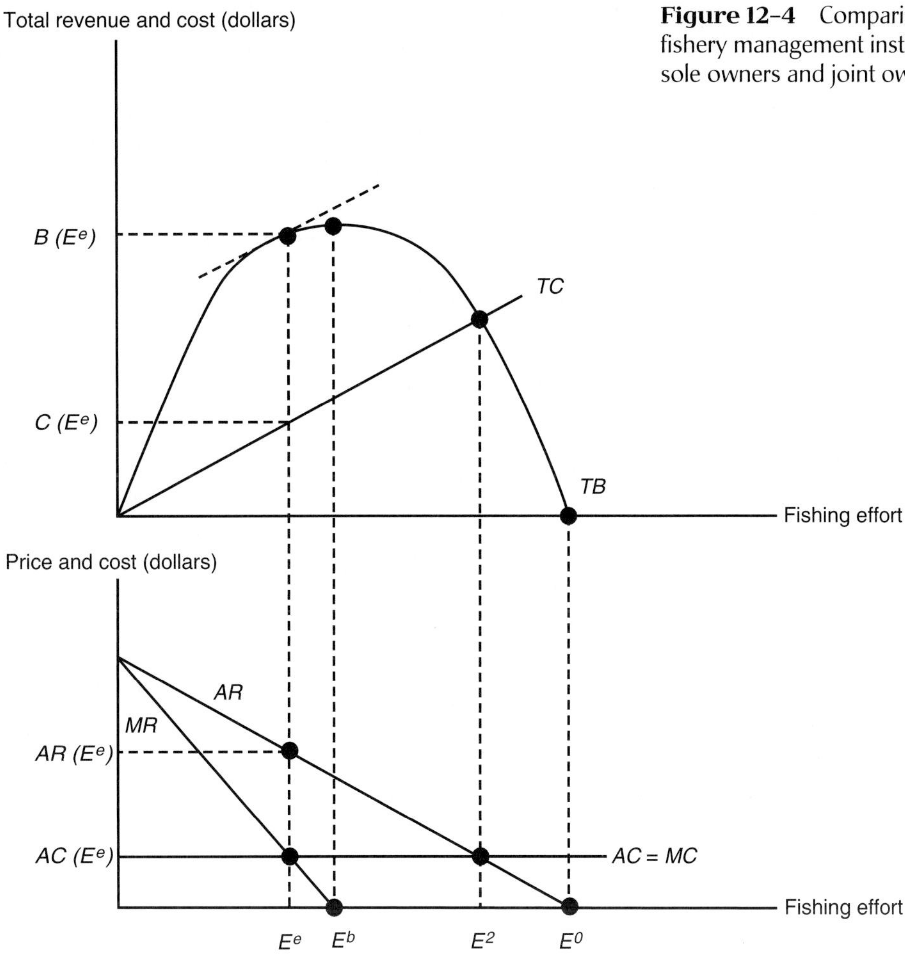

Figure 12-4 Comparison of fishery management institutions–sole owners and joint owners

Market Allocations

In the previous sections we defined an economically desirable harvesting pattern for a fish population and labeled it economically efficient. This information gives us the necessary tools to explore how well an unregulated market consisting of profit-maximizing fish harvesters achieves economic efficiency. If the results produced by profit-maximizing harvesters look different from those produced by economic efficiency, we are equipped to explore various policy measures for moving the market-determined path closer to economic efficiency.

Sole Owner

If the complete bundle of property rights to a fishery rests with a single person or interest, that fishery comes under control of what is called a sole owner. A sole owner of a fishery receives all the revenues and is responsible for all the costs. Assume the sole owner has clearly defined and enforceable property rights to catch fish and to exclude all other fishers from harvesting fish. Figure 12–4, which builds on Figure 12–3, permits us to analyze the sole owner's behavior. The upper and lower graphs have the same variable on the horizontal axis (fishing effort), so both graphs can be examined simultaneously to see what happens under several levels of fishing effort.

Presume the sole owner wishes to maximize profits. This fisher adds to profit through increased fishing effort as long as marginal revenue from the added effort exceeds marginal cost.

Effort expands until marginal revenue equals marginal cost.[5] Figure 12–4 shows that the effort level that maximizes profit occurs at level E^e. At that efficient level, profits are $B(E^e) - C(E^e)$. When the sole owner believes he has a secure property right to the fishery, he takes action that influences sustained profits; that is, he acts in a way to maximize the sustained income produced by the fishery. The sole owner's actions also have both social and private consequences, because society cares about sustained economic efficiency. The sole owner's actions promote the economically efficient use of the fishery as well as the resources used to harvest its fish. One of the earliest economists to reach this conclusions was Scott (1955).

Joint Owners: Open Access

Sole owners are rare in marine fisheries since it is hard to put a fence around one's private piece of ocean. Property rights to marine fisheries typically are not vested by single owners or even by a limited group of owners who band together to exclude others. History indicates that marine fisheries have been most commonly treated as an *open access* resource, which means that there is no limit to the number of fishers or fishing boats that can exploit a fishery. If people believe the fishing is good, there is nothing to stop someone with a boat and harvesting gear from getting into business.

Open access fisheries have major impacts on incentives. In years when harvesting technology was weak compared with the productivity of the oceans, incentives attracting new entrants into the industry rarely caused a problem. For example, suppose there is an open access fishery in which the fishing effort is currently at the economically efficient level, E^e. Think of the potential fisher who sees incomes being earned, adventure, and reward. She contemplates whether to buy and outfit a boat and start harvesting the fishery. Open access means that a potential fisher needs no one's approval to harvest fish, nor does she need to buy those rights from other fishers. The fish are there for the taking.

Since total effort is already at the efficient level, E^e, profits are being made and total effort is at a lower level than the maximum sustained yield. The newcomer compares the cost of the outfitted boat with the expected revenues. At effort level E^e, the average revenue per outfitted boat, $AR(E^e)$,[6] exceeds average cost, $AC(E^e)$. The open access status of the fishery continues to lure new fishers as long as there are profits to be made, and profits continue to lure new entrants as long as average revenue exceeds average cost. So new boats are bought and the fishery sees a steady stream of optimistic new entrants.

At what point does profit cease luring new entrants? Not until the attraction is gone. At what level of effort do profits vanish? Not until effort level E^2 is reached. At that level, the fishery produces zero net income. There are no market forces to restrain this increased effort until profits are reduced to zero at effort level E^2. The open access institution that governs the fishery sows the seeds of its own destruction, ultimately eliminating profit, wasting effort, and overexploiting the resource. At that effort level, economists say that unrestricted entry produced by open access causes *resouce rent dissipation.* Effort level E^2 does not reduce the stock below the critical level, but the critical level *is* reached when effort reaches E^0. So if the community of fishers is slow to reduce effort when it receives negative profit signals at effort levels exceeding E^2, there is a risk of effort expanding until the stock falls below the critical level. Open access fisheries are yet another example of the tragedy of the commons. See Example 12–2.

---

[5] This simplifies by assuming that the sole owner is only concerned about the current period's profit and ignores the future. A more realistic and complex model would assume that the sole owner wishes to maximize dynamic efficiency, that is, the discounted net present value of a stream of future incomes.

[6] If N is the number of existing fishers and each earns an average revenue of X, then even if the newcomer's catch comes entirely at the expense of existing fishers, the newcomer's expected revenue is $(NX/N+1)$, or about X, which in Figure 12–4 is $AR(E^e)$.

Example 12–2
Sustaining Europe's Fisheries

A 1997 study done by the European Parliament concluded that overfishing has several causes:

- Market failures: The price of fishing fails to send a signal of the economic cost of each new fisher's effort when access is unlimited.
- Free riders: Agreements are ineffective unless all fishery participants abide by them.
- Short-term success increasing long-term capacity: If a nonprice measure succeeds in conserving stocks, the now greater valued resource provides incentives to overfish through increased fishing capacity.
- Distributional impacts: Management systems that limit access by assigning property rights or by charging fishers for the right to fish can cause unfair redistributions of income and jobs.

Confronting the overfishing issue requires stating the objectives of fisheries management and establishing measures to carry them out. The objectives are correction of the problems of market failure, free riders, and undesirable distributional effects. These objectives can be accomplished by promoting stable, sustainable economic efficiency through accurate scientific assessment and effective enforcement of regulations. Measures to promote economic efficiency should include (1) definition of fishing rights and management rules, (2) creation of ways to allocate fishing rights, and (3) enforcement of fishing rights and management rules.

In many fisheries there is little relationship between this year's stock size and next year's growth. Identifying a level of fish catch that optimizes economic or physical yield is difficult because of fluctuations in growth that are hard to control. Even if an ideal level of fish catch could be identified, it would likely require a degree of government activity uncommon in Western democracies.

The individual transferable quota (ITQ) is a regulation intended to correct market failure and allow the market to adjust the amount of fishing to the opportunities available. An economic incentive is provided to eliminate overcapacity and encourage compliance. When ITQs or other fishing rights are initially given free, they tend to acquire considerable value to the recipients and make entry into the fishery more costly to prospective fishers. The European Parliament is optimistic about the potential for ITQs in which one group purchases from another exclusive right to a fishing area.

Source: European Parliament. 1997. *The common fisheries policy beyond 2002: Alternative options to the TACS and quotas system for the conservation and management of fisheries resources.* Available from *http://www.europarl.eu.int/workingpapers/fish/e7ensum_en.htm* [Accessed October 2, 2004].

The open access status of the fishery produces considerably more effort levels than are economically efficient. As illustrated in the article by Ricketts (Example 12–1), open access ultimately depletes the fish stock. And in the real world, open access has characterized most marine fisheries. Most government regulations have attempted to adjust the incentives facing fishers to remedy the overfishing problems caused by open access.

Is open access really different from sole ownership? Yes. Faced with the same level of profit as the total profits earned by joint owners, the sole owner avoids effort levels larger than E^e. Why? Because increasing effort beyond level E^e reduces profits, which violates the sole owner's economic self-interest. The incentive system facing the sole owner acts to protect the fishery, whereas the incentive system facing a group of joint owners does not protect the resource. The story has a different ending when all owners face unrestricted access under open access conditions. Each owner makes her own decisions, but the sum of actions taken by all owners expands effort beyond E^e, which reduces profits to the group. The cost of any single fisher's expanded effort is shifted mostly to other fishers. Each fisher is blind to the correlation between her own expanded effort and her own decline in income. Each fisher wishes that the others would reduce effort. In fact, each fisher may ask the government to regulate the others. Yet as Ricketts said, no fisher wants to be regulated by the same government that regulates everyone else.

Even though this analysis is theoretical, it has immense applications to the real world. Many fisheries currently are plagued by these problems: fisheries of the Black Sea (Knowler, Barbier, and Strand, 2001), Latin America (Ibarra, Reid, and Thorpe, 2000; Thorpe, Ibarra, and Reid, 2000), Malaysia (Tai, Noh, and Abdullah, 2000), the European Union (Bulte and Pennings, 1997), and Hong Kong (Lai and Yu, 2002). All of these studies identified significant overfishing. In all cases, if fewer fish are harvested, the same efficiency can be achieved with *fewer* boats used closer to each boat's capacity and *lower total cost.* For economically efficient levels of effort and harvest to occur in real-world fisheries, policies and institutions need to be designed so that incentives facing fishers produce efficient and sustainable outcomes.

Institutional Designs for Real-World Fisheries

Good policies for fisheries will promote efficiency, equity, and sustainability. Market solutions focus on private institutions to manage fisheries. Governments get involved through open access regulations that limit fishing practices and through limited entry regulations.

Market Solutions: Fish Farming and Fish Ranching

Inefficient and unsustainable fish harvests and fish stocks can occur when the resource is held as an open access resource. One policy solution is to design *private* property institutions to manage fisheries. Private markets work well when it is easy to limit or control the fish as private property, for example, when the fish do not move very far, when they can be limited by man-made barriers, or when they have a strong homing instinct to return to their birthplace to lay eggs.

Privatizing a fishery, when it can be done successfully, offers incentives to reduce overfishing. The incentives provided by potential profit encourage the private owner to invest in various measures to increase the fishery's long-run productivity and yield. Two examples of productive measures to increase a fishery's yield are controlling water temperature and increasing nutrients. Raising and harvesting fish in a controlled environment is known as *aquaculture,* a method that has seen some success (see Example 12–3). Difficulties with aquaculture often revolve around access to water, a resource that is managed as a common property in many cultures (e.g., Armitage, 2002; De Voe, 2000). Nevertheless, aquaculture has served as a major contributor to the word's food supply since 1985 (Alongi, 2002; Chopin et al., 2001; Naylor et al., 2000). See Table 12–1.

Economic theory predicts that most commerically successful aquaculture will occur where its profits are highest. These profits are likely to be highest where fish produced under common property conditions are the highest priced, either because of previous overfishing or because of high demand. Japan provides a good example of where both conditions are true. Japan has a large population that depends on fish for its protein. In recent years Japan has found that simply harvesting fish from the sea is inadequate to supply protein at affordable prices. For this reason, the country has become a major producer of fish from aquaculture; it has invested in some of the most modern aquaculture enterprises anywhere (Kishi, Uchiyama, and Iwata, 1994). The Japanese government has helped considerably by setting up private property rights for fishing in designated waters that were previously held in common. Local fishing associations then subdivide these designated areas and assign them to individual fishers for exclusive fishing. Treating the waters as private property allows each owner to invest in and improve the resource efficiently and sustainably.

Economic theory suggests that more aquaculture is likely in future years as it becomes more profitable in the face of overfished common property fisheries, whose prices will probably increase. Fish farming continues to make major contributions to the world's supply of harvested fish. By the mid-1980s, 8% to 10% of global fish consumption came from fish farms. By 2000, the figure had risen to 20% (FAO, 2002).

Fish ranching is another approach to aquaculture that harnesses the power of market incentives. Whereas fish farming is a highly controlled process for rearing a fish over its complete lifetime, fish ranching involves less control, less labor, and typically lower costs. Fish ranching capitalizes on the homing instincts in some fish, such as the Pacific salmon or ocean trout, which allow them to be captured at the time they return to their birthplace. Salmon and ocean trout eggs are hatched and kept in a confined area by the rancher for one to two years. When the fish are re-

Example 12–3
Common Property Resource Policy Challenges: Thailand

The competition to meet the world's growing demand for seafood has increased fishing pressure to the point that it threatens livelihoods of fishers in the Gulf of Thailand. Thailand's fishing grounds are fished heavily as a result of a growing global economy, changing markets, built-up fleets, and fast-moving development. A generation ago, fish were called the poor man's protein. But trends ranging from expanding international markets to health concerns in the United States and western Europe have helped increase seafood's popularity. The 200-mile exclusive economic zone set up through international treaties in the 1970s allowed nations to block out foreign fishing fleets from their waters and profit for themselves.

Property rights are a big part of the story. To make a profit, you must own the resources you use for production or rent them from someone who does. Fishers have no property rights in their catch until they actually make the catch and bring it into their boat. Historically, anyone with a boat, a crew, and fishing equipment can fish the ocean. With a small investment, skill, work, and luck, the returns can be attractive. Short-term economic gains look impressive because at the same time that fish populations are shrinking, demand and prices for fish increase. While fleet capacity was doubling, the catch increased by less than half, according to a study by the Food and Agricultural Organization (Poonnachit-Korsieporn, 2000). The world's fleet substituted commercially less valuable fish (e.g., pollock in Alaska). Meanwhile, government subsidies for fishers continue to be seen as entitlements, and boat owners have fought attempts to take them away.

Thailand is a classic example of a trend that continues to grow around the developing world: It has remade its coastline into a powerful economic resource, in both fishing and aquaculture, which helped it become the world's leading seafood exporter in 1994, passing the United States. Aquaculture is sometimes seen as a way to make up for growing prices of fisheries that rely on capture. Huge infusions of capital have changed the once quiet Gulf of Thailand coast into large numbers of shrimp farms. Environmental safeguards are rare. Pond sludge is set aside anywhere farmers can find room, which raises health and environmental concerns. Shrimp diseases can spread quickly through the congested rearing areas.

Meanwhile, the growing demand for fish in the United States, Europe, and Japan, and the developing world's desire to profit from fisheries, have combined to create a growing supply of seafood products from the developing countries, which have growing numbers of commercially motivated fishers. The aquatic resources center in Manila estimates that 50 million people worldwide are engaged in small-scale fishing through catching, processing, and marketing. Most live in small communities that until recently were insulated from those forces. In many cases these people are politically marginalized. They live with the changes established by consumers, producers, and governments thousands of miles away.

Sources: Barbier, E. B., and S. Sathirathai. 2001. Valuing mangrove conservation in southern Thailand. *Contemporary Economic Policy.* 19 (2): 109–122; McQuaid, J. 1997. Overfished waters running on empty. *The Times-Picayune,* New Orleans, LA. Available from *http://www.pulitzer.org/year/1997/public-service/works/1-2/* [Accessed October 2, 2004]; Food and Agricultural Organization. 1996. *The state of world fisheries and aquaculture.* Available from *http://www.fao.org/docrep/003/w3265e/w3265e00.htm* [Accessed October 2, 2004]; FAO. 1998. *The state of world fisheries and aquaculture.* Available from *http://www.fao.org/docrep/w9900e/w9900e00.htm*. [Accessed October 2, 2004]; FAO. 2000. *The state of world fisheries and aquaculture.* Available from *http://www.fao.org/docrep/003/x8002e/x8002e00.htm* [Accessed October 2, 2004]; Poonnachit-Korsieporn, A. 2000. *Coastal Fishing Communities in Thailand.* Food and Agricultural Organization. Available from *http://www.fao.org/documents/show_cdr.asp?url_file=/docrep/005/ac790e/ac790E00.htm* [Accessed March 8, 2005].

leased, they swim to the ocean. After they become mature adults, they return to the place of their birth where fish ranchers harvest and sell them. Some 1.7 million metric tons of salmon were harvested worldwide using ranching methods in 1999.

Aquaculture contributes more and more to the world's fish supply and could very well serve as an important backstop technology in the future for overfished natural fisheries. The success of aquaculture illustrates the considerable power of market incentives to supply an important and needed resource and substitute for depleted common property fisheries.

Table 12–1 World aquaculture and commercial catches, 1989–2003

	Aquaculture			Commercial catch			
	million metric tons						
Year	Inland	Marine	Total	Inland	Marine	Total	Grand total
1989	7.2	5.1	12.3	6.2	83.1	89.3	101.6
1990	7.7	5.4	13.1	6.4	79.1	85.5	98.6
1991	7.9	5.8	13.7	6.2	78.3	84.5	98.2
1992	8.9	6.5	15.4	6.1	79.2	85.3	100.7
1993	10.1	7.8	17.9	6.5	80.0	86.5	104.4
1994	11.7	9.1	20.8	6.7	84.7	91.4	112.2
1995	13.6	10.9	24.5	7.2	84.3	91.5	116.0
1996	15.4	11.4	26.8	7.4	86.0	93.4	120.2
1997	17.0	11.8	28.8	7.5	86.1	93.6	122.4
1998	18.1	12.7	30.8	8.0	78.3	86.3	117.1
1999	20.2	13.3	33.5	8.5	85.2	93.7	127.2
2000	21.3	14.2	35.5	8.7	86.8	95.5	131.0
2001	22.5	15.2	37.7	8.7	84.2	92.9	130.6
2002	23.9	15.9	39.8	8.7	84.5	93.2	133.0
2003	25.2	16.7	41.9	9.0	81.3	90.3	132.2

Source: *Adapted from Food and Agriculture Organization of the United Nations. 1996, 1998, 2000, 2002, 2004.* The state of world fisheries and aquaculture. *Available from http://www.fav.org/documents/show_cdr.asp?url_file-/docrep/007/y5600e/y5600e00.htm [Accessed March 8, 2005].*

Open Access Regulations

A major goal of fishery management is to bring fishing effort and fish mortality into line with what the resource can sustain on a continuing basis. There are two basic methods of achieving this balance: (1) open access regulations that limit fishing technology, fishing time, quantity or species of fish caught and landed, or fishing locations without reducing the number of fishers or fishing boats; and (2) limited entry, which restricts the number of fishers or fishing boats.

As fishing pressure mounts because of increased numbers of boats and/or more efficient harvesting technology, fish stocks are endangered. The natural tendency at this point is to find a way to regulate or control fishing pressure without actually reducing the number of fishers or boats. That is, there is a desire to regulate the gear, seasons, or species of fish without limiting who gets to fish. Open access regulations limit fishing *behavior* of fishers who participate in the fishery without directly limiting the number of fishers or fishing boats allowed to participate. So, although these regulations limit the effectiveness of access, they do not directly limit access to the fishery. Nevertheless, because open access regulations typically increase the cost of fishing, they can indirectly affect who participates in the fishery. Some fishers become unprofitable and they decide to leave the fishery.

Limiting access to marine fish populations can take many forms: fishing within set quotas, limiting certain kinds of fishing gear or the capture of certain species, limiting the size and horsepower of fishing boats, prohibiting fine-mesh nets so that smaller fish can continue to grow, and closing certain coastal waters, such as spawning grounds, during particular times. These are all forms of limiting effectiveness of access in order to reduce pressure on the fishery. Open access regulations have the intent of maintaining fish stocks at a target level. Theoretically that target level aims for maximum sustained yield. Many times however, the regulations were not enacted until the fish stocks had fallen dangerously below the level of maximum sustained yield.

How Fish Are Caught (Regulating Technology). Before the Industrial Revolution in the 1800s, fish stocks were virtually unaffected by fishing pressure. The harvesting power of a fishing boat, or even the larger power of a fleet of boats, on fishing populations was just too small. However, this has changed. The modern fishing fleet has the power to harvest huge numbers of fish quickly and efficiently. Sophisticated electronics are used to locate the fish. Fishing lines and fishing nets are measured in miles not feet. Huge engines quickly move boats to the fish soon after they are found. This impressive technology could produce sustained low fish prices for consumers and high incomes for fishers if the fisheries were privately owned. But when a fishery is subject to unrestricted open access without some kind of regulation, technology can decimate fish stocks, reduce incomes of fishers, and increase fish prices.

Which Fish Are Caught (Size and Weight Limits). Regulation of *which* fish may be caught typically sets a minimum size of a particular species of fish that is legal to catch and keep. The logic behind this type of regulation is to leave part of a fish stock in the sea to provide adequate breeding stock to guarantee future populations. Fishers typically deal with this kind of regulation by finding a net with a mesh size that allows the smaller fish to pass through while keeping the larger, legal fish.

When Fish Are Caught (Season Limits). Regulations that limit *when* fish may be caught aim to restrict harvests by restricting times during which fishing can occur. Occasionally regulations limit fishing to certain periods each day, but more commonly regulations close certain periods altogether each year. For example, the spawning (egg laying) season may be closed to fishing to avoid disrupting the spawn and/or to avoid fishing out the entire population, which may congregate in a particular area during that time.

Not all spawning fisheries are prohibited, because sometimes the spawning congregation is the economically most valuable part of the fishery. Caviar is the harvested, processed and salted eggs of certain large fish such as sturgeon.

Where Fish Are Caught (Vulnerable Populations). Some open access regulations limit where fish may be legally caught. These regulations aim to protect fish stocks where they concentrate in high numbers. For example, some kinds of commercial fishing is closed in locations where a population is so concentrated that much of it could be wiped out by overharvesting. These kinds of regulations also aim to protect fishing habitats from being lost to fishing. Pulling bottom-dragging nets through shallow sea areas can destroy valuable fish communities on the ocean floors.

How Many Fish Are Caught (Catch Limits). Some regulations consist of limits on the number, weight, or size of fish that can be taken in a given time period. These limits are sometimes called total catch quotas or total allowable catch (TAC). Total allowable catches give authorities a way to closely control yields. When a fish is very profitable (possibly because of high prices produced by low populations) and easy to harvest when found, its population can be decimated in a short period of time. Because the fish are so profitable, other kinds of open access regulation may be ineffective, so a limit of a small number of fish per boat per day might be established to limit exploitation. Roughgarden and Smith (1996) review scientific and policy evidence on the causes and remedies for collapse of fishers populations.

Economic Implications. A common feature of all open access regulations is an increase in the cost of fishing. By restricting inputs, seasons, boat sizes, and the like, regulators raise the cost of catching a given quantity of fish. Restrictions raise the total cost curve upward and to the left, giving it a steeper slope. In Figure 12–5, TC_1 is the total cost curve without regulations; a curve produced by a regulation might look like TC_2. After a regulation is in effect, under the same open access situation, market forces no longer lead to E^0 units of fishing effort but to a reduction of fishing effort at E^2 units. Because of this, regulators might make the regulation restrictive enough to rotate the cost curve to TC_3 so that it to intersects the maximum biological sustained yield level at E^b units of effort. If a regulation is made even more restrictive, the cost curve shifts up more, to TC_4, which reduces fishing effort to E^e units, approximately equal to the economically efficient level.

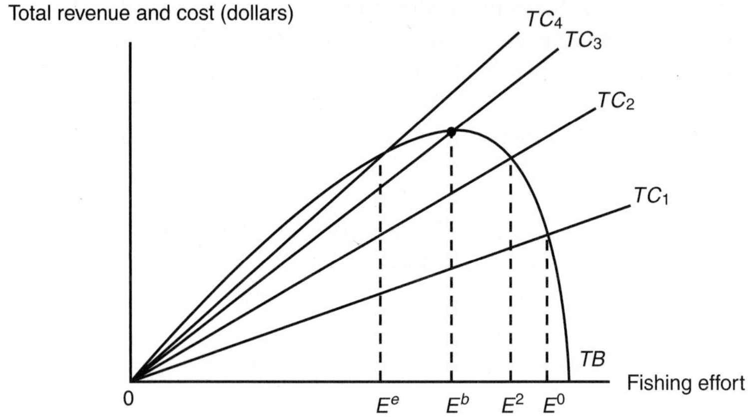

Figure 12–5
Reducing fishing effort by raising the cost of fishing

This regulatory, cost-increasing approach to open access fisheries poses two problems. First, it is possible to reduce levels of fishing effort through use of the profit motive by increasing harvesting costs and reducing effort levels, which increase stocks and ultimately raise future yields. But the cost of fishing is raised needlessly. *Economic efficiency means more than finding the optimal harvest and effort level.* It also means achieving this optimal level with the least cost of scarce resources. Policies may be *biologically efficient* but *economically inefficient*—and there is a difference. Biological efficiency means harvesting the quantity of fish equal to the maximum biological sustained yield. Economic efficiency means harvesting a quantity of fish that maximizes the net benefits to people, that is, benefits minus costs. Any policy that raises the costs of harvesting a given level of fish, such as gear or time restrictions, is economically inefficient. Costs matter to people. Gear or time restrictions mean higher costs, which mean lower incomes for fishers and higher prices for fish consumers.

The second problem with the regulatory, cost-increasing approach to fishing practices is that it sets up incentives for fishers to evade the regulations. When some fishing practices are constrained, fishers will attempt to substitute uncontrolled technologies for the controlled ones. For example, if authorities limit the kind of nets that can be used, fishers will substitute bigger nets, buy boats with bigger engines, and fish more hours per week. When one fishing area is closed, greater pressure is put on substitute areas.

It is still common for conservation policies of fisheries to be based on biological principles of achieving maximum sustained yield at any cost. Often these policies succeed in conserving the fish, but they hurt both fish consumers and producers by artificially restricting efficient fishing technologies. These policies represented in Figure 12–5 provide a clear contrast between the economic and biological approaches to fishery management. Figure 12–5 compares total cost of an economically efficient policy and the total cost after gear or time restrictions are established. The vertical distance between total benefit and total cost shows the net benefit (benefit minus cost) produced by the economically efficient policy at effort level E^e. After a gear or time restriction policy is implemented that raises costs, net benefits fall to zero, since $TB = TC_4$.

Society can do better. It can receive three kinds of increased net benefits: lower costs, higher fishing incomes, and lower fish prices. In fact, these potential net benefits are lost when a regulatory policy implements unnecessarily costly measures to catch a desired quantity of fish, such as using those represented in Figure 12–5. Rather than letting fishers use efficient technologies to reduce the cost of catching the desired number of fish, regulators prohibit those technologies and larger expenditures of all kinds are required to catch the same number of fish.

Time limits have a similar effect on raising the cost of fishing. Fishers want to use their equipment to fish in their preferred slice of time. When limits are set on fishing times (shorter seasons), fishers try to squeeze as many resources into that limited time as possible. They begin to substitute increased capital and labor for reduced time. Fishers feel compelled to buy big, expensive boats with costly harvesting gear so that they can catch as many fish as possible in the shortened season (this concept is known as overcapitalization). Without the time limit, the same total fish harvest would be accomplished with fewer, cheaper, smaller boats used to their full capacity. Again the effect is increased costs, lower fishing incomes, and higher fish prices.

Because of a biologically well-meaning but short-sighted aim of maximum sustained yield that ignores costs, society suffers three losses: higher fish prices, higher costs of buying fish, and lower incomes from fishing. Large numbers of big boats fishing at the same time in a short season is also dangerous, especially in bad weather. Policies that reduce fishing incomes through gear and time restrictions are also politically difficult to carry out, and they provide incentives for fishers to violate the regulations.

Limited Entry Regulations: Economic Analysis

Limited entry to marine fisheries means directly restricting the number of fishers and/or fishing boats allowed to participate in a particular fishery. In the United States, for example, limited entry policies reverse the long-standing practice of allowing fishers free access to the sea, a significant departure from the historical management of living marine resources. Limited entry controls access by assigning a finite number of individuals or fishing boats to operate in a given fishery. And because it is often a response to overcrowding, limited entry commonly leads to reducing the number of fishers or boats allowed to fish, either directly or through incentives.

Limited entry is a highly controversial way to manage marine fisheries. Many fishers feel that controlling the number of people who are permitted to fish removes their traditional rights and limits their historical fishing patterns. Nevertheless, a growing number of fishers as well as fishery managers and others recognize that limiting participation could be a significant measure in conserving and managing some marine fisheries (Hinman and Paulsen, 1993).

Individual Transferable Quotas. An *individual transferable quota* (ITQ) is an allocated privilege of landing a specified proportion of the total annual fish catch, which differs from the traditional, open access approach to commercial fisheries. ITQs divide the total annual catch quota into smaller individual portions. Typically, ITQs are transferable, which means owners of fishing vessels can sell their ITQ certificates, buy others' certificates, or in some cases lease their quota shares based on on how much (or whether) they wish to participate in the fishery.

A carefully crafted quota system on the number of fish that can be taken from the fishery has the potential to achieve economic efficiency and sustainability. Most quota systems, however are neither efficient nor sustainable so it is important to understand the meaning of *carefully crafted.*

An efficient and sustainable ITQ system[7] has five characteristics:

1. Each quota entitles its holder to catch a set weight of a particular type of fish.
2. The summed quantity of fish authorized by all quotas held by all fishers is equal to the economically efficient catch for the fishery.
3. Quotas are freely transferable among fishers, that is, they can be traded in the market.
4. Quota catch levels are enforced.
5. The total allowable catch per quota is adjusted with changing biological conditions.

Each of these five characteristics plays an important part in producing a sustainable and efficiently managed fishery. Suppose that the first condition is violated and that each fisher's quota is defined as the right to use one boat to harvest from the fishery (an effort quota), just like a parking sticker gives you the right to park one car in a lot. Fishing effort quotas are common. The effort

[7] An ITQ system is efficient and sustainable only if catch per unit-effort is independent of the fish stock density. If there are some periods of the year when fish harvest per unit-effort is high, an ITQ system would focus effort in that period with the risk that it could deplete the biomass.

quota is economically inefficient because it gives the fisher an incentive to buy a bigger boat, stock the boat with more equipment, hire a larger crew, and fish for more hours. Overall, it gives the fisher an incentive to expand the fishing capacity of that boat by substituting other things for the one-boat limit. By allowing fishers to substitute boat capacity for boat numbers, the number of fish caught per boat, and therefore the total number of fish caught, is too high. Sustainable economic efficiency requires limiting *catch* (output) not simply limiting *effort* (input).

Violating the second condition causes a simple arithmetic problem. Total catch per fisher times the number of fishers (quotas) fails to equal the economically efficient catch.

Condition three deals with transferability. If quotas are transferable, they will gravitate to the most efficient fishers simply because those fishers have the lowest costs per caught fish. Any fisher initially assigned a quota who has high production costs will find she can have a better income by selling the quotas than by fishing. The financial gain from selling the quota to a low-cost fisher is larger than the income she loses from leaving the fishery. Gains in trade go both ways. Quota buyers also benefit from transferability. Low-cost fishers who are assigned initial quotas discover they can make more income by buying added quotas than by fishing with only a single quota. Because of their low costs of production, the price at which they buy quotas (seller's income displaced) is less than the income they gain by added fishing rights produced by added quotas. Transferability also provides incentives that promote technical progress. Fishers who adopt cost-saving technology (e.g., more sophisticated gear) get two benefits: (1) a larger income on their current quotas and (2) increased profitability of new quotas bought from people who avoided the new technology. One can imagine the outcome of such a quota system. The economically efficient catch is harvested by professional captains, each running a large powerful boat with sophisticated gear and well-trained crews.

The fourth condition simply says that property rights require enforcement. Enforcement activities have to be established and financed so that fishers have incentives to follow the rules. Enforcing fishery regulations is typically expensive, because fishing is often done in isolated places and fishers can dump illegal fish overboard if necessary. ITQs are much cheaper to enforce than effort limits, as things like boat size and fishing gear are expensive to monitor and enforce.

The fifth condition deals with flexibility to change the allowed catch per quota in light of continually changing biological facts about the size of the fish stock. Fish stocks are expensive to measure accurately and sometimes they can change quickly and unexpectedly. If the existing quota permits catching too few fish, fishers suffer a lost income opportunity. But if the quotas are too optimistic about the safe stock size, the fishery risks being overfished unless each fisher's catch quota is quickly lowered. As better monitoring and measuring equipment provides new information about fish stock size and growth, flexibility allows incorporation of the latest information into each fisher's quota.

Distribution of ITQs. For each fisher, the initial quota is a property right. Two common approaches to distribute those initial quotas have received widespread attention. In the first approach government or other representatives of the people auction off the quotas. If the government keeps all the auction revenues, the auction will be unpopular among fishers, especially for a marine fishery that had been fished as an open access (free) resource for many years. It is hard to convince people to accept a program that charges for a right that has been free.

In the second approach government assigns quotas to the fishers for free, for example, in proportion to their historical catch. Afterwards, fishers can trade the permits among each other for cash until a market equilibrium is reached. All benefits of the quotas are kept by the current generation of fishers. If new fishers wish to enter the fishery, they must buy the quotas from the existing fishers who have the quotas. At that later time, competition among the quota buyers sets the price of a quota at a level equal to the discounted market value of future net incomes produced by a quota. Any future technology that increased the estimated income to be earned from having a quota (e.g., development of a new fish-finding technology) would increase a quota's price. This method of allocating quotas assigns the value of the quotas to the current generation of fishers. Future generations of fishers would pay the full value of the quota to have a fishing right; and to those future buyers, the price of the quota would be treated like any other cost of production.

Experience with ITQs. After the establishment and early successes of the New Zealand and Iceland ITQs in the 1980s and 1990s, several other nations implemented ITQ programs. In 2005 there were about 60 programs around the world in more than 15 countries. For data summaries, see Eythorsson (2000), Hannesson (2000), Repetto (2001), and Waitt and Hartig (2000). Despite the proliferation of ITQs, their future is uncertain because current programs continue to receive mixed reviews from fishers, governments, the fishing industry, and marine policy analysts. There are four widely publicized ITQ programs in the United States: the mid-Atlantic Surf Clam and Quahog fishery (beginning in 1990), South Atlantic Wreckfish (1992), Alaskan Halibut (1995), and Alaskan Sablefish (1995). Many environmental organizations see ITQ programs as the future of fisheries management.

Advantages of ITQs. A fishery managed under an ITQ system benefits in several ways. First, there are no time limits. People can fish when they want, and they do not need to race for fish. This improves safety[8] and profitability. Second, fishing rights are no longer tied to particular boats. Industry can consolidate its fleet and use only as many boats as are needed to catch the quota. This brings tremendous savings to the industry. Third, enforcement efforts can switch from costly operations at sea to comparatively cheap dockside enforcement as the fish are moved to processing plants. This saves taxpayers large amounts in Coast Guard cutter and aircraft time and other enforcement resources. Cheating falls for two reasons: (1) People are less likely to break the rules when they are able to make a good living from their work. (2) The rights to fish one's quota are valuable, and the quota holder is put at risk if caught cheating repeatedly. Hence, people will think twice before undertaking an illegal activity that, if detected and penalized, could result in the loss of their future income stream, which is what an ITQ represents.

Finally, regulators can stop micromanaging the industry. Instead of trying to regulate what goes *into* a fishery (e.g., when people fish, where people fish, how people fish, what gear people use, what size boat people use), regulators only need to specify an annual quota of what comes *out* of the fishery each year and leave the rest to creative enterprise. With an ITQ system, the fishery requires comparatively little government time to manage. Without ITQ management, regulators are tied up with costly and divisive details.

Potential Problems with ITQs. Despite their advantages, no ITQ system has yet fully resolved all of the potential problems. A few of the major problems are discussed next.

Distribution of Quotas. ITQs are a limited entry policy that attempts to reduce the fishing pressure on desirable overfished species. But how should that pressure be reduced? Somehow the total quantity of allowable catch from the ITQs must add up to less than the current catch, which means that (1) some current fishers will be denied entry, (2) all fishers will have to scale back their efforts, or (3) some combination of 1 and 2 must occur. How should the ITQs be distributed? Some may claim that newcomers are ineligible for the ITQs, for it is much harder to take a fishing livelihood away from a long-standing fishing family than to deny entry to a new fisher. The government could define the ITQ by cutting everybody's legal catch by a certain percentage, but that has problems. Economies of scale are common in fishing; that is, costs per caught fish fall with bigger boats, bigger nets, and so on. So reducing everybody's allowed catch by a certain percentage raises everybody's costs and passes on increased fish prices to consumers.

---

[8] Fishing at sea is the world's most dangerous occupation. Data provided from nations that keep accurate records show that occupational fatalities from sea fishing far exceed the overall national workplace averages. For example, in the United States the fatality rate among fishers is 25 to 30 times the national workplace average; in Italy it is more than 21 times the national average; and in Australia it is 16 times the national average (FAO, 2000).

A better approach from the view of costs is to reduce the number of boats harvesting fish without reducing the scale of each boat's fishing. This allows each vessel to be used at its efficient capacity without depleting the fish. Some fishers will have to forsake their livelihood and go into another business, but which fishers? One approach is to encourage the least efficient fishers to leave the industry with minimum economic suffering. For example, the government could buy fishing rights at least cost from those who are willing to sell them. Among current fishers, each one could state the lowest price that she would accept for leaving the industry. Then the government would buy the rights from the lowest cost bidders, thereby reducing the number of fishers by enough to reduce fishing pressure by the desired amount. The total price of the buyout could be financed out of the government's general revenues. Alternatively, it could be financed by a fee or tax on fishers that remain. The total number of permits retired would be increased until the fish population is protected. Because such a program is voluntary, those who leave the industry only do so if they feel adequately paid. If the buyout fee is paid by those who stay, they realize that their investment benefits them considerably in the future as the fish population increases.

Poor Data. Any fish population is hard to measure, especially a hypothetical population such as the one that produces maximum economic sustained yield. Biologists easily overestimate the size of the safe harvest. When that mistake occurs, there is a risk of issuing an unsustainably high level of quotas, thereby causing the stock to be at risk. Allocation of quotas is based on the biologists' estimate, and the practical implication is that an unsustainably high level of quotas can be issued. Therefore, the stock will be in jeopardy.

If the cost of buying back a large number of quotas is too large, the government might shift to allocating quotas by percentage share. Quotas would not give the right to an absolute number of fish but only to shares between 1% and 100% of a total allowed catch. (If biologists underestimate the safe catch, shares can later be adjusted.) If the allowed catch turned out to be twice too high to sustain the fishery, all quotas would be revised down to 50% of the original allowed catch. The total allowed catch could be determined annually by the government or other acceptable authority based on continually revised data, changing estimates of fish stock or changing technology for counting fish. The percentage share method is flexible because it allows the government to periodically adjust the total allowable catch, based on updated stock estimates, without the need to repurchase or resell large amounts of catch quotas.

Bycatch and High-Grading. Sometimes fish species are hard for fishers to target; the result is that they catch unintended species (called *bycatch*). If those unintended species are also regulated by an ITQ and the fishers have too few ITQs to cover the bycatch, they are landing unauthorized fish and they may be fined. A common method of dealing with bycatch is to dump it overboard. However the dumped fish typically die, so this is a two-edged sword: The bycatch stock is reduced and those that caught are not harvested. Bycatch is a major problem associated with ocean fisheries (Alverson et al., 1994).

Fish managers often have to deal with *high-grading,* which occurs when a quota defines the catch in terms of weight and not number of fish caught. Larger fish typically have a higher price per pound, so when a quota is defined in terms of a fixed total weight, fishers have an economic incentive to maximize the value of the quota by dumping overboard the smaller fish and continuing to fish for bigger, more lucrative ones. However, this action kills a high percentage of the smaller fish, so high-grading results in the same two costs as bycatch: smaller fish stocks and wasted harvests.

One way for fishery managers to deal with both the bycatch and high-grading problems is to make ITQs flexible. For example, ITQs could be set up to include credits for those with too many or too few fish. Fishers who catch too many fish for their quota could buy or lease credits from fishers who caught too few for their quotas, who would sell credits. If a legal way is established to lease credits, all fishers will have incentives to land and market fish legally, while protecting the fish stock. Some fishers may choose to underfish in hopes they can lease their shortages for a higher price than the profits lost by not filling their quota completely.

Overall Evaluation of ITQs. ITQ programs are intended to reduce overcapitalization, promote conservation of fish stocks, improve market conditions, and promote safety in the fishing fleet. Because ITQ programs guarantee a share of the catch, they generally slow or eliminate the race to fish and allow fishers flexibility over the rate and timing of their fishing.

ITQ programs have been criticized for increasing the incentive for fishers to file false catch reports and to high-grade their catches. In some cases, processors or wholesalers obtain monopoly control over catches. ITQs may discourage new entrants into a fishery because of the additional capital investment required to purchase or lease quota shares. In addition, ITQ programs may require additional enforcement expense and could cause substantial unemployment and socioeconomic dislocation in coastal communities. The fairness of current approaches to initial allocation of ITQ shares is questioned because they create wealth and windfall profits for the ITQ owner but they exclude processors and crew.

Knowledge and understanding of ITQ programs is evolving rapidly. Many of the early problems with ITQs resulted from program design and may not be inherent in the concept of ITQ management. However, few comprehensive analyses of ITQ program implementations have been conducted. How much of the criticism of ITQs is valid, and how much of the benefits claimed by proponents of ITQs might be realized, are yet to be determined.

Expanded Territorial Waters

The right of a coastal nation to complete control over a band of water along its shore (the territorial sea), has long been recognized in international law. In the eighteenth century the "cannon-shot rule" gained acceptance throughout Europe: Any coastal nation could control its territorial seas as far as a cannon ball fired from a cannon based on shore could reach. This was about 3 miles at the time, from which emerged the territorial sea limit of 3 miles, which was generally accepted until 1945.

In 1945, President Harry Truman proclaimed U.S. jurisdication over all natural resources on the U.S. continental shelf extending to 200 miles. This was the first major challenge to the internationally accepted freedom-of-the-sea doctrine that had been accepted for hundreds of years. Other nations soon followed by declaring that their ownership rights to fishing extend 200 miles from their shores. Later, a broader concept of a 200-mile exclusive economic zone was established. That distance was largely upheld for all coastal nations after intense negotiations that began in 1973 at the Third United Nations Conference on the Law of the Sea. In 1982 a constitution for the seas was adopted—the United Nations Convention on the Law of the Sea. Inside the 200-mile boundary, nations have exclusive control and responsibility to carry out effective fish management policies. Thus, lucrative fisheries in many nations' coastal waters can be protected from foreign nations, but there is still the internal problem of how to limit overfishing of a domestic open access resource (see Example 12–4). Fisheries in the open waters remain undecided until further international negotiations take place.

Enforcement Issues

A fishery regulation or policy will have no effect on people or fish without enforcement mechanisms. Unlike many regulations and laws, fishery policies are expensive to enforce. Oceans are wide and deep; sea coasts typically have many inlets and bays. Usually it is cheaper for fishers to violate the written laws than for authorities to enforce it.

The economics of law enforcement provides two insights into the design of efficient fishing regulations. First, the cost of enforcement has an important influence on the design of a policy. Second, an economically efficient fishery policy will maximize the present value of society's net benefits, including the benefits and costs of its enforcement. These two insights mean that the most efficient policy with no enforcement costs to fishers (or, voluntary compliance with the law) rarely produces the same behavior as the efficient policy with costly enforcement, to fishers. If policies are designed without incentives for a fisher to comply, the fisher will only comply

Example 12–4
The Cod Wars: Lessons for Fishery Policy

In a tense seven-month period between November 1975 and June 1976, Great Britain and Iceland confronted each other in a military crisis over Iceland's proclamation that it had authority over fishing rights up to 200 miles from its coastline. Although few shots were fired, several ships were rammed during the conflict. Declining fish stocks brought about Iceland's action, and since fishing is the main source of Iceland's income, that was a serious economic threat. Both sides agreed that stocks were declining but they could not agree on the cause or an allocation of rights to fish the remaining population.

The science of limiting fish harvest to conserve renewable resources for future fishers is poorly developed. Even less is known about the complex interaction between cod and the factors that influence cod (e.g., its prey species, water temperature breeding, and migration). Soon, after the cod wars, Iceland's cod stocks fell as too many cod and other commercial varieties of fish were caught, even though Iceland's fishers were in full charge. This happened partly because scientists were too optimistic in their calculations and partly because politicians felt pressure from voters.

Almost than 30 years later, Iceland is an outstanding example of how fish stocks can be managed even though considerable uncertainty clouds the links between science, economics, and politics. Each Icelandic fishing vessel has its unique fish quota, all catches are tracked, and fishers have incentives to play by the rules. Iceland's takes the precautionary approach: Any one year's total allowable catch is set at 25% of estimated stock, which leaves plenty to spawn. Icelandic scientists believe that by applying this rule, the chances of stock collapse are less than 1%. But Iceland's success came after bitter fight.

By the late 1980s debates on how to limit fishing effort in Iceland were influenced by economics principles discussed in this chapter, especially the use of individual transferable quotas (ITQs) to protect the fish resource and economy. The focus shifted from resource protection to the search for the least-cost method for achieving that protection. People recognized that under a system of ITQs, permanent and fixed quotas would help long-term planning. Market transferability would provide flexibility and efficient use of money, boats, and gear. Inefficient vessels (high cost per unit-catch) would be bought out, while efficient ones would remain to optimize their operations. In addition, the efficiency generated by ITQs could produce annual income for the government, which later became an important source of revenue in Iceland's fishing-dependent economy.

Iceland's 1990 Fisheries Management Act turned the country into a test site for a market-based fisheries management system. The act provided a basis for a tradeable quotas that continuously redistribute fishing rights among vessel owners, communities, and regions. The transformation of fishing rights into capital, represented by quota value, has been an important contribution to the growing economic strength of Icelandic companies with large holdings of quotas. In terms of export value of fish products and profits made in leading fishery companies, Iceland's ITQ program has been a success.

Sources: Brown, P. 2000. *Fighting cod war taught Icelanders the lesson of conservation. The Guardian.* Available from *http://www.guardian.co.uk/fish/story/0,7369,354910,00.html* [Accessed October 2, 2004]; Eythorsson, E. 2000. A decade of ITQ-management in Icelandic fisheries: Consolidation without consensus. *Marine Policy.* 24 (6): 483–492.

with the law if the benefits to her exceed the costs (see Example 12–5). Regulations that impose high costs of compliance on fishers are more likely to be ignored than those with low costs.

Efficient regulations impose costs on violators that are proportional to the damages to society caused by their violation.[9] Efficient regulations also contain provisions for dealing with noncompliance, for example, fines against violators. An efficient structure of fines will make the fisher's expected cost of a violation exceed its benefit.[10]

---

[9] This is an ancient subject. Blackstone, in the eighteenth century, wrote on the efficient design and enforcement of society's laws: "The end of punishment is to deter men from offending, it never can follow from thence, that it is lawful to deter them at any rate and by any means: . . . Where the evil to be prevented is not adequate to the violence of the preventive (price exceeds marginal cost), a sovereign that thinks seriously can never justify it" (Blackstone, 1942).

[10] The fisher's expected cost is measured as the probability of getting caught multiplied by the price of the fine. This probability increases as more enforcers are hired and decreases with the difficulty of detecting the violation. The fisher's benefit of the violation is defined as the increased net income from ignoring the regulation.

Example 12–5
Why Do Fishers Violate the Law in Southeast Asia?

An important question in fishery management is the extent to which fishers follow the law. A study in southeast Asia examined compliance by economically rational fishers whose behavior was motivated by various factors. Morality, legitimacy, and social influence were studied in addition to the conventional costs and revenues associated with illegal behavior.

The study's findings support the assumptions of traditional enforcement policy. A higher rate of detection and conviction arising out of enforcement activities discourages people from behaving illegally. Similarly, more expensive penalties and fines encourage fishers to comply with rules. What is interesting in this particular study area is that the probability of detection is low and violations are rarely detected, especially in Indonesia and the Philippines, given their geographic area and limited enforcement resources.

As stated in this chapter, theoretically, levels of compliance can be improved by increasing the probability of detection and conviction or penalty rate or both. However, large financial requirements are needed to test compliance theory. Instead, regulators might respond by enhancing enforcement resources and increasing the penalty rate to deter violators. Compliance theory predicts that the willingness to comply stems from moral obligations, and social influence is based on the perceived legitimacy of the authorities charged with implementing the regulations. Other evidence suggests that a major predictor of perceived legitimacy is the fairness built into the methods used to develop and carry out regulatory policy. Thus, enforcement authorities should try to determine what policies are judged fair by segments of the population subject to regulation. This may mean that civil penalties and other sanctions should be comparable in value to the larger harm done or gains realized. Therefore, to promote greater compliance, policymakers and enforcement authorities need to show violating fishers, and society at large, the extent of damage the violations cause so that they understand the legitimacy of the law and its penalties.

Source: Kuperan, K., and J. G. Sutinen. 1998. Blue waters crime: Deterrence, legitimacy, and compliance in fisheries. *Law and Society.* 32 (2): 309–337.

The enforcement economics just described bring out an important benefit produced to society when it regulates fisheries with private property institutions: The regulations are cheap. Aquaculturists or sole owners have little incentive to manage their fishery inefficiently, because they are the only ones who suffer. For them, little enforcement activity is necessary, other than the enforcement of their property rights to the fishery. In contrast, for any individual fisher who fishes in an open access situation, there may be considerable incentive to violate a regulation. This much larger incentive to violate the regulation raises the cost of enforcement.

■ Summary

Fisheries have provided a way of living for coastal communities for thousands of years. Currently they are the fifth-largest food resource and the world's major source of protein. A fishery—a biological population of fish—is a renewable resource, that is, its stocks can be replenished. Renewable resources can be perpetually replenished, but they also can be depleted through human actions. The growth of a renewable resource population typically depends on the population's size. If the population falls below a critical level, the population goes extinct.

Biologists have developed the concept of maximum sustained yield, which is the population of a fishery that permits the largest possible annual harvest, that can continue forever. If harvests continue to exceed annual growth, the population continues to fall over time, possibly to the critical level and even extinction.

The net benefits produced by the maximum sustained yield level of effort (biological efficiency) are lower than those at the effort level at which economic efficiency occurs. This result occurs because the *added* economic cost of the added fish is larger than the added economic benefit. So *net* economic benefits are lower under the biologically efficient level of effort than under the economically efficient level. The point of maximum biological efficiency incurs greater effort and produces more harvested fish than the maximum economic yield.

Where the sole owner of a fishery believes she has a secure property right to the fishery, she takes action that influences sustained profits. That sole owner acts in a way to maximize the sustained income produced by the fishery. The sole owner's actions have the potential to promote the economically efficient use of the fishery, as well as the resources used to harvest its fish. By contrast, the open access status of the fishery produces considerably more effort levels than is economically efficient. Unlimited access to commercially valued fisheries provides incentives to overexploit the stock. This excess effort is seen as too many fishers, too many boats, low incomes for fishers, and over the long run, depleted fish stocks. Where harvest costs are low, species extinction is possible. Where harvest costs are higher, economic extinction will occur long before biological extinction.

Considerable progress has been made in promoting sustainable fisheries programs. In countries where private property rights to fishing can be defined, successful aquaculture has developed. Creative methods for sharing the gains provided by greater economic efficiency could protect many marine fisheries from overexploitation. Individual transferable quotas (ITQs) are an intriguing method for preserving stocks while protecting incomes of fishers who rely on the fishery. Despite their considerable potential, no ITQ program has yet fully resolved the problems of the initial quota distribution, bycatch, high-grading, and honesty in reporting catches.

International treaties such as the United Nations Convention on the Law of the Sea have been signed to protect worldwide fish stocks, but it remains to be seen how effective they will be. In the future, we can expect to see individual fishers continuing to resist most forms of regulations until most see a convincing individual benefit that exceeds their cost. The extent to which such regulations can be designed and implemented while being compatible with the public interest, remains to be seen.

Review Questions

1. Why is the maximum sustained yield associated with more effort and higher fish harvests then the economically efficient sustained yield?

2. If a single sole owner had all rights to a fishery, could the economically efficient harvest exceed the biologically maximum sustained yield under very high discount rates? Explain.

3. Discuss both the efficiency and equity implications of assigning an exclusive license to harvest a fishery to one individual as a sole owner.

4. Would you expect a sole owner of a fishery to ever harvest the fishery to extinction?

5. Explain the difference between an open access fishery and a limited entry fishery.

6. Why is the overfishing problem made worse by advances in the efficiency of fishing vessels and fishing gear?

7. Discuss the advantages of fish farming and fish ranching as potential solutions to the overfishing problem.

8. Would you expect greater economic efficiency for a fishery under an individual transferable quota (ITQ) program than under a simple program of limiting total catch?

9. Does a nation's decision to establish a 200-mile limit for its fisheries remove the threat of overfishing inside that limit? Explain.

10. What are some of the limits of a policy of fishing technology that raises the cost of fishing as a solution to the overfishing problem?

11. Why will economically efficient regulations impose costs on violators that are proportional to the damages to society caused by their violation?

Problems

1. Quotas: Suppose a team of biologists and economists found that the economically efficient sustained yield of a fishery is 400 tons of fish harvested per year. The industry of fishers agreed to impose and enforce a catch quota of 400 tons harvested per year. This regulation is implemented by closing the fishery as soon as the 400 tons are harvested.

a. Explain why you would or would not expect the industry to harvest those 400 tons at minimum cost.
b. Suppose the industry assigns the quota among its members so that the total catch is split into shares, according to each member's average historical catch. A member who harvested 1% of the industry's historical catch would be entitled to 1% of 400 tons (4 tons allowed each catch). Will the industry harvest the 400 tons at minimum cost?
c. If the share quotas in part *b* can be transferred to others (i.e., they are ITQs) at a market price set by supply and demand, what do you think will happen to the total cost of harvesting the 400 tons? What will happen to the market value of the transferable quota in future years as fishing technology improves?
d. Suppose world fish prices are increasing due to population growth. If a current ITQ holder wants to sell an ITQ to a future fisher, how would you calculate the price the future fisher would be prepared to pay for it? How would that price change with a reduced interest (discount) rate?

References

Alongi, D. M. 2002. Present state and future of the world's mangrove forests. *Environmental Conservation.* 29 (3): 331–349.

Alverson, D. L., M. H. Freeberg, S. A. Murawski, and J. G. Pope. 1994. *A Global Assessment of Fisheries Bycatch and Discards.* Rome: Food and Agricultural Organization of the United Nations.

Anderson, L. G. 1977. *The Economics of Fisheries Management.* Baltimore: Johns Hopkins University Press.

Armitage, D. 2002. Socio-institutional dynamics and the political ecology of mangrove forest conservation in Central Sulawesi, Indonesia. *Global Environmental Change—Human and Policy Dimensions.* 12 (3): 203–217.

Beverton, R. J., and S. J. Holt. 1957. On the dynamics of exploited fish populations. *Fisheries Investigations Series.* 19 (2).

Blackstone, W. 1942. Commentaries on the Laws of England. In *Book 4, Of Public Wrongs.* Chicago: University of Chicago Press.

Bulte, E. H., and J. M. E. Pennings. 1997. Tradeable fish quota and futures markets. *European Journal of Law and Economics.* 4: 327–334.

Chopin, T., A. H. Buschmann, C. Halling, M. Troell, N. Kautsky, A. Neori, G. P. Kraemer, J. A. Zertuche-Gonzalez, C. Yarish, and C. Neefus. 2001. Integrating seaweeds into marine aquaculture systems: A key toward sustainability. *Journal of Phycology.* 37 (6): 975–986.

Ciriacy-Wantrup, S. V. 1952. *Resource Conservation: Economics and Policy.* Berkeley, CA: University of California Press.

Clark, C. 1990. *Mathematical Bioeconomics: The Optimal Management of Renewable Resources,* 2nd ed. New York: Wiley Interscience.

Clark, C. W. 1973. Profit maximization and the extinction of animal species. *Journal of Political Economy.* 81: 950–960.

DeVoe, M. R. 2000. Marine aquaculture in the United States: A review of current and future policy and management challenges. *Marine Technology Society Journal.* 34 (1): 5–17.

Eythorsson, E. 2000. A decade of ITQ management in Icelandic fisheries: Consolidation without consensus. *Marine Policy.* 24 (6): 483–492.

Food and Agriculture Organization. 2000. *The state of world fisheries and aquaculture.* Available from *http://www.fao.org/docrep/003/x8002e/x8002e00.htm* [Accessed October 2, 2004].

———. 2002. *The State of World Fisheries and Aquaculture.* Available from *http://www.fao.org/docrep/005/y7300e/y7300e00.htm* [Accessed October 2, 2004].

Gordon, H. S. 1954. The economic theory of a common-property resource: The fishery. *Journal of Political Economy.* 62: 124–142.

Hannesson, R. 2000. A note on ITQs and optimal investment. *Journal of Environmental Economics and Management.* 40 (2): 181–188.

Hinman, K., and C. Paulsen. 1993. *The crowded sea: An issue paper on limiting access to marine fisheries.* Available from *http://www.savethefish.org/publications2.htm* [Accessed October 28, 2004].

Ibarra, A. A., C. Reid, and A. Thorpe. 2000. Neo-liberalism and its impact on overfishing and overcapitalisation in the marine fisheries of Chile, Mexico and Peru. *Food Policy.* 25 (5): 599–622.

Kishi, M. J., M. Uchiyama, and Y. Iwata. 1994. Numerical-simulation model for quantitative management of aquaculture. *Ecological Modelling.* 72 (1–2): 21–40.

Knowler, D., E. B. Barbier, and I. Strand. 2001. An open-access model of fisheries and nutrient enrichment in the Black Sea. *Marine Resource Economics.* 16 (3): 195–217.

Lai, L. W. C., and B. T. Yu. 2002. The Hong Kong fish marketing organisation: A case study of the nature of the financial problems of a legal monopoly. *Pacific Economic Review.* 7 (1): 85–96.

Munro, G. R. 1982. Fisheries extended jurisdiction, and the economics of common property resources. *Canadian Journal of Economics.* 15: 405–425.

Naylor, R. L., R. J. Goldburg, J. H. Primavera, N. Kautsky, M. C. M. Beveridge, J. Clay, C. Folke, J. Lubchenco, H. Mooney, and M. Troell. 2000. Effect of aquaculture on world fish supplies. *Nature.* 405 (6790): 1017–1024.

Repetto, R. 2001. A natural experiment in fisheries management. *Marine Policy.* 25 (4): 251–264.

Ricketts, E. 1948. *Scientists report on sardine supply: Investigator blames industry, nature for shortages.* Available from *http://www.geocities.com/yosemite/gorge/5604/ed_ricketts1948sardine.htm* [Accessed October 2, 2004].

Roughgarden, J., and F. Smith. 1996. Why fisheries collapse and what to do about it. *Proceedings of the National Academy of Sciences of the United States of America.* 93 (10): 5078–5083.

Schaefer, M. 1957. Some considerations of the population dynamics and economics in relation to the management of the commercial marine fisheries. *Journal of the Fisheries Research Board of Canada.* 14: 669–681.

Scott, A. 1955. The fishery: The objectives of sole ownership. *Journal of Political Economy.* 63: 116–124.

Tai, S. Y., M. Noh, and N. M. R. Abdullah. 2000. Valuing fisheries depreciation in natural resource accounting. *Environmental and Resource Economics.* 15 (3): 227–241.

Thorpe, A., A. A. Ibarra, and C. Reid. 2000. The new economic model and marine fisheries development in Latin America. *World Development.* 28 (9): 1689–1702.

Waitt, G., and K. Hartig. 2000. Ecologically sustainable fishing in theory and practice: Individual transferable quotas in Australia's South East Fishery. *Australian Geographer.* 31 (1): 87–114.

Chapter 13

ENERGY

Outline

The Energy Resource 357
Energy Policy Objectives 359
Market Allocations of Energy 360
Institutional Designs for Renewable Energy Sources 380

Objectives

After studying this chapter, you should be able to

- Characterize the physical and economic nature of energy
- Identify desirable outcomes that guide the search for energy policies
- Describe how unregulated markets allocate energy
- Identify actions and government policies that influence outcomes in the energy market

Introduction

Life requires energy. The meat and the plants we consume depend, either directly or indirectly, on energy from the sun. On the Earth's surface and at shallow ocean depths, plant life depends on sunlight, and most animal life depends directly or indirectly on plants. The materials used to build our homes and make the products we need are mined from the earth, then transformed into finished goods through energy expenditure. Energy is essential for all economic systems and for life itself.

For the largest part of their energy needs, most industrialized countries depend on oil and natural gas, which are depletable, nonrecyclable energy sources. In less developed agricultural economies, energy dependence is more evident. People's basic need for calories is a need for energy. Conventional agriculture captures solar energy for human use, as solar energy captured in firewood serves the basic needs of home heating and cooking.

History shows that energy needs increase as economies develop. In ancient times, as supplies of firewood and other biomass proved inadequate to support growing populations, people turned to water power (a form of stored solar energy), then to coal, and later to oil and natural gas as major energy sources. Since the 1950s, nuclear power has entered the energy mix.

The twenty-first century will likely see continued transitions in energy sources. Economic analysis tells us the most important factor affecting the use of energy is its price. Reduction in the price of energy through government subsidy or technological advance, for example, increases energy use. Reduced energy prices increase standards of living for the energy user, but if increased use produces high social costs, the reduced prices may be undesirable for the larger society.

Because modern farms, homes, factories, and transportation systems rely on energy use, transitions away from dependence on fossil fuels (energy derived from coal, petroleum products, or natural gas) will incur large costs and new investments. The considerable economic and environmental costs of current energy use patterns and of changes in those patterns justifies a special focus on energy as a worldwide policy issue.

THE ENERGY RESOURCE

Throughout history people have developed new energy sources in the continued pursuit of comfort, enhanced security, and personal satisfaction. The domestication of animals freed up human labor, originally used for hunting and gathering, to be used for cultivation and crafts. Although captured war prisoners were sometimes forced to do the tasks nobody else wanted to do, nonliving energy sources turned out to be more productive for many applications. Water was a major energy source long before industrialization in the eighteenth century. Grain mills were located at waterfalls along major rivers to draw on the energy of falling water for turning waterwheels and mill machinery. Through the mid-nineteenth century, wood was the most important industrial fuel source, and throughout most of the late nineteenth century, wind was the main energy source for ships at sea.

History records three main energy eras: wood, coal, and oil, although there were no sharp breaks between the eras. The use of wood for heating predates written history and continued as a dominant source of energy for centuries. Coal was first mined on a large scale in Britain during the sixteenth century to replace shrinking forests as an energy source. The use of coal spread across Europe and into Asia where accessible coastal areas benefited from European coal exports. By the 1780s, the Industrial Revolution had ushered in a new age in which energy was used intensively in manufacturing. Coal, which had been coexisting with wood as an energy source, began to replace wood as demands for steam power grew in Britain. By the mid-nineteenth century, Britain's disappearing forests forced the nation to rely almost entirely on coal for steam and heat. Compared to wood, coal's considerably greater heat per unit of volume greatly reduced energy costs and helped the fuel gain acceptance in the marketplace.

Oil was discovered in the United States in the late 1850s, but it did not displace coal from the marketplace immediately. Instead, it became useful in new markets. Oil's capacity to reduce friction through lubrication was uniquely suited to the requirements of high-speed machinery. Many advances in metallurgy, manufacturing, and construction would have never occured without the qualities derived from oil. It was not until the development of the internal combustion engine[1] that oil displaced coal in the early twentieth century.

Natural gas began to compete with crude oil as a source of energy in the 1930s when the technology of long-distance, low-cost gas pipelines was developed. For a number of years gas consumption grew more rapidly than that of oil, and gas displaced oil in some markets because of its superior convenience, cleanliness, and economy. At present, oil and gas account for three quarters of U.S. energy consumption, and similarly high percentages are common in other industrialized nations.

Cheap, accessible energy has had a huge impact on productivity and standards of living around the world. The positive impact is illustrated by two significant inventions: electric power machinery and the internal combustion engine. Automobiles gave entire populations improved mobility and freedom. This mobility allowed increased flexibility in the location of industry and of peoples' homes. The internal combustion engine made it possible for fewer farm workers to turn out more food at lower prices. Farm vehicles quickly replaced animals as sources of power, further improving agricultural productivity. Cheap and widespread electrification advanced the development of new industry, which further improved the productivity of labor. The role of energy in improved standards of living may be seen most clearly in peoples' homes—refrigerator, stove, dishwasher, radio, television, air conditioner, and personal computer.

Measuring Energy Use

Comparing the amount of energy from its various sources is facilitated by a common denominator for measurement. Most of the time energy is measured by the heat content of the fuel (i.e., the amount of energy the fuel could provide if all of its heat content were converted into thermal

[1] In 1885 German mechanical engineer Karl Benz designed and built the world's first practical automobile to be powered by an internal combustion engine. On January 29, 1886, he received the first patent for a gasoline-fueled car. Benz & Cie, the company he started, became the world's largest manufacturer of automobiles by 1900.

Table 13–1 Selected energy conversion factors

Energy source	Measurement unit	BTU equivalents[a]
Electricity	Kilowatt hour (kwh)	3,412 BTU/kwh
Natural gas	Cubic foot (cf)	1,000 BTU/cf
Coal	Ton	22.14 MMBTU/ton
Crude oil	Barrel (b), or 42 gallons	5.85 MMBTU/b
Wood	Cord	22 MMBTU/cord
Uranium-235	Ton	7.0×10^{13} BTU/ton
Gasoline	Gallon	0.124 MMBTU/gallon

[a]MM = millions; k = thousands.

energy). The unit most commonly used in economic applications is the British thermal unit (BTU), the amount of energy required to raise the temperature of one pound of water by one degree Fahrenheit. Table 13–1 shows the BTU content of various energy sources.

Use of the BTU as a common denominator in comparing energy sources is probably the best simple measure available. For a comprehensive economic analysis of the costs and benefits of using different fuels, one must also examine engineering factors, such as conversion losses at various stages in moving from the energy content of raw fuels to the useful work performed for the final buyer. In addition, economic theory tells us that demand factors are important. Consumer preferences for certain types of energy sources are such that they are willing to pay higher prices for a BTU of energy supplied in one form rather than another. For example, electricity is often more expensive per BTU of heat than other energy sources, but its convenience, cleanliness, and flexibility are attractive. People are typically willing to pay a premium for electricity, especially when their incomes are high.

Energy and Poverty

Approximately 1.6 billion people, one quarter of the world's population, have no access to electricity and have limited access to cleaner, modern fuels such as kerosene, liquified petroleum gas, and natural gas (International Energy Agency, 2002). Most people who earn less than $2.00 per day rely on traditional biomass (fuelwood, charcoal, agricultural residues, and animal dung) to meet the bulk of their energy needs (Habtetsion and Tsighe, 2002; Jaafar, Kheng, and Kamaruddin, 2003; Pokharel, 2003). In the absence of new energy policies, 1.4 billion people will continue to lack electricity in 2030. Four of five people without electricity live in rural areas of the developing world, mainly in southern Asia and sub-Saharan Africa. Some 2.4 billion people rely on traditional biomass for cooking and heating; if there is no change in access to electricity, that number is likely to increase to 2.6 billion by 2030.

Lack of electricity worsens poverty and contributes to its perpetuation, as it blocks most industrial activities and the jobs it could create (Ezzati and Kammen, 2001). Access to affordable, modern energy services is a prerequisite for sustainable development and for alleviation of poverty (He, Zhang, and Ye, 1996). So energy policy takes on special importance for the developing countries of the world.

Need for an Energy Policy

Different nations and cultures have different objectives in their energy policies. But a common objective of government energy policies around the world is to ensure secure, diverse, and sustainable supplies of energy at competitive prices (Borthwick, 2000; Fielden and Jacques, 1998; Pandey, 2002; UK Government, 2002). Although private markets are a powerful institution for allocating energy to meet consumer needs (see Chapter 3), government intervention into private energy markets can be justified when any of the following market failures occur:

1. **Inefficiency:** Economic inefficiencies occur when decisions about private energy supplies do not produce maximum public benefit from available supplies (Brown, 2001). Four market failures associated with energy are common. First, some market failures arise occur because electricity and natural gas networks are natural monopolies. Monopolies limit the ability of suppliers to tailor products to meet the needs of individual consumers and the ability of consumers to communicate their wishes to suppliers. Second, significant environmental externalities often are associated with energy production and use; for example, the costs of environmental damage are not paid by the people who are responsible. Third, there are information asymmetries, particularly in households and small businesses, where consumers have much less information (e.g., energy saving possibilities) than suppliers. Fourth, the private market may supply too little innovation, especially in an unregulated market, because the private market cannot capture all the benefits of innovation that accrue to the wider society.

2. **Inequity:** Even when markets produce energy efficiency, outcomes can be socially undesirable for two reasons: First, when competition provides the most efficient outcome, markets are not necessarily equitable (e.g., the high environmental cost of certain fuels may impose a considerable burden on energy use by the poor). Second, equity considerations arise because energy is a necessity of life; everyone needs warmth and light to survive. In a modern society, access to electricity is essential to maintain a range of household needs. For example, refrigerators, televisions, and even access to the Internet are no longer regarded as luxuries, and in some countries their absence may increase social exclusion. Governments are concerned that many of the poor and vulnerable members of society cannot afford adequate supplies of energy and for that reason often examine measures to tackle fuel poverty.

3. **Insecurity:** Governments may intervene in energy markets for strategic reasons. Energy is essential to both households and businesses. Many energy systems rely heavily on two fuels, electricity and natural gas, for which energy storage is difficult. Electricity cannot currently be stored in significant quantities; natural gas also is difficult and costly to store except in large, specialized facilities. The difficulties in storing these fuels by individuals mean that security of supply is essential. Significant interruptions in supply can quickly escalate into a major political issue. The principles that govern intervention are normally established in relation to the objectives of a government's energy policy.

Energy Policy Objectives

Governments around the world strive to ensure secure, diverse, and sustainable supplies of energy at competitive prices. Specifically, they pursue the following policy goals:

- **Sustainable development:** Since the early 1990s, *sustainable development* has become an overarching energy policy goal for many governments. The concept of sustainability originated with environmental concerns; but the term is now defined in terms of balancing progress in economic, environmental, and social objectives.
- **Economic efficiency:** Allocating resources into energy programs that produce maximum net benefits in present value terms supports economic efficiency. Greater economic efficiency releases or creates resources that can be devoted to other goals, such as environmental protection.
- **Environmental quality:** The major goal of environmental protection involves several issues. Climate change is currently the most prominent, but there are many others: safeguarding individuals against poor air quality and dangerous chemicals, improving water quality, and providing lasting solutions to nuclear waste problems.

- **Equity:** Social issues influence the fair provision and use of energy. Currently, governments are confronting these issues: addressing fuel poverty; maximizing access to services that provide electricity; maximizing access to transport, especially in rural communities (Pokharel and Chandrashekar, 1998); minimizing employment impacts due to changes in the energy industry; and ensuring fair siting of energy infrastructure.[2]
- **Security:** Energy security can be viewed in terms of risk. The aims of energy policy are (1) to achieve an acceptably low level of the various risks of supply interruption and stress that confront the energy system and (2) to establish measures for managing the remaining risk. Risk can be managed with what amounts to an insurance premium. An energy policy that is secure often involves payment of a premium to insure against an adverse outcome. For that reason, more secure energy systems are likely to be more costly, other things remaining the same.

MARKET ALLOCATIONS OF ENERGY

Market forces have a major influence on the allocation of energy resources. To gain a better understanding of how those forces serve to allocate energy, we will develop a series of economic concepts applicable to energy, identify how those principles apply to various energy sources, analyze energy cartels, and examine the economics of alternative energy futures.

Economic Concepts

Decisions that affect current production and consumption of depletable resources (e.g., energy from fossil fuels) affect future supplies. A barrel of oil supplied today for today's consumption reduces the future resource base by the amount supplied. Facing that law of nature, the resource owner attempts to optimize production over many periods, recognizing that depletions in one period reduce opportunities in future periods. Because the major energy resources of the early twenty-first century—oil, natural gas, coal, and uranium—are depletable, it is valuable to understand how markets allocate nonrenewable resources over time.

Intertemporal Profit Maximization. The economics of nonrenewable resources is based on the idea that firms attempt to maximize long-run profits over time (intertemporally). When one period's produced output has no effect on resource supplies in future periods, competitive firms simply equate marginal cost with price in each period (see Chapter 3), which results in maximum profit for any period. The result is a production plan that produces an optimal stream of profits over time ($\pi_0, \pi_1, \pi_2, \ldots, \pi_n$), which maximizes the firms' long-run profits over time.

When a firm's product uses a nonrenewable resource, the firm's decision to produce and earn profits today from the resource necessarily blocks its ability to earn profits on that unit of resource in future time periods. Given the tension resulting from competition among time periods, how does the firm make production decisions when it is comparing profits from current and future periods? The answer is that it applies a discount rate to future profits, since a dollar's profit today is worth more than the same dollar's profit in some future period. Typically, the firm's discount rate closely approximates the market rate of interest, differing only by an adjustment for risk (see Chapter 6). So the interest rate sends a signal that guides the firm on the value of \$1 profit today earned from a unit of a mined resource to compare with a profit of \$$X$ from the same unit mined in Y years. If the market interest rate is 10% per year ($r = 0.10$), then a profit of \$1.10 that will not be earned until next year has a *present value* of \$1.00. The firm calculates the present value of a stream of future profits beginning in the current period 0 and ending at some period (n) in the future:

$$\text{Present value} = \pi_0 + \frac{\pi_1}{(1 + r)^1} + \frac{\pi_2}{(1 + r)^2} + \cdots + \frac{\pi_n}{(1 + r)^n} \tag{13.1}$$

[2] Van Horen and Simmonds (1998) describe conflicts between efficiency and equity for energy policy in South Africa.

Equation 13.1 guides the firm in regard to long-run *profit maximization.* By choosing a profit profile for all periods ($\pi_0, \pi_1, \pi_2, \ldots, \pi_n$) the present value of a stream of future profits can be measured and therefore maximized.

Market forces are at work to compel the firm to maximize the present value of future profits. If the firm finds itself in a competitive market, its survival depends on its ability to maximize profits. If the firm fails to select the optimal stream of profits, it becomes a candidate for a takeover. The presence of less than maximum profits provides an incentive for a new owner or manager to improve the profit performance.

User Costs. The *user cost* of a nonrenewable resource is defined as the opportunity cost of using the marginal unit for one period. The opportunity cost of using that marginal unit of a nonrenewable resource consists of the maximum profits foregone by taking the unit from another period and putting it to use in the period in question (Krautkraemer and Toman, 2003). User cost is a concept of considerable power and versatility, because it provides economic insight into the optimization problem faced by the owner of the nonrenewable resource.

Although Equation 13.1 correctly describes the present value calculation, alone it lacks the power of economic insight. In seeking that insight, we pose this question: Can the firm maximize the present value of profits from a nonrenewable resource by simply setting marginal cost equal to marginal revenue in each period? If the answer is yes, we arrive at a compelling result. If price equals marginal cost in each period, the competitive world oil price should be no more than $1.00 per barrel, because the marginal cost of producing oil is less than $1.00 per barrel in most Persian Gulf countries, home of the world's largest and cheapest oil reserves. So the answer must be no, since world oil prices are much higher and have been so for some time (Figure 13–1).

Therefore, for a nonrenewable resource, setting each period's price equal to each period's marginal cost fails to maximize profits. But why does this rule, which works so well for analysis of a single period, fail so conspicuously for a nonrenewable resource? The answer has to do with the special connection between time, the discount rate, and nonrenewable resources.

Where a firm uses *nondepletable* resource inputs, the decision to use today has no effect on the future cost of use. So marginal cost of the last unit produced today consists entirely of the marginal production cost (MC_p), the material, land, labor, and capital costs of producing today's last unit of output. But the case is different for nonrenewable resources. For example, the decision to produce a barrel of oil today blocks the possibility of producing that same barrel at some time in the future since that barrel, once produced, is gone forever. Each barrel produced today

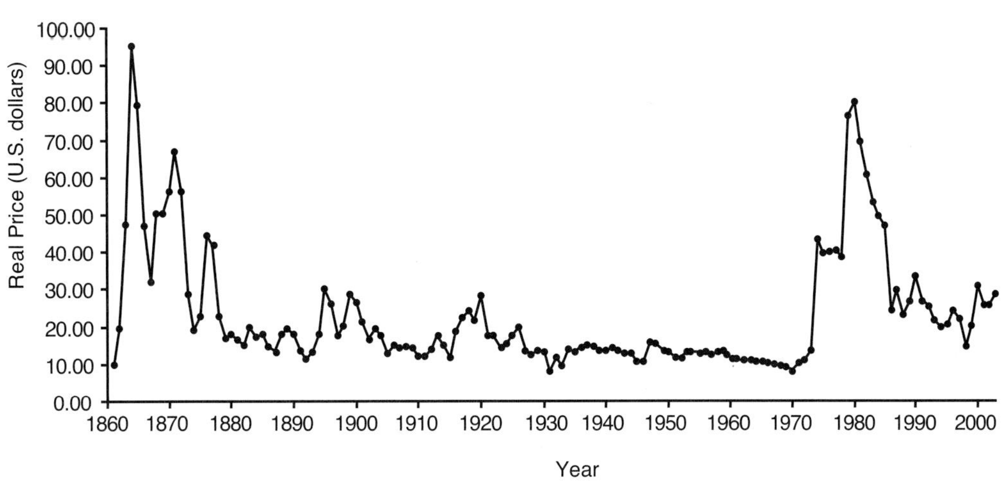

Figure 13–1 Real oil prices, 1861–2003
Source: British Petroleum. 2004. *Statistical Review of World Energy 2004.* Available from *http://www.bp.com/statisticalreview2004* [Accessed October 24, 2004].

is a barrel taken from the future. The resource's depletable nature requires that the owner compare marginal profits from producing the last unit now with the marginal profits of producing it in the most profitable future period. This mental exercise leads the resource owner to the heart of the principle of user cost: The user cost in any *i*th period is the maximum profit lost from taking a unit of a resource from any other period and producing it instead in period *i.* (User cost can also be defined for a renewable resource. Since that resource is renewed, there is no displaced profit attributable to using a renewable resource now. Its user cost is zero.)

Even though no accountant routinely collects these user costs, the firm still faces resource use costs (incremental profit) when it chooses to produce in period *i,* taking it from a future period's production.[3] So, for the nonrenewable resource, what is the total marginal cost of producing one more unit in period *i,* which we define as MC_i? It is the sum of conventional marginal production costs (MC_{ip}) and the user cost (U_i), that is,

$$MC_i = MC_{ip} + U_i \tag{13.2}$$

where U_i is the user cost by displacing the resource from production in another period and mining it in the *i*th period. The calculation of user cost rests on the principle of long-run profit maximization, which says that the competitive profit-maximizing firm sets P_i equal to total marginal cost, MC_i. Since user cost measures the opportunity cost of producing a marginal unit today, it is calculated by subtracting marginal production costs (MC_{ip}) from price (P_i):

$$U_i = P_i - MC_{ip} \tag{13.3}$$

We are now ready to apply the principle of user cost to predict how much of the nonrenewable resource will be used in each period. Since U_i represents the opportunity cost of selling a barrel in period *i,* the producer may elect to switch a marginal barrel from that period's production to some other period where it earns a higher marginal profit. What signal tells the producer to reallocate marginal barrels, and what signals the producer that profits are maximized and no more marginal barrels should be reallocated? As luck would have it, the same signal answers both questions.

Let us answer the second question first: How does the producer know when profits are maximized? If profits are in fact maximized, then no barrel of output reallocated among periods will increase the present value of profits defined in Equation 13.1. How does the producer know if that present value is being maximized? It is being maximized only if *discounted profits from all marginal barrels are equal in all periods.* The present value of profits is maximized when *discounted* user costs are equated in all production periods:

$$U_0 = \frac{U_1}{(1+r)^1} = \frac{U_2}{(1+r)^2} = \cdots = \frac{U_n}{(1+r)^n} \tag{13.4}$$

When the very special conditions of Equation 13.4 hold true, *undiscounted* user costs rise at the rate of interest:

$$\begin{aligned} U_1 &= U_0\,(1+r) \\ U_2 &= U_1\,(1+r) \\ &\cdot \\ &\cdot \\ &\cdot \\ U_n &= U_{n-1}\,(1+r) \end{aligned} \tag{13.5}$$

[3] This analysis of user cost assumes the firm has sufficiently secure property rights to treat losses in future profits as real costs. If property rights to the nonrenewable resource are insecure, the firm treats user cost as zero and produces where price equals marginal cost in every period until the resource is gone or until somebody claims it was not the firm's property in the first place.

Here is an example. If the interest rate is 10% ($r = 0.10$) and $U_0 = \$1$, then user cost in all future periods (U_i) must be \$1.10 in the first period, \$1.21 in the second, \$1.33 in the third, and so on. Only then will discounted user costs be equal over time. Here is another way to show the truth of that statement. Suppose user costs take on the following time path over periods 0 through 4 when the interest rate is 10%: \$1, 1.10, 2.00, 1.33. During period 1 undiscounted user cost rises exponentially at 10%, but for period 2 user cost is \$2.00. $U_2 = \$2$, so we know the producer has failed to choose a profit-maximizing production strategy. Why? Because the discounted value of the last barrel of oil produced in period 2 is about $\$1.65 = (U_2/(1.1)^2)$, but the discounted value for all other periods is \$1. To correct this mistake, the producer will increase the present value of the reserves by allocating more production to period 2 (reducing its user cost) and reducing production in the other periods (raising their user costs). Thus, the problem facing the wealth-maximizing producer (one who maximizes discounted profits) is to schedule production over time such that Equation 13.5 holds for the last barrel produced in any period (see Appendix I for mathematical details).

Hotelling's Rule. The link between user costs and prices is direct. If we assume that marginal costs of production (MC_i) are zero, then surprisingly, prices, like user costs, will rise at the rate of interest. In a famous article, Hotelling (1931) advanced this remarkable idea. Together with the assumption of perfect competition, one can substitute prices (P_i) for user costs (U_i) when marginal costs are zero from Equations 13.3 and 13.4 as follows:

$$P_0 = \frac{P_1}{(1+r)^1} = \frac{P_2}{(1+r)^2} = \cdots = \frac{P_n}{(1+r)^n} \tag{13.6}$$

To put it differently, when marginal costs are zero, prices will rise at the rate of interest:

$$P_1 = P_0\,(1+r)$$
$$P_2 = P_1\,(1+r)$$
$$.$$
$$.$$
$$.$$
$$P_n = P_{n-1}\,(1+r) \tag{13.7}$$

Hotelling's rule is simple, but it carries with it an underlying economic sophistication. It requires only three assumptions: (1) long-run profit maximization, (2) zero marginal production costs, and (3) perfect competition. These three assumptions predict the result that prices will rise in a smooth, predictable manner with the rate of interest. Hotelling's *r* percent growth rule makes economic sense: Deposits of oil, coal, or natural gas are irreplaceable assests, so luring producers into holding them for future periods requires that they receive a return of *r* percent per year by *not* producing everything immediately. If prices started temporarily rising at a faster rate than compound interest, producers would allocate more current production to future periods when prices were highest. This reallocation would reduce future prices and increase prices in the current period. This iterative process continues until the growth path of price returns to the equilibrium rate of interest, growing at *r* percent per year.

Prices must rise at *r* percent per year to establish a market equilibrium. But there are an infinite number of price paths that rise at *r* percent per year. For example, at $r = 0.10$, oil prices can move along one path of 1.00, 1.10, 1.21, 1.33, and so on, or along another path of \$300, \$330, \$363, \$399, and so forth. From this infinite number of price paths, how is the single, correct market-clearing price path determined and how does the market settle on that path?[4] In other

---

[4] The search for the market equilibrium price path is nothing more than a search for a resource price in the initial period, since we are assured that the price grows at *r* percent per year.

words, since we know that price in the initial period determines price in all remaining periods, what is the optimal price in the initial period?

To simplify, let us limit the number of time periods to two (current and the future), which makes a graphic solution possible. (The same principles can be used to find the mathematical solution for any number of periods; see Appendix I.) Suppose 50 barrels of oil are available for consumption, to be used in two periods, period 0 and period 1. Assume the demand curves in the two periods are given by the following two equations:

$$P_0 = 50 - 0.5Q_0 \text{ (period 0)} \tag{13.7a}$$

$$P_1 = 50 - 0.5Q_1 \text{ (period 1)} \tag{13.7b}$$

That is, both demand curves are linear, have an intercept (maximum price) of \$50, with a slope of -0.5. Each one barrel increase in output in a period reduces that period's price by \$0.50. Assume the interest rate is 0.10. To maintain the zero marginal production cost assumption, assume that the oil is sitting in open pools ready for competitive producers to haul off to market.

Figure 13–2 presents a graphic analysis of determining the optimal price path. The left-hand vertical axis measures price in the current period (period 0), while the right-hand vertical axis measures price in the second period (period 1). The horizontal axis measures quantity, for which the total of 50 barrels is to be allocated between the two periods. Beginning on the left vertical axis, note that the current period price is \$50 per barrel when current output is zero. The figure shows the demand curve, D_0, coming from the intercept, P_0 = \$50 with a negative slope of -0.5. Likewise, oil's future demand curve comes from the right-hand side vertical intercept, from P_1 = \$50, and also slopes negatively with the same slope equal to -0.5.

The two market equilibrium prices can be found by using the two demand schedules to compute the price for any possible output allocation between the two competing periods. For example, if $Q_0 = 40$ and $Q_1 = 10$, the two periods' demand curves tell us the prices in those periods will be as follows:

$$P_0 = 50 - 0.5(40) = \$30 \text{ (period 0)} \tag{13.7c}$$

$$P_1 = 50 - 0.5(10) = \$45 \text{ (period 1)} \tag{13.7d}$$

In summary, Figure 13–2 allows us to compute a set of two periods' prices for any allocation between the two periods that depletes the 50 barrels available.

Figure 13–2
Hotelling price path and production rates for an example of 50 barrels of oil to be used in two periods

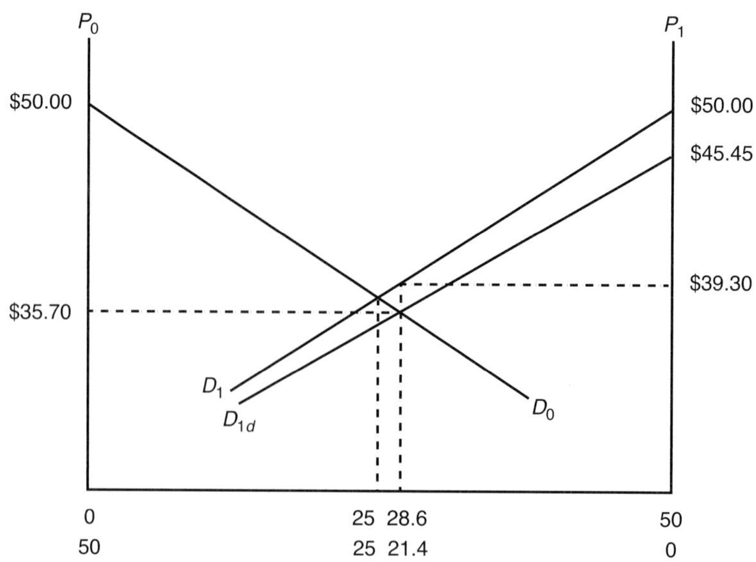

Next let us use the same figure to determine the user cost schedules, U_0 and U_1, for the two periods. Equation 13.4 tells us that when marginal extraction cost is zero, each period's user cost is identical to its price. For the current period, U_0, user cost is identical to Figure 13–2's demand schedule, D_0. Similarly, for the future period, U_1 equals demand schedule D_1. But recall the behavior of the resource owner who allocates production between the periods so that U_0 equals the discounted future period's user cost, $U_1/1.1$. That discounted future period's user cost is the schedule beneath D_1, labeled D_{1d}. The line D_{1d} is determined by discounting the intercept of D_1 ($50), which results in an intercept of $45.45. Similarly, the slope of D_1 is also discounted by 1.1, which results in a slope of approximately −0.45. So the line D_{1d} shows, for any output level in the future period, the present (discounted) value of its price equals its undiscounted market price in period 1, D_1, divided by 1.1.

According to Equation 13.4, the intersection of D_0 and D_{1d} (U_0 and $U_1/1.1$) is the graphic solution that equates *discounted* user costs. It indicates the optimal resource output allocation between the two periods. The only allocation of 50 barrels between the two periods that makes D_0 equal to D_{1d} is where the two lines cross. Figure 13–2 shows that the lines cross only when 28.6 barrels are produced in the first period and 21.4 barrels are produced in the future. At that *intertemporal allocation* of reserves, discounted user costs are equal in both periods.

Now how do we find the price in those two periods? We simply read off the correct price for each of those two quantities on each of the unadjusted demand curves, D_0 and D_1. A market price of $35.70 in the current period corresponds to 28.6 barrels in that period. Corresponding to the 28.6 barrels produced in the future period is a market price of $39.30 per barrel—a price just 10% percent higher than the current period price of $35.70. To put it differently, the discounted price of $39.30 in the second period equals the undiscounted price of $35.70 in the first period. No other output combination and price path will secure a long-run profit maximizing equilibrium. If you doubt this conclusion, consider an output level less than 28.6 barrels in the current period, which means something more than 21.4 for the future. At that output combination, U_0 exceeds $U_1/1.1$ because the first period's price exceeds $35.70 and the future period's price falls below $39.30. How will this combination of outputs affect market prices? It raises price in the first period and reduces price in the future. Now prices are growing slower than the interest rate. Holding oil in the ground for the future is a bad investment, considering the interest that could be earned by turning oil to money and banking the money. Producers will react to this slow growth in prices by increasing current output and reducing future output. The increase in current output lowers U_0 and raises $U_1/1.1$, so producers will find it profitable to expand current production. They will not stop expanding current production until $U_0 = U_1/1.1$. Using similar microeconomic thinking, you can show that if current production temporarily exceeds 28.6 barrels, market forces drive current production down to the optimal level.

Figure 13–2's graphic analysis emphasizes several determinants of the user costs and price path. As stated earlier, both future and current demand expectations are important. Marginal production costs are also important.[5] If oil reserves had been 100 instead of 50, user costs would be smaller, and the price path would start from a lower initial price. So as you would expect, discoveries of oil new reserves would reduce current and future oil prices.

Critics of Hotelling's model are quick to point out that observed oil prices do not move along a smooth price path, rising with the rate of interest. Figure 13–1, which shows the historical movement of real oil prices, indicates that the path of oil prices bears little resemblance to Hotelling's predictions. Before discarding Hotelling's simple model, however, recall its three assumptions: (1) long-run profit maximization, (2) zero marginal cost, and (3) perfect competition. Remember that any single price path comes from a unique set of expectations about present and future and demand and supply conditions. This alone can explain why historical oil prices have not moved along a smooth path. If producers continually adjust their expectations as they obtain new information, we might see a steady stream of jumping from one price path

---

[5] Results in the text would change only slightly for the case of positive marginal costs. In that case, net price (price minus marginal cost) must grow at the rate of interest (Solow, 1974).

Figure 13–3 Oil price path under several estimates of reserve size

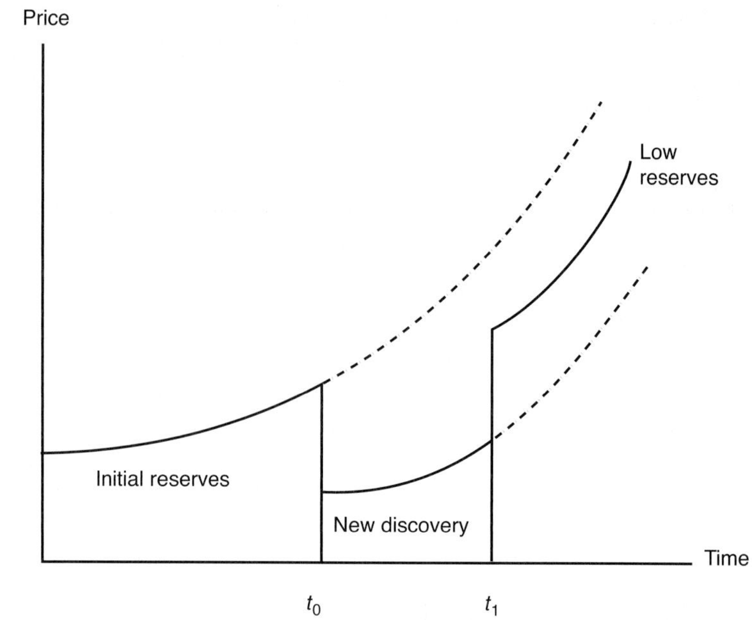

to another. Still other results will occur if the assumptions of zero production costs or perfect competition are relaxed. In any case, let us extend Hotelling's simple model by considering the effects of (1) the size of the resource base, (2) the presence of a backstop[6] fuel, and (3) the discount rate.

Size of the Resource Base. User costs play a major role in determining the price of oil (or other nonrenewable resources). User costs are based on expectations of *present* and *future supply* and *demand* conditions, so it is essential to consider what influences producers' perceptions of user costs. For a market equilibrium price path to emerge (e.g., for oil), producers must have a clear idea about the magnitude of the underlying reserve base. Knowledge of the size of existing oil reserves alone may be an inadequate predictor of the price path, because existing reserves are limited to actual reserves found to date. Investments in exploration have uncovered supplies in the past and explorations will likely do so in the future. If emerging new discoveries are consistent with producers' expectations, they need not revise their estimates of user costs, and the price path need not shift.

Figure 13–3 illustrates what happens to a competitive Hotelling price path with zero production costs after people change their beliefs about the size of reserves. As the initial period's oil prices are rising at the rate of interest, producers' expectations of user costs remain unchanged until period t_0. But at time t_0, geologists increase their estimate of the reserve base, perhaps prompted by an unexpected discovery or because a well-known discovery produces larger than expected reserves. The oil price path shifts downward because oil's scarcity value falls, so user costs are revised downward sharply. Following the revision, as long as expectations are unchanged from period t_0 to t_1, prices again rise at the rate of interest along the revised lower path produced by the new discovery. Now in period t_1 oil producers become convinced that the cumulative reserve base is smaller than they had previously thought, because new sampling on an existing field indicates smaller reserves than previous estimates. At t_1, prices immediately increase, defining a third higher equilibrium price path because of the reduced reserves; thereafter prices rise at the rate of interest until expectations are revised again.

---

[6] In baseball, the backstop is a wide structure that catches balls that get past the catcher. A backstop fuel is a substitute resource ready to supply society after its depletable resource is gone.

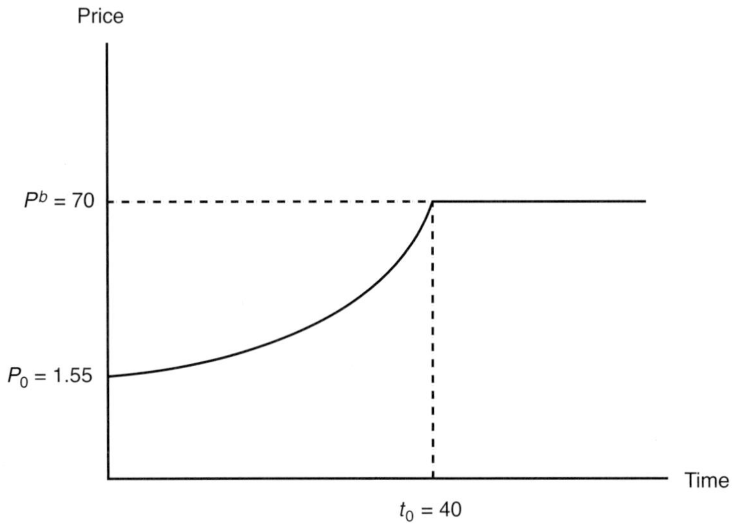

Figure 13–4 Oil price path when a backstop fuel is present

Figure 13–3 illustrates the important point that equilibrium prices need not follow a smooth price path. A steady stream of changing expectations cause switching to alternate equilibrium price paths. Thus, Hotelling's r percent model may be entirely correct but have only limited predictive power, because prices continually jump from one price path to another as a result of a steady stream of revised expectations and new discoveries.

Presence of a Backstop Fuel. In view of the huge potential supplies of crude oil alternatives from tar sands, oil shales, and coal, does economic theory have the power to predict the impact their development will have on the price of oil? The answer is yes. For clarity of exposition, assume that reserves of these oil substitutes become infinitely elastic at some price, P^b (i.e., the resource supply is so large that all current and future demands can be met at that constant price). These reserves are nonrenewable, but the reserve base may be so large that their user costs are effectively zero. (If each period's demand reaches zero at a finite price, then price can be made as close to zero as desired for a sufficiently high supply of a nonrenewable resource.) Alternatively, the backstop fuel may be a renewable energy source like solar energy.[7] In either case, virtually unlimited supplies are available. Figure 13–4 shows that the presence of a backstop fuel considerably alters oil's price path. No longer does the price continue to rise at the rate of interest for the entire planning period. Instead, the solid rising price path shows a world of perfect foresight. The price increases at rate r until it reaches P^b, at which time the backstop fuel enters the market at the price P^b to meet all demand. Presumably, producers watch the backstop fuel's price like a hawk and correctly anticipate that the backstop fuel production will be ready in year t_0 (the switch point) with sufficient capacity to meet expected demand that materializes at P^b. Moreover, oil producers plan to drill, produce, and sell all their oil reserves before t_0, because after that period, user cost and price cease their upward march.[8] After period t_0, oil prices remain at P^b for the indefinite future, or at least until a technological or institutional advance occurs that drives down the backstop fuel's price.

The presence of a backstop fuel waiting in the background to shine in its appointed day of glory has implications for today's price of oil. Suppose that coal liquefaction is too expensive to be

[7] Solar energy is an important backstop fuel on which many hopes around the world rest for promoting a long-term sustainable future. For example, Kaygusuz (2003) analyzes the use of imported fuels as a backstop for scarce domestic coal in Turkey.

[8] Conventional oil production may continue to occur in periods after t_0 because oil fields cannot be depleted instantaneously. Also, oil prices might overshoot the backstop fuel price, P^b, if in period t_0 the introduction of the backstop fuel is inadequate to meet market demand.

economically feasible currently, but it exists as backstop fuel for crude oil at a constant cost of \$70 per barrel (i.e., $P^b = \$70$). Suppose further that existing crude oil reserves are adequate for 40 years ($t_0 = 40$).[9] What is the optimal price of oil today? Applying Hotelling's model with $P_{40} = \$70$, we arrive at a current price in period 0 of \$1.55 for a 10% discount rate, using the following equation:

$$P_0 = P_{40}/(1.1)^{40} = \$70 / 45.26 = \$1.55 \qquad (13.8)$$

This equation says that the backstop fuel stands ready to meet all future demands after the age of conventional oil ends. For a discount rate of 5% rather than 10% the current price is $\$9.94 = \$70/(1.05)^{40}$. Calculations such as these show that even with highly conservative assumptions about the backstop fuel price and the years of available oil reserves remaining, world oil prices are well above levels implied by a competitive market.

Discount Rate. Inflation was not a problem in 1931, so Hotelling stated his results in terms of the nominal interest rate, unadjusted for inflation. Today, inflation rates are arguably more uncertain than oil prices, which causes many people to express oil price forecasts as real prices (i.e., in dollars of constant purchasing power). The forecast price of oil is future nominal price deflated by the future expected inflation rate (i). The real interest rate (r^*) is simply the nominal rate of interest (r) less the expected inflation rate (i):

$$r^* = r - i. \qquad (13.9)$$

For example, if the nominal interest rate is 10% (i = 0.10) and people expect a 7% inflation rate, then the real rate is 3%: 0.10 − 0.07 = 0.03. Hotelling's *r* percent price path model is correct even after accounting for inflation. We simply substitute the real rate of interest for the nominal rate of interest and interpret oil prices in dollars of constant purchasing power. Having defined the real rate of interest, we can ask whether that real rate, usually estimated from 2% to 4% per year, is the one oil producers use.

If holding oil reserves is a riskless way to hold wealth, private investors will use the real rate of interest, because it reflects the real return on a risk-free asset. In fact, oil producers commonly believe that geological and political risks of oil exploitation require the use of a considerably higher real discount rate to reflect the elements of uncertainty. Private investors believe that political risks are so large in the Middle East, for example, that a real discount rate of 9% may reflect a 3% real risk-free discount rate, plus a 6% risk premium.

Figure 13–5 shows the effects of changes in the real discount rate on price paths by showing price paths under three possible discount rates. Under the first high-risk scenario, producers use the highest real discount rate of 20% in anticipation of the host government nationalizing producers' assets. The result of this scenario is shown by the price path that begins lowest as large quantities of the resource are dumped on the market in expectation of future nationalization. Prices rise fastest since less is available for future supply, and the resource runs out earliest (labeled $r = 0.20$). Under that price path, prices grow at 20% per year, at which time the resource runs out in period t_0, and the backstop fuel supports remaining future demands. In the second scenario, a conservative, probusiness government is elected. Producers, therefore, believe the risk of nationalization is reduced, so the real discount rate is reduced to 5%. The resource price starts higher and grows more slowly, at 5% per year. Less is dumped on the market in early years and more is saved for later. The resource lasts a longer time and the backstop technology enters in period t_1 to support remaining demand. In the third scenario the government decides that the welfare of future generations should have nearly equal standing with that of current generations, and a 1% real discount rate is used. Now user costs are nearly constant over time. Initial prices begin at the highest of the three levels and grow at only 1% per year, reaching the point of depletion at

---

[9] More advanced analysis shows that the switch point is not predetermined (exogenous) as suggested. It is more accurate to think of P^b as the maximum price. Based on that price and on the size of known reserves, the initial price is determined as the level that permits price to grow at *r* percent, while depleting known reserves in the period in which price climbs to P^b. The number of years it takes for that to occur determines the switch point.

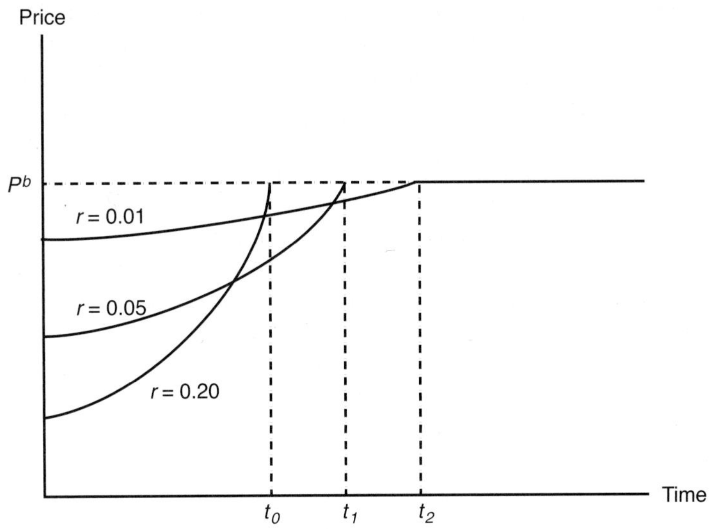

Figure 13–5 Price paths under varying discount rates

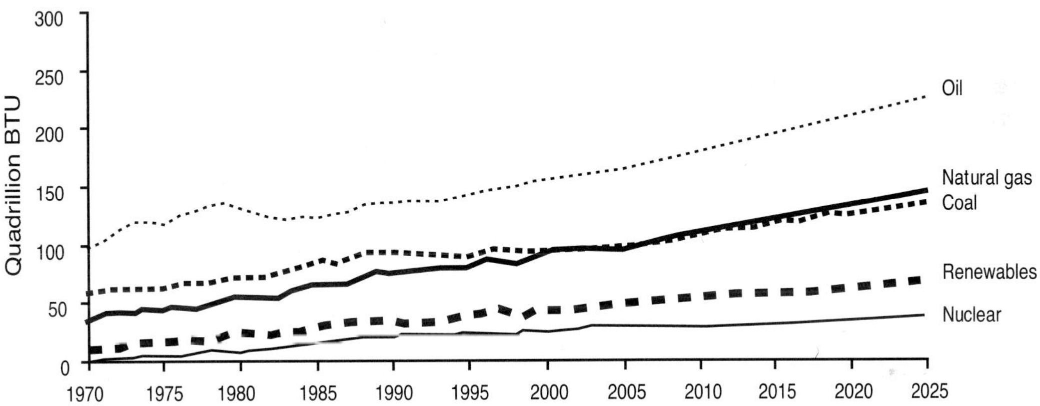

Figure 13–6 World energy consumption by source, 1970–2025
Source: U.S. Department of Energy. 2004c. *International Energy Outlook*. Available from *http://www.eia.doe.gov/oiaf/iev/world.html* [Accessed October 24, 2004].

the switch point in period t_2. After all, a 1% discount rate implies near indifference between a dollar today and a dollar in future years, when the backstop fuel is available.

Energy Sources

World energy consumption is projected to increase by 54% from 2001 to 2025, according to a 2004 forecast by the U.S. Department of Energy.[10] Much of the growth in energy use is expected to occur in the developing world. In contrast, slower growth in energy demand is expected for the industrialized world, averaging 1.2% per year. Most industrialized nations can be characterized as mature energy consumers with comparatively slow or zero population growth.

Fossil Fuels versus Renewables. Forecasts in Figure 13–6 show consumption of five major energy sources over the 2001 through 2025 period. The years prior to 2001 are shown as

[10] Source: Much of the analysis in this section relies on the U.S. Department of Energy (2004c).

actual consumption; later years are forecasts. Fossil fuel prices for electricity are expected to remain low relative to the costs of nuclear power and renewable energy sources. As a result, nonfossil fuels such as energy produced from *renewable resources* (renewables) are not expected to be economically competitive with fossil fuels in the near future.[11]

Oil is forecasted to remain the dominant energy source worldwide through 2025. Demand for oil is forecasted to increase by 1.9% annually, from 77 million barrels daily in 2001 to 121 million barrels daily in 2025. The United States, China, and other nations of developing Asia are expected to account for nearly 60% of the growth in oil demand. Coal is currently an important fuel for electricity production worldwide, and it is expected to dominate energy markets in developing Asia. World coal use has grown slowly since the 1980s, a trend that is expected to continue. Of all coal consumed, 64% is used for generating electricity. In almost every region of the world, power generation accounts for most of the projected growth in coal consumption. Natural gas is expected to be the fastest-growing primary energy source worldwide, maintaining an average growth of 2.2% annually over the 2001–2025 period.

Moderate growth, projected to average less than 2% per year, is forecasted for the world's consumption of hydroelectricity and other renewable energy resources. Much of the expected increment in renewable energy consumption in the industrialized world is projected to be in nonhydro renewables, including wind energy in western Europe and the United States. Lew (2000) describes the potential role of wind energy in China's energy future.

Nuclear Energy. Global warming, caused by carbon dioxide emissions (see Chapter 15), has rekindled interest in nuclear power, which emits no carbon dioxide. If new nuclear power plants are built, they will probably incorporate passive safety systems. Passive safety systems rely on the laws of physics, rather than mechanical systems or human judgment, to ensure that a nuclear reactor does not overheat or that nuclear reaction does not accelerate to unsafe levels. The laws of physics, unlike mechanical systems (e.g., safety valves and pumps), do not fail. Also, they permit less reliance on the judgment of people to make critical decisions.

One controversy over nuclear power is what to do with the nuclear wastes that result from the fission process. These wastes remain radioactive for more than 100,000 years. If the waste is not safely contained, it will contaminate groundwater and ecosystems and threaten human health. Existing and future nuclear waste from nuclear power plants is only a small part of the total problem; the majority of nuclear waste comes from obsolete nuclear weapons.

The future of nuclear power is highly uncertain. Critics oppose nuclear power because of its high cost, the threat of an accident that would release radioactivity, and the problem of storage of nuclear wastes (Radetzki, 2000). Supporters of nuclear power argue that safety can be improved by adoption of passive systems and by paying more attention to the role of people in safety. Supporters also argue that nuclear power is environmentally preferable, because it does not emit carbon dioxide or conventional air pollutants.

Electricity. Although renewable fuels and solar power ultimately will play an increasing role in the world's energy supply, most experts believe they will penetrate the market slowly. So for electric utilities, conservation has taken on an increasing role. The most important role for conservation is its ability to *defer capacity expansion.* Each new electricity-generating plant tends to cost more than the last, and frequently the cost increase is large. When new power plants are finished, prices of electric power are increased to pay for them. By reducing the demand for electricity, conservation delays the date when the new capacity is needed. So delays in the need to construct new plants translate into delays in rate increases.

---

[11] But there are exceptions: For example, Hillring (1999) describes the important role of wood for heating in Sweden. Furthermore, forecasts for fossil fuels are influenced by changes in government policies or programs. Renou-Maissant (1999) analyzes factors that influence substitution among energy-producing fuels. Nystrom and Cornland (2003) describe the relation between energy and forestry policy for renewable resources.

The dominant electricity pricing system, average cost pricing, is poorly designed to stimulate the efficient amount of conservation. Average cost pricing means that new, higher-cost supplies are averaged in with the lower-cost sources, which yields an average power price considerably lower than the incremental (marginal) cost of the new power. Because average cost pricing signals artificially cheap power, the consumer who is considering investing in conservation receives false signals. The consumer saves less money by conserving with average cost pricing than he or she would if the saved energy is priced at its true marginal cost. Consequently, average cost pricing fosters conservation practices at less than efficient levels.

Utilities are reacting to the problem of weak conservation price signals in a number of ways. One way is to look for methods to reward power buyers who invest in conservation measures, rather than in building expensive new power plants, when conservation is the cheaper alternative. Some programs have established rebates for residential customers to install conservation measures in their homes. Others have provided free home weatherization to qualified low-income homeowners. Still others have offered owners of multifamily residential buildings incentives for installing solar systems for heating water. Though the costs of these investments must be recovered from customers, utilities report that the savings have been large and customer satisfaction has been high. Recall from Hotelling's model that less power consumed (demanded) means non-renewable energy supplies last longer.

Total electric power demanded in a given year is not a utility's sole concern. The utility also wants to know how energy demand is spread throughout the year. The system's capacity must be high enough to satisfy demand even during peak demand periods. During off-peak periods, much of the instantaneous generating capacity remains underused. Demand during peak periods imposes two special costs on utilities. First, the peaking units, those electricity-generating facilities fired up only during the peak periods, produce electricity at a much higher marginal cost than do base-load plants (i.e., those operating nearly always). Second, growth in peak demand is what frequently triggers the need for capacity expansion. Slowing the growth in peak demand can delay the need for new expensive capacity expansion, and a higher proportion of the power needs can be met by the less-expensive base-load plants.

Utilities respond to the peak demand problem by adopting load management techniques to produce a more balanced use of this capacity over the course of each day and throughout each year. As described in Chapter 10, one load management technique is called peak load pricing. Peak load pricing attempts to charge the higher marginal cost of supplying peak power to people who use peak power by charging higher prices during the peak period. Although most utilities now use simple versions of peak load pricing, some are experimenting with more sophisticated applications. One innovative system, for example, transmits electricity prices every five minutes over regular power lines. In a customer's household, the lines attached to one or more appliances can be controlled by switches that turn the power off any time the prevailing price exceeds the customer's established limit. If you program your air conditioner to avoid the peak prices, it shuts down when the price reaches a critical level and turns back on when the peak passes.

One of the newest developments in the energy field is the movement to deregulate electricity production. In the United States, power deregulation began in 1992 when Congress allowed independent energy companies to sell power on the wholesale electricity market. Congress believed that requiring power generators to compete for customers would produce lower electricity bills for customers. Electricity deregulation also has raised environmental issues. Since electricity prices typically exclude the costs of environmental damages, power sources that can offer the lowest prices are often highly polluting. In this case, environmentally benign generation sources cannot compete on a level playing field. Some U.S. states have responded by adopting consumer right-to-know laws to offset, at least partially, this disadvantage. These laws force companies to list the types of fuels they use to generate power. The hope is that such green labeling will permit environmentally conscious power buyers to make choices based on environmental costs as well as on price.

The OPEC Cartel

The Organization of Petroleum Exporting Counties (OPEC) was formed on September 14, 1960, uniting five major oil exporters—Iran, Iraq, Kuwait, Saudi Arabia, and Venezuela—with the aim of restricting producer competition in the world oil market. As of 2004, OPEC's eleven members are Algeria, Indonesia, Iran, Iraq, Kuwait, Libya, Nigeria, Qatar, Saudi Arabia, the United Arab Emirates, and Venezuela. In the early years, OPEC had little effective economic power. If any OPEC member held out for higher prices or larger output levels, oil buyers could turn to other members or could rely on excess capacity in other countries. By 1970, however, conditions had changed radically. The most important single factor empowering OPEC was the disappearance of excess productive capacity outside the cartel.

By 1973, OPEC members watched rising market prices with great interest, and some of them conducted auctions to see just how high a price the market would support. When prices bid in these auctions exceeded \$15 (and in some cases even \$20) per barrel, some OPEC members thought they had at last found the true value of oil. The ease with which an unexpectedly high price increase was achieved by united OPEC action made the cartel ponder future actions. The cartel's first effort at monopoly pricing had succeeded. The critical question still facing OPEC was whether the price selected was the appropriate long-run, wealth-maximizing price path or whether it was too high or too low.

The period from 1981 through 1984 provided OPEC with its first major test. Cartel cohesion was rather simple to attain in the period from 1973 to 1980, as prices had increased tenfold and OPEC production was stable. In such a situation, everyone was aware of the benefits of collusion, and intercountry rivalries were suppressed. But the period from 1981 to 1984 brought home to OPEC the truth behind the laws of demand and supply. Simply put, the law of demand holds that as a consequence of the historically unprecedented rise in the price of oil during the 1970s, world consumption of oil decreased, particularly after consumers and competing producers were given sufficient time to adjust to the higher prices. See Example 13–1.

Figure 13–7 shows the economics surrounding a market consisting of OPEC producers combined with the rest of the world (non-OPEC producers). The law of supply predicts that the competitive fringe of oil producers (shown as $S_{\text{non-OPEC}}$), or the non-OPEC oil-supplying countries, will expand production in response to higher prices after a period of adjustment. The figure makes clear the economic model's predictions. The effects of OPEC raising the price from P^0 to P^1 are this: World oil consumption (D_{global}) decreases from q^0 to q^1 as oil buyers move back along their demand curve in the face of higher prices. Non-OPEC supply ($S_{\text{non-OPEC}}$) responds by increasing non-OPEC production from $q_n^{\,0}$ to $q_n^{\,1}$. The difference between the world demand and non-OPEC supply is the market that is filled by OPEC. The figure shows what you would expect: Higher prices reduce OPEC's sales from $q^0 - q_n^{\,0}$ to the considerably smaller $q^1 - q_n^{\,1}$. What happens to OPEC revenue? Under the low price, P^0, OPEC revenue is the area bound by P^0, q^0, and $q_n^{\,0}$. Under the high price, P^1, revenue is the possibly smaller area bound by P^1, q^1, and $q_n^{\,1}$.

Recall from principles of economics the distinction between the short-run and long-run demand schedules. The demand for oil is more elastic (price sensitive) in the long run than in the short run, because in the short run people are captives to a stock of energy-using equipment designed for cheap energy conditions. Ultimately, after an increased energy price, that stock of equipment is replaced by one designed for higher energy prices and the impacts of higher prices on reduced energy use are more fully realized. For example, from the early 1980s to the mid-1980s, more energy-efficient air conditioners, cars, and other energy-using equipment replaced old, more energy-inefficient counterparts. The short-run/long-run explanation also provides insight into explaining the rise in non-OPEC oil production over this period. Oil fields cannot be discovered and brought into production instantaneously. The percentage of non-OPEC production increased from 1977 to 1985 as these producers began moving up their supply curves in response to higher oil prices. See historical data on OPEC production in Figure 13–8.

Example 13–1
Renewable Energy Challenges in Brazil

Ethanol, the common alcohol in beer, wine, and other fermented and distilled liquors, has many uses. For example, it can be mixed in gasoline (petrol) as a transportation fuel. Ethanol is a clean-burning fuel made from corn, other grains, and dairy and other wastes. Renewable fuels like ethanol can provide environmentally clean, domestically produced energy, thereby contributing to many nations' independence from imported oil. In addition, farming communities benefit from the additional jobs created through the development of renewable fuel industry. More than two-thirds of the world's ethanol is consumed in Brazil.

In 1975, the Brazilian National Alcohol Program was established in response to the oil crisis of the 1970s. The program intended to reduce oil imports and limit any economic suffering that resulted from reduced imports. Brazil's program required that all oil be blended with ethanol; the ethanol share typically ranged from a fifth to a quarter. The Brazilian government also promoted the manufacture and sale of cars that ran exclusively on ethanol and enacted subsidies to increase domestic sugar cane production as well as construction of distilleries. As a result, sales of alcohol-powered vehicles reached a high of 96% of total sales by the mid-1980s. The news was not all good however. Remember, price exerts a powerful influence on energy's demand. A later decrease in oil prices, combined with Brazil's ethanol shortage in the late 1980s, reduced demand for all-alcohol cars; sales fell to less than 1% of all vehicle sales by 1997. But sales are increasing again in response to new tax breaks in alcohol-producing regions and government-required green vehicle fleets. As of 2004, Brazil still imports 40% to 50% of its consumed oil, but it is testing other combinations of fuels for blending, including methane, vegetable oils, and hydrogen. In addition, Brazil is developing a national biodiesel program to reduce dependence on imported diesel fuel and to limit air pollution.

Brazil is learning by doing. Since the introduction of its ethanol program in 1975, the cost of producing ethanol has declined in Brazil by 4% to 5% annually. Even though subsidies were eliminated in the late 1990s, ethanol is now a cheaper energy source than domestically produced gasoline from oil. The ethanol program has created more than a million jobs for 40% of the rural labor force and has reduced the nation's carbon dioxide emissions to 20% below what it would have been without the program. About 4 million cars run completely on alcohol, and sales of all-alcohol cars doubled in early 2002. Brazil's savings from avoided fuel imports over the past 27 years exceeds $52 billion, many times the total investments in the agricultural and industry sectors for ethanol production for vehicle use. Brazil now exports ethanol fuel.

Sources: Martinot, E. 2003. Renewable energy in developing countries: Lessons for the market. *Renewable Energy World.* 6 (4): 50–65; Martinot, E., A. Chaurey, D. Lew, R. Moreira, and N. Wamukonya. 2002. Renewable energy markets in developing countries. *Annual Review of Energy and the Environment.* 27: 309–348; Szklo, A. S., and M. T. Tolmasquim. 2003. Analysis of Brazil's cogeneration legal framework. *Energy Conversion and Management.* 44 (3): 369–380.

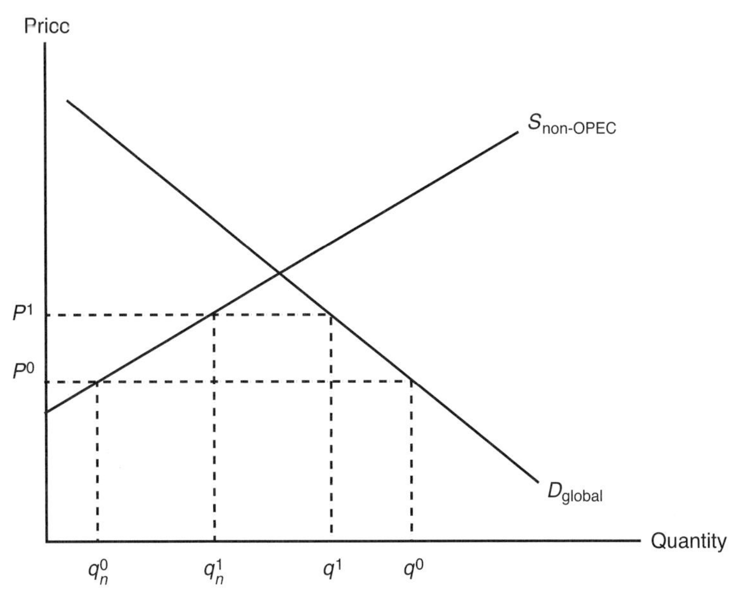

Figure 13–7
Economic determination of OPEC production

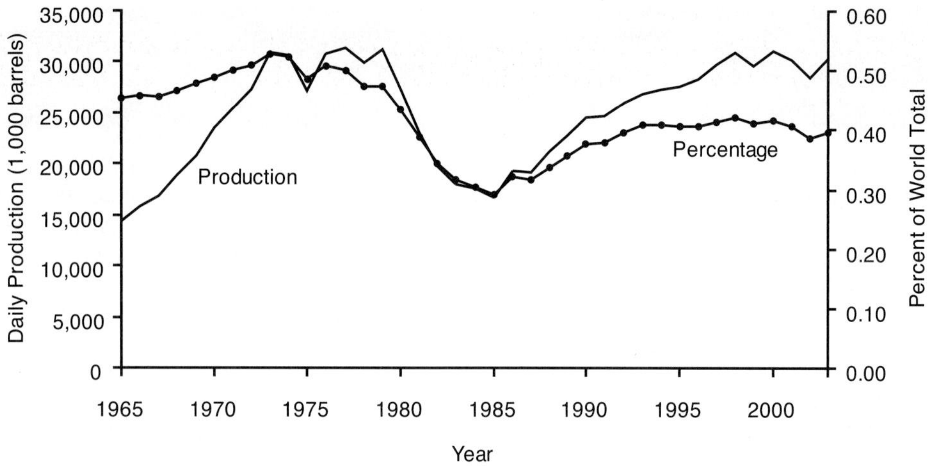

Figure 13–8 OPEC production by year
Source: British Petroleum. 2004. *Statistical Review of World Energy 2004.* Available from *http://www.bp.com/statisticalreview2004* [Accessed October 24, 2004].

Despite its stormy history, the OPEC cartel has been effective at achieving its aims.[12] Were the conditions that made it profitable special to the oil industry, or is the oil cartel the first of a growing list of other natural resource cartels? To answer these questions, let us examine the economic forces that enabled the cartel to occur. Four factors stand out from the rest: (1) shared interests among OPEC members, (2) supply response from competitors, (3) price elasticity of demand for OPEC oil, and (4) OPEC oil's income elasticity of demand.

Shared Interests. An important factor supporting a cartel in the natural resource market is the cartel's internal stability. With only a single seller (monopoly), that seller's aims can be pursued with little worry of alienating cooperating producers. If a cartel consists of several sellers, freedom of choice is more limited. As has been discovered with most multilateral treaties, the aim of each member and the aim of the group as a whole may be quite different.

Members of a cartel face incentives to cheat. Each member has an incentive to lower its price below the group's announced price, thereby greatly increasing its sales and stealing some of the market from the others. Each member is aware of the relevant economic theory: The price elasticity of demand facing an individual member is much higher than the elasticity for the group as a whole. Most of the increased sales by the potential cheater who cuts its price are sales taken from the other members. Starting from a given price and quantity, a higher price elasticity means that a lower price is required to maximize profits. Therefore, a successful and stable natural resource cartel must require the group to establish a mechanism for detecting cheating and for enforcing group agreements.

Cartels face another threat to their stability, namely, the extent to which members disagree on prices and output decisions. Each member has unique preferences for the group's price and output. Take the case of Saudi Arabia, which holds about one third of the OPEC total proven reserves (for emerging updates see Saudi Arabia Government, 2004), reserves exceeding those of any other member (Table 13–2). The importance of oil in Saudi Arabia's current and future economy (about 75% of its current income) means that it has a special need to sustain the value of its production and reserves. Setting OPEC prices too high reduces the world's future demand for its

[12] As of 2004, OPEC's aims were (1) coordination and unification of petroleum policies of OPEC countries and safeguarding their interests; (2) ensuring the stabilization of prices in international oil markets; and (3) securing a steady income to OPEC countries, a steady supply of petroleum to consuming nations, and an acceptable return to petroleum industry investors (Saudi Arabian Information Resource, 2004).

Table 13–2 World's largest oil reserves by country

Country	Reserves (billions of barrels)
Saudi Arabia	261.8
Canada	180.0[a]
Iraq	112.5
United Arab Emirates	97.8
Kuwait	96.5
Iran	89.7
Venezuela	77.8
Russia	60.0
Libya	29.5
Nigeria	24.0
United States	22.7
China	18.3
Qatar	15.2
Mexico	12.6
Norway	10.3
Algeria	9.2
Kazakhstan	9.0
Brazil	8.3
Azerbaijan	7.0
Oman	5.5
Angola	5.4
India	5.4
Indonesia	5.0
United Kingdom	4.7
Total	**1,168.2**
Total percent of world's reserves	**96.3**

[a] *World Oil* estimates Canada's reserves at a much more modest 5.5 billion barrels.

Source: *U.S. Department of Energy. 2004.* International Energy Annual 2002. *Available from http://www.eia.doe.gov/emeu/iea/pet.html [Accessed September 18, 2004].*

oil and increases the supply of non-OPEC oil. OPEC nations with more limited oil reserves recognize that their reserves may be gone soon, so they plan more for the nearer-term future.

Competitive Supply Response. An important factor in the ability of a cartel to exert control over a market is its ability to block competing suppliers outside the cartel from supplying to the market and reducing the price. Currently, OPEC produces about 40% of the world's oil. If non-OPEC producers have adequate resiliency to respond to higher prices by expanding capacity considerably, their action increases the quantity of oil supplies and causes the prices to fall, reducing OPEC's market share (Figure 13–7). As of 2004, only Russia appears to have sufficiently large reserves among the non-OPEC producers to make a significant difference in the world oil market. Since both the size of its reserve and its production capacity are highly uncertain, it is hard to assess Russia's ultimate impact on the future world market. Even though no single non-OPEC member is likely to have much effect on world oil prices, non-OPEC members collectively can have a major impact on price, as shown in Figure 13–7. OPEC must account for supply decisions of nonmembers when making its output and price decisions.[13]

---

[13] If OPEC has larger and lower-cost reserves than non-OPEC suppliers, a profit-maximizing strategy for OPEC may be to limit its own sales and production in early years, letting non-OPEC suppliers deplete their supplies. When the competition's supplies are gone, OPEC takes over the market as the sole seller, setting prices at whatever level maximizes the present value of its profits. Remarkably, the decision to form the cartel, by taking much OPEC production off the market in early years, permits non-OPEC suppliers to earn greater profits in the early years than if OPEC acted as a perfect competitor. This means that non-OPEC producers are an important force in the oil market in the early years, even if they control a minority of the world's oil production.

The impact of this competitive fringe on OPEC behavior was dramatically illustrated by events in the mid-1980s. In 1979, OPEC had accounted for approximately 50% of world oil production, whereas in 1986 this figure had fallen to approximately 30%. Total world oil production was down during this period more than 10% for all producers, and the pressures on the cartel mounted and prices ultimately fell. Between 1981 and 1986, the real price of crude oil fell to a third of its 1981 level (Figure 13–1). OPEC simply was not able to hold the line on prices because the necessary reductions in production were too large for the individual cartel members to sustain. Table 13–3 illustrates selected recent activities of OPEC.

Table 13–3 Selected OPEC decisions during 2003 and 2004

Date	Event
January 12, 2003	OPEC held its 123rd meeting to review oil markets. It raised the OPEC production quota from 23 million barrels per day to 24.5 million barrels per day, effective February 1, 2003, to ensure adequate supplies of crude oil.
March 17, 2003	The United Nations (UN) Secretary-General announced the temporary withdrawal of all humanitarian personnel from Iraq. The UN's independent oil overseers were withdrawn from the Iraqi oil platform in the south of Iraq, the only outlets for Iraqi oil exports allowed under the UN's Oil-for-Food Program.
March 19, 2003	War began in Iraq.
March 20, 2003	OPEC agreed to make up for any supply shortfall resulting from events unfolding in Iraq. Member countries pledged to use their available excess capacities to ensure supply.
April 24, 2003	OPEC held its 124th meeting and decided to reduce actual OPEC production to the new OPEC production quota of 25.4 million barrels per day, effective June 1, 2003.
May 22, 2003	The UN Security Council adopted Resolution 1483, lifting civilian sanctions on Iraq, ending UN monitoring of the export of crude oil and petroleum products from Iraq, and phasing out the UN's Oil-for-Food Program over the next six months.
May 27, 2003	The United States lifted most of its remaining sanctions on Iraq, thereby implementing UN Security Council Resolution 1483.
June 11, 2003	OPEC held its 125th meeting to review oil markets, leaving OPEC output quotas unchanged.
July 31, 2003	OPEC held its 126th meeting to review oil markets, leaving OPEC output quotas unchanged.
September 24, 2003	OPEC held its 127th meeting to review oil markets and decided to lower OPEC's production quota from 25.4 million barrels per day to 24.5 million barrels per day, effective November 1, 2003.
December 4, 2003	OPEC held its 128th meeting to review oil markets, leaving OPEC output quotas unchanged.
February 10, 2004	OPEC held its 129th meeting to review oil markets and decided to lower its production quota from 24.5 million barrels per day to 23.5 million barrels per day, effective April 1, 2004.
March 31, 2004	OPEC held its 130th meeting to review oil markets and reconfirmed the new production ceiling of 23.5 million barrels per day effective April 1, 2004.
June 3, 2004	OPEC held its 131st meeting to review oil markets and decided to increase its production quota to 25.5 million barrels per day effective July 1, 2004, and to 26.0 million barrels per day effective August 1, 2004.
September 15, 2004	OPEC held its 132nd meeting to review oil markets and decided to increase its production quota to 27.0 million barrels per day effective November 1, 2004.

Source: *U.S. Department of Energy. 2004.* Country Analysis Briefs, OPEC. *Available from http://www.eia.doe.gov/emeu/cabs/opec.html [Accessed September 18, 2004].*

Price Elasticity. The price elasticity of demand is an important factor in explaining a cartel's stability, because it reveals how sensitive quantity demanded is to price changes. When the price elasticity lies between 0 and −1.0, price increases lead to increased revenue (the closer elasticity is to 0 from −1.0, the more a given price hike increases total revenue). The amount by which revenue increases when price increases depends on the price elasticity of demand. A smaller absolute value of the price elasticity results in larger revenue gains from the decision to establish a cartel.

The price elasticity of demand for oil depends on the opportunities for conservation when energy price increases and on the price of and access to energy substitutes. Think of what needs energy, for example, transportation, space heating and cooling, and power production. If smaller cars can be produced that maintain desired convenience, or more energy efficient car engines can be built as oil prices increase, energy conservation will be inexpensive and reliable, and the price elasticity of demand for energy will be high. If good home insulation is cheap and effective, then outdoor temperatures extremes will not cause problems, because desired indoor temperatures can be maintained with less electric power or heating oil. The general principle emerging from this discussion is that a greater range of conservation opportunities and a smaller monetary expense needed to take advantage of them raise the price elasticity of demand for energy. Also, demand will be more price elastic in the long run when enough time has passed to permit adjustments, than in the short term when little adjustment is possible.

The price, access, and availability of OPEC oil substitutes is important because it limits the extent to which prices can be raised by the cartel. Abundant quantities of substitutes available at prices close to oil prices (e.g., backstop fuels described earlier) set an upper bound on OPEC's oil price. The ultimate backstop technology is solar energy, and it is this source that limits OPEC's ability to raise oil prices in the long run. Advances in solar technology, such as reductions in cost or increased reliability of solar photovoltaics, will place even greater constraints on OPECs ability to raise oil prices significantly. Nevertheless, solar energy is still, for the most part, considerably more expensive than the same energy content of oil. So the energy-consuming nations of the world will be waiting for some time before solar power is cost competitive with conventional energy.

Income Elasticity. The income elasticity of demand for oil is an important factor in predicting the stability of OPEC because it indicates how responsive the demand for oil is to growth in world economic activity. (*Income elasticity* is defined as the percentage change in demand resulting from a 1% change in income.) Even if oil's price is constant, as per capita income increases or total income grows through simple population growth, oil's demand will grow. Increases in demand due to increases in income support the power of OPEC to increase the price of its oil, thus adding strength to the oil cartel. Studies have found the income elasticity of demand for oil to be greater than 1 for developing (low-income) countries and to be between 0 and 1 for industrialized countries (for a recent analysis, see Atkins and Tayyebi Jazayeri, 2004).

The fact that income elasticity for oil is positive means that a given increase in OPEC's price will support more oil revenue growth when worldwide incomes are rising. The positive income elasticity of demand for oil was a major cause of the cartel weakening that occurred in 1983, when the world suffered a significant recession. That recession produced a large reduction in the demand for oil which placed pressure on the cartel to lower its price.

Market Outlooks for the Future

Over two-thirds of world energy consumption still comes from oil and natural gas. Even though real oil prices may decline for a time, over a longer horizon of 20 to 50 years, the real price of oil will likely rise, consistent with Hotelling's prediction. Recall that Figure 13–3 shows that the oil price will rise at rate r until it hits the backstop fuel price. Thereafter, the availability of the new backstop fuel constrains oil's price. Hotelling's model leads us to predict that conventional oil production ceases as soon as its price reaches the backstop fuel price, since user costs no longer rise at rate r.

Only one price will satisfy these two conditions: (1) oil prices must rise at rate r until reaching the backstop fuel price and (2) economic production of oil reserves will be exhausted at the point

Figure 13–9 Impact of reduced backstop fuel price

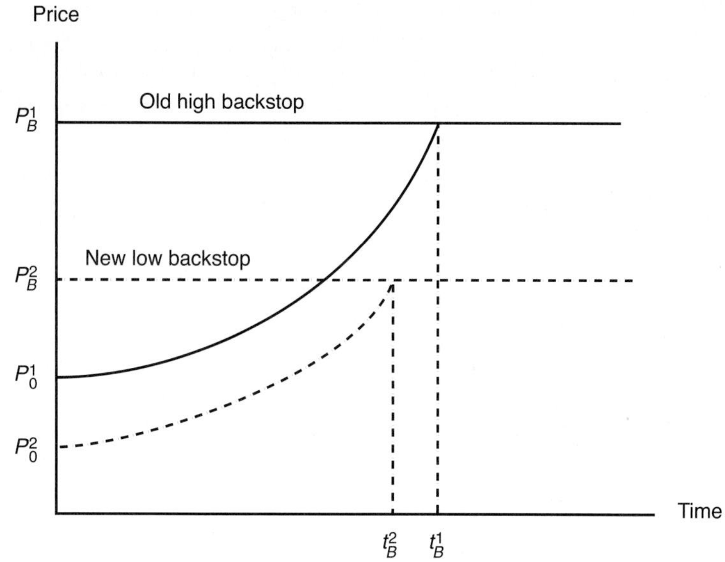

when the backstop price is reached. If the price path is too high, the backstop will be reached too early, and oil will be left over, because consumption rates at the higher prices are too low to use up conventional oil reserves. Conversely, if the price path is too low, the high consumption induced by the low price depletes conventional oil reserves well before the price path reaches the backstop.

Figure 13–9 shows the impact on the nonrenewable fuel price path of a technological breakthrough that reduces the backstop fuel price from P_B^1 to P_B^2. A good example is a breakthrough in solar *photovoltaic technology* that reduces its price from five times to twice the cost per kilowatt hour from conventional energy sources. See Example 13–2. The new backstop fuel price results in a new price path that reduces today's fuel price from P_0^1 to P_0^2. The new price path lies below the old price path everywhere. This occurs because a lower price path is required to keep the fuel price growing at the rate of r percent per year, while still depleting the unchanged reserves in the earlier period and producing a lower price guaranteed by the reduced backstop fuel price.

Under the original backstop fuel price, depletion was scheduled to take place in period t_B^1. However, now that the backstop fuel is cheaper, the conventional fuel reserve is scheduled for depletion in the earlier period t_B^2, the time at which the price climbs to intersect the price of the now cheaper backstop, P_B^2.

This discussion may seem too theoretical and distant from reality, but its implications are profound: Technological breakthroughs in backstop technology (e.g., a reduced price of solar energy) can significantly lower the current price path, even though the new backstop may not be reached until many years in the future. McVeigh et al. (2000) describe some institutional barriers to consumer acceptance of energy backstop technologies. The welfare gains from technological breakthroughs are potentially large, because price and consumer surplus are increased for all years into the future after successful reductions in the backstop price. For example, imagine the huge welfare gains that will be received by a grateful humanity of current and future generations if solar electricity can be produced around the world at prices less than \$0.05 per kilowatt hour. To the degree that new technology lowers the marginal social cost of conventional oil and of its backstop substitute fuel, society receives a major and permanent welfare gain. For these reasons, emerging backstop energy technologies are exciting subjects for analysis and debate.[14]

[14] This statement glosses over important policy questions being asked about transition fuels. The world's industrialized countries depend on oil and gas for most of their energy. The long-term future will see a transition to some renewable, nonpolluting energy source. What will happen in the intermediate period? Some optimistic forecasters hope that the transition to renewable, pollution-free energy will occur so quickly that no transition energy fuels will be needed, but most believe that transition fuels will be required (e.g., Dresselhaus and Thomas, 2001).

Example 13–2
A Renewables Success Story: Japan's Photovoltaics

Japan established net metering for photovoltaics (PV) in 1992, which required utilities to purchase excess power at the retail rate. Between April 2001 and March 2002 alone, Japanese electric power companies bought more than 124 gigawatt hours of surplus PV power. In 1994, Japan launched the Solar Roofs program to promote PV systems through low-interest loans, a comprehensive education and awareness program, and rebates for grid-connected residential systems that were provided in return for data about systems operations. Government promotion of PV has included publicity on television and in print media. The rebates declined gradually over time, from 50% of installed costs in 1994 to 12% in 2002, the year the program ended. In 1997, the rebates were opened to owners and developers of housing complexes, and Japan became the world's largest supporter of PV systems with a sevenfold increase in funding for the expanded program. The goal of Japan's PV program was to create market awareness and stimulate PV production, while reducing costs through economies of scale and improvements in technology. Ultimately, Japan wanted to foster large-scale power generation and export PV systems to the rest of the world.

Japan is now the world's leader in the manufacture and use of solar PV systems, having surpassed the United States at both in the late 1990s. When Solar Roofs ended, the program had exceeded its objectives: More than 420 megawatts of PV were installed between 1994 and 2002. Total installed capacity has increased an average of more than 42% annually since 1992. To keep up with demand, Japanese PV manufacturers have dramatically increased their production capacity by nearly 47% in 2002 alone. Japan was responsible for more than 43% of global PV production in 2002, and Sharp, a Japanese firm, is now the world's leading producer of solar cells. By some accounts, PV system costs in Japan have dropped at least 75% since the mid-1990s, far more rapidly than the decrease in average costs worldwide; the cost of residential, grid-connected PV systems in Japan declined by more than 40% between 1995 and 2003.

Sources: Haas, R. 2002. Building PV markets: The impact of financial incentives. *Renewable Energy World.* 5 (4): 195; Sawin, J. L. 2004. National Policy Instruments. Paper presented at the International Conference for Renewables, Bonn, Germany, January 2004, pp. 1–42. Available from *www.renewables2004.de/en/cd/default.asp*; Maycock, P. 2003. PV market update. *Renewable Energy World.* 6 (4): 84–101; Lamb, C. 2003. New laws reshape, fuel solar power growth. *Sacramento Business Journal.* October 20, 2003.

Two other factors may dramatically affect our energy future. On the demand side, major successes in energy conservation and increased efficiency in the end use of energy could considerably reduce current demand projections. On the supply side, the renewable energy sources that presently provide only a small part of the world's energy needs might assume a considerably larger role. The success of these two changes would require a combination of market price signals and government energy policies that promote energy efficiency and renewables.

Improved technologies for efficient energy use could achieve impressive reductions in energy use with little or no negative impact on living standards. In most industrialized nations, energy use per dollar of gross domestic product (GDP) has fallen steadily in recent years as technology improves and the economy becomes more service-based. Energy demand will continue to grow in developing nations, but the growth will not require heavy reliance on inefficient or environmentally damaging sources. Given such optimism, does anything stand in the way of a highly energy efficient future? The answer depends on price. The central economic fact pointing to heavy use of nonrenewable energy sources is the *low current price* for energy. Even though increasing energy efficiency by a factor of two or more may be technically feasible, it is not necessarily economically profitable. So long as energy prices remain low, energy-intensive production methods may remain the optimal economic strategy.

The low current price of fossil fuels is the prime barrier to developing nonfossil fuel energy sources. Remarkable advances in reducing costs of solar and renewable energy sources since 1980 have greatly shifted out their supplies and increased their quantities demanded (Figure 13–10). In fact, some alternative energy sources, like wind and biomass, have reached the threshold of competitiveness in the mass market, but in only a few cases have they crossed

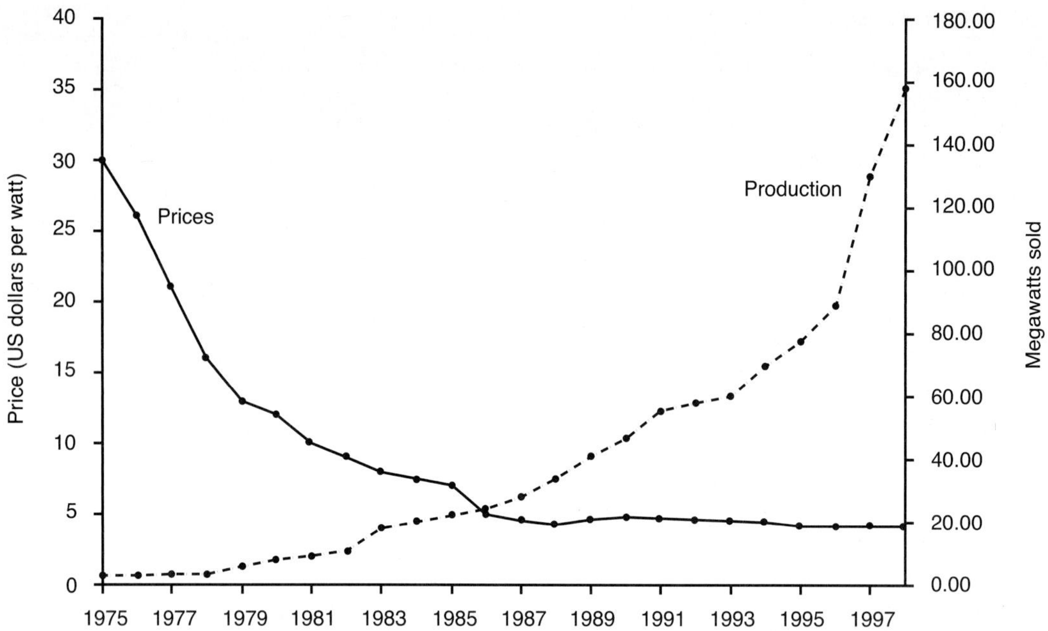

Figure 13–10 Photovoltaic production and prices (worldwide)
Source: U.S. Department of Energy. 2004f. *Solar Photovoltaics.* Available from *http://www.eia.doe.gov/cneaf/solar.renewables/page/solarphotv/solarpv.html* [Accessed October 24, 2004].

that threshold into large-scale commercial feasibility.[15] For example, solar power is competitive for water heating, but only rarely is it competitive for electric power supply.[16] In addition to price, other institutional issues have prevented alternative energy sources from gaining a major market share. Inertia plays a major role in slowing the transition to renewables. For example, consider the extensive infrastructure investment in fossil fuels, current power stations, machinery, and transportation systems; all support continued use of fossil fuels. Significant direct and indirect government subsidies that support fossil fuel industries also play a role. Worldwide, many governments enact policies that reduce energy prices to support industrial growth. Also, in many cases government authorities are committed to investment in standard coal- or oil-powered plants, so regulatory policy makes it hard for competitors to enter the market.

Institutional Designs for Renewable Energy Sources

Outcomes in the energy market will be influenced by government policies that affect the pace and level of development of renewables. Several key policy designs are discussed next: marginal cost pricing, environmental adders, energy and pollution taxation, green power, voluntary action, and incentives to promote energy conservation.

---

[15] A good Web page for staying abreast of the global solar electricity scene is U.S. Department of Energy, *Solar photovoltaic* (2004f).

[16] Without special government subsidy, solar electricity can be competitive in isolated sunny areas where there are no other commercially feasible power sources.

Marginal Cost Pricing

With all the barriers to obtaining increased energy supplied from renewables, what kinds of effective actions can be taken? Prices of fossil fuels often fail to fully account for the social costs of their environmental impacts, including their impacts on elevated greenhouse gasses (see Chapter 15). The disparity continues between the private costs of energy use produced through fossil fuels and their social costs (Romm and Ervin, 1996; Rubin and Leiby, 2000). As long as this disparity exists, there will be an inefficiently high level of environmental degradation associated with energy use and an artificially high level of energy use itself.

Two efficient measures for promoting greater use of renewables are a comprehensive series of marketable pollution permits and a system of per unit pollution taxes. Some combination of taxes and permits could be the most efficient method. The permit system might best be used for large, stationary sources of pollution (e.g., electric utilities and oil refineries), and taxes could be applied to mobile pollution sources (e.g., cars and trucks) as well as to small, stationary sources of pollution like heaters for privates homes and small businesses.

Renewable energy sources typically pollute less than fossil fuels. For example, if wood is burned to generate steam and produce electric power, the pollution from the wood is small compared with pollution from the oil or coal it displaces. Wood can be burned efficiently, which means that its combustion is nearly complete, leaving few unburned hydrocarbons in its emissions. Moreover, wood has essentially no sulfur content, so sulfur dioxide emissions (leading to acid rain) are greatly reduced. Finally, if the wood (or corn or other biomass fuel) is replanted after harvesting, then burning the biomass fuel does not contribute to global warming.

Global energy needs can be met with greater economic efficiency if all energy sources are priced at their marginal social cost, including the environmental costs produced by pollution, health risks, and global warming. Figure 13–11 shows how energy policies can adjust outcomes of competitive energy markets given the aim of economic efficiency and sustainability. The diagram shows the mix of a community's energy produced (e.g., in a single country) from two sources: (1) renewable sources, including biomass fuels, wind, and solar energy; and (2) fossil fuels, such as coal and oil. Price and quantity of the two sources are measured in a common denominator unit like BTUs. The diagram assumes that no pollution or greenhouse gasses are associated with the use of the renewable sources.[17]

The market's marginal private cost curve, which is equal to the supply curve summed over both renewable and fossil fuel sources, is constructed by horizontally adding the marginal private cost curve for renewable sources, MC_R, to the marginal private cost curve for fossil fuels, MC_{FP}. If D represents the total demand for energy use from all sources, then the unregulated market equilibrium quantity is Q_{TP} and the equilibrium price is P^0. At this point, A, the demand intersects that market supply curve, MC_{TP}. From total output, Q_{TP}, fossil fuels supply Q_{FP} units, which is shown by the intersection of MC_{FP} with price P^0. Renewable energy supplies the remaining $Q_R{}^0$, shown by the intersection of the renewable supply curve, MC_R, with price P^0 at point E. Notice that this competitive market solution maximizes total private benefits supplied by energy production, which you will recall from Chapter 3, is private consumer and producer surplus. That total private surplus equals area KJA. No other allocation of output between the two fuel sources produces a greater private surplus than the area KJA, the market allocation, produces.

Despite this idealistic story, your environmental instincts tell you something is wrong. Notice that the marginal social cost incurred by energy production from fossil fuels, MC_{FS}, is greater than its marginal private cost, MC_R. Each level of energy output supplied from fossil fuels incurs a marginal social cost described by the curve MC_{FS}, which includes the sum of marginal

[17] For an analysis of choices Dutch firms make in choosing energy sources, see DeGroot, Verhoef, and Nijkamp, 2001. For an example in Sweden, see Miranda and Hale, 2001.)

Figure 13–11
Economically efficient energy policy

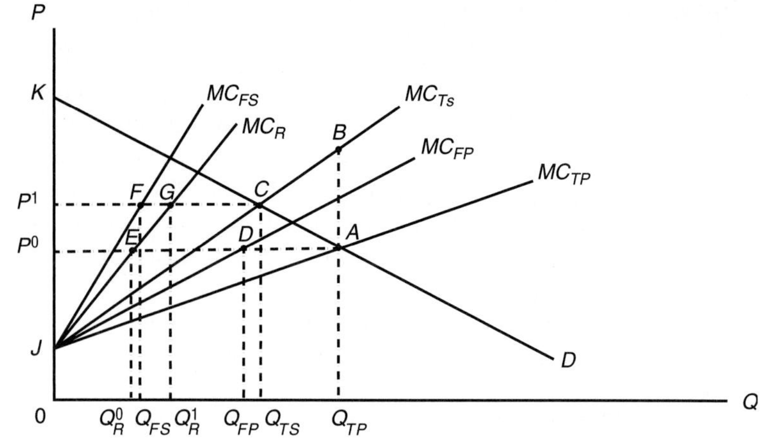

production costs paid to energy suppliers added to the marginal costs of pollution and of greenhouse gas accumulation. When all costs are counted, maximum economic efficiency occurs where the aggregate marginal social cost curve, MC_{TS},[18] intersects the demand function, D at point C. That socially efficient intersection point produces a lower equilibrium quantity, Q_{TS}, and higher equilibrium price, P^1, than the unregulated private market outcome, for which price is P^0 and quantity is Q_{TP}.

Can we say anything with confidence about the outcome of the unregulated competitive market described? Yes. Society suffers a welfare loss on the *last* unit of energy produced at output level Q_{TP}, equal to marginal social cost from all sources (point B on MC_{TS}) minus total marginal private cost from all sources (point A). The marginal loss equals the vertical distance BA. That distance consists of the external costs of pollution and greenhouse gas elevation suffered by the larger society not included in the power price, A. Where the efficient output level occurs at output level Q_{TS}, there is no welfare loss, since marginal social cost of the last unit, MC_{TS}, equals society's marginal benefit of the energy, point C on the demand curve D. At all output levels between the efficient level Q_{TS} and the unregulated competitive level Q_{TP}, the welfare loss is somewhere between zero and BA. The total welfare loss over all those units incurred by the unregulated market institution is the triangular area ABC. (If demand or marginal cost is nonlinear, the welfare loss figure is not a triangle.)

What can be done to give energy producers and consumers the incentive to account for these external costs? A policy decision could be made to increase the private cost of fossil-derived energy through an externalities tax, a tradeable pollution permits system, or an energy fuel tax. In that case, the marginal private cost for fossil fuels increases (pivots) from MC_{FP} to MC_{FS}. Raising the marginal private cost of fossil energy so that its actual costs are perceived by the market participants generates a reduced market equilibrium energy output of Q_{TS}, the economically efficient level. At output level Q_{TS}, the highest possible social consumer-plus-producer surplus is achieved, equal to area KJC. No other combination of production from renewables and fossil fuels produces a larger total net social benefit.

What does this efficient energy policy do to energy market conditions? The price of energy increases from P^0 to P^1, fossil energy use declines from Q_{FP} to Q_{FS} (point D to point F), and renewable energy use increases from $Q_R{}^0$ to $Q_R{}^1$ (point E to point G). Increasing the price of energy through the imposition of energy fuel taxes, tradable pollution permits, or pollution taxes accomplishes two goals in regard to efficiency: (1) It reduces pollution, since renewable

---

[18] The marginal social cost curve MC_{TS} is the horizontal summation of the marginal social cost of fossil fuel sources MC_{FS}, and the marginal cost of renewable sources, MC_R.

sources are generally cleaner; and (2) it reduces reliance on fossil energy, which will reduce pollution and make existing supplies last longer.

Environmental Adders

Environmental adders are estimates of the monetary value of the external damage imposed on society by each additional ton of a particular pollutant generated by an energy source (Eyre, 1997). Adders have been developed to provide incentives to change the choice among coal, natural gas, and other fuels in new electric power plants. These adders are intended to reflect damages done by emissions of pollutants from burning fossil fuels after existing required emission controls are implemented. For this reason, they are sometimes said to measure "residual environmental damage." When these residual damages are added to the direct cost of the energy source being considered, energy supplies fired by fuels with the lowest marginal social cost can be identified and built. Some U.S. state regulatory commissions, including the California Public Utility Commission, are requiring electric utilities to consider certain environmental externalities in evaluating the need for new generating units or for operating existing units. Utilities in those states are required to assign monetary values to selected environmental externalities and to add those values to the unit costs of each supply option responsible for the impact.

The principle of marginal cost pricing supports internalization of environmental externalities, particularly in connection with fossil fuels.[19] Renewable energy sources like wind and solar power have much lower (sometimes zero) external environmental costs than coal. So why does coal-fired electricity remain so cheap? It remains cheaper than power produced from renewables only because the true social costs of using coal are not reflected in its market price. The net result of the greater application of environmental adders would be increased power rates, particularly in coal-dominated regions, and the displacement of coal and other dirty fuels by more environmentally benign renewables. Overall power rates would increase; coal-fired power prices would rise the most and renewable-supplied rates would rise the least.

In the United States, environmental damage estimates have been used widely to determine environmental adders for regulation of the electricity industry. The use has been more for choice of electricity generation sources for newly constructed power plants than for choice of which existing plants to use for producing power at any point in time.[20] Despite the best intentions behind the use of adders, the values assigned to them have been quite variable and have not always reflected the best economic or physical science, especially in regard to health effects.

Energy and Pollution Taxes

Taxes on pollution that are caused by energy production or consumption are an important way to implement marginal cost pricing. Furthermore, estimates of the externality costs produced by the energy sector could be used to design these taxes. An analysis by Eyre (1997) in the United Kingdom reported two important results: First, electricity use has significant external environmental costs, which means that electricity is typically priced well below its full social cost. Second, energy taxes could drive behavior by electricity consumers in the direction of improved economic efficiency by providing incentives to block the inefficient use of energy and/or by discriminating among supply sources.

---

[19] Brown et al. (2001) present an optimistic scenario of how the United States could reduce most of its greenhouse gasses to 1990 levels. The scenario relies on the implementation of efficient measures for internalizing external costs of energy.

[20] The use of adders for new plants and not for the operation of existing plants might seem odd, because greater economic inefficiency gains arise from reduced use of older, dirty power stations with high external costs. However, in the case of U.S. electric power regulation, regulatory agencies have the most control in the choice of new plants.

Energy taxes pose political problems in countries that have experimented with them, such as the United States and the United Kingdom. Under some political conditions, charges on particular pollutants may be more politically workable than general energy taxes. Charges on particular pollutants can address the issue of technological dependence very directly, unlike a more general energy tax. Energy taxes may be a preferred choice, however, where there are many polluters (making command-and-control options hard to monitor) and where total damages to society are independent of the point of emission.

Carbon dioxide emissions provide an example. The problems of a carbon tax in which all affected countries agree to charge taxes to their own carbon emissions are mostly political (see Chapter 15). Where politically feasible, the value of an optimum carbon tax should be set at the marginal damage cost. However, it is not possible to identify even the approximate value of this number with any great confidence. Nevertheless, carbon taxes do have something to offer (Baranzini, Goldemberg, and Speck, 2000). In international negotiations, national targets are the most likely primary instrument for emission control. One possible approach would be to apply carbon taxation nationally or within a trading block, with the tax level determined by the cost of emission abatement. Of course, even national carbon taxation raises important issues of equity. If the overall economic benefit justifies a carbon tax, the political cost and will needed to deal with equity issues may be worth incurring.

Promotion of Renewable Energy

Renewable resources are increasingly recognized as important elements of a sustainable energy policy (Basosi, Maltagliati, and Vannuccini, 1999; Brophy, 1997; Charters, 2001; Dincer, 1999, 2000; Evrendilek and Ertekin, 2003). Table 13–4 shows the current worldwide status of energy supplied from renewable sources. Even though most renewable fuels themselves are inexpensive or free, efforts to promote renewable energy projects confront economic barriers. A number of factors combine to make most renewable energies more expensive than conventional energy (Bird, Aleagha, and Ferroukhi, 1995). Existing subsidies of fossil fuels and the failure to internalize all costs of nonrenewables erect high barriers to renewable energy. Financial barriers include the costs of renewable technologies, the need to achieve economies of scale in production, the lack of producer access to affordable credit, and the costs of connecting with the grid and transmission charges. Import duties on renewable technologies and components also make renewable energy more expensive. Two kinds of regulatory policies have been used to open the grid to renewables (Sawin, 2004). One guarantees price to the supplier (pricing laws); another ensures market share through government-mandated targets or quotas. See Example 13–3.

Pricing Laws. Pricing laws guarantee a price to renewable energy suppliers (Meyer and Koefoed, 2003). Under such laws, electric utilities are obligated to enable renewable energy plants to

Table 13–4 Global renewable resource use

Resource	Current use (exajoules per year)
Hydropower	10.0
Biomass energy	50.0
Solar energy	0.2
Wind energy	0.2
Geothermal energy	2.0
Ocean energy	0
Total	62.4

Source: *Johansson, T. B., K. McCormick, L. Neij, and W. Turkenburg. 2004.* The Potentials of Renewable Energy. *Paper presented at the International Conference for Renewables, Bonn, Germany, January 2004.*

Example 13–3
Power of Price in Promoting Renewable Energy in Germany

In 1990 Germany had little renewable energy industry, and few Germans believed the country would ever be a leader in renewables. Yet by 2000 Germany had become a renewable energy leader with a new, multibillion dollar industry supporting thousands of jobs. How did all this happen? Motivated by public concerns about nuclear power safety, energy security, environmental impacts, and global climate change, the German government enacted the Electricity Feed-in Law in 1990. The law required utilities to buy electricity generated from renewable technologies in their supply area and to pay whatever price was needed to guarantee that the best wind and solar energy sources were economically feasible. The law, which was inspired by similar policies that had worked in Denmark, traced its longer history to legislation in California in the late 1970s (see Example 13–5). After Germany's first energy pricing law was established, development of wind energy sustained steady growth. Farmers, small investors, and start-up manufacturers developed a new industry. Barriers to the industry's development were complex and numerous, but the German government was committed for the long run.

Among all policies promoting Germany's renewables, the pricing law has produced the greatest impact on Germany's renewable energy industries. The law ended economic uncertainties about whether producers could sell their electricity into the power grid and at what price. It also provided investor confidence and made it easier for producers to secure bank loans. Sustained investment produced improved technology, advanced experience, and scale economies that led to impressive cost reductions. The average cost of manufacturing wind turbines fell by 43% between 1990 and 2000, and wind capacity expanded from less than 60 megawatts in 1991 to 12,000 megawatts at the end of 2002—an expansion factor of more than 200. By 1997 Germany was the world's leading producer of wind energy. By the end of 2002, wind energy met 4.7% of the country's total electricity demand. Approximately 45,000 people worked in Germany's wind industry by early 2003. A sufficiently high number of Germans are employed in renewables industries, or own shares in wind turbines, solar, or other projects, to produce broad public support for renewable energy. Germany plans to reduce its carbon dioxide emissions 21% below 1990 levels by 2010 and expects to accomplish much of this through increased use of renewables.

Sources: Sanner, B., and W. Bussmann. 2003. Current status, prospects and economic framework of geothermal power production in Germany. *Geothermics.* 32: 429–438; Weiss, I., and P. Sprau. 2002. 100,000 Roofs and 99 Pfennig: Germany's PV financing schemes and the market. *Renewable Energy World;* French, I. 2003. Biogas in Europe: Huge potential in a growing market. *Renewable Energy World.* 6 (4): 120–131. For a summary of renewable energy programs in France, see Jean Baptiste, P., and R. Ducroux. 2003. Energy policy and climate change. *Energy Policy.* 31 (2): 155–166. For progress in Switzerland, see Jegen, M., and R. Wustenhagen. 2001. Modernise it, sustainabilise it! Swiss energy policy on the eve of electricity market liberalisation. *Energy Policy.* 29 (1): 45–54.

connect to the electric grid, and they must purchase any electricity generated with renewable resources at a minimum administered price. Guaranteed prices are typically set higher than the market-clearing price, and payments are usually guaranteed over a specified period of time. Normally, the price is set high enough to guarantee a sufficiently high return on renewables to encourage investment in them. See Example 13–4.

An early policy application of a guaranteed energy price to promote renewables occurred in California during the late 1970s. The U.S. Public Utilities Regulatory Act of 1978 (PURPA) required utilities to interconnect with and buy energy from qualifying facilities, including renewable energy plants, at the incremental costs of production saved (avoided). In California, implementation of PURPA involved standardized, long-term contracts with specified payments for all or part of the contract term. The contracts were financed through higher electric rates for consumers. This meant that consumers were charged a price increment in their power bills to pay for the additional cost of renewable energy compared to the cheapter price of nonrenewable energy. The time length of contracts (15 to 30 years for wind projects), combined with fixed energy prices, assured producers of a market for their product and also gave them an assurance they could take to the bank to obtain financing. Although most U.S. states saw little development in the use of renewable energy, California for a time became the world's leader in renewable energy development. See Example 13–5.

Example 13–4
Economic Incentives Promote Renewables in India

The Indian state of Tamil-Nadu took a serious look at its wind resource potential in the mid-1980s. The state electric utility started by building demonstration projects and the required grid infrastructure and by informing potential investors about good wind sites. The government opened the electric grid to private producers in 1991, encouraging investors to build and operate power plants and to supply power through long-term contracts with state electricity boards. In the spirit of California's clean energy legislation of the late 1970s, private companies received economic incentives to produce wind power from remote places with good wind and move it over transmission lines for sale.

Investment-based subsidies and a lack of turbine standards or production requirements led wealthy investors to use wind farms as tax shelters, and several early projects experienced poor performance. This occurred despite the significant technological advancements that had taken place since the early 1980s. Growth in wind energy slowed considerably when investment credits declined. Despite these bumps in the road, however, India now ranks fifth in the world for wind energy capacity, with 1,870 megawatts in 2003, about 2% of India's total power-generating capacity. India's turbine manufacturing capacity is about 500 megawatts a year, and 15 companies are contributing to the industry by building wind turbines and by creating domestic jobs.

India still faces huge energy challenges. India's opportunity for higher growth in gross domestic product will be stressed by sustained energy shortages because of the lack of far-reaching structural reforms in the fuel and power sectors. Nevertheless, India has made impressive gains in producing renewable energy. India has the world's largest decentralized solar energy program and the world's second largest biogas program. The Indian government established a number of financial incentives for photovoltaics (PV) in the 1980s and established local service centers as market volumes increased. Although government subsidies are still required for most PV installations, the nation ranks fifth worldwide for grid-connected PV capacity.

Sources: Sawin, J. L. 2004. National Policy Instruments. Paper presented at the International Conference for Renewables at Bonn, Germany, January 2004; Martinot, E., A. Chaurey, D. Lew, R. Moreira, and N. Wamukonya. 2002. Renewable energy markets in developing countries. *Annual Review of Energy and the Environment.* 27: 309–348; McPhee, M. 1996. Bringing solar light to the world. *The Energizer, Solar Energy Association of Oregon.* 16(3): 1, 8. Available from *http://www.solaror.org/ftp/autumn96.pdf.* Center for Strategic and International Studies. 2001. *Energy in India: An overview.* Available from *http://www.geocities.com/~dvadi/articles/india/energy_1999.html* [Accessed September 17, 2004]; Audinet, P., and E. Fages. 1998. Energy policy of India: Cost differences of two scenarios. *Energy Policy.* 26 (9): 669–686; Prasad, Y. S. 1996. Energy situation in India: Present situation, future outlook, and outlook for nuclear power. *Atw-Internationale Zeitschrift fur Kernenergie.* 41 (1): 11–13; Jagadeesh, A. 2000. Wind energy development in Tamil Nadu and Andhra Pradesh, India: Institutional dynamics and barriers–A case study. *Energy Policy.* 28 (3): 157–168.

Early pricing laws in Europe required that utilities provide access to the electric grid to small-scale wind producers and other private generators. The laws also guaranteed producers a minimum share of the retail trade in Denmark and in Germany. The German system was revised in 2000, and today most of its pricing laws provide a fixed payment for renewable electricity that varies by technology type, plant size, and location and is based on the costs of generation.

A variation on pricing laws, called net metering, permits consumers to install small, renewable systems in their homes or businesses and sell their excess power to the grid. This excess power, under the law, must be purchased at wholesale market prices by the utility. In some cases, producers are paid for every kilowatt hour they feed into the grid; in other cases they receive credit only to the point where their production equals their consumption.

Quotas. Pricing laws establish the price and let the market response determine capacity and generation; mandated quotas (targets) work in reverse. The government sets a target and lets the market determine the price. Typically, governments mandate a minimum share of capacity (or generation of electricity), or a percentage share of fuel to come from renewable sources. The share required often increases gradually over time, with a specific final target and an end date. The mandate can be placed on producers, distributors, or consumers.

The simplest form of quota system is one in which the government imposes a mandate on one energy supplier. For example, during the 1990s the Minnesota Public Utilities Commission

Example 13–5
What Caused California's Power Meltdown?

California increased its capacity to generate electric power in the 1980s because of 1970s legislation that guaranteed markets for sources that supplied renewable energy. By the early 1990s, the result was a significant excess of generating capacity. Then while California slept, its electricity demand grew much more quickly between 1996 and 2000 than anybody expected. The excess capacity gradually disappeared as demand for electricity grew and no new generating capacity was completed.

In 2000 higher demand shifted demand curves for power to the right and pushed production even further up the supply curve, thereby increasing competitive prices further. No new generating capacity had entered the market to shift the supply curve outward and to moderate price increases. By mid-December 2000, wholesale market prices surged to \$0.40 per kilowatt hour. With no retail price increases permitted by the California Public Utilities Commission, the state's major utilities were approaching insolvency as they found it hard to pay for the huge, uncollected, wholesale power charges. By early 2001, it became clear that California regulators would not permit raising retail power prices by enough to restore the utilities' credit.

In mid-January two major utilities (Pacific Gas and Electric, PG&E, and Southern California Edison, SCE) announced that they had too little cash to pay power suppliers. PG&E and SCE were effectively insolvent. Supply shortages soon followed. By late January 2001, it was clear that if the state took no action, the lights would go out as suppliers refused to generate electricity without guaranteed payment. Since retail rates were constrained by law, the state began to buy power to meet some of the utilities' financial shortfalls.

California made major errors in its power industry: experience teaches to avoid making its five major mistakes: (1) It refused to allow utilities with the responsibility of supplying retail demand to enter into forward contracts. (2) It ignored proposals for demand response programs that would allow customers to respond to wholesale price spikes by reducing consumption. (3) It misread false gains from cheap power available during temporary excess capacity. (4) It assumed that the situation of excess capacity would prevail long enough to defer reforming the institutions that support investment in new capacity. (5) It paid too little attention to the need for investing in new generation and transmission facilities.

Sources: Joskow, P. 2001. California's Electricity Crisis. *Oxford Review of Economics and Policy.* 17 (3): 365–388; Borenstein, S., J. Bushnell, and C. R. Knittel. 1999. Market power in electricity markets: Beyond concentration measures. *Energy Journal.* 20 (4): 65–88; Faruqui, A., H. P. Chao, V. Niemeyer, J. Platt, and K. Stahlkopf. 2001. Analyzing California's power crisis. *Energy Journal.* 22 (4): 29–52.

ordered the electric utility Northern States Power to install successive amounts of wind energy capacity, thereby helping to open up the wind market in that state. Several European countries now require that a specific share of diesel fuel contain biodiesel, and Brazil has become the world leader in ethanol production and use by requiring that ethanol make up a specified share of all fuel sold. As with the pricing law, the additional costs of renewable energy under quota systems are paid through a higher rate charged to all electricity consumers. Thirteen U.S. states, covering 30% of the U.S. power load, have mandated quotas. Quota systems are also in use in several other countries, including Japan, the United Kingdom, Italy, and Australia.

It is hard to control the costs of implementing energy pricing laws, but the size of subsidies can be controlled under a quota system. For example, if guaranteed renewable energy prices are set too high, they can encourage high levels of development in renewables and dramatically increase electric rates. If they are not high enough, the policy will attract little development. In 1998 the Danish government paid more than 100 million euros in subsidies to renewable energy, and additional costs in Germany totaled 200 million euros in 2000 (Menanteau, Finon, and Lamy, 2003). Quota systems are often credited with encouraging competition and driving down the cost and price of renewable energy.

Standards

Industry standards are an effective method for promoting renewable energy. There are several types of standards: technology standards and certification, project siting and planning standards, grid connection standards, and building codes. Standards can generate greater confidence in a product, thereby reducing risks, which is important for financing. Technology standards for wind turbines, for example, include everything from turbine blades, electronics, and safety systems to performance and compatibility with transmission systems. Denmark adopted wind turbine standards in 1979, largely due to pressure from wind industry leaders. The standards are now credited with playing a major role in Denmark's becoming a leading manufacturer of wind turbines.

Siting and planning requirements can reduce opposition to renewables if they address potential issues of concern, such as noise and visual or environmental impacts. Laws can set aside specific locations for development or restrict areas at higher risk of environmental damage or injury to animals such as birds. Both Germany and Denmark have required cities to reserve special areas for wind turbines and have set restrictions on the turbines' proximity to buildings and lakes. These policies have successfully reduced uncertainty surrounding the location of the turbines and about the planning process.

Renewables face two challenges when it comes to interconnection with existing power grids. First, unlike conventional power plants that have flexibility in siting, renewable plants must be sited where the resources are located (e.g., wind turbines do best in windy places). Second, some renewables, such as solar and wind, are intermittent. Both challenges can be dealt with through creation of connection standards that encourage access to the power grid.

The use of regulated standards to increase the use of renewables is a less efficient method than price incentives or transferable permits. Despite their weak efficiency properties, however, standards present some clear advantages. They are easy to administer and produce predictable results and they give renewable energy suppliers a clear set of rules for planning.

Green Power

Although renewable energy is typically more expensive than conventionally generated energy, public opinion polls internationally consistently reaffirm that consumers say they are willing to pay higher energy costs if doing so will improve the quality of the environmental. Accordingly, a number of power marketers in several U.S. states and in many countries worldwide (see Lipp, 2001, and Morthorst, 2000, for the United Kingdom and Maeng, Lund, and Hvelplund, 1999, for Denmark) have packaged *green power* plans made up mostly of wind-generated electricity. They have marketed those plans to ratepayers in places that give consumers the right to choose their mix of power sources. As an example, if a household is willing to pay an increased power bill of $50 per month to consume a power mix of 50% renewables over the baseline mix of no renewables, they would make that choice through their dollar votes, much like choosing to buy organically produced grapes in the produce section of their local grocery store.

Eighty utilities in 28 U.S. states offer special packages of renewable energy to ratepayers at a premium. In these plans, green power costs from $0.04 to $0.20 per kilowatt hour more than conventional power. Because of higher costs, no more than 1.5% of the retail customers in any state have signed up for the independently marketed programs to date. So despite the noblest of intentions, there seems to be a great difference between what people tell pollsters about their willingness to pay for environmental quality and their willingness to pay when presented with the opportunity in the marketplace. Until renewable-supplied electricity costs become price competitive to those of conventional energy, green marketing programs face major challenges in attracting large numbers of buyers.

Voluntary Action

Since the early 1990s, much attention has been focused on the potential for internalization of external costs without any form of government intervention (e.g., Welch, Mazur, and Bretschneider, 2000). Because of the significant public good produced by reduced energy pollution, voluntary action is important but accompanied by major challenges. The challenges come from the lack of economic incentive to reduce externalities without coercion by governments. Voluntary pollution control backed by the threat of social sanctions sometimes works. Remarkably, some long-established pollution behavior is controlled with minimum government action. Take the case of trash. People usually do not throw their trash in the streets, perhaps from fear of retribution by neighbors.

A few innovative private companies might be expected to adopt technologies that internalize external costs for green markets in advance of government coercion. But economic theory and casual observation agree that most people only internalize under conditions where economic incentives make it profitable. Without a more complete understanding of the relevance of external costs of energy, voluntary action on a large scale by individual companies in the energy sector is unlikely to occur. Voluntary agreements by groups of cooperating companies are more plausible than agreements by individuals, because group agreements risk less harm to an individual's competitive position compared to others. Voluntary action might be expected to occur if the potentially regulated industry expects regulatory action and, learning from earlier experiences, initiates action to achieve abatement in forms it finds most acceptable.

Conservation

Improved energy efficiency is a major factor in evaluating the effectiveness and cost of energy policies designed to reduce pollution or limit climate change. Technological improvements and conservation methods can play significant roles in reducing energy use; therefore, understanding the process of conservation is of central importance in the formulation of energy policy. On the surface, energy conservation is a simple idea, acceptable to all. Unlike increased energy production, which may be unattractive for environmental reasons, energy conservation seems like a completely costless policy. Just about any poll will show widespread support for energy conservation: Anyone against conservation must support waste (Clinch and Healy, 2002; Dincer, 2003). In contrast to the simplicity of energy conservation as a slogan, economics places a higher standard of rigor on the definition of energy conservation. In economics, energy conservation is any reduction in energy use for which the benefits exceed the costs. That is, conservation occurs only if the programs or investments that support energy conservation incur a lower cost than the benefits they produce.

Conservation Investment Decision. Suppose you wish to find out whether it is good economics for you to incur costs to reduce the use of natural gas consumed by your home's space heater. Your house consumes 125 million BTU of natural gas for a typical heating season. Using your hometown's prevailing price of $4.50 per million BTU, you calculate that you use $562.50 (see U.S. Department of Energy, 2004d) of natural gas per season. You have discovered that a $400 investment in improved insulation will last 15 years, and saves 20% of your seasonal use of natural gas (i.e., 25 million BTU or $112.50). What economic principles can you use to determine if the insulation is a good investment? You need to look no farther than the net present value formula presented in Chapter 5. If you use an interest rate of 10% to discount your future benefits (natural gas costs saved) and costs (your $400 cash outlay), you will reach the following conclusions:

- The present value of net benefits over the 15-year period is a positive $568.18.
- The cost per unit of energy conserved is about $2.10 per million BTU, much less that the price of gas at $4.50 per million BTU.
- The simple payback period (i.e., number of years before the present value of net benefits turns from positive to negative) is just over four years, considerably less than the expected lifetime of the insulation.
- The benefit-cost ratio, found by dividing the present value of gross benefits (discounted value of energy saved) by the present value of the cost of producing those savings is a positive and healthy 2.4 to 1.
- The internal rate of return from investing the $400, the discount rate at which the present value of net benefits is zero, is an impressive 38% over the life of the investment.

This example illustrates the calculations that homeowners and businesses make when deciding whether to invest in energy conservation.[21] If the present value of benefits exceeds the present value of costs, economic forces reward people who make investments in energy conservation. However, if the net present value of the investment is negative, economically motivated people will decline to invest in energy conservation. However, people do not always respond to economic opportunities. For example, Metcalf and Hassett (1999) found that energy consumers with long-established habits of high use often resist economically attractive conservation measures.

Let us now suppose that the incremental social cost of current levels of energy use exceed the incremental social benefits. This means that society thinks it is a good investment to conserve energy. There are a variety of possible policy instruments for conservation, each with different efficiency implications. The two discussed next rely on price methods or on nonprice methods.

Price Conservation Methods. ***Fuel Taxes.*** A distinctive feature of European energy policy is its reliance on fuel taxes such as taxes on gasoline (petrol) and heating oil. Taxes on gasoline exceed $2.00 per gallon (about $0.50 per liter) in many countries. The widespread use of fuel taxes is due to the low cost and simplicity of collecting them as well as the large potential for revenue. Government taxes are much lower in the United States; total state and federal taxes average about 27% of the retail price of gasoline (U.S. Department of Energy, 2004e).

A fuel tax is an efficient method for promoting a given amount of fuel conservation, because it reduces the lowest-valued uses of energy rather than taking high-valued uses away from consumers, which could occur under command-and-control regulations. Despite the efficiency of fuel taxation, political opposition to fuel taxes has typically focused on two factors. First, fuel taxes have large, often undesirable, effects related to income distribution. Second, fuel demands are so price inelastic (at least when the retail price is low) that fuel taxes have little effect on reducing use. These two sources of opposition are connected: A more price-inelastic demand schedule produces a larger income transfer for a given tax level while producing a smaller demand reduction.[22]

Why are fuel and many other energy demands so price inelastic? In the short run, the fixed, existing stock of energy-consuming equipment makes it hard to reduce energy use once the capital stock is in place. The equipment's efficiency characteristics are already established, so the consumer has only the opportunity to vary the equipment's use rate. For example, in early 1973, U.S. car buyers were saddled with their big gas guzzlers, so their only way to conserve gasoline was to drive less. Motor homes sat unused under shade trees waiting for cheaper gasoline. Economic studies continue to find that the short-run price elasticity of gasoline and other fuels is very low. This means that fuel tax increases needed to reach a given conservation target are high and transfer considerable income from fuel buyers to the government.

Taxes on Energy-Using Equipment. Although there is little argument over the importance of increased energy conservation in reducing pollution and limiting greenhouse gas emissions, there is intense debate about cost-effectiveness and about government policies that should be pursued to promote energy conservation. One method of reducing energy use is taxing energy-using equipment, for example, taxing the existing supply of equipment and/or equipment with low fuel efficiency. In Europe a common practice is to charge consumers a large excise tax at the time they buy a new car. Furthermore, some countries, like France, have encouraged the use of more fuel-efficient autos through a horsepower tax. The European Union has debated introducing a vehicle tax based on the vehicle's carbon dioxide emissions, a factor in global warming (Commission of the European Communities, 2002).

---

[21] Sowsinski, Rothwell, and Shirey (1995) describe several institutional constraints to energy conservation in Poland.

[22] Vehmas et al. (1999) describe political difficulties of taxing fuels in the Nordic countries.

Taxes on energy-consuming equipment, unlike fuel taxes, typically fail to minimize the cost in displaced social benefits for a given level of fuel conservation achieved. A fuel tax works through all three channels of substitution: stock, fuel efficiency, and use. However, a tax on fuel-consuming equipment works through no more than two of those channels. For example, a tax on cars raises the price of cars, thereby causing a reduced size in the stock (number) of cars. But the tax would not necessarily improve efficiency of the stock or the fuel use rate.

Nonprice Conservation Methods. ***Direct Quantity Rationing.*** Government regulations can promote energy conservation by rationing fuel among users. Fuel deliveries are simply reduced to those sectors that are presumed to be low priority. If such regulations are democratic, the cutbacks are assigned proportionally to some level of baseline use (e.g., historical) across all sectors (e.g., industries). Of course, there are advantages and limitations of quantity rationing.

Figure 13–12 shows a demand schedule for a given fuel. Quantity Q_{0f} is current consumption and Q_{1f} is the economically efficient consumption level based on setting marginal cost equal to the price, P_{1f}. The difference $Q_{0f} - Q_{1f}$ (equal to ΔQ_f) measures the desired reduction in demand needed to achieve economic efficiency. If prices are used to promote conservation of ΔQ_f, a tax can be enacted that raises the price from P_{0f} to P_{1f}. As a consequence of the increased price, all fuel uses with a marginal social benefit exceeding P_{1f} continue to be bought and used, while all fuel uses with social benefits between P_{0f} and P_{1f} drop out. Did the price mechanism eliminate the lowest-valued fuel uses? The answer is yes. The loss in social benefits from conserving ΔQ_f units of energy is the area $Q_{0f}Q_{1f}CD$, the right-hand, vertical trapezoid area shown in Figure 13–12. No other reduction of existing energy use from the current level, Q_{0f}, produces a social welfare loss that small.

Suppose now that some nonprice rationing mechanism (e.g., a priority allocation rule) selected the units lying between q_{0f} and q_{1f}, thus conserving an amount equal to Δq_f (equal to ΔQ_f). The total loss in energy-consuming benefits of the priority allocation rule is $q_{0f}\,q_{1f}AB$, a welfare loss trapezoid considerably higher than that resulting from rationing energy use through the price system. We can conclude with confidence that the price system identifies and excludes the units lying at the margin (between P_{0f} and P_{1f}) and, therefore, finds the conservation method

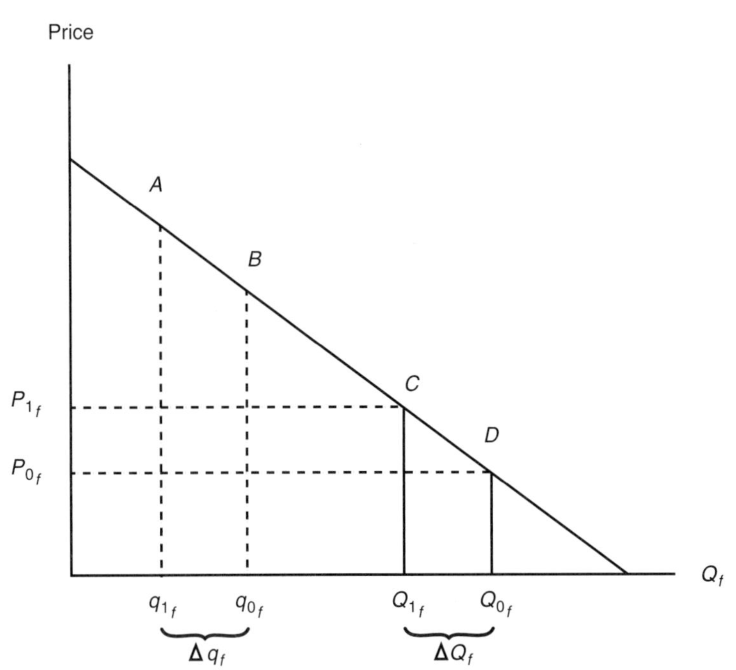

Figure 13–12
Competing energy conservation measures

that minimizes the loss in social benefits. So if the goal of conservation is to minimize the welfare loss of the amount of energy conserved, rationing the shortage through the price system is the most efficient method for doing it (Greene, 1998), because it minimizes the energy-consumption benefits that are lost.

Let us compare these results to a direct quantity rationing scheme. Could it also minimize the loss in social benefits from a given amount conserved? The answer is no, unless government planners have perfect knowledge and use it correctly. To ensure that only those uses with marginal social benefits of less than P_{1f} are rationed, planners must know each energy buyer's individual demand schedule. Knowing the overall aggregate market elasticity of demand is not enough, because quantity rationing occurs at the level of the individual buyer. Lacking this critical information, planners may try to simulate the market outcome by assigning certain fuel uses to be rationed as low priority, such as fuel for heating backyard swimming pools. (This happened in California in the 1970s.) Direct nonprice rationing assumes that the price elasticity of demand faced by all low-priority users is high, while price elasticities for all high-priority users are low. In fact, these elasticity conditions are unlikely to be true. Despite the inefficiency of nonprice restrictions on energy use, direct quantity rationing remains in widespread use because it avoids the politically undesirable income distribution impacts of raising energy prices. Consumers lucky enough to continue using the remaining quantity of fuel avoid paying the additional tax. Especially where the demand for energy is highly price inelastic (requires a large price increase to reduce use by a given amount), price rationing involves large income transfers from buyers to sellers.

Fuel Efficiency Standards. One popular nonprice method for energy conservation is a mandate that requires energy-using equipment to have certain minimum efficiency standards. For example, in the United States, laws were passed requiring auto manufacturers to sell a mix of cars that averaged 20 miles per gallon in 1980, gradually escalating to 27 miles per gallon by 1985. Similarly, a law could require home insulation standards to meet government requirements, thereby banning from the marketplace cheap, energy-inefficient home appliances such as air conditioners and dishwashers.

Governments enact fuel-efficiency standards for two reasons. First, like all nonprice energy conservation measures, standards avoid the politically unattractive income distribution effects of higher fuel prices. In addition, the standards correct market failures resulting from a lack of good information which encourages people to select equipment that is less fuel efficient. When appliances and cars that are low in fuel efficiency are banned from the marketplace, consumers are required by default to buy items that are fuel efficient. Supporters of free markets refer to such standards as paternalism, but supporters of the standards describe them as mandates for improved information. What do you think?

■ SUMMARY

Throughout history, humanity has developed new energy sources in the pursuit of comfort and productivity. Inexpensive, accessible, and reliable energy contributes to improved productivity and standards of living around the world. Lack of electricity and heavy reliance on traditional biomass are hallmarks of poverty in developing countries. Access to affordable, modern energy services is a prerequisite for sustainable development and for alleviation of poverty. A common objective of government energy policies around the world is to ensure secure, diverse, and sustainable supplies of energy at competitive prices.

Use of nonrenewable resources today for energy production reduces future supplies of those resources. The owner of a nonrenewable resource deals with this problem by optimizing production over many periods, recognizing that depletions in one period reduce opportunities in future periods. Firms attempt to maximize long-run profits over time. Maximizing those profits requires understanding user cost, which is the cost of using a marginal unit of a nonrenewable resource (i.e., maximum profit foregone by taking the resource from another period and using it now). The present value of a stream of future profits is maximized when *discounted* user costs are equated in all production periods. If marginal production costs are zero, Hotelling's rule states that the price of a nonrenewable resource must rise at a rate of r percent per year, where r is the discount rate. Hotelling's simple model can be extended to predict the price path of a nonrenewable resource according to (1) the size of the resource base, (2) the presence of a backstop fuel, and (3) the discount rate.

The Organization of Petroleum Exporting Countries (OPEC) was formed in 1960 with the aim of restricting producer competition in the world oil market. Although its early successes were limited, OPEC has been effective in achieving its goals since 1973. Four important economic forces have enabled the OPEC cartel to succeed: (1) shared interests among OPEC members, (2) supply response from competitors, (3) price elasticity of demand for OPEC oil, and (4) OPEC oil's income elasticity of demand.

More than two-thirds of world energy consumption comes from oil and natural gas. Over a horizon of 20 to 50 years, the real price of oil will likely rise, which will encourage greater production and consumption of energy from renewable resources. Nevertheless, reductions in the price of energy backstop technologies will reduce the price path of nonrenewable resources. Low current prices for nonrenewable resources—partly due to high supplies and partly due to government policies—are the greatest barrier to increased current use and increased development of renewables. Several policy designs will influence the pace and level of development of renewables: marginal cost pricing, environmental adders, energy and pollution taxation, promotion of renewable energy, standards, green power, voluntary action, and incentives for energy conservation.

Review Questions

1. Describe and define five energy policy objectives common to most countries.

2. Suppose you inherit an oil well containing reserves of 1 million barrels. You know with certainty that oil will be more valuable in 10 years when it will sell for $70 per barrel. Assuming the market rate of interest is 8%, for what price can you sell your reserves today?

3. Suppose in question 2 that you do not know future oil prices with certainty. To compensate for the risk attached to oil as an asset, you adopt a discount rate of 12%. In effect, if the risk-free return on treasury bonds is 8% you add a 4% risk premium. Now what is the discounted value of the oil well, applying a 12% discount rate?

4. Explain why a profit-maximizing producer must change the production time path until the present values of user costs are equated over time. To be more concrete, if $U_1/(1 + r) > U_0$ temporarily, how will economic forces change production, Q_0 and Q_1, in the two periods?

5. Do you think that Hotelling's rule can be applied to coal mining? Why or why not?

6. Suppose that expectations about future demand imply a $10 higher price at every output level than is currently the case. How would you find the new optimal output levels and price path?

7. Suppose the assumptions behind the numerical example in this chapter change. Instead of only 50 barrels of oil available for consumption, reserves are now 100 barrels. How would you solve for the new optimal output levels and the price path?

8. Suppose that the price intercept of the demand schedule in the numerical example in this chapter increased in the second period by 10% that is, from $50 to $55 (see Figure 13–2). What would this do to the equilibrium output path and to the equilibrium price path?

9. Explain why a conservation program that relies on price rationing of energy is more economically efficient than one relying on quantity rationing.

10. Explain how society's short-term and long-term economic welfare could be improved if it produced more energy from renewable sources, even if that action raised energy prices and reduced energy consumption.

11. Using Figure 13–7, explain what would happen to OPEC's output and to its total revenue if the cartel agreed to reduce the price from current P^1 to a P^2 that is higher than the current non-OPEC price, P^0. What would happen to OPEC's output and to its total revenue if the global demand for oil shifted out permanently from the current demand (*D* global) to a higher demand farther to the right? What would happen to the quantity of oil sold by non-OPEC producers?

12. Consider the example of testing the economic feasibility of investing in energy conservation that reduces the natural gas use of your space heater. How low does the price of natural gas have to fall per million BTU before the investment in conservation is no longer economically feasible? How high does the price of the investment itself have to rise before it is economically infeasible? Under what economic conditions can you expect people to make the largest voluntary investments in energy conservation?

13. What must occur for a program of public promotion of energy conservation to be successful? What are its relative strengths and weaknesses compared to direct quantity rationing and a fuel tax?

References

Atkins, F. J., and S. M. Tayyebi Jazayeri. 2004. *A literature review of demand studies in world oil markets.* Discussion paper, Department of Economics, University of Calgary. Available from *http://www.econ.ucalgary.ca/research/research.htm* [Accessed September 21, 2004].

Baranzini, A., J. Goldemberg, and S. Speck. 2000. A future for carbon taxes. *Ecological Economics.* 32 (3): 395–412.

Basosi, R., S. Maltagliati, and L. Vannuccini. 1999. Potentialities and development of renewable energy sources in an integrated regional system: Tuscany. *Renewable Energy.* 16 (1–4): 1167–1173.

Bird, F., A. V. Aleagha, and R. Ferroukhi. 1995. The economic-impact of subsidy phase-out in oil-exporting developing countries: A case study of Algeria, Iran and Nigeria. *Energy Policy.* 23 (3): 209–215.

Borthwick, W. K. D. 2000. The European Union approach to fuel cell development. *Journal of Power Sources.* 86 (1–2): 52–56.

British Petroleum. 2004. *Statistical Review of World Energy 2004*. Available from *http://www.bp.com/statisticalreview2004* [Accessed October 24, 2004].

Brophy, P. 1997. Environmental advantages to the utilization of geothermal energy. *Renewable Energy.* 10 (2–3): 367–377.

Brown, M. A. 2001. Market failures and barriers as a basis for clean energy policies. *Energy Policy.* 29 (14): 1197–1207.

Brown, M. A., M. D. Levine, W. Short, and J. G. Koomey. 2001. Scenarios for a clean energy future. *Energy Policy.* 29 (14): 1179–1196.

Charters, W. W. S. 2001. Developing markets for renewable energy technologies. *Renewable Energy.* 22 (1–3): 217–222.

Clinch, J. P., and J. D. Healy. 2000. Domestic energy efficiency in Ireland: Correcting market failure. *Energy Policy.* 28 (1): 1–8.

Commission of the European Communities. 2002. *Taxation of passenger cars in the European Union.* Available from *http://europa.eu.int/eur-lex/en/com/cnc/2002/com2002_0431en01.pdf* [Accessed October 24, 2004].

De Groot, H. L. F., E. T. Verhoef, and P. Nijkamp. 2001. Energy saving by firms: Decision-making, barriers and policies. *Energy Economics.* 23 (6): 717–740.

Dincer, I. 1999. Environmental impacts of energy. *Energy Policy.* 27 (14): 845–854.

———. 2000. Renewable energy and sustainable development: a crucial review. *Renewable and Sustainable Energy Reviews.* 4 (2): 157–175.

———. 2003. On energy conservation policies and implementation practices. *International Journal of Energy Research.* 27 (7): 687–702.

Dresselhaus, M. S., and I. L. Thomas. 2001. Alternative energy technologies. *Nature.* 414 (6861): 332–337.

Evrendilek, F., and C. Ertekin. 2003. Assessing the potential of renewable energy sources in Turkey. *Renewable Energy.* 28 (15): 2303–2315.

Eyre, N. 1997. External costs: What do they mean for energy policy? *Energy Policy.* 25 (1): 85–95.

Ezzati, M., and D. M. Kammen. 2001. Indoor air pollution from biomass combustion and acute respiratory infections in Kenya: An exposure-response study. *Lancet.* 358 (9282): 619–624.

Fielden, D., and J. K. Jacques. 1998. Systemic approach to energy rationalisation in island communities. *International Journal of Energy Research.* 22 (2): 107–129.

Greene, D. L. 1998. Why CAFE worked. *Energy Policy.* 26 (8): 595–613.

Habtetsion, S., and Z. Tsighe. 2002. The energy sector in Eritrea: Institutional and policy options for improving rural energy services. *Energy Policy.* 30 (11–12): 1107–1118.

He, J. K., A. L. Zhang, and Y. Ye. 1996. Technology options for CO_2 mitigation in China. *Ambio.* 25 (4): 249–253.

Hillring, B. 1999. The Swedish wood fuel market. *Renewable Energy.* 16 (1–4): 1031–1036.

Hotelling, H. 1931. The economics of exhaustible resources. *Journal of Political Economy.* 39: 137–175.

International Energy Agency. 2002. *World Energy Outlook: Energy and Poverty.* Paris, France: IEA Publications.

Jaafar, M. Z., W. H. Kheng, and N. Kamaruddin. 2003. Greener energy solutions for a sustainable future: Issues and challenges for Malaysia. *Energy Policy.* 31 (11): 1061–1072.

Johannson, T. B, K. McCormick, L. Neij, and W. Turkenburg. 2004. The potentials of renewable energy. Paper presented at the International Conference for Renewables, Bonn, Germany, January 2004.

Kaygusuz, K. 2003. Energy policy and climate change in Turkey. *Energy Conversion and Management.* 44 (10): 1671–1688.

Krautkraemer, J., and M. Toman. 2003. *Fundamental economics of depletable energy supply.* Discussion paper, Resources for the Future. Available from *http://www.rff.org/Documents/RFF-DP-03-01.pdf* [Accessed September 19, 2004].

Lew, D. J. 2000. Alternatives to coal and candles: Wind power in China. *Energy Policy.* 28 (4): 271–286.

Lipp, J. 2001. Policy considerations for a sprouting UK green electricity market. *Renewable Energy.* 24 (1): 31–44.

Maeng, H., H. Lund, and F. Hvelplund. 1999. Biogas plants in Denmark: Technological and economic developments. *Applied Energy.* 64 (1–4): 195–206.

McVeigh, T., D. Burtraw, J. Darmstadter, and K. Palmer. 2000. Winner, loser, or innocent victim? Has renewable energy performed as expected? *Solar Energy.* 68 (3): 237–255.

Menanteau, P., D. Finon, and M. L. Lamy. 2003. Prices versus quantities: Choosing policies for promoting the development of renewable energy. *Energy Policy.* 31: 799–812.

Metcalf, G. E., and K. A. Hassett. 1999. Measuring the energy savings from home improvement investments: Evidence from monthly billing data. *Review of Economics and Statistics.* 81 (3): 516–528.

Meyer, N. I., and A. L. Koefoed. 2003. Danish energy reform: Policy implications for renewables. *Energy Policy.* 31 (7): 597–607.

Miranda, M. L., and B. Hale. 2001. Protecting the forest from the trees: The social costs of energy production in Sweden. *Energy.* 26 (9): 869–889.

Morthorst, P. E. 2000. The development of a green certificate market. *Energy Policy.* 28 (15): 1085–1094.

Nystrom, I., and D. W. Cornland. 2003. Strategic choices: Swedish climate intervention policies and the forest industry's role in reducing CO_2 emissions. *Energy Policy.* 31 (10): 937–950.

Pandey, R. 2002. Energy policy modelling: Agenda for developing countries. *Energy Policy.* 30 (2): 97–106.

Pokharel, S. 2003. Promotional issues on alternative energy technologies in Nepal. *Energy Policy.* 31 (4): 307–318.

Pokharel, S., and M. Chandrashekar. 1998. A multiobjective approach to rural energy policy analysis. *Energy.* 23 (4): 325–336.

Radetzki, M. 2000. Coal or nuclear in new power stations: The political economy of an undesirable but necessary choice. *Energy Journal.* 21 (1): 135–147.

Renou-Maissant, P. 1999. Interfuel competition in the industrial sector of seven OECD countries. *Energy Policy.* 27 (2): 99–110.

Romm, J. J., and C. A. Ervin. 1996. How energy policies affect public health. *Public Health Reports.* 111 (5): 390–399.

Rubin, J., and P. Leiby. 2000. An analysis of alternative fuel credit provisions of US automotive fuel economy standards. *Energy Policy.* 28 (9): 589–601.

Saudi Arabia Government. 2004. *Saudi Arabia's petroleum policy.* Available from *http://www.mopm.gov.sa/html/en/oil_policy_e.html* [Accessed September 10, 2004].

Saudi Arabian Information Resource. 2004. *OPEC aims.* Available from *http://www.saudinf.com/main.d54.htm* [Accessed October 24, 2004].

Sawin, J. L. 2004. National policy instruments. Paper presented at the International Conference for Renewables, Bonn, Germany, January 2004.

Solow, R. 1974. The economics of resources or the resources of economics. *American Economic Review.* 64: 1–14.

Sowinski, J., G. Rothwell, and D. Shirey. 1995. Electric utility demand-side management in Poland. *Contemporary Economic Policy.* 13 (1): 84–91.

UK Government Performance and Innovation Unit. 2002. *The energy review: A performance and innovation unit report.* Available from *http://www.number-10.gov.uk/su/energy/1.html* [Accessed September 16, 2004].

U.S. Department of Energy. 2004. *Country analysis briefs, OPEC.* Available from *http://www.eia.doe.gov/emen/cabs/opec.html* [Accessed September 18, 2004].

———. 2004b. *International Energy Annual 2002.* Available from *http://www.eia.doe.gov/emen/iea/pet.html* [Accessed September 18, 2004].

———. 2004c. *International energy outlook.* Available from *http://www.eia.doe.gov/oiaf/ieo/index.html* [Accessed September 14, 2004].

———. 2004d. *Natural gas weekly update.* Available from *http://tonto.eia.doe.gov/oog/info/ngw/ngupdate.asp* [Accessed October 24, 2004].

———. 2004e. *A primer on gasoline prices.* Available from *http://www.eia.doe.gov/neic/brochure/oil_gas/primer/primer.htm* [Accessed October 24, 2004].

———. 2004f. *Solar photovoltaics.* Available from *http://www.eia.doe.gov/cneaf/solar.renewables/page/solarphotv/solarpv.html* [Accessed October 24, 2004].

Van Horen, C., and G. Simmonds. 1998. Energy efficiency and social equity in South Africa: Seeking convergence. *Energy Policy.* 26 (11): 893–903.

Vehmas, J., J. Kaivo-Oja, J. Luukkanen, and P. Malaska. 1999. Environmental taxes on fuels and electricity: Some experiences from the Nordic countries. *Energy Policy.* 27 (6): 343–355.

Welch, E. W., A. Mazur, and S. Bretschneider. 2000. Voluntary behavior by electric utilities: Levels of adoption and contribution of the climate challenge program to the reduction of carbon dioxide. *Journal of Policy Analysis and Management.* 19 (3): 407–425.

Chapter 14

POPULATION

Outline

Population and the Environment 397
Population and Demography 399
Macroeconomic Forces 400
Microeconomic Forces: Economic Theory of Fertility 407
Emerging Questions 413

Objectives

After studying this chapter, you should be able to

- Describe the relationship between population and the environment
- Identify demographic concepts for understanding population problems
- Describe macroeconomic connections between population growth and economic opportunities
- Show how the microeconomic theory of fertility explains desired family size
- Identify emerging population policy questions and knowledge needs

"Be fruitful and increase in number; fill the earth and subdue it."

The Bible, Genesis

"Population, when unchecked, increases in a geometrical ratio. Subsistence increases only in an arithmetical ratio. A slight acquaintance with numbers will show the immensity of the first power in comparison of the second."

Malthus (1798)

Introduction

Until about 1800, disease, war, epidemics, poverty, unclean living conditions, and a disordered society made many lives dangerous, risky, and often short. Because of resulting high death rates and low life expectancy, it took all of human history to reach the first billion people by the year 1800. Now it takes less than a decade for each added billion. Today, 90% of the current population increase takes place in developing countries, many of which cannot feed their own people. In 1993, Norwegian Prime Minister Gro Harlem Brundtland, in a lecture delivered to the United Nations, cautioned that these numbers may blind us to the fact that these people are individuals—people with rights, and inherent dignity—but who in many parts of the developing world lack access to health, education, and other resources to reach their full potentials (Brundtland, 1993). Increasing the number of people who face malnutrition and poverty deprives hundreds of millions of their full potential.

The biblical quotation suggests that a larger population is needed, even desired. People with a less optimistic view, such as Malthus, worry that a larger population aggravates crowding, stresses the environment, and adds to human misery. The debate about whether growing numbers of people are desirable is unlikely to disappear soon.

The pessimistic views of Malthus continue to be discussed by economists, sociologists, demographers, and political leaders. On one side of the debate, his critics believe that increases in human knowledge and technological innovation will enable the earth and its resources to provide rising standards of living despite growing populations. Optimists pointing to the notable successes in improving human welfare and the pessimists pointing to the pervasive poverty in developing countries and the continued stress to the environment.

■ Population and the Environment

The connection between population growth and environmental impact seems obvious at first glance: More people consume more resources, damage more of the earth, and generate more waste. As nations become larger or wealthier, consumption increases. This simple reasoning is true, but the larger picture is more complex. The connection between population and the environment is complicated and, for the most part, still poorly understood. Population and environment issues are based in different fields of study. Social scientists are overwhelmed by the encyclopedic knowledge they attribute to natural scientists, while natural scientists often have only a sketchy view of the social sciences, combined with a fear and mistrust of the political process. Few people specialize in studying population and the environment as an integrated discipline. Furthermore, people's use of the environment is an issue of major political and economic importance, and of considerable controversy (Davis, 1990).

Despite the complexities of integrating population and the environment, one simple identity introduced by Ehrlich and Holdren (1971) and currently used widely by many environmental scientists expresses a simple relation between environmental impact, such as pollution, and human population:[1]

$$\text{Environmental impact} \equiv \text{Population} \times \frac{\text{Income}}{\text{Population}} \times \frac{\text{Environmental impact}}{\text{Income}} \qquad (14.1)$$

$$\textit{dependent variable} \equiv \textit{term } 1 \times \textit{term } 2 \times \textit{term } 3$$

The triple equals sign means that the expression is an identity, not simply an equation. That is, the terms in the numerators and denominators cancel out, so the equation is always true. Since it always must be true, it cannot be used as a stand-alone test of theories. Even a completely false theory of environmental policy would look true if judged by this identity alone. Still, the expression offers a structure for thinking systematically about factors that affect the environment.

The term on the left, the dependent variable, is *total* environmental impact. It could be measured as the numerical value of some pollutant, such as sulfur dioxide released into the air, or the total water pollution in some region. Greater amounts of environmental impact occur with increases of any of the three terms on the right. The first term on the right is population, the second is average income per capita (affluence), and the third is environmental impact per unit of income produced (technology). Because the three terms are multiplied by each other, and not merely added, a proportional increase in any one of them (e.g., doubling) increases total environmental impact by that proportion. A doubling of all three terms on the right increases impact by a factor of eight. Thus, this equation shows the magnifying power on environmental impact resulting from simultaneous increases in all three right-hand terms.

Equation 14.1 provides a framework for analyzing the *environmental impact* of *population* growth because it separates effects of population growth alone from effects of the other two terms on the right. Three important influences on those two terms are economic changes, regulatory programs, and changes in technology.[2] Despite the fact that this equation is simply an accounting identity, it has three important uses: structural analysis, forecasting, and policy evaluation.

---

[1] Another way to express the identity is $I = PAT$, where I is impact, P is population, A is affluence, and T is technology. Environmental impacts can be expressed in terms of discharges, measured as a flow per unit time of an emission from a source. Another way to express environmental impact is in ambient condition of the emission that exists at a point in time, which is a stock. Other examples of flows are incomes earned per unit time and inflow of water into a reservoir per unit time. Stocks can be measured at a single instant in time, such as personal wealth or volume of water in a reservoir.

[2] Cramer (1998) shows how this equation can be used for environmental policy analysis when analyzing factors influencing air quality in California. Preston (1996) uses the equation to compare environmental impacts from agriculture and those of industry.

Structural Analysis

In structural analysis economists estimate the underlying relationships connecting population and the environment. One way to do this is to measure what factors influence each of the three right-hand terms in Equation 14.1. The size and sign of these influences are typically expressed as *elasticities.*[3] For example, if we can measure the sensitivity (elasticity) of population levels (term 1) with respect to the major factors that influence it, we know how each of these factors influences total environmental impact, other things being unchanged. If we are able to measure the elasticities of per capita income (term 2) with respect to education and natural resource supply, we know how a given percentage increase in either of these two will influence overall environmental impact. (For example, if the elasticity of per capita income with respect to education is 0.2, a 10% increase in average education level increases per capita income by $0.2 \times 10 = 2\%$.) If we can measure the elasticity of technology used (environmental impact per unit of income produced, (i.e., term 3) with respect to its determinants, we can infer how these determinants affect overall environmental impact. Structural analysis improves our insights into the factors that contribute to environmental impact.

Forecasting

In this chapter forecasts attempt to predict values of the dependent variable, total environmental impact, outside current experience. Typically, the forecast is a prediction based on other times, or places, or situations than have been experienced historically. We assume that the desired forecast is quantitative, explicit, and unambiguous, and therefore verifiable in the sense that there are possible outcomes that could validate or refute the forecast. Forecasted impacts help us prepare for the future or take the actions needed to accommodate these impacts.

Forecasters of future environmental impacts recognize that the three terms on the right-hand side of Equation 14.1 may have *compensatory* effects; that is, increases in one of the three right-hand terms may be offset by opposite decreases in one of the other terms. For example, if population in a given land area doubles (term 1), pollutants into the air and water may increase if there is no change in environmental technology (term 3). People may notice these increased impacts and demand that their elected representatives enact laws that require businesses and consumers to use more environmentally friendly technologies. The effect of these laws could be a reduction in the technology term (term 3) to half (or less) of its original level. So even though population doubles, people take control over their lives and change their environmental laws and institutions. The net effect of a doubling in the population (term 1) could be that pollutants per unit income (term 3) fall by half, thereby producing an overall impact of no net change in total pollutants.

Economists who use Equation 14.1 to forecast future impacts recognize that all three right-hand terms are controllable by people who are motivated to alter environmental changes produced by growing population. Moreover, the richer a society is, the more it is willing to invest in institutions, laws, and development or use of technologies that reduce the size of the technology term (term 3) to compensate for growing population. Many industrialized countries have found they can accommodate considerable population growth with little increase in total environmental impact.

Policy Evaluation

Perhaps the most important use of Equation 14.1 is for policy evaluation in a situation in which an analyst or decision maker must choose one policy, called a plan, from a given set of proposed policies. Policy evaluation is closely related to forecasting, and as in the case of forecasting, we

---

[3] Elasticity is the percentage change in the dependent variable from a 1% change in the independent variable. For example, suppose the dependent variable currently has a value of 100 and the elasticity equals 2. If the independent variable increases from 1,000 to 1,010 (a 1% increase), then the dependent variable will increase from 100 to 102 (a 2% increase).

assume in this chapter that the policy choice is quantitative, explicit, and unambiguous. A forecast is based in part on assumptions concerning the actions of decision makers. Similarly, policy evaluation is based in part on forecasts of the effects of each policy choice. The tasks of environmental impact forecasting and policymaking could be combined in an organization because both tasks involve making forecasts and evaluating proposed policies.

Assume that the aim of environmental policy evaluation is to design policies that mitigate negative environmental impacts of growing populations. Policies can be designed to alter any of the three right-hand terms of Equation 14.1 or they can be designed to alter all three, thereby lessening the overall environmental impacts of factors that would influence any of the three alone. For example, if population (term 1) and per capita income (term 2) are both growing due to underlying forces in the economy, policies could be enacted that reduced the environmental impacts of existing technologies (term 3). As an example of such a proactive environmental policy, consider California, where about one-third of all new U.S. migrants from foreign countries settle (see Cramer, 1998). Southern California's legendary smog became an urgent problem in the 1960s. Population growth in the state, mostly from immigration, had been high since 1940, quadrupling from 6.9 million in 1940 to more than 30 million in 1990. If air pollution per unit income (term 3 in Equation 14.1) had stayed constant, this huge increase in population combined with rising per capita income, would have made the smog intolerable. But stringent air quality regulations enforced by the state beginning in the early 1970s reduced term 3 considerably. By the 1990s the smog problem began to improve. The message is clear: Environmental policies can be designed to reduce the value of term 3 of Equation 14.1 enough that overall environmental impacts—even in the face of growing population—can be reduced, not increased.

■ Population and Demography

Much of the scientific work on population is done by demographers. This section introduces several terms demographers use to communicate their data. A key concept is *fertility rate*. Fertility is said to be at *replacement* when each generation of women exactly replaces the previous one. Typically, the replacement number is slightly larger than 2. The idea of replacement is a central factor in population projections, because it is the fertility level that, if maintained over time, produces zero population growth (assuming a constant level of mortality and no net migration). Positive or negative changes from replacement lead to population growth or decline, respectively. Currently, replacement fertility equals 2.4 births per woman in developing countries and 2.1 in industrialized countries. These levels are larger than 2 because children who die before reaching their reproductive years must be replaced with added births, and because the ratio of males to females born is slightly larger than 1. Figure 14–1 shows fertility trends for the United States and Figure 14–2 shows trends for selected other countries.

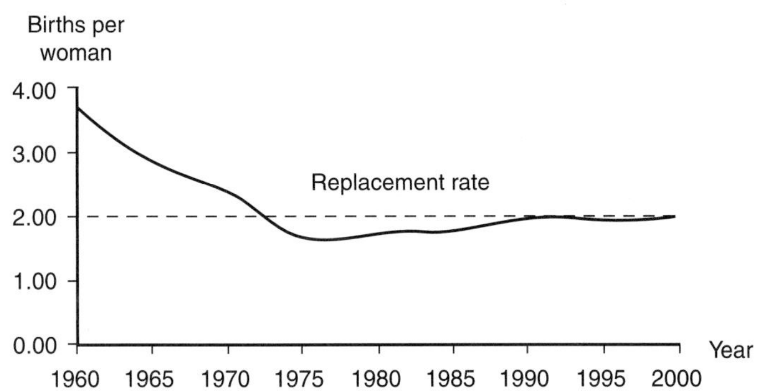

Figure 14–1 U.S. fertility rates, 1960 to 2000 Source: World Bank 2001. *World Bank Development Indicators.* Washington, D.C.

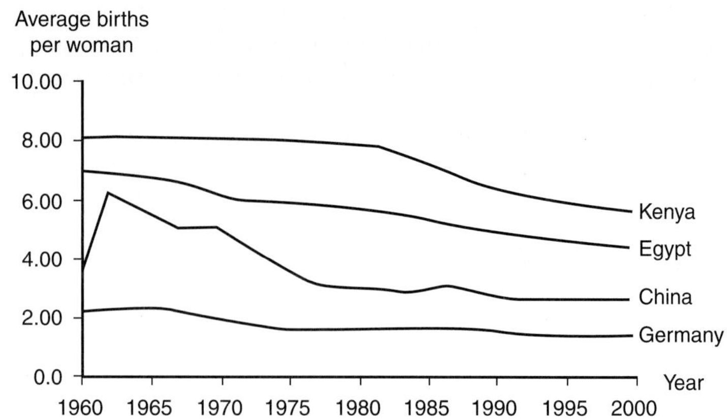

Figure 14–2 Selected national fertility rates, 1960 to 2000
Source: World Bank. 2001. *World Bank Development Indicators.* Washington, D.C.

Although there have been large declines in fertility rates in most countries since 1950, fertility rates in the year 2000 remained well above the replacement level in most developing countries, ranging from a high of 5.1 in Africa to 3.4 in south Asia and 2.7 in Latin America (World Bank, 2001).[4] These data indicate why fertility is a major force in developing countries. By contrast, as of the year 2000, fertility was below replacement in Europe (1.4), North America (1.9), and east Asia (1.8).

Table 14–1 shows population projections by the World Bank for the years 2000 and 2100 based on fertility and mortality projections of Bongaarts and Bulatao (1999). Worldwide population is projected to increase from 6.07 billion to 9.96 billion. Falling populations are projected for industrialized countries (Europe and North America); in contrast, the total population of developing countries is projected to increase from 4.89 billion to 8.86 billion.

Although the peak in absolute growth will likely occur sometime late in the twenty-first century or beyond, the rate of growth reached its maximum in the late 1960s and has been declining since then. A declining growth rate is consistent with a growing absolute gain in population because the base population to which the growth rate applies keeps growing. Population will likely more than double[5] before stabilizing, but the declining growth rate is consistent with claims that efforts to reduce population growth have had an impact (Bongaarts and Amin, 1997). Recent data about population trends suggest that the global population is going to decline in part because AIDS in Africa is slowing population growth there (United Nations, 2004).

Macroeconomic Forces

Population growth and economic growth *affect each other* (Cassen, 1993). Increases in population affect the volume and composition of output; and as economic growth takes place, changes occur that influence mortality and fertility. Nevertheless, for analytical purposes, we will consider the effect of economic growth on population growth separately from the effect of population growth on economic growth.

---

[4] For explanations, see Bongaarts and Watkins (1996) and Bongaarts and Bulatao (1999). Davis (1963) describes how the fertility rate fell in Japan from 2.7 in 1920 to 0.99 in 1959 in ways that mirrored decreases in western Europe and the United States. A revealing paper by the Chinese Government (1996) explains the Malthusian roots of its widely publicized one-child policy.

[5] The number of years it takes a population to double can be approximated by the famous law of 72, a law that says to divide 72 by the annual growth rate in percent terms. For example, the number of years to double a population growing at 6% per year is 72/6 = 12 years.

Table 14–1 World population projections for 2000 and 2100

	Projected population (billions)	
	Year 2000	Year 2100
East Asia	2.04	2.70
South Asia	1.48	2.69
Middle East and North Africa	0.33	0.73
Sub-Saharan Africa	0.67	1.98
Latin America and Caribbean	0.52	0.88
Northern America	0.31	0.37
Europe	0.73	0.61
Developing countries	4.89	8.86
Industrialized countries	1.18	1.11
Worldwide	6.07	9.96

Source: *World Bank 2001*. World Bank Development Indicators. *Washington, D.C.*

Effect of Economic Growth on Population Growth

A nation's economic growth can influence its population growth through various channels. This section examines some of the patterns linking economic growth and population growth seen around the world since the mid-nineteenth century.

The Demographic Transition. Most industrialized economies and more recently many developing economies, have passed through four stages of what is called the *demographic transition.* During the first stage, before the nation modernizes, birthrates are high and stable and somewhat higher than death rates, which ensures positive but small population growth.

In the second stage, the nation begins to modernize; death rates fall quickly and there is no reduction in birthrates. Death rates fall because of improvements in nutrition, hygiene, sanitation, and health care and as a result of the control of various causes of death, especially communicable diseases. Three main causes of decline in mortality in the twentieth century for developing countries are (1) reduced famines as a result of better agriculture and food distribution systems, (2) reduced death rates due to the control of disease, and (3) gradual improvement of general conditions of living and health (Cassen, 1976). The reduction in death rates produces significant increases in lengths of lives and a corresponding increase in the population growth rate. The second stage first occurred in western Europe and North America, where the period lasted about 50 years; other countries have since passed through this stage much faster.

The third stage involves significant declines in the birthrate. Recall that the birthrate was originally higher than the death rate, but in the third stage birthrate falls quickly and becomes nearly as low as the death rate toward the end of stage. (Montgomery, 1999, examines factors affecting the length of the lag between falling mortality rates and falling fertility rates.) Stage three results in continued increases in length of lives but lower population growth rates than stage two because birthrates are much lower. The fourth stage occurs when birthrates fall to the population replacement rate, that is, slightly more than two births per woman. Birthrates are slightly higher than death rates and both are stable. Ultimately, as the demographic transition runs its course, population stops growing and may actually fall.

Patterns of the Demographic Transition. The demographic transition is one of the best documented generalizations in the social sciences. According to the theory, societies that experience modernization change from traditional conditions of high fertility and high mortality to conditions in which both fertility and mortality are low.[6] Mortality decline since the late 1800s has been unprecedented in human history. Compared to fertility decline, reduced mortality is much easier to explain: reduction of epidemics through vaccinations and better hygiene, improved diagnosis and treatment of disease, efforts to reduce famines, fewer deaths from violence and civil wars, reductions in infant mortality, and improved standards of living. Fertility decline has also occurred, but an understanding of its underlying mechanism has proven much more elusive. Serious work began in the late 1920s by Thompson (1929), who divided countries into those with falling populations, rising populations, and rapidly rising populations, but he looked for few theories to explain the differences. Landry (1934) considered fertility reductions beyond explanation, except in the mechanical sense of increased use of contraception. He rejected the theory that the decline was due to physiological factors or to moral corruption, and he presented evidence that various forms of birth control were in general use in France as early as the late 1700s.

Notestein (1944) attempted a better understanding of the fertility decline in Europe. He observed that urban life took away from the family many functions in production, consumption, recreation, and education. In jobs and in society individuals were judged on their own accomplishments, not on those of the family. Newly mobile young people were not influenced as much by outside forces like the family and community. In a period of rapidly developing technology, new skills were needed and new opportunities arose for individual advancement. Education and a rational (nonsuperstitious) view of managing life became more common. The cost of child rearing increased and the economic contributions of children to the family declined. Falling death rates increased the number of family members who had to be supported and lowered the need for as many births. Women found new independence from household obligations and new economic roles less compatible with childbearing.

Despite many objections, qualifications, and doubts about the demographic transition just described, the force of the generalization remains. Its greatest strength is the prediction that the transition will occur in every society that is experiencing modernization. Its greatest weakness is its inability to forecast the threshold level of mortality or other social mechanisms that switch fertility to its downward path. Similarities between the European experience and modernization in many non-European countries can be explained by transition theory. In fact, the United Nations and the World Bank have based their population forecasts on the assumption of a standard transition.

When transition theory is applied to the experience of non-European countries, the patterns are impressive. In every instance mortality has declined first and has been followed by declining fertility, and the result in each case has been a considerable acceleration of population growth. The transition has become much more rapid and concentrated into shorter periods in non-European countries than it was in the European experience. A summary of what has been learned to date from experience outside Europe can be stated as follows (for more details, see Knodel and van de Walle, 1979):

- Fertility declines took place under a wide variety of social, economic, and demographic conditions.
- Family limitation was not practiced and was probably unknown among broad sections of the population before the decline in fertility began, even though a substantial proportion of births may have been unwanted.

---

[6] For an excellant review of the demographic transition, see Kirk (1996). He observed that the demographic transition is a matter of considerable controversy among demographers. For some, transition theory lies at the center of modern scientific demography; they recognize it as the central issue and theory of modern demography whose explanation has important significance for the future of the planet. To others, transition theory provides little theoretical insight and should be treated as an unproven generalization worthy of little serious discussion. Looking back to the theory's early development in the 1940s, Kirk notes that it is still unresolved whether the demographic transition is a theory, a generalized set of facts, a framework for analysis, or just an idea. The idea may have survived simply because no better theory has taken its place to explain demographic behavior in the modern world.

- Increases in the practice of family planning and the decline of marital fertility were essentially irreversible processes once under way.
- Cultural settings influence the onset and spread of fertility decline independently of socioeconomic conditions.

Several generalizations can be made about mortality and/or fertility as parts of the demographic transition toward modernization.

1. Mortality and fertility transitions have both experienced increasingly faster paces. For example, a mortality transition that took 75 to 100 years in northern Europe was achieved within 20 to 25 years in eastern Europe and within even shorter periods in less developed countries whose transitions occurred much later. The presence of an income effect on mortality has been shown, especially at lower income levels; that is, higher income reduces the death rate and prolongs life.

2. Some economists see reduced mortality as an absolute requirement of becoming modern. Views of life change and people begin to believe that it is possible to influence one's fate. People value individual achievement and a willingness to take personal risks. Historically, the decline in mortality came in three time periods (Gage, 1993): (1) reduced death rates as a result of greater public order and reduced violence beginning in the late eighteenth century, (2) medical discoveries and sanitary improvements up to World War I, and (3) the growth of antibiotics after World War II, initiated by the discovery of penicillin.

3. The dominance of economic factors was reflected in the early explanations of transition theory. A central theme in early discussions was that modernization changes the economics of childbearing and makes it seem economically disadvantageous. This led to the formulation of economic theory of fertility, which is discussed later in this chapter.

4. Government played a role of great importance in the decline of mortality and fertility. Increasingly active governments and the rise of the nation state brought public services that reduced mortality: provision of pure water, vaccination, control of epidemics and famines, and other public health measures. More widespread and sustained peace in the modern state reduced deaths from tribal and other warfare. Through better policing, government reduced violence and increased public safety.

5. Government influenced fertility levels. Compulsory schooling increased the cost of children, as did measures to outlaw child labor. The changing structure of the economy created more jobs suitable for women, including jobs in government.

No two countries have followed identical paths in demographic transition, because there are many possible combinations of marriage patterns, fertility, mortality, and migration at each stage of the transition. However, diversity is not irreconcilable with the universal truth of the transition. It is tempting to suggest that it was the decline of mortality that led to the fertility transition, but more than anything else reduced mortality reduced fatalistic attitudes, influenced the rise of the industrial economy, fostered the explosion in science, and continued belief in power of government to improve human lives.

Effect of Population Growth on Economic Growth

Policymakers pose many questions regarding the relation between the size of population and the size of the economy. Does a larger population improve or worsen the economic opportunities faced by a nation's people? Does the answer to this question depend on whether the nation is a rich or poor nation, crowded or sparsely populated, industrial or agricultural? Many nations that until recently have had high population growth are experiencing much slower growth rates. What is the effect of this change on their future economic growth and on their policies for influencing that growth? See Example 14–1.

Example 14–1
Controlling Population: Whose Welfare Counts?

Population researchers commonly assume that fertility control programs are desirable if they result in fewer people who individually are better off, but welfare economic theory questions that conclusion. Can we say with confidence that 99 people each enjoying real income Y is a better social state than 100 people each enjoying real income $Y - u$, where u is some positive number? Economics question the conclusion because two societies being compared have different members. However, in the society of 99 of 100, the 99 people would answer yes from their point of view. The problem lies with the 100th person: She may be very pleased indeed that she enjoyed $Y - u$ real income and would resent any statement that she should have never been brought into the world. This question is debatable in societies where Y is high and u is low, but it may be less debatable when the program at stake could alleviate human deprivation (low Y and a large u) on a large scale: If added people brought into the world without the program made the average situation much worse, few would consider it any improvement that the number of people enjoying this much lower state had gone up. And many of those added people compelled to receive the low income ($Y - u$) may themselves wonder about their fate. Concern for the lost welfare of averted births, as opposed to the welfare of those actually born in the past, present, and future, may be an unaffordable luxury in a poor, crowded nation.

Another welfare problem relating to population density occurs when the current generation shifts the high costs of increasing family size to the next generation. Present parents may benefit considerably from their children, especially in their later years; but the children themselves, especially the oldest, may wish they had fewer siblings and less income owed to the parents. Each member of the younger generation tolerates his current burdens as temporary and looks forward to his large number of children to compensate for those current burdens. One possible solution to the problem lies in the spirit of Kant's categorical imperative: Institutions should be designed that encourage each generation to procreate for the future as it would have wished the past generation had procreated for itself.

On the surface, this debate over population welfare seems like simple mental gymnastics. However, the debate is genuinely real for governments concerned about their future in the face of crowded conditions with rapid population growth. If the fertility of the present generation blocks a better life for future generations, to what extent should governments invoke the coercive power of the state in reducing current fertility?

Sources: Kant, I. 1785. *Fundamental Principles of the Metaphysic of Morals.* Translated by T. K. Abbott. Manchester, England: University of Manchester Press, 1955; Cassen, R. H. 1976. Population and development: A survey. *World Development.* 4 (10, 11): 785–830; Caldwell, J. C. 1978. A theory of fertility: From high plateau to destabilization. *Population and Development Review.* 4 (4): 553–577.

Let us begin with the simplest concept in the study of population—an idea going back as far as English economist Malthus—the idea of *diminishing marginal returns* to labor. With a given technology, fixed amounts of land and other natural resources, unchanged stocks of capital goods, and increased use of labor will lead to greater economic output, but after some point, diminishing increments of output will occur (Olson, 1998).[7] Still, the knowledge that a larger population produces more absolute output, a positive marginal product of labor, is not completely satisfying. The more demanding question is whether a larger population increases the average person's welfare. The answer to this question requires knowing whether the *marginal product of labor* (MP_L) from a growing population is larger than the *average productivity of labor* (AP_L). If $MP_L > AP_L$, then a growing population helps the average citizen economically; otherwise it does not. One easy and clear way to examine the impacts of growing population on a nation's economic welfare is to write a simple definition of the nation's output:

$$Y \equiv \frac{Y}{N} \times N \tag{14.2}$$

[7] The economic analysis of overpopulation rests on the law of diminishing returns: Without diminishing returns, the added labor produced by population growth would increase output in the same proportion. Even problems of pollution are special cases of the law of diminishing returns, since these problems are caused by diminishing returns to the environment's capacity to absorb wastes.

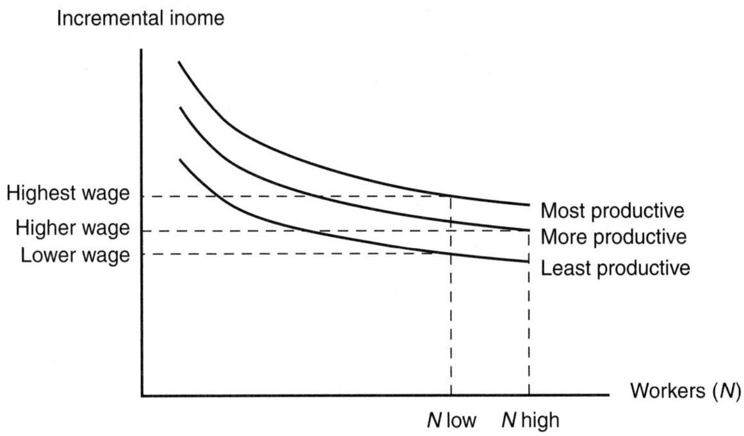

Figure 14–3
Workers, income, and technology illustrate how added capital shifts marginal productivity

This identity says that total output (Y) must equal output per worker (Y/N) times the number of workers employed in the nation (N). This equation can be changed to express the effects of a nation's population size by dividing both sides by total population (POP):

$$\frac{Y}{POP} \equiv \frac{Y}{N} \times \frac{N}{POP} \tag{14.3}$$

Equation 14.3 says that average output per capita (Y/POP) must equal the product of two important factors: output per worker (Y/N) and the share of the nation's population in the labor force (N/POP). Each of these two factors operates through a separate path in which increased population affects economic growth.

Output per Worker and Population Growth. Output per worker, Y/N in Equation 14.3, depends in a major way on a nation's stock of capital. In any one year, a nation is equipped with a stock of capital (i.e., man-made resources such as machines, factories, roads, dams, and irrigation facilities) that it has inherited from past investments. Each year's past investments were savings from that year's income, so for any time period

$$S = I$$

where S = savings and I = investment. Given the total inherited stock of capital at a point in time from all previous years, the current year's national output depends on the volume of labor employed (N) as well as the total stock of capital (C). Greater labor employed, even with the given stock of capital, will produce more output, but at a decreasing rate (see Solow, 1988). Equal successive increments of labor employed will generate successively smaller increments to output (Y). The smaller increments to output ($\Delta Y/\Delta N$), can be attributed to diminishing returns of labor applied to a fixed capital stock.

If a nation invests anything, its economy will have more capital at its disposal in the following period, since investment produces an equal increase in capital stock. In Figure 14–3, each of the three curves is a marginal product of labor for a different fixed capital stock. The middle curve relating output (Y) to employment (N), labeled as "more productive," lies completely above the bottom curve, labeled as "least productive." A vivid example of the power of added capital to shift up labor's marginal productivity to a higher curve is illustrated in the year-by-year increase in the power of personal computers. Each succeeding year a new person hired has a faster and better computer and hence can produce more output (Y) with the same human effort (N). The same volume of labor employed (N) will produce more output (Y) one year later simply because the previous year's savings was invested in new capital. Another example of capital improving the productivity of labor comes from agriculture. In a rural economy dependent primarily on farming, food production per worker can continue to grow if the quantity of tractors, roads, irrigation facilities, and other complementary capital increases at a greater percentage rate than the number of workers themselves.

The important macroeconomic principle at work here is that current savings provide the resources for current investment that adds to a nation's capital stock. For a given population and labor force, this gain in productivity expresses itself as added output per capita (*Y*/*POP*). Higher savings provide greater additions to the capital stock, and for that reason produce greater gains in per capita income.[8] Faster growing population rates have a retarding influence on savings, because they also have a higher percentage of people who are not employed in the labor force. Therefore, the population requires a higher percentage of income devoted to consumption and less devoted to savings. Because savings is a critical ingredient to a nation's economic growth, the relation between a nation's population's characteristics and its savings is important.

Age Structure of the Labor Force. Population growth has a direct effect on the percentage of a nation's population employed in the labor force (*Y*/*POP*) through what is called the *age structure effect* that results from changes in the age distribution of a growing population. Perform a mental experiment in which a nation's current population could grow at two possible rates, growing quickly and growing slowly. The faster growing population would have a higher percentage of younger people (couples are having larger numbers of children) and smaller percentage in the prime working ages. The higher population growth rate reduces per capita income (*Y*/*POP*) by decreasing the percentage of the population in the workforce (*N*/*POP*).

A study by Coale and Hoover (1958) explored the connection between the age structure of a population and its ability to accumulate capital and raise living standards for its citizens. Their findings support the theory that populations with lower percentages of people in the prime working ages save smaller percentages of their incomes, thus providing less capital stock for increasing future worker productivity. Figure 14–3 shows that a growing labor force produced by a growing population can still produce higher per capita incomes if the capital stock grows fast enough. But a stable population of zero growth fares even better. A no-growth population with a constant supply of labor will earn even higher per capita income, because a stable population saves a higher percentage of its income than a growing one. This higher percentage of income saved from a no-growth population produces even greater amounts of capital to complement the labor in production.

In summary, rising per capita income can accommodate a growing population and growing labor force as long as there is adequate savings to accumulate sufficient amounts of capital to shift up the demand for labor. This is shown by the higher wage in comparison to the lower wage. But a slower growing population will normally save higher percentages of its income, producing an even higher demand for labor yet, which results in the highest wage of all, shown as the top curve in Figure 14–3. The findings of Coale and Hoover have been tested several times in the past 50 years, and their original conclusions have been widely debated. Several recent studies support the original findings. Work by Kelley and Schmidt (1996) for developed and developing countries, Higgins and Williamson (1997) for 13 Asian countries over three, five-year periods, Lindh (1999) for 23 industrialized countries from 1950 to 1990, and Lindh and Malmberg (1999) found that reduced population growth increases standards of living and increases economic growth.

---

[8] But savings are not free; they come at the expense of current consumption. Every dollar saved is a dollar taken away from current pleasures. If savings produced no gain in future productivity or security, few would save, since the cost of lost current consumption is high and the gain in future consumption is limited.

■ Microeconomic Forces: Economic Theory of Fertility

The *economic theory of fertility* rests on the premise that couples make rational choices about family size. They choose to have smaller (larger) families as the benefits of added children are low (high) and the costs of rearing added children are high (low).[9] The economic theory states that population growth can be limited by (1) reducing the number of desired children and (2) providing access to contraceptives so couples can plan the size of their family. Desired family size depends on factors that influence the supply and demand for children. Figure 14–4 illustrates the economic theory of fertility; it graphs the economic forces that shape a couple's desired family size (sometimes it is not a couple's choice, but an individual's choice).

Demand for Children

People desire children for many reasons. Some people may be conscious of a psychological need to care for children in their own image (Kirk, 1991). Children may also satisfy other wants and needs:

- A wealthy family may wish to pass along its wealth and family name to children. The more children in the family, the higher is the probability that the bequest will be passed along.
- In a traditional agricultural society, children bring needed hands for farm chores.[10]
- Children may help support parents in their old age. This idea is especially relevant in countries with weak savings systems or no retirement pensions.
- In some cultures a woman's status and prestige is defined by the number of children she bears. For uneducated women with few opportunities to earn status outside the home, more children produce added prestige.
- High infant mortality reduces the probability of a child surviving to adulthood and increases the couple's demand for children.
- In cultures where a male child is highly valued, a couple may increase its family size until it produces at least one male child with a high probability of surviving to adulthood.[11]

The demand for children based on such factors is summarized by the downward-sloping line *MV* in Figure 14–4. The economic value of a child to the family is the present value of the stream of willingness of the couple to pay for consumptive value provided to it by that child and his or her descendants (Haveman, 1976).

Beginning at the origin when the couple has no children, the added value produced by the first child is *OB*. The second child brings *added* income, however, it is less than the increment produced by the first child. As the number of children increases, the line *MV* slopes downward because the incremental economic value produced by each added child decreases.

---

[9] Early microeconomic thinking regarding fertility decisions stated that reduced demand for children, as determined by income, prices, and tastes, was the driving force in the fertility transition to smaller family sizes (Schultz, 1974, and Becker, 1981). Kirk (1991) observed that for the Western high-income democracies, individual incentives for choosing family have become especially important since the 1960s. A new and vigorous individualism has emerged that takes a skeptical view of society's formal institutions and status conventions; it is expressed in the pursuit of higher personal status, higher real income, less deference to authority and social obligation, and greater freedom to exercise one's own judgment.

[10] Davis and Blake (1956) argued that high fertility was a rational response to economic conditions in a traditional agricultural society. (1) The extended family means that the cost and care of children are shared. (2) Children, once past young childhood, may pay for their costs, especially where there are cottage industries, but more generally in any farming situation. (3) Large families may bring economic strength through political strength in local organizations. Cassen (1976) analyzed conditions necessary for children to pay for themselves in India.

[11] Cassen (1976) cites Heer and Smith (1968) who computed that 6.8 children were required by a typical couple from India to ensure a 95% probability of producing one son surviving to the child's father's age of 65.

Figure 14–4
Economic theory of fertility: Demand and cost of children

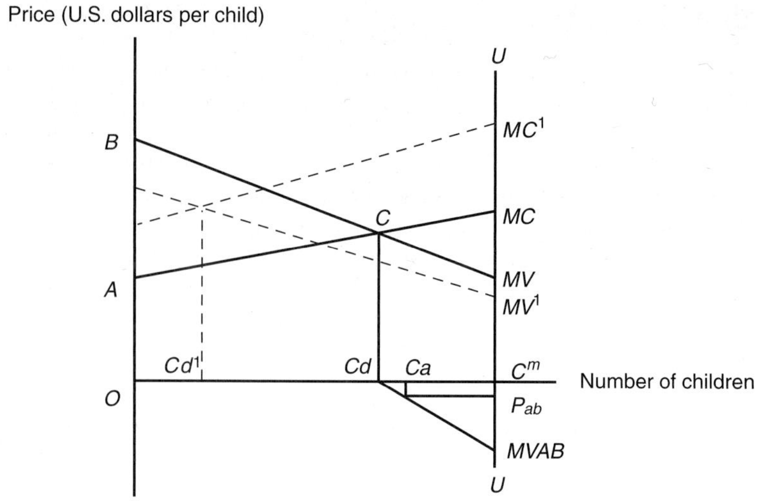

Cost of Children

Although children bring many benefits to the family, they also impose costs, which may limit the desired family size. The cost borne by the family of a child is the present value of the stream of consumption by or investment required to support the child and his or her descendants (Haveman, 1976). The marginal cost of children is summarized in Figure 14–4 by the upward-sloping line *MC*. The added cost of the first child is the vertical distance *OA*. Several factors influence the marginal cost of bearing and rearing a child:

- As the number of children increases, demands on the mother's time increase more than proportionally. If educational and status opportunities for the mother outside the home increase,[12] the opportunity cost of raising children is increased, which shifts *MC* up.
- As a culture urbanizes and industrializes, urban housing space becomes more scarce and the marginal cost of children increases compared to that in agricultural economies, which shift *MC* up, for example, MC^1. See Example 14–2.
- Costs of children increase with the rising costs and expectations the culture places on parents in regard to children's education.[13] Moreover, income lost to the family from children not working while in school is a major opportunity cost.
- Each extra child increases the family's cost of food, clothing, health care, shelter, and child care.
- Each added child imposes costs on existing siblings. When the family is already large, there are considerable incremental child-rearing and child-watching costs imposed on the added child's brothers and sisters. (Any adult who is a first born in a large family may remember these costs.)

---

[12] Davis (1984) describes how the percentage of U.S. women in the workforce increased from 4.6% in 1890 to 51.2% in 1982. This huge increase in the opportunity cost of the mother's time has doubtless shifted up the marginal cost of children, thereby contributing to reduced family sizes. Cassen (1976) describes the role of added education as affecting the cost of children. Education alters parents' perceptions of the advantages of smaller or larger families. The parents may acquire social or economic aspirations with which children would interfere; education may produce higher aspirations of the parents for their children, including more education for the children. Unless the parents are wealthy, these aspirations cannot be satisfied for more than a small number of children.

[13] For example, late in the nineteenth century in Europe, the United States, and Australia, as urban society developed, the middle classes and then the working classes sent their children to school because more occupations demanded a literate workforce and members of a richer society increasingly judged their neighbors by the way they schooled their children (Caldwell, 1999).

Example 14–2
Impacts of China's One-Child Policy

A commonly expressed rationale for family planning programs is that these programs attempt to bring together more closely the socially and privately desirable levels of fertility. Although the net social benefits of reducing births may be high, a program may have a very low ratio of private benefits to private costs and hence may not encourage voluntary participation. For this reason, according to some, the coercive power of the state may be the only way to enforce the desired cooperation.

China's one-child policy, in which couples cannot have more than one child without heavy penalties, illustrates the coercive power of state intervention in family planning. The state believed society receives an extenal benefit from a reduced family size that is not felt by the individual family. China's one-child policy was first expressed by social architect Deng Xiaoping in 1979 and was implemented in 1981. The Chinese government white paper defended its extreme policy: China was feeding 1.2 billion people, but cultivated land was only 5% of its territory. In contrast, U.S. cultivated land is 20% of its territory; combined with a much smaller population, the United States has nine times China's per capita farmland. China's farmers had to feed a growing population squeezed onto a shrinking land base, and this produced falling living standards for everybody.

China's policy has had several unforseen impacts. Chinese culture places heavy values on infant boys to continue family bloodlines. Many infant girls are abandoned or killed to make space for a boy, without exceeding the child quota. This strong cultural tie to boys skews the ratio of boys to girls, approaching 1.4 to 1 in some cases. In families with only one child, that child will support his parents alone; this development may increase pressure for publicly funded retirement pensions. Future high ratios of men to women will force many men to be bachelors, which may contribute to social tensions and a greater national focus on the military. Today's children, who receive sole attention from two parents and four grandparents, risk becoming spoiled–individually excelling but unable to work on teams.

China's one-child policy and its implementation illustrate the question that all governments of crowded, third-world countries face: What culturally compatible measures can control population, accommodate government budget constraints, and maintain basic human rights?

Sources: Chinese Government. 1996. White paper on family planning. *Population and Development Review.* 22 (2): 385–390; Johnson, K. 1996. The politics of the revisal of infant abandonment in China, with special reference to Hunan. *Population and Development Review.* 22 (1): 77–98; Johnson, K., H. Banghan, and W. Liyao. 1998. Infant abandonment and adoption in China. *Population and Development Review.* 24 (3): 469–510.

Unconstrained Optimal Family Size. Because each extra child brings both the added benefits and added costs described, the couple always faces a decision as to whether to expand the family further.[14] With complete and costless control over its desired family size, the economically motivated couple will expand its family size as long as the anticipated marginal benefit exceeds its marginal cost. This occurs for all family sizes for which the solid line *MV* in Figure 14–4 is larger than the solid line *MC*. The couple's welfare is maximized by a family size equal to the desired number of children, *Cd.* This occurs when the marginal value of an extra child just equals its marginal cost, shown by the intersection of the solid line *MV* and solid line *MC*. When the number of children is *Cd,* the couple's total economic welfare from its children is maximized by the triangular area *ABC*. Either a smaller or larger family size than *Cd* produce less total welfare.

[14] The couple's fertility decisions are the outcome of a negotiation between male and female. Rasul (2002) developed and tested a model of household bargaining over family size. He investigated the concept that when bargains are continually subject to renegotiation, the wife's bargaining power affects fertility. Understanding fertility as the outcome of a bargain between the male and female has two important advantages: It provides a way to introduce male-female conflicts over desired fertility and female empowerment can be understood as changes in female bargaining power. Using Malaysian household data on household fertility and male-female fertility preferences, Rasul concludes that husbands and wives are equally important in determining fertility.

Potential Family Size. Potential family size refers to the number of surviving children a household would have if fertility were not deliberately limited (Easterlin, 1975), which is shown by line *UU* in Figure 14–4. This maximum family size depends on the mother's natural fertility, her age at marriage and her health, and on the probability of a newborn surviving to adulthood. For a given level of natural fertility, an increase in the infant survival probability increases the potential family size. Mortality itself influences potential family size. Some diseases inhibit fertility and malnutrition affects the ability of the mother to bring a pregnancy successfully to term and the number of years in which she is fertile. Thus, the improvement of some health factors that influence mortality may increase potential family size. All the factors that affect potential family size vary widely by family, culture, and economic conditions. For example recently published data show *national average* fertility exceeding 10 children in Oman (World Bank, 2001).

Motivation for Fertility Regulation. The motivation for fertility regulation comes from the relative position of the potential family size (line *UU* in Figure 14–4) and the desired family size (*Cd*). Depending on the two positions, potential family size can be larger, smaller, or equal to its desired size. If the potential family size is less than the desired level, then line *UU* is to the left of *Cd,* and the demand for children exceeds potential family size. In this situation there is no desire to limit fertility and the couple looks for ways to expand its family size, such as adoption or fertility enhancement measures (Easterlin, 1975). If potential family size is much smaller than desired size, a couple may invest considerable time and money in expanding their family.

If potential family size exceeds demand, as shown in Figure 14–4, line *UU* lies to the right of *Cd.* At any point to the right of *Cd,* the cost of the added child exceeds its benefit to the family. In this situation, parents are faced with the possibility of having more children than they want and they are motivated to regulate their fertility. The situation of excess supply produces a demand for measures to limit fertility. Whether fertility control is actually used and the extent to which it is used depends on its cost, effectiveness, and acceptability.

Price of Fertility Regulation. The price of fertility regulation includes both the subjective drawbacks of fertility control, such as distaste for the idea itself or for particular techniques and objective costs, such as the time and money required to obtain and learn about the use of contraceptives. It also includes the price of access to contraception as well as its effectiveness (Easterlin, 1988). Three conditions must occur for actual fertility to fall below the potential family size: (1) fertility must be controllable through conscious choice; (2) reduced fertility must be perceived as advantageous; and (3) effective techniques of fertility reduction must be available, acceptable, and affordable (Coale, 1973). In other words, there must be motivation to regulate fertility and an affordable, available, and effective way to carry it out.

Without effective contraception measures, as is the case in many preindustrialized societies, the price of fertility regulation is high and the number of children is determined by a mix of social and biological factors.[15] Without effective, affordable, and acceptable contraception, couples do not reduce their actual number of children beneath the maximum potential; instead, they have

---

[15] Freedom of choice by couples is limited. Couples have the number of children that biology, culture, and society decide to give to them (Bourgeous-Pichat, 1967). Caldwell (1999) describes pressures that society placed on couples and that couples placed on themselves to discourage contraception in the late 1800s. Social pressures in Europe, the United States, and Australia limited middle-class couples' access to contraception until the last quarter of the nineteenth century. The central difficulty in achieving widespread fertility decline lay in an ironic relationship. As long as sex could not be discussed by either spouse without causing a loss of respect by the other, contraceptives remained primitive and sales outlets were generally thought disreputable. Better quality contraceptives sold through pharmacies could not occur until contraception became legitimate. Legitimization could be achieved only by a level of public discussion that convinced women and men that their spouses would not regard them as deviant if they raised such matters. Another example of these social pressures in Africa is provided by Bledsoe, Banja, and Hill (1998). The authors state that rural Gambian women who were asked how many children they wanted frequently responded by saying "whatever God gives me."

a family size defined by line *UU* in Figure 14–4. Cleland and Wilson (1987) point to the evidence that in traditional societies producing the desired number of children through conscious fertility control is difficult to achieve.

An averted birth is a baby not born that otherwise would have been born. In Figure 14–4, the first averted birth is the potential family size, line *UU,* minus one child. When the couple's actual family size is at line *UU,* which is to the right of desired family size, the first averted birth produces a greater benefit than cost. In Figure 14–4, this is shown as the vertical distance between *MC* and *MV* at line *UU.*

Actual Number of Children. As contraception becomes more widely available, effective, accepted, and affordable, couples in the situation of excess supply begin to see their ability to choose. In the terms of economics, they deliberately reduce their number of births to the left of line *UU* (Figure 14–4) by choosing larger numbers of averted births. They begin moving away from their potential family size and move down the line, reflecting the marginal value of an averted birth defined by the line *MVAB.* The marginal value of an averted birth (one child less) is equal to the marginal cost minus the marginal value of one child more. The vertical intercept of line *MVAB* on line *UU,* the marginal value of the first averted birth, is interpreted as the incremental value to the couple of the first averted birth below its potential family size.

As couples reduce the planned family size by increasing the number of averted births, they move down line *MVAB* and actual family size falls below the potential family size in its journey toward desired size. The process continues as family size is reduced from its maximum until *MVAB* is equal to the price of fertility regulation. If the price of fertility regulation is larger than the marginal value of the first averted birth, there is no fertility regulation and the couple's number of children stays at its maximum potential, *UU.* If the price of fertility regulation falls to *Pab,* the number of averted births increases; the number of actual children is *Ca* and the number of averted births is equal to *UU* minus *Ca.*[16]

If fertility regulation is so effective, acceptable, and available as to be costless, then the price of contraception is zero. The couple moves down its derived demand for averted births until the incremental value of an additional averted birth is zero. In this state of maximum effective choice, each couple has its desired number of children, *Cd.*[17] The desired number of children maximizes the couples' economic welfare, which is shown by the triangular area *ABC.*

Policy Options. The economic approach to population control is based on altering the couple's economic incentives by shifting the demand for large families to the left or shifting the marginal cost up. Various public policies could accomplish this aim. The underlying principle is to adopt policies that change the costs and benefits of child rearing so that more parents will perceive economic

---

[16] The actual number of children can be large or small, depending on the demand and cost of children, maximum potential family size, and price of fertility control. The analysis in Figure 14–4 does not assume a particular culture or context. The number of actual children minus the number of desired children is defined by population researchers as the "unmet need" for contraception (e.g., Bongaarts, 1991). Women who want to prevent pregnancy but do not practice contraception are said to have an unmet need for contraception. At least 120 million women in the developing world have this unmet need. These include women who want to postpone childbearing as well as those who want to limit their families. In terms of economic theory, the unmet need for contraception can be seen as the amount by which its demand exceeds its supply at a zero price. Despite the intuitive appeal of the unmet need idea, it may have problems. Many economists doubt the utility of the unmet need concept for population policy analysis, because there is an unmet need for every priced commodity in the marketplace. Each of us has an unmet need for everything and a greater unmet need for expensive things. If there is an unmet need for everything, policymakers may find it difficult to implement the concept of "unmet need" in distinguishing weak from strong programs. Carrying through with this logic, we are forced to conclude that since all proposed programs serve unmet needs, they should all be funded and carried out.

[17] Kirk (1991) observed that since fertility regulation in industrialized countries has become so widely accessible, acceptable, effective, cheap, and used, the fundamental demographic fact in Europe is that couples have moved closer and closer to having the number of children they actually want.

incentives associated with smaller families. The intent is to regulate population growth so that it meets national needs through policy actions that raise or lower costs and benefits of added children in individual families.

The most important government policies in the future will involve education, raising women's status, and improving child health. Such policies are valued in their own right, but education, especially of girls, has been shown to be the factor most closely related to fertility decline by delaying marriage and first births. Increasing equality between the sexes in legal, economic, and social affairs raises the cost of the mother's time with children, thereby raising the cost of rearing children to higher levels (e.g., from MC to MC^1 in Figure 14–4) by making roles other than childbearing more feasible and attractive to women. Reducing child mortality and adding measures to improve child welfare (e.g., compulsory schooling and child labor laws) reduce the economic value of children to the family to lower levels (e.g., from MV to MV^1 in Figure 14–4). Historically, these measures have been associated with fertility reduction everywhere, as shown in Figure 14–4 by higher MC^1 and lower MV^1 producing a lower desired family size of Cd^1.[18]

Education. In traditional societies with largely agricultural economies, children are a valuable source of labor. The provision of educational opportunities, which requires school attendance, lowers the labor value of children and raises their costs (e.g., books, uniforms, and school fees). Of all the social and economic factors that have been studied for their potential effect on family size, level of education, especially for girls, stands out most consistently. This finding is attributable to acceleration in cultural change and changes in values as a result of the introduction of mass schooling and enforcement of school attendance for both sexes. Demeny (1992) describes the influence on fertility rates worldwide produced by policies that introduce mass education with various distributions of the cost of education.

Women's Status. In societies where women's lives have long been confined to household matters, government policies that introduce changing educational and employment prospects for women are likely to be associated with reduced fertility. When offered the chance of a life that consists of something beyond childbearing, childrearing, and household chores, many women are likely to take the chance for themselves and for their children (Cassen, 1976). See Example 14–3.

Improvements in the economic, social, and legal status of women can reduce desired fertility in several ways. Such improvements raise the cost of children by making roles that do not involve childrearing more important. Empowering women is also likely to lead to reductions in (1) the dominance of husbands or other household members over women, (2) preference for male children, and (3) the value of and need for children as insurance against adversity in old age.

Child Mortality. A high death rate among children encourages high fertility for several reasons. (1) It makes the planning of families difficult because the number and timing of future deaths are unpredictable, and this contributes to superstition and fatalism. (2) It discourages investments in children's health and education. (3) It requires large numbers of births to ensure that the desired number of children survive to adulthood. All of these effects can be counteracted by the implementation of public health measures to reduce infant and child mortality. Historically, no population in the developing world has experienced a sustained fertility reduction without first having gone through a major decline in infant and child mortality. See Example 14–4.

---

[18] The higher MC^1 and lower MV^1 also shift out the marginal value per averted birth (not shown in Figure 14–4) and, for a given price of contraception, reduce the actual number of children by increasing the number of averted births.

Example 14–3
Benefits and Costs of Girls' Education in Africa

Education is expensive in sub-Saharan Africa, and the price of tuition is only a small part. Like other long-term investments, education carries many costs in the short run before the long-run benefits start to roll in. Despite steady progress in increasing school enrollment in sub-Saharan Africa since 1990, only 59% of children attended primary school from 1996 to 2002—the world's lowest rate of attendance. The heaviest burden falls on girls. For them, a mix of cultural beliefs, damaging poverty, and weak international educational aid produce even more dismal numbers. Only 57% of girls were enrolled in school during the same period. What may be most alarming is that the number of girls excluded from school in sub-Saharan Africa rose from 20 million in 1990 to 24 million in 2002. Both paid and unpaid costs of excluding girls from school are high. Uneducated girls are more likely to grow up to become poor as adults, die in childbirth, contract AIDS, and raise children who suffer the same fate.

School fees are a major obstacle to girls' education in sub-Saharan Africa. In societies where girls are considered an economically weak investment, many cash-strapped parents pay for their sons to attend school but keep their daughters at home. Many labor-intensive home tasks absorb girls' time, so there is a high opportunity cost of allowing the girls to leave home for education. Their home time involves activities such as helping mother fetch water and firewood, caring for younger siblings, cleaning the home, and cooking. Convincing mothers to give up that help by sending daughters to school presents a major economic burden to the family and a cultural challenge.

Several cultural factors block girls' access to education, particularly the practice of marrying girls as soon as they are physically ready. War and famine in Africa have also prohibited education for years. AIDS and other diseases have robbed families of breadwinners, leaving unmeasured numbers of children to care for their families.

Sometimes one person can make a difference. In west Africa, a turnaround for the village of Koutagba in Benin came in April 1997 with the arrival of one college-educated woman. She told villagers how they could benefit economically from sending their daughters to school. At first the village women avoided speaking to her, but soon she used economic incentives to attract the villagers' attention. She focused on the immediate economic value of education: Educated girls knew how to read a thermometer and could tell if a baby had a fever. Educated girls could calculate the costs of trading their produce at the local market. She finally won villagers' trust by offering a loan of $700 underwritten by the United Nations International Children's Emergency Fund (UNICEF). The mothers used the money to buy food from the nearest small town and bring it back to sell. In exchange, the mothers agreed to send a daughter to school. Economic incentives and individual will can play important roles ridding illiteracy from a family and a nation.

Sources: Sengupta, S. 2003. African girls' route to school is still littered with obstacles. *New York Times.* December 14, 2003. Available from *http://www.nytimes.com/2003/12/14/international/africa/* [Accessed December 15, 2003]; Bellamy, C. 2003. *The state of the world's children, 2004: Girls, education, and development.* UNICEF. December 11, 2003. Available from *http://www.unicef.org/publications/Eng_text.pdf* [Accessed October 3, 2004].

■ Emerging Questions

Many questions remain to be answered about government policies and gaps in our knowledge of the relationship between population and the environment:

- What will the demands of the growing world population mean for human use of the environment and natural resources?
- How will predicted increases in population affect the quality and quantity of the environment and natural resources throughout the world?
- What measures can be taken to maintain or improve living standards despite continued population growth, and do these measures change under various cultures and systems of government?
- What are the motivations for having children and how are these motivations altered by public policies?
- Where there is little mass schooling available, what other mechanisms for spreading cultural change, such as mass media, can influence fertility (Lloyd, Kaufman, and Hewett, 1999)?

Example 14–4
Women's Roles in Reducing Child Mortality

Sri Lanka, Costa Rica, and Kerala in India are poor countries, but they have all made impressive gains in reducing mortality rates. Falling mortality, especially reduced child mortality, is the first step toward reduced fertility and ultimately toward controlling population. Caldwell (1986) attempted to discover why those three countries reduced mortality so successfully with so few resources at their disposal. He discovered that the common denominator lay in women's freedoms. When it is socially acceptable for unmarried girls who have reached puberty to take responsibilities outside the home, or for older women to appear in public unescorted, several results follow, all of which reduce mortality. Girls stay in school longer, and mothers act on their own when there is a sick child or when they are sick themselves. When necessary, mothers travel to health clinics on their own, wait in mixed gender lines, and ask for clear explanations from male doctors. Women take responsibility for their own behavior rather than shifting it to a husband or male relative. The mother decides whether a sick child simply needs rest or whether to take the child to the doctor, she does not need to consult her husband or his family.

In societies where women take individual initiative, they are in a position to treat daughters equal to sons. In societies where women's premarital morality is less relevant to their status, women are likely to face less pressure to marry soon after puberty, and they may receive more and better education, training, and employment opportunities.

Women play a major role in making rural health care systems work, both in terms of staffing the centers and in using their services. Where women make more of their own decisions, girls are likely to become nurses. Where women make fewer decisions, parents worry about sending their daughters to medical training schools, working with male doctors or patients, or living without family or relatives far from home. When the culture permits many girls to be educated and parents are proud of their daughters who get jobs, it is easier to recruit women as health assistants or midwives. If such women can be recruited to work in their own home areas, such as they can in Sri Lanka, Costa Rica, and Kerala, they are more effective in house-to-house visits than health professionals from outside the community.

Source: Caldwell, J. C. 1986. Routes to low mortality in poor countries. *Population and Development Review.* 12 (2): 171–220.

- What do investments in education and improvements in income distribution, child health, and women's access to and control of resources contribute to voluntary fertility decline?
- What policy options can effectively reduce high demand for births in developing countries and reduce population momentum?[19]
- How are individuals' decisions on reproduction and family size affected by the knowledge, attitudes, and behaviors of others with whom they come into contact and by public policy (Montgomery and Casterline, 1998)?
- What are the effects on fertility of improved economic opportunities for women and how are these effects altered by government policies (Armin and Lloyd, 1998)?
- What is the least-cost mix of programs to reduce fertility and how does that program mix vary by culture, nation, and system of government? Knodel and Jones (1996) observed that the paths to fertility reduction are many and varied and no single path (e.g., sole focus on increased girls' schooling) should receive priority without analysis of many alternatives.
- How is contraceptive use altered by price and access to family planning programs (Easterlin, 1998)?
- How does family income alter benefits and costs of children?
- What influences a couple's maximum family size, desired family size, and actual family size?

[19] This question was posed searchingly by Bongaarts (1994). Long-term population growth is sensitive to small deviations from the replacement level. In India, for example, if fertility stabilized at the replacement level, the population would reach 1.6 billion in 2100 (Bongaarts and Amin, 1997). If fertility leveled off at a small 0.5 of a birth higher, the total would be 3 billion in 2100. This means that even modest programs to reduce fertility could have large absolute effects on population size.

Perhaps these questions will be answered as more nations move through their demographic transition. World population will stabilize or fall and stress on the environment and natural resources will be reduced. Evidence suggests that institutions that allow population and environment to coexist can work.

SUMMARY

The world's population is more than 6 billion and is growing by about 2% each year. Population growth can have a major impact on the environment and on the effectiveness of existing environmental policies. These impacts can be softened considerably, however, by judicious environmental policies. A simple formula allows for structural analysis, forecasting, and policy evaluation for a wide range of significant environmental issues.

The world continues to undergo a demographic transition that started in western Europe during the mid-nineteenth century. In the transition, a large number of children per family and high death rates are being replaced, first by falling death rates then by smaller family sizes. The transition continues to spread to Asia, Latin America, the Middle East, and slowly to Africa. The earth's population will continue to grow until the average number of children per family falls to the replacement rate of about 2.1, which many think will occur sometime between 2050 and 2100.

Increased population growth can reduce economic growth through two channels: (1) reducing the percentage of people in the workplace and (2) reducing savings rates, which reduces future productivity of labor, hence lowering per capita income.

The economic theory of fertility attempts to explain economic forces affecting a couple's choice of family size, as well as how government policies and social institutions alter those choices. This theory states that if a couple's desired family size is based on added benefits versus added costs of extra children, the couple will invest in more children as long as the added benefit to the family exceeds the added cost. According to the economic theory of fertility, a couple's actual number of children is influenced by (1) desired family size, (2) maximum number of potential children over the mother's reproductive years, and (3) price and effectiveness of contraception. If the maximum family size exceeds the desired size and if contraceptives are free, completely effective, and always used, a couple's actual number of children will equal the desired number.

Government policy can influence desired family size by reducing the marginal value of children (e.g., child labor legislation) and increasing the marginal cost of children (e.g., compulsory education requirements). Public policy also can influence a couple's maximum number of children by improving women's health and promoting changed attitudes toward family planning. Policies can be implemented that increase the number of averted births through actions regarding contraceptives: reducing price, increasing access, investing in research programs that increase effectiveness of contraceptives, and altering attitudes toward their use.

As nations move through their demographic transitions, average family sizes fall towards replacement rates. Girls receive more education, so women gain more freedom of choice for activities outside the home. A higher percentage of total children are wanted, and the actual number of children approaches desired family size. As social and cultural expectations rise regarding parents' child support responsibilities, the marginal cost of children shifts up and each child has access to more resources to support its rearing. As nations develop economically, fewer people will be employed in agriculture, so the demand for children as instruments for their parents' economic welfare will be reduced.

As more nations move through their demographic transitions, world population will stabilize or fall and stress on the environment and natural resources will be reduced. Evidence suggests that institutions that allow population and environment to coexist can work.

REVIEW QUESTIONS

1. Contrast environmental optimists and pessimists and explain the current debate over population growth.

2. What is the popular formula for analyzing environmental impacts of population growth? Give an example of how this formula can be used.

3. Explain the concept of demographic transition and provide an example.

4. What are the four phases of demographic transition? As a policymaker, what uses could you make of information on these four phases in connection with your environmental planning decisions?

5. Describe the effect of population growth on economic growth. What does this relationship mean for policies designed to reduce the population growth rate?

6. What is the fertility rate and why is it important for analysis of population growth?

7. What does the microeconomic theory of fertility attempt to explain, and why is it important to know what it explains?

8. What policy choices increase the marginal cost of additional children?

9. What policy choices reduce the marginal benefit of additional children?

10. Under what conditions do you believe the microeconomic theory of fertility predicts family size? Why?

11. Fertility rates vary widely among cultures globally. One hypothesis is that observed differences are due to different cultural beliefs; but another hypothesis says the differences are due to economic factors. How can economics be used to explain observed differences in fertility rate? Can you think of a test that would let you conclude which hypothesis is correct?

12. The microeconomic theory of fertility gives an opportunity to predict how public policies designed for purposes other than population issues could affect fertility rates. Describe some public policies that affect fertility rates, and explain whether they would tend to increase or decrease the rate.

13. Why is it important to know more about the motivations for having children and how these motivations are affected by public policies?

14. Identify at least three knowledge gaps that need to be filled before better policy choices can be made regarding global population problems. If you had a research budget, which of these knowledge gaps would you try overcome first? Why?

PROBLEMS

1. Suppose the equation for the demand for children per household in country *A* has been estimated to be as follows: children demanded = 10.0 − (0.001 * price), where price is the average dollar price per child annually. Now suppose the marginal cost of each extra additional child has been estimated to be $MC = 6{,}000$, where MC is the added dollar cost per additional child annually.

a. How would you interpret the three coefficients (numbers) 10.0, −0.001, and 6,000?
b. Compute the average number of desired children per household. (Hint: Find the number of children that makes $MC = P$).
c. A proposed education policy would raise the cost of educating each child so that the marginal cost per child equation is now $MC = 7{,}000$. Recompute the average number of desired children per household.

2. The demand for children per household in country *B* is estimated to be the same as in country *A:* children demanded = 10.0 − (0.001 * price). But now marginal cost is estimated at MC = 6,000 + (1,000 * "children") where "children" is the number of children.

a. How would you interpret the new coefficient, 1,000, that is multiplied by the number of children in the MC equation.
b. Compute the average number of desired children per household. (Hint: Solve the two equations simultaneously by finding the number of children that makes MC = price).
c. What economic conditions in country *B* that are absent in country *A* might give rise to the MC coefficient of 1,000?

REFERENCES

Armin, S., and C. B. Lloyd. 1998. Women's Lives and Rapid Fertility Decline: Some Lessons from Bangladesh and Egypt. New York: The Population Council. Policy Research Division Working Paper No. 117.

Becker, G. S. 1981. *A Treatise on the Family.* Cambridge, MA: Harvard University Press.

Bledsoe, C., F. Banja, and A. G. Hill. 1998. Reproductive mishaps and Western contraception: An African challenge to fertility theory. *Population and Development Review.* 24 (1): 15–57.

Bongaarts, J. 1991. The KAP-gap and the unmet need for contraception. *Population and Development Review.* 17 (2): 293–313.

——— 1994. Population policy options in the developing world. *Science.* 263 (5148): 771–776.

Bongaarts, J., and S. Amin. 1997. Prospects for Fertility Decline and Implications for Population Growth in South Asia. New York: Population Council. Policy Research Division Working Paper No. 94.

Bongaarts, J., and R. A. Bulatao. 1999. Completing the demographic transition. *Population and Development Review.* 25 (3): 515–529.

Bongaarts, J., and S. C. Watkins. 1996. Social interactions and contemporary fertility transitions. *Population and Development Review.* 22 (4): 639–682.

Bourgeous-Pichat, J. 1967. Social and biological determinants of human fertility in nonindustrial societies. *Proceedings of the American Philosophical Society.* 3 (3): 160–163.

Brundtland, G. H. 1993. Gro Harlem Brundtland on population, environment, and development. *Population and Development Review.* 19 (4): 893–899.

Caldwell, J. C. 1999. The delayed Western fertility decline: An examination of the English-speaking countries. *Population and Development Review.* 25 (3): 479–513.

Cassen, R. 1993. Economic implications of demographic change. *Transactions of the Royal Society of Tropical Medicine and Hygiene.* 87 (S1): 13–18.

Cleland, J., and C. Wilson. 1987. Demand theories of the fertility transition. An iconoclastic view. *Population Studies.* 41 (1): 5–30.

Coale, A. J. 1973. The Demographic Transition. *In Proceedings of the International Population Conference.* Leége, Belgium. Vol 1: 53–72.

Coale, A. J., and E. M. Hoover. 1958. *Population Growth and Economic Development in Low Income Countries.* Princeton, NJ: Princeton University Press.

Cramer, J. C. 1998. Population growth and air quality in California. *Demography.* 35 (1): 45–56.

Davis, K. 1963. The theory of change and response in modern demographic history. *Population Index.* 29 (4): 345–366.

———. 1984. Wives and work: The sex role revolution and its consequences. *Population and Development Review.* 10 (3): 397–417.

———. 1990. Population and resources: Fact and interpretation. *Population and Development Review. Supplement: Resources, Environment, and Population: Present Knowledge, Future Options.* 16: 1–21.

Davis, K., and J. Blake. 1956. Social structure and fertility: An analytic framework. *Economic Development and Cultural Change.* 4: 211–235.

Demeny, P. 1992. Policies seeking a reduction of high fertility: A case for the demand side. *Population and Development Review.* 18 (2): 321–332.

Easterlin, R. A. 1975. An economic framework for fertility analysis. *Studies in Family Planning.* 6 (3): 54–63.

———. 1988. The demand for family planning: A new approach. *Studies in Family Planning.* 19 (5): 257–269.

Ehrlich, P. R., and J. P. Holdren. 1971. Impact of population growth. *Science.* 171: 1212–1217.

Gage, T. B. 1993. The decline in mortality in England and Wales, 1861–1964: Decomposition by cause of death and components of mortality. *Population Studies.* 49 (1): 47–66.

Haveman, R. H. 1976. Benefit, cost analysis and family planning programs. *Population and Development Review.* 2 (1): 37–64.

Heer, D., and D. O. Smith. 1968. Mortality level, desired family size and population increase. *Demography.* 5 (1): 104–121.

Higgins, M., and J. G. Williamson. 1997. Age structure dynamics in Asia and dependence on foreign capital. *Population and Development Review.* 23 (2): 261–293.

Kelley, A. C., and R. M. Schmidt. 1996. Saving, dependency and development. *Journal of Population Economics.* 9 (4): 365–386.

Kirk, D. 1996. Demographic transition theory. *Population Studies.* 50 (3): 361–387.

Kirk, M. 1991. The problem of fertility 1936–1986. *Population Studies. Population Research in Britain Supplement.* 45: 31–47.

Knodel, J., and G. W. Jones. 1996. Post-Cario population policy: Does promoting girls' schooling miss the mark? *Population and Development Review.* 22 (4): 683–702.

Knodel, J., and E. van de Walle. 1979. Lessons from the past: Policy implications of historical fertility studies. *Population and Development Review.* 2 (2): 217–245.

Landry, A. 1934. *La révolution demographique. Etudes et essais sur les problems de la population.* Paris: Librairie du Recueil Sirey.

Lindh, T. 1999. Age structure and economic policy: The case of saving and growth. *Population Research and Policy Review.* 18: 261–277.

Lindh, T., and B. Malmberg. 1999. Age structure effects and growth in the OECD, 1950–1990. *Journal of Population Economics.* 12: 431–439.

Lloyd, C. B., C. E. Kaufman, and P. Hewett. 1999. The Spread of Primary Schooling in sub-Saharan Africa: Implications for Fertility Change. New York: The Population Council. Policy Research Division Working Paper No. 127.

Malthus, T. 1798. *An Essay on the Principle of Population.* London: Printed for J. Johnson in St. Paul's Churchyard.

Montgomery, M. 1999. Mortality Decline and the Demographic Response: Toward a New Agenda. New York: The Population Council. Policy Research Division Working Paper No. 122.

Montgomery, M., and J. B. Casterline. 1998. Social Networks and the Diffusion of Fertility Control. New York: The Population Council. Policy Research Division Working Paper No. 119.

Notestein, F. W. 1953. Economic Problems of Population Change. *In Proceedings of the Eighth International Conference of Agricultural Economics.* London: Oxford University Press.

Olson, M. 1998. Mancur Olson on the key to economic development. *Population and Development Review.* 24 (2): 369–379.

Preston, S. H. 1996. The effect of population growth on environmental quality. *Population Research and Policy Review.* 15 (2): 95–108.

Rasul, I. 2002. Household Bargaining Over Fertility: Theory and Evidence from Malaysia. Washington, D.C.: World Bank. World Bank Working Paper.

Schultz, T. W., ed. 1974. *Economics of the Family: Marriage, Children and Human Capital.* Chicago: University of Chicago Press.

Solow, R. M. 1988. *Growth Theory: An Exposition.* New York: Oxford University Press.

Thompson, W. S. 1929. Population. *American Journal of Sociology.* 34: 959–975.

United Nations. 2004. *World population to 2300.* Available from *http://www.un.org/esa/population/publications/longrange2/WorldPop2300final.pdf* [Accessed October 31, 2004].

World Bank. 2001. *World Bank Development Indicators.* Washington D.C.: World Bank.

Chapter 15

CLIMATE CHANGE

Outline

Scientific Evidence of Climate Change 419

Economic Analysis 421

Policy Options for Dealing with Climate Changes 427

International Agreements on Global Warming 434

Objectives

After studying this chapter, you should be able to

- Use economic analysis to understand causes and effects of global warming
- Show how economic analysis can be used to examine policy choices for controlling greenhouse gasses
- Identify major international agreements on global warming

Introduction

Since the late nineteenth century, growing use of fossil fuels in the industrialized world has provided the energy needed to substitute machines for human labor, increase productivity, and improve living standards. In the early 1990s, however, many scientists expressed concern about the evidence indicating that carbon dioxide (CO_2) emissions are warming the earth's climate, and some believe that the consequences of this warming will be serious (Cooper, 2000; Schelling, 1992). Increased atmospheric greenhouse gases (GHGs), such as CO_2, involve a time delay of many years after GHG emissions are released into the atmosphere. This delay occurs partly because oceans and trees moderate the reaction of climate to the increasing atmospheric concentrations. Currently, an estimated 700 billion tons of carbon remain in the atmosphere from past combustion of fossil fuels. Annual emissions globally are near 7 billion tons; a little more than half stays in the atmosphere, so the concentration increases by one-half percent per year. The threat of the greenhouse effect is a problem of the global commons.

Greenhouse gasses became a political issue in the late twentieth century because emissions into the atmosphere increased significantly. Burning fossil fuels, cutting forests, and operating motor vehicles create a layer of GHGs that trap enough heat to alter the earth's climate by raising the earth's surface temperature. Nations with the fewest resources have the least capacity to adapt to climate change and are the most vulnerable to the consequences (Albritton and Meira Filho, 2001; Alley et al., 2003; IPCC, 2001; Rosenweig and Parry, 1994; Wigley and Raper, 2001). In contrast, advanced market economies have considerable flexibility in adapting to climate changes.

A major question now focuses on the probable impacts of global climate change. Mendelsohn, Nordhaus, and Shaw (1994) Mendelsohn et al. (2000), Nordhaus (1993a), and Cline (1991, 1992) conclude that the effects will be modest, because agriculture and other climate-affected sectors will adapt to whatever changes occur, especially in industrialized countries. Economic benefits produced by measures taken to mitigate global warming accrue as a public good and pose the classic challenge of nations that would prefer to be free riders. Any nation that designs its own emissions abatement policy in isolation is likely to choose emission control targets that are too low (Poterba, 1993). Therefore, economically efficient policies to manage climate change are best made at an international level.

■ Scientific Evidence of Climate Change

The Intergovernmental Panel on Climate Change (IPCC), an international group of scientists, has made three major conclusions regarding impacts of climate change on temperature (IPCC, 2001):

1. The global average surface temperature has increased about 0.6°C (1.1°F) since the mid-nineteenth century, when reliable testing instruments became available.
2. Over the same period, the 1990s ranked as the warmest decade and 1998 was the warmest year.
3. The Northern Hemisphere's temperature increase in the twentieth century was the largest increase of any century of the past 1,000 years.

Table 15–1 shows estimated climate change outcomes, the likelihood of their occurring, and their impacts. Many of the forecasted changes point to more floods and droughts, with adverse impacts on ecosystems, economies, and human health. Table 15–2 shows the kinds of vulnerabilities that climate change could produce for different parts of the world.

Table 15–1 Climate change outcomes, probabilities, and impacts

Projected changes in twenty-first century	Likelihood	Impacts
Higher maximum temperatures; more hot days and heat waves over most land areas	Very likely	Increased death and illness in older age groups and urban poor Increased heat stress to crops, livestock, and wildlife Shift in tourist destinations Increased demand for electric cooling
Higher (increasing) minimum temperatures and fewer cold days, frost days, and cold waves over nearly all land areas	Very likely	Decreased cold-related human morbidity and mortality Decreased damage to some crops, and increased damage to others Extended range and activity of some pest and diseases Reduced heating demands
More intense precipitation events	Very likely	Increased flood, landslide, avalanche, and mudslide damage Increased soil erosion Increased recharge of some floodplain aquifers Increased stress on flood insurance systems
Increased summer drying and associated risk of drought	Likely	Decreased crop yields Increased damage to building foundations caused by ground shrinkage Decreased quantity of water resources Increased risk of forest fire Decreased hydropower protential in drought-prone regions
Increase in peak wind intensities of tropical cyclones	Likely	Increased risks to human life and risk of infectious disease epidemics Increased coastal erosion and damage to coastal buildings and infrastructure Increased damage to coastal ecosystems such as coral reefs and mangroves
Increased variability in Asian summer monsoon precipitation	Likely	Increase in flood and drought magnitude and damages in temperate and tropical Asia

Source: *Watson, R. T. et al., 2001.* Climate Change 2001: Synthesis Report. Summary for Policymakers. *Third Assessment Report of the Intergovernmental Panel on Climate Change. Available from http://www.ipcc.ch/ [Accessed July 21, 2003].*

Table 15–2 Climate change: Regional adaptive capacity and vulnerability

Region	Vulnerability
Africa	Overall: Adaptive capacity of human systems in Africa is low due to lack of economic resources and technology. Vulnerability high as a result of heavy reliance on rain-fed agriculture, frequent droughts and floods, and poverty. • Grain yields are projected to decrease for many scenarios, diminishing food security is predicted, particularly in small food-importing countries. • Increases in droughts, floods, and other extreme events would add to stresses on water resources, food security, human health, and infrastructures, and would constrain development. • Coastal settlements (e.g., in the Gulf of Guinea, Senegal, Gambia, Egypt, and along the Southeast African coast) would be adversely impacted by a rise in sea level through inundation and coastal erosion.
Asia	Overall: Adaptive capacity of human systems is low and vulnerability is high in the developing countries of Asia; the developed countries of Asia are more able to adapt and are less vulnerable. • Extreme events have increased in temperate and tropical Asia, including floods, droughts, forest fires, and tropical cyclones. • Rise in sea level and an increase in the intensity of tropical cyclones would displace tens of millions of people in low-lying coastal areas of temperate and tropical Asia; increased intensity of rainfall would increase flood risks in the same areas. • Climate change would increase energy demand, decrease tourism, and influence transportation in some regions.
Australia and New Zealand	Overall: Adaptive capacity of human systems is generally high, but there are groups in Australia and New Zealand, such as indigenous peoples in some regions, with low capacity to adapt and consequently high vulnerability. • Water is likely to be a key issue due to projected drying trends over much of the region and change to a more El Niño-like average state. • Some species with restricted climatic niches and with inability to migrate due to fragmentation of the landscape, soil differences, or topography, could become endangered or extinct. • Australian ecosystems that are particularly vulnerable to climate change include coral reefs, arid and semi-arid habitats in southwest and inland Australia, and Australian alpine systems. Freshwater wetlands in coastal zones in both Australia and New Zealand are vulnerable, and some New Zealand ecosystems are vulnerable to accelerated invasion by weeds.
Europe	Overall: Adaptive capacity is generally high in Europe for human systems; southern Europe and the European Arctic are more vulnerable than other parts of Europe. • Summer runoff, water availability, and soil moisture are likely to decrease in southern Europe, thereby widening the difference between the north and drought-prone south. Increases in precipitation are likely in winter in the north and south. • Upward and northward shift of biotic zones will take place. Loss of important habitats (wetlands, tundra, isolated habitats) would threaten some species.
Latin America	Overall: Adaptive capacity of human systems in Latin America is low, particularly with respect to extreme climate events, and vulnerability is high. • Loss and retreat of glaciers would adversely affect runoff and water supply in areas where glacier melt is an important source of water. • Floods and droughts would become more frequent; floods would increase sediment loads and degrade water quality in some areas. • Increases in intensity of tropical cyclones would alter the risks to life, property, and ecosystems from heavy rain, flooding, storm surges, and wind damages. • Yields of important crops are projected to decrease in many locations in Latin America, even when the effects of CO_2 are taken into account. Subsistence farming in some regions of Latin America could be threatened.
North America	Overall: Adaptive capacity is generally high in Canada and the United States for human systems, but is less in Mexico. • The rate of biodiversity loss would increase. • Some crops would benefit from modest warming accompanied by increasing CO_2, but effects would vary among crops and regions, including declines due to drought in some areas of Canada's prairies and the U.S. plains.

Table 15–2 Continued

Region	Vulnerability
North America *continued*	• Rise in sea level would result in increased coastal erosion, coastal flooding, loss of coastal wetlands, and increased risk from storm surges, particularly in Florida and much of the U.S. Atlantic coast. • Weather-related insured losses and disaster relief payment in the public sector in North America have been increasing. Insurance sector planning has not yet systematically included climate change information, so there is potential for surprise.
Small island states	Overall: Adaptive capacity of human systems is generally low in small island states and vulnerability high. Small island states are likely to be among the countries most seriously affected by climate change. • The projected rise in sea level of 5 mm per year for the next 100 years would cause increased coastal erosion, loss of land and property, dislocation of people, increased risk from storm surges, reduced resilience of coastal ecosystems, saltwater intrusion into freshwater resources, and high resource costs to respond to and adapt to these changes. • Islands with very limited water supplies are highly vulnerable to the impacts of climate change on the water balance. • Limited arable land and soil salinization make agriculture of small island states, both for domestic food production and crop exports, highly vulnerable to climate change. • Tourism, an important source of income and foreign exchange for many islands, would face severe disruption from climate change and rise in sea level.

Sources: *Watson, R. T. et al., 2001.* Climate Change 2001: Synthesis Report. Summary for Policymakers. *Third Assessment Report of the Intergovernmental Panel on Climate Change; Kelly, P. M., and W. N. Adger. 2000. Theory and practice in assessing vulnerability to climate change and facilitating adaptation.* Climatic Change. *47 (4): 325–352; Smith, B., I. Burton, R. J. T. Klein, and J. Wandel. 2000. An anatomy of adaptation to climate change and variability.* Climatic Change. *45 (1): 223–251.*

■ Economic Analysis

Policymakers have a keen interest in the CO_2 problem because current and future policies affect the amount by which CO_2 builds up. The costs of any CO_2 emission abatement program are likely to be large because the total volume of CO_2 emissions from all sources around the world is so huge.

Popular treatments of the global warming issue typically assume that if society discovers that global warming is the result of human activity, we should stop the warming at all costs. This attitude reflects the common view that environmental issues involve moral rights and wrongs and that environmental policy should eliminate contamination regardless of the price. By contrast, the discipline of economics views the global warming issue by asking these questions: What are the benefits and costs of various measures for doing something about global warming? How are those benefits and costs distributed among the various countries whose actions could reduce warming? See Example 15–1.

The climate change problem is unique for this reason: Actions taken involve large costs that are incurred *long before* benefits can be realized or measured with confidence (Weyant, 1993). The stakes are huge; possible investments and/or economic impacts range in the order of hundreds of billions of dollars a year.

Role of Technological Change

People are flexible. The power of people to develop and use technology that adapts to possible greenhouse warming should not be underestimated. Economic impacts of global climate change are much smaller today than they would have been in the year 1900. Efforts to project economic damages of climate change in 1900 would have missed the mark badly (Schelling, 1992). There would have been no way to forecast the impact of greenhouse warming as a result of air travel, high-speed communication, miles of highways, and undreamed-of medical advances. Anticipating

Example 15–1
How Does the Public Perceive Climate Change?

Few citizens anywhere are likely to admit that they do not care about climate change, as this would reveal both ignorance and insensitivity. But does the average person care enough about global warming to make major commitments? Bord, Fisher, and O'Connor conducted a survey in the United States to determine the importance of global warming compared with issues like pollution, national parks, poverty, the federal deficit, and a strong military. You might be surprised at what they found. Global warming came in last.

Global warming is low on the priority lists of U.S. citizens for all the reasons stated in this chapter that make it hard to negotiate international agreements to limit global warming. Global warming has all the characteristics of a public good in which everyone has an economic incentive to free ride on other people's investments to solve the problem. In addition, global warming is a complex issue that is poorly understood even by scientists. Global climate change is far removed from most people's personal experience. Most of us experience weather changes but we do not experience global warming.

Bord, Fisher, and O'Connor asked if we should spend less, the same, or more on seven environmental problems and global warming came in *last* in terms of preferred spending. They found that for actions like replacing old appliances, buying a fuel-efficient car, and using less cooling and heating, Americans took the moral high ground, as long as it was not inconvenient.

The authors concluded that the public, both in the United States and elsewhere, share these characteristics:

- Strong environmental awareness and support for environmental goals
- Awareness and concern for global warming
- Poor understanding of the causes of global warming
- Perceived significant threat from global warming, but at levels lower than threats from other environmental and social problems
- Limited willingness to pay to control global warming

All other things being equal, most of the world's citizens will support national and international initiatives designed to limit global warming, as long as these initiatives do not demand a major change in their lifestyles. Most people everywhere are environmentally aware but poorly informed about global warming. Nevertheless, it is likely that public awareness will translate into support for policy initiatives in connection with global climate change, if policy designs and implementation limit impacts on individual lifestyles.

Source: Bord, R. J., A. C. Fisher, and R. E. O'Connor. 1998. Public perceptions of global warming: United States and international perspectives. *Climate Research.* 11: 75–84.

the automobile, economic forecasters might have been worried about the impact of GHG on flooding and droughts due to more mud and dust, not imagining that virtually all roads would be paved. The impact of climate warming on health would have been considered without antibiotics or other medical advances. The lesson is that so we need to guard against inflating forecasted economic impacts of future climate change. The human impacts of GHG warming will almost certainly be smaller by 2100 than they are now.

Economic Impacts

Climate change is likely to have different effects on different economic sectors and in different nations. The sectors of the economy that depend heavily on uncontrolled ecosystems (e.g., naturally occurring rainfall, stream flow, or temperatures) will be most influenced by climate change. Agriculture, forestry, outdoor recreation, and coastal activities are good examples. Industrially developed regions, such as Taiwan, Japan, western Europe, and North America, are somewhat insulated from the impacts of climate change, whereas developing parts of the world, such as Africa, Latin America, and most of Asia, are more vulnerable (see Table 15–2).

CO_2 emissions have an economic impact in two ways: (1) they change incomes and unpriced values produced by climatic changes and (2) they add costs for emission abatement. An economically efficient policy selects a mix of emission reduction and adaption at which the marginal benefit of reduced GHG concentrations equals their marginal cost. Such a policy would also find the least-cost way to reduce GHG emissions (Chapter 5).

Analytical Framework

Two concepts provide powerful insights into the benefits and costs of policies designed to mitigate greenhouse gasses: the *greenhouse damage function* and the *greenhouse abatement cost function* (Nordhaus, 1991a). The greenhouse damage function measures the costs to society from incurring damages produced by a changing climate. The function can be used to measure the *benefits* of policies that *reduce* damages from greenhouse warming. That is, any policy that slows GHG emissions would produce a benefit equal to the reduction in greenhouse damages compared to a base policy of no change. Damages included in this function include things like impacts of changing crop yields, land flooded by rising sea levels, increased costs of air conditioning, and other impacts shown in Table 15–2. The greenhouse abatement cost function measures the costs incurred by policies that slow global warming, for example, the costs of switching from fossil fuels to hydroelectric power, planting trees to absorb more CO_2, building structures to contain sea floods, and so forth. We know something about the size of abatement costs, but we know little about the greenhouse damage function.

In addition to these two functions, the familiar concept of economic efficiency is part of the framework for analyzing benefits and costs of policies to mitigate GHGs. The economically efficient level of GHG reductions occurs where the present value of the *marginal cost* of reducing GHG emissions equals the present value of the *marginal benefits* produced by those lower emissions. This also occurs where the present value of marginal *net* benefits equals zero. This economically efficient level of reduction depends on four factors: (1) cost of controlling GHGs, (2) economic damages from greenhouse warming, (3) the discount rate, and (4) the length of time it takes for climate to react to CO_2 emissions. The economically efficient level of GHG control would be nearly zero (no need for a GHG control policy) if there were high costs, low damages, high discounting, and long delays between emissions and climate change. For the opposite case of low abatement costs, high damages, no discounting, and quick response of climate change to emissions, the efficient degree of control is nearly one-third of GHG emissions (Nordhaus, 1991a, b).

Figure 15–1 shows the steps required to proceed from a proposed GHG policy to its economic impact:

- Step 1 (the far left column) defines a pair of policies: the baseline policy of no change and a proposed new policy. Two policies are the minimum required for analysis; at least one policy must be compared to a baseline policy to permit an evaluation. The proposed policy could be an international treaty or a national, regional, or local policy that restricts or otherwise limits its GHG emissions.

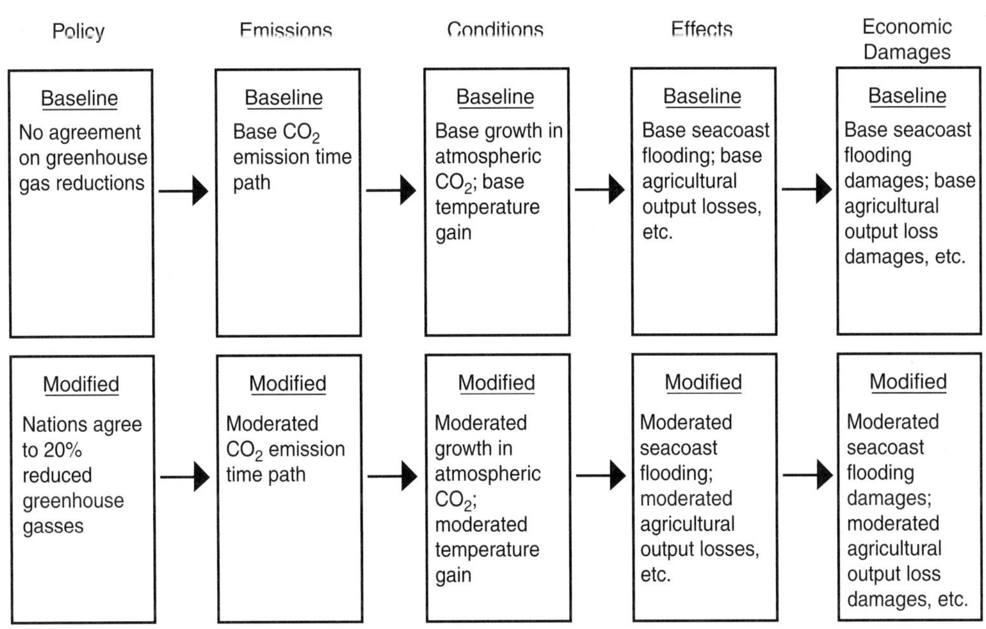

Figure 15–1 Steps required to proceed from a proposed greenhouse gas policy to determination of the policy's economic impact
Source: Adapted from Fisher, A. C., and F. M. Peterson. 1976. The environment in economics: A survey. *Journal of Economic Literature.* 14 (1): 1–33.

- Step 2 (column 2) estimates the impact of the proposed policy on GHG emissions in current and future years.
- Step 3 measures the effect of GHG emissions on the current and future conditions of the atmosphere. Two important dimensions of atmospheric CO_2 are concentrations and temperature.
- Step 4 estimates base and proposed policy impacts on physical or biological effects.
- Step 5 measures economic damages of both the base and the proposed policy, including marketplace efforts as well as nonmarket effects.

Completion of these five steps permits the estimation of benefits and costs of any proposed GHG control policy compared with a baseline policy of no change. Figures 15–2 through 15–4 illustrate important parts of this five-step analysis.

Figure 15–2 CO_2 emissions with and without an agreement to reduce greenhouse gas emissions

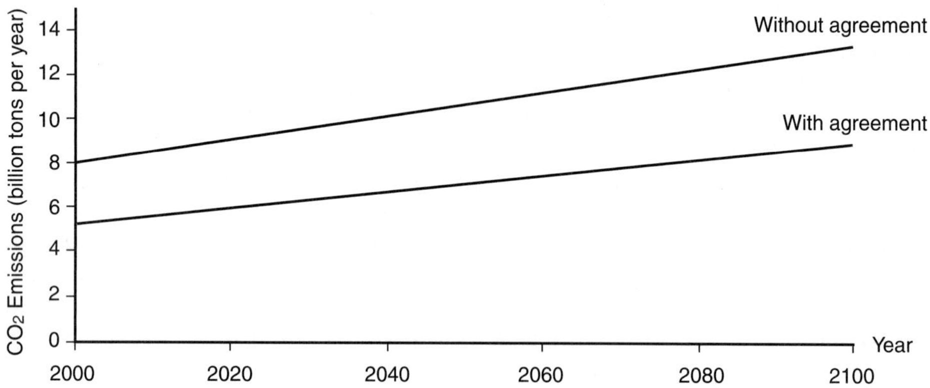

Figure 15–3 Atmospheric CO_2 with and without an agreement to reduce greenhouse gas emissions

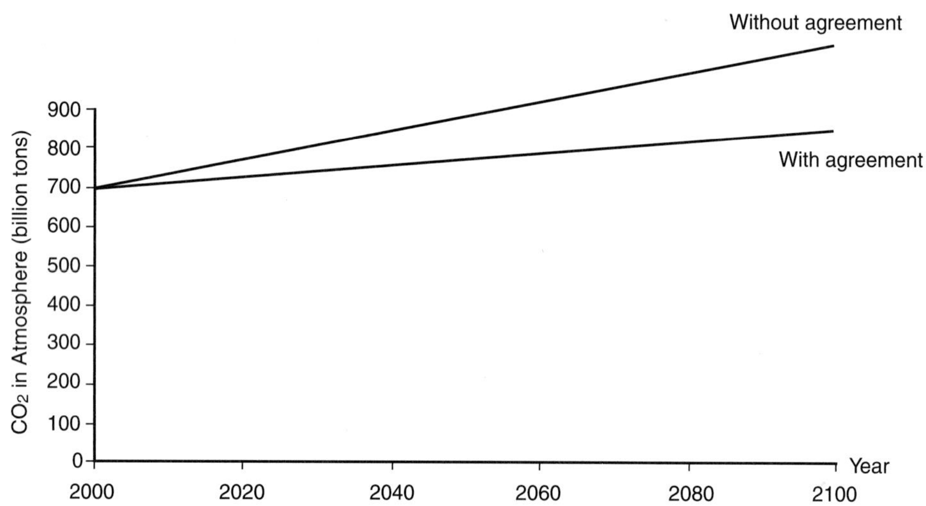

Figure 15–4 Global temperature increase with and without an agreement to reduce greenhouse gas emissions

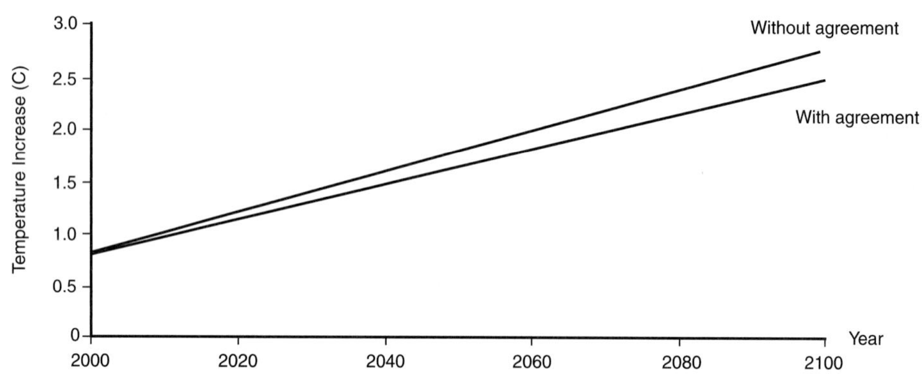

Figure 15–2 shows the kinds of CO_2 emission forecasts needed for step 2. It shows GHG emissions in the years 2000 to 2100 *with* and *without* an international agreement that would limit GHG emissions by 34%. Data for this example are based on the physical science climate models described by Nordhaus (1991a). Figure 15–3 shows plausible atmospheric CO_2 concentrations from 2000 to 2100 with and without the same agreement. Figure 15–4 continues the example by showing forecasted global temperature change with and without the agreement. Notice the very long delay in impact on temperature from an agreement, which is consistent with the global climate models developed by most climate specialists. Rodhe, Charlson, and Anderson (2000) and Stott and Kettleborough (2002) describe some of the uncertainties surrounding climate change models.

Benefits of Controlling Greenhouse Gases

The benefits of controlling GHGs are measured subtracting the greenhouse damages that occur with a proposed level of control from the greenhouse damages that occur with no control. Because CO_2 fertilizes trees and other plants and because warm climates may be attractive, the greenhouse effect itself may produce an economic advantage, that is, negative "benefits" for GHG reductions. Of all the major industries, agriculture is most likely to be affected by climate change (Mendelsohn, Nordhaus, and Shaw, 1994) and would likely receive the greatest benefits from mitigating climate change. Rising sea levels would impose heavy costs on coastal areas, but the impact would likely occur slowly and over many years. Construction, energy, recreation, and water systems would be affected but less so than agriculture.

It is important to guard against inflating the benefits of GHG controls. For example, international trade among nations provides a significant buffer to deal with climate change. If each nation specializes in areas of commodity production and sector development that are least affected by its own climate change, then trade among nations softens the global impact of climate change. Reilly and Hohmann (1993) estimated that trade reduces the economic impacts of climate change by a factor from 5 to 10, because greater specialization decreases impacts of climate change on both production and consumption. The market has the flexibility to adapt to outside events like climate change; failure to be aware of this power exaggerates losses from climate change.

The benefits of reducing emissions occur when lower current emissions reduce future damages that would have been caused by climate change. The methods shown in Figure 15–1 let us translate emission reductions into a marginal benefit function by tracking impacts of a policy reduced emissions to changes in ambient conditions, then into *physical effects,* then into annual *economic impacts,* and finally into the *present value of the economic impact.*

Costs of Controlling Greenhouse Gasses

Costs of controlling GHGs, typically defined as abatement costs, include all costs incurred to limit greenhouse gasses. Projecting the costs of reducing carbon emissions requires making many assumptions about what will happen to the world over very long time periods, both with and without a GHG control program (Shogren and Toman, 2000; Weyant, 1993). Estimates of the GHG abatement costs vary widely. Some studies conclude that the United States could meet its Kyoto Protocol target at almost zero cost (the Protocol is discussed later in this chapter); others claim that the United States would lose between 1% and 2% of its GDP each year. The debates about GHG abatement costs reflect different beliefs about three important assumptions that affect the costs of a GHG policy:

1. **Level of abatement:** Higher costs result from a greater required abatement brought on by an ambitious target level, or if baseline emissions of GHG are high, costs of GHG control depend on the speed of control as well as its scale.
2. **Flexibility of policy instruments:** Most long-term target GHG concentrations can be achieved at much lower costs if abatement is increased gradually over time. Cost savings well

in excess of 50% could be achieved by using a cost-effective strategy for meeting a long-term concentration target instead of a path that mandates more aggressive early reductions.

3. **Potential for new technology:** Some engineering studies conclude that 20% to 25% of current carbon emissions could be eliminated at nearly zero cost if people changed to new technologies (e.g., fluorescent lightbulbs, improved insulation, efficient heating and cooling systems, and more energy-efficient appliances). In addition, future GHG abatement costs may be lower than now because of continued trends toward increased energy efficiency. These trends can be improved by policies that deliver economic incentives for technology that reduce GHGs.

Benefit-Cost Analysis

Few benefit-cost analyses of greenhouse warming have been attempted because of heavy data requirements, major uncertainties in the basic sciences, unexpected institutional innovations that could occur, unknown future adjustments, and unknown scientific or technical advances for reducing emissions. Nevertheless, Nordhaus engaged in such analyses. In one study, he found that at the efficient control level of GHGs, the estimated total global annual costs of reducing emissions was about \$68 billion in 1990, while the total global annual benefits were about \$180 billion (Nordhaus, 1991b). The marginal damage of greenhouse gases was estimated to be about \$5.3 billion per year, or about \$75 per ton of carbon equivalent[1] in the CO_2, which would produce an efficient reduction of 34% of total GHG emissions (the high-damage case of Nordhaus; Tol (2002) and Unruh (2000) present more recent estimates).

Figure 15–5 shows the total global benefits and total costs of various possible percentages of GHG reductions from levels of use in 1990. Nordhaus (1991b) found that the economically optimum level of GHG control will produce a net benefit, compared to no control measures, equal to an annual value \$112 billion (\$180 billion − \$68 billion).

Marginal Benefit. The marginal benefit of reducing CO_2 is the decrease in damages resulting from reducing GHG emissions one more percentage point. The horizontal axis of the marginal benefits function shown in Figure 15–6 measures GHG emissions as a percentage of the uncontrolled quantity. The uncontrolled quantity takes on a value of zero percent when GHGs are not controlled, that is, when GHG emissions operate in an environment without regulations or international agreements.

Marginal Cost. The marginal costs measure the added (opportunity) cost of reducing GHGs by 1%. Economic theory tells us two important things about the shape of the marginal abatement cost function. First, the marginal cost equals zero when there is no GHG control. This means that the first units of GHG reduction can be had at virtually no cost, because at the first unit GHG emissions are unpriced. That is, each emitter continues adding CO_2 emissions until the marginal value of added emissions is zero. Therefore, the value displaced by eliminating that last unit (marginal cost) must be zero. Second, marginal abatement costs increase as the level of abatement increases; that is, greater cost is placed on polluters who reduce GHG emissions by one extra percent when they have already reduced GHG emissions by a high amount than when they have not reduced emissions at all.

Figure 15–6 shows marginal benefits and marginal costs of various possible levels of GHG reductions (Nordhaus, 1991b). The economically efficient level of reduction is 34% compared to baseline levels based on assumptions of high GHG damage. Also shown are reduction levels that

[1] The marginal cost of achieving a 1% reduction in GHGs can be expressed in billions of dollars per year (BDY) or dollars per ton (DPT) of carbon. The marginal cost in DPT equals marginal cost in BDY times one divided by 1% of current carbon output in billions of tons. Since 2002, carbon emissions are about 7 billion tons, the marginal cost of emissions in dollars per ton = \$75 per ton = \$5.3 billion divided by (1% of 7.0 billion).

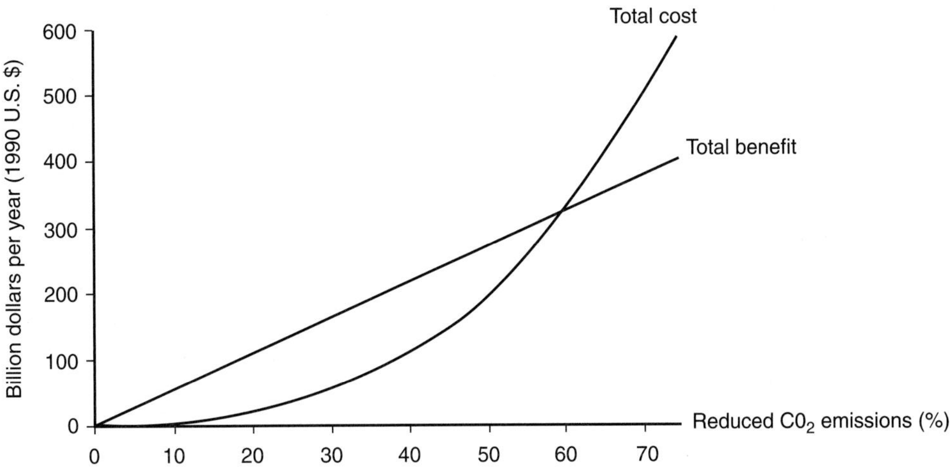

Figure 15–5 Total annual global benefits and costs of an agreement to reduce greenhouse gas emissions

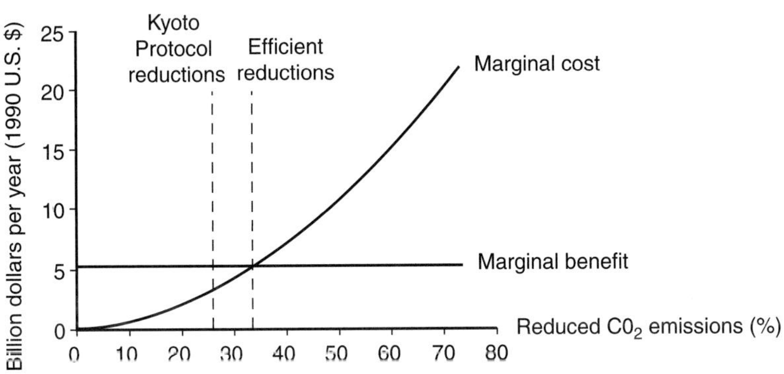

Figure 15–6 Marginal benefit and marginal cost of reducing greenhouse gas emissions

would result from the 1997 Kyoto protocol, in which average global reductions to 5% below 1990 levels have been negotiated.

Policy Options for Dealing with Climate Changes

An important question with major economic implications asks what kinds of policies can be designed to control changes in global climate. Four possible strategies are adjustment, mitigation, prevention, and planet insurance. Under option 1, people *adjust* to the warmer climate if it occurs. This option could take place without any formal public policy or international cooperation among nations. Through the ordinary responses of people, institutions, and private markets, climate gradually warms and sea levels increase. If some areas become less productive, economic resources will leave the areas and migrate to more productive regions. If ocean levels rise, people will gradually leave coastal areas unless explicit public policy decisions are made to protect the areas. This option is much like adjustments made by people who leave flood-prone areas when no flood control structures are built. Under option 2, society attempts to *mitigate* global warming through measures such as planting trees and climate engineering. One estimate found that 100,000 tons of carbon can be neutralized by 1 ton of trace iron particles in the oceans. Little is known with certainty about the effectiveness of any of the mitigation proposals (Nordhaus, 1993b). Option 3 is *prevention,* which receives the most political and popular attention. Most policy debates focus on lowering energy consumption or substituting energy sources that are not based on fossil fuels. Option 4 is *planet insurance.* If society chooses the fourth option, it simply pays whatever abatement costs are needed to ensure that there will be no severe damages from climate change. An efficiency-minded society will design institutions that mix these four options to produce the greatest possible discounted net benefits of GHG control.

Adjustment

In industrialized countries most people's work, income, and leisure do not directly depend on outdoor temperature. In these countries several kinds of adjustments can be made to reduce the impact of global warming, for example, spending more time indoors during the hotter months and more time outdoors during the colder periods. Other adjustments could capitalize on the benefits of higher temperatures, such as taking advantage of longer seasons for outdoor leisure activities.

Mitigation

Trees absorb CO_2 and are fertilized by it. Greater levels of CO_2 are absorbed if more trees are planted, fewer trees are cut, fewer settlement schemes are undertaken that require clearing forests, and tree harvesting is delayed as described in Chapter 8.[2] Despite the beneficial effects of these actions, typically the benefits do not accrue to the individual organizations or businesses that would plant the trees or delay the harvest; instead, the benefits accrue to other people, nations, or generations. Since market forces produced by outcomes of individual decisions produce too little GHG mitigation, some kind of public action is necessary. See Example 15–2. One option is to offset the global warming through climate engineering (Jamieson, 1996; Nordhaus, 1993a), for example, using measures that would increase reflected sunlight or absorb more carbon emissions before they reach the atmosphere. Scientists have suggested two possibilities: launching reflective devices into space and seeding the oceans with iron particles to absorb CO_2. Regardless of what mix of mitigation policies is proposed, economics emphasizes policies that produce the maximum reduction in climate damages for a given level of spending. See Example 15–3.

Prevention

Combustion of fossil fuels produces CO_2. Any strategy for controlling GHGs needs to address this energy combustion. CO_2 emissions can be reduced if producers or consumers use less energy or switch to energy sources other than fossil fuels, such as wind, photovoltaic, hydroelectric, or nuclear power. Carbon taxes and emission trading are two policy measures for reducing CO_2 emissions.

Carbon Taxes. A *carbon tax* could be imposed on atmospheric carbon emissions from fossil fuels like gasoline (petrol), jet fuel, or coal-fired electric power plants. Imposing this tax would raise the price of CO_2 emissions from zero to a level nearer marginal cost. (Baranzini, Goldemberg, and Speck, 2000; Greenaway, 1991). Carbon dioxide is one of many GHGs. To avoid rewarding people who simply switch from CO_2 to another GHG, an efficient tax would be based each GHG's contribution to global warming.

Advantages and Disadvantages. There are five major advantages of a carbon tax (Goulder and Mathai, 2000; Pearce, 1991). First, carbon taxes produce the double dividend of pricing GHGs efficiently while reducing damages to incentives caused by taxes: (1) Remember that most taxes damage incentives by inflating commodity prices and damaging work incentives. However,

---

[2] Trees are examples of carbon sinks, which are reservoirs that absorb carbon. Soil and trees act as natural sinks for carbon; each year hundreds of billions of tons of carbon in the form of CO_2 are absorbed by oceans, soils, and trees.

Example 15–2
New Hampshire's Experience with Climate Change

On May 9, 2002, New Hampshire Governor Jean Shaheen signed HB 284, the New Hampshire Clean Power Act. The law made New Hampshire the first U.S. state to enact a reduction in four common pollutants from power plants, including carbon dioxide. (Massachusetts and California have also enacted greenhouse gas control legislation.) A major partner in passing the Act was the Public Service Company of New Hampshire (PSNH), owner of the state's three fossil fuel power plants. By 2007, PSNH agreed to reduce sulfur dioxide emissions, the chief component of acid rain and airborne soot, by 75% below federal requirements that took effect in 2000. Also by 2007, nitrogen oxide emissions, which cause smog and acid rain, will be reduced by 70%, thereby reducing that pollutant to 90% beneath 1990 levels. Trading emission credits will play a major role in achieving goals. If PSNH cannot meet the new standards with actual emissions reductions, it can buy pollution credits from utilities in other states that have already reduced their emissions.

Tradeable emission permits are a classic free-market solution to the problems caused by negative externalities. Recall from Chapter 4's discussion of market failures that greenhouse gas pollution is a negative externality in which polluters create a larger cost to society than to themselves. Because the polluter does not count the pollution as part of its costs of production, it ignores those costs in its pollution emission decisions. New Hampshire adopted one method of dealing with this problem by relying on economic incentives. Tradeable emission permits allow the government to give companies licenses to pollute at a certain level. Companies can buy, sell, and trade these permits on the market.

Is this license to pollute immoral? Maybe not: The New Hampshire license comes at a price. Even with the license, it is in the selfish interests of companies to pollute as little as possible. If they pollute at a level higher than their permit allows, they have to buy permits from another company. If they pollute less than they are allowed to, they can sell their permit. So the incentives work in the direction of reducing GHG emissions. In many parts of the world, most environmental laws regulate or limit how much polluters can emit, which effectively *give them a permit to pollute at that level.* This command-and-control method damages incentives to reduce pollution, and it raises the cost of complying with the rules if the polluter currently exceeds those standards. Tradeable pollution permits are a less expensive and more effective way to deal with the externality problem.

Source: University of New Hampshire. 2002. *Climate, air quality, and weather stories.* Available from *http://unfccc.int/resource/kpstats.pdf* [Accessed October 4, 2004].

Example 15–3
Engineering Approaches to Climate Change versus Behavioral Approaches

People who believe most strongly that climate change is occurring often resist engineering approaches (EAs) to reverse it, for example, planting trees, fertilizing the oceans, and even unconventional measures like launching mirrors into space. One argument against EAs is based on the idea that it is morally wrong to interfere with natural processes: Nature knows best. Another argument holds that two wrongs do not make a right: Rather than righting a wrong, EAs add a second wrong on top of the first. Indeed, many say the second wrong of engineered compensations correcting for human-induced climate change is worse than the first wrong of inadvertently changing climate, even if the second wrong reclaims the original climate.

Jamieson (1996) concludes that changing human behavior may be economically cheaper and more effective than altering the climate through engineered methods. Human behavior has a long history of adaption to a wide range of external changes. People adapt quickly to emerging conditions, especially when institutions like markets and price systems reward their flexibility. This does not mean that changes in human behavior are simple to implement; but, in principle, human behavior *can* be modified or reversed. Flexibility is not as common in physical systems. Technology can change physical systems so much that it is not economically or technically possible to recover the original states. This is clear in the case of nuclear weapons and nuclear power, but it also could be true in the case of climate change. If we change climate, we may not be able to reverse the change at any economically feasible cost. Since human behavior is changeable through learning, behavior adjustment may be the economically cheapest response strategy for addressing many environmental problems.

Sources: Jamieson, D. 1996. Ethics and intentional climate change. *Climatic Change.* 33 (3): 323–336; Ludwig, D., M. Mangel, and B. Haddad. 2001. Ecology, conservation, and public policy. *Annual Review of Ecology and Systematics.* Palo Alto, CA: Annual Reviews.

a carbon tax improves incentives by requiring producers and consumers of GHG fuel to pay the cost of damages imposed on society by using the fuels. (2) Governments that enact a carbon tax can use the tax revenues to reduce other taxes that damage incentives, such as income taxes.

Second, a carbon tax offers political power by helping to forge international agreements for GHG reduction. Given the widespread use of fossil fuels and their important role in supporting economic development, few developing countries are likely to agree to reduce fossil fuels without some kind of financial incentive. Each country has an incentive to sign up to a GHG reduction agreement only if its net benefit of contributing to GHG control exceeds zero. Without compensation, most developing countries will typically perceive the costs as simply too high compared with the benefits.[3] The use of carbon tax revenues to reduce other taxes internationally could encourage noncooperating nations to cooperate.

Third, carbon taxes could reduce compliance costs for industry, and hence for consumers who bear a part of the tax. High-cost emitters of CO_2 will pay the tax rather than decreasing emissions. Low-cost emitters (those who can reduce emissions at low cost) will invest in carbon emission control because it is cheaper than paying the tax.

Fourth, carbon taxes provide a sustained incentive for CO_2 emitters to invest in carbon-conserving technology. Command-and-control standards are typically linked to a particular technology, so CO_2 emitters who switch to the mandated technology have no incentive to adopt even cleaner technology when it becomes available. But a tax, which always rewards investments in cleaner technology by lowering the cost of the tax whenever CO_2 emissions are reduced, provides sustained incentives to reduce carbon emissions.

Fifth, carbon taxes can be adjusted easily and quickly based on new information. Regulations, by contrast, are much more difficult to adjust. Carbon taxes offer policymakers an instrument for responding to a steady stream of unexpected changes in data, science, technology, and institutions.

There are two disadvantages of carbon taxes. First, it is hard to identify the carbon tax that reduces CO_2 emissions to an economically efficient level. Tax policies will miss the efficient percentage reduction in CO_2 emissions unless the elasticities of marginal benefit and marginal cost (slopes and levels of both functions), such as shown in Figure 15–6, are known accurately. Second, carbon taxes would affect low-income households most severely. This argument has been a central part of resistence to higher gasoline taxes in the United States. The ratio of fossil fuel to income declines as household income increases; therefore, a higher percentage of a poor family's income would be spent on a carbon tax.

Gasoline Taxes. Governments in both the United States and Europe have taxed gasoline (petrol) for many years, typically to finance new roads, and fossil fuel taxes are accepted widely. Few nations, though, have imposed carbon taxes.[4] Because each gallon of gasoline consumed in a vehicle contributes about the same amount to greenhouse warming, the price of gasoline with an added carbon tax should be the same per unit volume everywhere.[5] An economically efficient carbon tax would raise the price of all CO_2 emissions to their real marginal cost. Real marginal cost means the value of all opportunities displaced by the added CO_2 emissions. If each emitter is charged the full marginal cost, the carbon tax produces an incentive for both producers and con-

---

[3] Results of the 1992 Earth Summit and its follow-up conventions (described later in the chapter) have brought out the political and economic importance of the uneven distribution of costs and benefits of controlling GHGs. That uneven distribution continues to be a major stumbling block in international negotiations.

[4] Finland, in 1990, was the first European nation to impose a carbon tax. Though it generates a little less than 1% of total tax revenue, the tax, based on the carbon content of fuels, encourages switching to fuels such as natural gas, which emit less CO_2 per unit energy output when burned.

[5] Gasoline taxes in Europe are more than five times higher than taxes in the United States, and have been this way for years. Higher European gasoline prices use reduce per capita consumption to less than half of U.S. levels. In Europe, cars are smaller, driving distances are shorter, and use of public transportation is more widespread.

sumers to take action such as these: reducing CO_2 emissions, switching from a private car to public transport, changing technologies from coal to hydroelectric or nuclear electric power.

An efficient carbon tax would price CO_2 emissions the same everywhere for the reasons just described. However, gasoline taxes are charged to produce revenues to build roads, and these taxes vary widely among nations. So while these two taxes are a step in the right direction, they could be more economically efficient. The low gasoline tax in the United States sends a signal to the gasoline buyer that the environmental cost of driving in the United States is less than in Europe. The low U.S. price signals U.S. drivers that gasoline is cheap and rewards behavior that involves heavy use of gasoline (e.g., taking long motor vacations, commuting with few passengers per car, and substituting driving for other forms of work and leisure). Higher European gasoline taxes reward the opposite behaviors.

Environmentally sustainable energy systems will likely require technologies that produce fewer GHGs. One way to encourage sustainability is to design better economic incentives. Policies need to motivate people through the price system to alter their behavior and emit fewer GHGs, both in producing and consuming energy. Carbon taxes will produce revenue, and revenue can be used in many ways: reducing other taxes that damage work incentives, paying for costs of controlling emissions (e.g., reforestation), subsidizing solar- or wind-powered electricity, subsidizing development or manufacture of electric cars, or providing tax breaks for solar heating or cooling.

Emission Trading. *Emission trading* refers to a program in which a private property right to emit pollutants is auctioned off by a responsible organization to the highest bidder. The total supply of marketable GHG emission permits is established at a certain level. One widely discussed way to implement an emission trading program would allocate an initial amount of permits (without charge) to each nation. After the initial allocation, each nation could trade permits with other nations, depending on whether it had greater need for the cash or for the permit to emit. For details of emission trading, see Chapter 16.

The trading process could begin by allocating a limited quantity of transferable permits to each nation. The initial allocation might be based on current or forecasted population or CO_2 emission levels, or any other method with political support. Each nation's permitted allocation would allow it to produce up to a given maximum of allowed emissions quantity per unit of time (e.g., per year). After the initial allocation, a nation could incur costs to reduce its emissions below its permitted amount. These costs could be paid for by selling or renting its unused emission permits to other nations that need additional permits to stay in compliance. A trade of cash for permits between two nations lets the buying nation increase its GHGs by no more than the same amount as the selling nation reduces GHGs. All GHG emissions would require a permit.

Some internationally negotiated enforcement organization would be required to ensure that every nation followed the agreed-on rules and that emissions occurred only if there were supporting permits. Even as population or economic growth occurred, the total permit supply, and therefore the total initial amount of GHG emissions allowed globally, would not exceed the initially set allowances. (Total permit numbers allocated could change as scientific information emerged or economic conditions changed.) Emissions trading offers advantages to sellers and buyers of permits:

1. **Reduced cost of emission control:** The tradeability of permits makes the total cost of complying with reduced emission allowances considerably lower than costs if the allowances were not tradeable. A global market in allowances facilitates reduction by nations that can reduce emissions at least cost. The potential to reduce the cost of controlling GHG emissions through market forces is an attractive feature that may increase political support of GHG controls. Seller nations would be those that could easily and cheaply reduce emissions, for example, nations that can substitute well-developed public transport for private transport and water-rich nations that can develop nonfossil power at low cost.

Buyer nations would be those for which reduction in GHG emissions would be expensive or impossible. For example a developing country may use a CO_2-emitting technology like coal-fired power plants because it has no other affordable way to produce power. And in some nations with high levels of private ownership of vehicles, people cannot or will not switch to public transport.

2. **Reduced GHG emissions over time:** One way to reduce emissions over time is to limit the sales of excess emission rights. For example, no more than 90% of unused permits can be sold on any given trade. The allowable percentage could vary based on the need to limit emissions and based on discoveries in the basic climate sciences.
3. **Payment separate from implementation:** Emission trading lets each nation specialize in its comparative advantage. Some nations can contribute to GHG controls by buying a permit rather than paying for installing more expensive controls. Other nations will contribute by using revenue from permit sales to finance less-expensive measures of emission control.
4. **Flexibility in allocation:** Flexibility in allocating initial allowances offers considerable hope for political success. One allocation method is to distribute permits to each nation based on current or forecasted population. This method would allocate more allowances to heavily populated countries, such as China. A moral argument can be made that population is the only legitimate basis for permit allocation: Everybody is treated equally and has an equal right to benefit from control of GHG emissions.

   Heavily populated nations with low GHG emission would benefit. Their current atmospheric conservation would be rewarded in the marketplace, because they would be able to sell large allowances to other nations.

   Another allocation method is based on current or forecasted GHG emissions. This method would assign larger numbers of permits to nations that produce greater current GHG emissions, that is, nations consuming large amounts of energy or fossil fuel power (e.g., China or the United States). Allocating permits based on the criterion of current GHG emissions rewards nations that are heavy emitters and punishes countries that have conserved the global atmosphere.
5. **Banking and innovation:** Some nations may wish to place their emission allowances into a bank (i.e., avoid using them) by simply not selling them, even though they were currently emitting less than their permitted level. *Emission banking* would keep open their option for future permit use. The option provides an insurance policy that allows an emitter to buy the permits back at an affordable price if needed for its future development. One nice outcome for nations who bank permits is that total global current emissions fall in proportion to the total percentage banked.

A permit trading program rewards nations that develop new technologies for reducing emissions. For example a country with a sunny climate (e.g., Egypt or Italy) has a potential comparative advantage in producing photovoltaic power, so it could produce electric power directly from the sun. The climate might attract companies that produce low-cost solar panels for obtaining electric power. An emissions trading program would provide low-cost market incentives for the country to reduce its dependence on fossil fuels. Furthermore, the financial rewards of switching from being a buyer of a GHG emission permits to a seller could help finance the solar panel development.

Designs of emission trading programs must include the following elements for a program to work:

- Clearly defined property right: All program participants must be free to choose what they want to do with the allowances.
- Exclusivity: All benefits must accrue solely to the property right owner and to nobody else.
- Transferability: Permits must be voluntarily marketable, not at a controlled price but at whatever price the market will bear.
- Security: The property right must be secure from seizure by others.

Market forces of supply and demand would determine permit prices, and the price would vary widely and often based on economic and political conditions in each nation. However, trades could only take place after the international authority assigned to carry out the program was confident that (1) the permit seller reduced emissions by at least the permit-stated amount and (2) the permit buyer increased emissions by no more. That is, after the trade occurs, there is no net gain in emissions.[6]

Planet Insurance

Although global warming has received much attention from scientists and politicians since the early 1990s, nobody knows for sure whether global warming is a serious problem. If it poses no problem, then the benefits of policies that deal with it will be zero but the costs will be in the trillions of dollars. Two types of information are essential to the development of a reasoned climate policy: the gains from reducing GHG emissions and the costs of attaining those reductions. Currently there is considerable uncertainty about that information, which can lead to errors in policy development. The cost of errors from overinvesting in GHG control, should it prove a serious issue, are smaller than the costs of underinvesting in it should it prove benign.[7] Governments can guard against this asymmetry of errors by investing in what could be called planet insurance. One way of doing this is to make decisions that would reduce the after effects of greenhouse warming if it occurs. Insurance would not reduce damages produced by climate change; it would, however, arrange for those who are hurt to receive compensation after the change occurs.

In valuing environmental resources such as current climate conditions, irreversibility is an important factor. Once altered, climate cannot be restored to its current condition, at least on a relevant time scale (Chichilnisky and Heal, 1993; Heal and Kristom, 2002). The value of avoiding destruction of a unique resource, such as the current climate, can be considerable. Preserving the resource in its current state lets us change our minds later if we need to. Altering the resource irreversibly throws away that option. So the decision to preserve the resource can be credited with an *option value,* the value of having an option to reconsider the decision at some time in the future.

Statisticians say we make a type I error when we *accept a false hypothesis.* For example, we accept the hypothesis that global warming is serious but it turns out not to matter much. The cost of making this kind of error is measured by the expense of overinvesting resources in protecting the atmosphere or investing resources too soon. Current and future generations pay a needlessly high price for energy and lose real incomes that could have been used for other programs. Planet insurance costs, however, could be large but not ruinous.

We make a type II error when we *reject a correct hypothesis.* For example, we reject the hypothesis that global warming is serious but later events prove it to be so. The cost of making this kind of error is the cost of failure to recognize irreversible ruin from damages to the planet imposed on future generations.

Table 15–3 shows one economist's view of potential policy responses to global warming. Note how the four possible strategies of policy design are applied.

---

[6] For many years the western United States experienced a similar problem with "paper water." The owner of a water right sells a right to a stream that has not been used for many years to a buyer who diverts the water from the stream immediately. Total stream depletions increase, even though there is no added stream flow. Some perfectly legal holder of a water right then faces a dry stream (i.e., has paper water, but no wet water).

[7] Some environmental advocates believe the risks to the planet posed by climate change are so large that some action should be taken now. They see the cost of that action as relatively low compared with the alternative of doing nothing about GHG warming then finding out later that it is a serious problem. In contrast, many business interests view the risks of climate change as too uncertain and the costs too high to justify drastic action. They prefer a wait-and-see policy or, at most, a limited expenditure on GHG emission reductions.

Table 15–3 Potential policy responses to global warming

Adjustment:	Adjust to greenhouse warming
	Market adjustments
	People and capital move to more temperate zones
	Farming areas migrate north
	Government policies
	Build seawalls to protect coastal areas
	Regulate land use
	Support research on flood- and drought-tolerant crops
Mitigation:	Offset effects of greenhouse warming
	Send particles into the atmosphere
	Fertilize ocean with trace iron
	Plant trees
	Reduce timber harvests
Prevention:	Slow greenhouse warming
	Reduce energy demand
	Reduce GHG emissions per unit of energy supply
	Substitute to fuels that produce less CO_2
	Block CO_2 from getting into the atmosphere
Insurance:	Allow enough abatement costs to insure against a catastrophe

Source: *Nordhaus, W. D. 1991a. A sketch of the economics of the greenhouse effect.* The American Economic Review. *81(2): 146–150.*

■ INTERNATIONAL AGREEMENTS ON GLOBAL WARMING

Understanding the problem of global warming and trying to determine what to do about it may have been the most important environmental issues of the late twentieth century. In 1988, as a response to the emerging body of scientific evidence suggesting that man-made emissions of certain gases contributed to global warming, a group of scientists and environmentalists organized to seek an international plan that would slow or reverse global warming. The group recommended holding a framework convention on climate change that would establish a system to promote international cooperation to resolve the scientific and political issues involved in slowing global warming. That year, the United Nations Environmental Program (UNEP) set up the Intergovernmental Panel on Climate Change (IPCC) to make scientific assessments of global warming and to identify the extent to which individual countries would likely be affected.

United Nations Framework (1992)

In 1990, the United Nations General Assembly created the Intergovernmental Negotiating Committee to negotiate a treaty on global warming. At the 1992 Earth Summit in Rio de Janeiro, Brazil, 153 nations signed the resulting United Nations Framework Convention on Climate Change (UNFCCC). That document (United Nations, 1992) called on industrial nations to reduce emissions of carbon dioxide and other heat-trapping gases to 1990 levels by the year 2000. It was the first formal, international statement of concern and agreement to do something about global warming.

The UNFCCC proposed three goals:

1. To stabilize atmospheric GHG concentrations to a level that would avoid dangerous interference with climate systems in a time period that would permit natural ecosystems to adapt to climate change
2. To ensure that such climate changes would not threaten food production
3. To promote sustainable economic development

Although all signatories committed themselves to cooperate in reducing greenhouse emissions, there was a clear understanding that the developed countries had both a greater responsibility and greater capability to reduce emissions. About 75% of emissions were coming from developed countries. The principle of common but differentiated responsibilities gave the developed world some responsibility for helping the developing world lower its emissions without lowering its development.

The developed countries that agreed to assume emission limits and reductions were listed in Annex I of the UNFCCC and are refered to in official documentation as Annex I parties. (Annex I countries include all the states in the Organization of Economic Cooperation and Development plus central and eastern European states with economies in transition.) The UNFCCC committed some developed countries to provide new and additional technological and financial resources to developing countries so they could establish emissions inventories and begin developing clean technologies. Developing countries were scheduled to assume emissions reductions at a later time, contingent on whether the developed countries met their responsibilities.

The UNFCCC was a step toward the objective of a global warming treaty: stabilization of heat-trapping gas concentrations in the atmosphere at a level that would prevent dangerous impacts on the climate. Table 15–4 summarizes key accomplishments at international conferences that have been convened since then and some are described next in more detail.

Berlin (1995)

Beginning in 1995, summits of all the signatories of the UNFCCC, called Conferences of the Parties (COPs), have been held each year in various locations. At COP1, signatories set up a process to identify quantitative targets and timetables for Annex B countries[8] to reduce CO_2 emissions to 1990 levels. This aim is known as the Berlin Mandate, and for the next two years that mandate established and distributed emissions limits and reduction aims. Several issues were involved, including (1) which gases to include in the targets; (2) what starting year to use for comparing later emissions; (3) by which target years emission reductions should occur, (4) by how much countries should reduce emissions by those target years; and (5) the distribution of targets and timetables among all Annex B countries. Overall, the issue was how to efficiently, fairly, and sustainably share the costs of reduced emissions.

The U.S. delegation acknowledged the need for a post-2000 aim for global warming, but it was not prepared to commit to any particulars. The United States called for expanding the effort to combat global warming to include actions by large, rapidly industrializing, developing countries (e.g., China and Indonesia). Developing countries feared that the United States wanted to impose emissions commitments on them that would seriously impede their ability to industrialize. The Berlin conference addressed this issue by calling on developing countries to advance implementation of their existing commitments under the 1992 treaty, but required no binding emissions targets or timetables.

The United States and other industrial countries advanced the principle of joint implementation, which would allow industrial countries to meet part of their commitment for emissions reduction through public or private sector investments in developing countries. They argued that such investments are often less expensive and more effective at reducing GHGs than investments in the home country. For example, a utility building a coal plant in the United States may find it cheaper to offset CO_2 emissions from that plant by jointly investing in energy efficiency measures in Poland or reforestation projects in Central America. In an effort to

---

[8] Annex B countries are those developed countries that agreed to set a target for their GHG emissions. The list of Annex B countries is similar but not identical to Annex I countries.

Table 15–4 Outcomes of international climate change negotiations

1992: Earth Summit, Rio de Janeiro, Brazil
Resulted in the United Nations Framework Convention on Climate Change (UNFCCC), an agreement whose aim was to prevent dangerous man-induced interference with the earth's climate system. Industrial countries included in the treaty's Annex B were to enact policies aimed at reducing their emissions to 1990 levels by the year 2000. However, no particular policies were required, and Annex B countries (similar to Annex I countries) obligated themselves only to try to reduce their emissions not to succeed. UNFCCC was signed by 153 countries and entered into force on March 24, 1994. The United States ratified it in October 1992.

1995: COP1, Berlin, Germany
Resulted in the Berlin Mandate, a statement that the UNFCCC would have little effect on reducing greenhouse gas emissions unless Annex B countries were held to quantified limitations and reductions within specific time frames. Later, this approach became known as setting targets and timetables for emissions reduction.

1996: COP2, Geneva, Switzerland
Called for establishing legally binding emissions targets as earlier proposed at COP1. Rejected the COP1 plan that uniform policies should be imposed so that Annex B had countries flexibility to develop their own policies.

1997: COP3, Kyoto, Japan
Resulted in the Kyoto Protocol, in which most Annex B countries were assigned emissions targets to be achieved by 2008–2012. The average target was about 95% of the country's 1990 emissions. Many of the details of implementation were left for future negotiations.

1998: COP4, Buenos Aires, Argentina
Adopted a two-year plan to design measures for implementing the Kyoto Protocol. Issues that were discussed included financial transfers and a clean development mechanism (CDM) participation by developing countries. Also discussed were measures for incorporating carbon sinks.

1999: COP5, Bonn, Germany
Established methods to monitor progress on the program of work agreed on at COP4.

2000: COP6, The Hague, Netherlands
Intended to complete details on carrying out the Kyoto Protocol but negotiations ended with no agreement. Many issues were left unresolved: how the protocol's mechanisms would operate, what measures would be applied to enforce compliance, the size of allowances for sinks that remove atmospheric CO_2, and whether there would be limits on use of the protocol's flexibility mechanisms.

2001: COP6 followup, Bonn, Germany
Continuation of COP6 following the unresolved issues of The Hague. In March 2001 U.S. President George W. Bush declared that the United States would not participate in the Kyoto Protocol. Other Annex B countries agreed to proceed without U.S. participation. Major sink allowances were given to Canada and Japan. Meeting produced a series of recommendations on implementing the Protocol to be discussed at COP7.

2001: COP7, Marrakesh, Morocco
Formally adopted most of the recommendations of COP6. Completed rules for use of flexibility mechanisms, particularly the CDM. Also established a compliance committee to facilitate, promote, and enforce compliance with the Protocol. In the event of noncompliance, the enforcement branch of the compliance committee may subtract 1.3 times the amount of the violation from the violator's emissions allowance for the subsequent commitment period. Also finished the accounting procedures to be used for sinks.

2002: COP8, New Delhi, India
Formally announced the link between sustainable development and climate change and established the importance of carrying out commitments of previous conferences. Promoted development and use of renewable energy resources and technologies that reduce greenhouse gas emissions. Implemented the CDM by channeling private investment into emissions-reduction projects in developing countries, while offering industrialized countries credits against their Kyoto emission targets.

2003: COP9, Milan, Italy
Announced that climate change remains the most important global challenge to humanity and that its adverse effects are already a reality in all parts of the world. Raised questions about methods for developed countries to finance greenhouse gas reductions in countries most vulnerable to climate change.

2004: COP10, Buenos Aires, Argentina
Discussed the Kyoto Protocol on global warming, which went into force on February 16, 2005, after Russia's ratification in November 2004. Russia's entry raised the worldwide total of ratifying nations to exceed the 55% threshold of existing GHG emissions required for the Kyoto Protocol's implementation. The Protocol will require industrial nations to make substantial cuts in their emissions of greenhouse gases. A goal of the conference was to draw back the United States, which withdrew in March 2001, into discussions about ways to mitigate climate change after 2012, when the Kyoto agreement expires.

Sources: *McKibbin, W. J., and P. J. Wilcoxen. 2002. The role of economics in climate change policy.* Journal of Economic Perspectives *16 (2): 107–129; United Nations. 2003. Ninth Session, Conference of the Parties, Accomplishments; United Nations. 2004. Preliminary Agenda. Tenth Session Conference of the Parties.*

build confidence in joint implementation, the Berlin conference established criteria for joint implementation projects to be initiated during a pilot phase.

Geneva (1996)

The second COP meeting (COP2), which was held in Geneva, Switzerland, in 1996, moved the goal of a strong international global warming treaty closer to reality. In a significant breakthrough, most of the world's environment and foreign ministers approved a strong declaration in favor of legally binding targets to reduce emissions of heat-trapping gases. A major shift in the U.S. negotiating position helped promote support among industrialized countries. The Geneva Declaration advanced the treaty process beyond the 1995 Berlin Mandate in three ways. First, it endorsed the conclusions of the second assessment report of the IPCC, a report that had been criticized by industry groups and OPEC countries since its being released in 1995. The Geneva Declaration stated that the IPCC report work is the most complete and authoritative assessment of the science of climate change and that it provides a scientific foundation for action to control emissions of GHGs. Second, the Geneva Declaration accepted the IPCC's assertion that the continued rise in atmospheric concentrations of heat-trapping gases will result in dangerous interference with the climate. This position is stronger than that of the original treaty signed in Rio. Third, the declaration called on negotiators to speed their work to complete a binding protocol and to include provisions for legally binding objectives and significant reductions (i.e., binding targets and timetables) in GHG emissions. The declaration called for a binding protocol to be signed at the third COP that would occur in 1997.

Kyoto (1997)

A major step forward in climate change negotiations occurred when the Kyoto Protocol was drafted at COP3 in 1997 in Kyoto, Japan (United Nations, 1997). Annex B countries (including but not limited to all Annex I countries of the original 1992 Earth Summit agreement minus Turkey) agreed to establish total emissions thresholds for six gases during varying commitment periods. Table 15–5 shows that the protocol establishes different levels of reductions for each nation, depending on its economic conditions. Reduction targets would start taking effect in 2008 and would be achieved over various five-year periods. Industrialized countries of the world agreed to reduce their emissions of heat-trapping gases by an average of 5.2% below 1990 levels for the first period, which runs from 2008 through 2012. The U.S. delegates agreed to an average reduction of 7% below 1990 levels for this period (implementation requires ratification by the U.S. Senate). European Union countries agreed to jointly meet a target of an 8% reduction from 1990 levels in the first period. Other industrialized countries, such as Russia, received more lenient targets, which makes the average target for industrialized countries about 5% below 1990 levels. If implemented successfully, these targets would reduce annual CO_2 tonnage from 1990 levels of about 8 billion tons per year to about 7.6 billion tons per year. Figure 15–6 shows the difference between the economically efficient level of reductions, about 36% of CO_2, and the reductions agreed to under the Kyoto Protocol, about 27%, compared with the growth rate that would have occurred.[9]

Flexibility Mechanisms. The set of targets and timetables did not resolve the issue of actual emission reductions, because it was clear that several Annex B countries would not meet their

[9] The growth rate of CO_2 emissions without the Kyoto Protocol is based on the assumption that per capita CO_2 emissions would have stayed at the 1980–2000 average of 1.12 tons per capita worldwide, and that population would have grown at the forecasted rate of about one-half a percent per year from the years 2000 to 2100. This produces forecast CO_2 emissions of 10.4 billion tons per year produced by a population of 9.96 billion without the agreement.

Table 15–5 Emissions commitments under the Kyoto Protocol for 2008 through 2012

Country	Emissions target (from 1990 baseline)
European Union	Reduce by 8%
United States	Reduce by 7%
Japan	Reduce by 6%
Russia and some transition countries	Stabilize at 1990 levels
Australia	Increase by 8%
Iceland	Increase by 10%

Source: *Data from United Nations. 1997.* Kyoto Protocol to the United Nations Framework Convention on Climate Change. *Available from http://unfccc.int/resource/docs/convkp/kpeng.pdf [Accessed October 4, 2004].*

agreed-on commitments. Therefore, the Kyoto Protocol laid the foundation for mechanisms by which developed countries could be flexibile in meeting their emission commitments. Essentially these mechanisms permit one country to receive credit for reduced emissions in another country. Three such mechanisms were seriously examined: emission trading, joint implementation, and a clean development mechanism.

Emission Trading. The Kyoto Protocol allows nations with emission targets to trade greenhouse gas allowances. As described earlier in this chapter, emission trading involves issuing of CO_2 credits to each country and trading them according to whether it is cheaper to buy credits or reduce the CO_2 emissions. The advantage of this mechanism is that it is market driven and thereby self-regulating within set parameters such as legislated reduction targets and number of credits issued. How the credits are allocated was not decided. They could be auctioned or they could be allocated to firms on the basis of historical emissions. Within countries, emission taxes (e.g., carbon taxes) may be used rather than emission trading, but the taxes must be linked to the price of permits. The attractiveness of tradeable permits is that any national scheme can function internationally. Under an emission trading system, countries or companies can buy cheaper emission permits from countries that have an excess supply of permits because they met their targets without fully using their permits. Structured effectively, emission trading provides a powerful market incentive to cut emissions while allowing flexibility in taking cost-effective actions.

Joint Implementation. In joint implementation, groups of developed countries work jointly to reduce emissions or enhance sinks (carbon-absorbing sources), thereby lowering the total cost of reducing emissions. Countries with emission targets may obtain credit toward their targets through jointly organized, project-based emission reductions in other such countries. The private sector could make a major contribution in these activities.

Clean Development Mechanism. The Kyoto Protocol established a clean development mechanism (CDM) that allows developing countries to voluntarily participate in projects that will reduce their greenhouse gas emissions. The mechanism promotes two types of projects. The first, which is energy related, is designed to transfer technology to developing countries so that they can achieve economic growth without depending on the fossil fuels as industrialized countries do. The technologies include energy-efficient methods and equipment and renewable energy technology (e.g., wind and solar power). The second type of project is designed to restore and protect forests in developing countries. Forestry projects have the potential to provide many related benefits, such as conserving biodiversity and protecting watersheds and other ecosystems. Nations or companies that make investments in forestry would receive emission credits that could be used to comply with their treaty commitments.

The CDM supports the mechanism of joint implementation for credit in developing countries. Developed countries can consider emissions reductions from project activities in developing countries as contributions to their compliance targets. Thus, companies in the developed world can participate in cooperative projects to reduce emissions in the developing world. The companies will reduce emissions at lower costs than they could in their home country, while the developing countries will receive the kind of technology that allows them to grow more sustainably. Under the CDM, companies can choose to make investments in projects or to buy emission reductions. In addition, companies are encouraged to set aside a small part of proceeds to help particularly vulnerable developing countries, such as island nations subject to coastal flooding, adapt to the environmental consequences of climate change.

Treaty Accomplishments. The Kyoto Protocol is a major step in ongoing international efforts to limit global warming. The developing countries have insisted that industrial countries make the first reductions in emissions, but most have agreed that all countries ultimately need to accept limits on emissions to protect the planet from global warming. Two accomplishments of the treaty are noteworthy:

- The Protocol embraces the principle of international emission trading. This would allow countries that achieve greater reductions (beyond their targets) to trade its credits to other countries.
- The Protocol allows some counting of sinks toward a country's target for emission reduction. Sinks are CO_2 storehouses; the main example is forests. Forests absorb CO_2 from the air as they grow and reduce the speed of its buildup in the atmosphere. As yet there is no consensus on the best method for assessing sinks.

Coalitions. Three main coalitions have emerged in global climate negotiations. The *Umbrella Group,* which was formed at the talks in Kyoto, includes the United States, Australia, Canada, Russia, Ukraine, New Zealand, Norway, and Japan. All are industrialized nations outside the European Union, and they favor maximum freedom in how countries are allowed to reduce their emissions. The *G-77 and China* (G-77/China) includes more than 130 nations, most of which want a treaty. This coalition also includes oil-producing nations, such as Kuwait, which fear revenue losses if a treaty takes effect and reduces global demand for oil. The *European Union* consists of 25 European nations working together to set a single global warming policy for the group.

Buenos Aires (1998)

The first major meeting to follow the Kyoto conference was held in Buenos Aires, Argentina, in November 1998 (COP4). The meeting resulted in the Buenos Aires Plan of Action, which produced major accomplishments: First, it strengthened implementation of major provisions of the original 1992 Rio treaty on technology transfer and assistance to developing countries in adapting to the impacts of climate change. Second, it set out a scope of work and a deadline for decisions on unfinished business from Kyoto. It established that a treaty will be in force when two conditions are met: 55 countries sign and ratify the treaty and those countries represent at least 55% of the emissions from industrialized countries. This has made it difficult for the treaty to be in force without U.S. ratification.

Marrakesh (2001)

The seventh COP was held in November 2001 in Marrakesh, Morocco, and included 172 governments. The aims of the conference were to complete agreements on the details for carrying out commitments on reducing GHG emissions under the 1997 Kyoto Protocol. See

Example 15–4
Climate Change Negotiation in Brazil

Brazil has a major stake in the climate change issue, because many of its economic interests require the growth of its energy sector and development of its Amazon forest region. Brazil could contribute to achieving an international agreement that will slow climate change and promote its own national interests, but the country faces conflicting economic incentives when deciding its role.

The size and scale of its tropical forests makes Brazil's forestry decisions important to it and to the rest of the world. Deforestation of the Amazon poses a double threat for global warming: Burning of forests releases large amounts of carbon into the air and that burning robs the burned area of a productive carbon sink. Brazil has critical interest in how changes in forests are figured in emission calculations because of the huge size of the Amazon, as well as the importance of the Amazon in its development plans. The decision to cut more or fewer trees in the Amazon presents a very different stream of benefits and costs for Brazil than it does for the rest of the world.

Brazil's negotiating stance, therefore, is a product of the kind and amount of its emissions, which come from two sources: energy use and changes in land use changes, mostly deforestation. Like most developing countries, Brazil's energy emissions are low partly because incomes are so low that each household can afford little energy. Deforestation of the Amazon, most of which lies in Brazil, is the major contributor to the country's greenhouse gas emissions. According to Brazil's own estimates, about 12% of the Amazon has been deforested. Deforestation rates continue to increase each year. Why does Brazil continue to lose trees? The economic answer is that Brazil's benefits from harvesting Amazon trees exceed its costs. The deforestation occurs in several ways: clearing of trees for farming and ranching, logging, flooding caused by building hydroelectric dams and reservoirs, and infrastructure and industrial development (e.g., building roads and mines).

Any future Brazilian commitments to reduce greenhouse gas emissions will require controlling its deforestation, which has proven elusive to date. Daunting obstacles block Brazil's efforts to control deforestation:

- Immense size of the Amazon.
- Importance of harvesting the Amazon's trees as a resource for economic and social goals.
- Weak federal coordination among the agencies charged with protecting the forest and its people.
- Developed countries that have little economic incentive to join an international green timber certification system.

Brazil also has considerable interest in maintaining sovereignty over its natural resources, particularly the Amazon forest. The Brazilian government considers the Amazon to be a natural resource to be used in the country's quest for economic growth and social development. There is also a national security component to its interest: Brazil's military has argued for years that control of the Amazon frontier is essential to the nation's defense. Here is the challenge to the international community interested in slowing global warming: Help Brazil control deforestation through economic incentives so that the control produces benefits to Brazil that exceeds its costs. Designing and financing these incentives are a major challenge for the world's environmental future.

Sources: Johnson, K. 2001. Brazil and the politics of the climate change negotiations. *Journal of Environment and Development.* 10 (2): 178–206; Intergovernmental Panel on Climate Change 1995. *Second assessment synthesis of scientific technical information relevant to interpreting Article 2 of the UN Framework Convention on Climate Change.* Available from *http://www.ipcc.ch/pub/sarsyn.htm* [Accessed October 4, 2004].

Example 15–4. The European Union and the G-77/China were determined to organize and promote a deal that would ensure sufficient ratifications of the Kyoto Protocol. A deal was struck and ratification of the Protocol became a more realistic possibility. In making this deal, the EU and the G-77/China conceded to many of the demands of key Umbrella Group countries. See Example 15–5. Knowing that their participation was essential for the Protocol to be ratified, the Russian Federation, Japan, Australia, and Canada used their leverage to weaken the conditions needed to achieve ratification. The extent of their bargaining power, as well as the level of their determination to protect national interests (as predicted by the economic the-

Example 15–5
What Climate Change Means for the Dry Country of Egypt

Four thousand years ago, the prophet Joseph warned Egypt's king of seven years of abundance and seven years of famine. Egypt's king knew his people would not plan for the future while times were good, so he put Joseph in charge of Egypt's food policy. Joseph decided that during the seven years of abundance, when the Nile River flowed full and produced bountiful crop yields, he would collect and store quantities of grain so that food would be available for the coming bad times. Sure enough, when famine spread over Egypt, Joseph took grain from storage and sold it all over the country. He also earned gold for Egypt by selling grain to other countries.

Joseph avoided a devastating famine through a series of brilliant food storage and water management actions, and he brought to Egypt a surplus in gold as well as a food management plan that was the envy of the ancient world. Today, Egyptian planners lack the luxury of perfect forecasts. Instead, they rely on understanding and analysis of physical, economic, and cultural systems. The scientific understanding of these systems allows for better planning to meet local and regional climate change, unexpected events, and uncertainty surrounding the economy and culture. The real possibility for climate changes due to increased greenhouse gas emissions highlights the importance of development plans.

Egypt depends on water supplies that are sensitive to climate change. A large part of productive farmland in Egypt's Nile Delta would be greatly affected by increases in sea levels caused by climate change. Farmers need water from the Nile for irrigation, a water supply which is itself vulnerable to precipitation and temperature changes inside the entire Nile basin, which starts as a few drops of water high in the mountains of central Africa.

Egypt has no economically feasible water supply options other than the Nile itself. Climate change could affect flow in the Nile and therefore Egypt's water supply. Furthermore, Nile stream flows will likely be influenced by upstream development in countries to the south of Egypt such as Sudan and Ethiopia. Increased temperatures would increase water use for crops and reduce crop yields, thereby increasing food prices and reducing real incomes throughout Egypt. Like other developing countries whose futures are tied to climate change, Egypt takes a special interest in global negotiations to limit greenhouse gasses.

Source: Yates, D., and K. Strzepek. 1998. An assessment of integrated climate change impacts on the agricultural economy of Egypt. *Climatic Change*. 38 (3): 261–287.

ory) became most apparent. See Example 15–6. Although the deal struck satisfied everybody, the European Union and G-77/China gained the most with almost all of their demands adopted.

A key issue that will make or break compliance with the agreement of COP7 relates to the legal nature of decisions adopted by the enforcement branch of the compliance committe established by COP7. Some observers argue that there are two reasons why measures that ensure consequences of decisions will be legally binding. First, although it was agreed that only parties to the Kyoto Protocol will decide on the methods of adoption of decisions, adoption is expected to be made by an amendment, which enables the enforcement branch to decide legally binding consequences. Second, by providing that a party's eligibility to participate depends on its compliance procedures, the agreement provides incentives for everybody to ratify the compliance amendment.

The value of the Kyoto Protocol may lie not in its direct impact on limiting greenhouse gas levels, but rather in serving as a precedent for concerted and coordinated global action on climate change. In this regard, the Marrakesh negotiation sends the right signal, as few parties have an excuse not to ratify the Protocol. Furthermore, the Marrakesh accomplishments may provide the United States with an acceptable legal structure to join the process of combating global climate change.

Example 15–6
Economic Challenges of a Climate Change Treaty

National sovereignty presents a problem in the search for a greenhouse gas (GHG) treaty. Any GHG treaty that constrains a nation's sovereignty is likely to be rejected. Even if it is ratified in the spirit of good citizenship, it may be ignored later by citizens who are asked to pay for its implementation. A national government can coerce individual citizens into following the law through use of its police powers (if supported by the majority), but the power of coercion breaks down at the international level. No international organization designed to date can coerce individual sovereign countries to carry out a climate change agreement that the country finds incompatible with its own interest. This is an example of the classic public goods problem: Each nation has an incentive to contribute nothing to reduce GHGs and to free ride on other nations' contributions.

The Kyoto Protocol to establish emissions thresholds for six gases was signed in 1997. The Protocol specified that two conditions must be met before it became binding: It must be ratified by at least 55 countries and it must be ratified by countries accounting for at least 55% of Annex B emissions. By August 2004, the Protocol had been ratified by 64 countries contributing to 44.2% of Annex B emissions. The United States, acting on its perceived self-interest, withdrew from negotiations in March 2001, which made the second condition hard to satisfy. The United States alone accounts for about 33% of 1990 Annex B emissions.

Designing a workable international GHG agreement requires making it economically attractive for governments to address domestic costs of people and firms who comply with GHG reductions. This economic principle has been noticed. One important policy instrument of the Kyoto Protocol is a system of internationally tradeable emission permits. But *international* permit trading runs the risk of being inefficient, given uncertainties in the marginal cost of abating greenhouse gas emissions. International permit trading also has a potentially serious political flaw: It may generate large transfers of wealth between countries. Moreover, each individual government has a selfish motive to avoid policing the agreement. Monitoring carbon dioxide polluters is costly, and punishing violators charges domestic residents a price in return for global climate benefits that will accrue mostly to foreigners and mostly to future generations. So each government has a strong incentive to ignore the illegal behavior of domestic firms that exceed their emission permits. Like any other public good, for a GHG treaty to work, each participating country needs to have confidence that all other participants enforce their commitments.

Sources: McKibbin, W. J., and P. J. Wilcoxen. 2002. The role of economics in climate change policy. *Journal of Economic Perspectives.* 16 (2): 107–129; United Nations Framework Convention on Climate Change. 2004. *Kyoto Protocol, status of ratification.* Available from *http://unfccc.int/resource/kpstats.pdf* [Accessed October 4, 2004].

■ SUMMARY

The problem of climate change is unique because the large costs of actions are incurred long before benefits can be realized or measured with confidence. Climate change and the expenses of its control are likely to have different effects on different economic sectors and in different nations. Sectors that rely heavily on uncontrolled ecosystems, which depend on naturally occurring rainfall, runoff, or temperatures, are most influenced by climate change.

Two functions explain the benefits and costs of policies designed to mitigate greenhouse gases. The greenhouse damage function measures the costs to society from incurring damages produced by a changing climate (e.g., impacts of changing crop yields, land flooded by rising sea levels, increased cooling costs). It can be used to measure the *benefits* of policies that *reduce* damages from greenhouse warming. Any policy that slows GHG emissions would produce a benefit equal to the reduction in greenhouse damages compared with a base policy of no change.

The abatement cost function measures the costs incurred by policies that slow greenhouse warming (e.g., the cost of switching from fossil fuel to other energy sources like hydroelectricity or solar electricity, planting trees or crops to absorb more CO_2, building to contain sea floods). We know something about the size of abatement costs, but we know little about the greenhouse damage function. The economically efficient level of GHG reductions occurs where the present value of the marginal cost of reducing GHG emissions equals the present value of the marginal benefits produced by those lower emissions. The benefits of controlling GHGs are measured this way: greenhouse damages with no control minus greenhouse damages with a proposed level of control. Marginal abatement costs measure the added (opportunity) cost of reducing GHGs by 1%.

Four kinds of policy options deal with climate change. Under the adjustment option, people adjust to the warmer climate if it comes; no formal public policy response is required. Society can also mitigate GHG warming (the second option) through measures such as planting trees and climate engineering. Little is known about the effectiveness of these actions. The third option is prevention through measures such as reducing energy consumption or substituting nonfossil sources of energy. The fourth option is planet insurance in which society simply pays whatever abatement costs are required to ensure there are no severe damages as a result of climate change.

International negotiations on climate change policy began seriously in the early 1990s at the Earth Summit in Rio de Janeiro organized by the United Nations. The outcome of the summit was the United Nations Framework Convention on Climate Change (UNFCCC), which was signed and ratified by most countries of the world. The goal of the UNFCCC was to stabilize emissions of greenhouse gases at 1990 levels by the year 2000 through voluntary actions taken by individual countries.

The convention established a mechanism by which negotiations could continue at periodic Conferences of the Parties (COPs). Between 1995 and 2004, there were several COPs, commonly referred to by number as COP1, COP2, and the like. The main decision reached at COP1 in 1995 in Berlin was that for UNFCCC to have a significant effect on reducing greenhouse gas emissions, individual countries had to be held to quantified limitation and reduction aims within specified time frames, an approach later described as setting targets and timetables for emissions reduction. All subsequent COPs have aimed at designing an international treaty in which participating countries would agree to achieve specific targets for greenhouse gas emissions by a set date. The resulting Kyoto Protocol, initially adopted at COP3 in 1997, has been revised and refined in subsequent meetings.

Even after the Kyoto Protocol was ratified by a sufficient number of nations in 2004, national sovereignty is an issue. Any treaty that constrains a nation's sovereignty risks being not accepted by that country. Even if it ratifies the treaty, the country may ignore it later. A single nation's government can coerce individual citizens into following its laws through use of its police powers, but no international organization designed to date can coerce sovereign countries to carry out a climate change agreement that the country finds incompatible with its own interest. Designing an effective cooperation mechanism in which all countries benefit from the actions of each nation's effort to limit their GHG emissions presents major research and policy challenges.

Review Questions

1. Why it is hard for nations to reach a global agreement about the economically efficient amount of climate change mitigation to undertake?

2. Climate change policymaking requires assessment of the benefits and costs of climate change mitigation policies. How are the benefits measured?

3. Explain the double dividend from carbon taxation when carbon tax revenues are recycled back into the economy.

4. Using the standards of economic efficiency and administrative ease, compare the advantages and limits of using carbon taxes and tradeable permits to control carbon dioxide emissions.

5. In the future, a number of developing countries will likely contribute far more greenhouse gases to the atmosphere than they do at present. What regions pose the greatest threats and why would they wish to avoid promoting policies to limit their GHG emissions?

6. Large differences between developing and developed countries underlie much of the international conflict over climate change policy. Discuss major differences in ways countries approach development of global warming policies.

7. Why is the Kyoto Protocol a significant, though limited, international agreement?

Problems

1. (requires calculus) Suppose that the cost, C, of reducing GHG emissions increases with higher reductions as well as with higher existing emission levels, E. The emission reduction for the ith country is defined as R_i. Suppose the formula is

$$C_i = \delta_i E_i R_i^2$$

where δ_i is a number unique to the ith country. Suppose also that the benefits, B_i, of emission reduction for the same country depend on the emission reduction undertaken by *all* countries including itself:

$$B_i = \alpha_i \sum_j E_j R_j$$

where E_j and R_j refer to emissions and reductions of the jth country. The term α_i is a number unique to the ith country and is larger for countries that suffer more from greater temperature increases. The benefits formula says that total emission and emission reductions from all countries affect the benefits received by each country.

a. Calculate emission reductions if each country acts alone to maximize $B_i - C_i$.

b. Calculate efficient reductions if all countries cooperate to maximize global net benefits.

c. Calculate the difference in global net benefits for the two cases.

d. Is global cooperation worthwhile to reduce GHG emissions?

e. Is it worthwhile for any nation to act on its own to reduce GHG emissions?

f. What actions would produce more global cooperation to reduce GHG emissions?

2. (calculus not required) Use the same benefit and cost formulas from problem 1 but assume there are only two countries in the world.

	δ	α	E
Country 1	1	2	100
Country 2	1	2	200

a. Calculate emission reductions if each country acts alone to maximize net benefits. (Hint: Calculate each country's net benefits, $B_i - C_i$, for emission reduction levels equal to 0, 1, and 2.)

b. Calculate the efficient emission reductions if both countries cooperate to maximize global net benefits. (Hint: Calculate the nine combinations of emission reductions, $R_1 = 0$, 1, and 2, and $R_2 = 0$, 1, and 2. Select the one combination of R_1 and R_2 that maximizes global net benefits.)
c. What is the difference in global net benefits for the two countries?
d. Is it worthwhile for both countries to cooperate to reduce GHG emissions?
e. Is it worthwhile for either country to act on its own to reduce GHG emissions?
f. What actions would bring about the kind of global cooperation needed to reduce GHG emissions to the globally efficient level?

References

Albritton, D. L., and L. G. Meira Filho. 2001. *The scientific basis, technical summary.* Working Group I to the Third Assessment Report of the Intergovernmental Panel on Climate Change. Available from *http://www.ipcc.ch/* [Accessed July 21, 2003].

Alley, R. B., J. Marotzke, W. D. Nordhaus, J. T. Overpeck, D. M. Peteet, R. A. Pielke, R. T. Pierrehumbert, P. B. Rhines, T. F. Stocker, L. D. Talley, and J. M. Wallace. 2003. Abrupt climate change. *Science.* 299 (5615): 2005–2010.

Baranzini, A., J. Goldemberg, and S. Speck. 2000. A future for carbon taxes. *Ecological Economics.* 32 (3): 395–412.

Chichilnisky, G., and G. Heal. 1993. Global environmental risks. *Journal of Economic Perspectives.* 7 (4): 65–86.

Cline, W. R. 1991. Scientific basis for the greenhouse effect. *The Economic Journal.* 101 (407): 904–919.

———. 1992. *The Economics of Global Warming.* Washington, D.C.: Institute for International Economics.

Cooper, R. N. 2000. International approaches to global climate change. *World Bank Research Observer.* 15 (2): 145–172.

Goulder, L., and K. Mathai. 2000. Optimal CO^2, abatement in the presence of induced technological change. *Journal of Environmental Economics and Management.* 39 (1): 1–38.

Greenaway, D. 1991. Economic aspects of global warming. Editorial note. *The Economic Journal.* 101 (407): 902–903.

Heal, G., and B. Kristrom. 2002. Uncertainty and climate change. *Environmental and Resource Economics.* 22 (1–2): 3–39.

Intergovernmental Panel on Climate Change. 2001. *IPCC Third Assessment Report: Climate Change 2001.* Available from *http://www.ipcc.ch/* [Accessed October 4, 2004].

Jamieson, D. 1996. Ethics and intentional climate change. *Climatic Change.* 33 (3): 323–336.

McKibbin, W. J., and P. J. Wilcoxen. 2002. The role of economics in climate change policy. *Journal of Economic Perspectives.* 16 (2): 107–129.

Mendelsohn, R., W. Morrison, M. Schlesinger, and N. Andronova. 2000. Country-specific market impacts of climate change. *Climatic Change.* 45 (3–4): 553–569.

Mendelsohn, R., W. D. Nordhaus, and D. Shaw. 1994. The impact of global warming on agriculture: A Ricardian analysis. *The American Economic Review.* 84 (4): 753–771.

Nordhaus, W. D. 1991a. A sketch of the economics of the greenhouse effect. *The American Economic Review.* 81 (2): 146–150.

———. 1991b. To slow or not to slow: The economics of the greenhouse effect. *The Economic Journal.* 101 (407): 920–937.

———. 1993a. Optimal greenhouse-gas reductions and tax policy in the "DICE" model. *The American Economic Review.* 83 (2): 313–317.

———. 1993b. Reflections on the economics of climate change. *Journal of Economic Perspectives.* 7 (4): 11–25.

Pearce, D. 1991. The role of carbon taxes in adjusting to global warming. *The Economic Journal.* 101 (407): 938–948.

Poterba, J. M. 1993. Global warming policy: A public finance perspective. *Journal of Economic Perspectives.* 7 (4): 47–63.

Reilly, J., and N. Hohmann. 1993. Climate change and agriculture: The role of international trade. *American Economic Review.* 83 (2): 306–323.

Rodhe, H., R. J. Charlson, and T. L. Anderson. 2000. Avoiding circular logic in climate modeling. *Climatic Change.* 44 (4): 419–422.

Rosenweig, C., and M. L. Parry. 1994. Potential impact of climate change on world food supply. *Nature.* 367: 133–138.

Schelling, T. 1992. Some economics of global warming. *The American Economic Review.* 82 (1): 1–14.

Shogren, J., and M. Toman. 2000. *How much climate change is too much? An economics perspective.* Resources for the Future. Available from *http://www.rff.org* [Accessed July 21, 2003].

Stott, P. A., and J. A. Kettleborough. 2002. Origins and estimates of uncertainty in predictions of twenty-first century temperature rise. *Nature.* 416 (6882): 723–726.

Tol, R. S. J. 2002. Estimates of the damage costs of climate change, part 1: Benchmark estimates. *Environmental and Resource Economics.* 21 (1): 47–73.

United Nations. 1992. *United Nations Framework Convention on Climate Change.* Available from *http://unfccc.int/resource/docs/convkp/conveng.pdf* [Accessed October 4, 2004].

———. 1997. *Kyoto Protocol to the United Nations Framework Convention on Climate Change.* Available from *http://unfccc.int/resource/docs/convkp/kpeng.pdf* [Accessed October 4, 2004].

———. 2003. *Ninth session, Conference of the Parties, Accomplishments.* Available from *http://unfccc.int/cop9/* [Accessed October 4, 2004].

———. 2004. *Preliminary agenda, Tenth session, Conference of the Parties.* Available from *http://unfccc.int/cop10/* [Accessed October 4, 2004].

Unruh, G. C. 2000. Understanding carbon lock-in. *Energy Policy.* 28 (12): 817–830.

Weyant, J. P. 1993. Costs of reducing global carbon emissions. *Journal of Economic Perspectives.* 7 (4): 27–46.

Wigley, T. M. L., and S. C. B. Raper. 2001. Interpretation of high projections for global-mean warming. *Science.* 293 (5529): 451–454.

Chapter 16

Environmental Quality and Pollution

Introduction

Rachel Carson was born in 1907 on her parent's farm in Springdale, Pennsylvania. She spent much of her childhood on the farm exploring the outdoors and writing, and she published her first work at the age of 10 in a children's magazine. She graduated in 1929 from what is now Chatham College. In 1936 she passed the Civil Service exam and was hired by the U.S. Bureau of Fisheries. In 1951, Rachel Carson published *The Sea Around Us,* which appeared on the *New York Times* bestseller list for a year and a half. After the commercial success of her book, she quit her job and devoted her energy to full-time writing, completing *The Edge of the Sea* in 1955.

In the mid-1940s Carson expressed growing concern about the use of newly invented pesticides, especially DDT.[1] The more she learned about the use of pesticides, the more she worried about their environmental costs. She began research on what would become her most famous and lasting work, *Silent Spring,* which focused largely on the environmental costs of pesticides (Carson, 1962). Carson also was aware of the benefits produced by pesticides, namely, higher crop yields and increased standards of living that cheaper food prices bring to poor people everywhere.

Today, more than 40 years after *Silent Spring* was published, environmental realities such as air and water pollution, use of toxic substances, global warming, and their impacts on human health and welfare, still present hard choices. Large-scale abatement of environmental pollution will reduce our access to cheap energy, food, transportation, housing, and a host of other valued services. But the consequences of ignoring pollution problems are dangerous to our health and welfare. Economic analysis has a vital role to play in the hard choices involved in designing and implementing plans for dealing with the quality of the environment.

[1] Dichlorodiphenyltrichloroethane (DDT) is a pesticide used to control mosquitoes and other insects.

Outline

Physical Basis of Environmental Pollution 446

Desirable Outcome Criteria 448

Unregulated Market Outcomes 461

Institutional Designs for Controlling Pollution 461

Objectives

After studying this chapter, you should be able to

- Describe the physical nature of environmental pollution
- Identify desirable outcomes for policies that improve environmental quality
- Analyze environmental quality outcomes resulting from an unregulated market
- Evaluate policy designs that influence the quality of the environment

■ Physical Basis of Environmental Pollution

The physical aspects of environmental pollution can be analyzed through four dimensions: (1) absorptive capacity, that is, stock versus flow; (2) spatial scale, that is, local versus regional versus global; (3) origin, that is, point versus nonpoint sources; and (4) occurrence, that is, continuous versus episodic.

Absorptive Capacity: Stock versus Flow

One simple and important dimension of environmental pollutants is whether they accumulate over time (*stock pollutants*) or tend to dissipate soon after being emitted (*flow pollutants*). Stock pollutants are those for which the environment has little or no absorptive capacity. Over time, these pollutants accumulate as emissions continue to enter the environment. Glass bottles and metal cans lying about the countryside are stock pollutants, as are heavy metals like lead that accumulate where they are emitted in soils. Policies dealing with stock pollutants must consider their allocations over multiple time periods; these pollutants cannot be analyzed in a one-period (static) framework.

Flow pollutants are those for which the environment has some absorptive capacity. These pollutants, which tend to dissipate over time, do not accumulate if their emission rate is below the environment's carrying capacity. A classic example of a flow pollutant is noise; as long as the source produces noise, it is emitted into the environment. As soon the source shuts down, the noise stops. Flow pollutants include a range of organic materials released into an oxygen-rich water body and altered by local bacteria into less harmful, inorganic material. Carbon dioxide is absorbed by plant life and the oceans, at least up to a point. Flow pollutant policies can be dealt with effectively in a single-period analyis; that is, an analysis of a single period's benefits and costs can provide a framework for analyzing flow pollutants.[2]

Stock pollutants accumulate in the environment in nearly the same amounts as they are emitted. Radioactive waste, for example, decays so slowly that for all practical purposes it will be with us permanently. Plastics are another example: They decay very slowly by human standards, so what we emit into the environment will be with us almost permanently. For both stock and flow pollutants, the economic policy problem is the same: trying to compute the environmental damages and comparing those damages to the costs of reducing emissions. Although the problem sounds the same for both kinds of pollutants, the task is much harder for stock than for flow pollutants. With flow emissions, ambient concentrations in the environment depend only on current emissions, so shutting off the emission source soon produces zero ambient concentrations. With stock emissions, the relationship is more complex. Accumulation of a pollutant over time in the environment breaks the link between current emissions and current damages. It also complicates the science needed to remediate the problem. The cause-and-effect between emissions and damages is harder to identify when there is a long (and sometimes unknown) time period between the two.

Stock pollutants also present a human problem: Principles of long-term economics (see Chapter 6) remind us that it is hard to get people to focus on long-run damages produced by today's emissions. If today's emissions produce large current benefits (in low product prices and high living standards), and if the costs of abating those emissions are high, history indicates that people will not worry much about the future. Many people believe that future generations will be richer than they are today, and this belief diminishes the incentive for policymakers to worry about the welfare of future generations.

---

[2] Stock and flow pollutants are polar extremes. Between those two poles, there are many kinds of emissions that are partly, but not completely, stock pollutants. For example, consider the case of organic wastes emitted from municipal waste treatment plants. Once emitted the emissions are subject to various chemical processes that break down the organic material into more benign material. Most water bodies have a natural assimilative capacity that allows them to accept some organic materials and render them more benign. When the water body's assimilative capacity is not adequate in a particular case, the emission source can be shut down; in a fairly short period of time, water quality will return to normal levels. But once emissions exceed that carrying capacity, the emissions become stock pollutants.

Spatial Scale: Local versus Regional versus Global

Some emissions produce impacts in restricted, local regions, whereas others produce impacts over wider regions, including the global environment. Emissions of noise and a degraded visual environment typically produce only local impacts; damages produced by any particular source may be limited to small numbers of people in a well-defined region. Even if the damages are great to that region, those damages are limited in their geographic sphere. If there are similar emissions in various places around the world, each pollution source can be dealt with independently and by local people who are most affected.

Other pollutants, such as acid rain, produce widespread effects over larger regions. For example, common technology used by many electric utilities in a large, upwind area of the United States or Europe may cause damages to ecosystems in a wide swath downwind of the utilities, hundreds or even thousands of miles away. Consider another example: Ozone-depleting effects produced by chlorofluorocarbon (CFC) emissions from countries all over the world can produce chemical changes in the earth's stratosphere, resulting in global impacts.

Other things the same, local environmental problems are easier to remedy than regional or national problems. Say I pollute my neighbors by burning trash for free instead of spending $5 to haul it to the dump. We may be able to design a solution among ourselves, or we can appeal to our local political institutions to reach an agreement. But if my behavior causes more distant or more delayed pollution, and if my behavior is hard to isolate from the behavior of others, solutions will be harder to produce. When each person's (or country's) actions make a small contribution to a large problem, especially if the problem is international in scope (e.g., global warming), we are a long way from having effective means of responding. The precise nature of the physical impacts is hard to isolate, and international political institutions for reaching satisfactory environmental agreements are still rudimentary.

Origin: Point versus Nonpoint Sources

Pollution sources differ by the cost of identifying discharge points. It is not hard to identify the points at which sulfur dioxide emissions leave a power plant; they come out of the plant's smokestacks. Municipal wastewater treatment plants (sometimes called sewage treatment plants or water pollution control plants) remove some of the worst pollutants from used water before it is discharged into local waterways. Normally these plants have a single point from which all of the wastewater is discharged. Both of these examples are *point-source pollutants*. For many pollutants, however, there are no well-defined discharge points. Some water pollutants come from several sources and combine; these are called *nonpoint-source pollutants*. As water flows, it picks up and carries contaminants. This polluted water ends up in streams, lakes, and oceans by flowing directly or by going through untreated storm drains. Water also carries pollutants into underground drinking water as it soaks into the ground. Agricultural chemicals, for example, typically run off the land in a diffuse pattern, and while they may pollute well-defined streams or aquifers, there may be no single pipe or obvious source from which these chemicals are emitted. In urban areas, runoff from storm water—containing oil, hazardous toxins, and disease-causing agents—also is a nonpoint-source pollutant problem. In comparison to nonpoint-source pollutants, point-source pollutants are easier to measure and monitor and easier to understand in terms of the connection between emissions and impacts.

Occurrence: Continuous versus Episodic

Emissions produced by power plants and city sewage treatment plants are essentially continuous—they occur all the time. In this situation, the policy challenge is to manage the rate and level of the discharges. Clear comparisons can be made between various control programs and the associated emission rates of each. The fact that emissions are continuous does not mean that damages are also continuous. Meteorologic and hydrologic events can turn continuous emissions into uncertain damages. Control programs are often easier to carry out when emissions are not subject to large-scale fluctuations.

Many pollutant emissions occur on an episodic basis, such as in accidents. Accidental chemical or oil spills are classic examples. Think of the large number of accidental releases near the U.S. Louisiana-Texas border in connection with the petroleum/chemical industry releases to the air associated with startups, shutdowns, equipment failures, and process upsets (U.S. EPA, 2002). Since we never know when an accident will happen, the policy challenge is to design, implement, and evaluate a system to reduce the probability of accidental discharges into the environment. With an episodic pollutant, there may be nothing to measure 99% of the time. Even though there are no large-scale emissions, a pollution problem remains if damaging releases that are preventable at low cost take place often. Where recurring accidents are a real issue, we enter the world of statistical probability. Measuring the probabilities of episodic emissions requires factual data on actual occurrences over a period of time. The policy question then centers on the economic value of added insurance against these events versus the cost of paying for that insurance through more reliable systems and better trained operators.

Desirable Outcome Criteria

There are many kinds of environmental pollutants, media in which the pollution occurs, and cultural contexts. Each condition has special characteristics that tell us whether a policy response is more likely to succeed in some circumstances and fail in others. In evaluating the effectiveness of a policy for addressing any problem of environmental pollution control, it is important to have a unified set of evaluation criteria. This avoids the tendency of being swept away by the special characteristics of any single situation. Three classic criteria for dealing with pollution policy are economic efficiency, equity, and sustainability.

Economic Efficiency

Economic efficiency is a democratic standard for making decisions. As discussed in more depth in Chapters 3 through 5, efficiency makes no distinction among people. A dollar of net benefits to one person resulting from controlling pollution is considered equal to a dollar produced for anybody else at any other source. However, applying this standard may result in unfairness. For example, economic efficiency tells us that $100 of benefits received by one rich person offsets $1 of cost imposed on each of 100 poor people because the algebraic sums are equal. Most of us view an outcome that produces benefits for rich people at the cost of poor people as unfair; nevertheless, an economically efficient policy produces no guarantees that its outcome will be equitable. An efficient environmental pollution policy is one for which total benefits of reducing the pollution exceed total costs, regardless of who benefits and who pays the costs. An efficient pollution control policy moves toward the point where the benefits of reducing the pollution damages exceed the costs of abating the pollution.

Pollution Damages. Pollution damages refer to all negative effects that users of the environment experience as a consequence of that environment's degradation. The many kinds of these damages vary from one environmental resource to another. For example, the effects of river pollution include (1) inconveniences for vacationers who are unable to use the river for a previous purpose such as boating, fishing, or swimming; (2) health problems for people who have a higher probability of contracting waterborne diseases; and (3) costs incurred by people in nearby cities whose water bills increase because of greater treatment costs required to make the water drinkable. Damages produced by air pollution also vary.[3] Increased deaths and sickness from diseases

---

[3] For example, the U.S. EPA (2004c) summarizes effects of six major air pollutants: ozone, particulate matter, carbon monoxide, nitrogen oxide, sulfur dioxide, and lead. Bityukova and Argenbright (2002) describes economic impacts of air pollution in Moscow, Russia. Chay and Greenstone (2003) analyze impacts of air pollution on infant mortality.

such as chronic bronchitis, emphysema, and lung cancer are related to increased levels of pollutants, such as radon, particulate matter, and asbestos fibers. Air pollution can also cause damages by degrading various materials,[4] and environmental destruction has significant impacts on elements of the natural ecosystem. Some of these impacts, such as destroyed genetic information in a rare or endangered species, may produce important implications for people.

Damage Functions: Concept. As a general principle, large amounts of any pollutant produce greater damages and increase the value of programs that restore those damages or prevent those damages from occurring. A damage function describes the relationship between the quantity of a pollutant and the resulting damage it produces. Two kinds of *damage function methods* are in widespread use: (1) emission damage functions, which show the connection between the quantity of a pollutant emitted from a source and the resulting damage it produces; and (2) ambient damage functions, which show the connection between concentration of a particular pollutant and the damages that concentration produces. The concept of marginal damage function can be applied to the two functions just described. A marginal damage function indicates the additional damages produced by a one-unit change in emissions or ambient concentration. Sometimes, the marginal damage function provides insights into total damages, based on the idea that the areas under marginal damage functions are total damages. The structure of any marginal damage function depends on the pollutant, the culture, and the economic conditions of the people affected by the pollutant. See Example 16–1.

Figure 16–1 presents four marginal damage functions. The top row shows two functions for marginal emission damage. The x-axis measures the quantity of some pollutant emitted into the environment during a particular time period. The units of measurement (e.g., pounds, tons) per unit time (e.g., day, month, week, year) depend on the particulars of the pollutant and culture involved. The y-axis measures environmental damages in monetary terms. Any single example of environmental pollution typically produces many kinds of impacts, the particulars of which depend on the special pollutant involved and the culture, time, and place where it is emitted. Navrud (2001) illustrates how a given kind of emission produces very different values in different cultures and countries by comparing U.S. and European impacts. One way to treat these impacts comprehensively is to reduce them to a common denominator, such as through a monetary scale (see Chapter 7).

The marginal emission damage function in panel a (Figure 16–1) shows marginal damages increasing only slightly at low emission levels, but more rapidly as emissions become progressively larger. Empirical analysis by economists and environmental scientists has shown that this may be a typical shape for many pollutants. At low emission levels, marginal damages are small; ambient concentrations in the environment are so low that only the most sensitive people in the area are significantly affected. As emission levels increase, however, damages increase; at still higher levels of emissions, incremental damages produced by further increases become more pronounced as environmental damages increase in scale, amount, and intensity. Figure 16–1b shows a marginal emission damage function with a shape similar to Figure 16–1a, but the function starts out much higher on the y-axis and rises more quickly. It could indicate a dangerous (toxic) material with known deadly impacts even at low emission levels.

The bottom row of Figure 16–1 presents two marginal ambient damage functions. The y-axes still measure damages in monetary units, but the x-axes now measure ambient concentrations (e.g., parts per million). Figure 16–1c shows an ambient damage function that increases quickly at low concentrations, levels out until very high concentrations are reached, then increases quickly. This function could apply to an air pollutant that produced significant damages among

[4] Several important outdoor sculptures from Renaissance Italy have been moved inside to protect them from air pollution. Hundreds of monuments are threatened throughout Italy, and unlike humans, monuments do not regenerate. Air pollutants cause a black crust to form on the surface of outdoor monuments and churches. Over time, the crust thickens and the stone or marble underneath weakens, cracks, and even breaks off (taking down a nose or an arm). Given that the art work may not be replaceable, the economic cost of the damages produced by pollution presents itself as an important question, with major policy implications for the value of culture itself and for policies designed to promote its preservation. See Accardo, Giani, and Giovagnoli (2003).

Example 16–1
Water Pollution Policy in China

China has an equal amount of water as Canada, but China supports 100 times the population. Per capita water reserves in China are 2,500 cubic meters, one-fourth the world's average. Water pollution accidents, usually triggered by heavy rainfall, floods, or drought, often overwhelm the inadequate wastewater storage and treatment infrastructure. Such events can release industrial wastewater and sewage overflows and reduce the supply of fresh water for drinking or pollution dilution. As of the late 1990's only 5% of municipal wastewater and 17% of industrial discharge received any treatment before being discharged into lakes, rivers, irrigation ditches, or coastal waters. Few of China's major rivers meet government standards for primary drinking water supplies. The degraded condition of much of China's surface water and a lack of coordination between public health and water quality pose serious human health threats. And chronic health consequences that result from changes in environmental conditions only begin to appear in health and epidemiologic records after a decade or two.

Another problem is pesticides. In China, the pesticide market is dominated by organophosphate insecticides, which are toxic to aquatic fauna and are hard to control. Fish and aquatic organism kills have been reported, even though their causes have not been established. Some of these chemicals are volatile; therefore, they pose a risk to human health.

What does China's future hold? A more affluent and growing population, more intensive use of agrochemicals, and continued industrial expansion will increase the amount of human, animal, and industrial waste entering surface and groundwater streams. Economic impacts of water pollution are stressing the Chinese government's capacity of managing water resources in the traditional fashion. The World Bank estimates the impact of water pollution on human health is approximately $3.9 billion per year (41.73 billion yuan). In sum, evidence indicates that water pollution is playing a significant role in shaping China's environmental risks. Attention to and correction of water pollution problems would result in many public health benefits.

Sources: Wu, C., C. Maurer, Y. Wang, S. Xue, and D. L. Davis. 1999. Water pollution and human health in China. *The Sino Sphere Journal.* 5: 10–15; Calamari, D., and L. P. Zhang. 2002. Environmental risk assessment of pesticides on aquatic life in Xiamen, China. *Toxicology Letters.* 128 (1–3): 45–53.

Figure 16–1 Typical marginal damage functions

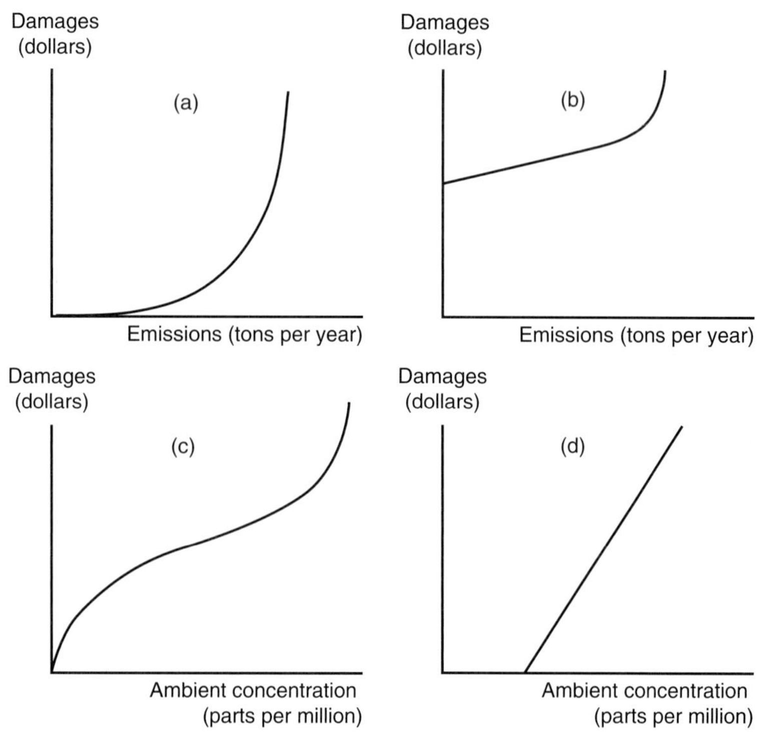

unusually sensitive people at low concentrations and among everybody at high concentrations, but did not produce much marginal increase in the midranges. Figure 16–1d shows an ambient marginal damage function that stays at zero until threshold effects occur at rather high concentrations, at which point the function increases linearly with growing ambient concentrations.

Figures 16–1a and d illustrate the controversial characteristic of *thresholds;* that is, values of emissions or ambient concentrations below which marginal damages are zero. Thus, the pollutant can increase to these threshold levels without causing any increase in damages. As will be seen in chapters to come, the assumed existence or nonexistence of a threshold in damage functions for particular pollutants has had important impacts on environmental control policies. There have been long, vigorous arguments about whether the damage functions of certain types of pollutants do or do not have thresholds.

Damage Functions: Analysis. The concept of the damage function is important because it is used to express and analyze a variety of different types of pollution problems and public policy approaches. Figure 16–2 shows two marginal emission damage functions. Like demand and supply curves, these functions are time specific: They show emissions and marginal damages for a particular period of time. There are a couple of ways of thinking about this. One is to assume, for purposes of simplicity, that the function refers to a strictly noncumulative pollutant. Thus, all damages occur in the same period as emissions. A more complicated assumption is that for a pollutant that accumulates over time, the damage function shows the total value that people place on current and future damages.

Consider first the marginal damage functions labeled MD_a. In previous chapters we discussed the relationship between marginal and total quantities, such as the relationship between marginal and total costs. We have the same relationship here. The height of the marginal damage curve shows how much total damages would change with a small change in the quantity of emissions. When the effluent level is at the point marked E_2, marginal damages are \$10. That is to say, if emissions were to increase by 1 ton per year from point E_2, the damages experienced by

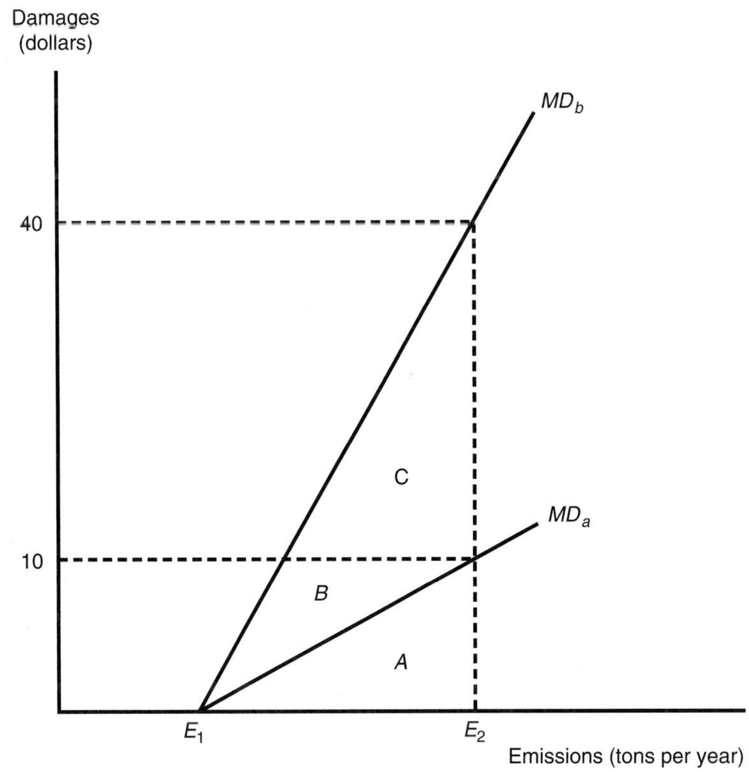

Figure 16–2
Structure of a marginal damage function

people exposed to those emissions would increase by \$10. By the same token, if emissions decreased by 1 ton per year beginning at point E_2, total damages would be reduced by \$10. Because the height of the curve, as measured on the y-axis, shows marginal damages, the area under the curve between the point where it is zero at point E_1 and some other point, like the one labeled E_2, shows the total damages associated with that level of emissions. In the case of the marginal damage function MD_a, and point E_1, total damages are equal to the monetary amount expressed by the triangular area bounded by the x-axis, the curve MD_a, and the effluent quantity E_2 (area A).

Notice that MD_b always lies above MD_a. What factors could account for the difference, even if the two figures refer to damages from the same pollutant? For any given level of emissions, marginal damages are larger for MD_b than for MD_a. At emission level E_2, a one-unit increase in emissions raises damages by \$10 if the marginal damage function is MD_a, but it increases damages by \$40 if it is MD_b. One explanation is that MD_b occurs when many people are affected by an emission, as might happen in a large city, while MD_a applies to a more sparsely populated rural area where fewer people affected produces smaller incremental damage. Therefore, an important factor shifting marginal damage functions up and to the left is an increase in the number of people exposed to the pollutant in question.

Could those two functions refer to the same group of people? The answer is yes. The two functions could apply to the same people facing emissions in different time periods. Think of a city that experiences air pollution episodes because of atmospheric inversions. The marginal damage function MD_b would occur when there is a temperature inversion that traps the pollutant over the city and produces relatively high ambient concentrations. When normal wind patterns blow the pollutants out of the area, the damage function is MD_a. So the same emission levels at two different time periods (even if the two periods are quite close together) could produce significantly different damage levels due to the changing structure of the environment itself. Since MD_b lies above MD_a, it produces both higher marginal damages and higher total damages. At emission level E_2, total damages produced by MD_a are area A, but total damages are area $A+B+C$ when the damage function is MD_b.

We may be tempted to ignore costs and to conclude that the damage function tells us everything we need to know about pollution control decisions. Perhaps we are tempted to say that a civilized society should aim for emission levels at or below point E_1 where marginal and total damages are zero. However, to analyze emission targets comprehensively, we need to know something about abatement costs.

Abatement Costs. *Abatement costs* are the costs of controlling the level of emissions released into the environment or the costs of reducing ambient concentrations. These costs can be borne by business, government, consumers, or a combination of them. For example, a coal-fired power plant has emissions that blow downwind. In its course of operation, the power plant produces a considerable quantity of sulfur dioxide. If the plant has free access to the region's ordinary air waste dispersal and dilution services, the cheapest disposal method is to simply release its untreated emissions into the air. The plant could incur some costs to treat these emissions, to render them more harmless, but that treatment adds something to costs and nothing to revenues. The costs incurred to render emissions more harmless are called abatement costs, because they are costs paid to abate, or limit, the quantity of residuals emitted into the air.

Abatement costs depend on a number of factors. Costs of abating sulfur dioxide emissions from an electric power plant will be different from the costs of reducing the organic load from a paper mill (Stiffler, 2004). Even in the case of several sources producing the same effluent, the costs of abatement often vary widely because of differences in the technological features or age of the source. One source may have been installed recently, using modern, environmentally friendly production technology, whereas another may have been installed years ago and may use more heavily polluting technology. In the remainder of this chapter, abatement costs include the large number of measures for limiting emissions: changing inputs, altering production technology, treatment of potential emissions, recycling of emissions, changing locations of the source, and the like.

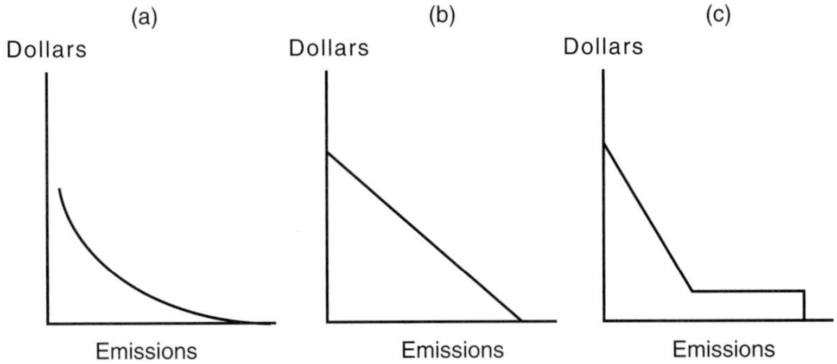

Figure 16–3 Typical marginal abatement cost function

Abatement Cost Functions: Concept. Figure 16–3 shows the principle of abatement costs graphically, using the notion of a marginal abatement cost function. The units on the x-axis and y-axis are the same as those for the damage functions, namely, monetary value of costs on the y-axis and quantities of emissions on the x-axis. The marginal abatement cost measures the incremental cost of decreasing emissions by one unit. Another interpretation is abatement cost savings from reducing abatement (increasing emissions) by one unit. The x-axis shows that marginal abatement cost functions start at the far right, at what are called uncontrolled emission levels (emission levels before abatement begins). Where emissions are measured on the x-axis, abatement (subtraction of emissions) is typically read from right to left. From this rather unusual origin, marginal abatement costs indicate the incremental cost of producing one more unit of reduced (abated) emissions. For these reasons, the marginal abatement cost function rises from right to left (falls from left to right), indicating an increasing marginal cost of producing abated emissions.

Figure 16–3 shows three unique but typical marginal abatement cost functions. The function in Figure 16–3a shows marginal abatement costs rising very slowly as emissions are first abated, but later rising more quickly as abatement increases and emissions fall off. Figure 16–3b shows marginal costs rising quickly and steadily as soon as abatement starts. Figure 16–3c shows a marginal abatement cost curve that has an initial finite but flat phase, followed by increasing marginal costs. This could represent a situation in which slight emission reductions can be handled by technical measures that require a large up-front investment; but for larger reductions, the marginal abatement costs steadily increase. The marginal abatement cost functions shown in the three parts of Figure 16–3 are taken to be minimum costs of abating emissions, but cost minimization is not a trivial task. Costs of any activity such as pollution abatement depend on available technology as well as on managerial and technical skills. Use of the wrong technology, or application of the correct technology improperly, will produce needlessly high abatement costs.

Abatement Cost Functions: Analysis. Figure 16–4 presents two marginal abatement cost curves. The high curve, MAC_b, starts out at the uncontrolled emission level, E_2, and slopes up to the left. Starting at E_2, the first units of abatement can be accomplished with little marginal cost. Think again of the coal-fired power plant. This first, small decrease in emissions might be produced with the addition of a more efficient coal-burning technology, but as emission levels are reduced further, the marginal cost of achieving additional reductions increases. For example, to reduce emissions by 30%, the plant's owners may have to invest in a new type of technology that burns coal more efficiently. A 70% emission reduction might require considerable amounts of new treatment technology in addition to the step taken previously; a 90% abatement might require costly equipment to recycle nearly all of the production residuals produced in the plant. For these reasons, a common pattern is that larger emission reductions require higher marginal costs of producing further reductions.

There is an upper limit on abated emissions and an associated upper limit on the marginal costs of abating those emissions. A plant owner's most extreme (and painful) choice is to close

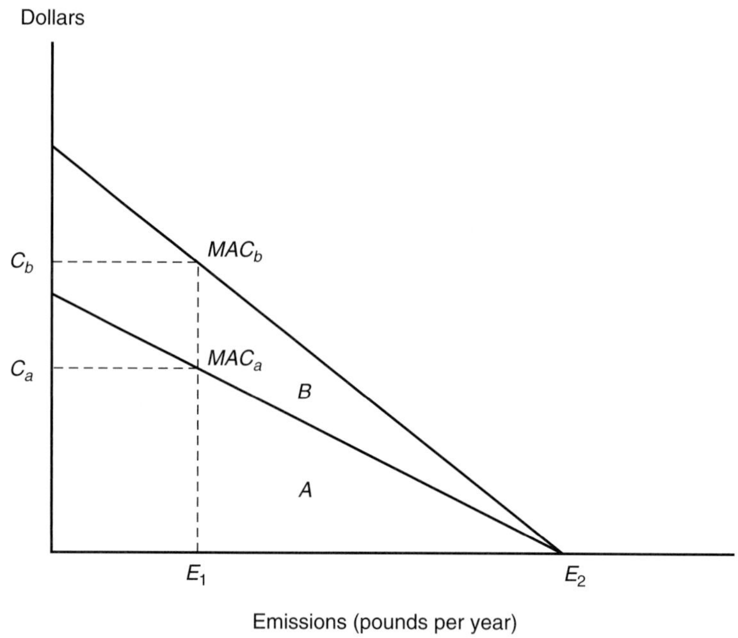

Figure 16–4 Structure of a marginal abatement cost curve

down the business and cease operations,[5] which produces zero emissions by default, (however, processing of certain toxic materials produces residual leakages into the environment long after the firm shuts down). Marginal graphs such as Figure 16–4 can be used to show both marginal and total values. If the source currently emits at E_1 tons per year, the number on the vertical axis shows the marginal cost of abating one extra unit of emissions. The area beneath any marginal abatement cost curve, between its origin at point E_2 and any particular emission level, is equal to the total costs of abating emissions to that level. For example, with the curve labeled MAC_b, the total abatement cost of achieving an emission level of E_1 tons per year is equal to the area under the curve between E_2 and E_1, the area $A + B$. Remember we are reading the graph from right to left.

Now look at the lower marginal abatement cost curve shown in Figure 16–4, MAC_a. It lies below MAC_b, which means that it represents conditions where marginal abatement costs for any level of emissions are lower than costs of MAC_b. For example, at E_1 pounds per year of emissions, the marginal cost of abating one more pound is only c_a, considerably lower than c_b, the marginal abatement costs for MAC_b at the same level of emissions. Even if we are dealing with identical pollutants in both cases, what conditions could produce this difference? One possibility is that each applies to a different emission source. For example, MAC_b refers to abatement costs from an old plant built with old control technology, while MAC_a refers to abatement from a newer plant using more modern technology. The newer plant produces lower cost abatement.

Even if MAC_a and MAC_b are abating the same pollutant at the same source, the difference could be explained by abatements in different time periods. The bottom function (MAC_a) could represent the situation after installation of sophisticated new pollution-control technology, while the top one (MAC_b) applies before the change. In other words, technical change in pollution abatement reduces the marginal abatement cost curve for a given pollutant. The same graph can be used to measure the cost that this source saves by installing the new technology when the emission rate is a constant E_1.

[5] The costs of this extreme abatement depends on the conditions. If the emitter is a lone, small plant in a major industry made up of similar plants, shut-down costs may be modest. That plant's closure may produce little distinguishable impact on the worldwide price to consumers of the firm's product (e.g., paper), even though the impact on local income and jobs may have high economic and political costs. However, if the abatement is a worldwide response, for example coal-fired electric power worldwide responding to an international greenhouse gas mitigation agreement, the option of zero emissions produces huge costs.

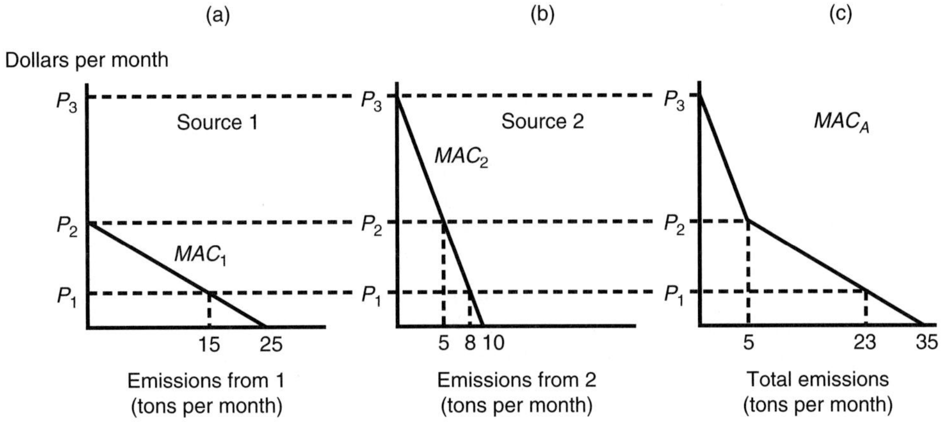

Figure 16–5 Aggregate marginal abatement costs

Prior to adopting the more advanced technology, the source total cost for abating emissions from E_2 to E_1 equals $A+B$ per year. Afterward, total abatement costs fall to A. So abatement costs saved by the technological change are B. Analyses like this take on considerable importance when evaluating programs for pollution abatement, because an important standard for judging abatement policies is the size of cost-saving incentive the programs provide for firms to conduct research and development to improve methods of pollution control.

Distribution of Marginal Abatement Costs. The previous analysis considered the marginal abatement cost function as it applies to only a single emission source or firm. But in the world of policy, analysts often think about impacts of proposals on the behavior of a group of firms in an industry, economic sector, region, or nation. National environmental policies often focus on abating emissions from groups of similar pollution sources, not isolated single polluters. Andreoni and Levinson (2001) and Ansuategi and Perrings (2000) describe conditions under which a nation's economic growth affects its costs of abating pollution.

National environmental policies rarely apply to a series of identical firms. In any large-scale environmental policy analysis, considerable heterogeneity is likely among firms that fall under a given set of regulations. Therefore, individual marginal abatement cost functions typically differ widely from one regulated source to the next. For example, the control of organic pollutants in any heavily industrialized polluted river basin requires abating emissions from a large variety of different sources in different regions with varying pollution control technologies. For these reasons, a single, *aggregate* marginal abatement cost function is needed for a set of regulated firms to aggregate all relevant individual marginal abatement cost functions. The challenge posed by the need for aggregating individual behavioral responses to a proposed policy leads to an important principle underlying the design of effective environmental policy: a single, aggregate abatement cost function that minimizes the cost of complying with a given regulation. Properly constructed, that function reveals the least costly way to achieve emission abatements, and it can apply to various operations of firm or to a group of many firms.

Beginning from the left, Figure 16–5 shows two different sources of a given pollutant, labeled source 1 and source 2, each with its own marginal abatement cost function. Figure 16–5a shows that MAC_1 begins from the right at 25 tons per month of emissions and rises steadily as emissions are reduced. Figure 16–5b shows that MAC_2 starts from the right at the uncontrolled emission level of 10 tons per month, but rises much more quickly. The two functions are different because source 2 may be an older plant with fewer technical or managerial choices for controlling emissions. Another explanation is that the two sources, while emitting the same material, manufacture different products or use different production methods: or the two firms could be in different regions, operate in different industries, or face different weather or climate conditions.

Figure 16–5c shows the aggregate marginal abatement cost curve, MAC_A, which is an aggregation of the two individual relationships on the left. But aggregation is tricky, and there are many traps to avoid. Since the two individual marginal abatement cost curves are different, the method chosen for adding the two curves together has a major influence on the outcome, and on environmental policies to be recommended based on that aggregation. With two or more emission sources with different abatement costs, aggregate costs of abatement depend strongly on how the total emissions are allocated among the sources. So what principle should be used in choosing an aggregation method? Add the two (or more) individual abatement cost functions in whatever way produces the least possible aggregate marginal abatement costs. To do this, add the functions horizontally. Choose a starting level of marginal abatement cost, for example, point P_1 on the y-axis. We know that P_1 is the marginal abatement cost produced by reducing effluent by 1 ton at levels of 15 tons per month from source 1. It is also the marginal cost of reducing by 1 ton an effluent level of 8 tons per month from source 2. So on the aggregate curve shown in Figure 16–5c, a marginal abatement cost of P_1 is associated with the horizontal addition of emissions; in this case, emissions equal 15 tons + 8 tons = 23 tons per month.[6] Each of the other points on the aggregate marginal abatement cost curve, MAC_A, are found in that way, namely, by summing *horizontally* over all individual marginal abatement cost functions.

Although this procedure seems rather mechanical, it has important economic underpinnings. Once again we are encountering the *equimarginal principle,* an idea developed and applied at several previous points in this book. *Minimizing aggregate abatement costs for a given emission level requires allocating aggregate emissions among all sources so that they all have identical marginal abatement costs.* For example, begin in Figure 16–5c on the aggregate curve MAC_A at the point of 23 tons per month. This 23-ton total can be distributed among the two sources in any number of ways: half from each source; all from one and none from the other; 15 tons from source 1 and 8 from source 2, and so on. Only one allocation, however, will produce the lowest aggregate abatement costs;[7] this is the allocation that leads the different sources to the point at which they have *equal marginal abatement costs.* We will return to this point many times in this chapter.

Efficient Pollution Abatement. At this point, we have separately analyzed two important concepts—the marginal damage function and the marginal abatement cost function—in connection with some given pollutant discharged at some time and place and under some known condition. Now we are ready to synthesize those two building blocks. Figure 16–6 brings together the concepts of damages avoided and costs required to avoid those damages. It shows a typical set of marginal abatement cost and marginal damage functions, respectively, *MAC* and *MD*. They show that marginal abatement costs are zero at the uncontrolled emission level E^2, while marginal damages are zero at the threshold emission level E^1.

You would probably guess that the economically efficient level of emissions occurs at the level at which marginal damages are equal to marginal abatement costs, and you would be right. The logic behind this has been stated many times in this book: It is the logic of trade-offs that are inherent in the analysis of pollution control policy. Greater emissions expose the affected society to greater damages produced by more environmental damage. Reduced emissions require somebody to commit more resources to incur larger costs for emission abatement actions. The economically efficient emission level (as well as the economically efficient abatement level) occurs where these two kinds of costs offset each another. That circumstance occurs only at one point, namely, where the incremental costs of abating one more unit equals the incremental benefits of reducing the costs of damages. This occurs at emission level E^* in the Figure 16–6. Marginal damages and marginal abatement costs are equal to each other and both are equal to P^* at this level of emissions.

---

[6] The horizontal addition of P_1 has an important economic interpretation. When total abatements over all sources are found from the cheapest source, the 23rd ton per month can be abated at a marginal cost of P_1.

[7] This statement assumes that at least two sources have different marginal abatement costs schedules. If all individual MACs are identical, all abatement allocations produce the same aggregate abatement cost. In that special case, any abatement-sharing policy, such as an equal sharing of abatement responsibilities, produces the same aggregate cost.

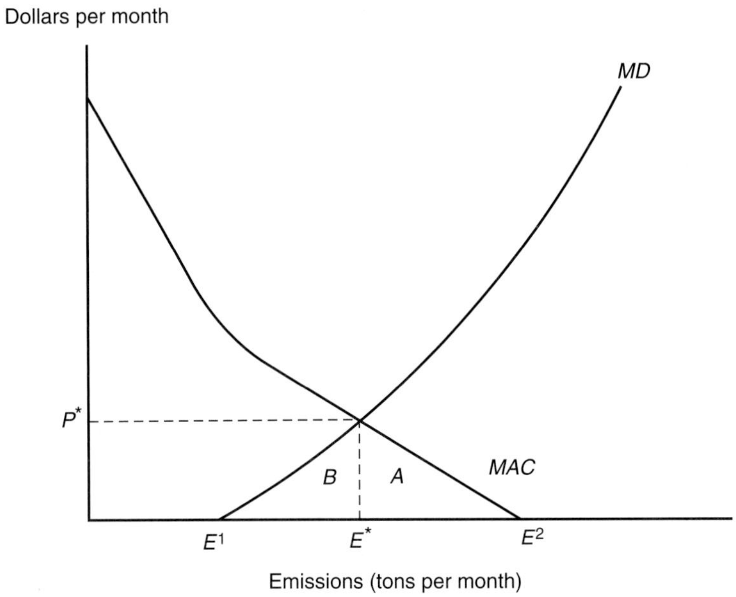

Figure 16–6 Efficient level of emission abatement

You also can think of the efficient emission level in terms of totals rather than marginals by remembering that totals are always the area under marginal functions. In Figure 16–6, the triangular area *B* shows total damages produced by emission level E^*, while triangular area *A* shows total abatement costs at the same emission level. The sum of these areas, $B+A$, measures total costs to society produced by E^* tons per month compared to a situation of no damage and no abatement cost. The emission level E^* is the one point that minimizes the sum of total abatement costs and total pollution damages. Try any other emission level and you will see that total damage plus abatement costs are higher. Every emission level produces its own aggregate costs to society, but point E^* is the cheapest way to confront the environmental problem.[8]

The intent of this graphic analysis and marginal discussion is to develop a unified conceptual way of coping with trade-offs. In reality, each pollution problem has its special local flavor, special characteristics, and unique political and cultural environment. Therefore, an overarching framework is essential for understanding the major choices available and consequences of those choices without being locked into the details of each special case. Marginal analysis supplies a unified method of framing the problem, and that method can be adapted to the particulars of any given environmental pollution problem.

Efficient Emission Levels: Concept. Figure 16–7 shows three different environmental pollution situations. Although each of the three graphs looks rather different, economic efficiency is always analyzed using the same framework. In all three cases, E^* shows the efficient level of emissions and P^* presents marginal damages and marginal abatement costs produced at the efficient emission level.

- Figure 16–7a shows a pollutant for which the efficient level, E^*, is zero. Marginal damages, MD_1, for this condition are high, as are marginal abatement costs, MAC_1. But marginal damages are always higher than marginal abatement costs. This rather unusual looking situation may be

[8] Don't be taken in by the following false impression: On the surface, emission level E^* looks like it produces high emission levels, since it is far to the right on the x-axis. So some might conclude that economically efficient emission always produces high emissions and imposes high costs to society for coping with them. That is a mistake. In fact, the economically efficient emission level could be zero for a dangerous emission or could produce zero damages for a harmless emission. Watch out for the "fallacy of misplaced concreteness," which falsely concludes general truths from a small set of special numeric examples.

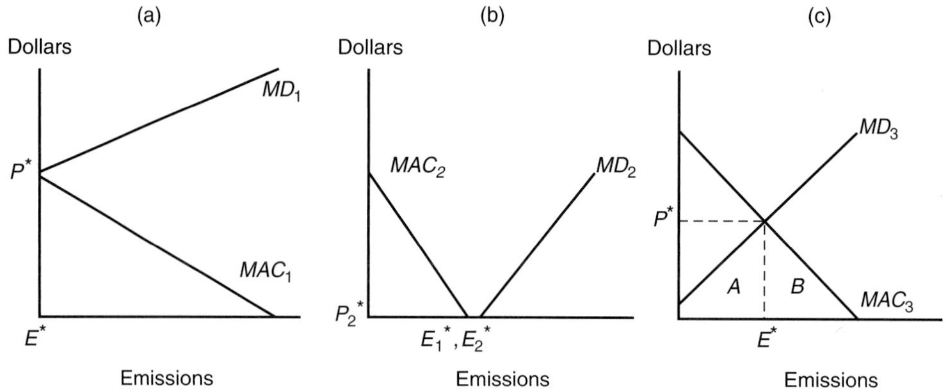

Figure 16–7 Efficient abatement by pollutant

caused by emissions of a toxic pollutant (e.g., dioxins, benzene, mercury, or the pesticide DDT), where marginal abatement costs are always lower than marginal damages avoided by the abatement.

- Figure 16–7b shows a situation where the marginal abatement cost function (MAC_2) begins at zero, stays unchanged at zero, then rises when emissions fall below E_1^*. The marginal damage function (MD_2) also begins at its zero origin and does not rise until emission level E_2^* is reached. In this case, both E_1^* and E_2^* (where $E_1^* < E_2^*$) exceed zero; P_2^*, which equals zero, is well below what it was in Figure 16–7a. So Figure 16–7b shows that as long as actions can be implemented that keep emissions between E_1^* and E_2^*, both marginal abatement costs and marginal damages stay at zero. An example of this set of conditions is relatively harmless pollutants in rural areas. In fact, conditions shown in Figure 16–7b explain why legislated pollution abatement programs with heavy control expenses typically receive little political support in many rural areas of the world. The abatement would be nice, but it may produce little in added benefit compared with the added cost.
- Figure 16–7c shows a pollutant for which E^* is significantly larger than zero. Marginal damages at this point are large and so are total damages and abatement costs, as shown by the large size of areas *A* and *B* corresponding to these values. These could be the conditions for a pollutant where both marginal abatement costs and marginal damages increase steadily. For such circumstances, significant abatement levels are required to most efficiently improve society's welfare at emission level E^*.

Efficient Emission Levels: Analysis. Things change, but usually they change for a reason. This is especially true in the world of environmental policy. Last year's efficient level of emissions may be different from the efficient level this year, and both levels can be different from next year's level. When any of the forces that influence the marginal abatement cost function or the marginal damage function are altered, the functions themselves shift to the right or left. Therefore, the economically efficient emission level will change, and the environmental policies that produce that efficient emission level also may change. This discussion presumes that the economically efficient emission level (i.e., the one that balances damage costs with abatement costs) is an ethically defensible and desirable aim for environmental policy. Nothing guarantees that the level of emissions that actually occurs will be economically efficient, either before or after shifts in the marginal damage or marginal abatement cost functions. Still, knowing what happens to the efficient emission level is important in the formulation of environmental policies based on efficiency.

Figure 16–8 analyzes several factors that can change the efficient emissions level, E^*. Figure 16–8a shows the case of a left shift in the marginal abatement cost function, from MAC_A to MAC_B. The main reason for such a change is improved pollution control technology (see Industry Canada, 2004). The large part of abatement costs depend on technology available for re-

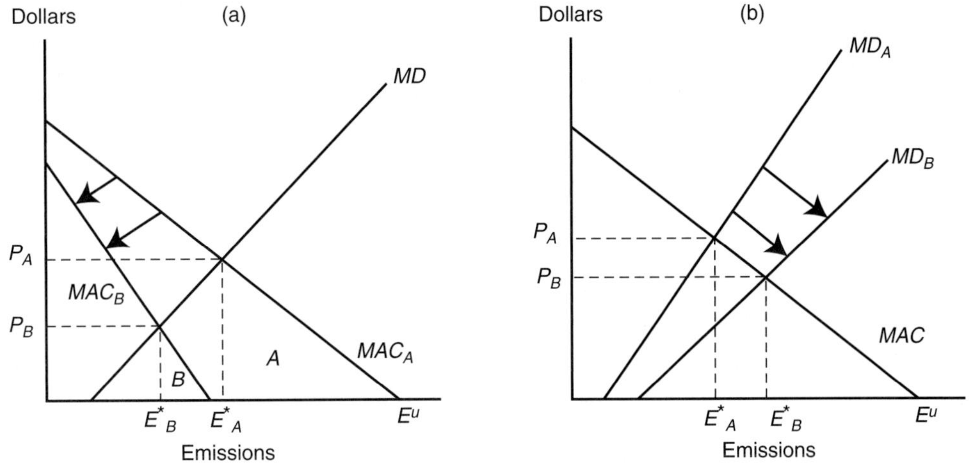

Figure 16–8 Factors changing the efficient abatement level

ducing emissions, for example, recycling technology, treatment technology, and substitute fuel technology. These new technologies are based on hard work, motivation, time, and the potential for reward. They come from training, energy, money, and resources put into research and development, sometimes because the technology is required to meet emission control requirements. In other words, the shift of the marginal abatement cost down and to the left comes from a new recycling or treatment method produced by investments in new technologies.

As you would expect, the downward, left shift of the MAC_A to MAC_B reduces the efficient level of emissions, as shown by the change from E^*_A to E^*_B. You might guess that the reduced cost of abating pollutants would surely reduce the cost of abatement, but that is not necessarily true. Surprisingly, a left shift in MAC_A can lead to either an increase or reduced total cost of abating emissions under the new efficient level E^*_B. Before the change, total abatement costs equal the area between the uncontrolled level E^u and the efficient level E^*_A, or area *A*. After the change, for the same reasons, total abatement costs equal area *B*. Although, area *B* is smaller than area *A* in this diagram, it could be larger depending on the shapes of the damage and abatement cost curves. Knowing whether total abatement costs at the efficient emission level increase or decrease depends on the comparative size of areas *A* and *B*. The relative size of those two areas depends on the shapes of the damage and abatement curves and the size and nature of the shift in the marginal abatement cost curve.

Figure 16–8b illustrates a shift downward and to the right for the marginal damage function, from MD_A to MD_B. The two causes for the shift are population migration out of the area and a growth in per capita incomes that reduces vulnerabilities to pollution exposures. Fewer people exposed or higher incomes that insulate people from the dangers of exposure mean that a given amount of effluent will cause less damage. This leads to a commonsense conclusion: Decreases in the marginal damage function increase the efficient level of emissions from E^*_A to E^*_B. The same logic indicates what the opposite will produce: An increased marginal damage function shifts it to the left and, under the economic efficiency criterion, encourages us to press more resources into pollution control, and fewer to other higher-valued uses.

Equity in Pollution Control

Equity, or fairness, is another important criterion in evaluating environmental pollution policy. Equity is, first and foremost, a matter of morality[9]; equity involves how the benefits and costs of

[9] In common law, equity means fairness or justice. It is the name given to the part of the law that deals with disputes between individuals when neither has broken the law but there is a conflict between their rights or claims.

Example 16–2
Controlling Environmental Pollution in India

Much recycled waste material from industrialized countries is shipped to poor countries like India and Bangladesh. In India, the trader who buys the material often auctions it to backyard recyclers. From there, along with locally generated plastic waste, it travels to urban slums, where poor people often breathe in toxic fumes emitted during the recycling process. India's government values the economic benefits produced by this recycling activity. Importers in India have been allowed to import plastic waste as a business simply because the material has not as yet been classified as hazardous waste under international standards. So industrialized countries are shifting their environmental risks to the poorest countries of the world.

Poor people who handle waste engage in a toxic and risky livelihood. But like other risky ventures, the returns can be high. In 1997, India's Supreme Court ordered that all polluting industries in New Delhi be relocated outside the city. This action placed thousands of migrant industrial workers at economic risk. These workers, mainly daily wage earners, stood to gain little from the compensation package, as they had never registered on official rolls and, therefore, did not exist.

New technology has created major environmental justice issues in India. In early 2000, farmers who had lived and carried out agriculture for decades in what is now the green belt of New Delhi, protested against establishing a high-tech hazardous waste landfill in their neighborhood to accommodate the city's waste, generated mostly by its middle class. Groundwater, on which they farmers relied heavily for drinking, would be contaminated. More generally, technology can shift environmental risks to people who are the least empowered. Indian cities are seeking measures to cope with growing mounds of waste and are proposing the installation of waste incinerators, to be located in the poorest urban neighborhoods. There seems to be little concern that those who live in the vicinity will be exposed to toxins that enter the food chain through the environment.

The conflict between jobs and the environment is an old one, but it is increasing with the changing power balance between those who represent both interests. India stands to benefit considerably by defining common views and creating partnerships. One place to start would be to allow and encourage workers to be recognized as key stakeholders for environmental change, thereby reducing conflict between labor and environmentalists in what could be a common challenge rooted in shared values.

Source: Agarwal, R. 2002. Shifting environmental risks in India. Available from *http://www.indiatogether.org/environment/opinions/raviag02-1.htm* [Accessed March 22, 2005.]

environmental improvements ought to be distributed among members of society. Fairness is also important from the standpoint of policy effectiveness, because proposed policies may not be supported enthusiastically in the political arena if they are considered inequitable. Nevertheless, there is much debate on how much weight should be put on the two objectives of efficiency and equity (see Chapter 18). See Example 16–2.

Equity considerations also are paramount in the design of international environmental pollution policy. As discussed in earlier chapters, countries at different stages of development embrace different views on how the burdens of international pollution-control programs are best distributed. These views may be driven by considerations of what is fair given great economic disparities.

Environmental Sustainability

The third criterion for dealing with pollution policy is sustainability. Sustainable environmental and natural resource activities are desirable for three reasons. The first is economic. Sustainable activities are those for which benefits exceed costs over the long run. Unsustainable behavior, by assigning low weights to the welfare of future generations, could be expected to produce higher costs than benefits. Any plan that maximizes social welfare over time is likely to require sustainable activities. The complete truth of this statement hinges on how sustainability measured and on what discount rate is used. Still, it is hard to imagine any economically efficient growth pattern over several time periods that avoids leaving to future generations much of what is desirable. The second reason sustainability is desirable rests with its morality. The current generation has a moral duty to all generations that follow. Accepting this duty implies actions that avoid limiting opportunities to future generations. The third reason for desiring sustainability is ecological: Diverse, intact, and resilient ecosystems are desirable by all generations. Any policy, activity, or institution that threatens to limit such ecological outcomes is undesirable. Sustainability of environmental and natural programs is discussed in more depth in Chapter 19.

Unregulated Market Outcomes

Now that we have defined and provided examples of economic efficiency in environmental policy, we are ready to ask whether an unhindered market system produces economically efficient outcomes. If we rely on the interaction of market demand and supply to determine the level of environmental emissions and abatement, will the market settle on the level of output for which marginal benefit equals marginal cost? Recall from Chapter 3 that despite its limits, the market system contains certain incentive structures that have the potential to be harnessed toward the aim of improved environmental quality. Two important incentives are cost reduction in competitive situations and rewards produced from taking the initiative to seek out cheaper technical means of production. The power of incentives embedded in the market system can be harnessed in the search for efficient environmental policies. It may be more effective to alter existing incentives so they better reflect society's environmental values than to scrap the market system altogether and establish a whole new set of institutions.

In addition, recall from Chapter 12 that unregulated markets can promote economic efficiency in the use of environmental resources in the special case where those resources are managed by sole owners. When a fishery is operated by a sole owner, that owner tends to take care of it so that it produces a maximum economically sustainable income. The same is true for environmental resources generally. Sole owners of forests, recreation areas, streams, and lakes tend to manage them efficiently and sustainably, because it is in their best economic interest to do so. Nevertheless, sole ownership has limits. Natural environments that absorb residual emissions produced by ordinary economic activities rarely affect only a single sole owner. Weak property rights and the associated open-access status of most natural environments typically produce considerably more use of those resources than is economically efficient. All the market failures described in Chapter 4 (e.g., common property, public goods, and externalities) regularly appear in natural environments. Because the sole owner of a natural environment is rare, a range of institutions play a role in bringing about economic efficiency for environmental resources.

Institutional Designs for Controlling Pollution

Efficient, equitable, and sustainable policy solutions are highly desired ways for approaching environmental pollution. Several concepts for finding these solutions are discussed next.

Decentralization

Decentralized environmental policies are defined as policies that let involved stakeholders design their own solution to the environmental problem. Suppose a clean lake has two manufacturing plants ringing its perimeter. For a fish-canning plant, the lake's water is packaged in its canned fish. For a chemical company, the lake's water is used as a garbage pit for disposing its waste (for example of chemical pollution, see United Nations, 2001). Obviously, these two firms face a major conflict over how the lake should be used. Is there any way the two can work out a solution on their own without resorting to formal public action? Is there any way to balance the pollution damage suffered by the fish cannery with the increased abatement costs of the chemical company in working out that solution? A decentralized method for finding the efficient ambient water quality levels in the lake would let the two plants work it out on their own.

Decentralized approaches offer two important advantages. First, because the stakeholders involved are the ones earning revenue and paying costs for the environment's use, they have strong incentives to design their own solutions. Second, those same stakeholders may know most about the environmental damages as well as the costs of abatement. So, by virtue of their special positions, stakeholders may be best equipped to design and carry out efficient actions. Two institutions contribute to the decentralized solution: liability laws and assignment of property rights.

Liability Laws. *Liability laws* hold individuals responsible for the consequences of their behavior. One way to do this is to require a party to pay compensation for damages caused to others. So using liability laws, society could require polluters to be liable for damages they cause. This approach serves two purposes: (1) parties who are injured receive compensation (equity), and (2) potential polluters avoid actions that produce damages that exceed the costs of avoiding those damages (efficiency).

Efficiency of Liability Laws. Figure 16–9 shows the power of liability laws to promote economic efficiency in connection with the environment. The conventional model of environmental pollution is shown as containing marginal abatement costs (*MAC*) and marginal damages (*MD*). Both of these functions are related to the emission rate of a pollutant. Starting from the right side, the uncontrolled emission rate begins at E^1, significantly more emission than the efficient level E^*. From point E_1, society passes a liability law requiring polluters to compensate parties damaged by an amount equal to the damages caused by the pollution. The law's effect is to require the polluter to confront and take account of damages to society. The law internalizes the environmental damages that were merely external to the polluter prior to the law; now these costs, shown as *MD,* are costs that polluters are liable to pay. That liability gives the polluter an economic incentive to pay attention to marginal damages when deciding on an emission rate. At the uncontrolled emission rate, E^1, total damages are a monetary amount equal to the area under the marginal damage curve $C+D+B$.

The process by which society moves from the passage of a liability law to the law's outcome in promoting economically efficient emissions goes something like this: Beginning at emission level E^1, the polluter thinks of available choices because it realizes its legal liability can be reduced by limiting emissions. The action of emission reduction comes at the price of positive and increased marginal abatement costs. Still, as long as incremental abatement costs are smaller than marginal damages, the polluter sees an economic incentive to reduce its emission rate, moving left on the figure. In principle, a liability system has the potential to lead this polluter to emission level E^* with no special need for a centralized authority to mandate emission abatement. All it requires is a set of liability laws that allows parties who suffer damage from pollution to seek and get compensation for their damages. The liability law approach (1) provides needed incentives to motivate people to fully account for potential environmental damages and (2) measures the compensation levels required for parties who suffer damages. Both of these outcomes could deal with the problem of finding the location of E^* on the x-axis. Depending on a country's legal traditions, a liability system could be implemented through common law procedures or by formal statutory law enactments.

Common Law. One way to structure an environmental liability system is to rely on common law. Common law, which originated in the unwritten laws of England, is based on judicial precedents rather than statutory laws. Judgments normally are based on precedent established from similar cases in the past. The critical question is whether common law institutions can promote an emission level approaching the efficient level, E^*. Although this level might be approximated by the courts of law, acting on the evidence presented by polluters and injured parties, many factors influence the final outcome.

One factor is called the *burden of proof.* Where the burden of proof lies and what is required to establish that proof are important elements of any liability system. For example, in the United States a person who believes he or she has been damaged by pollution must file an action in a set

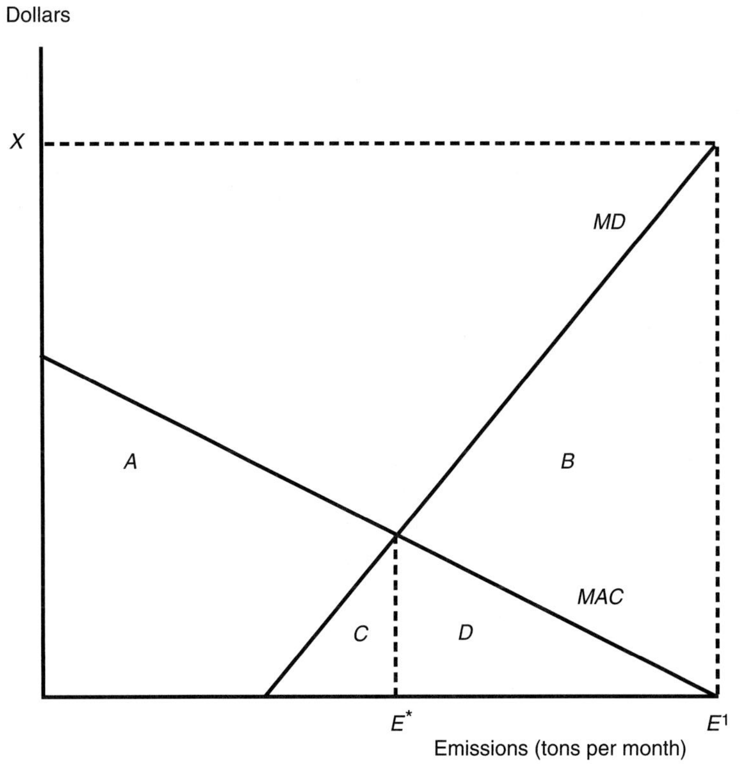

Figure 16-9 Liability laws and economic efficiency

period of time, often within two to three years.[10] Then that party must provide a direct and clear causal linkage between the pollution and the damage suffered. Establishing the connection between the defendant's pollution and the party's injury is no easy task; two possibly expensive steps are involved.

1. The plaintiff (person initiating the lawsuit) must show the pollutant was a direct cause of his or her injury. The standards of proof required by the courts may be more than science can deliver. Causality can be hard to prove: merely because *B* followed *A* does not mean *B* was caused by *A*. The best a plaintiff can hope for is to implicate a pollutant on a probabilistic basis. For example, exposure to a toxic substance produced an increased probability of acquiring a disease, but not a certainty. Even that probability may be hard to demonstrate and may require access to statistical expertise and technical information.

2. It must be shown that the pollutant to which the plantiff was exposed came from the defendant's source. Sometimes this proof is easy to establish; for example, it is hard to doubt that the oil on a shoreline came from a nearby, highly publicized oil spill. In many cases, however, direct linkage cannot be established with any confidence. If you are an urban resident in a polluted city, how are you going to convince a judge that a particular industrial plant or source produced the sulfur dioxide molecules that you breathed? There could have been hundreds of sources. If a polluting substance cannot be traced to a particular defendant, people who have been damaged may not be able to receive compensation.

Transaction costs, which refer to the costs of reaching and enforcing agreements, are high in liability systems where opposing parties compete in court to detemine liability and agree on compensation levels. If the case is simple, with only a single party on each side and a clear case

---

[10] For recent debates on upgrading environmental liability laws in the European Union, see European Parliament (2002).

of damage, the liability system may work well. For example, low transaction costs may produce an outcome that is close to an efficient emission level. In the case of the two factories on a small lake, the two can go to court and argue about the economic values to each of them of using the lake for their purposes. Because these values are comparable, it presumably would not be too difficult for a judge to determine the extent of the damages that one firm is inflicting on the other. In contrast, where the scale of damages is high, with large numbers of parties polluting and being damaged, transactions costs are often high.[11] Complex scientific questions of fact are typically involved, and even well-trained scientists or statisticians disagree on the type and size of causality. With large numbers of parties involved, and with many channels of influence connecting emissions and damages, it is difficult to discover or agree on efficient solutions.

Liability arrangements through common law are reliable in bringing about efficient emission levels when there are few parties, clear causal linkages, and easy to measure damages. Such conditions could be met for some local, small-scale cases of pollution; but for larger, more complex interdependencies, these conditions do not occur. We must look for other methods to connect polluters and the people who experience the damages.

Statutory Law. Statutory law, which consists of laws enacted by the legislative branch of government, are another way to organize a liability system for environmental improvements. For example, a legislative body could enact a law requiring polluters to pay compensation when they cause damage. Several countries have passed laws explicitly describing liability of their oil companies for damages from spills in their coastal waters. For a U.S. example, see U.S. EPA (2004b) and for a European example, see SETAC Europe (2004).

A prime example of the potential role that statutory law can play in promoting economically efficient environmental decisions involves oil spills by tankers. Because oil spills are hard to monitor, they present special challenges. The spills only occur with the rare accident, so there is no sustained emission flow to measure. The equivalent of abated emissions is reduced probability of spills, which depend on shipping practices (e.g., tanker maintenance, crew training) as well as on the design of the tanker itself. Even though behavior is expensive to monitor, society wants to know that potential polluters have incurred all appropriate expenses to reduce spill probabilities to suitably small levels. Potentially, statutory liability laws can contribute to structuring economically efficient incentives, but how well they succeed depends on the formulas for assigning liability. The laws provide the right incentives only if the compensations required for the injured parties have an expected value equal to damages produced. If the law requires compensations that are too low, potential polluters invest too few resources to reduce accident probabilities.

Assignment of Property Rights: Who Is Damaging Whom? Recall from Chapter 4 the importance of property rights in promoting efficient use of the environment. Consider again the case of a lake that one firm uses for waste disposal and another firm uses for a water that it adds to its canned fish. To simplify, assume that these two firms are the only stakeholders; nobody else is affected by the actions of either firm. Which firm causes damages and which one suffers damages? This may seem like a silly question, because our instincts tell us that the waste disposer is the cause of the problem. But it can be argued that if the waste-disposing firm is required to clean up its pollutants, the presence of the food company imposes economic damages on the polluter by making it invest in emission abatement. Who causes damages and who suffers injury depends on who owns what property. Property ownership assigns a certain bundle or rights within which the owner is free to make decisions about resource uses. When a resource lacks an owner, nobody has a clear economic incentive to care for the resource.

---

[11] Montero, Sanchez, and Katz (2002) show the importance of transaction costs in making it difficult to reduce air pollution in Chile.

Rights to Pollute. Let us begin with a situation where neither firm owns the lake. So ownership rights could be assigned either to the polluting firm or to the firm using it for a water supply. Will the assignment of those rights affect the lake's pollution level? Surely it would lead to zero emissions if owned by the water supply user and to uncontrolled emission if owned by the polluter, wouldn't it? Maybe not, if the lake owners and its nonowners are permitted to negotiate under the law. Remember what it means to own property: The owner decides how the resource is used and may block any unauthorized use. The owner also can negotiate a contract with anyone else who wishes to use the resource, (in much the same way a homeowner can rent a room to a tenant).

Take another look at Figure 16–9. Suppose that the effluent emitter, firm 2, owns the lake and *MD* refers to marginal damages suffered by firm 1, the water supply user. Does firm 2 have any incentive to even think about damages suffered by firm 1, that is, about the shape and location of *MD?* Or will firm 2 just continue to pollute? Firm 2's marginal abatement cost for reducing effluents into the lake is shown as *MAC.* Remember that the property rights say firm 2 owns the lake, so it may use it any way it pleases. Uncontrolled emissions are at E^1. Firm 2 commits no resources whatsoever to abating its emissions. As long as things are unchanged, Figure 16–9 shows that marginal damages from emission E^1 are $X, even though marginal abatement costs are zero. If both firms are free to negotiate, firm 1 has every incentive to offer firm 2 a payment to reduce its emission stream; for the first ton of abated emissions, any payment between zero and $X improves the welfare of both firms. The two firms have every incentive to continue bargaining over the marginal emission units as long as the curve *MD* is larger than *MAC.* And as long as *MD* is greater than *MAC,* firm 2 is better off by abating its pollutant for any monetary payment that exceeds its *MAC.* Any payment less than *MD* improves the welfare of firm 1. Under this continued stream of bargaining, we would see movement towards point E^*, the only the point at which *MAC* equals *MD.* This is likely to be what happens when firm 2 owns the lake.

Rights to No Pollution. Suppose, by comparison, that firm 1, the firm suffering damages from the polluter, owns the lake. Under these conditions, firm 1 would permit no damages to its property, which means that emissions of zero damages would be allowed, emissions would be no larger than where *MD* crosses the emissions axis. If zero damage emissions are what the property owner allows, will emissions stay there? Not if bargaining is permitted. If bargaining is allowed, firm 2 would have incentives to buy rights from firm 1 to discharge wastes into the lake. Any price for this dumping right less than the polluter's *MAC* and higher than the water user's *MD* improves the welfare of both parties. And so the bargaining goes. As emissions move up toward E^*, the polluter's *MAC* starts to fall and the water user's *MD* continues to rise. The forces for this movement to the right continue as long as *MAC* is greater than *MD.* Only after the efficient level of emissions E^* is reached can the adjustment come to an end. At emission level E^*, further adjustment stops because the polluter's *MAC* is no higher than the water user's *MD,* and there is no further basis for trade. (Antweiler, Copeland, and Taylor, 2001, describe why greater international trade can be good for the environment.)

This example shows that if property rights for the use of environmental assets are well defined, and if property owners and all potential users are allowed to negotiate an allocation of the property, the efficient emission level can occur, regardless of who is assigned the property right. This striking result is the Coase theorem, as described briefly in Chapter 4. Thus, defining property rights combined with the legal right to transfer them among owners sets the stage for automatic decentralized achievement of efficient levels of environmental quality. Another appealing aspect of the institution of property rights is that the parties who are involved in the bargaining (stakeholders) may know the most about the comparative values of abatement costs and environmental damages. Therefore, the correct efficiency point may be achieved with no need for centralized, often expensive, bureaucratic authority. The appeal to the power of free negotiation, combined with private property, has led many to propose converting environmental resources to private ownership as a way to promote their efficient use.[12]

Command-and-Control Approach

The command-and-control approach relies on rules or standards to bring about environmental improvements. For emissions, a standard is a maximum emission rate allowed under the law. The idea behind such a standard is this: If society wants its members to avoid doing something that damages other members, it passes a law that outlaws the damaging action then finds a means to enforce the law.

Figure 16–10 shows marginal damages and marginal abatement characterizing the rate at which an emission is discharged into the environment. The uncontrolled emission level is at E_u, a rate significantly above the efficient emissions rate, E^*. Authorities can mandate the emission rate E^* by establishing an upper limit (standard). They can enforce the standard by hiring and fielding the enforcement authorities needed to discover potential violations. When violations are uncovered, the polluter is fined something that bears some relation to the damages produced by those emissions. To the extent that a firm limits emissions to comply with the standard, it incurs abatement costs. Total abatement costs of area *A* measure the cost of complying to meet the mandated standard.

Standards are popular with lawmakers and members of the public as a method for controlling pollution because they are simple and direct. By setting clear and visible emission targets, standards appeal to the universal need to confront environmental pollution and do something about it; they support the ethical belief that pollution is bad and should be controlled. The legal system provides a convenient method for confronting the problem.

Economics of Standards. A big challenge facing the legal system is where to set a standard. Standard-setting authorities may debate the question of whether they should account for damages alone or both damages and abatement costs. It may come as little surprise that one approach has been to set standards by reference only to the marginal damage function; see *MD* in Figure 16–10. Some environmental laws have relied on the principle of setting the standard at what is called zero-risk level, that is, the level that protects everyone from damage, irrespective of their vulnerabilities. Following the principle of zero risk means setting emission standards at the threshold level, labeled E_T. This principle works if a threshold can be found. But as we saw earlier in this chapter, the question of thresholds is a matter of considerable debate. Some physical scientists who study environmental damages believe that some pollutants have no threshold level at all, which means that the marginal damages function is positive for any emission level larger than zero. For highly toxic materials, the zero level is certainly worth aiming for; but the zero level would be an expensive goal, indeed, to accomplish for all emissions. See Example 16–3.

Another possibility is to set the standard at a level that tolerates a small level of damages, for example, emission level E_0 in Figure 16–10, at which marginal damages are still significantly below marginal abatement costs. Still, this method sets the standard with no consideration of the costs of abatement. The principle of economic efficiency tells us to set the standard

---

[12] Recall the assumptions required for the Coase theorem to produce economically efficient use of an environmental resource: Property rights must be defined, transferable, and enforceable. There are three reasons why bargaining attempts can fail even if the benefits of the bargaining exceed the costs: (1) If transactions costs are too high, it may be too expensive for the two sides to meet. For example if there is a city of 100,000 residents who uses the lake for recreation and 10 polluting firms, the cost of getting the stakeholders together may be too high to produce efficiency. (2) If either side lacks information about the benefits or costs of limiting pollution, an inefficient result could occur, because neither party would know what to offer the other party. (3) If either party engages in strategic behavior, an economically efficient agreement may be blocked.

Example 16–3
Experimenting with Environmental Regulations in Chile

Developing countries often apply environmental regulations that were originally established, developed, and tested in industrialized countries. Even if the benefits of a regulation exceed its costs, implementation may require costly abatement technology that places stress on a poor country's financial capacity. So a highly debated question in policy circles is this: Should developing countries import environmental standards designed for developed countries, or should they invest resources to design their own? Those who support the theory of importing believe that doing the basic research needed to discover and establish environmental regulations is expensive. They recommend that developing countries borrow and adapt findings from their industrialized counterparts. Critics of imported standards respond that such standards risks being irrelevant to the culture, conditions, and technologies of poor countries. Each view point claims that the benefits of carrying out its theory exceeds the costs of the importing country.

For example, Chile reaps a considerable economic benefit from its copper exports. Imported environmental regulations on the mining and processing of copper may be misunderstood and misapplied without considerable support from local researchers in adapting regulations to local needs. Foreign environmental standards may not even be applicable to Chile because of geographic differences.

Environmental regulators in developing countries are rarely informed about the methods used for designing environmental standards or the associated probabilities. Moreover, environmental criteria are often copied from regulations already established in developed countries and presumed to be accurate and relevant for local conditions. Environmental regulations on trace copper in seawater present a good example. Research by Chilean scientists is showing a higher resistance to copper by northern Chile marine algae compared with algae elsewhere. Perhaps in Chile, algae have evolved to tolerate naturally high levels of copper in northern Chile coastal waters: If so, setting environmental standards in Chile for copper in seawater ignores historically adapted biological realities and produces little or no economic or biological benefit for the costs incurred in lost copper income and jobs displaced.

Source: Castilla, J. C. 1997. Environmental regulations and the hazards of risk assessment: Attitudes in developing countries. *Human and Ecological Risk Assessment.* 3 (5): 659–664. Much of this example was researched by New Mexico State University environmental science student Hilary Brinegar.

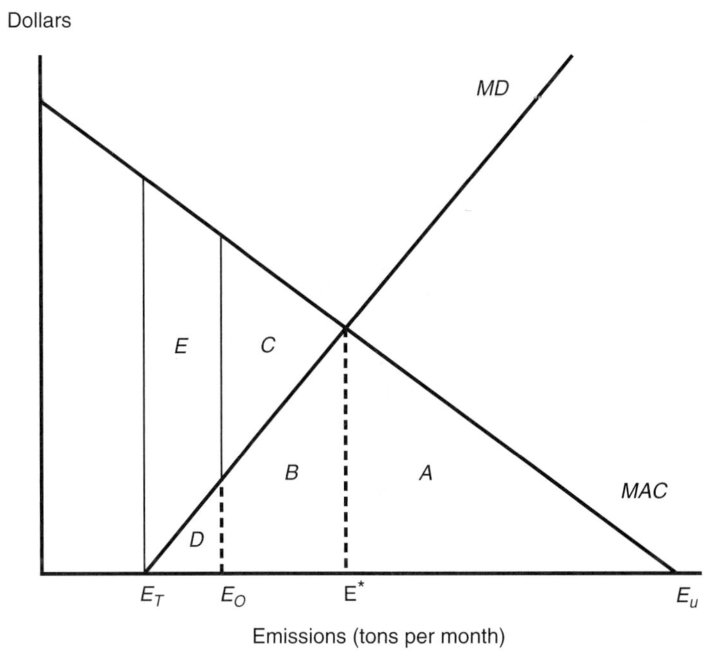

Figure 16–10
Efficiency of emission standards

so as to balance marginal damages with marginal abatement costs, namely, at our favorite point E^*.[13] If authorities restrict emissions below the economically efficient standard set at E^*, at a point like E_0, costs exceed benefits by the amount labeled *C*. Further tightening the standard to the zero-risk level (E_T) produces a further incremental net loss, labeled *E*.

Standards' Uniformity. In setting standards policymakers usually must decide whether a standard should be applied uniformly or be adapted to various conditions. Relevant conditions could be affected populations, emission control technologies available, type of industry and the like. One important condition is location: spatial uniformity versus spatial flexibility. A problem in applying standards uniformly is that different locations may differ widely in their factors influencing abatement costs and damage functions. One set of standards applied the same everywhere may come at a high cost of lost economic efficiency.

Figure 16–11 shows two marginal damage functions, one for each of two regions, and the same marginal abatement cost, *MAC*, for each. Suppose that MD_{ua} applied to an urban location and MD_{ra} applies to a rural area. MD_{ua} lies above MD_{ra} because larger numbers of people live in the urban location, so any given level of emissions affects the health and welfare of more people in the urban area than in the rural area. Marginal damages produced by emissions are considerably higher in the urban location, so the economically efficient emission level is lower in the urban area; the efficient level is E_{ra} in the rural area and E_{ua} in the urban location. As another example, compare industrial and residential areas. If few people live in the industrial area, we would expect marginal damages from pollutants to be higher in the residential area, which calls for lower economically efficient emissions there.

A policy that applies a single, uniform standard is unlikely to be economically efficient everywhere. Setting a single standard at E_{ua} is too strict for the rural area and produces a welfare loss for the rural area of L_r. Applying the standard at E_{ra} is too lax for the urban location and produces a welfare loss there of L_u. The only way to avoid this loss in efficiency in both places is to implement the policy by setting separate standards unique to each area's conditions. Political constraints might require that a single standard be applied everywhere. In that case, minimum damage is done to economic efficiency if the standard is chosen to minimize the size of total efficiency losses. That means looking for the one standard that minimizes the summed areas of the triangles L_u and L_r.

Standards' Cost-Effectiveness. If marginal abatement costs vary from one location to the next, setting the standard efficiently requires that the total abatement costs of meeting any given standard are minimized. In other words, marginal abatement costs should be equal at each location. This is yet another example of the equimarginal principle: Producing the largest abatement in total emissions for a given total abatement cost requires controlling emissions everywhere, so that all sources have equal marginal abatement costs. Sources with varying emissions are controlled by different amounts, depending on the each source's unique marginal abatement cost function. Identical standards are cost-effective only in the special case that all emission sources have equal marginal abatement cost functions.

Figure 16–12 and the accompanying table illustrate marginal abatement cost functions for two unique sources, each discharging a material that comes under the same regulation. You can see that the two sources' marginal abatement cost functions vary. They increase more slowly with reduced emissions for source 1 compared to source 2. There are several possible reasons for the differences. The sources could produce and sell different final products using different technologies. If source 2 is an older plant, it may have less flexibility; therefore, its cost of abating emissions is more than that of source 1, which is a newer plant with newer equipment. As a rule,

---

[13] Balancing marginal damages and marginal abatement costs assumes the each regulated polluter (Yates and Cronshaw, 2001) and the government regulator have accurate information (Hoel and Karp, 2001; Huesemann, 2001). It also assumes that the regulator wants to maximize society's net benefits, an assumption questioned by Levinson (2000).

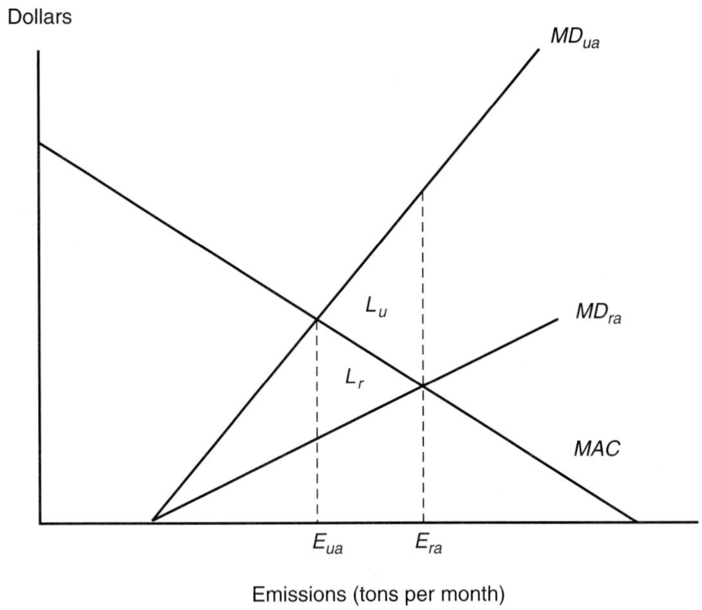

Figure 16–11
Geographic variation in abatement efficiency

expect considerable variability in abatement costs among firms, even if they discharge the same pollutant.

Suppose that without regulatory standards emissions start at the uncontrolled level of 10 tons per week for *each* source, for a total of 20 tons. Authorities wish to reduce *total* emissions (summed over both sources) to 10 tons per week by setting emission standards. How should they apply the standards? The procedure that seems most fair is to apply the same standard to each source; in this case, 5 tons per week. This certainly looks fair enough, since that method requires reductions of the same proportion from current levels for each source. However, not all sources are created equal economically. Each has unique marginal abatement costs. The application of uniform standards to economically different sources violates the equimarginal principle for economic efficiency. Uniform standards produce a considerable loss in abated emissions for the abatement costs incurred. Let us look at the numbers to see why this is true.

At a 5-ton emission level, source 1 incurs marginal abatement costs of $500 per ton, while source 2 incurs the much higher marginal abatement costs of $2,000 per ton. Recall that total costs are the marginal costs summed. So the total compliance costs are $1,500 for source 1 and $6,000 for source 2, for a grand total compliance cost of $7,500.[14]

How much could be saved by applying the equimarginal principle to setting the standard? The table shows that the total emission reduction can be achieved at considerably reduced compliance costs by satisfying the equimarginal principle. Application of that principle indicates that source 1 should cut its emissions to 2 tons per week (abate 8 tons) and source 2 to 8 tons (abate 2 tons). At these emission levels both sources' marginal abatement costs are equal ($800 per ton) and total abatement costs are $4,800 ($3,600 for source 1 and $1,200 for source 2). This application of the equimarginal principle produces a 36% savings in total costs from the uniform standards case. We also can think of it as producing greater abatement for a given abatement cost: For the $7,500

[14] The exact total cost is the area under the marginal cost curves. For the linear marginal cost curves shown, that area is $1,250 for source 1 and $5,000 for source 2, or a grand total of $6,250 when both sources abate five units. The discrepancy lies in integer approximations of total costs described in the table. Total costs are the sum of the previous marginal costs over increments that approach zero. The table's integer approximations affect neither the outcome nor the conclusions for policy.

Figure 16–12
Standards when marginal abatement costs vary

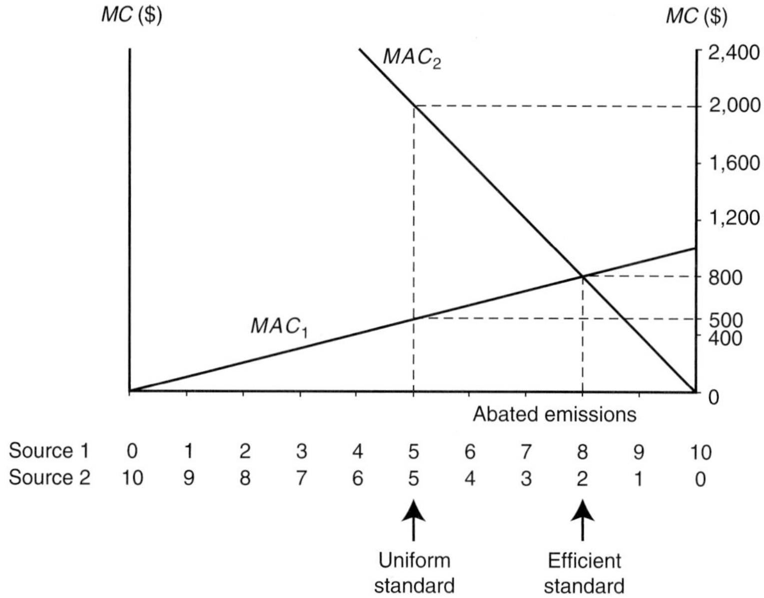

Abatement (tons per week)		Marginal abatement cost (dollars)		Total abatement cost (dollars)		
Source 1	Source 2	Source 1	Source 2	Source 1	Source 2	Total
0	10	0	4,000	0	22,000	22,000
1	9	100	3,600	100	18,000	18,100
2	8	200	3,200	300	14,400	14,700
3	7	300	2,800	600	11,200	11,800
4	6	400	2,400	1,000	8,400	9,400
5	**5**	**500**	**2,000**	**1,500**	**6,000**	**7,500**
6	4	600	1,600	2,100	4,000	6,100
7	3	700	1,200	2,800	2,400	5,200
8	**2**	**800**	**800**	**3,600**	**1,200**	**4,800**
9	1	900	400	4,500	400	4,900
10	0	1,000	0	5,500	0	5,500

abatement cost incurred under the uniform standards case, we can produce a cleaner environment, by limiting sources unequally, according to the equimarginal principle. To illustrate the point, cutting source 1 all the way to zero emissions (total compliance cost = \$5,500) and reducing source 2 to emissions of 7 tons per week (total compliance cost = \$2,400) produces total compliance costs about equal to the case of identical standards, but with much lower total emissions (7 tons per week rather than 10). So the equimarginal rule can be used to save costs for regulated sources, or it can be used to produce a cleaner environment, or both.

How does this simple example generalize to the real world? When marginal abatement costs are different for different sources, implementing uniform (identical) standards produces smaller emission reductions for a given total cost. Larger differences in marginal abatement costs across sources means that equal standards do progressively worse compared with application of the equimarginal principle. For a public agency to vary the application of standards in accordance with the equimarginal principle, it needs data on the marginal abatement costs for each regulated source. These data are typically expensive and/or hard to find. If the data are provided by the regulated industry, do not expect them to be perfectly objective. The general message of Figure 16–12 and the accompanying table is this: Uniform standards are inefficient, but nonuniform standards can be costly to implement.

Incentives Produced by Standards. Holding all firms to the same emission target can be an expensive way to meet pollution control policy goals. Although standards may effectively limit emissions, they usually impose high costs by forcing some firms to undertake unnecessarily expensive means of controlling pollution. Since the costs of limiting emissions may vary widely among firms—even among various emission sources within a single firm—the most cost-effective technology in one situation may be too expensive in another. Costs can vary because of the design, age, and configuration of a firm's production facility. Worse, command-and-control regulations may freeze the development of technologies that could produce greater amounts of control. Thus there is little economic incentive for firms to exceed their emission control targets while command-and-control standards discourage adoption of new technology. How standards might provide incentives in the short run and in the long run is discussed next.

Short-Term Incentives. Short-term incentives encourage firms to use their existing plant to abate emissions as cheaply as possible. The standards approach to controlling emissions is weak in this respect because standards discourage firms from looking for incremental savings. Standards are either met or not met. If firms are already meeting the standards, they have no special incentive to do better, even if the costs would be small and benefits of abated emissions high. Similarly, polluters must meet the standards, even if the last few units of abatement are more expensive then the value of potential damages avoided.

Long-Term Incentives. The long run is defined as the period of time in which all costs are variable, that is, the length of time in which a firm can make all needed adjustment to meet standards. Long-term incentives provide the length of time required to seek out the cheapest methods to reduce future emissions. A firm needs such incentives to make cost adjustments to current emission standards, even though the required technologies may not be in place.

In the long run, efficient policies of pollution control produce incentives to search for the kinds of technical and organizational changes needed to reduce future costs of emission abatement. How well do standards provide these long-term incentives? Figure 16–13 shows the evidence. It shows a firm's marginal abatement costs under two conditions: MAC_A refers to these costs prior to a given technical improvement and MAC_B refers to these costs after the firm invests in research and development (R&D) to achieve better emission abatement. With no pollution control standards, E_3 is the firm's uncontrolled and unregulated pollution level. There is no incentive to incur costs on R&D to limit emissions. Now suppose standards are established that the firm must meet at the efficient emission level E_2. When the firm uses existing abatement technology, its total annual costs for compliance is shown as area $D+E$, the area under the marginal abatement cost function starting at the uncontrolled level. If the firm responds to those externally imposed standards by investing in R&D, compliance costs of reducing emissions to E_2 decrease to area E, the area under the lower marginal abatement cost function. The cost difference, D per year, is the amount by which compliance costs are reduced and represents the *payoff incentive* for making the R&D investment.

Figure 16–13
Abatement standards and technology

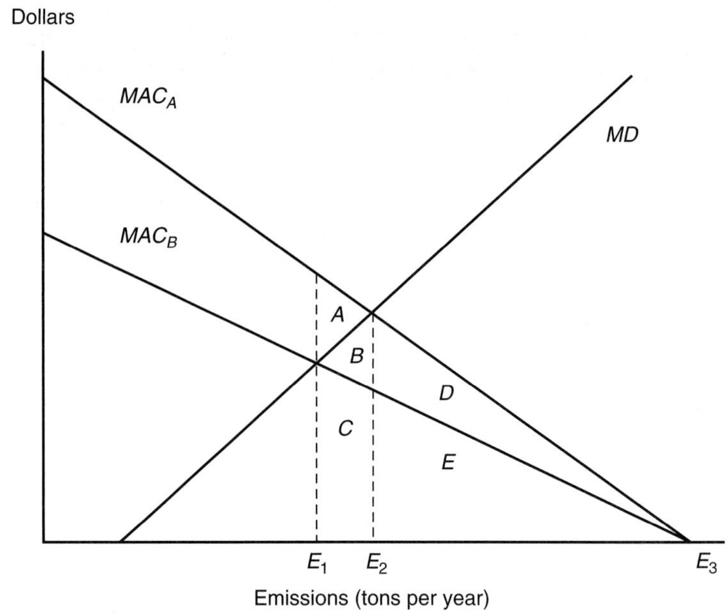

Despite this optimistic picture, however, standards can also provide disincentives for long-term investments in technical advancement. Remember, standards are established by regulatory authorities, then implemented by polluters. Interaction between the regulator and the regulated can lead to patterns of negotiation that significantly affect outcomes.[15] Imagine that a regulatory authority begins by setting a standard at something like the efficient emission level, E_2, and that a regulated firm responds by finding an abatement technology that reduces the marginal abatement cost curve to MAC_B. Even though this innovation will reduce the cost of complying with E_2, there is another impact developing behind the scenes. The regulatory authority notices that the reduction of marginal abatement cost from MAC_A to MAC_B reduces the efficient emission level to E_1. That is, the reduction increases the abatement level (reduces the emission level) at which benefits of abatement exceed the costs by the greatest amount.

The authority is not blind to the impact of its regulations on the firm's incentives to innovate. After seeing the success of its regulation in promoting cost-saving innovation, the regulatory authority might ask, "Why not let society share the gains of this magnificent cost savings? Let's again tighten the regulatory standards to emission level E_1 to reflect the new lower costs." Suppose the authority does that. Now the firm's cost of complying with the still stricter regulations is $C+E$ per year, the appropriate area beneath MAC_B. The net change in total abatement cost produced by this second set of standards, displacing the first set, is the area $C-D$. For the regulated firm this means that its cost savings are significantly reduced compared from the savings when the standard stayed at E_2. Compliance costs will be *higher* than before the R&D program, if area $C-D$ exceeds zero.

The regulated firm can see the impact of its successful innovations on the regulator's incentive to re-regulate. In fact, the regulated firm is likely to anticipate that its successful innovations, through their impact on future standard tightening, could make it worse off. This anticipation could encourage the firm to spend money on hiring lobbyists and influencing public opinion to resist any regulations in the first place (e.g., Damania, 2001). Such reactions have detracted from the incentive to produce and implement new technologies for pollution control. The series of events just described demonstrates what are called perverse incentives, which work against the

[15] Berman and Bui (2001) describe this interaction for an example from the South Coast Air Basin in southern California. A competing hypothesis is advanced by Khanna (2001) and Maxwell, Lyon and Hackett (2000), who argue that voluntary compliance can be cheaper than waiting to be regulated.

Table 16–1 Notable environmental incentive mechanisms outside the United States

Mechanism	Places applied	Comments
Environmental labeling	Most industrialized and several developing countries	Government programs to promote environmentally friendly products. Credited with decreasing Volatile Organic Compounds and other emissions in Germany and increasing recycled paper use in Korea.
Noise pollution charges	Belgium, France, Germany, Japan, the Netherlands, Norway, and Switzerland	Not used in United States
Effluent charges	Several countries	Evidence of incentive effects in Germany, the Netherlands, and Malaysia
Water depletion charges	Several countries	Incentive effects noted in several Asian countries and projected in the Netherlands
Carbon/energy taxes	Denmark, Finland, the Netherlands, Norway, and Sweden	Not used in the United States. Incentive effects noted in Sweden and Finland and projected effects in the Netherlands.
Deposit systems	Numerous countries, including Austria, Greece, Korea, Norway, and Sweden	Korea requires deposits from producers; links refunds to proper postconsumer disposal, motor oil, other products. Austria includes lightbulbs. Greece, Norway, and Sweden include car hulks; incentive effects noted.
Fertilizer and pesticide charges	Austria, Belgium, Scandinavian countires	Higher than fees in U.S. states. Incentive effects noted in Austria and Sweden.
Other product charges	Various countries (e.g., Belgium's disposables tax, Italy's plastic bag tax, Germany's industry-imposed packaging fees)	More products than in United States. Effects noted or projected for some of these charges.
Livestock and manure charges	Belgium and the Netherlands	Little use in United States
Motor vehicle quota system	Singapore	Not used in United States
Congestion pricing	France, Norway, United Kingdom, and Singapore	Use just beginning in United States. Incentive effects noted in all three countries
Motor fuel taxation	Various countries	U.S. credit approach. Sweden taxes diesel emissions
NOx emissions charges	France and Sweden	Little used in the United States outside California. Incentive effects noted in Sweden
SO_2 emissions charges	France, Japan, other countries	Differs from U.S. approach of tradable allowances under acid rain program

Source: *Adapted from U.S. Environmental Protection Agency. 2001.* The United States experience with economic incentives for protecting the environment. *Available from http://yosemite.epa.gov/ee/epa/eermfile.nsf/vwan/ee-0216b-01.pdf/$file/ee-0216b-01.pdf [Accessed October 6, 2004].*

aims of the environmental regulation. Perverse incentives, in connection with standards, may discourage long-run advances in pollution abatement technology.

Harnessing Market Incentives

Market-based incentives are instruments that regulate behavior through market signals rather than through explicit rules or standards regarding pollution control levels or methods. Two important types of market incentive policies for controlling pollution are emission charges and emission trading. Table 16–1 shows several environmental economic incentive mechanisms in use outside the

United States. Both types of policies have the potential to harness market forces by encouraging firms to undertake pollution control efforts compatible with their own interests.[16] Both require centralized public action to start up but rely on the incentive-harnessing power of the market to carry out efficient pollution control. For emission charges, firms are encouraged to respond in whatever way best serves their self-interest (sometimes called actions of the "rational polluter," Spence, 2001) in the face of what amounts to a new price for using the environment's services. Emission trading is even more flexible and is set up to work with a minimum of government intervention.

Emission Charges. A pollution charge assesses a fee or tax on the quantity of pollution that a source produces. In practice, this approach makes it economically worthwhile for the source to reduce its emissions to the point where the marginal abatement cost equals the tax rate. If regulators wish to encourage economic efficiency in pollution abatement, the charge they impose will be set equal to the marginal benefits of cleanup, and that charge will encourage the efficient level of cleanup. This is a theoretical ideal, but policymakers more commonly think in terms of a desired level of cleanup, partly since methods to measure many of the benefits of environmental cleanup are still being developed.

Under a system of emission charges sometimes called emission taxes[17], regulatory authorities tell polluters: "You may release whatever level of emissions you wish, but we will measure your emissions, and we will charge you a certain amount for every unit (e.g., pound) that you discharge." When authorities set up an emission charge, they remove the moral stigma of pollution and replace it with a business culture. Aronsson (2001) describes how an emissions charge can influence technological change and Ono (2003) describes how it can affect long-run economic growth. Table 16–2 shows some examples of emission charges in European countries. Essentially, an emission charge requires polluters to pay for the emission discharge services provided by the environment, in much the same way we pay for our trash-hauling services. The charge effectively says that sources responsible for emission will be charged for the services of the environment—waste dilution, chemical decomposition, and the like. In bringing the services of the environment under the umbrella of a business culture, the services of the environment are treated like other priced inputs. Since business responds to the incentive to conserve scarce land, labor, capital, and materials, the emission charge provides an incentive to conserve the use of the environment's resources. Again like labor and other inputs, the highest-priced inputs are conserved most intensively. To conserve their use of the environment, firms are free to choose whatever methods save them the most costs. Sometimes the cheapest method is to move to a "pollution haven," that is, a country where emission taxes are lower (Eskeland and Harrison, 2003; Xing and Kolstad, 2002). Table 16–3 compares U.S. and European taxation rates on gasoline (petrol) and diesel fuel.

A culture of regulation commonly surrounds environmental control, so the approach just described may sound unusual. However, freedom to choose is the main advantage of emission charges. By removing the stigma of pollution and giving polluters freedom to choose their best way to abate emissions, charges are designed to harness the creative forces released by the desire to maximize profits. The emission abatement choices on the menu can be any combination of treatment, technical change, changes in the mix of inputs, emission reuse, products that pollute less, and the like. The strength of emission charges is that they provide a powerful incentive for polluting firms to seek out the cheapest methods for abating emissions. The government is not involved in telling firms how to abate emissions.

---

[16] In contrast, conventional approaches to regulating the environment (e.g., command-and-control regulations) allow little flexibility in the methods for achieving society's policy goals. Command-and-control regulations typically meet the policy aims by setting uniform standards for firms. The two most common are technology-based standards and performance-based standards. Technology-based standards specify the method, or even the equipment, firms must use to comply with a given regulation. A performance standard sets a uniform control target for firms, while giving them a choice in how they meet this target.

[17] Fullerton and West (2002) describe how a tax on gasoline can have effects on reducing air pollution that are similar to the effects of emission taxes.

Table 16–2 Selected European effluent fees

Effluent[a]	Country	Price (highest marginal, U.S. dollars)	Use of fee revenues
CO	Czech Republic	33/ton	Environment
	Estonia	1.36/ton	
	Lithuania	1.75/ton	
	Poland	22/ton	
	Russia	0.09/ton	
	Slovakia	20/ton	
CO_2	Denmark	42/m^3, diesel, kerosene, gas oil 38/ton, coal 0.02/kwh, electricity	General budget
	Finland	38–43/m^3, gasoline, diesel, kerosene 39/ton, coal 0.003–0.006/kwh, electricity	
	Netherlands	45/m^3, gasoline, diesel, kerosene 0.02/kwh, electricity	Income tax relief
	Norway	59/ton, coal	General budget
	Sweden	106–131/m^3, gasoline, diesel, kerosene 114/ton coal 0.02/kwh, electricity	
SO_2	Bulgaria	0.02/kg	Environment
	Czech Republic	45/ton	
	Denmark	1.60/kg	General budget
	Estonia	95/ton	Environment
	Finland	30/m^3 diesel or oil	General budget
	France	32/ton direct emissions	Research
	Hungary	2.40/ton	Environment
	Italy	62/ton direct emissions	
	Lithuania	46/ton	
	Norway	0.01/liter of fuel/0.25% sulfur content	General budget
	Poland	83/ton	Environment
	Russia	6.10/ton	
	Slovakia	33/ton	
	Spain	39/ton	General budget
	Sweden	3.33/m^3 of liquid fuel/0.1% sulfur content	
NOx	Bulgaria	0.05/kg	Environment
	Czech Republic	45/ton	
	Estonia	216/ton	
	France	27/ton	Research
	Hungary	4/ton	Environment
	Italy	123/ton	
	Lithuania	67/ton	
	Poland	83/ton	
	Russia	5.08/ton	
	Slovakia	27/ton	
	Sweden	5/kg	

[a]CO-carbon monoxide; CO_2-carbon dioxide (greenhouse gas); SO_2-sulfur dioxide; NOx-nitrogen oxide.

Source: *Adapted from Stavins, R. N. 2001. Experience with market-based environmental policy instruments. Washington, D.C.: Resources for the Future Discussion Paper 01–58. Available from http://www.rff.org. Reprinted with permission from Elsevier.*

Table 16–3 Tax rates on motor vehicle fuels in the United States and Europe

Country	Gasoline (euros per 1,000 liters)	Country	Diesel (euros per 1,000 liters)
United States	53	United States	70
Romania	204	Romaina	112
Latvia	211	Lithuania	129
Estonia	224	Bulgaria	135
Lithuania	243	Estonia	166
Bulgaria	267	Latvia	198
Slovakia	268	Slovakia	209
Czech Republic	304	Czech Republic	229
Poland	314	Poland	237
Greece	321	Portugal	246
Luxembourg	347	Greece	247
Portugal	349	Luxembourg	253
Hungary	358	Spain	270
Slovenia	369	Austria	283
Spain	372	Belgium	290
Ireland	374	Slovenia	290
Austria	408	Hungary	308
Switzerland	467	Ireland	325
Belgium	494	Finland	325
Denmark	519	Denmark	346
Sweden	529	Netherlands	347
Italy	542	France	367
Finland	561	Sweden	373
Germany	562	Germany	378
France	586	Italy	403
Netherlands	592	Switzerland	487
Norway	651	Norway	519
United Kingdom	801	United Kingdom	801

Source: *Speck, S., J. McNicholas, and M. Markovic. 2002.* National experience with pollution charges and energy taxes in central and eastern Europe. *Available from http://www.devtechsys.com/CASE022.pdf [Accessed October 6, 2004].*

Economic Concepts. Figure 16–14 and the accompanying table illustrate the significant points of an emission charge; the numbers in the table refer to one source (i.e., one firm) of a given pollutant. Suppose the efficient level of emissions has been found to be E^* (see figure), or 7 tons per year (see table). Suppose two possible emission charges (taxes) of \$300 per ton or \$500 per ton are being debated. The two possible tax rates are shown as T^* and T^1, respectively. Column 3 in the table shows the firm's marginal abatement costs, *MAC* in the figure, and the column 4 shows total abatement costs, which is the area under *MAC* starting from point E^u at the unregulated level. Columns 5 and 6 show the firm's marginal tax cost under each of the two tax rates. Columns 7 and 8 show the total annual tax bill the firm pays at the \$300 and \$500 per ton tax rate. Columns 9 and 10 show total costs, under each of the tax rates, which consist of abatement costs plus the tax bill.

Under the \$300 tax rate, the firm begins by considering all possible emission levels, ranging from 0 to 10 tons. The firm recognizes that it can minimize its total costs, at \$2,700, by emitting 7 tons per year (abating 3 tons per year). It knows that 3 tons will minimize costs by comparing its marginal abatement costs to its marginal tax bill saved. Suppose the firm starts by considering emitting 10 tons per year. From the 10-ton level, it recognizes that cutting emissions to 9 tons *cost \$100 more in abatement costs,* but it saves *\$300 in tax charges;* so being a good citizen also reduces its total cost. Based on this reasoning, the firm realizes it could reduce costs by continuing

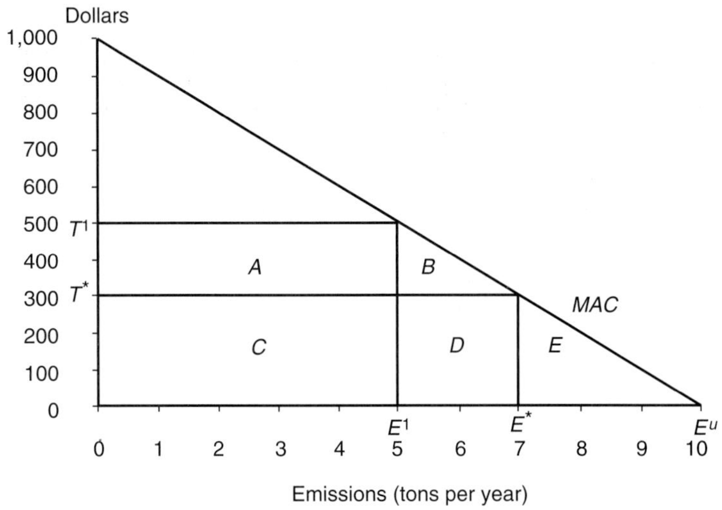

Figure 16–14 Impact of an emission charge

1	2	3	4	5	6	7	8	9	10
				Marginal tax bill		Total tax bill		Total costs	
Abatement	Emissions	Marginal abatement costs	Total abatement costs	$300 per ton	$500 per ton	$300 per ton	$500 per ton	$300 per ton	$500 per ton
		dollars							
10	0	1,000	5,500	300	500	0	0	5,500	5,500
9	1	900	4,500	300	500	300	500	4,800	5,000
8	2	800	3,600	300	500	600	1,000	4,200	4,600
7	3	700	2,800	300	500	900	1,500	3,700	4,300
6	4	600	2,100	300	500	1,200	2,000	3,300	4,100
5	**5**	**500**	**1,500**	**300**	**500**	**1,500**	**2,500**	**3,000**	**4,000**
4	6	400	1,000	300	500	1,800	3,000	2,800	4,000
3	**7**	**300**	**600**	**300**	**500**	**2,100**	**3,500**	**2,700**	**4,100**
2	8	200	300	300	500	2,400	4,000	2,700	4,300
1	9	100	100	300	500	2,700	4,500	2,800	4,600
0	10	0	0	300	500	3,000	5,000	3,000	5,000

to reduce emissions as long as the tax rate (marginal tax bill saved) exceeds its marginal abatement costs. The firm's decision rule is this: Continue to reduce emissions until marginal abatement costs equal the marginal tax bill saved (emission tax rate). This is shown in Figure 16–14 as the firm moves from the uncontrolled emissions level, E^u (10 tons), to the efficient level E^* (7 tons) required to respond to the $300 tax.

The secret to predicting the polluter's actions is to recognize that it wishes to minimize total costs and that these costs equal the sum of abatement costs and the emission tax bill. After the firm reduces its emissions to 7 tons under the $300 tax, its total annual tax bill is $2,100. Annual abatement costs are $600. Graphically, total abatement costs correspond to the area under the marginal abatement cost function, labeled E in Figure 16–14. The total tax bill is equal to emissions times the tax rate, or the rectangle labeled $C+D$. Under the emission charge program, a polluting firm's total costs are its pollution abatement costs plus the tax payments made to the authority: $2,700, or area $C+D+E$.

We might ask if the emission tax really has any influence on the firm's polluting behavior. What is to stop the firm from simply ignoring the charge and continuing to pollute at the unregulated level of E^u? In fact, there is nothing to stop the firm from doing just that. If it continues emitting 10 tons of emissions, its total costs are $3,000 per year, consisting completely of a tax bill. But the emission charge harnesses the firm's incentive to reduce its costs by reducing its emissions. No emission reduction at all produces higher total costs than the $2,700 the firm can achieve by reducing emissions to 7 tons. The firm's response to an emission tax depends on many factors, but one factor that can be controlled by the regulator is the level of the tax. A higher tax produces greater reduced emissions. Figure 16–14 and table show that a tax of $300 encourages the source to emit 7 tons, whereas a $500 tax produces a cutback to 5 tons per year.[18] A higher tax rate reduces emissions by more. A steeper sloped marginal abatement cost function produces a smaller reduced emission for any given tax.

Now let us compare the emission charge (tax) approach with the emission standard approach. With the $300 tax, the firm pays total tax and abatement costs equal to $2,700. Of course, regulatory authorities could have decided on emission standard equal to 7 tons per year. Suppose the authorities did just that and relied on the 7-ton standard to limit the firm's emissions to 7 tons per year. This regulatory action would cause the firm's total costs to equal $600, all of which are abatement costs with no tax costs. This means that the emission charge program costs the polluting firm more than the standards program. A regulatory standard produces abatement costs that are equal to those of the emission charge, but with the standard the firm still has access to the environment's disposal services for free. Under the emission charge program, the firm must pay for the environment's services. Therefore, you can expect polluting firms to prefer standards instead of charges. Emission charges can produce the same efficient level of emissions as regulations, but they come at a higher cost to the regulated industry, which is why industry typically resists charges (see Wolff, 2000).

Cost-Effectiveness. Despite resistence by regulated firms, a strong argument can be made for supporting a program of emission charges; namely, emission charges can control *multiple* emission sources at minimum cost. Figure 16–15 and the accompanying table show the effects of an equal emission tax rate applied to different sources that have varying marginal abatement cost functions. If each source follows its economic incentives and minimizes its own costs, each reduces emissions until its marginal abatement costs equal the tax. Since each source equates its own marginal costs to the common tax rate, the outcome is equal marginal abatement costs over all sources.

Suppose there are two sources of a pollutant, source 1 and source 2. The two marginal abatement cost functions are the same ones used in Figure 16–12. Marginal abatement costs of source 1 increase more slowly with abated emissions than do costs of source 2. If regulatory authorities impose an effluent fee of $800 per ton on each source, source 1 reduces its emissions to 2 tons per year (abates 8) and source 2 emits 8 tons (abates 2). Each response is an independent, cost-minimizing decision. Any other behavior produces higher total costs. After the two cost-minimizing reductions, both sources have equal $800 marginal abatement costs. Total emissions abated are 10 tons per month, so the effluent charge has accomplished a remarkable task. Through no special effort on the part of the gov-

[18] At a $500 tax ($T^1$), the firm sets its marginal abatement cost equal to the $500 tax and emits 5 tons (point E^1), pays abatement costs of $1,500 ($B+D+E$), and pays taxes of $2,500 ($A+C$). When the firm takes the $500 tax as a fixed price, beyond its control, its emission of 5 tons minimizes total costs at $4,000 ($A+B+C+D+E$). Other emission levels at that tax rate produce an equal or greater total cost.

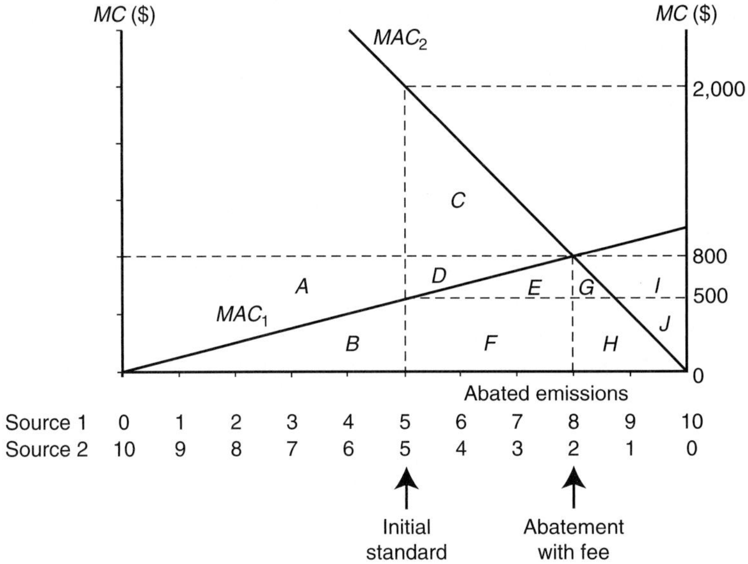

Figure 16-15
Emission charges and cost-effective abatement

Institution	Emitter	Units abated	Marginal abatement costs	Total abatement costs	Fee per unit	Units emitted	Emitter fees paid	Taxpayer fees received	Net costs
			← dollars		→		← dollars		→
Without fees	Source 1	5	500	1,500	–	5	–	–	1,500
	Source 2	5	2,000	6,000	–	5	–	–	6,000
Total		10		7,500		10			**7,500**
With fees	Source 1	8	800	3,600	800	2	1,600	1,600	3,600
	Source 2	2	800	1,200	800	8	6,400	6,400	1,200
Total		10		4,800		10	8,000	8,000	**4,800**

ernment, the emission fee program has effectively distributed abatements between the two firms consistent with the equimarginal principle. Source 1's total abatement costs are $3,600 (area $B+E+F$), the area under its marginal abatement function MAC_1. Source 2's abatement costs are $1,200 (area $G+H$), the area under its marginal abatement curve, MAC_2.

Now, who pays how much in fees? Source 1 pays effluent fees of $1,600 ($G+I+H+J$), while source 2 pays $6,400 in fees ($A+B+D+E+F$). The sum of effluent fees both sources pay, $8,000, are costs to the firms, but taxpayers receive *offsetting* tax revenues of $8,000. So, society sees effluent tax revenues as neither a net cost nor a net benefit (but transfer payment only).[19] Therefore, total net costs under the $800 emission fee are the costs of abatement only, $4,800 ($B+E+F+G+H$).

[19] This discussion ignores the double-dividend debate, that is, whether revenues from pollution taxes can help the environment while providing a financing source for additional, efficient environmental and economic programs. Bosquet (2000) surveys the recent evidence for environmental taxes in particular. Burtraw et al. (2003) analyze the evidence for carbon taxes, and Gielen and Chen (2001) analyze double dividends for environmental taxation in China.

An emission tax avoids equal abatement by exploiting cost advantages of more efficient pollution abaters. The tax produces a larger proportional emission reduction from sources with lower marginal abatement costs. Firms with steeper (less efficient) marginal abatement costs reduce emissions proportionally less. Therefore, an emissions tax is more cost-effective than proportionate emission reductions. To see this, suppose that regulatory authorities rejected the emission charge and instead enacted a proportional abatement with the intent that all polluters should receive equal treatment. For example, assume that the two sources are required to abate emissions by 50% (from 10 to 5 tons per year). Figure 16–15 and the accompanying table show that this "equal treatment" produces different marginal abatement costs for the two sources. Total abatement costs for source 1's 5-ton emissions are $1,500 (area *B*), and for source 2, they are $6,000 ($C+D+E+F+G+H$), for a total net cost of $7,500 ($B+C+D+E+F+G+H$) which is $2,700 higher than costs produced by the fee program. The $2,700 welfare loss, which is shown as area $C+D$, is called a *deadweight loss* produced by an inefficient policy.

Implementation of an emission charge poses several challenges and costs of enforcement. Any publicly administered charge system requires reliable facts on what is being taxed. Emissions must be measurable cheaply and reliably; that is, emissions must be concentrated in a finite number of recognizable flows to be monitored. So an emission program is hard to administer for most nonpoint emission sources, because those sources are spread over too wide an area to be measurable. Emissions from agricultural runoff and other diffuse sources are especially hard to measure reliably, so these would not be good candidates for an emission fee program.

Emission Trading. Although pollution charges can encourage economically efficient emission patterns, they have two disadvantages: (1) it is hard to predict the abatement a given charge will produce; and (2) the charges impose high costs on industry, so industry, particularly in the United States, rarely supports them. For major environmental policies, abatement targets are often set in terms of reduced percentages of current amounts. An important question is how to best achieve a specified abatement target while also achieving economic efficiency. One approach is to set up an *emission trading program,* also known as a *transferable emission permit system.*[20] Under a transferable permit system, an allowable, overall level of pollution is established and allocated among firms in the form of permits. Firms that keep their emission levels below their allocated level may sell their unused permits to other firms or use them to offset excess emissions in other parts of their facilities. Here is how they work: First a target level of abatement is set, which leads to a target set of emissions. Then a total number of permits is set to match that target. The permits can be distributed to affected firms in any way that is politically acceptable, for example allocating based on historical emissions or selling the permits at an auction.

Once allocated, the permits are fully transferable (tradeable). Anybody can buy the permits—other firms or environmental groups that wish to reduce emissions. Firms are free to make their own choice on whether to abate emissions or buy permits for the pollution they emit or plan to emit. The total volume of emissions discharged by all affected firms in the affected area cannot exceed the maximum amount established by the total supply of permits. This system has the added attraction of giving any stakeholder interested in abating emissions the opportunity to pur-

[20] Two highly publicized examples of emission trading are (1) sulfur dioxide trading under the 1990 U.S. Clean Air Act and (2) the Kyoto Protocol for dealing with global warming. In the first case, the acid rain program, (established under Title IV of the Clean Air Act) sets a cap on total sulfur dioxide emissions. Emission allowances, each representing one ton of sulfur dioxide, are distributed to companies affected by the cap, based on previous emissions. To avoid federally mandated reduction measures, affected firms are permitted to choose their own method of sulfur dioxide abatement. If a firm reduces emissions below its allowance, it can sell the surplus permits, which gives it an added incentive to lower emissions. The ability to trade encourages abatements to occur where abatement is the least expensive. Firms with low marginal abatement costs can reduce emissions and sell the excess permits to firms with high marginal costs. This trading leads to potential cost savings. For details, see U.S. EPA, Emissions Trading (2004a) and Colby (2000). In the second case, the Kyoto Protocol introduced the idea of its parties trading greenhouse gas emissions to help reduce the cost of meeting their overall emission objective.

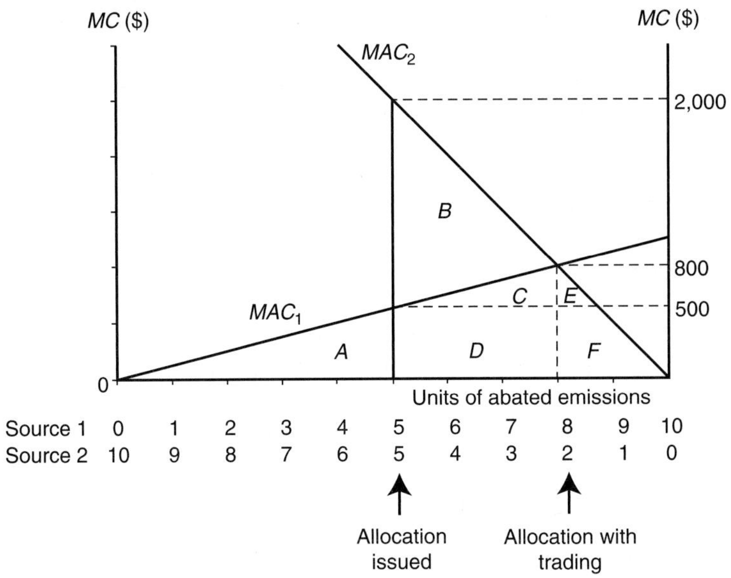

Figure 16–16 Cost savings from emission trading

Institution	Emitter	Permits issued	Units abated	Abatement costs	Trade	Units traded	Permit price	Permit value	Net costs
				dollars			←	dollars	→
Without trades	Source 1	5	5	1,500	–	–	–	–	1,500
	Source 2	5	5	6,000	–	–	–	–	6,000
Total		10	10	7,500	–	–	–	–	**7,500**
With trades	Source 1	5	8	3,600	Sell	3	800	2400	1,200
	Source 2	5	2	1,200	Buy	3	800	2400	3,600
Total		10	10	4,800					**4,800**

chase permits and permanently retire them, which reduces total pollution below the original target level. One variation on this theme is for permits to be issued for a limited time; after this time, smaller numbers of permits are issued, thereby reducing overall emission levels.

Figure 16–16 and the accompanying table present an economic analysis of the cost savings produced by emission trading. This highly simplified figure presumes only two sources, each currently emitting 10 units of pollution, for a total of 20 units. The presumed policy aim is to reduce emissions by an aggregate of 10 units over the two sources, which means there can only be 10 units of emission between the sources. The figure and table show all possible ways in which a total abatement of 10 units can be distributed between the two sources. The marginal abatement cost curves for the two sources are plotted in different directions on the same axis, with pollution abatement by source 1 going from left to right for source 1 and from right to left for source 2. As demonstrated earlier in the chapter, this method of analysis simplifies discovery of the point at which the equimarginal principle is satisfied, which is the point where both firms' marginal abatement costs are equal.

Before the emissions trading program, the two firms together emit 20 units of pollution. To achieve the goal of 10 units abated, only 10 emission unit permits are issued. One possible allocation of that property right is to allocate 5 to each firm. If permits are not tradeable, each firm reduces its emissions from 10 to 5, a total reduction of 10, as shown at the midpoint of the figure. At this point, marginal control costs are \$500 for source 1 and \$2,000 for source 2. This permit allocation achieves the policy goal in reduced emissions, but is it cost-ineffective? Each firm's total control cost can be seen on the graph as the area under its unique *MAC* curve. Source 1's total cost for pollution control is area *A* (\$1,500), and the total cost for source 2, \$6,000, is area *B*+*C*+*D*+*E*+*F*. The combined cost to achieve 10 units of pollution reduction is *A*+*B*+*C*+*D*+*E*+*F* (\$7,500).

Emission trading, however, produces the following opportunity: *Both* firms can improve their own positions, and overall cost-effectiveness, by *trading permits*. It will be to the advantage of source 1, which has lower abatement costs, to reduce pollution by 3 additional units, for a total of 8 reduced. Source 1 thus uses only 2 permits and sells the extra 3 permits to source 2. Source 2 has high abatement costs, so it finds it profitable to buy 3 permits from source 1. This purchase permits source 2 to reduce pollution by only 2 units instead of its original required 5. The equilibrium price for the permits is \$800 per permit. That price represents the marginal control costs for both firms at the point where source 1 reduces emissions by 8 units and source 2 reduces by 2 units. At the posttrade equilibrium, total control cost for source 1 is area *A*+*C*+*D* (\$3,600), up \$2,100 from the pretrade level; total cost for source 2 is area *E*+*F* (\$1,200), down \$4,800 from the pretrade level. The emission reduction goal of 10 units abated has been achieved at minimum cost. Area *B* (\$2,700) represents the net savings from this more efficient approach; that area is the deadweight loss avoided by trading compared with requiring both firms to abate 5 units. Figure 16–16 and the table show costs with and without trading. *Emission trading reduces total costs for each firm as well as reducing the combined costs.*[21] Table 16–4 shows several cases where economic incentives can reduce the cost of complying with environmental standards compared with the traditional regulatory approach.

Emission trading combines the precision of regulation with the price incentives offered by emissions charges. It permits regulatory authorities to establish a well-defined limit on total pollution levels (which charges cannot do), while harnessing market incentives to minimize the cost of abating pollution (which regulations cannot do). It is economically attractive for the firms involved for it allows a given amount of pollution abated at minimum total economic cost. Furthermore, other stakeholders in a cleaner environment can contribute to abated emissions by buying and retiring permits, which helps clean up the environment progressively more in future years (Malik, 2002). Table 16–5 shows some of the tradeable permit systems in various countries around the world. See Example 16–4.

Enforcement

A typical pollution control program, whether it is based on standards, emission fees, or emission trading, calls for some level of reduced emissions from uncontrolled levels. Since abatement costs can be expensive and since no firm loses customers by not abating, there is a natural incentive to avoid limiting emissions, no matter what the law says. You might expect that penalties written into the law for violating program abatement would produce complete compliance, but this rarely occurs. The search for the self-regulating pollution control program has been long and so far unsuccessful. Like other laws, pollution control laws require monitoring and enforcement (e.g., Dasgupta, Hettige, and Wheeler, 2000 and Dasgupta et al., 2001), and these tasks require resources. A common challenge is finding adequate resources for monitoring and enforcement to

[21] Not all emission trading systems are equal. Trading systems can be characterized by several important details: (1) Scope: Is trading restricted to averaging emissions within a single facility, among facilities owned by the same firm, or among firms under different ownership? (2) Cap: Is there a limit on total emissions or on effluents? (3) Commodity being traded: How will the commodity be defined? (4) Distribution of tradeable permits: What method is used to allocate initial permits? (5) Trading ratio: Is the required trading ratio 1:1 or some greater ratio, such as 2:1, meaning the buyer would need to buy two permits to have the right to continue emitting one unit? Higher trading ratios are designed to reduce a region's total emissions over time, (6) Banking: Can tradable certificates be banked or otherwise reserved for future use? (7) Monitoring: How is credit generation and trading monitored (See U.S. Environmental protection Agency, 2001)?

Table 16–4 Cost savings from pollution control through incentives

Pollutant controlled	Study source, year	Location	Traditional regulatory approach	Ratio of costs: Traditional approach versus incentive approach
Hydrocarbons	Maloney and Yandle (1984)	DuPont facilities in United States	Uniform percent reduction	4.15
Nitrogen dioxide	Seskin et al. (1983)	Chicago	Proposed reasonably available control technology (RACT) regulations	14.4
Nitrogen dioxide	Krupnick (1986)	Baltimore	Proposed RACT regulations	5.9
Total suspended particulates (TSP)	Atkinson and Lewis (1974)	St. Louis	State implementation plan (SIP) regulation	6.0
Particulates (TSP)	McGartland (1984)	Baltimore	SIP regulations	4.18
Particulates (TSP)	Spofford (1984)	Lower Delaware Valley	Uniform percent reduction	22.0
Particulates (TSP)	Oates et al. (1989)	Baltimore	Equal proportional treatment	4.0
Reactive organic gases and nitrogen dioxide	SCAQMD[a] (1992)	Southern California	Best available control technology	1.3
Sulfur dioxide	Roach et al. (1981)	Four Corners Area	SIP regulation	4.25
Sulfur dioxide	Spofford (1984)	Lower Delaware Valley	Uniform percent reduction	1.78
Sulfur dioxide	ICF resources (1989)	United States	Uniform emission limit	5.0
Sulfates	Hahn and Noll (1982)	Los Angeles	California emission standards	1.07
Chlorofluorocarbons	Palmer et al. (1980); Shapiro and Warhit (1983)	United States	Proposed emission standards	1.96
All regulated air pollutants	Bates et al. (1994)	Poland	European Community and German standards	1.1 to 1.2
Ozone	Hahn (1995)	United States	Vehicle mandate in California and northeastern United States	1.3 to 2.0
Nitrogen oxides	Krupnick et al. (2000)	Eastern United States	EPA SIP provisions	1.8 to 2.0

[a]SCAQMD = South Coast Air Quality Management District.

Source: *Adapted from U.S. Environmental Protection Agency. 2001.* The United States experience with economic incentives for protecting the environment. *Available from http://yosemite.epa.gov/ee/epa/eermfile.nsf/vwan/ee-0216b-01.pdf/$file/ee/0216b-01.pdf [Accessed October 6, 2004].*

produce efficient levels of compliance. Heyes (2001) describes the importance of having monitoring resources that accurately measure emissions.

Economics, as you have seen, deals with questions about resource allocation, and this is certainly true for enforcement of environmental laws. Enforcement presents a classic efficiency trade-off between resources used, which have alternative uses, and benefits produced by greater compliance. Inclusion of enforcement costs in analysis of environmental policy raises the question of whether programs should be designed (e.g., standards set) with enforcement costs included. Tougher standards require higher enforcement costs because they require larger abatement costs from emission sources (costs the sources would prefer to avoid). More lax standards can be achieved with lower levels of enforcement resources. In fact, it is possible that larger emission abatement levels

Table 16–5 Tradeable permit systems in selected countries

Country	Program[a]	Traded commodity[b]	Application period	Effects
Canada	ODS Allowance Trading	CFCs and methyl chloroform	1993–1996	Low trading volume
		HCFCs	1996–present	Low trading volume
		Methly bromide	1995–present	High trading volume
	PERT	NOx, VOCs, CO, CO_2, SO_2	1996–present	Pilot program
	GERT	CO_2	1997–present	Pilot program
Chile	Santiago Air Emissions Trading	Total suspended particulates emission rights traded among stationary sources	1995–present	Low trading volume
European Union	ODS Quota Trading	ODS production quotas under Montreal Protocol	1991–1994	Faster ODS phaseout
Singapore	ODS Permit Trading	Permits for use and distribution of ODS	1991–present	Permit price increase
United States	Emission Trading Program	Criteria air pollutants under the Clean Air Act	1974–present	Savings of $5 to $12 billion
	Leaded Gasoline Phasedown	Rights for lead in gasoline among refineries	1982–1987	Faster leaded gasoline phaseout; $250 million annual savings
	Water Quality Trading	Point-nonpoint sources of nitrogen and phosphorous	1984–1986	No trading occured; ambient standards not binding
	CFC Trades for Ozone Protection	Production rights for some CFCs, based on depletion potential	1987–present	Environmental targets ahead of schedule
	Heavy Duty Engine Emissions Trading	Averaging, banking, and trading of credits for NOx and particulate emissions	1992–present	Standards achieved
	Acid Rain Reduction	SO_2 emission reduction credits; mainly among electric utilities	1995–present	SO_2 reductions achieved ahead of schedule; $1 billion per year savings
	RECLAIM Program	SO_2 and NOx emissions among stationary sources	1994–present	Unknown as of 2000
	Northeast Ozone Transport	Mostly NOx emission by large stationary sources	1999–present	Unknown as of 2000

[a]ODS = ozone depleting substances; PERT = pilot emission reduction trading; GERT = greenhouse gas emission reduction trading.

[b]CFC = chlorofluorocarbons; HCFCs = hydrochlorofluorocarbons; NOx = nitrogen oxides; VOCs = votatile organic compounds; CO = carbon monoxide; CO_2 = carbon dioxide; SO_2 = sulfur dioxide.

Source: *Adaped from Stavins, R. N. 2001. Experience with market-based environmental policy instruments. Washington, D.C.: Resources for the Future. Discussion paper 01–58. Available from http://www.rff.org. Reprinted with permission from Elsevier.*

could be achieved by using more lax standards (lower regulated emissions) that are cheap and easy to enforce than by using tougher standards that require greater costs of enforcement.

Figure 16–17 presents a simple model of pollution abatement in which the costs of enforcement are intergrated into the search for the economically efficient program. The bottom function shows marginal abatement costs excluding enforcement costs, labeled *MAC*. Adding marginal enforcement costs to *MAC* produces a marginal cost function that includes enforcement costs labeled $MAC + F_1$. The vertical distance between *MAC* and $MAC + F_1$ at any given emission level indicates marginal enforcement costs. The graph is drawn with the built-in assumption that incremental enforcement costs are proportional to emissions abated. Therefore, the more pollution is abated, the more costly it is to enforce further reductions. One factor reducing the cost of enforcement is incentives provided by punishment potential. Most pollution control regulations include provisions for fines or jail terms that may be imposed on violators if and when they are caught and prosecuted. Polluters and regulators know that if punishments are too small for the

Example 16–4
How Can Economic Incentives Control Pollution in Central Asia?

Many reformers in the New Independent States (NIS; also called Eastern Europe, Caucasus, and Central Asia, EECCA, countries) of the former Soviet Union recognize that environmental regulations are ineffective. For this reason, economic incentives are being scrutinized for their potential flexibility and cost-effectiveness. The use of economic instruments in NIS dates back to the period long before the breakup of the Soviet Union in 1991.

All NIS countries have legal systems of environmental liability, or damage compensation, implemented in part with pollution charges and natural resource payments. Polluters and users of natural resource can become liable for damages to the environment in cases where license or permit license provisions are violated. However, enforcement has been weak. Various NIS countries have experimented with taxes on certain waste products. As in the United States, these attempts have been mostly unsuccessful because of resistance from industry. Some of the economic instruments applied in the Soviet Union have disappeared since independence (e.g., the deposit refund system). Pollution tax rates and tax revenues vary widely among the NIS countries, but one thing they share is that the system of pollution charges functions badly.

Where the benefits of environmental improvement cannot be measured, cost-effectiveness analysis encourages available resources to be allocated in a way to achieve as much environmental improvement as possible. Through use of cost-effectiveness, the selection and design of environmental policy instruments can be based on comparison of the costs they impose and the expected environmental improvements they are expected to produce.

Enforcement presents the same problems to the NIS as they do to other countries. Measures to improve compliance absorb resources, so they need to be considered carefully. The ultimate effect of economic instruments depends on effective enforcement, including the level of collection of environmental taxes and charges. Compliance levels by emitters depend on good monitoring as well as on credible and consistent punishments for violation. One way to ensure better compliance by emitters is to involve industry in designing and establishing realistic targets as well as in regular and transparent reporting during implementation. Another is to hire honest regulatory authorities, pay them well, and promote a culture of honesty in which bribes by emitters are not accepted. (Heyes, 2001, describes why it might not always be economically efficient for corporate polluters to be honest.) But scarce funds with which to pay authorities present challenges. Special problems in using economic incentives to clean up the environment occur in the NIS because market culture and institutions are still largely absent.

Source: TFIEAP. 2003. *The use of economic instruments for pollution control and natural resource management in EECCA.* Centre for Cooperation with Non-Members Environment Directorate. Available from *http://www.oecd.org/dataoecd/37/18/26732337.pdf* [Accessed October 6, 2004].

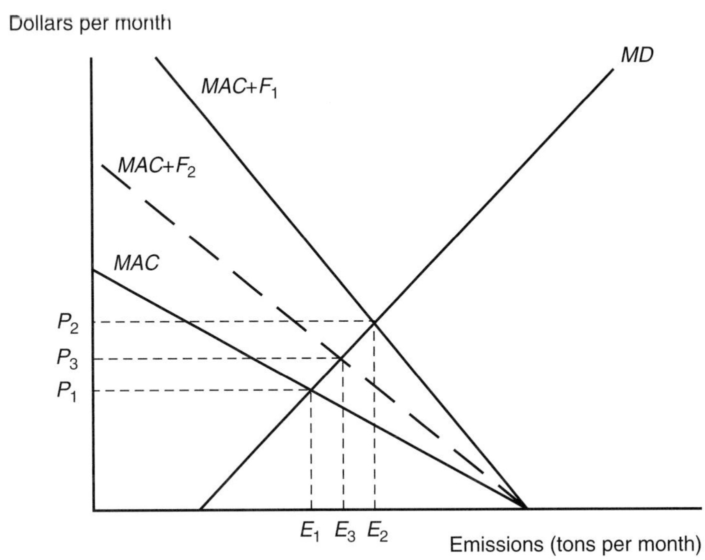

Figure 16 17
Enforcement costs affect the efficient abatement level

potential size of marginal damages inflicted on the public, there are strong incentives to ignore the law. Punishments proportional to marginal damages caused by emissions promote economic efficiency in the allocation of both abatement and enforcement resources.

Enforcement costs effectively raise the economically efficient emission level compared with emissions produced under zero enforcement costs (or perfectly law-abiding polluters). Let us analyze how this makes economic sense. If the marginal cost of reducing emissions requires adding expenses of monitoring emissions and enforcing the program, we would expect to see the economically efficient emissions level increase. In the extreme case, if a pollution control law is unenforceable at any cost (e.g., certain nonpoint pollutants), then no amount of enforcement will bring about the efficient emission level and actual emissions will remain at their unregulated level. This analysis illustrates the importance of having effective enforcement technology: In Figure 16–17, reduced marginal enforcement costs shift $MAC + F_1$ down to $MAC+F_2$, closer to MAC. This increase in enforcement efficiency reduces the economically efficient emission level. In the extreme case, if the marginal costs of enforcing additional abatement are zero (e.g., if a self-enforcing program could be designed), then economically efficient emissions are reduced all the way to E_1. For this reason, technical advancement in environmental enforcement produces the same changes (qualitatively) on the efficient abatement level as technical improvement in abated emissions. This is why better enforcement technology (e.g., improved monitoring equipment) tends to reduce emission levels.

Summary

An important physical basis of environmental pollution is whether pollution accumulates over time (stock pollutant) or tends to dissipate soon after being emitted (flow pollutant). Some emissions produce impacts in restricted, local regions, whereas others produce impacts over wider regions, including the global environment. Pollution sources differ by whether they come from point or nonpoint sources. Emissions can be continuous or episodic.

Three classic criteria for dealing with pollution policy are economic efficiency, equity, and sustainability. An efficient environmental pollution policy is one for which total benefits of reducing the pollution exceed total costs, regardless of who benefits and who pays the costs. An efficient pollution control policy moves toward the point where the benefits of reducing the pollution damages exceed the costs of abating the pollution. An important method for measuring the economic efficiency of proposed policies is the use of pollution damage functions and abatement cost functions. Equity, or fairness, is an important criterion for evaluating environmental pollution control policy. Equity is concerned with how the benefits and costs of environmental improvements ought to be distributed among members of society.

An important policy question is whether unregulated market incentives produce adequate amounts of pollution control. The question is this: If we rely entirely on the interaction of demand and supply expressed through the market to determine the level of environmental emissions and abatement, will the market settle on the level of output for which marginal benefit equals marginal cost? Natural environments that absorb residual emissions produced by ordinary economic activities rarely affect a single organization that has the property rights necessary to use market incentives to control those emissions. Weak property rights and the associated open-access status of most natural environments typically produce considerably more use of those resources than is economically efficient. All types of market failures (e.g., common property, public goods, and externalities) regularly appear in natural environments.

Because private property rights associated with the control of pollution are so rare, a range of institutional designs are available to bring about economic efficiency for environmental resources. Several are relevant in controlling environmental pollution efficiently, equitably, and sustainably: decentralization, command-and-control approaches, harnessing market incentives through emission charges and emission trading, and enforcement of society's pollution control laws.

Review Questions

1. How do stock and flow pollutants differ? Distinguish pollutants from point sources versus nonpoint sources.

2. Explain the three classic criteria for designing environmental pollution policies: efficiency, equity, and sustainability. Can you think of a policy that is efficient but not equitable and not sustainable?

3. What is a marginal damage function and how is it used?

4. What are thresholds in marginal damage functions and why are they controversial?

5. How do you interpret the statement that marginal damage of one more unit of a pollutant is $10? Why is that useful information for a policymaker to know?

6. What is a marginal abatement cost schedule?

7. Why does a marginal abatement cost function rise from right to left?

8. How are aggregate marginal abatement cost schedules constructed from individual ones?

9. Explain how to find the efficient level of emission abatement and what occurs at that point. Use words, tables, graphs, mathematics, or any combination.

10. Explain why economically efficient abatement levels can be high for some pollutants, low for others, and zero for others.

11. What kinds of factors can change the economically efficient level of abatement for a given pollutant?

12. Explain how liability laws influence the search for efficient pollution abatement levels.

13. Why are emission standards unlikely to be efficient?

14. What must occur for emission standards to produce economic efficiency?

15. How can an emission charge lead to efficient emission levels?

16. Explain why polluting industries typically prefer emission trading programs to emission charges.

17. Why is enforcement usually required for economically efficient emissions to occur?

PROBLEMS

1. Cost-effective emission standards: The accompanying table shows marginal abatement cost functions for two sources. Note that the two cost functions are different. Regulatory authorities want to reduce total emissions to 10 tons per week by setting emission standards. They decide to apply the same standard to each source, 5 tons per week.

Abatement (tons per week)		Marginal abatement cost		Total abatement cost		
Source 1	Source 2	Source 1	Source 2	Source 1	Source 2	Both
		←		dollars		→
0	10	0	8,000	0	44,000	44,000
1	9	200	7,200	200	36,000	36,200
2	8	400	6,400	600	28,800	29,400
3	7	600	5,600	1,200	22,400	23,600
4	6	800	4,800	2,000	16,800	18,800
5	5	1,000	4,000	3,000	12,000	15,000
6	4	1,200	3,200	4,200	8,000	12,200
7	3	1,400	2,400	5,600	4,800	10,400
8	2	1,600	1,600	7,200	2,400	9,600
9	1	1,800	800	9,000	800	9,800
10	0	2,000	0	11,000	0	11,000

a. What is source 1's marginal abatement cost? How about source 2? What are the total abatement costs of each source? What are the total compliance costs for both sources?
b. What allocation of abatement could the authorities secure by applying the equimarginal principle? What are the total compliance costs produced by that allocation of abatement? How much compliance cost is saved by applying the equimarginal principle?

2. Impact of an emission charge: The accompanying table shows the impacts of two possible emission charges for a single source of a given pollutant. Suppose the efficient level of emissions has been found to be 6 tons per year. Two possible emission tax rates are being debated, \$100 per ton and \$200 per ton. The table's columns show the same concepts presented in the text's discussion of emission taxes.

Impact of an emission charge

Abatement (tons per week)	Emissions (tons per week)	Marginal abatement cost (dollars)	Total abatement cost (dollars)	Marginal tax bill		Emission tax bill		Total costs	
				\$100 per ton	\$200 per ton	\$100 per ton	\$200 per ton	\$100 per ton	\$200 per ton
10	0	500	2,750	100	200	0	0		
9	1	450	2,250	100	200	100	200		
8	2	400	1,800	100	200	200	400		
7	3	350	1,400	100	200	300	600		
6	4	300	1,050	100	200	400	800		
5	5	250	750	100	200	500	1000		
4	6	200	500	100	200	600	1200		
3	7	150	300	100	200	700	1400		
2	8	100	150	100	200	800	1600		
1	9	50	50	100	200	900	1800		
0	10	0	0	100	200	1000	2000		

a. What is the firm's level of emissions when there is no tax at all? (Hint: Find the emission level for which total abatement costs are minimized.)
b. What is the firm's emission level under a \$100 per ton emission tax? (Hint: Find where the marginal tax bill equals the marginal abatement cost.)
c. What is the sum of total abatement costs and emission tax bill at that emission level?
d. What is the firm's emission level under a \$200 per ton tax?
e. At what level should the tax be set to secure an emission level equal to the efficient level?

3. Emission trading: Two polluters can abate their emissions at different marginal costs (*MC*) and different total costs (*TC*).

Polluter 1	Polluter 2
$MC_1 = 100\ Q_1$	$MC_2 = 400\ Q_2$
$TC_1 = 50\ Q_1^2$	$TC_2 = 200\ Q_2^2$

Q_1 and Q_2, respectively, are the quantity of *emission abated* by the first and second polluter. Suppose with zero emission abatement, each firm emits 10 units of emissions, for a total of 20 units. Describe the effects of each of the following programs in terms of abatement costs for each firm: total abatement costs, fees paid, government revenues, total net costs for each firm, total net costs, and total quantity of abated pollution.

a. A regulation requiring each polluter to abate its pollution by 5 units.
b. An emission fee of \$800 per unit of pollution emitted. (Hint: Treating the fee as a price received (tax saved) by each unit abated means each polluter will set that price equal to the marginal abatement cost.)
c. An emission trading system in which the rights are issued to emit 10 total pollution units, 5 to each firm. No trading is allowed.
d. Firms are allowed to trade their initial permit allocation in a competitive market. (Hint: You can find the price at which the permits trade by solving for the value of Q_1 and Q_2 at which the two marginal costs are equal. Remember that $Q_1 + Q_2 = 10$.) After you solve for Q_1 and Q_2, insert each quantity back into the respective marginal cost equation; that gives you the equilibrium price of the permits.
e. How much lower is the total cost of abated emissions for each firm under emissions trading than under the emissions fee?

■ REFERENCES

Accardo, G., E. Giani, and A. Giovagnoli. 2003. The risk map of Italian cultural heritage. *Journal of Architectural Conservation.* 9 (2): 44–49.

Andreoni, J., and A. Levinson. 2001. The simple analytics of the environmental Kuznets curve. *Journal of Public Economics.* 80 (2): 269–286.

Ansuategi, A., and C. Perrings. 2000. Transboundary externalities in the environmental transition hypothesis. *Environmental and Resource Economics.* 17 (4): 353–373.

Antweiler, W., B. R. Copeland, and M. S. Taylor. 2001. Is free trade good for the environment? *American Economic Review.* 91 (4): 877–908.

Aronsson, T. 2001. Green taxes and uncertain timing of technology change. *Resource and Energy Economics.* 23 (1): 41–62.

Berman, E., and L. T. M. Bui. 2001. Environmental regulation and productivity: Evidence from oil refineries. *Review of Economics and Statistics.* 83 (3): 498–510.

Bityukova, V. R., and R. Argenbright. 2002. Environmental pollution in Moscow: A micro-level analysis. *Eurasian Geography and Economics.* 43 (3): 197–215.

Bosquet, B. 2000. Environmental tax reform: Does it work? A survey of the empirical evidence. *Ecological Economics.* 34 (1): 19–32.

Burtraw, D., A. Krupnick, K. Palmer, A. Paul, M. Toman, and C. Bloyd. 2003. Ancillary benefits of reduced air pollution in the U.S. from moderate greenhouse gas mitigation policies in the electricity sector. *Journal of Environmental Economics and Management.* 45 (3): 650–673.

Carson, R. 1962. *Silent Spring.* New York: Houghton Mifflin.

Chay, K. Y., and M. Greenstone. 2003. The impact of air pollution on infant morality: Evidence from geographic variation in pollution shocks induced by a recession. *Quarterly Journal of Economics.* 118 (3): 1121–1167.

Colby, B. G. 2000. Cap-and-trade policy challenges: A tale of three markets. *Land Economics.* 76 (4): 638–658.

Damania, R. 2001. When the weak win: The role of investment in environmental lobbying. *Journal of Environmental Economics and Management.* 42 (1): 1–22.

Dasgupta, S., H. Hettige, and D. Wheeler. 2000. What improves environmental compliance? Evidence from Mexican industry. *Journal of Environmental Economics and Management.* 39 (1): 39–66.

Dasgupta, S., B. Laplante, N. Mamingi, and H. Wang. 2001. Inspections, pollution prices, and environmental performance: Evidence from China. *Ecological Economics.* 36 (3): 487–498.

Eskeland, G. S., and A. E. Harrison. 2003. Moving to greener pastures? Multinationals and the pollution haven hypothesis. *Journal of Developmental Economics.* 70 (1): 1–23.

European Parliament. 2002. *Proposal for a directive of the European Parliament and of the Council on Environmental Liability.* Commission of the European Communities, Brussels. Available from *http://europa.eu.int/eur-lex/en/com/pdf/2002/en_502pc0017.pdf* [Accessed October 6, 2004].

Fullerton, D., and S. E. West. 2002. Can taxes on cars and on gasoline mimic an unavailable tax on emissions? *Journal of Environmental Economics and Management.* 43 (1): 135–157.

Gielen, D., and C. H. Chen. 2001. The CO_2 emission reduction benefits of Chinese energy policies and environmental policies: A case study for Shanghai, period 1995–2020. *Ecological Economics.* 39 (2): 257–270.

Heyes, A. 2001. Honesty in a regulatory context—good thing or bad? *European Economic Review.* 45 (2): 215–232.

———. 2002. A theory of filtered enforcement. *Journal of Environmental Economics and Management.* 43 (1): 34–46.

Hoel, M., and L. Karp. 2001. Taxes and quotas for a stock pollutant with multiplicative uncertainty. *Journal of Public Economics.* 82 (1): 91–114.

Huesemann, M. H. 2001. Can pollution problems be effectively solved by environmental science and technology? An analysis of critical limitations. *Ecological Economics.* 37 (2): 271–287.

Industry Canada. 2004. *Technology roadmap: Lumber and value-added wood products.* Available from *http://strategis.ie.gc.ca/epic/internet/infi-if.nsf/en/fb01315e.html* [Accessed October 6, 2004].

Khanna, M. 2001. Non-mandatory approaches to environmental protection. *Journal of Economic Surveys.* 15 (3): 291–324.

Levinson, D. J. 2000. Making government pay: Markets, politics, and the allocation of constitutional costs. *University of Chicago Law Review.* 67 (2): 345–420.

Malik, A. S. 2002. Further results on permit markets with market power and cheating. *Journal of Environmental Economics and Management.* 44 (3): 371–390.

Maxwell, J. W., T. P. Lyon, and S. C. Hackett. 2000. Self-regulation and social welfare: The political economy of corporate environmentalism. *Journal of Law & Economics.* 43 (2): 583–617.

Montero, J. P., J. M. Sanchez, and R. Katz. 2002. A market-based environmental policy experiment in Chile. *Journal of Law & Economics.* 45 (1): 267–287.

Navrud, S. 2001. Valuing health impacts from air pollution in Europe: New empirical evidence on morbidity. *Environmental and Resource Economics.* 20 (4): 305–329.

Ono, T. 2003. Environmental tax policy and long-run economic growth. *Japanese Economic Review.* 54 (2): 203–217.

SETAC Europe. 2004. *Introduction to natural resource damage assessment and restoration-based compensation for environmental damages, short course, Prague, Czech Republic.* Available from *http://www.zuova.cz/setac2004/body/courses.php?x=9* [Accessed October 6, 2004].

Spence, D. B. 2001. The shadow of the rational polluter: Rethinking the role of rational actor models in environmental law. *California Law Review.* 89 (4): 917–998.

Stiffler, L. 2004. *Our troubled sound: Pipeline will mitigate paper-mill pollution.* Available from *http://seattlepi.nwsource.com/local/109818_kimberly24.shtml* [Accessed October 5, 2004].

United Nations. 2001. *Environment, mineral resources forum.* United Nations News Archive. February. Available from *http://www.mineralresourcesforum.org/news_events/archive/news0102.htm* [Accessed October 6, 2004].

U.S. Environmental Protection Agency. 2002. *Episodic Release Reduction Initiative (ERRI).* April 2002. Available from *http://www.epa.gov/earth1r6/6en/a/erri07-5fin.pdf* [Accessed October 6, 2004].

———. 2004a. *Emissions trading.* Available from *http://www.epa.gov/air/trading.html* [Accessed October 6, 2004].

———. 2004b. *Oil Pollution Act frequently asked questions.* Available from *http://www.epa.gov/oilspill/opafaqs.htm* [Accessed October 6, 2004].

———. 2004c. *What are the six common air pollutants?* Available from *http://www.epa.gov/air/urbanair/6poll.html* [Accessed October 6, 2004].

Wolff, G. H. 2000. *When will business want environmental taxes?* Available from *http://www.rprogress.org/publications/etr_business/etr_business_execsum.html* [Accessed October 6, 2004].

Xing, Y. Q., and C. D. Kolstad. 2002. Do lax environmental regulations attract foreign investment? *Environmental and Resource Economics.* 21 (1): 1–22.

Yates, A. J., and M. B. Cronshaw. 2001. Pollution permit markets with intertemporal trading and asymmetric information. *Journal of Environmental Economics and Management.* 42 (1): 104–118.

Chapter 17

ENVIRONMENTAL RISK

Introduction

In the 1850s steamboat pilots who navigated North America's Mississippi River faced considerable danger. Their boats could be torn open by sharp underwater rocks, stuck in shallow water banks, or blown to bits by exploding boilers. In his book *Life on the Mississippi,* Samuel Clemens offered revealing insights on how a group of steamboat pilots reduced their environmental risks by forming a labor union. Through organization and cooperation, these "association pilots" informed each other of ever-changing river conditions, while hiding that information from pilots excluded from the association (referred to as outsiders). This exclusion made outsider piloting riskier and more treacherous. Outsider pilots soon began grounding and sinking steamboats. Finally, the day came when every captain of every steamboat was told by steamboat insurers to fire the outsiders and hire association pilots in their place.

Sickness, accidents, premature death, and other unexpected bad things have always injected unwanted risk into people's lives. Centuries ago, some smart organizer invented the idea of insurance as an institution for coping with certain risks. As long as people are willing to pay a small premium to insure against the risk of a catastrophe, they are guaranteed damage reimbursement if the bad thing comes to pass.

Take fire insurance as an example. Suppose the risk of fire causing \$100,000 damage to your house is 0.0001 (one in ten thousand) in any particular year. For any year, the expected value of this risk (expected damage) is \$100,000 × 0.0001 = \$10. If you fail to insure your house, in most years there is zero damage and you are lucky. Every once in a while, however, a major loss occurs. Suppose 10,000 people organize themselves, and each contributes \$10 annually to a fund, producing \$100,000 each year. The fund can expect about one fire per year on average, with some variability from year to year. Those fires cost the fund 1 fire × \$10,000 = \$10,000 in the typical year, but sometimes more and sometimes less. As the number of people in the group expands, the variability of the annual loss falls. With 1 million houses in the pool, revenues are \$10 million, but it is almost certain that there will be something close to 100 houses burned each year at a total cost of about \$10 million. Expanding the

Outline

Nature of Risk 492

Desirable Outcomes 495

Unregulated Market Outcomes 500

Institutional Designs for Coping with Environmental Risks 504

Objectives

After studying this chapter, you should be able to

- Characterize the nature of risk
- Identify desirable outcomes that guide policies to control environmental risk
- Explain environmental risks that result from unregulated market activity
- Identify institutions for effectively controlling environmental risk

number of people insured moves from a situation of few to no losses most years to a situation with about 100 losses every year with smaller variability. Each person pays a small premium, but each is covered in case of a catastrophic loss.

Every day individuals and communities make choices based on their understanding of the likely consequences: sometimes they accept uncertainty, sometimes they take a known risk, and sometimes they face hazards deemed acceptable while avoiding others deemed unacceptable. They make these choices based on their perception of the costs and consequences of those choices. Policymakers need information on environmental risks and on its costs and consequences to frame effective legislation to protect human health and environmental quality. This chapter explores how the concepts and methods of economics can be used to understand, assess, and control environmental risk.

Nature of Risk

Risk, the chance of something bad happening, is part of all our lives. Coping with risks involves two tasks: (1) identifying the presence and seriousness of risks and (2) responding to them. Most of us deal with these tasks, although rarely in systematic way. Often our perception of risks is not based not on factual analysis but on instinct. For example, we look for a designated driver on nights when we drink too much. Perceptions of risks and reactions to them, while rarely scientific, are what determines how each individual deals with life's risks. A person may choose to accept a risk as is, find ways to reduce the risk consistent with available resources, or try to avoid the risk entirely.

Risk analysis is also a central part of public policy analysis, because an important element of public policy is formulating an organized response to society's risks. Since governments are responsible for designing risk responses for an entire society, they are expected to do more than simply account for individual risk perceptions that are intuitive, subjective, and rarely scientific. Instead, public policy analysts and policymakers turn to a more systematic analysis of risk before designing a policy response. Nevertheless, because possibilities rather than certainties are involved, *environmental risk* is harder to measure, assess, and manage than deterministic (nonrandom) environmental pollution problems. Cause-and-effect connections are unclear, and exposure of people, ecosystems, or property to a pollutant does not guarantee damages, only an increased probability of effects. Furthermore, effects from the exposure may be so small that no objective method of analysis can measure the danger produced by the risk. Although it may be common to see an adverse effect of a pollutant in controlled laboratory conditions, it is much harder to show it in real-life conditions where there can be hundreds of things happening to each of us, our property, and our ecosystems daily.

Complexity

Think of the complex path of causality associated with environmental risks. Take, for example, the case of disposing of a dangerous pesticide in a carefully built landfill. If everything goes according to plan, the pesticide remains in the landfill, contained where it is supposed to be. However, there is always some probability that the landfill will leak, but is a leak always bad? Maybe and maybe not. If the landfill leaks, the pesticide may remain in that part of the landfill that does not leak. If the pesticide does leak, it might or might not get into the groundwater. If it does get into groundwater, it might get into the local drinking water. If the public drinking water supply is contaminated, there may or may not be an adverse health effect for a typical person who drinks the water. This example shows the complex series of events between introducing an environmental hazard (the pesticide) and potential dangerous health impacts. Some of the process is unknown but possibly knowable, such as the quantity of dangerous material from the pesticide in the food supply. Other parts of the causal chain may have a considerably larger element of randomness, such as whether one person develops cancer from eating food with a given amount of pesticide residue and others experience different effects.

Causality

Once an adverse effect on human health is observed, it is hard to work backward to ascribe the effect to a well-defined cause. The considerable unknown link between the cause(s) and observed effects leads to much uncertainty. Take air pollution. Causal biological links between health effects and air pollution are weakly understood. Exposure to air pollution has been associated with reduced immune defense mechanisms of the lungs. But tobacco smoking has also been shown to cause similar problems. For some air pollutants, breathing particulate matter is known to produce an inflammatory response that affects the bone marrow, and that response is thought to produce various kinds of heart and breathing morbidity. Certain exposures to oxides of nitrogen have been associated with increased susceptibility to bacterial and viral infections. Carbon monoxide combines with hemoglobin to reduce the oxygen-carrying capacity of the blood, which can contribute to anemia and adverse pregnancy outcomes (e.g., miscarriage, stillbirth, low birth weight, and early infant mortality). When the biological pathways of effect are unknown, the issue is how to establish causality. In some cases where damage has occurred, such as somebody contracting cancer, several parties might be responsible or may share responsibility. Maybe several companies dumped old gasoline on the ground when cleaning up; which accident caused the cancer? Maybe the individual with cancer contracted it by being exposed to bad drinking water and dangerous food. Or maybe the individual contracted cancer from a normal biological process related to aging.

Time delays present special problems in untangling complex causality. With environmental hazards, long time delays are common. In some cases it can take decades for an ordinary process to turn into a hazard and, ultimately, to become a health risk. After building an environmentally safe landfill, it may take years for a leak to occur, more time for the leak to get to the groundwater, and more time yet for the contaminated groundwater to reach somebody's drinking water pumped from the polluted aquifer. After all that, it may take 20 more years for somebody to become sick from drinking the well water.

Voluntary versus Involuntary

The importance of an environmental risk depends in part on whether a person's exposure is voluntary or involuntary (see Example 17–1). Many economic studies have found that people will *voluntarily* take on greater risks if they receive a higher wage as compensation. However, the additional compensation is typically smaller than what people require for a similar *involuntary* risk they face, such as the added risk of having a toxic dump site installed near their homes. Most of us routinely take on added risks if there is a benefit to go with it. When we are exposed to risks without our consent, however, we typically demand a higher level of safety or more compensation.

A *voluntary risk* is one taken through an individual's conscious choice. Each of us elects to make personal choices that add or reduce risks compared with the risks assumed by doing nothing. Since we make these choices voluntarily, we respond to the risks assumed by adjusting our personal exposure to the relevant hazard. Suppose a business person living in Miami, Florida, wants to attend a meeting in Seattle, Washington. She can fly by commercial airline, getting there in half a day; take trains, which will take several days; or stay at home, reducing the exposure of hazards to zero. Each possible choice brings different risks and different benefits. If the meeting is important enough, if jet travel is fast enough, and if risk is perceived as suitably low, she flies; otherwise, she does something else.

Example 17–1
What Makes Environmental Risks Unacceptable?

Some risks trigger much more alarm, anxiety, or outrage than others, even though they may have the same expected costs or damages. The United Kingdom Department of Health found these risks to be of great concern:

- Risks that are involuntarily imposed (e.g., pollution) as compared to voluntary ones (e.g., smoking)
- Activities that pose a threat of a dreaded form of death, injury, or illness, such as cancer
- Risks that are inescapable even by taking personal precautions
- Risks that arise from unfamiliar or novel sources
- Man-made risks such as pesticides and nuclear power plants, compared to natural ones like floods and earthquakes
- A risk that produces a single, large-scale consequence (e.g., airline accident), compared to ones that produce several small-scale consequences (e.g., car accidents)
- A risk that damages identifiable victims, compared to those that harm anonymous victims
- Risks for which consequences of exposure are delayed and which cause hidden or irreversible damage
- An unfair distribution of risks and benefits as a result of a particular activity
- Activities that pose a risk to certain groups such as children and future generations
- Risks that are subject to contradictory statements from responsible sources, or, even worse, from the same source

Source: United Kingdom Department of Health. 1998. *Communicating about risks to public health: Pointers to good practice.* Available from *http://www.dh.gov.uk/assetroot/04/03/96/70/04039670.pdf* [Accessed October 9, 2004].

Because an individual's voluntary risks are largely influenced by that individual, the government typically plays only a small role in controlling those risks. When the public sector does take action, it is usually limited to getting information on poorly understood dangers and communicating that information to the public. Food labeling and public service announcements are two examples of risk communication activities performed by the government.

Sometimes people are exposed to hazards that are beyond their control. One characteristic of *involuntary risk* is that individual responses are limited or expensive. For example, to reduce the risk of disease from polluted air, a person can avoid living in industrial cities. The individual's risk of harm is reduced but not eliminated. Involuntary risk is expensive for private individuals to control. Governments often attempt to control society's exposure to certain involuntary risks beyond what an individual can afford to do with his or her own resources. Although the dividing line between voluntary and involuntary risk is not always sharp, it is there. The point at which that line is crossed is when governments might take action. Consider the case of cigarette smoking and secondhand smoke. Cigarette smoking is an individual decision that produces voluntary risks. In the mid-twentieth century, there were believed to be few risks to society beyond what the individual accepted by smoking. Government action in connection with smoking and health risks was minimal, limited to things like public service announcements alerting smokers to the dangers of smoking. However, in more recent years, research found that secondhand smoke produced risks to the larger society, not just to smokers. As passive smoke became recognized as a public problem, communities started taking a more active role. Today, many communities have enacted laws that prohibit or greatly control smoking in public buildings because of the involuntary risks imposed on nonsmokers.

Perceived versus Objective

Table 17–1 lists several risks, all of which are estimated to involve a risk of loss of life of one in a million per year. You might be surprised at the magnitude of many of these risks; although the activities described are familiar, most of us have a hard time assessing the risks. Psychologists have found that most of us tend to overestimate the probability of low likelihood events and underestimate the probability of high likelihood events. The term *objective risk* deals with the sta-

Table 17–1 Activities carrying a one in a million risk of death

Whole body exposure to 0.1 mSv of ionizing radiation[a]
Smoking 1.5 cigarettes
Traveling 50 miles (80 km) by car
Rock climbing for 1.5 minutes
Being a man aged 30 for 9 hours
Being a man aged 53 for 53 minutes

[a]The maximum safe dose allowable per year of ionizing radiation is 0.1 millisieverts.

Source: *Pochin, E. E. 1974. Occupational and other fatality rates.* Community Health. *1 (July–August): 2–13. Adapted and reproduced with permission from the BMJ Publishing Group.*

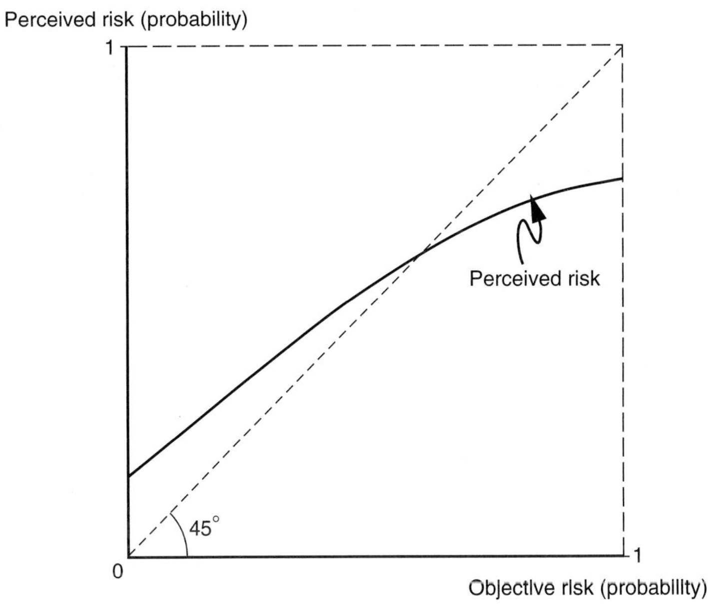

Figure 17–1 Perceived versus objective risk

tistical relation between exposure and adverse effects. By comparison, the term *subjective risk* means an individual's perception of his or her objective risk.

Figure 17–1 shows the problem in which people often overestimate low-probability risks and underestimate high-probability risks. The 45-degree line indicates that a person's subjective risk exactly equals the objective risk, defined by objective measures. The flatter, sloped line represents evidence from a range of experiments and surveys identifying how people subjectively rank the threats posed by objective risks. The results: People typically inflate low risks over which they have little or no control (e.g., global warming) and understate high risks that they can control to some degree (e.g., walking across the street). Often people worry more about where a risk comes from than its size (e.g., synthetic versus natural carcinogens).

Some risks are simply more tolerable than others. Those who routinely take on the risk of driving a car at high speeds may be unwilling to accept risks produced by a nearby toxic waste dump. People are more willing to accept voluntary risks they believe they can control than involuntary risks they consider beyond their control. Analysis has found that space, time, familiarity, dread, anxiety, regret, and perception of control drive the idea of voluntary risk (Viscusi, 1992).

■ Desirable Outcomes

Economic efficiency, equity, and sustainability are discussed next. All three are desirable outcomes for judging programs that would reduce or limit environmental risk.

Economic Efficiency

Environmental decision makers—including political leaders, professional managers, and the general public—realize that controlling environmental risk is expensive. They have a natural interest in measuring the effectiveness of those expenditures compared with other uses of the same resources. The economic value of various measures of controlling environmental risks is an important element of many environmental policy decisions.

Finding a defined value of life has been a pursuit of governmental agencies around the world for many years. It is regarded as an essential element of benefit-cost analysis to guide public policy in regulating environmental risk. Saving one life is correctly seen as a moral imperative on which no monetary value can be placed, but prevention of every possible accidental death would incur larger costs of resources than are available in public and private hands. So some measure of the value of saving a life is necessary to ration resources among life-saving measures and other ends desired by society. To deal with this issue, the term *value of a statistical life* (VSL) is used to emphasize that people are seen putting a value not on a particular life, but on safety measures that reduce the statistically expected number of accidental fatalities by one. Being able to measure that value is made difficult because markets rarely exist that enable measurement of the reductions in risks achieved through proposed environmental regulations. This lack of direct information on the market value of environmental risk reduction requires consideration of indirect methods.

Valuation Methods. Economists commonly use several methods to value reductions in environmental risk: human capital, willingness to pay, and expected utility.

Human Capital. *Human capital* refers to a person's skills that are economically productive. It includes the education, special training, and experience embedded in a person that gives him or her greater productivity on the job than would occur without those assets. For example, think of a woman's human capital as the investments she has made in herself that lets her command a higher salary. The human capital approach values risk reductions by examining the gain in a person's prospective earning power. According to this approach, the value of controlling risks by enough to save a life is calculated by what the person contributes to society through the increased net present value of future earnings. One great advantage of the human capital approach is that it is actuarial, which means it measures observed data on income earned according to people's age, education, and other characteristics. A major problem with the approach is that it assigns reduced values to the saved lives of women and minorities and zero value to retired people. Even though those reduced values are supported by income data, their use presents an environmental justice problem; that is, investments made in reducing the environmental risks of these groups produce a lower economic value. Another problem with this approach is that it violates economic theory whenever people are found willing to pay more for reduced risks than their gain in earning power.

Willingness to Pay. The willingness-to-pay approach for valuing risk reduction is based on the principle that people value risk reduction if it leads to a greater level of welfare. The welfare produced by a given risk reduction, from R_1 to R_2, is measured by the maximum the person in question is willing to pay to reduce the risk. (Another measure is the minimum compensation a person is willing to accept for tolerating the status quo at risk level R_1 for being unable to face the reduced risk level, R_2.) By watching people pay for greater safety to reduce risk (or accepting payment to compensate for no reduction in risk), economists can see what people are actually paying to save lives and reduce sickness. Three empirical methods for measuring the willingness to pay for risk reduction are in widespread use: wage-risk trade-offs, averting behavior, and stated preferences.

Wage-risk trade-offs are based on the concept that the marketplace rewards people who take on greater risks in their jobs by paying a wage premium. Many people will accept a higher wage or salary for accepting higher job risks. Therefore, we should be able to measure the added wage or salary the market pays for people who accept greater risks. By watching how wages are progressively raised in the face of incremental increases in risk, we can develop a schedule of the marginal cost of various elevated risk levels. Of course, our real interest here is the economic value of reducing risk. The value of this risk reduction is the incremental willingness to pay (in reduced wages) for the attribute of job safety. Economists have measured the risk premium required to compensate for riskier jobs and have found that the VSL from reduced risk is between \$700,000 and \$16,300,000, in 1997 dollars (U.S. EPA, 2003). One limitation of this method is that people who do not work show no willingness to pay for increased job safety, so the value of a saved life for this group cannot be measured.

The averting behavior method estimates willingness to pay for reduced risk by watching what people actually pay to protect their families and themselves by averting risk. Many of us avert risk at a price when we buy things like smoke detectors, fire alarms, flu shots, seat belts, bicycle helmets, and bottled water. According to this method, estimates of the VSL range from \$700,000 to \$4,800,000 in 1997 dollars. A debated assumption is whether using a VSL is justified for valuing public programs that reduce *involuntary* risks to people based on what they pay to voluntarily limit their risks.

Stated preferences methods (contingent valuation) ask people directly, through a survey or interview, how much they would be willing to pay to reduce risk. The approach constructs a hypothetical market in which a person buys or sells safety and attempts to determine a person's willingness to pay for risk reduction. The challenge is to make these hypothetical markets realistic and relevant to people. As described in Chapter 7, many economists and others criticize contingent valuations that produce VSLs in which a respondent answers a hypothetical question without enforcing his actual budget constraint. One estimated range of the VSL is \$1,500,000 to \$4,600,000 in 1997 dollars (U.S. EPA, 2003).

Expected Utility Theory. Expected utility theory is based on the principle that each of us has a well-defined and rational understanding of risk and of the effectiveness and cost of measures to control it. The use of expected utility theory presumes that individuals have clear preferences for risk that influence their choices. When they confront a risk, they can evaluate probabilities and consequences systematically and predictably. Consider the following example: Suppose an individual receives a single good (income), C, but faces the risk of being unsure how much she will consume of that good in the future. Two outcomes can result, but she does not know which one will occur. With probability p_A (a number between 0 and 1), event A will occur. With probability p_B, event B will occur. If event A occurs, the amount of C (income) she receives is C_A; if B occurs, she receives C_B. Let us call label A and B "states of the nature" and C_A and C_B payoffs under each state of nature. Each state of nature occurs with a known probability. To keep things simple, suppose that the two states of nature are mutually exclusive and exhaustive (both cannot occur together, but one must occur) and only those two outcomes are possible: $p_B = 1 - p_A$.

We begin with an ordinary welfare function, $W(C)$, defined for certain payoffs, then extend it to account for risky outcomes. From that, we define the welfare (utility) produced by a choice made under risk as the expected utility produced by the various payoffs in different states of nature. That expected utility is

$$W(C_A, C_B, p_A, p_B) = p_A\, W(C_A) + p_B\, W(C_B) \tag{17.1}$$

where $W(C_A)$ is the welfare received if event A occurs, and $W(C_B)$ is welfare if event B occurs. That means that total expected welfare is computed as the sum of the probabilities of each event occurring times the welfare received under each outcome. A good way to see what is going on is to substitute numbers for C_A, C_B, p_A, and p_B. Assume that C_A is $-\$10{,}000$ (a loss), C_B is \$6,000, (a gain) and p_A and p_B are each 0.5. What would you expect this risk to produce on average, in monetary terms, if you repeated it many thousands of times? You could write a computer program that tells you what would happen, but it is clear that the expected monetary value comes from the formula in Equation 17.1: 0.5 times $-\$10{,}000 + 0.5$ times $\$6{,}000 = -\$2{,}000$. So, taking a chance in which you might lose \$10,000 or might win \$6,000, on average produces an expected loss of \$2,000.

Figure 17–2 presents a graphic analysis of these choices and outcomes. The welfare function always increases at a decreasing rate, indicating decreasing marginal gains from added

Figure 17–2 Expected utility analysis

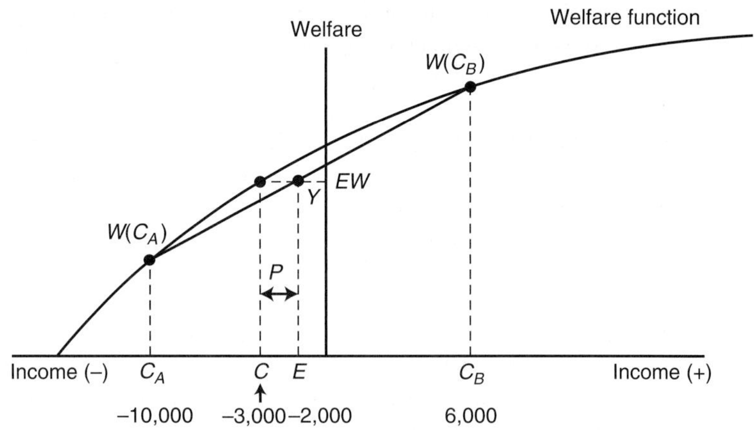

income as income increases. In other words, when you are poor, you receive more welfare from an added dollar of income than when you are rich. The horizontal axis shows payoffs from the two events, *A* and *B*. Each event produces the welfare shown on the graph: Event *A* produces C_A (−\$10,000) with welfare $W(C_A)$, and event *B* yields C_B (+\$6,000) with welfare $W(C_A)$. The expected income produced by this risk is *E* (expected payoff), computed as $p_A C_A + (1 - p_A)\, C_B$. In this example, $E = -\$2{,}000$.

But the story is not over. Expected income is not the same as expected welfare. Expected welfare produced by this risky situation is $p_A\, W(C_A) + (1 - p_A)\, W(C_B)$, shown as *EW* (expected welfare). Point *Y* is located between points $W(C_A)$ and $W(C_B)$ in proportion to p_A and $(1 - p_A)$, in this case halfway between the two points. (*Y* is a weighted average welfare of $W(C_A)$ and $W(C_B)$ with the weights being p_A and $(1 - p_A)$. *Y*'s horizontal coordinate is *E* and its vertical coordinate is *EW*. The welfare function's diminishing marginal effects (curvature) has important implications. It means there is some level of *certain* (nonrisky) income (less than *E*) that produces an equal amount of welfare as produced by the risky situation with two possible outcomes. Economists usually label that amount of income the *certainty equivalent* of the risk. That certainty equivalent, *C*, is found by looking for the income that produces a welfare level for which $W(C) = EW$. Notice that *C* is less than *E*, which means the individual in this example is indifferent between having *C* for certain (a bird in the hand) as opposed to *E* on average (two in the bush). The difference between these two incomes, $E - C$ is called the *risk premium* (*P*). *P* is the added amount of income needed for the individual to voluntarily take a risk instead of accepting the risk-free outcome. Figure 17–2 shows that *E* is −\$2,000 and C is −\$3,000. The individual is indifferent between losing \$3,000 for sure versus accepting the risk, even though on average only \$2,000 will be lost. Her attitude comes from the fear of possibly losing \$10,000. The difference between the two numbers, \$1,000, is the risk premium, *P,* which can also be interpreted as the amount of income the individual is willing to pay to convert the risk to a certain average value of *E*.

Not all welfare functions are equal. The one in Figure 17–2 is for somebody who is risk averse. Imagine a welfare function that is less curved (i.e., closer to a straight line). As the welfare function approaches a straight line, the individual's risk premium declines. When the welfare function is a straight line, there is no risk premium at all. People with that kind of welfare function are called *risk neutral.* People who are risk loving have a welfare function that is opposite the curvature of the risk averter; that is, the function shows increasing marginal benefits from higher income levels.

Expected Damages Saved and Risk Neutrality. Multiplying the probability of an accident by damage to the victim should an accident occur, yields the *expected damage* of the accident. For instance, if there is a 10% probability of an accident that causes \$10,000 of damage, the expected damage is \$1,000. The use of expected damages allows us to reduce the uncertain event to something certain. Then we can think of measures for reducing the expected damages, which economists call *expected damages saved.* Expected damages can be reduced by anything that reduces the probability of the accident or anything that reduces the damages the accident produces, or both. So, total expected damages in a risky situation are measured as

$$ED = pD \tag{17.2}$$

where ED is expected damages, p is probability of accident, and D is damages produced by the accident. The expected value of damages saved through precautionary measures can be found by determining what kinds of things change Equation 17.2. Changes are measured as

$$EDS = -\ \Delta ED = -\Delta pD - p\ \Delta D + \Delta p\ \Delta D. \tag{17.3}$$

EDS is expected damages saved and Δ is change (reduction) in either the probability of an accident or in damages produced should it occur. Laws requiring seat belts to be worn in cars are a good example of a public action taken to reduce expected damages from car accidents. Even though a seat belt law has little influence of the probability of an accident occurring ($\Delta p = 0$), by successfully saving lives when an accident does occur ($\Delta D < 0$), the law results in a positive EDS by producing a large value for the term $-p\ \Delta D$.

Use of the EDS formula has major implications for formulating or evaluating public policies that control environmental risk. With the formula, expected damages saved can be computed for any policy being considered that would reduce damages from an environmental accident (or exposure) or reduce the probability of such an accident occurring, or both. The use of expected damages saved to value environmental risk-reducing programs requires that people who suffer these damages are *risk neutral;* that is, it must be assumed that they are willing to pay an amount to reduce an environmental risk equal to the expected damages saved. Although this assumption is unlikely to be true, factual data on the willingness to pay for reducing risk are often unavailable; so the *assumption* of risk neutrality is often made when performing analyses of risky situations. Assuming risk neutrality is also attractive because it establishes a lower bound on the true willingness to pay where people are risk averse. More applications based on risk neutrality are described later in this chapter.

Using Value Information. Economically efficient choices for reducing risk are those for which the marginal social benefits of risk reduction exceed or equal the marginal social costs. Efficient choices are based on considering the projected costs of implementing an environmental improvement program, together with the corresponding benefits of expected improvements in the health of the affected population. This approach provides risk managers with the ability to evaluate and select the best available environmental control strategies from a range of proposed options. Efficient choices are based on assigning monetary values to quantify health benefits gained from the reduced incidence of disease (morbidity) and reduced risk of premature death (mortality). Several institutions will be evaluated later in this chapter based on their likelihood for promoting efficiency as defined in this way.

Equity

Environmental equity refers to fairness in the distribution of environmental risk (see Gardashuk, 2003). According to many environmental justice advocates (see Chapter 19), inequitable (unjust) distribution of the costs of environmental risk resulting from the application of technology has been common in industrial societies since the early twentieth century. Environmental risks are often not shared equitably in the United States or in other parts of the world. For example, an epidemiological study undertaken by the World Health Organization in Manila in the Philippines (United Nations, 2000) showed that the hazards of air pollution are particularly felt by those who make their living in the streets, (e.g., street workers, peddlers, mass transport drivers, and traffic police). Several policy approaches for dealing with the region's inequity have been proposed:

- Establishing laws that take into account indicators of air quality management
- Programming targets by establishing clear indicators of progress and setting achievable targets, such as phasing out lead, reducing key pollutants like particulate matter and sulphur

oxides, introducing cleaner fuels and promoting cleaner engines, and developing a fuel-efficiency labeling program

- Reforming institutions that respond to air pollution issues, with special emphasis on supporting research for improving capacities for setting standards
- Combining market-based approaches with command-and-control enforcement procedures
- Mobilizing civil society by encouraging public participation in programs aimed at the reduction of air pollution

Sustainability

Environmental risk management is all about promoting sustainable actions. As described in greater detail in Chapter 19, the safe minimum standard (SMS), or *precautionary principle*, is one mechanism for promoting sustainability where a small probability of a serious outcome is possible. A comprehensive account of the SMS involves establishing constraints on environmental pollution and other risky activity so that high-cost risks are made suitably small. A workable version of the SMS suggests that under reasonable uncertainty allowances, threats to humanity or valuable ecological systems are made suitably small, provided that these do not involve uneconomical costs. Of course, for real policy decisions to be carried out using the SMS standard, somebody needs to make informed judgments about what is reasonable, valuable, and uneconomical.

Three conditions justify arguments for a SMS of risk avoidance. The first is the presence of uncertainty; often there is incomplete knowledge of the range of consequences of current human decisions. Major habitat changes could lead to widespread loss of species or considerable but unknown risks to human health and lives. Second, environmental and human health consequences produced by many economic activities are often irreversible. Third, losses associated with these irreversible actions can be large, although clouded by ignorance because of uncertainty. When all three of these conditions seem present, it is desirable to adopt a risk-averting, sustainable approach. Even if an action does not look dangerous on the surface, building in safeguards may be justified to reduce the risk or cost of serious damages, unless the cost of doing so is intolerably high.

■ Unregulated Market Outcomes

Unregulated markets have dealt with risks through the use of insurance for a long time. People seek insurance to reduce their own risk by transferring it to a larger group of people who can bear it more easily. Their utility can be increased if the small probability of a large loss is taking away from the injurer (the one who potentially injures the public, typically a polluter) in exchange for the certainty of a small loss. The costs of compensating an individual (the insured) who suffers a large loss are spread across a large group of individuals. The average loss in an insurance pool becomes more certain (reduced variance) as more independent and identically distributed exposure units are added in the pool. The insured pays a premium to the insurer, fixed in advance, for this transfer of risk and for a guarantee that should a loss occur, the insurer pays for the loss incurred. The price of the premium exceeds the expected value of the loss; the difference, on average is equal to the insurer's income. Risk creates disutility for people and organizations who are averse to risk. For risk-averse organizations, the maximum premium they are willing to pay exceeds the expected value of their loss.

The risk-averse injurer also has a demand for insurance. He perfers the certainty of a small loss (the payment of the insurance premium) in exchange for shifting the financial responsibility of a catastrophic loss to the insurance company (see Faure, 2001) which increases the utility of the injurer by reducing his financial responsibility for the loss.

Environmental risks typically involve a small probability of accidents that produce catastrophic damages. Two examples are the risks associated with global warming and hazardous waste disposal. Environmental insurance has several advantages; arguably the most important is the guarantee that if an accident occurs, victims are compensated. Another advantage is a reduction in the haunting fear that an injurer may face bankruptcy. A third advantage is that insurance companies that have invested years in developing expertise on risks and damages can supply informed advice to potential injurers without government intervention. The prospective injurer who uses this infomation to control risks is rewarded by lower insurance premiums. The potential gains to both provide an incentive for the two to cooperate. A fourth advantage is that through insurance the injurer's risk exposure is reduced and spread among many similarly situated organizations. So, if a potential injurer is risk averse regarding large environmental accidents (and most firms are), insurance provides a way of turning a risky situation into a sure bet. If private insurers are willing to provide coverage for environmental risk, the public may pose several questions: Will adequate amounts of insurance emerge from the private sector without need for government action? Are there barriers to the private supply of insurance? If there are such barriers, can public action remove or reduce them?

Choices under Risk

We influence many of the risks we confront. We move or reduce physical activities when air pollution becomes bad. We buy bottled water if we worry that our drinking water is polluted. We fasten our seat belts when driving on a crowded highway. An individual can buy a water filter, join an exercise club, run, eat healthful food, and take other actions to reduce risk. How an individual invests time and money to reduce the risk of bad things happening depends both on her attitude toward risk and her ability to reduce it.

Observed risks depend on both natural events and individual self-protection decisions. How people make decisions about risk differs across individuals and situations, even though the natural phenomena that trigger these efforts apply equally to everyone. Therefore, assessing risk levels solely in terms of natural science can be misleading. Relative prices, incomes, and other economic and social parameters that influence any person's self-protection decisions affect risk. Just as public–policy–based economics requires an understanding of the physical and natural phenomena that underpin choices, public–policy–based natural science requires an understanding of the economic phenomena that affect risk.

Environmental Risk Insurance

For environmental insurance to work, the risk must be attractive to insurers and premiums must be attractive to potential injurers (see, for example, Janssen, 2001; Pearce, 2000; Wong and Dutschke, 2003). Several conditions must be met before insurers are willing to provide insurance policies against an uncertain environmental loss.

Verifiable Loss. Many losses can be anticipated, and enforceable insurance policies can be written against these losses. Economists call these losses verifiable (see Doherty and Muermann, 2002). Insurers establish a relationship to pay for the verifiable losses when the insurance contract is written. Some losses are nonverifiable. For example, it is hard to insure against mental anguish produced by an environmental accident because nobody can measure the severity of mental anguish. The injured may feel the anguish, but it cannot be easily verified. In addition, losses that are hard to verify are often easy to fake.

Many losses produced by an environmental hazard are hard to verify. For instance, suppose somebody proves that company *Y* contaminated the groundwater aquifer used for city *X*'s drinking supply with a known carcinogen. Now the city's ratepayers are exposed to a risk. Has there been a loss? Is exposure to the risk the loss, or must somebody contract cancer for there to be a loss? Has anyone who contracts cancer for any reason suffered that loss? Remember, cancers come from many sources and it may be hard to show that a particular cancer is produced by the contaminated water. Insuring against environmental risks requires that potential losses be verifiable by independent third parties. Where this cannot be done, it will be hard to insure environmental risks privately.

Risk Pooling. In environmental insurance, loss must be capable of being *pooled;* that is, the risk must be shared across a large number of people. The larger the number of people over which the risk is pooled the better. Since the insurance company collects premiums from all who are insured, the impact of losses incurred by the few is paid for by the premiums. Also, risk pooling reduces the variability (variance) of the risk. The law of large numbers tells us that as the size of the insured group rises, so does the accuracy with which the size and the frequency of the loss can be estimated. It is hard to estimate the chance of any one organization having an accident, but it is much easier to estimate the chance that an accident will occur within a larger group. For example, there may be a 0.0001 chance that an accident will occur, but nobody can say that the accident will happen to individual *X*.

For risk pooling to work, it is important that the risks faced by any one individual are *uncorrelated* with risks faced by risks of others. If not, a significant part of the insured group will be making simultaneous claims that cannot be paid by premium revenue. If the environmental resource being protected has public good properties, then its loss for one person will be a loss for others too. The requirement of uncorrelated risks fails. This suggests that the public good aspects of environmental protection in many forms may not be easily insurable.

Risk pooling is based on various demographic variables such as age, gender, and the value of the asset insured, all of which can affect future outcomes. By analyzing these variables and characteristics of a large population over time, insurers can better predict risk and reduce uncertainty over prospective claims. This, in turn, helps consumers by ensuring more economic efficiency in the premiums they pay. Responding to the competitive pressures of the marketplace, insurance companies invest in the development and use of different methods to improve their identification and classification of people of low risk. This analysis enables insurers to offer policyholders who are better risks a lower premium and attract them away from their current insurers.[1]

Premium Measurability. Computing the premium to charge for insurance is not a minor detail, because a mistaken premium can cause the market to fail. If the premium is too large, no insurance will be sold. If it is too small, insurers will lose money or possibly go bankrupt. If an insurer has experience with a loss, it is possible to calculate a premium based on the historic frequency of that loss. For instance, because there is a great deal of information on house fires, the probability that a randomly chosen house will burn down in a year can be computed, and this is the information needed to compute a premium.

Damages caused by weather are insurable, because long-run weather information is available. However, climate change might not be insurable. Climate change can be thought of as the long-run trend of weather events, so it may seem that climate should be insurable. However, climate change produced by greenhouse gasses may combine features of temperature levels and changes in those levels for which there is no reliable historical information. If insurers have no experience with a loss, they must guess on the loss's frequency and size. If uncertainty clouds that risk, which typically is the case for environmental risks, the insurer needs to err on the high (safe) side in estimating frequency of damage or loss to avoid catastrophic losses.

---

[1] The opposite of risk pooling is risk spreading. Risk spreading occurs when the varying risks of individual policyholders are combined and spread among all policyholders of a given population. For example, community rating mandates that premiums be assessed according to a community's characteristics, not individual characteristics. Low-risk groups are charged higher premiums so that high-risk groups can pay lower premiums. Because the premiums mandated by community rating represent an underpriced bargain for high-risk individuals, low-risk individuals find little value in the overpriced insurance and often decide to opt out of the market. Insurers are left with the high-risk policyholders taking advantage of the mandated low rates. To hold down rates for this subsidized group, insurers must continue to raise prices on those low-risk subscribers who choose to stay in the market, which further increases the dropout rate. Ultimately, fewer and fewer people can afford insurance; healthy, low-income individuals and families are especially affected. Moreover, insurance companies often find that the revenues collected from premiums cannot cover increasing claims costs. Rather than face eventual insolvency, many insurance companies leave the market.

Limited Loss Exposure. Many environmental risks have long gestation periods. If company *X* dumps toxic chemicals on the ground, it may take years for the chemical to leak into the water table. If the chemical wastes leak into the aquifer and get into the drinking water supply, it may take years of drinking contaminated water for people to receive a dangerous exposure to the chemical. Some illnesses themselves have long gestation periods, with delays years and even decades between exposure to a toxic substance and onset of a disease such as cancer. This means that several decades may pass between the dangerous activity (dumping the chemical) and the damages produced (cancer). It is hard to insure a risk for which the damage may occur at any time in the future and with no time limit. As time advances, information changes. It is hard to estimate potential damages from an oil spill over the next year, let alone over 20 years. In 20 years, the institutional environment (society's rules), as well as knowledge of the physical science, may have altered considerably. This is especially true with environmental risks. For an insurer to offer a policy over a very long time period, the premiums charged will likely be so high that potential injurers will be unwilling to buy them.

Moral Hazard. Moral hazard means that an insured's risky behavior changes after that person takes out an insurance policy designed to compensate for costly outcomes produced by the behavior. For example, an insured homeowner takes fewer fire precautions after buying a fire insurance policy. The insurer needs to be able to predict the behavior of the insured; typically this is done by looking at past data on frequency and cost of accidents. If the insured changes her behavior after she is insured and if this behavior cannot be predicted, premiums will be set either above or below expected actual damages. If the insured's behavior changes to increase the expected frequency of adverse events, premiums will be set too low. Eventually, over large numbers of policies issued, the insurer will go broke.

So the presence of significant moral hazard reduces the likelihood that insurance will succeed. Insurance companies, though, can confront moral hazard. Sometimes they monitor the behavior of the insured and raise premiums if the insured is seen doing things that increase risk. Companies may lower rates if the insured is seen as investing in added precaution. If you get a speeding ticket, your car insurance rates may be raised. If an oil company uses a double-hulled tanker to transport oil instead of a single-hulled one, its insurance premiums protecting against ecosystem damages from oil spills might be reduced. Moral hazard is a problem of *asymmetric information*. The insurance company has access to some information, but the insured has information of which the insurer is not aware. Changing asymmetric to symmetric information, in which the insurer knows what the insured is doing, is, therefore, a way to deal with moral hazard.

Incentives matter. People take greater precaution when there is a greater risk of uninsured loss. If you do not take out car insurance because you cannot afford the premiums, you are likely to drive more carefully. If the insurer removes the risk of loss, the incentive to be careful is reduced. If the insurer can monitor the amount of care, the problem of moral hazard is reduced. If there are significant activities that are hidden to the insurer, the insurer must assume the worst. This raises premiums, often to the point at which they become unattractive to the insured. There are other ways of dealing with moral hazard, including writing a policy with a deductible, that is, a sum that the insured pays toward the cost of compensation if an accident occurs. When the insured must pay a part of the accident's damage, the insured has an incentive to take efficient precautions. Typically, deductibles are a fixed sum (e.g., the insured pays the first $250 of loss), but they can also be a fixed percentage of the loss. Luckily for the environment, many environmental risks are produced by large companies. When insuring a large organization, it is cost-effective for the insurer to monitor the activities of the insured through audits, inspections, and the like. This tends to reduce the potential damages produced by moral hazard.

Adverse Selection. Adverse selection refers to the process by which risks not calculated by the insurer become insured. It refers to the tendency for people who are above-average risks to enroll in an insurance plan for which the expected value of their loss exceeds their premium. In the world of environmental risk, where one injurer is more likely to incur a large loss, the insurance company needs information about above-average risks to calculate its premiums, since premiums are set by expected losses. If the insured has a characteristic that affects the chances and size of a loss, but it is unobservable to the insurer, the latter must assume the worst and charge a high premium. This high premium is unattractive to potential injurers who are better risks, so these better risks opt out of the insurance. This sets up a vicious cycle in which the only potential injurers who buy insurance are bad risks, which further increases rates. The outcome may be no market at all, because the insurance company will go broke or leave the market.

The solution rests with finding the needed information and adjusting the premiums, which is a common practice in insurance. One important method for getting the relevant information is to audit the insured, either directly or by requiring the insured to submit to an independent audit. Although the verdict is still out, adverse selection may not be a big problem for environmental risks because the insured may have a hard time hiding important information. Insurance companies can often find out whether a company is more likely to have an accident based on past performance or facility characteristics.

Contract Enforceability. Insurance works only if the insured pays the premium and the insurer honors the obligation to pay out for proven damages. There must be some underlying legal force relating to an insurance contract, that is, there must be a right of an injured party to claim damages from the individual or agent causing the damage. If this legal force is lacking, either party could walk away from the contract whenever it is convenient, which means that the environment receives little effective protection.

Insurability and the Market

In addressing the question of whether environmental risk is insurable, remember that the core business of insurance companies is to provide financial protection against future loss arising from a defined set of causes. There are constraints on the types of risks that can be insured and on the size of risks that can be transferred to insurance markets. Attempts to identify standards for insurability of environmental risk point to factors such as expected frequency and severity of losses, maximum total loss, degree of moral hazard, adverse selection, and other influences (see Dahlstrom, Skea, and Stahal, 2003; Skea, 2004). In general, low-frequency/high-severity events, like nuclear power incidents, are not insurable, whereas high-frequency/low-severity events, such as car accidents and fires, represent ideal risks for private insurers.

An important question for public policy is whether unregulated market outcomes, like the earlier account of the steamboat pilot's association, control environmental risk efficiently. Because it is hard for us as individuals to effectively control personal exposure to environmental hazards, we often turn to public safeguards. These could be government environmental regulations backed by the force of law or private-sector guidelines that require its members to voluntarily adhere to them. Circulating information on the dangers posed by exposure to a particular hazardous substance is a public good. Each of us may have an incentive to free ride on the discoveries of others. Each of us, as individuals, is unlikely to have enough incentive to take on the personal cost of doing the needed research to discover the degree of risk. Thus, it may be that none of us as individuals is expected to produce the efficient amount of information, discover the size of risk, or act on it. As a result, governments may need to play a substantial role in stimulating research on the nature of hazards, providing for the delivery of information to people likely to be affected, and, in some cases, regulating those hazards.

Institutional Designs for Coping with Environmental Risks

Risk assessment and risk management are two important means for coping with environmental risks and for bringing public resources to bear on managing risks more efficiently, fairly, and sustainably. To implement these means, regulators use mechanisms such as liability rules, regulations, incentives, and others.

Risk Assessment

Environmental *risk assessment* involves examination of environmental risks to human health or the ecology. It characterizes the potential adverse health effects of human exposure to environmental hazards. As shown by the formula developed for expected damages (Equation 17.2), risk assessment contains two elements: the probability an accident (unfavorable event) will occur and the damages produced by the accident if it occurs. The intent of environmental risk assessment is to improve understanding of the likelihood and damages of various environmental risks as well as to use that knowledge to make better choices for controlling the risks. Finding and organizing the relevant data can help focus attention on sources of risk that, if controlled, produce the greatest public health and ecosystem benefits. In the United States, the EPA uses a four-step risk assessment approach first presented by the U.S. National Academy of Sciences (1983, 1984).[2] The four steps are hazard identification, dose-response analysis, exposure analysis, and risk characterization.

Hazard Identification. *Hazard identification,* the first step in risk assessment, is the process of determining whether exposure to an agent can cause an increase in the incidence of a health condition (e.g., abortion, birth defect, etc.). It involves characterizing the nature and strength of the evidence of causation and identifying the kinds of toxic effects (e.g., damage to the nervous system, cancer, etc.) that the pollutant causes. Hazard identification also summarizes the degree of certainty with which toxic effects observed under specified conditions hold (transfer) for the population whose risks are being analyzed. The aim of this step is to decide if a hazard needs the attention of a more detailed quantitative assessment.

Hazard identification is used to develop a weight-of-evidence ranking of potential adverse health effects from exposure to a hazard. That ranking comes from two major information sources: long-term animal studies and human epidemiological studies. Hazards that produce evidence of adverse effects in long-term animal studies are typically seen as high risk, unless there is significant contradictory evidence. Although *epidemiology* is an important source of information, epidemiological studies are not experimentally generated and are only capable of detecting relatively large increases in risk.[3] Therefore, the inability to detect effects does not rule out potential health risks. Hazard identification provides the qualitative information needed for the analyst to weigh the available information to determine whether a more detailed quantitative assessment is warranted.

Dose-Response Analysis. The second step of risk assessment involves the *dose-response relationship.* This step aims to measure how health responds to *changes* in exposure levels. Dose refers to the amount of exposure to a hazard, and response refers to the percentage of the population exposed suffering death or illness. Dose-response analysis is the process of characterizing the relation between the dose of an agent administered or received and the incidence of an adverse health effect as a function of the exposure to the agent. It takes account of intensity of exposure, age pattern of exposure, and other factors that might affect response (e.g., gender and lifestyle). A dose-response assessment often extrapolates from animals to humans.

Threshold exposure levels, if they exist, are dose levels below which only minimum health risks exist. Two methods are in widespread use for estimating dose-response functions: human-to-animal comparisons and epidemiological studies.[4] Human-to-animal comparisons use animal

---

[2] For environmental risk assessment in the United Kingdom, see Green, Short, and Taylor (2000).

[3] Epidemiology studies the distribution and determinant of diseases and injuries in human populations. It has five components: (1) hunch that some factor may influence occurrence of disease, (2) formulating a hypothesis based on suspicions concerning influence of some factor on disease occurrence; (3) testing the hypothesis to see if there is a statistical association between factors and disease occurrence; (4) assessing validity of association by seeing if the observed association really exists; and (5) making a conclusion.

[4] Sometimes clinical human studies are used in which human subjects are exposed to various ambient levels of the hazard; then health effects are determined. Because human subjects are involved, these studies are expensive, controversial, and rare. They seldom provide sufficient information for a risk assessment.

Figure 17–3 Dose-response relationships

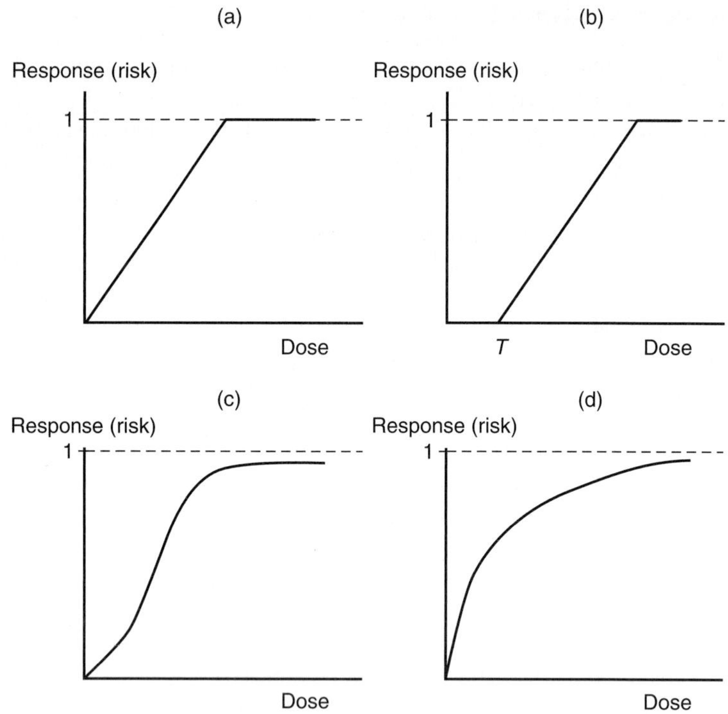

bioassays to determine the dose-response function. A bioassay is a test used to determine the impact of a hazard by comparing the hazard's effects on a test organism with the effects of a baseline, standard preparation. The aim is to extrapolate the high-dose responses of experiments performed on animals to infer lower dose responses on humans. The epidemiological method estimates dose-response functions by observing human health in nonexperimental (observed) surroundings. Raw data on individual health are derived from surveys or from population-wide health information. After collecting and processing raw data, econometric models are developed that characterize the dose-response function. The dependent variable (Y) is typically some measure of human health that is related to several independent (X) variables. The X variables typically include the dose of the contaminant in question along with anything else that would cause Y to vary among observations, such as lifestyle characteristics.

Figure 17–3 shows four hypothetical dose-response functional forms. Each graph shows a positive relationship between dose level and response. What varies is the rate at which the response changes with the dose. Figure 17–3a, for example, shows a linear dose-response relation that is bounded at the top by 100% certainty. It means that the rate of increase between dose and response is constant. Notice also that this relation starts at the origin. Some positive response to the dose is expected no matter how small the dose; to put it differently, no threshold level has been observed. Now compare Figure 17–3a with the relationship shown in Figure 17–3b. In this case, there is a linear effect of dose and response, but that effect occurs only beyond dose level T. Up to and including point T, there is no response at all, which indicates that the hazard shown by this relation has an identified threshold at dose level T. Figure 17–3c shows an S-shaped relationship drawn from the origin. Here, the response level increases at an increasing rate, and then increases at a decreasing rate. In Figure 17–3d, the relationship increases at a decreasing rate throughout its range. Results produced by hazard identification and dose-response analysis reveal something about the risks of an environmental hazard based on a known population being examined.

Exposure Analysis. *Exposure analysis,* the third step in risk assessment identifies the population at risk and the likelihood of exposure to the hazard. Exposure analysis begins with the existence of a person and a contaminant in the same environment at the same time (Jantunen and Finland, 2002). It is the process of measuring the intensity, frequency, and duration of human exposures that could arise from the release of hazards into the environment. In its most complete form, exposure analysis describes the magnitude, duration, schedule, and route of exposure; the size, nature, and kinds of populations exposed; and the uncertainties in all estimates.

Exposure analysts typically start by finding information about a suspicious hazard, including its molecular structure and physical and chemical properties. Then they study the locations at which the hazard enters the environment, which lets them describe its exposure pathways and its environmental fate. The analysts also assess a hazard's potential mobility, its reaction to other compounds, and how long it stays in the environment over time and space (temporal and spatial persistence). Then they measure the hazard's concentrations and identify the exposed population by location, habits, and size. Typically, exposure analysts connect concentration estimates with the exposed population to develop an exposure profile, which describes the size of the population exposed, exposure routes, and the exposure's intensity, duration, and frequency.

Risk Characterization. The fourth step of risk assessment is called *risk characterization,* which is the process of integrating information from the first three steps to develop qualitative and quantitative estimates of the likelihood of an adverse effect. The assessment of the severity of outcome is an important part of this process. The quantitative component of risk characterization identifies the magnitude of the risk and provides a way to compare one risk with another. Risk can be measured as a probability that an event will occur; a numerical value quantifies the likelihood of occurrence in some time period. Some probabilities are based on what are called actuarial risks, those determined from factual data. Actuarial risk is found by calculating the number of victims of a given hazard relative to the total number exposed. Probability measures, such as cancer risks associated with chemical exposure, are often based on inferences derived from animal bioassays or epidemiological studies. For example, the risk of getting cancer in a year from drinking chlorinated water has been estimated to be 0.0008%, or 8 in 1 million persons exposed. At this fourth step, risk assessment estimates risk thresholds or safe exposure levels below which further analysis is not needed.

Synthesis. Environmental risk assessments serve several purposes: (1) to wisely use resources by setting priorities among risk-reduction programs; (2) to decide which hazards should be regulated; and (3) to identify hazards that require further study. The four-step framework described can be applied to all types of health threats that arise in the environment. For environmental risk assessment methods used by the U.S. EPA, see Abernathy and Roberts, 1994. The methods and assumptions involved in each of the four steps vary according to the particular environmental pollutant, the strength of available scientific data, and our understanding of the pollutant's effects on human health. The framework provides the method by which available knowledge about a particular environmental risk is organized and evaluated by the risk assessor (Figure 17–4). Once the framework is set up, systematic measures can be developed to identify conditions that may create excessively high risks and to evaluate the comparative effectiveness of various forms of intervention. Intervention includes everything from regulations to warnings to education to economic incentives to law enforcement to medical measures to political pressures. This is the territory of risk management. See Example 17–2.

Risk Management

Governments acting on behalf of the public's welfare face a variety of opportunities to reduce risk. In the United States, fire protection in airplane cabins costs $200,000 per life saved; automobile side-door protection standards save lives at $1.3 million each; Occupational Safety and Health Administration (OSHA) asbestos regulations save lives at $89.3 million each; EPA asbestos regulations save lives at $104.2 million each; and a proposed OSHA formaldehyde

Example 17–2
Environmental Risk Assessment: Uses and Abuses

Quantitative risk assessment (QRA) is a process that measures the likelihood that adverse ecological effects may occur or are occurring as a result of exposure to one or more stressors. Several U.S. government agencies started using QRA in the 1970s and by the mid-1980s most U.S. environmental organizations had experience with QRAs. As the practice became integrated into the EPA's procedures for implementing environmental statutes, environmentalists expressed skepticism about its merits.

In the late 1990s, Tal surveyed 17 U.S. national environmental groups and 16 local ones to assess their views on the merits of risk assessment. More than 85% of the groups surveyed were skeptical. Many environmentalists consider it immoral to assign people and ecosystems to intolerable environmental fates without their consent. A consensus of the groups surveyed agreed that if preventing environmental hazards is the ultimate public policy objective, risk assessment will not promote that objective. By investigating the *magnitude* of the risk instead of how to *eliminate* it, most respondents felt that risk assessment distorts the environmental debate. All respondents surveyed supported the following view: Risk assessment should not replace hazard elimination as the operating paradigm for environmental regulations.

Environmentalists generally do not believe QRA methods sufficiently characterize the danger of environmental hazards to humans and ecological systems. They believe that too much energy goes into quantifying risks and too little is done to reduce or eliminate them. Almost unanimously, environmentalists dislike the exclusionary nature of risk assessments that undermines democratic participation in local environmental decisions. Environmentalists do not believe QRA methods characterize the full danger of environmental hazards to humans and ecological systems. It is widely believed that too many government and industry resources go into quantifying risks and too little is done to find ways to reduce or eliminate them. The environmental movement enjoys broad public support and has enormous influence over environmental debates. Although improved risk communication techniques may contribute to greater comfort with risk-related issues, QRA has a long way to go to reduce the doubts held by most environmental groups.

Source: Tal, A. 1997. Assessing the environmental movement's attitudes toward risk assessment. *Environmental Science and Technology.* 31 (10): A470–A476.

Figure 17–4 Risk assessment

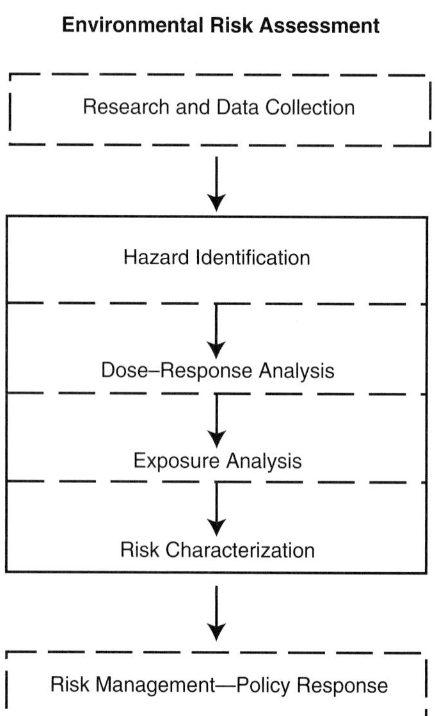

Example 17–3
Genetically Modified Food: Risks and Returns

Many of the processed foods available to U.S. buyers contain at least one ingredient that has been genetically altered. Certain genetic materials are "cut" from the DNA of another organism and "pasted" into another. Typically these changes are made to improve what consumers want from food, for example, a long shelf-life or bright color. Genetic modification (GM) has successfully defended crops against pests, reduced land area needed for a given plant yield, and allowed reduced water requirements for irrigation. But GM critics raise important questions: How safe are the modified foods? Have the health effects been tested? What are the risks compared with the benefits?

Modern agricultural research boasts impressive improvements: (1) grains and cereals that tolerate frosts, acid soils, and drought; (2) bananas, strawberries, and other fruits that stay ripe for days longer than their unmodified relatives; (3) knowledge to grow plants that can absorb toxic metals from the soil, thereby enabling land polluted by lead, for example, to have a new life in farming. Critics point out that many such breakthroughs come at the cost of unforeseen consequences. Detractors worry that too little testing has been done on the long-term human health impacts of genetically modified foods. Suppose part of the DNA of an apple is embedded into the genetic code of a peanut. Would people with a severe peanut allergy suffer by eating that apple? Should the apple be required to have a warning label? Because of such concerns, critics of GM have urged that governments follow the precautionary principle, that is, better safe than sorry.

Proponents of GM remind us that there is never 100% certainty about a new innovation's safety. If we routinely apply the precautionary principle, we will always be safe but never advance. Proponents of GM apply a modified version of the precautionary principle to genetically modified foods, a principle adjusted to the problems posed by current dynamics of the world population. Rapidly growing populations in poor countries with scarce farmland means more mouths to feed and little land for providing the food. For environmentalists who want to preserve the few remaining untouched lands (e.g., wilderness areas), it is simply too costly *not* to look for new ways like GM to grow more food on less land. Compared to embracing GM, *failure* to embrace the land-saving technologies of GM risks producing greater and more dangerous environmental disasters. If science cannot discover how to produce more food on fewer acres, it will not be the industrialized countries' rich activists who suffer. Suffering will fall on the world's poor. Supporters of GM say that the least ethical course of action is to avoid genetically modified foods, because by so doing we fail to feed the world's poor.

Source: Balko, R. 2004. *Genetically modified food: An environmental risk or millions' best hope for survival?* Available from *http://www.aworldconnected.org/article.php/483.html* [Accessed October 10, 2004].

---

standard costs $72 billion per life saved. Risk management deals with the questions of which policies should be pursued and which provide benefits that are commensurate with their costs (Viscusi, 1992). Environmental risk management, which is the process of deciding what should be done about a hazard, has the following characteristics:

- It is the process of assessing measured levels of environmental risk and making decisions to reduce risk levels.
- It is typically carried out by regulatory agencies under various legislative mandates.
- It takes into account political, economic, and engineering information along with risk-related factors to choose the appropriate regulatory response to a potential health hazard.
- It refers to measures for allocating resources to reduce risks in proportion to the toll they take on human health and welfare (Rodricks, 1994).
- It involves decisions that often require value judgments on questions such as these: What level of risk is acceptable? What level of expenditure is reasonable for controlling the risk? See Example 17–3.

Determining Acceptable Risk and Policy Instruments. All risk management strategies aim to reduce risk. For each policy proposal, however, somebody must decide how much risk reduction is worthwhile. That knowledge dictates how stringent public policy must be. In setting the level of acceptable risk, the policymaker faces a difficult situation.[5] If the acceptable risk level is set at zero, the policy must find a way to reduce society's exposure to zero if there is not a threshold. Although this action succeeds in taking away the health damages, it may do so at a politically unacceptably high cost. For example, if the hazard is a chemical used in an industrial process, reducing exposure to zero means the government must ban its use, thereby displacing jobs, reducing consumer choices, and increasing prices.

One guide that is used to help policymakers choose an acceptable risk level is the principle of *de minimis risk* (is sometimes taken to be the risk of a natural hazard, like an earthquake). This is a baseline risk level, a level so low that further reductions cannot be justified by their cost. Once that baseline is established, the decision maker can use comparative risk analysis to determine how variations in that risk level compare to other risks society faces. Table 17–2 lists and describes some of the most important U.S. environmental legislation drafted to cope with various environmental risks.

Once the degree of acceptable risk has been identified, the second task confronting policymakers is to choose the policy instrument for effectively achieving that risk level. Options are likely to include a mix of market-based, voluntary, and command-and-control approaches. In carrying out the decision, the risk manager thinks about the size of risks, benefits, and costs associated with each available control measure. See Example 17–4.

Economic Analysis. Carrying out the two risk management tasks, determining the acceptable risk level and choosing the appropriate policy instrument, requires making good choices from the menu of options. From an economic view, three considerations are the most important:

- Level of risk established
- Benefits society receives from adopting the policy
- Costs of policy implementation

Over time, several risk management strategies have been developed. Two dominant ones are cost-effectiveness analysis and benefit-cost analysis.

Cost-Effectiveness Analysis. Cost-effectiveness analysis compares the relative costs of a range of proposals for controlling environmental risk, all intended to meet the same objective. The analysis requires choosing the central objective that the reduced risk will achieve. Once a clear objective is identified, several control measures can be identified and developed. Although all control measures may meet the same objective, they may use a wide range of methods to accomplish that end. Cost-effectiveness analysis seeks the cheapest method to achieve the desired objective. For cost-effectiveness analysis to work, all measures for controlling risk must provide *equally effective* levels of environmental protection. If that assumption holds true, than the most cost-effective method is the one that meets the objective at minimum cost, in present value terms. In the United States, use of cost-effectiveness analysis is required by some environmental laws such as the Toxic Substances Control Act and Federal Insecticide, Fungicide, and Rodenticide Act.

Cost-effectiveness analysis typically calculates health benefits as a reduction in health risk. As in any other cost-effectiveness analysis, the monetization of health risk is not attempted. Cost-effectiveness analyses typically attempt to define the relation between a given risk reduction and the cost of environmental control measures necessary to reduce that risk. Increased marginal costs as a result of reducing risks are common; that is, an incremental reduction in health risk typically

---

[5] Slater and Jones (1998) summarize the challenges faced by the United Kingdom Environmental Agency in its aims of protecting and enhancing the environment. Achieving those aims requires a scientific understanding of the pressures, conditions, and responses of the environment. The authors see a pressing need for an accountable and open way of dealing fairly with environmental issues and other factors such as benefits, costs, uncertainties, and resource constraints. They describe the significant role played by methods of environmental economic analysis described in this chapter when they explain that the agency is developing strategic risk assessment methods that enable it to do three things: (1) compare the severity of risks from a wide range of sources or pressures; (2) allocate resources to areas where the greatest level of risk reduction may be achieved; and (3) determine which policy option for any given scenario is likely to result in the greatest benefit.

Table 17–2 Major U.S. environmental legislation

Short title	Year enacted, amended	Aims	Provisions
Clean Air Act	1970, 1977, 1990	Achieve national ambient air quality standards (NAAQS) (1990 amendments dealt with acid rain, ground-level ozone, stratospheric ozone depletion, and air toxins)	Regulates air emissions from area, stationary, and mobile sources. EPA establishes NAAQS to protect public health and the environment.
Clean Water Act	1972, 1977, 1981, 1990, 2002	Regulate discharges of pollutants into U.S. waters	EPA implements pollution control programs such as setting wastewater standards for industry.
Federal Insecticide, Fungicide, and Rodenticide Act	1996	Control pesticide distribution, sale, and use	Requires certification of pesticide users.
Occupational Safety and Health Act	1970	Ensure workplace safety	Ensures that employers provide a workplace free from recognized safety and health hazards.
Oil Pollution Act	1990	Control catastrophic oil spills	Requires oil storage facilities and ships to submit plans detailing how they will respond to large discharges.
Resource Conservation and Recovery Act	1976	Control hazardous waste	EPA regulates generation, transportation, treatment, storage, and disposal of hazardous waste.
Safe Drinking Water Act	1974	Protect U.S. drinking water quality	Requires public water systems to comply with health-related standards
Comprehensive Environmental Response, Compensation, and Liability Act (Superfund)	1980, 1986	Clean up uncontrolled or abandoned hazardous waste sites	EPA has power and budget to seek out parties responsible for release and ensure their cooperation in cleanup. EPA cleans up orphan sites when potentially responsible parties cannot be found or when they fail to act.
Toxic Substances Control Act	1976	Track 75,000 industrial chemicals currently produced or imported into United States	EPA screens and requires reporting or testing of chemicals that pose an environmental or human health hazard. It can ban manufacture and import of chemicals that pose unreasonable risk.

Source: *U.S. EPA. 2004.* Major environmental laws. *Available from http://www.epa.gov/epahome/laws.htm [Accessed October 10, 2004].*

Example 17–4
Risks and Benefits of DDT in Africa

Malaria is a serious and sometimes fatal infectious disease transmitted by infected mosquitoes as they suck blood from people and leave parasites in the blood stream. Malaria is accompanied by recurring fever and can be fatal. Approximately 300 to 500 million cases of malaria are reported annually worldwide and 90% of these are in sub-Saharan Africa; health care costs and lost income exceed $2 billion annually. The Greek historian Herodotus, in fifth century B.C., described the use of bed nets by Egyptians for protecting against mosquitoes:

> Against the mosquitoes that abound (in Egypt), the following have been devised by them: those who dwell higher up than the marshy country are well served by the towers where they ascend to sleep, for the winds prevent the mosquitoes from flying aloft; those living about the marshes have a different recourse, instead of the towers. Every one of them has a net, with which he catches fish by day, and at night he sets it around the bed where he rests, then creeps under it and sleeps. If he sleeps wrapped in a garment or cloth, the mosquitoes bite through it; but through the net they absolutely do not even venture. (*Book 2*)

Today, malaria can be prevented by a variety of methods such as reducing the population of mosquitoes, reducing the extent of swampy ground on which they breed, and spraying with pesticides. No effective vaccine for malaria has been developed. Attempts to sterilize mosquitoes and to breed noninfective mosquitoes have failed. The most cost-effective control measures vary by location, but the common feature of success is sustained investment, systematic management, continual evaluation of services and their impact, and vigilance.

Eradication of malaria has been successful in many countries, including large parts of India, Sri Lanka, and Latin America, using the pesticide DDT as part of a prevention program.

The World Health Organization (2000) considers DDT one of the most cost-effective insecticides available for controlling malaria. In this context, cost-effectiveness is defined as the cost of averting one disability-adjusted life year (DALY). Assessments in Mauritius show that the cost per DALY averted is about $4.20 (U.S. dollars). Programs elsewhere in Africa using other methods (e.g., bed nets) have been assessed at between $19 and $85 per DALY averted. Today, in much of sub-Saharan Africa DDT seems to produce a greater DALY than all its substitute measures.

Source: Roberts, J. L. 2002. *Malaria control: A case study from the Republic of Mauritius.* United Nations. Available from *http://africa.unep.net/casestudies/malaria-mauritius.asp* [Accessed October 9, 2004].

requires a greater incremental rise in the cost of control when health risks are lower than when they are higher. The cost of control for an initial reduction of health risk at high levels of exposure will be much less expensive than a further reduction in remaining amounts of risk incurred at already lower exposure levels.

Figure 17–5 and the accompanying table illustrate the cost-effectiveness idea with hypothetical data. They show potential environmental standards and control costs for two programs: increased water contaminant control and increased particulate emission control. In each program, higher environmental standards produce reduced concentrations, that is, fewer parts per million (ppm). Those reduced concentrations require polluters to incur higher control costs, as they are required to implement more extensive control measures. When control costs are low, further increases in control costs increase environmental safety quickly. In contrast, added gains in safety are much more modest when control costs are high, after the cheap and easy controls have already been installed. As control costs are increased and safety improves, greater reductions occur in mortality risk.

The top table shows choices available for program 1. Reduced mortality risk[6] increases from a low of zero with unregulated controls (100 ppm) to a high of 0.00220 (reduced probability of premature death) when the control standard is set to 9 ppm. For an affected population of 1 million, increased costs of complying with controls saves more lives, as greater reductions in mortality risk

[6] For several ways of expressing those mortality risks, see Slovic (1999). A few examples are deaths per million people in the population, deaths per million people within *X* miles of the exposure source, deaths per unit of concentration, deaths per facility, and deaths per ton of toxin released.

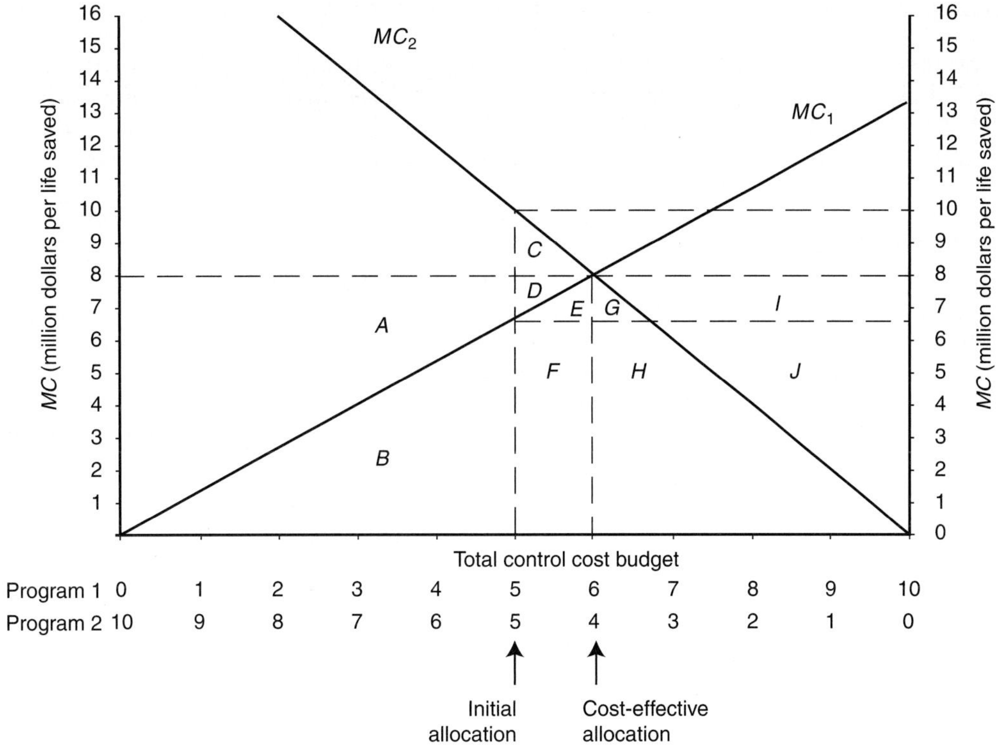

Figure 17-5 Cost-effective risk reduction

Program 1, increased water contaminant control[a]

Potential standard (parts per million)	Annual control costs (billion dollars)	Total reduced mortality risk	Total saved lives	Incremental lives saved (lives per million dollars)	Incremental cost per life saved (million dollars per life)
100	0	0	0	–	0.00
50	1	0.00075	750	0.75	1.33
33	2	0.00113	1,125	0.38	2.67
25	3	0.00138	1,375	0.25	4.00
20	4	0.00156	1,563	0.19	5.33
17	5	0.00171	1,713	0.15	6.67
14	**6**	**0.00184**	**1,838**	**0.13**	**8.00**
13	7	0.00194	1,945	0.11	9.33
11	8	0.00204	2,038	0.09	10.67
10	9	0.00212	2,122	0.08	12.00
9	10	0.00220	2,197	0.08	13.33

Figure 17-5
Continued

Program 2, increased particulate emission control

Potential standard (parts per million)	Annual control costs (billion dollars)	Total reduced mortality risk	Total saved lives	Incremental lives saved (lives per million dollars)	Incremental cost per life saved (million dollars per life)
100	0	0	0	–	0.00
44	1	0.00050	500	0.50	2.00
25	2	0.00075	750	0.25	4.00
16	3	0.00092	917	0.17	6.00
11	**4**	**0.00104**	**1,042**	**0.13**	**8.00**
8	5	0.00114	1,142	0.10	10.00
6	6	0.00123	1,225	0.08	12.00
5	7	0.00130	1,296	0.07	14.00
4	8	0.00136	1,359	0.06	16.00
3	9	0.00141	1,414	0.06	18.00
2	10	0.00146	1,464	0.05	20.00

[a]Both programs in the table are based on an affected population of 1 million.

Source: *Adapted from U.S. Environmental Protection Agency. 1997.* Presidential Commission on Risk Assessment and Risk Management. *National Center for Environmental Assessment. Available from http://cfpub.epa.gov/ncea/cfm/pcrarm.cfm [Accessed Ocotober 10, 2004].*

are multiplied by 1 million. The most demanding pollution standard of 9 ppm incurs compliance costs of $10 billion and saves 2,197 lives. Incremental lives saved from increased compliance costs (the fifth column) can be thought of as the marginal product of control cost resources. That marginal product is measured as the gain in total lives saved divided by the additional compliance costs needed to save those added lives. The far right column, marginal (incremental) cost per added life saved, is the reciprocal of the marginal product shown in the fifth column. That marginal cost is interpreted as the incremental control costs required to save one more life. Notice that the highest level of environmental control at 9 ppm incurs control costs of $10 billion and has a marginal cost per life saved of $13.33 million. Other things being equal, life-saving programs with low marginal costs are preferred to ones with higher marginal costs, because programs with low marginal costs free up resources to save more lives.

The bottom table shows similar choices for program 2, in which greater control costs are incurred to meet regulations requiring reduced concentrations (parts per million) of air contaminants. Tighter regulations produce higher compliance costs and a corresponding increase in lives saved. The hypothetical numbers in the table show that incremental costs per added life saved range from a low of zero when emissions are uncontrolled to a high of $20 when control is maximized at 2 ppm at a compliance cost of $10 billion.

Suppose that a proposal would set the regulations at 17 ppm in program 1 and 8 ppm in program 2, respectively. Polluters incur an equal $5 billion in total compliance costs for each program. As a result, total lives saved are 1,713 for the water pollution control program and 1,142

Example 17–5
Controlling Carbon Monoxide Risks in Korea

North Korea has a long history of carbon monoxide (CO) poisoning, which reduces the blood's oxygen-carrying capacity. It is especially damaging to people with diabetes, respiratory diseases, hypertension, and other conditions in which the heart has an impaired capacity to meet oxygen demand. The CO poisoning rate in North Korea is among the highest in the world. Poisoning occurs largely from heating and cooking systems that use coal briquettes. The problem stems from North Korea's lack of energy sources; historically, the country imports more than 90% of its energy, so safer energy sources are expensive.

Anthracite coal, which is used nationwide as briquettes for cooking and for heating houses in the winter is a cheap fuel. Homes are heated by smoke from the stove, which is conducted underneath rooms to provide heating before it goes up the chimney. Floors are often made of dried mud covered with specially coated paper. In many homes it is hard or expensive to prevent chimney gases from seeping into rooms where the family sleeps. CO poisoning in North Korea experienced huge growth in the 1960s and 1970s and by the early 1980s, more than a million people were being treated in hospitals annually.

Prevention approaches often fail to deliver economically efficient or cost-effective solutions. Furthermore, where there are many contributing factors, it is hard to disentangle the contributions each make to CO poisoning. Still, there have been some successes. Death rates from CO poisoning have been decreasing gradually as a result of a government initiative to find the most cost-effective measures by which people could more easily substitute dirty fuels with clean ones.

In 1995, the Korean Ministry of the Environment revised the nation's ambient air quality standards for CO, establishing levels not to be exceeded more than three times per year. Results of several studies conducted on indoor and ambient air quality suggests this policy has done some good, since indoor CO levels in newer homes and offices are not significantly different from those found outdoors in many cities.

Source: Baek, S. O., S. M. Hwang, and Y. H. Moon. 1999. Carbon monoxide pollution in Korea: Public health implications. *Indoor and Built Environment.* 8: 156–167.

---

for the air emission control program, for a total of 2,855 lives saved. However, a comparison of the marginal cost per life saved through each program shows that total lives saved can be increased by reallocating total compliance costs. When each program absorbs $5 billion, the incremental number of lives saved per million dollars is approximately 0.13 for the water contaminant program (1,838 − 1,713) / ($6,000 − $5,000) and 0.08 for the air emissions control program (1,225 − 1,142) / ($6,000 − $5,000). Since the marginal number of lives saved is much lower for the air program, reallocating control costs from air to water will increase the total number of lives saved with no increase in total costs.

For a total compliance cost of $10 billion, the *cost-effective* set of regulations occurs when $6 billion is allocated to water (14 ppm) and $4 billion is allocated to air (11 ppm). Those regulations produce the maximum number of lives saved (2,880) for the $10 billion compliance costs. The cost-effective set of regulations can be found by looking for the allocation of scarce control costs for which marginal cost per life saved is equal for both programs, shown by the intersection of the two marginal cost schedules in Figure 17–5. The cost-effective set of regulations is that for which this equimarginal rule holds: *Marginal cost per effect averted is equal for all programs.* (For this example, the effect averted is one life saved.) When this equimarginal rule is followed, total lives saved are maximized for the given compliance cost incurred. Indeed, Figure 17–5 and the tables show that for the $10 billion compliance cost, marginal cost per life saved is $8 million under both programs. No possible reallocation of control costs can increase total lives saved. For the water program, the cost-effective control standard is 14 ppm; for the air quality program, it is 11 ppm. One especially attractive feature of cost-effectiveness analysis is that it permits risk managers to define an acceptable level of environmental risk without requiring the conversion of human health benefits into monetary measures. The relationship between a level of control and an acceptable level of risk is estimated directly. Cost-effectiveness analysis is the method of choice for many risk management evaluations. See Example 17–5.

Environmental laws often establish the risk reduction (or risk level) to be achieved, which means that the level of associated benefits is set by the reduction in risk required. In such cases, the risk manager can turn to cost-effectiveness analysis, since the measurement of benefits is not an issue. Thus, economics can contribute to the goal of selecting a policy instrument that meets the legislated risk objective at minimum cost. The advantages of cost-effectiveness analysis are that its calculations are straightforward and that it can help find the cheapest control measure from a range of options, with no information required on benefits. However, the method is limited in scope. Only one objective can be chosen, and the objective chosen at the beginning of the analysis is not always seen as desirable after the study is finished.

Benefit-Cost Analysis. Benefit-cost analysis (sometimes called risk-cost benefit analysis) is a specialized method for conducting economic analysis of risk management choices. It compares the estimated costs of implementing an environmental improvement program with the estimated benefits produced by its anticipated improvements in human health (or reduced damages from sickness or premature death). BCA provides risk managers with the ability to evaluate and select the most economically efficient environmental risk control strategy from a range of proposed options. It involves assigning monetary values to quantify health benefits gained from the reduced incidence of disease (morbidity), lessened risk of premature death (mortality), or reduced risk of ecosystem damage. When the monetary present value of net benefits is larger than zero, projected human health or ecological benefits outweigh the control costs. This finding, if it occurs, tells us that the environmental risk control program in question makes good economic sense. Similarly, when the net present value is negative, the environmental risk control strategy is not defensible in economic terms (see Goldman, 2000).

In the United States, President Reagan's 1981 Executive Order 12291[7] is largely responsible for the increased use of benefit-cost analysis in designing environmental policy.[8] In 1993, President Clinton's Executive Order 12866 continued to support this analysis of benefits and costs when formulating risk management strategies, and still later the Bush administration has kept the principles of both executive orders.[9] BCA sets an "acceptable" risk level using the standard of economic efficiency. For proposed actions that would reduce risks incrementally, the decision maker would compare the monetary value of social benefits with the associated costs to find the efficient risk level. That efficient risk level occurs where the *marginal social benefit* (MSB) equals the *marginal social cost* (MSC) of risk reduction. Another way of saying the same thing is that the efficient risk level corresponds to the risk level that maximizes the difference between total social benefits (TSB) and the total social costs (TSC).

---

[7] It is the first such order to designate net benefits as the standard for assessing proposed regulations. In the words of the executive order, "regulatory objectives shall be chosen to maximize the net benefits to society" and "regulatory action shall not be undertaken unless the potential benefits to society from the regulation outweigh the potential costs to society." The order assigns major importance to economic feasibility in the setting of U.S. environmental control. Details are at U.S. EPA (2004).

[8] Ball and Goats (1996) describe a framework role for benefit-cost analysis for environmental risk control in the United Kingdom. They describe BCA as aiming to achieve a reasonable balance between the costs of risk control measures and the benefits gained, with both are expressed in monetary terms so far as it is possible.

[9] BCA is not the only method of analysis using economic principles for controlling environmental risks. Compare it with the precautionary principle. The precautionary principle says it is better to be safe than sorry. That says we are ignorant about many important aspects of the environment and human health. It acknowledges scientific uncertainty and guides our actions in response to it. It says when an activity raises threats of harm to human health or the environment, precautionary measures should be taken even if some cause-and-effect relationships are weakly established scientifically. In this context, the proponent of a potentially risky activity, rather than the public, should bear the burden of proof. It says that aggressive regulation is justified even in the face of scientific uncertainty—even if it is unclear that environmental risks are serious. By contrast, the use of benefit-cost analysis says it is only justifiable to undertake more aggressive regulation if defended by an assessment of the benefits and costs of regulatory action.

BCA can be used in the policy appraisal stage much in the same way it is used to find an acceptable risk level.[10] At this stage, economic efficiency is used to evaluate the effectiveness of an ongoing initiative. In practice, estimates of the MSB and MSC at the risk level specified by policy are compared to see if they are equivalent. If not, the risk manager knows that the acceptable risk level needs to be reduced if MSB of risk reduction is greater than MSC, and increased if MSB is less than MSC.

The BCA approach compares the estimated costs of improved environmental controls against the expected benefits of the reduced likelihood of illness or death associated with various exposure reduction scenarios. A monetary *benefit-cost ratio* found to exceed one means the projected health benefits outweigh the costs of control. Economists normally conclude that an environmental control program makes good economic sense when expected benefits outweigh costs. Conversely, a benefit-cost ratio less than one means an environmental risk control strategy is not defensible in strictly economic efficiency terms, although this does not necessarily disqualify it from further consideration on noneconomic grounds.

Acceptable use of BCA depends on the correct identification of suitable control options. If risk control options are poorly conceived, with inadequate accounting for best technology or best practice, then many of the proposed policy options will show poor benefit-cost ratios and all might be rejected as infeasible. One way to guard against the mistake of excluding workable control measures is to assure that knowledgeable stakeholders contribute to the development of control measures, as well as to the estimation of costs associated with each. Business stakeholders often have technical and cost information unavailable to the regulators—information needed to design the most efficient control initiatives. However, much of that information is closely guarded against competing firms.

How does BCA compare with a cost-effectiveness analysis? A cost-effectiveness analysis asks how a fixed amount of total compliance cost should be allocated to maximize effectiveness in averting undesirable effects, such as reducing the number of lives lost from environmental hazards. A BCA does that, too, but goes one step further. It assumes that the economic value per undesirable effect averted can and should be measured. BCA asks *to what level should total compliance costs be expanded, and how should those costs be allocated* to maximize the net benefits of the averted effects. A BCA finds an *economically efficient* set of regulations to manage environmental risks. Any economically efficient allocation of regulatory resources is cost-effective, but not all cost-effective allocations of regulatory resources are economically efficient. Achieving economic efficiency requires following this equimarginal rule: *Marginal cost per effect averted (e.g., per life saved) is equal for all programs and is equal to their common marginal benefit.* In the case of the two programs shown in Figure 17–5 and the accompanying table, if the marginal benefit is $12 million per statistical life saved, then economically efficient regulations occur when $15 billion of total compliance costs are allocated to the two programs. The efficient allocation of that $15 billion occurs when $9 billion is allocated to the water quality program and $6 billion to the air quality program. At that allocation of compliance costs, the marginal cost per saved life for each program is $12 million and is equal to the marginal benefit per saved life. The regulations that produce that economic efficiency in allocation of compliance costs is 10 ppm for the water quality program and 6 ppm for the air quality program. Figure 17–6 shows the total gain in economic efficiency produced by tightening environmental regulations and expanding polluters' associated compliance costs from $10 billion (point *C*) to $15 billion (point *B*). Regulations are tightened from the (cost-effective) 14 ppm to the (economically efficient) 10 ppm for the water quality program, and similarly from 11 ppm to 6 ppm for the air quality program. The gain is area *ABC*. That area has an important economic interpretation. It is net gain in value of lives saved over the added compliance costs incurred by tightening the regulations. An

[10] Strictly speaking, the acceptable risk level is not a fixed standard. Where changes in technology, institutions, or other factors affecting the stream of benefits and costs occur, a new acceptable risk level will be economically efficient. For example, in the face of population growth, more people are exposed to a given hazard, so a lower level of acceptable risk becomes economically efficient.

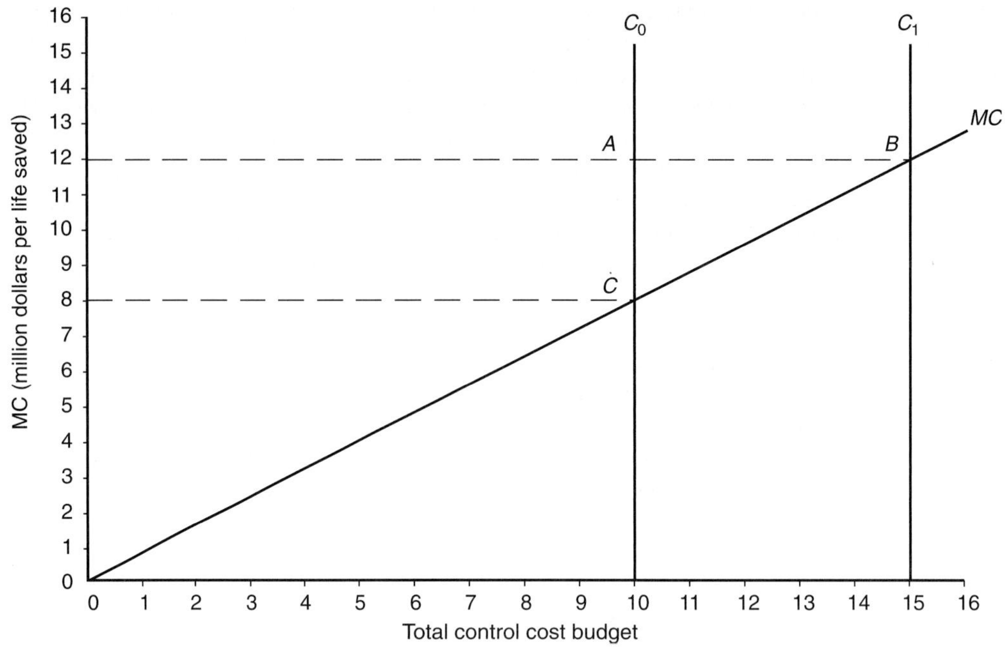

Figure 17–6 Economically efficient risk reduction

economically efficient tightening of environmental restrictions means that environmental control resources are moving from a lower-valued to a higher-valued end use. Area *ABC* is society's net gain produced by reallocating those resources.

One of the greatest debates over BCA involves its use of monetary benefits for projected human health or ecological benefits. The use of monetary values for measuring benefits gained or lost sometimes makes this kind of analysis hard to defend for environmental risk evaluation. As stated earlier in this chapter, a reliable estimate of the cost of an illness is a major barrier to the use and political acceptability of BCA.

Risk Management Institutions

Regulators have access to several methods for controlling environmental risks. They can establish liability rules, enact regulations, modify prices through taxes and subsidies, promote creation of markets, and promote better risk communication. Driving a car, hiking the wilds, hiring people to work for you, and even taking your family on a camping trip, can lead to accidents. In light of the constant potential for accidents, society wants to reduce risks to acceptable levels; it wants to know that its members take adequate precaution when driving, hiking, building, or doing anything that is risky. Incentives matter and they affect people's risky behavior. One way to take advantage of incentives is to design laws that influence and alter choices people make in connection with their risky activities. For example, most countries have a large number of laws influencing the driving of motor vehicles.

Liability. Liability law has developed over hundreds of years as a way of dealing with risky activities. The basic idea is that anyone engaging in an activity that imposes risks to other people may have to pay compensation to the victim in case of an accident. When an accident causes a victim to be injured, the victim may recover damages by suing the injurer. Liability law provides incentives to potential injurers to take precautions when they do things containing an element of risk to others. Think of liability law as serving to charge a price to injurers for the costs they impose on others. Without liability law, the accident damage to others would be an externality and would be ignored by the potential injurer. In other words, liability law has the effect of internalizing the accident damage into the benefit-cost calculation made by the person undertaking the risky activity. So liability law, in principle, can promote the economically efficient amount of risk in people's activities by providing incentives to invest in the efficient amount of precautionary behavior.

Risk Allocation under Liability Laws. In considering how a liability law influences risky behavior, think of an organization that undertakes risky activity as the "injurer" and the person or group who stands to be damaged if an accident occurs as the "injured victim." Recall from Equations 17.2 and 17.3 how to measure expected damages produced by risky activities. The injurer can take steps to reduce the probability of an accident occurring or to reduce the damages it produces, or both. To simplify, any action that reduces expected damages incurs a certain level of "precaution." So the injurer can invest in some amount of precaution, at a cost, to reduce expected damages to the victim. But more precaution is not free, it comes at a cost. These costs increase with greater levels of precaution, much like a rising supply curve. And these costs must be absorbed by the injurer. When investments made in precaution are low, marginal cost of precaution will also be low. The first steps taken to improve precaution are the cheapest. However, after those steps have been taken and precaution levels rise, there are fewer and fewer opportunities for cheap improvement, so the marginal cost of more intense precaution increases.

Does the injurer pay any attention to the liability laws? Do liability laws affect the injurer's precautionary behavior? The answer is yes to both questions. A lack of liability laws means that economic incentives are signaling injurers to take no precautions. Any precaution is costly, but the benefits accrue to the victim and not the injurer. Now watch what happens when a liability law is introduced. The law says that the costs of any accident to the victims are borne by the injurer. So the injurer faces total costs equal to the costs of precaution plus expected damages from Equation 17.2. The injurer who faces both those costs, will take actions that minimize the sum, which happens when marginal costs of more precaution equal the marginal value of expected damages saved (Equation 17.3). At this point, the injurer has acted in self-interest by making investments in precaution that produce that equimarginal condition.

Negligence versus Strict Liability. Two important kinds of liability law are *negligence* and *strict liability* (Coleman, 2003). Strict liability means that if the actions of the injurer (the party allegedly responsible for the accident) cause an accident, the injurer must compensate the victim, whether or not the injurer is at fault.[11] No matter how little the victim invested in attempting to avoid the accident, the injurer must pay compensation. The entire burden rests with the injurer. Negligence liability law assigns some responsibility in avoiding accidents to the victim. Only if the potential injurer has taken less than the economically efficient level of precaution will the injurer be liable for damages to the victim from an accident. The standard of liability, the Hand formula, applied by U.S. courts, is an economic one.[12] That test says that the defendant is guilty of negligence if the loss caused by the accident, multiplied by the probability of the accident, exceeds the cost of undertaking the precaution needed to avoid the accident. When the test is applied correctly, it is precisely the economic efficiency standard. So in setting up incentives for avoiding accidents, negligence law is compatible with economic efficiency. In 1947 Judge Learned Hand wrote that the injurer's liability for the damages from an accident depend on (1) the probability of the accident multiplied by the (2) damages produced should the accident occur compared to (3) the cost of undertaking precautions to avoid the accident (*United*

---

[11] Alberini and Austin (2001) found that requiring strict liability for the cost of cleaning up contaminated sites increased the level of care taken by the firm to avoid uncontrolled releases of pollutants into the environment.

[12] The standard of liability (dividing line between liable and not liable) is the test applied by the courts in deciding whether the defendant has exercised adequate care (invested in suitable levels of precaution). Judge Learned Hand (1872–1961), was an influential state and federal judge in the first half of the twentieth century.

States v. Carroll Towing Co.).[13] Hand's reasoning can be sharpened by extending Equation 17.1. The defendant is guilty of negligence if and only if the following inequality holds true:

$$C < pD \tag{17.4}$$

where C is the cost of precaution (burden of prevention), and as defined earlier, p is the probability of the accident, and D is damages the accident produces. This formula has come to be known as the Hand formula. Hand went on to conclude that, taking into account the surrounding circumstances, the expected damages from the accident outweighed the cost of precaution to avoid the accident, so he held the defendant liable. Presumably, if an accident occurs in which the cost of precaution that could have been taken to avoid the accident is larger than the expected damages to the victim, the injurer will be found not liable.[14]

The Hand formula is attractive because of the correct incentives it offers injurers to take actions to avoid accidents. Assuming that the level of precaution required by the courts is equal to the economically efficient level of precaution, the injurer will normally take the efficient level of care to avoid being held liable of negligence. Will the victim become careless, knowing that the injurer must either take efficient precautions or pay for any injuries caused by failure to do so? Not likely, because not all losses can be compensated with money. In addition, victims are often risk averse with respect to large losses and tend to prefer uniform levels of utility over time. A risk-averse victim would prefer a constant utility of 10 over time to a utility of − 10 (due to injury) in one period and a utility of 35 (including compensation for the previous injury) at a later time. For these reasons, in a negligence legal system, victims as well as injurers are likely to continue to take precaution, and injurers have incentives to take efficient precaution levels. Because of the potential of a lawsuit lying behind every negligent act, the aim of a negligence system is to encourage all potential injurers to invest in the efficient amount of precaution without writing special rules to govern all actions. By requiring us to face the marginal cost of all our actions, negligence law is a form of marginal cost pricing; therefore, it is a method of enforcing actions that promotes and rewards an efficient amount of precaution.

Under a strict liability rule, the injurer pays for the damages his conduct causes to the victim regardless of whether or not he was negligent (i.e., regardless of whether or not he incurred the economically efficient level of precaution). The injurer pays for negligent and non-negligent damages. If no precaution by the victim could reduce the expected damages produced by an accident (i.e., if expected damages cannot be influenced by the victim), then the only relevant costs of precaution are those of the injurer. In that special case, a defendant (injurer) who is held strictly liable, will take the economically efficient level of care to minimize his expenses. If his precaution costs are less than the expected damages of the accident, the injurer will take the precaution rather than pay for the injury. If the cost of the precaution is more than the damages saved in reduced injury, it is cheaper for the injurer to pay for the resulting damages than to take

---

[13] Hand's reasoning is based on benefits and costs: Any preventive action taken will have both benefits and costs. We want people to take all of those precautions for which the benefits outweigh the costs, but only those precautions. Speed limits are an example of benefit-cost calculations. An international speed limit of 35 miles per hour undoubtedly would decrease the number of deaths on the world's highways, but we would all go mad on long trips.

[14] Presumably, the cost of precaution refers to the least-cost combination of actions that could have been taken by the injurer and the victim. When the victim could have avoided the accident at a cost of $100, but the injurer could have done the same at a $150 cost, then the victim's $100 cost is the relevant cost of precaution. Also, the entire question of who the victim is goes back to the presumed distribution of property rights. Is the plaintiff who suffers damages produced by the accident the victim, or is the defendant who constrains activities because of the presence of the plaintiff the real victim? If property rights are well defined (to take risks with impunity or not to have one's safety impaired by others), some of the conclusions of the Coase theorem (Coase, 1960) may be applied. It suggests that decentralized measures produce an efficient allocation of precaution, thus reducing court costs. Notice that the application of the Hand formula requires the court to find facts that allow computation of (1) a schedule of marginal costs of precaution, (2) another schedule for marginal benefits of precaution, and (3) whether the precaution level produced by their intersection exceeds the actual precaution level taken by the injurer. Requirements of these computations place a heavy burden (cost) on the court. They also put a burden on the injurer who must implicitly make the same computations in deciding how much precaution to invest in to avoid accidents.

the added precaution. In this special case, the injurer will take the efficient level of precaution. The problem with strict liability is that victims have no incentive to undertake their own precaution to reduce accident damages. For example, take the case of the rancher who knows a railroad will be held strictly liable for all livestock its trains kill. The rancher receives a higher net income by letting more of her livestock wander near the tracks rather than incurring the precautionary expense of keeping them penned. In this case, it is economically rational to take no precaution. In situations where controlling expected damages requires both the injurer and the victim to undertake precaution, a strict liability rule is inefficient. Nevertheless, strict liability has a saving grace: The danger of discouraging a victim from taking precaution may be mitigated if the victim is averse to risk, especially if the victim stands to suffer uncompensated physical injury.

Joint and Several Liability. Joint and several liability is a legal doctrine that makes each of the parties responsible for damages awarded in a lawsuit. Since the 1970s in Europe and the United States, joint and several liability has had important significance for environmental damages. A long, complex, and uncertain chain of events connects hazardous landfill waste and contaminated groundwater. This complexity presents a major problem for proving causation, which typically must be done to collect damages under the two forms of liability described. During long time lags between accidents and damages, ownership of businesses changes, some companies go bankrupt, and records are lost. Ultimately, liability of any kind is hard to prove, and damages are hard to measure and collect. Remember that a good liability law is supposed to encourage potential injurers to take precautions when they engage in risky activities. If a company is unlikely to be sued, even if it behaves negligently, then the incentive effects of liability are weakened, and there is too little precaution for economic efficiency.

One solution to this problem is the concept of joint and several liability. For example, under the U.S. Comprehensive Environmental Response, Compenstation, and Liability Act, the principle of joint and several liability establishes liability for site cleanup and other costs on the part of more than one potentially responsible party. If there were several owners or users of a site that became contaminated over the years, they could all be considered potentially liable for cleaning up the site. Take the case of the bankrupt landfill. Injured parties may choose to sue the richest of several companies that contributed waste to the landfill *for all damages.* Under joint and several liability, the defendant (the injurer) could then attempt to recover damages from other potentially responsible parties. The idea is that somebody has to absorb the costs of the accident, and it may as well be one of the contributing parties rather than the victim. If the victim was forced to recover from each potentially responsible party in bite-sized chunks, the lawsuit would be less likely to ever take place, so the incentive properties of liability would be undermined.

Regulations versus Liability. In some situations, environmental regulations may be preferable to liability laws. If hazards are widespread and risks large, it may be more efficient for a government regulator to define the level of precautionary behavior than to ask the courts to shoulder that burden for huge numbers of lawsuits. Regulations may simply be a cheaper way to get the job done. Liability also presents the problem that victims do not always sue, even when justified. Bringing a lawsuit is costly and inconvenient. If an injurer is not held accountable for legitimate damages, the power of liability laws to promote efficiency is weakened. Bankruptcy could be an even larger problem with liability laws. Nobody can sue a polluter for more than the firm's insured assets; the firm will simply declare bankruptcy. That potential outcome means that injurers will take too little precaution to avoid an accident with high expected damages, so liability provides too few accident-avoidance incentives.

Incentives. Market-based incentives rely on the self-interest provided by economic motivations to encourage cost-effective environmental protection (see U.S. Council of Economic Advisors, 1996, and U.S. EPA, 1997). Consider the use of tradeable sulfur dioxide emission allowances to reduce acid rain. This program, established under the 1990 amendments to the U.S. Clean Air Act, permits electric utilities to reduce their emissions of sulfur dioxide, a contributor to acid rain, *below* maximum allowed levels (overcomply) and sell the unused emission allowances to companies whose abatement costs are too high for them to comply (see Chapter 16). The program caps aggregate sulfur dioxide emissions well below historical levels. It also allows emission reductions to be achieved more cost-effectively instead of requiring every company to install the most expensive technology for controlling sulfur dioxide emissions.

Alternative compliance provides greater flexibility to industry by allowing choices in achieving emission or risk reduction specifications. This option is designed to achieve higher levels of environmental protection at lower cost and to foster integration of local concerns in environmental risk management decisions. Substantial savings can result for industry, communities, or any regulated entity that participates. For example, EPA's Project XL (U.S. EPA, 2002b), launched in 1995, allows six private firms and two government agencies to experiment with different strategies for improving environmental protection. Government also can provide greater flexibility in compliance for companies that are attempting to use innovative pollution reduction and control technologies. For example, some companies have experimented with the bubble concept, in which a well-defined geographic area is established, then the cheapest way to reduce a pollutant inside that "bubble" is determined. This method has reduced the costs of complying with environmental regulations.

Consensus, Mediation, and Dialogue. Negotiated rule-making and dialogue projects, such as EPA's Common Sense Initiative (U.S. EPA, 2002a), offer opportunities for stakeholders to design new standards and solutions that protect human health and the environment more reliably and with greater cost-effectiveness and public acceptance. In that program, industry-specific strategies were explored to promote cleaner and cheaper environmental performance. The EPA convened consensus-oriented teams of stakeholders to look for measures to turn complicated and inconsistent environmental regulations for six major industries—automobile manufacturing, computers and electronics, iron and steel, metal finishing, petroleum refining, and printing—into comprehensive strategies for environmental protection that are compatible with a sector's unique requirements. For example, several pilot projects in Michigan, Rhode Island, and Connecticut have been testing innovative, cost-effective ways to help the metal-finishing sector improve environmental performance. The EPA, metal finisher trade associations, and interested stakeholders have developed a strategic goals program that seeks a voluntary commitment from metal finishers to make a number of improvements in pollution prevention and efficiency. Benefits to metal finishers participating in the program include demonstration of environmental leadership, potentially greater operational flexibility, technical assistance in cost-effective pollution prevention, and research and technology development.

Monitoring, Research, and Surveillance. Monitoring of emissions and ambient conditions helps to keep track of progress on environmental quality improvement. Monitoring also serves as a useful tool in risk management when a community is skeptical about the effectiveness of risk management actions, by making alternatives to traditional command-and-control programs more acceptable. Research can be an important management option when lack of knowledge about the source of a problem or its impacts make a course of action unclear. EPA's cooperative effort with scientists to identify and design appropriate research projects on hormonally active agents, as directed by the U.S. Congress, is an example of research to inform future decision making. Health surveillance is an underused way to observe effects of pollution and control measures on human health. Although the incidence of cancer from environmental toxins may be so small as to be unobservable, surveillance of other health effects such as asthma or heart attacks may lead to identification of problems that need additional risk reduction.

Risk Communication. Risk communication and informed consent provide a major benefit: With more information, people can make informed choices based on preferences toward risk rather than simply follow government regulations. Risk managers are challenged to present information to consumers so as to produce more informed and accurate decisions. For example, the language on hazard warnings should advance the primary objective of informing consumers and enabling them to make better decisions. By paying attention to the economic concepts of decision making under risk, warnings can better convey the information consumers need to make sound choices regarding risks and precautions. For example, the EPA recently published a booklet showing a variety of data visualization tools (e.g., maps, color coding, icons, graphs, and simulations) to show, interpret, and communicate various elements of environmental risk (U.S. EPA, 2002c).

Summary

Risk refers to the chance of something bad happening. To minimize or avoid risks requires identifying, recognizing, and understanding them. Causality presents a difficult problem for environmental risks. It is hard to work backward, that is, to assign to an observed effect a particular cause and a person or organization responsible. It is especially hard to establish causation in the case of environmental hazards because of the considerable time delays involved.

The importance of an environmental risk depends partly on whether a person's exposure is voluntary or involuntary. People are willing to bear significant risks for a very small incremental benefit when they voluntarily make a choice. When people are exposed to involuntary risks, they typically demand a considerably higher level of safety. Perceived risks are often much different from objectively measured risks. Most people consistently overestimate the probability of a low likelihood event and underestimate the probability of a high likelihood event.

The economic value of reducing environmental risks is an important part of many environmental policy decisions. Measuring that value is complicated because markets that directly value the reductions in risk rarely exist. Economists commonly use indirect methods to value environmental risk reductions: human capital, willingness to pay (which includes wage-risk trade-offs, averting behavior, and stated preferences), and expected utility theory. Regardless of how risk reduction is valued, the use of economic efficiency in environmental risk decisions is based on recommending actions for which the marginal social benefits of risk reduction equal or exceed their marginal social costs.

Equity refers to fairness in the distribution of environmental risk. According to many advocates of environmental justice advocates, unfair distributions of the costs of environmental risk have been common in industrial societies since the early twentieth century. A growing issue is the search for policies that produce more fair distributions of risks caused by various environmental stressors. Environmental risk management is about promoting sustainability. The precautionary principle (better safe than sorry) is one mechanism for promoting sustainability where a small probability of a serious outcome is possible.

Insurance is the classic response through which private markets control environmental risks. However, because market insurance requires stringent conditions, private markets will likely provide inadequate incentives for efficiency when any of the conditions fail to occur. Because of inadequate incentives for controlling risk efficiently, governments often play a role in setting the boundaries on ethical responses to risk, in funding research on the nature of hazards, and in informing affected parties.

Because individuals find it difficult to control their personal exposure to environmental contaminants, they turn to community safeguards. Such safeguards could be government environmental regulations with the force of law or voluntarily established guidelines in the private sector based on voluntary compliance by business organizations. Risk assessment and risk management are two important means in widespread use for coping with environmental risks and for bringing public resources and regulations to bear on managing risk more efficiently, fairly, and sustainably. Important risk management institutions include liability laws, regulations, economic incentives, research, and risk communication.

Review Questions

1. Why is it hard to assign an observed adverse environmental effect to a particular cause and an organization responsible for the effect?

2. Why are perceived risks typically much different from objectively measured risks?

3. Why is the economic value of environmental risk reduction an important part of many environmental policy decisions?

4. Explain the following methods economists use to value environmental risk reductions: (1) human capital, (2) willingness to pay, (3) wage-risk trade-offs, (4) averting behavior, (5) stated preferences, and (6) expected utility theory.

5. Explain why economically efficient choices in environmental risk decisions are based on seeking actions for which the marginal social benefits of risk reduction equal or exceed their marginal social costs.

6. Why are private markets expected to provide inadequate incentives for efficiency in managing environmental risks?

7. Give four reasons why enviornmental insurance is desirable as a way to cope with environmental risks. Why might we expect too little environmental insurance to be supplied by the private sector?

8. Explain risk assessment and risk management as means of coping with environmental risks.

9. Explain the strengths and limits of the following institutions for managing environmental risks: liability laws, regulations, economic incentives, research, and risk communication.

PROBLEM

1. Cost-effective environmental risk management: Two environmental programs reduce mortality risks (save lives). One controls drinking water contaminants and the other controls air particular emissions. For each program, stricter regulations incur greater costs but by reducing exposure to hazards, each program saves more lives by reducing the risk (probability) of death. Weaker regulations cost less, but save fewer lives. The accompanying table shows that the water pollution program currently costs $2 billion per year and saves 1,100 lives by reducing mortality risks by 0.0011. The air emissions program also costs $2 billion, but saves 1,800 lives by reducing mortality risk by 0.0018.

	Increased water contaminant control				Increased particulate emission control			
Total program budget (billion dollars)	Reduced mortality risk	Total lives saved	Marginal lives saved (lives per billion dollars)	Marginal cost per life saved (billion dollars per life)	Reduced mortality risk	Total lives saved	Marginal lives saved (lives per billion dollars)	Marginal cost per life saved (billion dollars per life)
0	0	0	–	–	0	0	–	–
1	0.0006	600			0.0010	1,000		
2	**0.0011**	**1,100**			**0.0018**	**1,800**		
3	0.0015	1,500			0.0024	2,400		
4	0.0018	1,800			0.0028	2,800		
5	0.0020	2,000			0.0030	3,000		
6	0.0021	2,100			0.0030	3,000		
7	0.0022	2,200			0.0030	3,000		

a. Using the objective of total lives saved, conclude whether or not the current $4 billion is well spent by allocating $2 billion to each program. (Hint: For each program, use the equimarginal rule for cost-effectiveness to compare the marginal lives saved per added dollar.)

b. Suppose a reliable source finds that the economic value of a statistical life is $2.0 million (for a range of values of a statistical life, see Viscusi, 1993). If people believed that value, how much cost will be incurred in each program to maximize economic efficiency? Explain how you would defend your request to increase your total budget by the amount needed to produce economic efficiency.

c. Environmental justice advocates observe that low-income communities receive fewest resources to clean up environmental hazards, because their low incomes produce a low value of a statistical life. Therefore, they conclude that the use of economic efficiency for controlling environmental risks perpetuates environmental injustice. Do you agree? Can you think of an analytical approach that promotes efficiency in the use of regulatory resources while improving environmental justice?

REFERENCES

Abernathy, C. O., and W. C. Roberts. 1994. Risk assessment in the Environmental Protection Agency. *Journal of Hazardous Materials.* 39 (2): 135–142.

Alberini, A., and D. Austin. 2001. Accidents Waiting to Happen: Liability Policy and Toxic Pollution Releases. Washington, D.C.: Resources for the Future. Discussion paper 01–06.

Available from *http://ww.rff.org/documents/rff-dp-01-06.pdf* [Accessed October 10, 2004].

Ball, D., and G. Goats. 1996. Towards a coherent industrial safety and environmental risk management philosophy in the United Kingdom. *International Journal of Environment and Pollution.* 6 (4-6): 397–414.

Clemens, S. 1983. *Life on the Mississippi:* Bantam Classics.

Coase, R. H. 1960. The problem of social cost. *Journal of Law and Economics.* 3: 1–44.

Coleman, J. 2003. *Theories of tort law. Stanford Encyclopedia of Philosophy.* Available from *http://plato.stanford.edu/entries/tort-theories/* [Accessed October 9, 2004].

Dahlstrom, K., J. Skea, and W. R. Stahel. 2003. *Innovation, Insurability, and Sustainable Development: Sharing Risk Management Between Insurers and the State.* London: Blackwell.

Doherty, N. A., and A. Muermann. 2002. *Brokers and the insurance of non-verifiable losses.* Available from *http://fic.wharton.upenn.edu/fic/papers/04/muermann%3bdoherty.pdf* [Accessed October 9, 2004].

Faure, M. 2001. *Environmental damage insurance in theory and practice.* University College London Centre for Law and the Environment. Available from *http://www.cserge.ucl.ac.uk/faure.pdf* [Accessed October 9, 2004].

Gardashuk, T. V. 2003. *Environmental justice and environmental policy in the modern world.* National Ecological Center of Ukraine. Available from *http://www.fondazionelanza.it/epa/abstract/gardashuk_full.pdf [Accessed May 25, 2005]*

Goldman, B. 2000. An environmental justice paradigm for risk assessment. *Human and Ecological Risk Assessment.* 6 (4): 541–548.

Green, E., S. Short, and M. Taylor. 2000. *Guidelines for environmental risk assessment and management.* Available from *http://www.defra.gov.uk/environment/risk/eramguide/index.htm* [Accessed October 9, 2004].

Herodotus. 1995. *Book 2.* Available from *http://www.perseus.tufts.edu/greekscience/hdtbk2.html* [Accessed October 10, 2004].

Janssen, J. 2001. Risk Management of Investments in Joint Implementation and Clean Development Mechanisms Projects. Ph.D. dissertation, Universitat St. Gallen, Switzerland.

Jantunen, M., and K. Finland. 2002. The Role of Exposure in the Chain of Events from Sources to Health Outcomes: Dermal and Aggregate Exposures. In *Proceedings of the International Conference on Occupational and Environmental Exposures of Skin to Chemicals: Science & Policy.* Available from *http://www.cdc.gov/niosh/topics/skin/conference/s1t1.html* [Accessed October 9, 2004].

Pearce, D. 2000. *The insurance industry and the conservation of biological diversity: An analysis of the prospects for market creation.* OECD Working Group on Economic Aspects of Biological Diversity. Available from *http://www.cserge.ucl.ac.uk/oecd_insurance.pdf* [Accessed October 9, 2004].

Rodricks, J. V. 1994. Risk assessment, the environment, and public health. *Environmental Health Perspectives.* 102 (3): 258–264.

Skea, J. 2004. *Insurance, technological risk and emerging science and technology policies.* Available from *http://www.psi.org.uk/research/project.asp?project_id=11* [Accessed October 9, 2004].

Slater, D., and H. Jones. 1998. Risk assessment and environmental protection. *Risk Decision and Policy.* 3 (1): 53–66.

Slovic, P. 1999. Trust, emotion, sex, politics, and science: Surveying the risk-assessment battlefield. *Risk Analysis.* 19 (4): 689–701.

United Nations Economic and Social Commission for Asia and the Pacific. 2000. *Ministerial Conference on Environment and Development in Asia and the Pacific.* Available from *http://www.unescap.org/mced2000/so5.htm* [Accessed May 25, 2005].

U.S. v. Carroll Towing Co. 159 F2d 169, 174 (2d Cir 1947).

U.S. Council of Economic Advisors. 1996. *Economic Report of the President.* Available from *www.gpoaccess.gov/usbudget/fy96/pdf/econ_rpt.pdf* [Accessed October 10, 2004].

U.S. Environmental Protection Agency. 1997. *Presidential Commission on Risk Assessment and Risk Management.* National Center for Environmental Assessment. Available from *http://cfpub.epa.gov/ncea/cfm/pcrarm.cfm* [Accessed October 10, 2004].

———. 2002a. *Common Sense Initiative.* Available from *http://www.epa.gov/history/topics/reinvent/10.htm* [Accessed October 9, 2004].

———. 2002b. *Learn about Project XL.* Available from *http://www.epa.gov/projectxl/learnabt.htm* [Accessed October 9, 2004].

———. 2002c. *Risk communication in action: Environmental case studies.* Available from *http://www.epa.gov/ord/nrmrl/pubs/625r02011/625r02011.html* [Accessed October 9, 2004].

———. 2003. *Guidelines for preparing economic analyses.* National Center for Environmental Economics. Available from *http://yosemite.epa.gov/ee/epa/eed.nsf/webpages/guidelines.html* [Accessed October 9, 2004].

———. 2004. *Historical background.* National Center for Environmental Economics. Available from *http://yosemite.epa.gov/ee/epalib/ee222.nsf/0/6c836c88fc105951852567a500651399?opendocument* [Accessed October 9, 2004].

U.S. National Academy of Sciences. 1983. *Risk Assessment in the Federal Government: Managing the Process.* Washington, D.C.: National Academy Press.

———. 1994. *Science and Judgment in Risk Assessment.* Washington, D.C.: National Academy Press.

Viscusi, K. 1993. The value of risks to life and health. *Journal of Economic Literature.* 31 (4): 1912–1946.

Viscusi, W. K. 1992. *Fatal Trade-offs.* Oxford, UK: Oxford University Press.

Wong, J., and M. Dutschke. 2003. Can Permanence Be Insured? Consideration of Some Technical and Practical Issues of Insuring Carbon Credits from Afforestation and Reforestation. Hamburg Institute of International Economics. Discussion paper 235. Available from *http://www.hwwa.de/hwwa_engl.html* [Accessed October 9, 2004].

World Health Organization. 2000. *DDT use in malaria prevention and control.* Available from *http://www.who.int/inf-pr-2000/en/note2000-15.html* [Accessed October 10, 2004].

Chapter 18

ENVIRONMENTAL JUSTICE

Outline

What Is Environmental Justice? 527
Measuring Environmental Justice 529
Economic Theory and Policy Analysis 530
Applications to Achieve Environmental Justice 534
Evidence of Environmental Justice 540

Objectives

After studying this chapter, you should be able to

- Describe and define environmental justice
- Analyze economic concepts that guide the search for environmental justice
- Apply economic concepts in understanding and improving environmental justice
- Interpret evidence on extent of environmental justice

Introduction

In 1978, a young assistant professor of sociology at Texas Southern University, Dr. Robert D. Bullard, fresh out of his graduate studies, was asked to study the history and pattern of placement of waste facilities in Houston. The study was prompted by a lawsuit filed by African American residents of a northeast Houston suburban neighborhood who wished to block a garbage dump from being put in their community. Bullard completed his study in 1979 and published it in 1983 as "Solid Waste Sites and the Black Houston Community" (Bullard, 1983). In it, he documented Houston's history of placing waste facilities in largely African American communities. All five city-owned garbage dumps, six of the city's eight incinerators, and three of the four privately owned landfills were sited in African American neighborhoods. Later, Bullard helped organize the first national People of Color Environmental Leadership Summit, which was held in Washington, D.C., on October 17, 1991 (see Example 18–1). From this summit emerged 17 principles of environmental justice that laid the framework for how environmental justice organizers would work together and how they would relate to other national environmental groups.

Due in large part to Bullard's work, the Clinton administration raised environmental justice to a high level of international visibility by expanding the U.S. EPA's Office of Environmental Justice and by appointing the National Environmental Justice Advisory Council. On February 11, 1994, President Clinton signed Executive Order 12898, which states that federal activities should not produce disproportionately high and adverse impacts on minority and low-income populations. It also says that the U.S. government will not deny these populations the health, safety, and other benefits of environmental improvements or exclude them from participation in programs that would produce those benefits. The executive order requires federal agencies to include environmental justice issues in their research and data-gathering activities. The order requires that public documents be easily accessible and understood through measures such as language translations.

In 2002, an 11-year follow-up conference was held, Leadership Summit II. More than 1,200 delegates from grassroots and community-

Example 18–1
Principles of Environmental Justice

The following statement was adopted at the People of Color Environmental Leadership Summit held on October 17, 1991, in Washington, D.C.:

> We the people of color, gathered together at this multinational People of Color Environmental Leadership Summit, to begin to build a national and international movement of all peoples of color to fight the destruction and taking of our lands and communities, do hereby re-establish our spiritual interdependence to the sacredness of our Mother Earth; to respect and celebrate each of our cultures, languages and beliefs about the natural world and our roles in healing ourselves; to insure environmental justice; to promote economic alternatives which would contribute to the development of environmentally safe livelihoods; and, to secure our political, economic and cultural liberation that has been denied for over 500 years of colonization and oppression, resulting in the poisoning of our communities and land and the genocide of our peoples, do affirm and adopt these Principles of Environmental Justice.

To review the 17 principles, go to the EJNet Web site.

Source: EJNet Web Resources for Environmental Justice Activists. 2003. *Principles of environmental justice.* Available from *http://www.ejnet.org/ej/principles.html* [Accessed October 10, 2004].

based organizations, faith-based groups, organized labor, civil rights, youth, and academic institutions made their way to the U.S. capital. The environmental justice issue had come of age, and economics has much to offer in the design, implementation, and evaluation of actions affecting environmental justice.

■ What Is Environmental Justice?

In 1997, the U.S. Environmental Protection Agency established the following definition of environmental justice, a definition that continues to guide U.S. federal environmental policy (U.S. EPA, 2003).

> Environmental Justice is the fair treatment and meaningful involvement of all people regardless of race, color, national origin, or income with respect to the development, implementation, and enforcement of environmental laws, regulations, and policies. Fair treatment means that no group of people, including a racial, ethnic, or a socioeconomic group, should bear a disproportionate share of the negative environmental consequences resulting from industrial, municipal, and commercial operations or the execution of federal, state, local, and tribal programs and policies.

These noble principles led to many questions about what kind of environmental policies are considered fair and just.

In the search for ways to evaluate the extent to which environmental justice occurs, let us start with the idea of justice itself. John Rawls' famous book, *A Theory of Justice* (1971), has had a major impact on how we think about justice. Rawls' lifelong interest in justice developed out of his concern, while a soldier during World War II, with the question of why there is injustice in the world. As he started his inquiry as to whether a just society is possible, Rawls rejected utilitarianism. *Utilitarianism* says that laws and other public actions are just when they promote the greatest possible algebraic sum of happiness, added up over all members of society, even if the worst-off members of society suffer. Issues of fairness have always been a problem for utilitarian philosophy, which accepts sacrifices of individual and minority welfare for the sake of gains by the majority. To develop an alternative, Rawls turned to the social contract traditions of the philosophers Locke, Rousseau, and Kant. The social contract views society as a cooperative venture from which everyone should expect to benefit. To benefit, each person requires access to opportunity. One's opportunities for success in life are heavily shaped by luck. In turn, each person's luck is shaped in part by institutions designed by those who participate in the political, social, and economic systems.

From his research, Rawls derived two key principles to define a just society:

1. Each individual has a right to the greatest liberty compatible with a like liberty for all (the liberty principle).
2. Social and economic inequalities are justified only if they benefit everybody (the difference principle).

According to Rawls, the first criterion about basic liberties takes precedence over the second. That is, gains accomplished when judged by the second principle are not considered just if they cause losses measured by the first principle. Rawls' difference principle is especially important for the study of environmental justice. It holds that differences in wealth, status, power, and other desirable resources are just only if they create a system of market forces and incentives whose productivity gains make everybody better off than they would be otherwise, especially the worst-off members of society. For example, paying high salaries to doctors is just if everybody's health care improves, especially people who have the least access to health care.[1] In effect, the difference principle says that if you want to assess how just a society is, look at how its worst-off members are doing.

Based on Rawls' two standards for justice, how should institutions be organized to produce the just society? Rawls imagined an "original position" in which we exist as self-interested individuals who have no knowledge of our standing; that is, each of us is unaware of the socially significant facts about ourselves. Each of us has no knowledge of our generation, inherited characteristics (e.g., race, gender, beauty), wealth, access to environmental resources, or of the position we would occupy in any jointly chosen social structure. Furthermore, each of us is free of any biases acquired by living in any special set of circumstances. From behind Rawls' "veil of ignorance," we come together and select rules to live by, without knowing whether we will be prosperous or destitute in the society governed by those rules. Rawls concluded that as a result of not knowing our position in society, each of us is driven by an equal concern for the fate of all. Our priorities are not only preserving personal and political liberty but also reducing economic inequality. For fear of finding ourselves at the bottom of the pile, we reject laws and policies based on the utilitarian standard that maximize average *utility*. Because of the veil of ignorance, these are no incentives to design self-serving programs; instead, objectivity characterizes the debates leading to establishment of a just social contract. Participants in the original position are forced, even though self-centered, into designing institutions that produce a just society. Any principles that emerge are bound to be just because all incentives for any person to contribute to an unjust system are neutralized.[2]

Rawls believed that the incentives faced by all *from* the original position (i.e., *behind* the veil of ignorance) lead to equal justice and equal opportunity. Society permits unequal distributions of resources only if such distributions are to the advantage of the least favored. Desirable principles of environmental justice are those we would all agree on if each of us were in a position to rationally, impartially, and freely deliberate just arrangements.

---

[1] Higher payments made to any person pass Rawls' test of justice if those payments have the effect of bringing forth that person's added output so that the fruits of that output accrue as welfare gains for the worst off. If high payments to auto executives cause more and better cars to be produced so that the poorest have access to better transportation than they would have without those high payments, then the high payments pass Rawls' justice standard.

[2] This discussion assumes that human justice is the sole purpose of environmental justice. In fact, many environmentalists believe that nonhuman environmental entities have moral standing; that is, independent of the benefits to humans, people have a moral responsibility to treat the environment with respect.

Practically, it may be difficult to design an environmentally just society. Still, policies can be enacted that improve environmental justice compared with the status quo. So it is really *improvements* in environmental justice that we wish to achieve.[3] Rawls' principles underlie the definition established by the U.S. EPA, which opened the door to U.S. federal responsibility for environmental justice.

■ MEASURING ENVIRONMENTAL JUSTICE

Trying to determine the presence or absence of environmental justice involves the problem of measuring environmental risk across groups of people. Physical measures are often chosen to characterize risk such as pollution or hazard, for example, ambient concentrations of a pollutant or nearness to a toxic landfill. At least four physical measures have been used: (1) emissions, (2) ambient concentrations, (3) exposure, and (4) health risk. Each measure can lead to different conclusions about a society's level of environmental justice.

1. A region's emissions equal actual emissions, minus emission exports to other regions, plus imports from elsewhere. Emissions are a flow; that is, they are measured as an output per unit of time. Income is another example of a flow that is measured as an amount earned per hour, per day, or per month. Emissions do not exist at a point in time. To know something about the path from any emission's source to its impact on people, we need to know about its dispersion. Dispersion determines what percentage of the regional population is affected by the emissions. Emission releases and dispersion are important to track because any region can produce significant pollutants yet not impose risks to local people if emissions are dispersed or exported. For example, people who live close to an air emission source may have little health risk if prevailing winds take the emissions in the opposite direction. Likewise, a region importing pollution could face considerable health risks even though that region has no emission sources. Still, even after imports and exports are accounted for, emissions fail to account for ambient concentrations, and they ignore characteristics of affected populations that could increase or reduce the risk of harm.

2. Ambient concentration of a pollutant indicates concentration at a single instant in air, water, or land. Ambient concentration is a stock just as the amount of money in your savings account or the water in a reservoir are stocks, because they exist at a single point in time. Ambient concentrations, measured as parts per million or milligrams per cubic meter, are a better indicator of environmental risk than emissions because they have greater measurable health implications. Ambient concentrations quantify the density of what people will be exposed to in the medium.

3. Exposure connects ambient concentration to human behavior; it allows for the behavior of the population that is at risk. For example, if people with low income tend to work outdoors and people with high incomes work indoors, lower income groups will be exposed to more outdoor pollution, even if the ambient concentration of an outdoor pollutant is the same for both groups. Income also can be used to insulate against ambient pollution. If higher income groups buy indoor air filters, then their exposure to pollution will be less than that of people who cannot afford the filters. Exposure is a better measure than ambient concentration because it allows for differences in behavior among groups of people, such as timing or location of employment. Although exposure comparisons among

[3] These improvements may be expensive, and their high price may have major implications for other forms of justice. For example, declaring that everybody has a right to clean water and a healthy environment appears to pass the test of environmental justice. But a majority of fully informed voters might vote against exercising the right to that environmental justice. And it would be rational (and possibly just) to vote against it if the cost of achieving that justice, through reallocated public budgets, sacrificed programs that produce an acceptable level of nutrition, adequate medical care, or decent housing for people who have limited access to those resources.

populations can be a good indicator of potential environmental justice, typically few comparisons are made because the data are expensive to find or simply do not exist.

4. Health risk is an improvement over exposure as an indicator of environmental risk because it allows for health characteristics of the people exposed to a risk. A person's health risk depends on exposure plus personal characteristics such as nutritional habits, overall preexisting health, and ability to afford health-improvement resources (e.g., specialized physicians, fitness centers, and athletic clubs). For equal levels of exposure, the poor are likely to be at greater health risk than the rich.

How environmental risk is measured has a great impact on conclusions about the presence or absence of environmental justice. Some measure of exposure or health risk is the most accurate way to judge the risk of these hazards, but usually ambient concentrations—or worse emissions—are the best means available. An indicator like emissions will likely produce seriously misleading implications about the environmental justice of policies.

Another important element in reaching conclusions on the presence and extent of environmental justice is the geographic scale of analysis, which can range from large units (e.g., entire regions) to small ones (e.g., individual addresses). The main problem with analyzing very large numbers of very small units is the cost of getting the data. In addition, there may be too little variation in pollution from one unit to the next to draw conclusions about its importance. A problem with analyzing small numbers of large units is that much important detail is lost in averaging; more localized inequities will be missed. The growing use of geographic information systems is permitting better analysis of data of all sizes than was previously possible.

Economic Theory and Policy Analysis

Economic theory provides insights into why specific locations are selected for hazardous waste treatment, storage, and disposal facilities (TSDFs). A population's economic vulnerability may play a role in influencing locations. No potential receiving community finds TSDFs inherently desirable. Although there may be a best location for a TSDF from a national or global view, the search is really for a "least bad" location. Virtually any fully informed community will express the "not in my backyard" (NIMBY) sentiment, unless there is some kind of offsetting compensation scheme. Understanding the economic forces underlying the TSDF site location requires recognition of the incentives facing the owners of the facility and the receiving community.

Facility owners or developers of TSDFs typically want to maximize their net income, so economic theory predicts their behavior: They seek a site location that allows them to process or dispose of hazardous wastes at least cost to them. Since these activities usually require much land, owners look for locations with low land costs. A second cost element is the cost to the business of having the host community regulate its hazardous activities, for example prescribing methods of shipping waste to the site and setting up various safety regulations for handling hazardous materials. The owners look for communities that have low costs associated with these regulations. A third element of cost is the risk of lawsuits: Owners want to find site locations where the character of the local community poses a small risk of those costs. In short, facility developers of TSDFs are like any other business owner in that they search for the least-cost way to conduct their business, for a given level of revenue.

TSDF-receiving communities face very different incentives from those of TSDF owners or developers. Any potential community wants to maximize whatever net benefits it could receive from being chosen as a site location. To do this, it wants to be sure that the site is safe for employees as well as for local residents. But no community would accept a TSDF without added compensation, such as increased tax revenues, new jobs, new schools, added public services, and the like.

Economic theory predicts that low-income communities will be the most likely places to attract a TSDF, because of incentives to facility owners as well as to the community itself. Land prices are typically lowest in low-income communities because there are few income-generating activities to bid up land prices. Environmental regulations and the risk of lawsuits are smallest because low-income communities have the greatest need for added economic activity. Because poor communities have the least income, they have the greatest need for more: They are likely to place the smallest demands for compensation by way of jobs, tax revenues, and community services. For these reasons, lower income communities (in cities, counties, nations) may expect to see more TSDF facilities. In fact, this economic logic lies behind a famous World Bank memo, leaked in 1991, written by Lawrence Summers, the bank's chief economist at that time. That following memo argues that a policy of exporting pollution to third-world countries is economically beneficial and rational. (Summers, 1991):

> **1.** The measurements of the costs of health-impairing pollution depends on the foregone earnings from increased morbidity and mortality. From this point of view a given amount of health-impairing pollution should be done in the country with the lowest cost, which will be the country with the lowest wages. I think the economic logic behind dumping a load of toxic waste in the lowest wage country is impeccable and we should face up to that.
>
> **2.** The costs of pollution are likely to be nonlinear as the initial increments of pollution probably have very low cost. I've always thought that underpopulated countries in Africa are vastly *under*-polluted, their air quality is probably vastly inefficiently low compared to Los Angeles or Mexico City. Only the lamentable facts that so much pollution is generated by nontradeable industries (transport, electrical generation) and that the unit transport costs of solid waste are so high prevent world welfare enhancing trade in air pollution and waste.
>
> **3.** The demand for a clean environment for aesthetic and health reasons is likely to have very high income elasticity. The concern over an agent that causes a one in a million change in the odds of prostate cancer is obviously going to be much higher in a country where people survive to get prostate cancer than in a country where under 5 mortality is 200 per thousand. Also, much of the concern over industrial atmosphere discharge is about visibility impairing particulates. These discharges may have very little direct health impact. Clearly, trade in goods that embody aesthetic pollution concerns could be welfare enhancing. While production is mobile the consumption of pretty air is a nontradeable.

Summers said that his memo was intended to sharpen the debate on important issues by intentionally taking as narrow-minded an economic perspective as possible. What Summers described as that economic perspective is an example of the economic efficiency argument described throughout this book. The economic efficiency position is reviewed next; you be the judge of whether it is morally acceptable.

Figure 18–1 shows that if the incremental damages (marginal costs) from toxics or other pollutants increase with increasing levels, then a given amount of added pollution in a clean part of

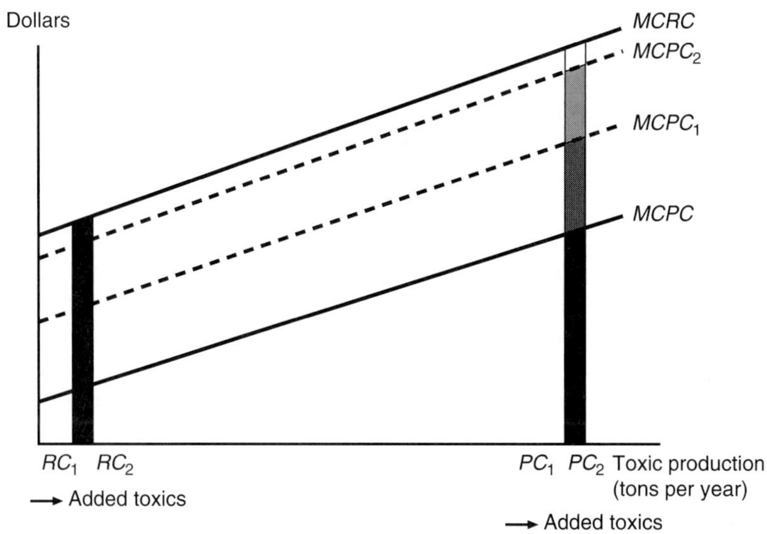

Figure 18–1 Economic efficiency and economic justice in locating toxic sites

the world (or clean neighborhood) will cause less economic damage than comparable incremental emissions in an already dirty community. This is the efficiency logic behind the statement of Summers about underpopulated countries in Africa being underpolluted compared with cities like Los Angeles or Mexico City.

When Summers said that the demand for a clean environment for aesthetic and health reasons is likely to have very high income elasticity,[4] he was appealing to the efficiency logic: People value a clean environment less in poorer communities. Concerns over the odds of getting prostate cancer are much higher in countries where people survive to get prostrate cancer than in a country where more people die at a young age. Summers' memo indicates that exporting waste to poor communities reduces physical damages to the environment and produces much lower damage measured as the willingness of the community to accept payment in exchange for wastes.

Figure 18–1 is a graphical treatment of Summers' concept applied to the economic efficiency question of where to place a new TSDF facility. The top solid line represents the marginal cost associated with increasing toxic production in a rich community (*MCRC*). That marginal cost schedule is high because incomes are high, real estate prices are high, and people who live in that community are likely to have good information on the health and property impacts of the proposal TSDF facility. Simply put, the top line says it would take a very high payment to compensate those people for the damages associated with placing the TSDF facility in their community. The siting would increase toxics from the current low level of RC_1 to RC_2, increasing the rich community's damages by the area in the black vertical column. That area is the amount of payment required to compensate the rich for damages.

The bottom solid line shows a much lower marginal cost associated with increasing toxics in the poor community (*MCPC*). Costs are lower because in the poor community there are lower incomes, higher unemployment rates, fewer public facilities, poorer access to running water, and little specialized knowledge of the environmental impact of the TSDF. So the poor community has a much smaller willingness to accept payment in exchange for accepting the facility in its backyard. The poor community's added damages from locating a TSDF facility equals the area of the smaller all-black-shaded column bound by toxic levels PC_1 and PC_2 and the bottom marginal cost schedule (*MCPC*). Summers' pure efficiency argument states that society as a whole suffers smaller damages if the TSDF is located in the poorer community. The poor community would suffer damages equal to the solid area beneath the bottom solid line (*MCPC*) between PC_1 and PC_2; the rich community would suffer damages equal to the larger solid area beneath the top solid line (*MCRC*) between points RC_1 and RC_2.

To analyze how environmental justice is connected with the efficiency argument just described, let us pose a compelling question: Are the top and bottom marginal cost schedules in Figure 18–1 set in stone or can they be affected by policy? In other words, what kinds of policies could be enacted to make the two marginal cost schedules move closer to each other so that society is not always best served by placing new TSDFs in poor communities? Start with income. Other things being equal, if policies are enacted that give the poor community better job skills, higher incomes, and a better tax base, the bottom marginal cost curve begins to shift upward. Higher incomes shift it up until it reaches the bottom dashed line at $MCPC_1$. Next, add improved infrastructure, public facilities, and technical environmental impact knowledge, as well as the time to acquire these, and the curve shifts up more—to the top dashed line $MCPC_2$ which is close to or equal to the rich community's marginal cost curve *MCRC*.

So the issue of environmental justice is this: As policies are enacted to improve the lives of vulnerable people (and nations), the marginal damages from TSDF sites increases. Notice in Figure 18–1 that when the poor community's marginal cost schedule reaches $MCPC_2$, it is no longer more economically efficient to locate the marginal site in that community. The poor community's damages produced by the site are now equal to the black area plus the two lighter areas in the right-hand column. The preexisting level of justice in a society has major implications for

[4] Recall that the income elasticity is defined as the percentage change in demand resulting from a 1% change in income.

Example 18–2
Environmental Justice in the United States

In 1993, Bullard and Wright issued a call to action for improved environmental justice in the United States. Their call was based on analysis of who gets what environmental justice, why, and in what amounts. They found that working-class people, people of color, and people with low incomes were exposed to increased environmental hazards in their work, where they live, and in places where their children play. The authors believed that this situation has less to do with science and more to do with institutional arrangements. They found that locally unwanted land uses and health risks were not randomly distributed but were located partly because the people of the community offered the least resistance.

Bullard and Wright found major inequities in exposure to environmental hazards. Race and income were related to the distribution of air pollution, municipal landfills, incinerators, and pesticide exposure. For example, they found that in the Los Angeles air basin, more than 71% of African Americans and 50% of Latinos lived in areas with the most polluted air, whereas only 34% of whites lived in these places. The dirtiest zip code in California (90058), a square mile section of Los Angeles County, was filled with waste dumps, smokestacks and wastewater pipes from polluting industries. In 1989, 18 industrial companies discharged 33 million pounds of chemicals into the environment. Population in this neighborhood was 59% African American and 38% Latino. Bullard and Wright also found injustice on American farms. Most U.S. migrant and seasonal farmworkers are Latino. More than 40% of all U.S. farmworkers have been sprayed or exposed to drifting spray. Despite federal legislation, migrant farmworkers were rarely informed about what hazardous substances are used at their workplace or how to find protection from those hazards.

Health risk inequities were also found. For example, African American children had greater rates of lead poisoning than white children. Nearly half of African American children living in cities experienced increased lead levels compared with 16% of white children living in the same places. Even at similar income levels, middle-class African American children were three times more likely to be poisoned by lead than middle-class white children. Bullard and Wright found similar evidence of environmental injustice in West Dallas; Louisiana's Cancer Alley (the name given by the media to one of the most heavily industrialized and polluted areas in the United States, along the Mississippi River between Baton Rouge and New Orleans); East St. Louis, Illinois; the Carver Terrace subdivision in Texarkana, Texas; and Chicago's south side.

Overall, Bullard and Wright concluded that the nation's poorest are being poisoned at an alarming rate, and many of the families have little or no access to health care. In their call to action for greater environmental justice for all Americans, the authors made several national policy recommendations: soliciting input from affected populations; developing pollution prevention models, data sets, and government agencies that incorporate an environmental justice framework; obtaining community information through development of health databases, and educating family physicians in small towns to recognize signs and symptoms of environmental injustices.

Source: Bullard, R. D., and B. H. Wright. 1993. Environmental justice for all: Community perspectives on health and research needs. *Toxicology and Industrial Health.* 9 (5): 821–841.

the design of economically efficient policies. When a society already possesses a just distribution of all resources, as well as TSDFs, it is unlikely to be economically efficient to add TSDF sites to the most vulnerable populations. That is, it is only economically efficient to add more environmental insults to the most vulnerable populations if environmental *injustice* already exists. You can see the vicious circle: Injustice combined with an efficiency standard for environmental policy choices perpetuates and expands injustice. See Example 18–2.

Now look at what happens when a low-income community that lacks environmental justice has a low willingness to accept payment in compensation for accepting a TSDF. That community's acceptance of environmental damages produced by placement of a TSDF can be bought with lower tax revenues, wages, and salaries compared with what would be necessary in high-income communities. Thus, Kuehn (1997) found that the hazards faced by people of color and people with low incomes and other vulnerabilities are likely to be given inadequate attention by analytical methods like benefit-cost analysis:

> If minority and poorer communities suffer from disproportionate pollution because they have the fewest resources and can offer the least resistance to proposals to impose risks

on their communities, an expert-driven, resource-intensive process that is subject to the influences of the individual risk assessor and the power and money of the risk producer will still leave these communities at the mercy of government decision makers and risk producers (p. 272).

Does economic theory tell us that location of TSDFs in economically disadvantaged communities is an obvious case of environmental injustice? Maybe, but not necessarily. For example, if placing a TSDF depresses already low land prices, home ownership may become more affordable to families with low incomes. Some families, even with full information about the TSDF, may voluntarily move into the area, accepting the environmental hazard in exchange for affordable home ownership and improved public services. Some argue that any place where a TSDF is located will experience a reduction in real estate prices. That price reduction can be expected to bring in more lower income people who otherwise could not have afforded the land. So seeing lower income people located nearer TSDFs poses the difficult statistical question of attempting to separate overt discrimination based on income and its correlates such as race, from the ordinary dynamics of land markets. More research needs to be conducted by economists and others[5] about the structure of land and real estate markets to obtain a better understanding of situations in which certain groups of people are unfairly exposed to environmental risks. See Example 18–3.

Applications to Achieve Environmental Justice

Environmental justice focuses on how the benefits and costs of environmental programs are distributed among society's members. From the view of policy design, justice is important because democratically elected politicians resist supporting policies that are seen as unjust.

Benefits and Costs of Controlling Environmental Degradation

One way to design more environmentally just programs is to compare the distribution and level of benefits and costs of various environmental proposals. We want to see how the benefits and costs of a particular environmental policy proposal (net benefits) are distributed among all society's groups. A more environmentally just program tries to spread out the benefits and costs evenly among society, which could lead to a more environmentally just society. Such a society will enact policies that assign higher costs to groups that receive higher benefits.

The hypothetical numbers in Table 18–1 illustrate the challenges facing environmental policymakers. Think of these numbers as the benefits (summarized as net benefits) and costs of eight competing programs for controlling a region's air pollution, water pollution, or toxic wastes.[6] The first three columns are the program, total regional population, and the (present value of) net benefits, respectively. Suppose that group *Y* and group *Z* refer to a more and less vulnerable population, respectively. Vulnerability could be measured, for example, by gender, race, income level, religious preference, or beauty. Vulnerability could also occur by location, such as nation. It could even refer to position in time: For a depletable resource policy, the vulnerable population could be a future generation that risks having nothing of the resource remaining (see Chapter 19). See Example 18–4.

[5] Various noneconomic forces are also at work that determine observed levels of environmental justice. A rather cynical view is stated by Buckingham-Hatfield and Evans (1996): People with economic and political power tend to secure the best environments and are usually prepared to tolerate environmental exploitation and degradation so long as it is confined to localities they do not need, value, or frequent.

[6] A label need not be limited to a single program; it could refer to a mix of environmental programs. For example, label *A* could refer to a program of establishing three specified wilderness areas, regulating vehicle emissions in specific air shed, and enforcing existing water quality regulations. Label *B* could refer to a program mix that establishes one wilderness area, has no regulations on vehicle emissions, and establishes and enforces new water quality regulations. The labels could also refer to programs that are not explicitly environmental in their intent but produce significant negative or positive environmental impacts. Energy, transportation, food, and water policies typically produce major environmental impacts.

Example 18–3

Environmental Justice in Drinking Water Regulations

In connection with drinking water, environmental justice is all about environmental protection for poor, minority communities. More than a third of all U.S. households with incomes below $10,000 cannot meet at least one basic need. In 1995, 20% of Americans lived in households that had at least one difficulty in meeting a basic need such as safe drinking water.

Higher prices for drinking water have resulted from increased costs associated with meeting new safety regulations, which place added economic burdens on disadvantaged communities. This statement assumes that vulnerable communities who receive safer drinking water are obliged to pay for the added cost of the water. In fact, no such thing need occur. The costs of improved drinking water can be borne by whomever the political system says should pay. Increased public health benefits from the added safety regulations may be offset in a disadvantaged community by the hard choices households must make to pay higher water bills. Paradoxically, substituting money from other needs to pay for the cost of complying with a new drinking water regulation could reduce already low health in these communities. A decent supply of safe drinking water is an essential requirement, however, to sustain welfare anywhere. Also, safe drinking water attracts business, industry, jobs, and tax revenues.

The most important question facing policymakers dealing with safe drinking water is this: Are the benefits from setting tougher regulations to produce more healthful drinking water worth the added monthly water bills faced by disadvantaged households who must pay those increased bills? To put it more constructively, the challenge in producing improved environmental justice through safer drinking water is to design regulations for which the benefits to the disadvantaged communities exceed the cost of higher water bills. In fact, implementing environmental justice through guaranteeing that a group's benefits exceed its costs may require that the most vulnerable communities receive large enough subsidy so that they pay no costs whatsoever for their safer water.

Safe drinking water remains inaccessible for 1.1 billion people in the world. Another important policy question is what is the best way to get safe drinking water to poor people around the world. A classic method, based on the assumption that safe drinking water is simply too expensive for the average family, is for the government to build and run water treatment facilities–charging nothing for water but getting water to almost noone. This method may pass the test of environmental justice but deliver little safe drinking water. Another method is to let the private sector build and run water treatment facilities, charging the full cost of water, which would require very poor people to pay up to 25% of their income for safe water. In either case, how does one charge a high enough price for water to produce adequate revenue to pay for a water utility without keeping water from the poor who cannot afford the price of water? Some studies have shown that even if a family makes only a dollar a day, they may be willing to pay up to a quarter of that for safe drinking water. This finding suggests that private responsibility for safe drinking water can work while being environmentally just.

Source: Pontius, F. 2000. Environmental justice and drinking water regulations. *Journal of American Water Works Association.* 92 (3): 14.

Table 18–1 Environmental justice: Net benefits of controlling environmental degradation by program and group

			Group *Y*		Group *Z*			
Program	Population	Net benefits (dollars)	Percentage of population	Percentage of net benefits	Percentage of population	Percentage of net benefits	Justice[a]	Efficiency[b]
A	100	1,000	20	50	80	50	U	I
B	100	1,000	20	20	80	80	J	I
C	100	1,000	20	10	80	90	U	I
D	100	1,000	20	−10	80	110	U	I
E	100	4,000	20	50	80	50	U	E
F	100	4,000	20	20	80	80	J	E
G	100	4,000	20	10	80	90	U	E
H	100	4,000	20	−10	80	110	U	E

[a] *U* = unjust; *J* = just.

[b] *I* = inefficient; *E* = efficient.

Example 18–4
Siting Waste Treatment Facilities and Environmental Justice

Typically people recognize the need for developing a new waste treatment facility, because waste must be treated. But no community wants the facility near it. So how should a site be chosen and how can principles of environmental justice help in making a choice?

Minehart and Neeman proposed an auction-like method for selecting one site among an array of competing possibilities. Principles of economic efficiency say that a site should meet two standards: (1) construction and operation costs should be minimized consistent with the requirements of the facility and (2) the welfare loss (disutility) to the host community should be minimized. The second standard causes the most controversy, for selecting a site that minimizes disutility is subjective, hard to measure, and brings great emotional outpouring. The subjectivity of each potential host community's welfare loss allows each to manipulate other communities' perceptions. For example, any community might be motivated to overstate its suffering by a factor of 10, 100, or more if such an overstatement can reduce its chances of being the host.

In the authors' proposed system, each of several communities submits a bid equal to its total cost for hosting the treatment, storage, and disposal facility (TSDF), equal to construction costs plus disutility. The cost of building and operating the facility at each site is presumed to be known. The facility is located in the community with the lowest total cost. All potential host communities who entered the pool pay transfer payments that compensate the chosen host. That is, the communities share a payment to the host community equal to the second lowest total cost, allocated proportionally according to some fair standard, such as the quantity of waste each community generates.

This proposal of Minehart and Neeman satisfies several important standards: First, it is easy to apply to many real-world siting situations. Second, the siting method balances the budget, because it does not require outside subsidy and it does not produce a surplus. Third, it also encourages voluntary participation by all communities, which have incentives to abide by the outcome of the procedure even after the bids become public information. Most importantly, the community that incurs the least total cost ends up hosting the facility.

Still, the proposed method may raise issues of environmental justice issues. It is possible that poor and minority communities may have a lower disutility for hosting a TSDF because they have more to gain from the tax and job benefits the facility could attract. Economic efficiency may result in poor communities hosting TSDFs more often than richer communities. If environmental justice is a separate issue from economic efficiency, it may be necessary to give up some efficiency for a more equal distribution of TSDFs.

Source: Minehart, D., and Z. Neeman. 2002. Effective siting of waste treatment facilities. *Journal of Environmental Economics and Management.* 43: 303–324.

Programs *A, B, C,* and *D* have the same total net benefits of 1,000, but the distribution of those net benefits is just (*J*) for program *B*, but unjust for programs *A, C,* and *D* (*U*).[7] Program *B* is just, for under that program, the incidence of group Y's net benefits is equal to the percentage of base population (20%). Programs *A, C,* and *D* can be considered unjust because their population proportions do not equal their net benefits proportions. The goal of environmental justice can conflict, however, with the goal of economic efficiency. Programs *E, F, G,* and *H* produce a net much higher benefit of $4,000, compared with the much lower net benefit of $1,000 so programs *A* through *D* have an inefficient label (*I*) in the table, whereas programs *E* through *H* receive an efficient (E) label. Program *F* gets a yes vote for environmental justice, and programs *E, G,* and *H* fail on justice for the same reasons that applied to programs *A* through *D*.

[7] Justice is defined in the sense of Rawls' use of the term: A just program is supported by a vote of people located in the original position, behind a veil of ignorance. If nobody knew whether they were a member of group *Y* or group *Z*, the environmentally just program (*J*) would receive voter support, while the unjust one (*U*) would lose it. One could argue that program *A* is just, for under program *A* total net benefits are distributed in higher percentage to the more vulnerable group *Y* (50%) than is that group's percentage of the total population (20%).

Example 18–5
Environmental Justice in Nepal's Community Forestry

Environmental justice is a major issue in Nepal. Lama and Buchy (2002) explored the ways in which women and other disadvantaged people are excluded from effective participation in community forestry. Participation in programs varies widely based on gender, class, and caste. Rich and upper-caste men control most decisions. People in lower castes are entitled to use forests according to the community's constitution, but they usually do not claim their rights. They are socially disadvantaged and typically own little private land.

Many disadvantaged say that their voices are ignored. Even if they are heard, they are unlikely to cause any action. The poorer people claim that upper-caste people abuse their power by relying on community resources when they are in a position to use resources from their own larger private lands. Lower-caste people depend considerably more on the community forests, since they have little land of their own and little knowledge of specialized forestry. Few women attend community meetings because they feel they have nothing to gain. Fear of intimidation by rich and upper-caste people is a major obstacle to participation by the disadvantaged.

Economic theory predicts what has occurred in Nepal: People participate in group forestry planning only when they perceive a personal benefit of participating that exceeds the cost. People who think their costs exceed their benefits do not get involved in the group's decision making. The rich and powerful perceive their benefits of participating as exceeding their costs; the poor and disadvantaged do not. The environmental justice challenge for Nepal is to design institutions that encourage participation in group decision making, so that each person perceives a benefit exceeding the cost.

Economic theory also predicts the behavior of the forestry department staff. They contact the people they consider to be the most important people in the village, because those contacts produce greatest impact per unit of budget spent. Women and lower-caste people have less contact with staff and are less targeted in government extension programs. The disadvantaged are further marginalized in the process because the field forest staff feels more at ease and able to communicate with the powerful. Low salaries and scarce travel budgets further damage environmental justice by discouraging forestry staff from contacting the least approachable (highest cost) people. Exclusion of the disadvantaged may reduce the sustainability of community forest programs, because poor people may be forced to break the rules (e.g., stealing from protected forest areas). Lama and Buchy conclude that environmental justice is seriously undermined in Nepal's community forestry program because the program ignores the needs of the most vulnerable.

Source: Lama, A., and M. Buchy. 2002. Gender, class, caste, and participation: The case of community forestry in Nepal. *Indian Journal of Gender Studies*.9 (1): 27–41.

If there are only resources available for one of these eight programs, which one should be selected? The choice of the best program depends on the objective of the electorate. If pure efficiency is the only objective (zero weight assigned to justice), then programs E though H are equally preferred, and each is preferred to any of the first four programs (A through D). Each produces because $3,000 more in net efficiency benefits. For this simple case, program F may be the best overall choice because it passes the test of justice while also producing the highest overall economic efficiency.

It is hard to generalize about what projects actually get selected in the real world. Depending on the country, region, culture, and context, varying weights are assigned to efficiency versus justice. A common approach in the United States is to look for programs that have a mix of acceptable efficiency and justice (i.e., each objective has a weight between zero and one and the weights add to one). For example, federal water programs are required by law to pass the efficiency test of benefits exceeding costs, but the most efficient program is rarely selected by Congress or federal agencies. More often, the selected program is one for which net benefits are positive to all easily identifiable groups while producing overall benefits that exceed costs. See Example 18–5. However, often the most vulnerable groups are the most difficult or expensive to identify.

Table 18–2 Distribution of net benefits of an environmental proposal

Net benefits category (i)	Population share (p)	Total net benefits share ranked from worst to best (n)	Cumulative population share (P)	Cumulative net benefits share (N)
First (lowest)	0.20	0.05	0.20	0.05
Second	0.20	0.11	0.40	0.16
Third	0.20	0.17	0.60	0.33
Fourth	0.20	0.24	0.80	0.57
Fifth (highest)	0.20	0.43	1.00	1.00
Total	1.00	1.00		

Preferences for justice versus efficiency vary widely by elected officials and others. For this reason, environmental policy analysts can make an impressive contribution to the policy debate by presenting results of program proposals in terms of benefits, costs, and net benefits to all relevant identifiable groups that time and budget permit. Availability of all of the data allows environmental decision makers to establish their own rankings of each group's importance, and at the same time permits the power of rational science to contribute to the policy debate.

Quantifying Environmental Justice

Perfect environmental justice can be defined as complete proportionality in the distribution of net benefits among the population of all affected groups. The actual level of environmental justice can be measured and compared across proposed programs to find the most just program. To measure relative equality, we begin with its opposite. Inequality is a property of the distribution in a population of some valued resource such as income, wealth, or the net benefits of environmental programs. The distribution of the net benefits of environmental programs is often highly skewed, with the most vulnerable people receiving the fewest benefits and incurring the most costs. One way to conceptualize inequality in the distribution of environmental program net benefits is with the Lorenz curve. All in the population are ranked by net benefits from the lowest to the highest, as in Table 18–2.[8]

The Lorenze curve (L) shown in Figure 18–2, is the plot of the cumulative share of net benefits (N) against the cumulative share of population (P). Notice that the 45-degree diagonal line is the line of perfect equality, E, in which all groups receive an identical share of net benefits. Line E is the line of complete environmental justice, consistent with the U.S. EPA's definition. A popular measure of inequality is the *Gini coefficient.*

This coefficient can be used to measure inequality for a large number of groups when each has its own distribution of population as well as net benefits. The Gini coefficient (G) is calculated from the Lorenz curve shown in Figure 18–2 as this ratio:

$$G = \text{Area } A \,/\, (\text{Area } A + \text{Area } B) \tag{18.1}$$

[8] All groups are assumed here to receive positive net benefits. In fact, many groups incur net costs in which their net benefits are less than zero. One way to deal with a negative algebraic sign for net benefits is to add in an amount of the worst-off group's negative net benefits to all groups, thus reassigning zero net benefits to the worst-off group. This method inflates the economic impact of the policy to all receiving groups but maintains the correct distribution of relative environmental justice.

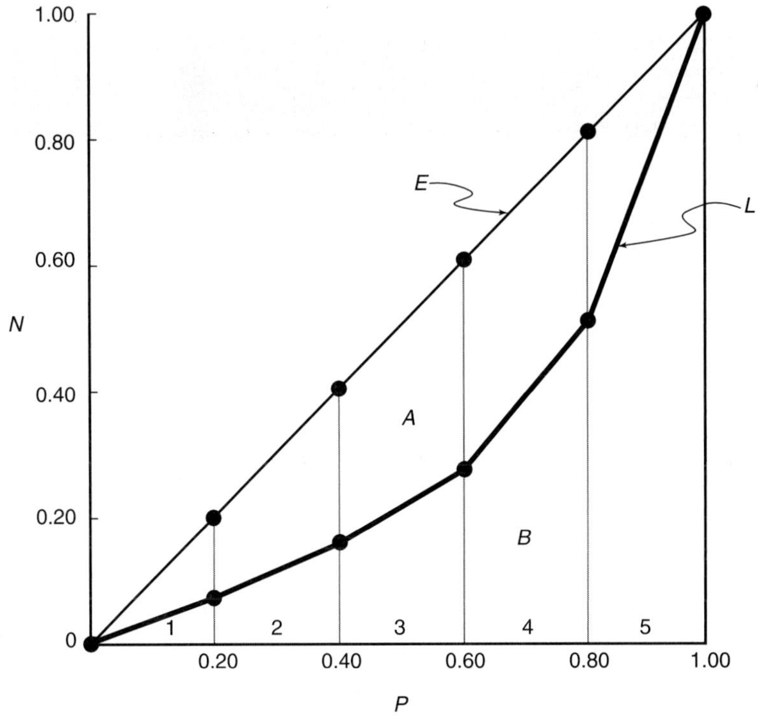

Figure 18–2 Lorenze curve (L) and Gini coefficient: $G = A/(A+B)$

Table 18–3 Gini coefficient calculation according to the geometric trapezoid method

	Base * average leg height	Trapezoid area
Area *A* + Area *B*	1.00 * 1.00/2	**0.500**
Area 1	0.20 * 0.05/2	0.005
Area 2	0.20 * (0.05 + 0.16)/2	0.021
Area 3	0.20 * (0.16 + 0.33)/2	0.049
Area 4	0.20 * (0.33 + 0.57)/2	0.090
Area 5	0.20 * (0.57 + 1.00)/2	0.157
Area *B*		**0.322**
Area *A*	0.500 − 0.322	**0.178**
Gini coefficient	0.178/0.500	**0.356**

Remember from high school geometry that (Area A + Area B) is the area of a triangle, given by 1.00 * 1.00/2 = 0.500. Calculation of the Gini coefficient for the data in Table 18–3 is shown by the steps shown in Table 18–3, which are based on calculating a sum of vertical trapezoids. (A trapezoid's area is the base times the average of the two unequal sides; note that a right triangle is a trapezoid with one side of length zero.) The Gini coefficient in this example is 0.178 divided by 0.500, which equals 0.356.[9] Depending on the distribution of net benefits, the Gini coefficient

[9] The calculated Gini coefficient (G) follows the spirit of the EPA's definition of environmental justice. However, notice that the coefficient ignores the relation between existing vulnerability levels and proportions of net benefits received from an environmental program. It assumes that designing a program with a higher proportion of net benefits to a worse-off group is equally as unfair as giving a higher proportion to a high-income group. Many would debate this assumption.

Table 18–4 Gini coefficient calculation, formula method

Net benefits category (i)	Population share (p)	Total net benefits share (ranked low to high) (n)	Cumulative population share (P)	Cumulative net benefits share (N)	$P_i - P_{i-1}$	$N_i + N_{i-1}$	*RHS*	*G*
First (lowest)	0.20	0.05	0.20	0.05	0.20	0.05	0.010	
Second	0.20	0.11	0.40	0.16	0.20	0.21	0.042	
Third	0.20	0.17	0.60	0.33	0.20	0.49	0.098	
Fourth	0.20	0.24	0.80	0.57	0.20	0.90	0.180	
Fifth (highest)	0.20	0.43	1.00	1.00	0.20	1.57	0.314	
Total	1.00	1.00					0.644	**0.356**

Formula: $G = 1 - \sum_i (P_i - P_{i-1})(N_i + N_{i-1})$

The formula method works best for calculating the Gini coefficient for large numbers of groups. A group's population share (p) is its population divided by total population in the program's affected area. A group's net benefits share (n) is its net benefits from the environmental program divided by total program net benefits summed over groups. Its cumulative population share (P) is the sum of population shares (p's) of all groups below and including itself. A group's cumulative net benefit share is calculated using the same logic as its cumulative population share. A group's $(P_i - P_{i-1})$ is its population share (p). Its $(N_i + N_{i-1})$ is its cumulative net benefit plus those of the next lower group. Its product is called the right-hand side (RHS) of the sum (Σ) term. It is $(P_i - P_{i-1})$ times its $(N_i + N_{i-1})$. Finally, the Gini coefficient is one minus the sum of RHS elements.

ranges from zero (perfect justice) to one (perfect injustice). In the Lorenz curve, the 45-degree line is an environmental policy of perfect justice. In general, the Gini coefficient is smaller when the Lorenz curve is closer to the line of perfect justice.

A simplified formula for the Gini coefficient is given by the following equation:

$$G = 1 - \sum_i (P_i - P_{i-1})\,(N_i + N_{i-1}) \tag{18.2}$$

where $P_i - P_{i-1}$ means that for each population group i, you need to calculate the cumulative population share in its group minus its share in the next lower group. For each (ith) population group, the term on the left is $P_i - P_{i-1}$. That difference is multiplied by the term on the right, $N_i + N_{i-1}$, which is the cumulative net benefits proportion in the given group plus its proportion in the next lower group. Table 18–4 has more details.

Evidence of Environmental Justice

Much of the scholarly work on environmental justice looks for patterns to answer this question: What is the distribution of environmental hazards among races and income levels? Typically, such analyses require cross-sectional data in which different geographical locations contain different kinds of people who face different kinds of hazards at some given point in time.[10] Cross-sectional data have an advantage because they can indicate large changes in en-

[10] Time series (longitudinal) data could also be used, in which behavior patterns could be observed at one or more places as environmental hazards change over time. One disadvantage of this method is that the hazards usually change slowly and/or by small amounts, so it is difficult to discern behavioral responses to those changes.

vironmental hazards from one place to another, and behavioral adjustments to those patterns can be easily discerned. Three classic studies by (1) Bullard (1983), (2) the United Church of Christ (1987), and (3) the U.S. General Accounting Office (1983) attempted to answer the distribution question.

- Study 1: Bullard's previously described study examined an attempt by an African American community in Houston, Texas, to block the placement of a hazardous waste site in their community. Bullard concluded that race, as well as income status, were probably factors in decisions about locating hazardous waste sites.
- Study 2: In 1987 the United Church of Christ examined the problem of hazardous waste siting from the U.S. national view. This study came to some remarkable conclusions, including that race was the most important predictor of location of hazardous waste sites. The findings also showed that communities with the largest number of hazardous waste facilities had the highest percentage of racial and ethnic residents. In communities containing the nation's five largest landfills, the minority population was more than three times larger than in communities that lacked those facilities.
- Study 3: The U.S. General Accounting Office examined characteristics of the four hazardous waste sites in Region 4 of the U.S. EPA, (which consists of eight southeastern states). The results showed that three of those sites were in communities with mostly African American residents and the fourth was in a community with residents of below average income.

Since those three classic studies were published, many more studies have followed. Findings have been mixed. Some of the studies found strong correlations between hazards and income or race. A few have not. An analysis by Turner and Wu (2002) found that studies varied widely in unit of analysis, geographic scale, hazard examined, time period, and concept of environmental justice. In short, the studies are hard to compare. Moreover, studies that focus only on existing distributions of environmental hazards typically ignore the historical events or institutions that created them. Pearce (2003) reviewed more than 40 studies linking environmental hazards and race/income. Results generally (but not always) found that existing distributions of environmental risk for emissions, concentrations, and health risks for land, water, air, and hazardous waste sites are biased against the nonwhite and the poor.

Hird and Reese (1998) examined relationships between the distribution of 29 air, water, and hazardous waste pollutants and the racial, political, and socioeconomic characteristics of affected communities. Their analysis of U.S. data showed that race and ethnicity are strongly associated with poor environmental quality. Both nonwhite and Hispanic populations experienced significantly higher pollution levels, and this uneven distribution was a national rather than a regional finding. Surprisingly, higher income was associated with poorer environmental quality, but the reasons for this were not clear. Overall, Hird and Reese found that race as well as many other factors are correlated with pollution, for example, political mobilization, population density, manufacturing activity, and income.

Few studies of environmental justice have been conducted in Europe. However, one study (Coenen and Halfacre, 1999) examined differences in pollution across urban neighborhoods in the Netherlands. Little environmental injustice was found to be present in low-income and ethnic minority areas. This seems remarkable given the fact that ethnic minorities make up about 10% of the Dutch population.

Boer et al. (1997) and Carlson and Zasloff (2003) looked at data from southern California to see whether certain racial groups are disproportionately affected by air toxics. Their findings showed that for African and Asian Americans located near southern California air toxic facilities, there appears to be no significant environmental justice issue. Asian Americans are

Example 18–6
Environmental Justice: Human versus Animal Welfare

Schools of large, mature, yellowfin tuna congregate and swim beneath large groups of dolphins in the eastern tropical Pacific. In the 1950s U.S. purse seiners (fishers who use very large nets) began to exploit this association by circling the dolphins with large nets to capture the tuna beneath them. By the early 1970s, U.S. purse seiners were believed to cause the deaths of more than 300,000 dolphins per year. The U.S. Congress enacted the Marine Mammal Protection Act (MMPA) in 1972 to limit the killing or injuring of dolphins as part of commercial fishing. The law limited the number of incidental killings of dolphins by U.S. purse seiners and banned imports of fish caught with commercial fishing technology dangerous to dolphins.

Key issues were how to resolve the interests of the competing actors—people, dolphins, turtles, and tuna—as well as identify the principles of justice underlying these choices. Congressional debates in the early 1970s and again in 1996 and 1997, which resulted in the U.S. International Dolphin Conservation Program Act of 1997, illustrate the challenges in answering the question of how we should rank the interests of competing species. People's interest lie in fishing, both as a livelihood and as a supply of economically affordable protein. It is not too far-fetched to assume that individual tuna have an interest in surviving, and the facts indicated that the tuna could have been threatened with extinction if certain common fishing methods developed in response to the MMPA had continued. We can also assume that individual dolphins would prefer to survive and to avoid human-induced suffering. Turtles, sharks, and other affected species also likely have an interest in individual survival.

Congressional debates about the 1997 act highlighted the relative values placed on different species: The rank appeared to be human interests, dolphin interests, turtle interests, and last with no interest, tuna. In these debates competition was often set up in terms of two competing human interests: economic concerns and species stewardship. Several debates also relied on the humanlike qualities of the dolphin, such as native intelligence, beauty, humor, gentleness, and ability to communicate.

Never seriously debated was that purse seine fishing itself was the underlying problem or that it possessed too much destructive power to be ethically acceptable. Wastefulness of dolphin deaths was an issue, but only when there were economically feasible alternatives to that waste. In the end, the 1997 act was passed as a compromise. Under that compromise, dolphin suffering did not necessarily rank ahead of the potential extinction of a species: Dolphin net settings were defined as dolphin safe if no dolphin deaths occurred.

Source: Cheyne, I. 2000. Law and ethics in the trade and environment debate: Tuna, dolphins and turtles. *Journal of Environmental Law.* 12 (3): 293.

actually slightly underrepresented in TRI[11] neighborhoods compared with their representation in Los Angeles County. Latinos, however, fared worse. Neighborhoods near the top 100 TRI facilities contain a significantly larger percentage of Latino residents than similar neighborhoods throughout Los Angeles County. The Latino population in Los Angeles County is 44%; however, 60% of residents living near the top 100 TRI facilities are Latino. Even when controlling for income, Latinos are still significantly overrepresented among people affected by air toxins (similar findings are reported by Morello-Frosch, Pastor, and Sadd, 2002). The authors found that income was at least as important as race in southern California. Neighborhoods around the top 100 TRI facilities are typically poorer than the county overall, measured by median gross rent per month and median housing value. The authors concluded that southern California has a major environmental justice problem in which people with low incomes are considerably more affected by air toxics than others. See Example 18–6. A similar study examined evnironmental justice for water rights in south Asia. See Example 18–7.

[11] U.S. federal law requires firms that handle large amounts of chemicals report quantities of each chemical they release into the land, air, and water. The federal government assembles the reported information into the Toxics Release Inventory (TRI). The authors took the reported data in the TRI for the 100 firms in Los Angeles county that release the largest amounts of air toxics. They analyzed locations by surrounding population, race and ethnicity, and housing value, and examined population groups within a half mile radius of each TRI facility.

Example 18–7
Environmental Justice in Water Rights: South Asia

In much of south Asia, men own and control most irrigation water rights. The little control that women have over water may be an important indicator of gender inequality as well as contributing to it. Zwarteveen and Meinzen-Dick (2001) found that not only do gender differentials in water rights indicate environmental injustice, but they also may lead to inefficiencies in total water use, and threaten household welfare and food security. That is, environmental injustice may lead to economic inefficiency by reducing total economic output from water used.

According to the authors, institutions play a major role in assigning rights and responsibilities for managing water between men and women in south Asia. They give an example from the *Chhattis Mauga* system in Nepal, where male farmers do not want their wives or daughters to participate in ditch maintenance activities to avoid having them work alongside male strangers who might make embarrassing jokes. Such works may also dictate longer travel distances than are considered right for women.

Microeconomic theory predicts that the benefits and costs of participating in water users' organizations and meetings are different for women than for men. Women's heavy domestic workloads raise the opportunity costs of women's time in attending meetings. Moreover, their benefits may be lower, because the effectiveness of formal meetings as a way to defend their interests may be lower for women. Public meetings in south Asia are historically viewed as a man's domain. Economic theory predicts what has occurred: Few women participate in planning meetings. A woman's ability to participate in irrigation management decisions depends on the freedom to interact with men. But in much of south Asia, women's effective participation is limited by *purdah*, the practice of prohibiting women from public observation by having them wear concealing clothing and by using walls, curtains, and screens inside the home.

Zwarteveen and Meinzen-Dick illustrate how women manipulate their status to get what they need. A widow in the Buttala irrigation system in Sri Lanka refused to clean her share of the community ditch. When she was criticized about violating the laws, she convinced the authorities that she was innocent. Because she was an old woman, she said, she could not be expected to do her share of the work, and she could not make her sons help. Further inspection showed her ability to do all the farming activities herself; her alleged vulnerability was simply a clever way to avoid cleaning the ditch.

Environmental justice is desirable on its own, but improved environmental justice may also lead to greater economic efficiency. As long as women in South Asia can escape sanctions for breaking the rules of the system, excluding them from the system weakens and damages the effectiveness of water use organizations.

Source: Zwarteveen, M., and R. Meinzen-Dick. 2001. Gender and property rights in the commons: Examples of water rights in South Asia. *Agriculture and Human Values.* 18: 11–25.

Summary

According to the U.S. Environmental Protection Agency, environmental justice is the "fair treatment and meaningful involvement of all people regardless of race, color, national origin, or income with respect to the development, implementation, and enforcement of environmental laws, regulations, and policies." John Rawls' famous book, *A Theory of Justice,* has had a major impact on how we think about environmental justice. According to Rawls, a just society could be designed if each of us were in a position that would allow us to rationally, impartially, and freely deliberate just arrangements. Rawls' principles underlie the definition of environmental justice established by the U.S. EP Agency, which opened the door to U.S. federal responsibility for environmental justice.

At least four measures of physical environmental risk influence environmental justice: emissions, ambient concentrations exposure, and health risk. How environmental risk is measured has a major impact on the conclusions reached about the presence or absence of environmental justice. Conclusions on the presence and extent of environmental justice also depend on the geographic scale of analysis, which ranges from entire regions to individual addresses.

Economic theory provides insights into why specific locations are selected for hazardous waste treatment, storage, and disposal facilities (TSDFs). TSDF developers typically want to maximize their net income. So economic theory predicts they will seek a location that allows them to process or dispose of hazardous wastes at minimum cost to them. TSDF-receiving communities want to maximize whatever net benefits they could receive from being chosen as a TSDF location. No community would accept a TSDF without added compensation such as increased tax revenues or new jobs. Economic theory predicts that low-income communities will be the most likely places to attract a TSDF because of incentives facing facility owners and those facing the community itself.

The argument of pure economic efficiency states that society as a whole suffers smaller damages if TSDFs are located in poorer communities. This argument was used in an internal memo of the World Bank that was leaked to the public. The memo caused considerable controversy because locating polluting facilities in poor countries is likely to fail the test of environmental justice. Few of us would vote for such an unjust scheme if there was a significant chance we would end up living in such a community.

Environmentally just programs can be designed by analyzing the benefits and costs accruing to the most vulnerable people in society from various proposals for locating hazardous waste TSDFs. Democratically elected bodies often resist supporting policies that seem unjust. The Gini coefficient, which ranges from a value of zero to one, can be used to measure the degree of observed environmental justice in the distribution of net benefits of proposed environmental programs.

Much scholarly work has attempted to analyze the distribution of environmental hazards among races and income levels. Many, but not all, of these studies have found strong correlations between location of environmental hazards and income or race. The growing use of geographic information systems will bring continued improvements in analysis and understanding of environmental data about hazardous waste sites as well as in the design of environmentally just policies.

Review Questions

1. How can a system of environmental justice designed and enforced by the majority do anything but favor the majority's welfare?

2. In many cities, environmentally undesirable facilities are located where the most vulnerable people live. What economic forces might be causing this to happen? Describe what methods you would use to find out if this has happened in your town.

3. How would you apply economic principles to design a more environmentally just society?

4. Propose your own definition of environmental justice. What economic principles would you use to discover whether a proposed environmental policy is environmentally just?

5. Describe the economic principles that guide the placement of TSDFs. How can those principles be used to ensure that locations of TSDFs are environmentally just?

6. What does the evidence in this chapter tell you regarding the presence or absence of environmental justice? What implications for environmental policy reform does this evidence suggest?

7. How would you decide whether a proposed environmental reform would create a more environmentally just society?

8. Describe the economic forces behind the levels of environmental injustices you see.

Problem

Distribution of net benefits of an environmental program

	Group *A*		Group *B*		Group *C*			
Program	Population	Net benefits	Population	Net benefits	Population	Net benefits	Gini coefficient	Justice ranking
1	200	100	100	100	200	100		
2	200	50	100	200	200	250		
3	200	400	100	1,200	200	400		
4	200	400	100	200	200	400		

1. Use the accompanying table to compare environmental justice across programs.[12] First compute each program's Gini coefficient. The first step in calculating the Gini coefficient for any program is to sort the groups of people by proportion of net benefits received per unit population, from worst to best. Then use the following formula to compare perfect justice and actual justice for each group

$$G = 1 - \sum_i (P_i - P_{i-1})(N_i + N_{i-1})$$

where $P_i - P_{i-1}$ means that for each population group i, calculate the cumulative population share in its group minus its share in the next lower group. That left-hand share difference is multi-

[12] A perfectly just program occurs when each group's percentage of net benefits equals their percentage of the base population. More just programs produce coefficients closer to these values.

plied by the right-hand term $N_i + N_{i-1}$. That right-hand term is the cumulative net benefits proportion in the ith group plus its proportion in the next lower group. Your task is to use the formula to rank the four programs on their environmental justice performance.

References

Boer, J. T., M. Pastor, J. L. Sadd, and L. D. Snyder. 1997. Is there environmental racism? The demographics of hazardous waste in Los Angeles County. *Social Science Quarterly.* 78 (4): 793–810.

Buckingham-Hatfield, S., and B. Evans. 1996. *Environmental Planning and Sustainability.* London: John Wiley.

Bullard, R. D. 1983. Solid waste sites and the Black Houston community. *Sociological Inquiry.* 53: 273–288.

Carlson, A. E., and J. Zasloff. 2003. *Environmental justice.* Available from *http://www.ioe.ucla.edu/publications/report01/* [Accessed October 10, 2004].

Coenen, F. H. J. M., and A. C. Halfacre. 1999. *Environmental risk perception, income, and ethnicity: Does the Netherlands have an environmental justice problem?* Available from *http://www.utwente.nl/instituten/cstm/publications/* [Accessed October 10, 2004].

Hird, J. A., and M. Reese. 1998. The distribution of environmental quality: An empirical analysis. *Social Science Quarterly.* 79 (4): 693–716.

Kuehn, R. R. 1997. An analysis of the compatibility of quantitative risk assessment with the principles of environmental justice in the United States. *Risk Decision and Policy.* 2 (3): 259–276.

Morello-Frosch, R., M. Pastor, and J. Sadd. 2002. Integrating environmental justice and the precautionary principle in research and policy making: The case of ambient air toxics exposures and health risks among schoolchildren in Los Angeles. *Annals of the American Academy of Political and Social Science.* 584: 47–68.

Pearce, D. 2003. The Distribution of Benefits and Costs of Environmental Policies: Conceptual Framework and Literature Survey. Prepared for the OECD Environment Directorate Workshop on the Distribution of Benefits and Costs of Environmental Policies.

Rawls, J. 1971. *A Theory of Justice.* Cambridge, MA: Belknap Press of Harvard University.

Summers, L. 1991. *Internal memo.* Available from *http://www.sustainableworld.org.uk/summersmemo.htm* [Accessed October 10, 2004].

Turner, R. L., and D. P. Wu. 2002. Environmental Justice and Environmental Racism. An Annotated *Bibliography and General Overview, Focusing on U.S. Literature, 1996–2002.* Institute of International Studies, University of California. Available from *http://globetrotter.berkeley.edu/envirpolbib/B07-turnerwu.pdf* [Accessed October 10, 2004]

United Church of Christ. 1987. *Toxic Wastes and Race in the United States: A National Report on the Racial and Socio-economic Characteristics of Hazardous Waste Sites.* New York. United Church of Christ, Commission for Racial Justice.

U.S. Environmental Protection Agency. 2003. *Environmental justice.* Available from *http://www.epa.gov/compliance/environmentaljustice/index.html* [Accessed October 10, 2004].

U.S. General Accounting Office. 1983. *Siting of Hazardous Waste Landfills and Their Correlation with Racial and Economic Status of Surrounding Communities.* Washington, D.C.: U.S. Government Printing Office.

Chapter 19

SUSTAINABLE DEVELOPMENT

Outline

Deciding What to Sustain 547

Measuring Progress of Sustainability 553

Instruments for Implementing Sustainability 555

Policy Applications 560

Objectives

After studying this chapter, you should be able to

- Identify what could be sustained in the search for a sustainable future
- Examine ways to measure progress in the search for a sustainable future
- Identify economic concepts that support the search for sustainability
- Describe policy applications for promoting sustainability

Introduction

Gro Harlem Brundtland was born in Oslo, Norway, in 1939. When she was 10, her family moved to the United States. As a young adult she studied medicine in the United States and completed her training in Norway. In 1974, Brundtland gained national attention in Norway by campaigning for health reform; as a result of her efforts, she was asked to join the Norwegian government where she soon discovered her political talents. In 1981, at age 42, Brundtland was the first woman elected as Norway's prime minister.

In 1983 Brundtland accepted an invitation from the United Nations to chair the World Commission on Environment and Development to examine global environmental and developmental problems and propose solutions. In 1987 the commission drafted a proposal to address the world's environmental and economic problems. The report, *Our Common Future* (also called the Brundtland Report), had an immediate influence on how scientists, governments, and citizens thought about the concept known as sustainable development. The report defined sustainable development as "development that meets the needs of the present without compromising the ability of future generations to meet their own needs" (Brundtland, 1987, p. 51).

A proposal in the report that called for a conference on development and the environment led to the Earth Summit in Brazil in 1992, which attracted leaders from approximately 200 countries.

Sustainable development focuses on the connection between a sustainable human society and the environment. Economic principles offer powerful tools that can be used to gain insights into this connection. In addition, economic principles help in the development of sustainable paths for a modern economy. In turn, better understanding of sustainable paths lays the foundation for designing sustainable policies and for determining what sorts of outcomes to expect. This chapter shows how economic concepts can help us think coherently about sustainable development. It also describes several challenges that lie ahead in the search for sustainable development.

■ Deciding What to Sustain

At its simplest level, sustainability is the ability to maintain something undiminished over some time period. Three defining questions are (1) what is to be sustained, and at what scale and in what form; (2) over what time period and with what level of certainty; and (3) through what social process and with what trade-offs against other social goals (Lele and Norgaard, 1996). Sustainability means more than a generally comfortable feeling that resources will be available to provide for future generations. Something must be sustained for the very long run, and it is important to agree on what that something is.[1]

Most of us feel that certain unique and irreplaceable natural or cultural assets, such as the Grand Canyon of Arizona, the pyramids of Egypt, or the Great Barrier Reef of Australia, should be preserved for their own sake. But should that sort of preservation be required for all elements of the natural or built environment? Probably not. It would be neither possible nor desirable to leave the natural or built world as it is now, or require that it be returned to the condition in some previous period. Most environmental resources are desirable to people not for what the resources are, but for what they do. It is the capacity of natural resources to provide human welfare that gives them value.

In this chapter a sustainable path for any economy is defined as one that allows every future generation the option to be as well off as previous ones. By accepting the duty of designing and carrying out sustainable policies, one generation gives the next generation a future with the resources needed to achieve a standard of living at least as good as its own, and to treat the subsequent generation similarly. As Nobel laureate Robert Solow showed in several papers in the 1980s and 1990s, for any generation the mission is to avoid depreciating its bequest to future generations.

The standard of living achievable for any generation depends on that generation's inherited endowment of environment, resources, technologies, and knowledge—everything that could affect current or future capacity to produce well-being. That includes depletable resources, like fossil fuels, as well as renewable resources such as water, forests, and fisheries. It also includes technical knowledge and the general level of education and supply of human talents, as well as environmental amenities and the environmental life support system.

A sustainable path for the economy is not necessarily a path that preserves intact every single thing or even *any* single thing. It is a path that replaces whatever it takes from its inherited endowment of environmental, natural, human, and capital resources. What matters is not the special form the replacement takes, but only the capacity to produce what posterity will enjoy.

Reasons for Sustainability

Sustainability can be considered desirable for three reasons. The first is economic; that is, sustainable activities are those for which benefits exceed costs (see Chapter 5). Sustainable activities are likely to produce benefits that exceed costs in present value terms. Unsustainable behavior, by assigning low weights to the welfare of future generations, likely produces higher costs than benefits. Any plan that maximizes social welfare over time is likely to require sustainable activities. The complete truth of this statement hinges on how we measure sustainability and on what discount rate we use. Still, it is hard to imagine any economically efficient growth pattern over time that does not leave to future generations much of what is desirable. The second reason sustainability is desirable is its morality. The current generation has a moral duty to all generations that follow (the idea of moral duty underpins the Brundtland Report). Accepting this duty implies acting in ways that avoid limiting opportunities to future generations. The third reason for desiring sustainability is ecological. The ecological principle rests on the idea that diverse, intact, and resilient ecosystems are desirable by all generations. Any policy, activity, or institution that threatens to limit such

[1] Much of this section relies on Solow (1992).

ecological outcomes is undesirable. Public institutions and private behavior should be organized to avoid irreversible ecological disruption (i.e., to sustain ecological desirability).

Definitions of Sustainability

The Brundtland Report raised the principle of sustainability to a high level of visibility and importance worldwide. Since the early 1980s, hundreds of economists, ecologists, and political leaders have offered definitions of sustainability, all of which can be seen as attempts to apply the principles of the Brundtland Report. Several such definitions are discussed next; they range from definitions based on consumption and resource use to definitions that remove humanity from center stage.

Nondeclining Consumption per Capita. Some economists have defined sustainability based on the principle that welfare per capita (or what is usually more easily measured, consumption per capita) should be constant or growing over the future. That definition began with John Hartwick (1977) and Robert Solow (1974) in the mid-1970s. They view the ultimate end of economic activity as consumption or utility, and they consider environmental and natural resources, as well as capital and human resources, as inputs to that process. What society desires to sustain is human welfare, and it cares little about the means by which that welfare is produced.

Production Opportunities. In terms of production opportunities, sustainability means actions that maintain for future generations the opportunity to produce adequate output for their needs.[2] Production potential at any time period depends on the stock of productive resources available for use. Productive resources can consist of human labor and capital, if capital is defined broadly enough. Capital defined broadly consists of the stock of useful resources (with the exception of human labor):

- *Physical capital* (K_p): factories, machines, infrastructure, and buildings, all of which come from devoting part of each period's total output to investment
- *Intellectual capital* (K_i): the stock of useful knowledge, also called the state of technology
- *Human capital* (K_h): the stock of learned talents residing in each person. This stock improves the potential of that person to apply his or her labor productively.
- *Natural capital* (K_n): the stock of natural resources and the environment, including water resource systems, aquifers, minerals, forests, fisheries, fertile land, and the earth's environmental life support system. Natural capital includes exhaustible and renewable resources.

Based on this definition of capital and labor, the economy's total production potential is expressed as the following aggregate production function:

$$Q = Q\,(K,L) \tag{19.1}$$

where Q is the economy's total production potential made possible by useful application of the two aggregate resources, L, human labor, and K, total capital consisting of a simple sum of its four parts:

$$K = K_p + K_i + K_h + K_n \tag{19.2}$$

More output is produced by increases in either total K or L or both. Each input is subject to diminishing marginal returns (i.e., increases in each increases output at a decreasing rate). Many studies have found that aggregate production functions have approximately constant returns to scale; that is, increasing both inputs by a given percentage increases output, Q, by the same percentage.

---

[2] Solow's famous cake-eating parable sheds light on the importance of production opportunities. The parable poses the question of how to best allocate some fixed quantity of a nonrenewable resource (a large cake) among future generations. The answer is easier than you think; you can have your cake and eat it too. In fact, it isn't cake we want at all, but the productive resources to make cake, like land and water to grow wheat and produce sugar. As long as we can sustain cake-making output for future generations, we have solved the problem of having and eating cake.

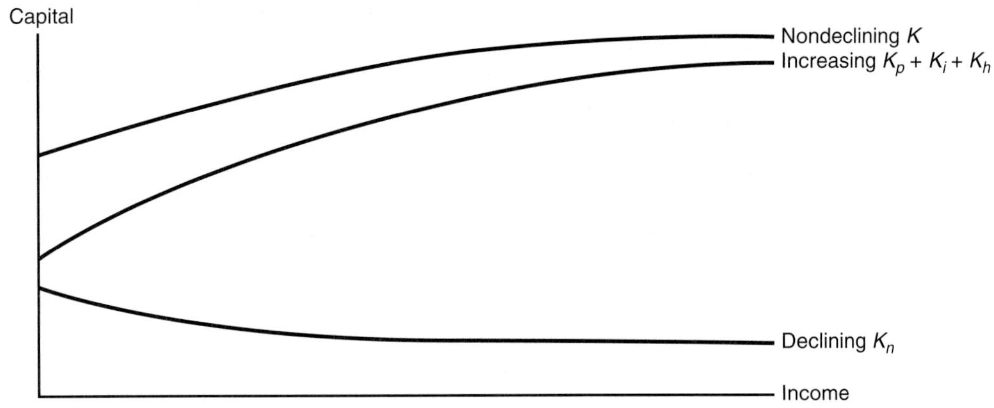

Figure 19–1 Production opportunities and natural capital

Figure 19–2 Weak sustainability

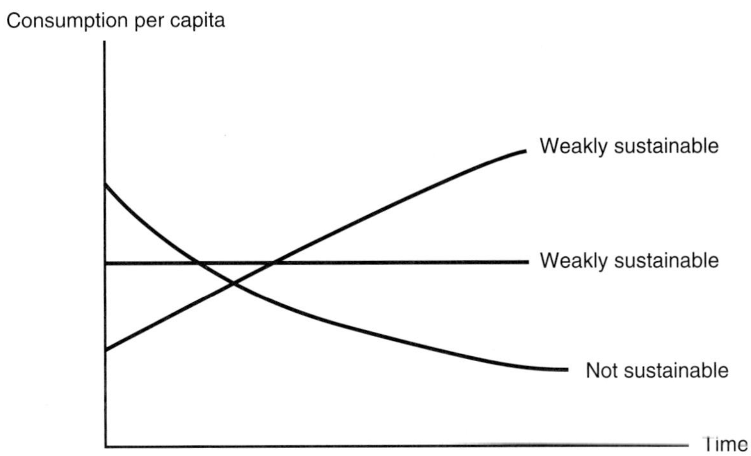

Maintaining production potential for a given population means ensuring that total capital, K, is constant or increasing over time. If none of the four elements of capital can be substituted for each other in production, and if each is required for production, then productive potential can be maintained only if each of the four elements is constant or increasing over time. But substitution can mean that if any of the four elements of capital (e.g., K_n) is depleted through human action, total capital[3] and therefore, total productive potential can be maintained by increasing any one of the other three elements.[4] If the four kinds of capital can indeed substitute for one another, there is no reason why all of the four elements have to be nondecreasing as long as the total is maintained (Figure 19–1). This concept of sustainability is sometimes referred to as *weak sustainability* (Figure 19–2). Thus, under weak sustainability productive potential bequeathed to future generations can be guaranteed even if natural capital decreases if those decreasing supplies can be substituted by increasing quantities of physical, intellectual, or human capital.

[3] A related issue is how those forms of capital should be compared to each other. How should barrels of oil be added to acres of old growth forests, or how should remaining numbers of an endangered fish be added to numbers of an endangered wolf. If weights (e.g., monetary values) could be assigned to each kind of capital, we could add them up to a meaningful total value, which would permit common denominator comparisons.

[4] Perfect (infinite) substitution among the four sources of capital occurs if total capital (K) is a linear function of each of its four parts. Zero substitution means that total capital (K) falls by X percent whenever its most limiting part falls by X percent. So if natural capital (K_n) is the limiting resource, human depletions of natural capital by 20% reduce overall capital by 20%, no matter what happens to the other three sources.

Resource Use. Some economies that are based on natural resources (e.g., those of Iceland, Norway, Russia, the state of Oregon) rely heavily on access to its renewable resources like agriculture, water, forests, and fisheries. If an economy is poor and has little human-produced capital or exhaustible resources, it may depend heavily on its renewable resource base. A good example is an irrigation economy in a developing country that depends on surface water from rivers. For this kind of economy, sustainability may depend more on sustained use of renewable resources with less immediate concern for maintaining natural capital. For such economies, the concept of *sustained yield* is important. A sustained yield describes a condition in which policies are enacted to keep constant the stock of a renewable resource, while producing a constant flow of services indefinitely, (Think of living off interest while leaving principle untouched.) One example is a fishery. When harvested at suitably low rates, a fishery can deliver a steady and sustained flow of services in the form of harvested fish without ever depleting the principle.

The maximum sustained yield of a renewable resource is the highest possible flow of services that can be sustained from an ecological system. Although the maximum sustained yield typically does not produce maximum net present value from the system, it is, nevertheless, sustainable. For that reason maximum sustained yield may be politically attractive in some conditions. For such economies, sustainable policies would focus on maintaining sustained yields (harvests). Pezzey (1992a, b) provides details.

Nondeclining Natural Capital Stock. A definition of sustainability based on nondeclining natural capital places emphasis on natural capital assets and requires policies to guarantee that these assets do not decline over time. This definition says that after a certain critical level is reached, no increase in other kinds of capital or human labor can substitute for further decreases in natural capital. If natural capital is essential to either production or consumption, and other productive resources or outputs cannot substitute for it, then preventing a decline in natural capital is a required condition for sustainability. People who prefer this definition of sustainability, sometimes known as strong sustainability, believe that once the critical level is reached, reduced stocks of natural capital cannot be made up with increased labor or other types of capital (Figure 19–3).

Krutilla (1967) provided one of the earliest explanations of why nondeclining natural capital is required for sustainability to occur, independent of its value as an input in a productive process. He based his idea on the concept of uniqueness and examined unique (nonsubstitutable) attributes of nature such as the Grand Canyon, an endangered species, an ecosystem, or a biotic community required for a species' survival. He analyzed the time stream of benefits from activities that require preserving that natural capital (e.g., environmental life support) and compared it to the benefits of using the natural capital for extractive activities (e.g., logging or mining) that have adverse consequences on the value of the preserved natural capital.

Figure 19–3
Sustainability of natural capital stock

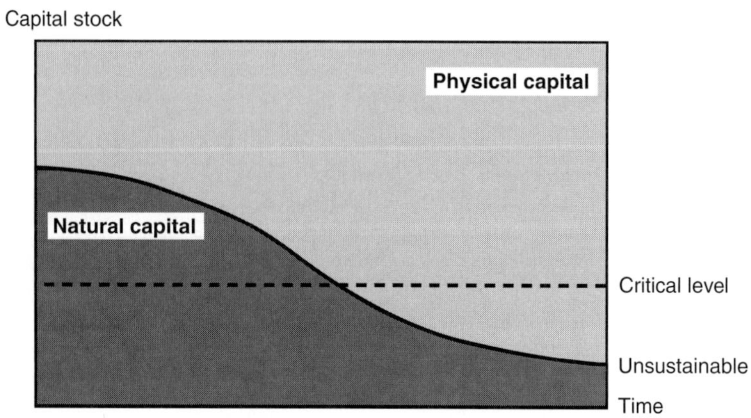

Defining sustainability as nondeclining natural capital is based on Krutilla's concept of uniqueness, which means the capital in question has no effective substitutes. For that special kind of capital, sustainability requires protecting it. Even if history provides evidence of successful substitutions, past successes do not guarantee that a trend will continue. When various functions of the natural environment are examined it appears that many of them can only be performed by stocks of natural capital, functions for which no substitution is possible.

Ecosystem Stability and Resilience. Another concept that defines sustainability requires guaranteeing a minimum stability and resilience of ecosystems over time. This definition does not focus on humanity; rather, it looks at sustainability from the view of the ecological system of which humans are one part. Sustainability is defined by the extent to which the structure and function of the ecosystem can be maintained. The ecological concept of stability is the tendency of a population to return to an equilibrium condition after an outside disturbance. Resilience refers to the tendency of the ecosystem to retain its original structure and function after an outside disturbance. An ecosystem can be resilient without all of its constituent populations being stable. The resilience of an ecosystem is measured by its ability to maintain its organization without sustaining catastrophic change even if it experiences disruptive outside shocks. Resilience tends to be higher when the extent of connectedness or complexity of the ecosystem is greater. If we accept the premise that ecological sustainability is a desirable objective, then policies should be developed to minimize the probability of disturbances that threaten the resilience of the whole ecosystem.

Risk Avoidance. Ciriacy-Wantrup, a founding father of environmental and resource economics who taught in the Department of Agricultural and Resource Economics at the University of California Berkeley from 1938 to 1980, formulated the principle of risk avoidance as a way to promote sustainability. He coined the term *safe minimum standard* (SMS) as a conservation target of environmental and natural resource policy (Ciriacy-Wantrup, 1968).[5] It is achieved by avoiding what he termed the critical zone, namely, the physical conditions brought about by human action that make it uneconomical to halt and reverse depletion. Some economists consider the SMS synonymous with more the widely used term *precautionary principle.*

When an environmental or natural resource has a high economic value because of its uniqueness or its potentially great but unknown future productivity, we may wish to avoid outcomes in which the resource is driven to extinction or to a worthless state. In that case, a policy that constrains a resource's harvesting or use may be desirable even if its expected benefits are less than its expected costs. The safe minimum standard of conservation is a risk management device for coping with a small probability of a potentially catastrophic outcome. A comprehensive version of the SMS involves establishing constraints on resource harvesting, pollution, or other risky activities so that all risks to the survival of the renewable resource are made suitably small. A more workable application of the SMS suggests that under reasonable uncertainty, threats to the survival of valuable environmental systems are made suitably small, provided that the costs to do this are economically acceptable. For policy decisions to be carried out using this workable SMS standard, informed judgments must be made about what is reasonable, valuable, and economical.

---

[5] Born in Germany, Siegfried Von Ciriacy-Wantrup completed his undergraduate work at the University of Berlin, the University of Vienna, and the University of Bonn prior to receiving a master's degree from the University of Illinois in 1930 under an international exchange program. In 1931, he obtained a doctorate from the University of Bonn and remained there as a lecturer until 1936. Confronted by Nazi repression, he moved to the United States and joined the Berkeley faculty in 1938. He devoted his professional life to making economic analysis useful for lay people, policymakers, and his students. He was a pioneer in the economics of natural resources, environmental issues, and values.

Three conditions justify arguments for a SMS of conservation, originally developed by Ciriacy-Wantrup in connection with analysis of biodiversity:

1. Uncertainty is present. Typically all consequences of decisions are not known (e.g., widespread loss of species arising from major habitat change).
2. The environmental consequences produced by economic activity, particularly resource harvesting or extraction, are irreversible.
3. The irreversible losses can be large, although unknown.

When all three of these conditions are relevant, a risk-averting approach is desirable. Even if an action does not seem dangerous on the surface, we may wish to build in safeguards in our behavior to reduce the risk or cost of serious damages even further, unless the cost of doing so is intolerably high.

Coevolutionary Development. Coevolutionary development (Norgaard, 1984) is based on the ecological principle in which several species coevolve based on the reciprocal response of each to the other. Coevolutionary development extends the idea of ecological coevolution to include social as well as ecological systems. If economic development is sustainable (not simply one group exploiting another or this generation living better at the expense of the future), institutions must be designed that allow economic systems to develop. Conventional economic thinking bases the potential for sustainable development on the ability to design technologies and institutions that improve the quantity of capital and labor. It is hoped that this continued improvement in capital and labor will substitute for exploitation of lower grades of exhaustible resources and reduced qualities of ecosystems and environmental assets.

In contrast to conventional economic thinking, the coevolutionary development school views successful sustainable development as development in which there is harmony with nature. Where social and natural systems depend on each other, coevolution sees development of a social system and its natural environment in a way that is nondestructive to both. In this school of thought, human institutions and technologies are designed to be flexible enough to smoothly adapt to unexpected changes in nature.[6] Sustainable social institutions anticipate and adapt to unexpected changes in the natural world. Sustainable development depends on knowing there is a need to adapt to changes in the natural environment as well as investing the resources needed to carry out plans for adapting. A culture survives and flourishes if it adapts to the limits imposed by nature, rather than erecting structures and institutions that insulate against nature. To take a simple example, the coevolutionary view says it is cheaper to allow our tastes to evolve so that we are comfortable in a wide range of indoor environments, rather than to build, operate, and maintain expensive heating and cooling systems that fix indoor temperature at a certain constant temperature year round.

Species Existence. Chapter 2 showed that several nature-centered ethical systems hold that nonhuman species have an inherent right to a sustained existence, independent of their instrumental value to humans. In contrast, an anthropocentric view assigns primary standing to humans, human needs, and human satisfaction and lower standing to needs of nonhumans. The nonhuman part of the environment is assumed to be an instrument to be used by people as people see fit, which means that the environment is defined in terms of the benefits it provides to people. Although there is no single, accepted *biocentric* view, most who accept its principles view the preferences of all natural organisms, species, and ecosystems as equally important. The biocentric view recognizes that human desires and human values are important, but it considers them as a part of a larger biological system. Living nature in its various forms has moral standing on its own apart from its instrumental value to people.

---

[6] Adaptive development is related to coevolutionary development. Adaptive development is a way to carry out a sustainable plan when future benefits and costs of that action are uncertain because of either inadequate knowledge or because outcome depends on the future state of nature itself. By watching the result of a first stage, information can be discovered and processed so that a sustainable plan can be carried out in future stages.

Nonhuman species' sustainability is not yet a foundation of any nation's environmental policy. It may take widespread increases in education, per capita income, and personal security among the masses, for whom personal struggles with nature's constraints need to be reduced considerably, before national decision makers base environmental policies on principles of all species' sustainability.

MEASURING PROGRESS OF SUSTAINABILITY

Every community wants to know how it is progressing over time. To measure a community's progress, *national income accounts* were developed by Nobel laureate Simon Kuznets in the 1930s. Before that time, policymakers had no access to timely macroeconomic data, so they could not take policy actions in response to current economic trends. Because they could not measure national welfare, they did not know the nation's economic status on if their policy actions were helping or hurting. A major part of national income accounting focuses on net domestic product (NDP), which measures the market value of all goods and services bought by consumers, firms, and government agencies, minus the value of the capital stock depreciated by producing that output. Because the NDP gives an overall snapshot of the state of any nation's economy, development of it and the other national income accounts rank among the top inventions of the twentieth century. NDP per person, called per capita NDP, is the average person's share of the economy's NDP. A major economic policy goal in most countries is to promote the highest growth rate for per capita NDP consistent with low inflation and other national objectives.

Despite the importance of the national income accounts, they can give misleading indications of sustainable economic development. NDP fails to account for nonrenewable and renewable resource depletion as well as environmental degradation. This is not simply an intellectual criticism; rather, it strikes at the heart of economic policy and can be especially important in nations and cultures that depend heavily on natural resources and clean environments. If a nation cuts its forests, depletes its soil fertility, and pollutes its air and water, it can become poorer despite an increasing NDP. National income accounts simply record the value of harvested trees, overfished fish stocks, and depleted oil as net contributions to NDP. The numbers may lead economic policymakers to interpret the nation's falling output as growing output. By masking environmental and natural resource depletion, an increasing NDP might block attempts at corrective environmental action. If we are measuring economic growth with the wrong yardstick, we can hardly expect people to make the right decisions for either the economy or the environment.

Green-Adjusted NDP

One method for revising national income accounts to keep track of the potential of sustainable development is to estimate a *green-adjusted NDP*, which accounts for renewable and nonrenewable resource depletion as well as for environmental damages from pollution. Three major modifications to the conventional NDP have been proposed:

1. For nonrenewable resources, subtract from NDP an amount equal to the change in stock of nonrenewable resources multiplied by the difference between their price and marginal cost.
2. For renewable resources, subtract from NDP a quantity equal to the population's annual harvest minus annual growth (change in stock). That quantity is also multiplied by the difference between its price and marginal cost.
3. For the case of pollution, subtract an amount equal to the change in stock of pollutants (emissions minus environmental assimilative capacity). That quantity is multiplied by the difference between its price and marginal abatement cost.

These three adjustments produce the following green-adjusted NDP:

$$\text{Green-adjusted } NDP = NDP - (P_D - MC_D)\Delta D - (P_R - MC_R)\Delta R - (P_E - MC_E)\Delta E \quad (19.3)$$

where P_D and MC_D are the price (marginal benefit) and marginal (opportunity) cost, respectively, of depletable resources; ΔD *is* the net change in stock (depletion minus discoveries) of depletable resources; P_R and MC_R are the price and marginal cost of renewable resources; ΔR is the net change in stock of renewable resources; P_E and MC_E are the price of added emissions (abatement cost saved) and marginal cost of added emissions (pollution damages incurred); and ΔE is the excess of pollution emissions over local environments' assimilative capacity. The equation's three adjustments correspond to the three modifications listed. They adjust NDP for changes in depletable resources, renewable resources, and pollution emissions. This equation applies the sustainability concepts of Solow and Hartwick as well as implements the first three of Daly's operational principles which will described subsequently. Nobel laureate John R. Hicks had the concept of sustainable income in mind when he showed how measured income should be adjusted for depletion of an exhaustible resource (wasting asset):

> If a person's receipts are derived form the exploitation of a wasting asset liable to give out at some future date, we should say that his receipts are in excess of his income, the difference between them being reckoned as an allowance for depreciation. (Hicks, 1946, p. 187)

Equation 19.3 attempts to adjust NDP to account for wasting environmental and natural resources assets. The use of the equation has major implications for measuring sustainability, as well as for policies that could be enacted to carry it out. If the marginal benefit equals the marginal cost of any of the three terms (ΔD, ΔR, or ΔE), then that term effectively takes nothing away from NDP. For example, think about the term ΔE. If policies have already been enacted that efficiently manage a pollutant's emission, then the marginal abatement cost saved (P_E) is equal to the marginal pollution damages incurred (MC_E) for increases in emissions. In that case, no subtraction from NDP is required for proposed policies that affect that emission, because that kind of pollution is already managed efficiently. However, if price is higher than marginal cost for any of the three terms, further depletions incur large subtractions from NDP. For example, suppose high levels of oil production cause a large reduction in the net stock of oil (large ΔD). If the price of that oil is high compared with its extraction cost ($P_D >> MC_D$), then correct sustainability accounting requires a subtraction from unadjusted NDP equal to $(P_D - MC_D)\Delta D$.

Equation 19.3 not only reveals when things are being done sustainably, but it provides insights into the design of more sustainable policies. For example, to keep unadjusted per capita NDP from falling, sustainable policies need to manage all three terms so that all three prices equal their respective marginal costs. When that occurs, all three terms are zero and green-adjusted NDP equals its conventional counterpart. In that case, policies that promote sustainable unadjusted NDP will also promote sustainable green-adjusted NDP. Of course, carrying this out is not easy. Considerable investments may be required to measure all those prices and marginal costs. Valuing pollution damages incurred by emissions (MC_E), for example, has far to go before it passes the test of scientific credibility.

Conditions for Sustainability

Two conditions form the foundation for the sustainable economy just described. First, at each point in time, NDP indicates the largest consumption level that can be allowed this year while guaranteeing no decrease in future consumption. That is, NDP measures the maximum current level of consumer satisfaction that can be sustained forever. Second, this year's NDP is this year's interest on society's total stock of capital. Capital means everything, tangible and intangible, in which the economy can invest or disinvest, including knowledge. This capital stock should be evaluated at the right prices.

The first condition tells us that NDP in any period is a measure of the highest sustainable income achievable, given the total stock of capital available in that period. The second condition tells us that NDP at any instant can be thought of as that same stock of capital multiplied by the discount rate. The goal of economic policy is to make investment and depletion decisions this year in a way that does not erode sustainable income. Those decisions should not permit the aggregate capital stock to fall; preserving sustainability requires maintaining society's capital intact.

Any generation can take from the future. If decisions of the current generation result in high current consumption, the next generation will be forced to start with a lower stock of capital. This is equivalent to saying that the new sustainable level of income is lower than the old one. The high-consumption generation fails to achieve the goal of sustainability. A generation can also give to the future. A generation that consumes little leaves behind a larger stock of capital than it inherited. That generation has increased the sustainable income level and has done it at the expense of its own consumption. Sustainability requires enough investment to maintain the complete stock of capital intact. It does not require keeping intact the stock of everything in the mix; trade-offs and substitutions are allowed.

Each generation depletes a part of the earth's original resource endowment. So what should each generation return in exchange for depleted resources if it wishes to keep its contract with future generations? Each generation should add to the total capital stock in other forms by enough to maintain the aggregate resource stock intact. In other words, each generation should replace the depleted resources with other assets of equal value. The quantity deducted from conventional NDP will give a truer NDP if it takes account of resource depletion. A society that invests what it depletes, possibly in substitute forms, preserves its capacity to sustain a constant level of consumption. This principle can be applied to environmental amenities and the environmental life support system as well as being applied to renewable and depletable resources.

Instruments for Implementing Sustainability

This section describes several principles for implementing sustainable development. Once these principles have been identified, we can discuss a range of possible measures that show whether an economy is growing sustainably. We also will consider the data required to carry out such measures.

Finding Right Prices

Environmental resources are overused because the price attached to their use is typically too low or in some cases is zero. If a resource price is too low, the resource is priced at less than its marginal (opportunity) cost, which damages incentives to use the resource efficiently. For example, a chemical plant pays no price to use the nearby river for waste disposal; thus the plant faces inadequate incentives to reduce its pollution. One significant barrier to attaining sustainable development is that too large a burden is placed on ecosystems by economic activities—too much pollution, too many fish harvested, and the like. Finding policies that put the correct price on various environmental resources, through measures like ecotaxes, would move us closer to sustainable development, since heavy use of ecosystems for current economic activity would likely be reduced. See Example 19–1.

The Hartwick-Solow Rule

Hartwick (1977) and Solow (1974, 1992) have shown that an economy requiring a nonrenewable resource as a production input can guarantee, at a minimum, constant consumption levels over future years if it follows this simple rule: Reinvest all rents (e.g., the difference between a mineral's price and its marginal extraction cost) earned by exploiting the resource into other forms of capital, for example, human capital. As stated previously, this zero net depreciation results in nondeclining consumption through substitution. For example, as the stock of fossil fuels is depleted, society continues to produce and consume electric power through developed solar photovoltaic electric cells.

Example 19–1
Sustaining Development in Poor Cities in Africa

The concept of sustainable development has a strong environmental theme, and the continent of Africa has enviable environmental quality. Yet most people in African cities live in poverty without the most basic human needs like food, clothing, and shelter. Housing shortages in African cities are staggering. In Egypt, for example, overcrowding and poverty drive more than a million people to make their homes in Cairo's cemeteries. In Ghana, housing shortages have forced garages and kitchens to be converted into bedrooms. Lack of basic infrastructure in African cities requires poor people to adopt unsustainable lifestyles. Energy, water, and basic services like garbage collection are often unavailable. People routinely choose between survival and the environment. Usually, the need to survive overrides environmental considerations.

Africa's lowest per capita energy use is good for the environment, but is it good for human living standards? Electricity supply interruptions and fuel shortages dramatically reduce industrial output which makes it hard to overcome poverty. In Africa's urban areas, people use fuelwood for cooking and space heating. With few other reliable energy sources, growing urban populations are placing growing pressure on the urban ecosystem. Indoor and outdoor air pollution occur as a result of carbon dioxide emissions from burning fuelwood. Moreover, the depletion of forested areas near cities can be measured by the long and growing distances traveled to fetch fuelwood. This burden is borne mostly by children and women, who suffer lost education and lost opportunities for viable economic activities.

Despite these problems, potential exists for sustainable development efforts in Africa's cities. An optimist sees a window of opportunity for Africa to save its natural environment by learning from experiences of developed countries. In addition, Africa is environmentally sustainable for economic reasons. Many environmentally polluting activities (e.g., use of agricultural chemicals and fertilizers, household chemicals and transportation vehicles), are expensive and Africa's poverty prohibits their use.

Africa faces a difficult dilemma. Balancing development with ecosystem conservation is an enormous task. It is made harder by low and falling prices of Africa's exports because of deteriorating terms of trade, which results in inability for Africa to earn foreign exchange to buy what it needs. Until this dilemma is resolved, sustainable economic and environmental development in African cities will be difficult to achieve.

Sources: Ebohon, O. J. 1996. The scope and limits of sustainable development in Africa's built environment sector. *International Journal of Sustainable Development and World Ecology.* 3 (1): 1–12; Hope, K. R., and M. H. Lekorwe. 1999. Urbanization and the environment in southern Africa: Towards a managed framework for the sustanability of cities. *Journal of Environmental Planning and Management.* 42 (6): 837–859.

Three assumptions are required for this rule to work: First, utility (total welfare) depends only on consumption priced in the marketplace. Second, the environment is only valued as an imput to production and has no value as a stand-alone producer of utility. Third, all forms of capital (natural, human, intellectual and physical) can be substituted for each other in production. That is, losses of natural capital can be substituted by gains in human capital with no reduction in consumption (technically, the elasticity of substitution is larger than zero). All of the three assumptions may be true in special cases, but they have not been shown to be universally true. For that reason, an important area of scientific, economic, and policy research is the testing of conditions under which any or all these assumptions hold true. If the rule is followed and the assumptions hold true, per capita consumption will remain constant or grow over time, no matter what happens to natural capital.[7]

Critical Natural Capital

If it is required to keep some amount of the existing stock of natural capital to permit future generations to have the same standard of living as the current generations then a rule for sustainable development is required to maintain that natural capital. If we could measure natural capital, a

[7] Howe (1979) approaches this measure of sustainability from a slightly different way but ends up with the same conclusion. He aims to sustain the "effective resource base," by which resource depletion of exhaustible resources is balanced with capital accumulation and technical progress.

strong sustainability requirement says that sustainable development requires a nondecreasing stock of natural capital. If natural capital stocks (e.g., fisheries, wetlands) are maintained, then no decision that depletes the net stock of any component of natural capital would be allowed, regardless of how high its benefits are. One way to maintain the stock of natural capital is through substitution. For example, an action that depleted the stock of wetlands would require an offsetting replacement by creating a new wetland. Applying the principle of maintaining natural capital raises many difficult questions. How much of that natural capital stock should be held constant? What kinds of substitutions among the various elements are permitted? How should we aggregate spatially? Within which geographic area should we hold those stocks constant? If, for example, a 1,000-year-old redwood forest is to be harvested for new homes, does a newly planted redwood forest of the same acreage (or 10 times the acres) at a different location on a different kind of land at another place count as a substitute? Many environmental assets (e.g., the Grand Canyon) are unique—no physical substitute exists. Does the uniqueness of an asset mean that any development that threatens its uniqueness be prohibited? What characteristics of an environmental asset are required to make it unique?

Safe Minimum Standard

The potential of irreversible species loss and uncertainty associated with the value of biodiversity led Ciriacy-Wantrup and his followers to advocate the safe minimum standard (SMS) of conservation (described earlier) as an alternative or addition to formal benefit-cost analysis. History teaches that it is wise to establish safeguards that guard against catastrophic crashes such as irreversible declines in fish stocks, droughts, or ozone depletion. A crash in the earth's life support system and, therefore, in the economy that depends on it, is like a sudden onset of illness or serious injury in a person. Hoping the event will not happen is a poor way to plan; instead, it is best to guard against catastrophic crashes through appropriate action. For example, a current reduction of carbon dioxide emissions from coal-fired power plants guards against risks of a crash caused by the unknowable effects of future global warming. A future catastrophe in an ecosystem carries a large cost to society. If that large cost has a positive probability of occurring, then current actions that reduce that probability or cost will reduce those costs. A crash such as extinction of a species is irreversible. Prudent actions taken to avoid it increase society's options for the future. The more costly the anticipated crash or the higher is its probability of occurring, the more society is willing to pay to avoid it.

Given uncertainties about environmental valuation techniques, unknowns about the evolution of society's environmental values, and knowledge gaps on requirements for ecosystem viability, SMS supporters argue it is best to err on the safe side. As originally conceived by Ciriacy-Wantrup, the SMS principle focused on irreversible species extinctions. In this case, application of the SMS attempts to keep intact the minimum viable population of a species, unless the cost of doing so is unacceptably high. The SMS begins by identifying the smallest viable population or habitat requirement for that population. Management actions that could safeguard that minimum are identified and estimates are made of those actions' costs. If the policy proposal threatens the SMS, society is presumed to rule against it, unless the costs of the safeguarding action are judged to be too high.[8]

One difficulty with the SMS principle is the problem of identifying how society should decide conditions under which the cost of setting the SMS is too high. Another problem with implementing the SMS concept involves its extension to elements of natural capital other than various wildlife populations and their habitats. Still another problem is measuring minimum population or habitat sizes, when these levels could change over time or with new scientific knowledge.

---

[8] This process of implementing the safe minimum standard is close to the methods by which the U.S. Endangered Species Act (ESA) and its amendments have been implemented. The ESA prohibits federal agencies from actions that jeopardize the continued existence of any endangered species.

Example 19–2
Sustaining Development after Natural Disasters in Nicaragua

In 1998 Hurricane Mitch pounded Nicaragua and caused widespread damages. This tragedy illustrates the vicious cycle of unsustainable natural resource depletion combined with a natural disaster. Because of deforestation, forest fires, and deterioration of hillsides that occurred throughout Nicaragua prior to Hurricane Mitch, what could have been minor damages caused by a hurricane were dangerously elevated. Years of poverty in Nicaragua, including little available land for agriculture and widespread shortages of necessities, caused both urban people and farmers to relocate to locations vulnerable to flooding, like riverbanks, lakeshores, and mountain slopes. The only shelters these people could afford were cheap, flimsy shacks that were vulnerable to heavy rain, mudslides, and flooding.

Unsustainable policies adopted previous to the hurricane compounded the damages. Years of poverty had encouraged international lending agencies to promote economic policies that sacrificed a safe environment for short-term economic growth. The World Bank and International Monetary Fund encouraged policies in which Nicaragua would reduce imports and increase exports, thereby earning more foreign exchange to buy a wider range of foreign goods. Nicaragua could increase its exports only by unsustainably increasing natural resource harvesting and production. Nicaragua made many structural adjustments during the early 1990s that increased its dependence on export agriculture and placed small subsistence farmers in a dangerous position. The adjustments also led to increased clearing of forests and more agricultural land cultivation, so Nicaragua became even more vulnerable to natural disasters.

Now Nicaragua has few sustainable policy choices. For example, a common policy proposal is for governments to offer property rights to poor farmers to give them incentives to practice sustainable use of resources. But this policy would be hard to implement in Nicaragua because of a lack of cultivatable land due to severe soil erosion. After Hurricane Mitch, even less cultivatable land remains. Another policy proposal would tax farmers who practice unsustainable development to offset damages to the environment and to future generations. This presents a problem of environmental justice: Most of the unsustainable land use takes place in the hands of the rural poor, who have little choice but to overexploit natural resources available to them for survival. Taxing these farmers would worsen their situation considerably.

Hurricane Mitch teaches important lessons about sustainable environmental and economic planning. By focusing on the long-term outcomes of policy decisions on society, countries can protect themselves better against negative impacts that are outside their control, such as natural disasters.

Source: Lane, H. 2000. Sustainable development versus economic growth: A case study on natural disaster in Nicaragua. *Journal of Environmental and Development.* 9 (2): 175–182.

There is a major difference between implemention of the SMS and the critical natural capital method: The SMS for any resource in question is allowed to be violated if society believes the (opportunity) cost of setting the SMS is unacceptably high. According to the critical capital principle, the critical capital level is safeguarded at any cost. Both the SMS and critical capital method leave unanswered the question of who pays the costs of safeguarding the environment. Many believe that the brunt of preservation efforts falls disproportionally on poor developing countries. As shown in several examples in this chapter, developing countries feel considerable pressure to use their minerals, land, fisheries, water, and rain forests for current economic benefits. See Example 19–2.

Avoiding Irreversibility

If environmental and natural resource policy actions are reversible, then the sustainability argument loses much of its force and power. If society discovers that current behavior is unsustainable, it simply changes that decision at whatever time period the altered decision produces benefits that exceed costs by the greatest amount. Reversibility means that nothing is lost forever. Bigger challenges are posed by irreversible decisions. Use of an exhaustible resource is irreversible in the sense that after it is depleted, that quantity is gone forever. We may regret our choice to use it, but the decision cannot be reversed. Irreversibility may be even more important for renewable resources. If a current harvesting rate leads to a fish stock falling below the critical level, the species will become irreversibly extinct.

Another kind of irreversibility affects decisions that significantly change the use of an environmental resource. Once developed, a wilderness area cannot be returned to its original state in a timescale relevant to normal human actions. This situation poses a classic asymmetry: A decision to avoid development can be reversed sometime in the future. A decision to develop can happen only once; it cannot be reversed. Carrying out the development option closes off the preservation option forever.

The argument surrounding irreversibility was stated by Krutilla (1967). He found that at any point in time where the environmental footprint left by technology is getting smaller, there will be a greater current conversion of a natural environment into an industrial resource than would occur in the future with more advanced technology. Furthermore, with growing public value of an undeveloped natural environment, its conversion into industrial products will proceed even further than if that conversion were reconsidered in the future. If these natural environment conversions are irreversible, it is impossible to achieve a level of well-being in the future that would have been possible had that natural environment conversion been delayed. So we are faced with carrying out a current action with considerable ignorance about future tastes, future technologies, and remaining supplies of unique environmental assets. Although formal benefit-cost analysis may suggest that environmental development for industrial uses produces a higher net benefit than preservation, current development may produce negative net benefits compared to development in the future. Where there is a finite risk of a high cost from that kind of future regret occuring, setting a SMS may be an attractive method to avoid incurring that cost.

Steady-State Principles

Daly (1990) proposed four principles that, if followed by policymakers, would give rise to sustainable development. The first principle applies to renewable resources such as forests, fish, game, rangelands, and renewable water. For those resources, policies should be enacted that set harvest levels at no more than the population's growth rate. In no case should the harvest rate exceed the maximum sustained yield, which you will recall is the largest possible annual sustainable harvest.

The second principle deals with nonrenewable resources such as minerals, or nonrechargeable aquifers. For these resources, he proposes an investment of enough of current returns from the mined resource so that in the future, when the resource is economically depleted, renewable substitutes produce as much productive capacity as existed when extraction began. This is another way of stating the Hartwick-Solow rule, which suggests regulating the extraction of natural resources so that renewable supplies are available by the time the natural resources are economically depleted. For example, Daly suggests that groundwater pumping should be regulated so that renewable surface water supplies are available by the time the aquifer is economically depleted. A pumping permit issued for a depletable aquifer would require an offset, which could be carried out by building an aquifer recharge (reinjection) area near a river to accept water recharge in wet years. As another example, *severance taxes* could be placed on coal or oil deposits that finance the development of universities, high-tech industries, or even interest-bearing financial instruments that can substitute for future depleted deposits. See Example 19–3.

Daly's third principle applies to pollutants: Emissions should not exceed the receiving environment's capacity to assimilate them in any given time period. This principle raises the challenge of how to measure an environment's capacity to assimilate wastes. It also presents a challenge in a situation where assimilative capacity varies widely across time periods or across locations. In situations where the environment has zero capacity to accept a pollutant, the correct policy is to set discharge limits at zero. A river's capacity to assimilate discharges may be high during the spring runoff or in the mountains near its origin, for example, but that capacity may be nearly zero in late summer when flows dwindle or in downstream locations where most of the river is depleted.

Example 19–3
Sustainable Water Programs in Malawi

Many rural areas of the world lack piped water, and policymakers face the question of how to best supply it. Malawi, a small landlocked nation in southeastern Africa, has experimented since the 1960s with programs to bring piped water to rural areas. Malawi's water supply programs provide a historical record for determining what is best in rural areas.

In 1968 the Malawi government built its first gravity-fed water project to supply drinking water to rural populations. By 1980 activities had expanded to the construction of 32 such projects. A notable feature in these programs was the absence of local cost-sharing, so local people had an economic incentive to request such programs. Another feature was comparatively little central government involvement; most of the operation and maintenance was done by local water-using communities. In 20 years the programs supplied water to 13% of the nation's population. Postconstruction studies showed that more than 80% of the taps still had running water.

Despite many successes, maintenance problems pointed to unsustainable future water supplies for nearly half the projects. The fact that water was free was a major cause of poor maintenance. Consumers were reluctant to pay regularly toward a maintenance fund when everything seemed to be working just fine. If the price of free water is an unsustainable water system caused by inadequate revenues to keep the system going, then free water is unsustainable. Securing a stream of revenues to keep up the system is a big part of system maintenance. Obtaining that revenue is considerably more important than preserving the water supplier's good relationship with consumers. That steady stream of revenues would go a long way toward repairing pipes and conducting other system maintenance. Another problem produced by free water is the necessity of asking for large amounts of in-kind contributions, especially volunteer labor. Skilled labor, developed and supplied at a premium wage, keeps the system running much better than free, unskilled volunteer labor.

Overall, Malawi's piped water programs have worked well to date. The smallest and newest rural programs are working best, but nearly half perform badly. Two policy actions could improve performance: charging local water buyers for most or all the water they use and building smaller systems. The participatory approach works well for a community organization that can manage small, rural, piped, gravity-fed programs, but it works poorly in meeting the needs of more complex schemes, which local, unskilled labor cannot maintain.

Source: Kleemeier, E. 2000. The impact of participation on sustainability: An analysis of the Malawi rural piped scheme program. *World Development*. 28 (5): 929–944.

Daly's last principle deals with placing limits on the overall scale of human activity, including use of depletable resources. Daly views the increased scale of people and their related density as a pervasive source of human unhappiness.[9] To implement his fourth principle, Daly recommends more restrictive controls, for example, on pollution emissions, energy use, depletable resource use, and most politically controversial, on population. Presumably, these controls should be set to maximize the present value of future net benefits. But if future population is precluded by those controls, how does one deal with the present value of their unrealized net benefits (see Chapter 14)?

■ Policy Applications

The purpose of understanding sustainability is to enact policies that make it happen. Suppose some government agency publishes data on the economic value of each year's resource depletion. Based on the data, we know that we owe to the future a quantity that compensates for depletion of this year's inherited stock. We have taken a known quantity whose economic value is X, and

[9] Daly's point is illustrated by many who wrote about the delights of living in uncrowded, sunny, temperate, pollution-free California before 1940. Although California was still a desirable place to live for many of its 35 million plus residents in 2004, most residents would find living there more desirable if population could be reset to 1940 levels of 7 million. But what about the 28 million Californians who would have to leave? Would they be better off?

we should provide a replacement, possibly in another form, but at least as large as *X*. If, for example in a frontier economy people enter a virgin forest and harvest trees worth *X* to future generations, those people can keep their contract with the future by using those harvested trees to build houses, factories, farms, and technical knowledge worth at least *X*. Borrowing from future generations may have happened in the United States in the late nineteenth century. Depleted natural and environmental resources were traded for a growing capacity for food and industry as well as newly developed centers of knowledge.

A sustainable policy produces an economically equivalent amount of net investment, large enough to maintain society's capital stock. A commitment to sustainability translates into a commitment of making a measurable amount of productive investment. This kind of policy would likely have a high priority in developing economies that produce primary products, economies that supply the advanced industrial world with resource-based products like oil. To keep their contract with future generations, these developing countries would channel their resource rents (differences between market price and extraction costs) into productive investment. They might even want to invest more, because mere sustainability fails to help future generations in poor countries improve their living standards. Several oil-producing countries, in fact, have pursued a sustainable path by channeling oil revenues into factories, farms, and universities. Some have squandered part of their oil revenues into luxury consumption.

Environmental preservation can be handled in much the same way, but the concept of sustainability is much more difficult to apply to the environment for many reasons. Some environmental assets may confer a large benefit to society, not as an input to further commodity production, but as a final product with value in and of itself. Unique and irreplaceable natural treasures are examples. If some feature of the natural environment or environmental life support system is claimed to irreplaceable it is not open to substitution by human production without severe loss in welfare. Although this claim can be disputed, it is certainly true in some cases. When such resources are found, failure to protect them for future generations makes sustainability an expensive if not impossible proposition. The cost of such failure depends on the nature of utility functions for unique environmental assets and the extent to which substitution occurs in those functions between environmental and man-made assets.

Loss of environmental assets poses a special problem in developing countries. For these countries, a dilemma arises when they discover that adoption of environmental standards set by developed countries makes local industries uncompetitive in global markets. The poor countries then face a choice between cooperating to degrade their own environment to gain material affluence and keeping their natural environment while continuing in economic poverty. Resolving this choice in a country's best interest lies in making known trade-offs between an immediate loss of environmental amenity and a gain in future economic well-being. Temporary acceptance of poor environmental conditions can be made more acceptable if policies are established to translate incomes into productive investment. A portion of the population's higher incomes could be spent on environmental repair, as long as the total stock of valued resources is maintained. When poor countries in search of their own economic goals contribute to global environmental damage, such as India's and China's and contribution to global warming by burning coal for power, much more difficult policy questions emerge. See Examples 19–4 and 19–5.

Institutional Design

Market forces provide strong incentives for environmentally safer production methods as prices and profits respond to reduced environmental quality. Some people feel that market forces occur too slowly, too late, or not at all. Government interventions by way of regulations, taxes, subsidies, and direct spending on environmental research and development may be necessary.[10] Let us consider several such sustainable, environment friendly institutions.

---

[10] The British government has made plans to implement sustainability (United Kingdom Government, 2001).

Example 19–4
The Search for Sustainable Environmental Policies in India

India's large population, widespread poverty, and sensitive natural environment make it an excellent case study for examining the impacts on various income groups of economic and environmental policy reforms. Also, India's growing use of polluting energy (e.g., coal) increases the importance of good environmental policies and raises the costs of poor ones. In a poor developing country like India, failure to integrate environmental and economic policies has had several undesirable effects, including damaging the environment, reducing the level of income, and increasing income inequality.

Price controls and subsidies are widespread in India. Keeping prices of essential consumer goods like food, water, and energy low was historically justified on equity grounds, as lower prices help the poorest people. Similarly, government price control on inputs like fertilizers, irrigation water, and energy is designed to increase industrial production and food output, both of which help the poor. Despite the best intentions, however, many of these subsidies have high environmental costs. For example, agricultural chemicals have damaged groundwater aquifers and energy production has contributed to global greenhouse gases.

India's wide gap between rich and poor threatens sustainable production opportunities because the poor typically rely more heavily than the rich on natural resources like topsoil, water, and minerals. So the poor may be depleting natural resources faster than they would if they had higher incomes and could afford to use nonpolluting resources. This phenomenon of greater environmental damages produced by falling incomes is known as the poverty-environment hypothesis. If the hypothesis is true, degraded environments increase poverty while greater poverty also degrades the environment. Therefore, environmental and economic policies in India need to account for the potential circular link between poverty and the environment in which causality moves in both directions.

Environmental and economic policies are formulated by different people in different organizations with different backgrounds, skills, and beliefs. To show how more integrated policies could help the people of India, Nugent and Sarma developed a model in which all major environmental resources, economic sectors, and income levels are linked together. Their model keeps track of environmental impacts of economic policies and economic impacts of environmental policies. They conclude that under the right conditions, better policy integration can be achieved in such a way to achieve economic growth, better environmental quality, reduced poverty, and greater equity—all at the same time. They suggest an integrated policy to simultaneously reduce energy, fertilizer, and irrigation susbidies. Such integrated polices could increase overall income, produce greater equity, and reduce environmental damages.

Source: Nugent, J. F., and C. V. S. K. Sarma. 2002. The three E's—efficiency, equity, and environmental protection—in search of win-win policies: A CGE analysis of India. *Journal of Policy Modeling*. 24: 19–50.

Extended Producer Responsibility. Extended producer responsibility (EPR) is a policy measure in which laws are passed that require producers to be financially responsible for their products after those products' useful lives are over. EPR programs are growing in acceptability in some Asian and European countries. EPR requires that producers reclaim used products and deal with them through remanufacture, recycling, or reuse. Costs are financed by the producer and, ultimately by the consumer through higher product prices. The programs are based on the principle that ultimately consumers are responsible for paying environmental costs of the product, including downstream waste management and recycling, and for pollution emissions through remanufacturing, recycling, and reuse. Requiring producers and consumers to be responsible for all environmental costs of a product provides two important incentives for sustainably protecting the environment:

- Producers are rewarded with higher profits if they use product designs that are made to ease environmental damage, for example, products that use fewer materials, require less packaging, are easier to fix when broken, and are easier to recycle.
- Consumers prefer to buy more environmentally safe products because the price is lower. That is, when two products are identical except for environmental costs, the products with lower environmental costs also have a lower market price, thereby encouraging consumers to buy that product.

Example 19–5
Sustainable Energy Programs in China

China's remarkable economic growth has had negative environmental impacts because of the growth in energy use, which increased at an annual rate of 5.3% from 1980 to 1991. Energy-intensive economic growth in China has been accompanied by growing air quality problems. Dirty air is visible in many large cities where large amounts of fossil fuels are consumed to produce power to heat homes and commercial buildings.

China's recent history is one of the most energy-intensive among Asian countries. Although energy intensities vary in China, on average it requires three to four times as much energy input per unit of output as developed countries require. However, globalization means that China must compete with production in other countries that use much less energy per unit output. For it to compete in the long run, China must decrease energy intensity before other countries exploit their comparative energy efficiency advantage to capture emerging markets. Energy intensity among industrialized, high-income countries has continued to decline through aggressive pursuit of energy-efficient production since the energy crisis of the 1970s. High levels of air pollution in many of China's cities make reduced energy intensity even more important. Three of the 10 large cities with the world's highest monitored sulfur dioxide concentrations are in China (Chenyang, Guangzhou, and Bejing).

Careful development and use of renewable resources, such as solar, wind, and geothermal energy, could improve China's overall efficiency in two ways. First, renewable resources reduce the risk of escalation of fuel costs. Second, renewable sources typically do not require large-scale operations for maximum efficiency. Instead, they can be scaled to current demand needs and later be expanded to meet new demand as it occurs. The most important contributions of sustainable renewable energy sources are their considerable health and environmental benefits.

Choosing an energy development path that emphasizes renewable sources may be in China's sustainable economic and environmental interest. China could move in this direction by investing in efficiency and renewability quickly, because it does not have an already built energy system based on large-scale, depletable, polluting resources like many industrial economies have.

Sources: Byrne, J., B. Shen, and X. G.Li. 1996. The challenge of sustainability: Balancing China's energy, economic and environmental goals. *Energy Policy.* 24 (5): 455–462; Byrne, J., B. Shen, and W. Wallace. 1998. The economics of sustainable energy for rural development: A study of renewable energy in rural China. *Energy Policy.* 26 (1): 45–54.

Globally, EPR programs have been initiated in 20 countries in the Organization for Economic Cooperation and Developement (OECD) and in some non-OECD countries. Despite some successes, however, the concept of EPR remains controversial. To determine if an EPR program is environmentally effective and economically efficient, the benefits of the cleaner environment must be measured and compared with the costs of the higher product prices that EPR brings about.

Tax Reform. A common policy of many countries is to tax destructive environmental activities. *Environmental tax reform* proposals are based on the idea of raising the price through taxation of activities like pollution and exhaustion of depletable resources. Taxes on environmentally damaging activities raise the relative price of environment-damaging or resource-depleting activities, and reduce the relative price of activities that are resource-conserving and environmentally benign. Here are two examples of tax reform:

1. A policy enacted by the U.S. government in the late 1970s was intended to promote greater use of renewable solar energy. Under the program a percentage of the cost of installing passive solar energy equipment on homes could be written off as a tax credit. Each dollar of the tax credit was one dollar in reduced tax liability, so the tax credit for investments in passive solar energy offered a much more powerful financial incentive than an equivalent tax deduction. Many citizens who lived in sunny climates, who might have otherwise rejected passive solar energy, received the double benefit of reduced tax liability and a lower winter heating bill by installing solar energy equipment.

2. In the late 1970s the state of California enacted a reform to reduce environmentally destructive production of electricity (e.g., burning coal). California's approach was to establish legal mandates requiring utilities to buy power at their avoided costs from environmentally cleaner sources (e.g., solar energy, hydroelectric power, or natural gas). As a side effect of this policy, California power buyers had to pay much higher prices than the rest of the nation, nearly 50% above the national average (see Chapter 13).

The challenge posed by such policies is to continue using tax codes and regulations to promote environmentally friendly programs as long as the benefits of the programs exceed their costs. The question remains, however, of whether the environmental benefits of cleaner but more expensive power exceed the costs. To formulate economically effective environmental tax reforms, policymakers must be able to credibly measure the benefits of a higher-quality environment so that those benefits can be objectively compared with the higher price of environmentally regulated activities.

Ecolabels. Ecolabel programs are designed to promote sustainable production and consumption by passing some environmental costs to consumers. *Ecolabels* provide a signal to consumers of the environmental impacts of producing or using a product. They are voluntary, but in some markets they are becoming an important competitive factor. Although there are many different labeling programs run by governments, private companies, and nongovernmental organizations, there are only three types of labels. The Geneva-based International Standardization Organization is creating standards for each.

Type I labels compare products with others within the same category. Labels are assigned to products that are environmentally preferable through their complete life cycle. The criteria are set by an independent body and monitored through a process of certification. Labeling products in this way requires difficult judgment calls. Consider two products that are identical, except that one pollutes air and the other pollutes water. Which is superior? Presumably superiority is based on the damages produced by each of the two kinds of pollution, but those damages are hard to measure, verify, and defend. Perhaps ecolabels will help promote sustainable exports for Tanzania, as discussed in Example 19–6.

Type II labels are environmental claims made about a product by its manufacturer, importer, or distributor. The claims are not independently verified, do not use predetermined and well-established standards for reference, and provide the least information of the three types of environmental labels. Here is an example: A manufacturer's label states the product is biodegradable without defining the term.

Type III labels list a product's environmental impacts throughout its life cycle. These labels are similar to nutrition labels on food products that summarize fat, sugar, or calorie content. The information categories can be set by a trade group or by independent bodies. In contrast to type I labels, type III labels do not compare the environmental costs of products, leaving that decision to product buyers. Critics of type III labels believe few consumers have the time or will to judge whether, for example, emissions of sulphur are more costly than emissions of carbon dioxide.

Not all proposals for ecolabels are voluntary. For example, the European Union (EU) wants World Trade Organization (WTO) rules changed so that trade can be restricted on the basis of the environmental impact of the way in which products can be produced, processed, and distributed throughout the product's life cycle. The WTO position on ecolabeling has been that it does not discriminate among different producers based on environmental impacts of products. Under the EU proposal, however, required ecolabel certification would show that all regulations are being followed throughout the product's life cycle. In addition, the manufacturer would be obligated to meet the cost of disposing of the product at the end of its life cycle. To ensure that imported products that avoid these life cycle costs have no cost advantage over environmentally responsibly produced domestic products, the imported product would be excluded from sale without the ecolabel. By requiring this ecolabel as a condition to sell products in the EU, importers would have an economic incentive to buy only environmentally safe products. The WTO has made no decision on whether ecolabels are compatible with WTO's more limited mission of helping trade flow smoothly, freely, and predictably.

Example 19–6
Sustainable Exports in Tanzania

After the Earth Summit of 1992, Tanzania (a country in southeast Africa) decided to set up its own sustainable development program. Tanzania found that two sectors of its economy, mining and tourism, were major income earners, but both acted unsustainably, thereby risking the sustainability of the country's environmental and economic future. Both sectors are heavy users of environmental and natural resources, so a significant effect of their expansion has been a reduced quality of the environment. Of special concern was deforestation and biodiversity loss caused by these economic activities.

Land clearing at a mining site is where the environmental problem starts. Mining activities bring concentrations of people, who clear the land of trees for settlement puroses, building materials and fuel, and subsistence agriculture. This deforestation, in turn, leads to soil erosion and reduced water quantity and quality. Coastal tourism has similar ripple effects that damage the environment. Often tourism begins with destruction of mangrove forests (coastal rain forests that protect shorelines from erosion and wave damage), estuaries, beaches, and coral reefs, which are important bird and fish habitat and breeding areas. This can have a negative effect on a region's fish stocks and related industry. Many structures serving tourists (e.g., hotels) in Tanzania have been built with mangrove poles, and demands for wood is increasing due to human population and tourism growth.

Kulindwa analyzed the environmental and economic impacts of a range of possible mining and tourism policy reforms and found that several measures could be taken to direct sustainable economic activity. As seen from this chapter's several definitions of sustainability, sustainable development requires integrated actions in many economic and environmental sectors. Well-coordinated implementation poses special challenges to a poor country like Tanzania. For one thing, sustainable development is hard to sell to policymakers who are constrained by immediate problems like political and economic survival. Another challenge is that government agencies often have little incentive or money to cooperate and coordinate sustainable development plans. Kulindwa found that better training for civil servants could help, because these people have a major influence on carrying out sustainable policies. For example, they interact with communities in sustainable programs like wildlife conservation, forestry, and fishing. Another important way to facilitate a sustainable development path is through educational programs in environmental protection and sustainable technologies. Kulindwa recommends that a Commission for Sustainable Development be established in Tanzania to oversee, coordinate, and facilitate the integration of policies, strategies, and activities for achieving a sustainable development path.

Source: Kulindwa, K. 2002. Economic reforms and the prospect for sustainable development in Tanzania. *Development Southern Africa.* 19 (3): 389–403.

ISO 14000. Not all environmental regulations have to be set by government legislation to be effective. Groups of business may voluntarily agree to regulate themselves. Since the late 1940s, the International Standardization Organization (ISO), a worldwide group that sets national standards, has provided voluntary manufacturing standards. These standards are established through a voluntary consensus approach and are held in high esteem by most of the world's manufacturers.

Environmental regulations influence comparative costs and trading patterns among trading partners. Growing globalization gives rise to greater opportunities for businesses to engage in international trade. More widespread international trade can make it more difficult to enforce consistent international environmental standards. A common argument made by businesses is that domestic firms will lose business to foreign nations that can undercut them in price by manufacturing under less strict environmental standards. For example, a foreign country that requires no environmentally friendly packaging can sell its packaged products more cheaply than another nation which does require it. Domestic firms that comply with the regulation, producing their product in an environmentally responsible way, could be driven out of business by foreign competitors.

Following the establishment of the WTO in 1986, a need was identified for developing international environmental standards to deal with the political problems of inconsistent environmental standards. The proposed standard ISO 14000 is expected to provide consistent international certification for manufacturing and other businesses in an environmentally sustainable way. Although much remains to be done regarding certification requirements, some progress has been made toward establishing uniform international standards. The following working groups have been established by the ISO:

- Environmental Performance Evaluation: Provides guidelines on defining baselines and measuring performance
- Life Cycle Assessment: Establishes principles and methods for the conduct of a life cycle assessment
- Environmental Management Systems: Sets up principles and guidelines for organizing high-quality products consistent with minimum environmental damage
- Environmental Auditing: Identifies methods and standards for audit of an environmental management system
- Environmental Labeling: Establishes a framework for wording and methods for identifying environmental impacts of a product

ISO 14000 was set up so that becoming certified under it would benefit domestic business, international business, worldwide consumers, and the environment. Supporters of the ISO 14000 process hope that firms that comply with its certification requirements will realize reduced costs of waste management, lower energy costs, a better public image, and a way to help them formulate environmental plans and identify their environmental performance. Supporters hope that ISO 14000 will actually provide firms a competitive advantage as well as a method for identifying progress in their environmental programs.

Global Actions

Several global actions have been initiated to promote a sustainable future. Some of the most important are discussed next.

Conference on the Human Environment (1972). The late 1960s witnessed tremendous progress by the then-new environmental movement in North America and Europe. In 1972, the international community met in Stockholm, Sweden, to examine for the first time global environmental and development issues. This was the United Nations Conference on the Human Environment. It was the first world conference on the environment and resulted in a number of important decisions, including the Stockholm Declaration on the Human Environment and the creation of the United Nations Environmental Program. Hundreds of other conferences have been organized since then; a few of the most important ones are described next.

The World Commission on Environment and Development (1987). By 1987, the World Commission on Environment and Development had been established by the United Nations to study the process of global environmental degradation and examine ways to ensure the long-term viability of human society. As noted at the beginning of this chapter, the commission was chaired by Norway's prime minister Brundtland. Its report, *Our Common Future,* has become the benchmark for understanding the global environment.

Rio Summit (1992). The 1992 United Nations Conference on Environment and Development (also known as the Rio Summit and as Earth Summit 1992) brought environment and development issues to front and center stage. This summit was the largest-ever gathering of heads of state, senior government officials, and stakeholder representatives. It raised global awareness of the importance of sustainable development and emphasized the need to integrate environmental and economic development issues, which typically were viewed as independent issues.

The Rio Summit is noted for several accomplishments. The Rio Declaration on Environment and Development reaffirmed the Stockholm Declaration, provided several guiding principles for actions, and contributed to the development of general principles of international law. Agenda 21 was an action plan for sustainable development; it summarized actions, roles of specific groups, and methods of implementation, to achieve sustainable development. The United Nations Commission on Sustainable Development (CSD) was created to protect the implementation of Agenda 21 and encourage dialogue and partnerships between governments, the international community, and nine major groups. The Framework Convention on Climate Change established commitments among nations for developing national inventories of emissions and formulating programs to limit climate change. The Convention on Biological Diversity was established for conservation and sustainable use of biodiversity by promoting scientific and technical cooperation, establishing protected areas and eradicating undesired plant species, and providing financial support.

To follow up on the achievements of the Rio Summit, the CSD was to track and publicize implementation of the summit's agreements. After five years a meeting was held to review progress made since the Rio summit, but the meeting did not achieve significant movement forward for sustainable development.

Kyoto Protocol (1997). At a 1997 meeting dealing with sustainability in Kyoto, Japan, participants agreed to a protocol to reduce emissions of heat-trapping gases that cause global warming. The protocol established different levels of reductions for individual nations, depending on their economic and national situations. The Kyoto Protocol purposed that the reduction targets begin to take effect in 2008 and achieved over various five-year periods, (details are discussed in Chapter 15).

Johannesburg Summit (2002). Since the Earth Summit in 1992, there has been some progress in promoting the ideals of sustainable development. Despite high hopes for improvements, the facts show continued poverty for a growing world population, inadequate health care, and degradation of the environment. The issues are compounded by a lack of funding and a complex of factors such as poor and ineffective implementation and efforts, strategy gaps, and lack of overall coordination and agreement.

In Johannesburg, South Africa, in 2002, the World Summit on Sustainable Development began with hopes for major accomplishments. The conference provided an opportunity to reenergize and refocus the global commitment to the achievement of sustainable development. Some of the decisions reached by the countries participating in the summit are listed next:

- To halve by 2015 the number of people around the world who lack clean water and basic sanitation. Worldwide about 1.1 billion people lack access to adequate drinking water. Developing proper sanitation would significantly reduce diseases such as cholera.
- To take action to help the poor gain access to affordable energy. Participating countries failed, however, to agree on specific targets to boost the share of energy produced from renewable sources like solar or wind power. Globally, 2 billion people lack access to electricity. The EU wanted targets but the United States and some oil-producing nations opposed targets.
- To aim to make health care affordable and consistent with basic human rights and cultural and religious values.
- To reduce the speed at which threatened plants and rare animals become extinct. The plan did not set specific targets, and the wording of the plan does not limit countries from pursuing development projects. The World Wildlife Fund criticized the plan for providing little protection for rare species.

The Kyoto Protocol on global warming received major support at the summit when Russia and Canada announced that they would ratify the Protocol. Their support means that enough of the major producers of greenhouse gas have signed to bring the treaty into effect by early 2005. Nevertheless, the Kyoto Protocol suffered a setback when the United States announced it would not ratify it. Negotiators also agreed on giving precedence to environmental regulations over trade and focused on the need to fight corruption and promote democracy and the rule of law, but they did not make aid conditional on democratic governance.

■ SUMMARY

The concept of sustainability received worldwide attention after the publication of a 1987 report by the United Nation's World Commission on Environment and Development. The report, titled *Our Common Future* (the Brundtland Report) defined sustainable development as development that meets the needs of the present without compromising the ability of future generations to meet their own needs. Any generation wishes to give to its posterity not any particular thing, but rather a future endowed with the resources needed to achieve a standard of living at least as good as its own and to look after the next generation similarly.

Economists have offered various definitions of sustainablity, all of which are attempts to apply the principle of sustainability consistent with the Brundtland Report: Some definitions, for example, are based on nondeclining consumption per capita, production opportunities, use of resources, nondeclining natural stock of capital, ecosystem stability and resilience, risk avoidance, coevolutionary development, and species existence.

One method to measure the potential of sustainable development is to estimate green-adjusted net domestic product (NDP). This method modifies conventional national income accounts to keep track of the long-term impact of renewable and nonrenewable resource depletion, as well as to accounting for environmental damages from pollution.

Several principles have been proposed for implementing sustainable development. These include finding the right prices; the Hartwick-Solow rule, which ensures zero net depreciation of the capital stock; defining and enforcing critical levels of natural capital; implementing the safe minimum standard; avoiding irreversibility, and carrying out Daly's steady-state principles.

Regulatory institutions are being developed to bring about a more sustainable future. Among those in use internationally are extended producer responsibility, environmental tax reforms, ecolabels, and ISO 14000, an international environmental standard proposed by the International Standardization Organization. Finally, several important global actions have been initiated to promote a sustainable future: the United Nations Conference on the Human Environment (1972), the Brundtland Report (1987), the Rio Summit (1992), the Kyoto Protocol (1997), and the Johannesburg Summit (2002).

■ REVIEW QUESTIONS

1. Sustainability requires thinking about efficiency and equity. Do you agree? Why or why not?

2. Define and explain the use of each of the following concepts: weak sustainability, strong sustainability, and natural capital depreciation.

3. Suppose sustainability is defined as maintaining production opportunities for the future. Explain why and how those opportunities can be guaranteed even if supplies of natural capital decrease.

4. Critique this statement: A morally acceptable contract with the future requires avoiding any damages to the natural environment.

5. Under what conditions does a sustainable future require maintaining the current natural capital stock?

6. Describe green-adjusted NDP, why there is a need for it, and what desirable policies could result from its measurement and use.

7. Explain the safe minimum standard and some of the difficulties involved in designing programs that carry it out.

8. Every part of a nonrenewable resource used today blocks its use by future generations. Therefore, the only ethically defensible policy for the current generation is to develop and use only renewable resources. Do you agree? Why or why not?

9. Analyze and discuss the following: Future generations neither vote in current elections nor buy in current markets. Therefore, current market decisions will always ignore the interests of future generations.

10. Critically analyze the following statement: If well-enforced regulatory structures do not exist, no individual or community will take the needed steps to achieve a sustainable future.

11. Explain the strengths and limits of using international environmental standards among trading partners as a way to promote sustainability.

12. Explain the economic incentives that influence private businesses to invest in measures to reduce pollution. How might public policies take advantage of these incentives?

13. Extended producer responsibility (EPR) presumes that all players in the product cycle have a role to play in promoting sustainability. What is the role of the consumer (product buyer)?

14. What are ecolabels and how do they give consumers an opportunity to register support for a sustainable future? Do you see any difficulties in relying completely on voluntary purchases of ecolabeled products as a way of promoting a sustainable future?

15. What are the goals of ISO 14000? What methods would ISO 14000 use to improved environmental quality worldwide? Do you see any problems in relying exclusively on voluntary participation in ISO 14000 by private businesses?

Problems

1. Accounting: Explain, using the green-adjusted NDP formula (Equation 19.3), how unadjusted NDP can increase while green-adjusted NDP falls over time. What kinds of policies could be enacted so that green-adjusted NDP increases over time?

2. Policy analysis: Can green-adjusted NDP be increasing over time while mineral depletion occurs? Use the green-adjusted NDP formula to show what is required for sustainable economic development to occur when mineral resources are being depleted.

References

Brundtland, G. H. 1987. *Our Common Future.* New York: Brundtland Commission, United Nations.

Ciriacy-Wantrup, S. V. 1968. *Resource Conservation: Economics and Policies.* 3rd ed. Berkeley: University of California Agricultural Experiment Station.

Daly, H., 1990. Toward some operational principles of sustainable development. *Ecological Economics.* 2: 1–17.

Hartwick, J. M. 1977. Intergenerational equity and the investing of rents from exhaustible resources. *American Economic Review.* 67 (December): 972–974.

Hicks, J. R. 1946. *Value and Capital.* 2nd ed. Oxford UK: Clarendon Press.

Howe, C. 1979. *Natural Resource Economics.* New York: John Wiley & Sons.

Krutilla, J. V. 1967. Conservation reconsidered. *American Economic Review.* 57 (September): 777–786.

Lele, S., and R. B. Norgaard. 1996. Sustainability and the scientist's burden. *Conservation Biology.* 10 (2): 354–565.

Norgaard, R. B. 1984. Coevolutionary development potential. *Land Economics.* 60 (2): 160–173.

Pezzey, J. 1992a. Sustainability: An interdisciplinary guide. *Environmental Values.* 1 (4): 321–362.

———. 1992b. Sustainable development concepts. Washington D.C.: World Bank Environmental paper no. 2.

Solow, R. 1974. Intergenerational equity and exhaustible resources. *Revew of Economic Studies.* Symposium: 29–45.

———. 1992. *An Almost Practical Measure of Sustainability.* Washington D.C.: Resources for the Future.

United Kingdom Government. 2001. *Resource productivity: Making more with less.* Available from *http://www.number-10.gov.uk/su/resource/default.htm* [Accessed October 12, 2004].

Appendix A

Supply and Demand Mathematics

This appendix uses calculus to show how to analyze the impacts of supply and demand changes on price and quantity in a single market. Those impacts are caused by changes in supply, demand, and environmental policies. Equilibrium price and quantity are determined by the intersection of supply and demand schedules.

Demand is influenced by price, various demand shifters, and sometimes by environmental policies. The term α appears in the demand curve to represent the impact of any demand shifter. Examples of α are tastes, substitute prices, income, and the like. Supply is also influenced by price, several supply shifters, and various environmental policies. The parameter β appears in the supply curve to represent impacts of any nonprice supply shifter such as resource prices, technology, and seller numbers. Environmental policy decisions are represented by the term E. It can appear in *both* the demand and supply curve (but it does not have to). It represents the impact of an environmental policy variable that could shift either demand or supply.

Supply and demand are each represented by one equation. Each equation contains two unknown variables (equilibrium price and quantity). The two equations also contain three parameters, factors that can shift either schedule to the left or right: α, β, and E. The two equations are

$$Q - D(P, \alpha, E) = 0 \text{ [demand function]} \tag{A.1}$$

$$Q - S(P, \beta, E) = 0 \text{ [supply function]} \tag{A.2}$$

where

Q = equilibrium quantity

P = equilibrium price

D = demand equation

S = supply equation

α = demand shifter

β = supply shifter

E = environmental policy factor that shifts demand or supply or both

Demand equations are downward-sloping (inverse) functions of price, and supply equations are upward-sloping (direct) functions of price. When this is true, a solution for both can be found. The solution is called a reduced form, and it expresses quantity and price as separate functions of the three parameters α, β, and E. That reduced form solution can be expressed as

$$Q = f_1(\alpha, \beta, E) \text{ [quantity]} \tag{A.3}$$

$$P = f_2(\alpha, \beta, E) \text{ [price]} \tag{A.4}$$

where f_1 is the equation for directly finding quantity for known values of α, β, and E; and f_2 is the function used for directly finding equilibrium price.

We are most interested in knowing how equilibrium quantity and price change with changes in any of the three parameters, α, β, or E; that is, we want to know (3 parameters) × (2 variables) = 6 impact relations. Assume that the impact of each parameter on demand and/or supply can be measured and is known. Assume that increases in α increase demand, $\frac{\partial D}{\partial \alpha} > 0$; increases in β increase supply, $\frac{\partial S}{\partial \beta} > 0$; and that environmental policy, E, shifts increases both demand and supply, $\frac{\partial S}{\partial E} > 0$ and $\frac{\partial D}{\partial E} > 0$. So our aim is to find the impact of each of the three parameters on equilibrium quantity, where that quantity is defined by Equation A.3. Mathematically, we are looking for $\frac{\partial Q}{\partial \alpha}, \frac{\partial Q}{\partial \beta}$, and $\frac{\partial Q}{\partial E}$.

We also want to find the impact of each of the same three parameters on equilibrium price, where price is shown in Equation A.4. Mathematically we are looking for $\frac{\partial P}{\partial \alpha}, \frac{\partial P}{\partial \beta}$, and $\frac{\partial P}{\partial E}$.

A calculus-based method for finding these six economic effects is to differentiate the two supply-demand relations, Equations A.1 and A.2, with respect to each of the three parameters whose impact we want to predict. We could find the first two desired relations, $\frac{\partial Q}{\partial \alpha}$ and $\frac{\partial P}{\partial \alpha}$, by differentiating Equations A.1 and A.2 with respect to the demand shifter parameter, α, to get the following two equations. Both are linear in the two unknown variables $\frac{\partial Q}{\partial \alpha}$ and $\frac{\partial P}{\partial \alpha}$.

$$\frac{\partial Q}{\partial \alpha} - \frac{\partial D}{\partial P}\frac{\partial P}{\partial \alpha} - \frac{\partial D}{\partial \alpha} = 0 \tag{A.5}$$

$$\frac{\partial Q}{\partial \alpha} - \frac{\partial S}{\partial P}\frac{\partial P}{\partial \alpha} = 0 \tag{A.6}$$

Equations A.5 and A.6 may look strange, but they are nothing more than two linear equations in our two unknowns, $\frac{\partial Q}{\partial \alpha}$ and $\frac{\partial P}{\partial \alpha}$. They can be solved using whatever method you like best for solving linear equation systems (e.g. substitution, subtraction, or matrix algebra). Results are as follows:

$$\frac{\partial Q}{\partial \alpha} = \frac{\frac{\partial S}{\partial P}\frac{\partial D}{\partial \alpha}}{\left[\frac{\partial S}{\partial P} - \frac{\partial D}{\partial P}\right]} \tag{A.7}$$

$$\frac{\partial P}{\partial \alpha} = \frac{\frac{\partial D}{\partial \alpha}}{\left[\frac{\partial S}{\partial P} - \frac{\partial D}{\partial P}\right]} \tag{A.8}$$

Equations A.7 and A.8 indicate that we can find the effect on a market's equilibrium quantity (Equation A.7) and price (Equation A.8) from changes in any of the three parameters: α, β, and E. For example, the effect on equilibrium quantity from an outward shift in demand, $\frac{\partial Q}{\partial \alpha}$, is positive. In fact, if demand increases at all (shifts out), the only way there could be no impact on equilibrium quantity $\frac{\partial Q}{\partial \alpha} = 0$ is if the supply schedule is a vertical line, that is, $\frac{\partial S}{\partial P} = 0$, which makes the numerator and hence the entire Equation A.7 zero.

Table A–1 Impact of changes in supply, demand, and environmental policy on market price and quantity

Impact	of increased demand	of increased supply	of policy that increases demand and supply
On quantity	$\frac{\partial Q}{\partial \alpha} = \frac{\frac{\partial S}{\partial P}\frac{\partial D}{\partial \alpha}}{\left[\frac{\partial S}{\partial P} - \frac{\partial D}{\partial P}\right]}$	$\frac{\partial Q}{\partial \beta} = \frac{\frac{\partial D}{\partial P}\frac{\partial S}{\partial \beta}}{\left[\frac{\partial D}{\partial P} - \frac{\partial S}{\partial P}\right]}$	$\frac{\partial Q}{\partial E} = \frac{\partial D}{\partial E} + \left[\frac{\partial D}{\partial P}\right]\left[\frac{\frac{\partial D}{\partial E} - \frac{\partial S}{\partial E}}{\frac{\partial S}{\partial P} - \frac{\partial D}{\partial P}}\right]$
On price	$\frac{\partial P}{\partial \alpha} = \frac{\frac{\partial D}{\partial \alpha}}{\left[\frac{\partial S}{\partial P} - \frac{\partial D}{\partial P}\right]}$	$\frac{\partial P}{\partial \beta} = \frac{\frac{\partial S}{\partial \beta}}{\left[\frac{\partial D}{\partial P} - \frac{\partial S}{\partial P}\right]}$	$\frac{\partial P}{\partial E} = \frac{\left[\frac{\partial D}{\partial E} - \frac{\partial S}{\partial E}\right]}{\left[\frac{\partial S}{\partial P} - \frac{\partial D}{\partial P}\right]}$

A similar kind of differentiating of Equations A.1 and A.2 can be used to find the last four impacts, $\frac{\partial Q}{\partial \alpha}, \frac{\partial P}{\partial \alpha}, \frac{\partial Q}{\partial E}$, and $\frac{\partial P}{\partial E}$. Their values are summarized in Table A–1. These formulas let you predict impacts on equilibrium price and quantity from any change in a supply shifter, a demand shifter, or an environmental policy.

Appendix B

TOTAL BENEFIT, MARGINAL BENEFIT, AND CONSUMER SURPLUS

In everyday life we say we get value for our money when we buy something at a good price. What we are really saying is that *value* and *price* can be different, and that when the price is *low* we get more *value* for the price we pay. Value is the benefit we receive and price is what we pay. In economics the ordinary idea of value is called *marginal benefit,* which we measure as the maximum price somebody is willing to pay for one more unit of a good.

■ DEMAND AND MARGINAL BENEFIT

The demand curve for bottled water in Figure B–1 shows the relation between value and price. When the price per bottle of water is $1.50, Julie's quantity of bottles demanded (bought) is four bottles per week. Her demand curve is the same as her marginal benefit. The figure's demand curve shows the maximum price Julie is willing to pay for any given quantity. It also shows that when four bottles a week are available, the most Julie is willing to pay for the fourth bottle is $1.50, shown at point *B*. Her marginal benefit from the fourth bottle is $1.50, also shown as point *B*.

■ CONSUMER SURPLUS

Sometimes we are lucky enough to be charged a lower price than the maximum we are willing to pay. When we do not have to pay as much as something is worth to us, we receive a *consumer surplus.* For the last unit bought, consumer surplus is the marginal benefit minus the actual price paid. For all units bought, consumer surplus is the total benefits minus the total amount paid. Julie's demand curve for bottled water tells us the number of bottles she plans to buy at each possible price, which is the same as her marginal benefit from bottled water at each quantity. If the price of water is $1.00 a bottle, Julie buys six bottles a week. She spends $6 total on water ($1 per bottle times six bottles), which is shown by the area 0*AHG*.

The demand (marginal benefit) curve has four interpretations, all of which are correct. (1) It is the marginal benefit of the last unit supplied: $MB = f(\text{supply})$. (2) It is the maximum price willingly paid for the last unit supplied: $P = f(Q)$. (3) It shows the competitive market clearing price if all units are supplied: $P = f(\text{supply})$. (4) It shows the quantity demanded at each price: $Q = g(P)$. The letters *f* and *g* refer to "functions of."

To calculate Julie's consumer surplus for all six bottles, we need to find her consumer surplus on each bottle and add these surpluses together. For the sixth bottle, her marginal benefit equals $1 and she pays $1, so her consumer surplus on this bottle is zero. For the fourth bottle, her marginal benefit is $1.50. So on this bottle, she receives a consumer surplus of $1.50 minus the actual price of $1.00, which is $0.50, or *DE,* shown as the vertical distance between the arrows. That illustrates the principle that consumer surplus for any given unit is the marginal

Figure B–1 Total benefit, marginal benefit, consumer surplus

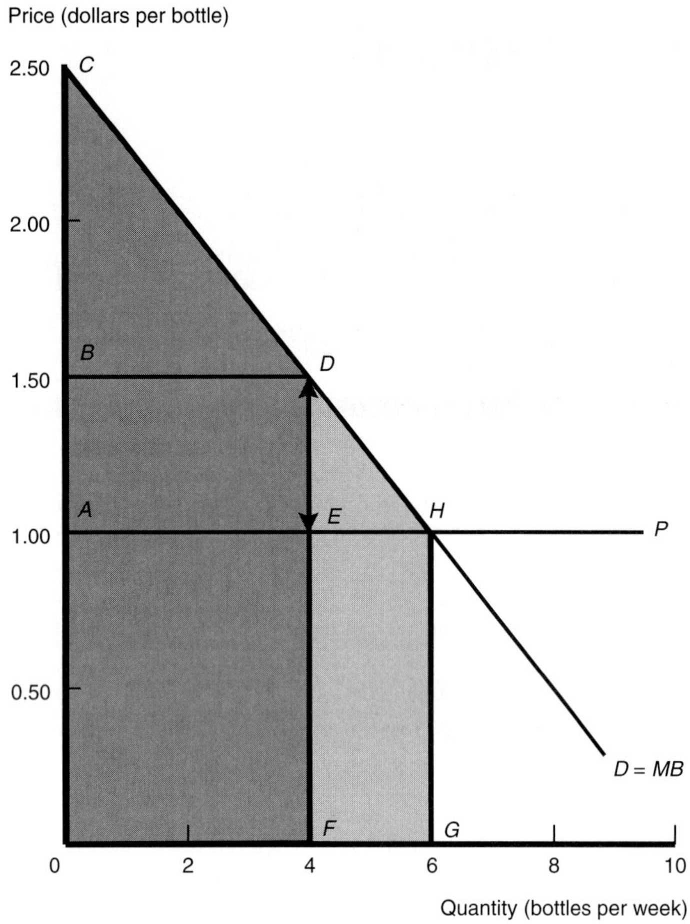

benefit minus the price. For the first bottle, Julie's marginal benefit is nearly \$2.50, although her price is still only \$1.00. So she receives a consumer surplus of \$1.50; (\$2.50 − \$1.00).

Julie's total consumer surplus when the price is \$1.00 for all bottles is the sum of her consumer surplus on all six bottles. It is \$4.50 a week, which is shown by the area *ACH.* Total consumer surplus is the area of the triangle. Its base is 6 and its height is \$1.50, so its area is 6 × \$1.50 ÷ 2 = \$4.50.

Total Benefit

Julie's total benefit for all six bottles bought at the price of \$1.00 per bottle is the amount she pays, equal to the rectangle 0*AHG* (\$6.00) plus her consumer surplus. Her consumer surplus is equal to the triangle *ACH* (\$4.50). So her total benefit is \$6.00 plus \$4.50, or \$10.50. If the price per bottle is raised to \$1.50, Julie's total benefit falls to the area 0*CDEF,* which equals \$8.00. Her consumer surplus falls to area *BCD* (\$2.00) and the amount she pays at the new price of \$1.50 stays the same at \$6.00, as it was at a price of \$1.00, although its area changes from 0*AHG* to 0*BDF.* Here is an important point: When the price increases (decreases) total benefit falls (rise), consumer surplus falls (rises), and the total amount paid can increase, decrease, or stay the same. The light-shaded area *FDHG* equal to \$2.50, shows the amount by which Julie's total benefit falls when the price per bottle increases from \$1.00 to \$1.50. It is also the incremental benefit lost when a policy raises the price from \$1.00 to \$1.50.

Appendix C

Use of Programming Methods to Value Public Actions

Mathematical programming methods can be used to derive the value of a publicly supplied program input when that input is used by a commercial business. The model defines the increment in net producer income resulting from adding the publicly supplied input to a production process as the producer's willingness to pay for the added input supply.

For example, let I_j refer to each jth production input and Q_i refer to each ith output. The subscripts 1 and 0 attached to all variables refer to values with (1) and without (0) the public program's input. The input quantity supplied by the environmental program is defined as the quantity I_1^1 *with* the program and I_1^0 *without* the program. If input prices (C_j) and output prices (P_i) are unaffected by the public program, the change in net income (ΔY) associated with the program's added input supply per unit of time is

$$\Delta Y = Y^1 - Y^0 = \left(\sum_{i=1}^{n} P_i^1 Q_j^1 - \sum_{j=2}^{m} C_j^1 I_j^1 \right) - \left(\sum_{i=1}^{n} P_i^0 Q_i^0 - \sum_{j=2}^{m} C_j^0 I_j^0 \right) \qquad \text{(C.1)}$$

To find the value per unit of the program input, C_1, divide the expression in the equation by the incremental quantity of program input (i.e., ΔI_1). Mathematical programming is one method used to find this value. Its use involves assuming that the representative producer (e.g., a farmer or power supplier) maximizes net income by adjusting all nonproject inputs and outputs *with* the environmental program as well as *without* it.

Measuring benefits in this way assumes that producers attempt to maximize income. In fact, various institutional constraints may operate so that the producer is unable or unwilling to maximize net income. In that case, Equation C.1 should not be maximized; instead, it is set up to replicate conditions that would occur with and without the proposed program. In any case, whether or not the equation is optimized, its use assumes that input and output prices correctly reflect social values, and that all inputs and outputs are correctly measured.

Appendix D
IMPROVING COST-EFFECTIVENESS OF ENVIRONMENTAL PROGRAMS

Cost-effectiveness is a method of analysis that is used to find the least-cost method or methods to reach a particular environmental target. It can also be used to indicate how many extra costs are incurred by failing to implement cost-effective progams (e.g., lives not saved) whose output is measurable. Using the example of human lives saved, cost-effectiveness tells us the added costs to save a given number of lives or the added lives saved for a given total cost. Consider the following formula, from production economics theory, for measuring the added quantity of a program's output produced by incurring additional costs:

$$dQ = \sum_i \frac{P_i\,dX_i}{MC_i} \tag{D.1}$$

where dQ is the added program output (e.g., lives saved), P_i is the price per unit of the ith program input, dX_i is the added quantity of the ith program input (e.g., intensity of a regulation), and MC_i is the marginal cost of producing one more program output from the ith program input (e.g., lives saved from increasing the intensity of a regulation). The equation indicates the amount by which output can be increased or total program costs can be reduced, when existing marginal costs of all program inputs are not equal.

Equation D.1 says that for a fixed total cost of the program (i.e., $\sum P_i\,dX_i = 0$) and for a proposal that would shift program inputs to change output, if marginal costs of increasing program outputs are equal across all programs, then dQ must be zero, which means there is no possible increase in output. However, when marginal costs are different across the various methods of producing output (MC_i not equal), we can do better: Output can increase for a fixed total program cost ($dQ > 0$) or equal output can be had for less expenditure, that is, $\sum P_i\,dX_i < 0$. The farther apart marginal costs are to start with, the greater potential there is for the program's output to increase for given costs, or for total costs to fall for a given output, or both.

Appendix E

Preferences, Demand, and Benefits

This appendix illustrates how welfare measures and their approximation can be estimated from observable data. Assume that a policymaker wants us to find the benefits produced by an action that would produce a price change. Illustrating with the Cobb-Douglas preference ordering, we specify the consumer's utility as

$$U = \prod_i X_i^{B_i} \tag{E.1}$$

in which U = ordinal utility, X_i = number of units consumed of each ith commodity, and B_i are parameters, which we typically assume to sum to 1. The Greek symbol (capital pi) means terms to the right should be multiplied by each other. The Marshallian demand equation is found by maximizing utility subject to a constraint that the sum of the X_i multiplied by their prices cannot exceed monetary income. That equation turns out to be

$$X_i = \frac{B_i Y}{P_i} \tag{E.2}$$

This demand equation predicts the consumption of each ith good, based on known values for income and prices. The expenditure function measures minimum expenditure required to hold ordinal utility constant at its level reached under *initial* prices (P^0) and money income when the consumer faces new prices (P). That expenditure function is given by the following formula:

$$e(P, P^0, Y) = Y\left(\frac{\prod_i (P_i)^{B_i}}{\prod_i (P_i^0)^{B_i}}\right). \tag{E.3a}$$

This expression has the important property that it consists entirely of terms that are observable. Note that when the new prices equal initial prices, minimum expenditure equals actual income. A second similar expenditure function can be defined as referenced to *new* prices. It measures the minimum expenditure required to hold ordinal utility constant at its level reached under *new* prices (P^1) and money income when the consumer faces any given prices (P).

$$e(P, P^1, Y) = Y\left(\frac{\prod_i (P_i)^{B_i}}{\prod_i (P_i^1)^{B_i}}\right). \tag{E.3b}$$

These two expenditure functions can be easily coded into a spreadsheet to perform the calculations required, which are described in more detail next.

Welfare Measures

From the two expenditure functions, we can compute welfare measures associated with price changes in one or more goods. The compensating variation (CV) and equivalent variation (EV) produced by price changes in one or more goods are

$$CV = Y - e(P, P^0, Y) \tag{E.4a}$$

$$EV = e(P, P^1, Y) - Y \tag{E.4b}$$

Both the *CV* and *EV* can be measured entirely from observable data, that is, from the parameters estimated from the Marshallian demand as well as observed prices, quantities, and income. The consumer surplus approximation of the two exact welfare measures is the area under the Marshallian demand schedule (Equation E.2) bound by old and new prices. See Chapter 7.

Compensated Demands

The compensated (Hicksian) demand function is found by writing the Marshallian demand Equation E.2 with the income effect deleted. Its formula is the same as Equation E.2 with one of the expenditure functions, Equation E.3a or Equation E.3b, inserted for actual money income (*Y*). That substitution produces

$$X_i^*(P, P^0, Y) = \frac{B_i e(P, P^0, Y)}{P_i} \tag{E.5a}$$

where the asterisk superscript indicates demand is compensated to indifference with conditions achieved under *old* prices. The compensated demand described by Equation E.5a and the Marshallian demands are equal when prices are set at *old* levels. The compensated demand required to produce indifference with *new* price conditions is

$$X_i^*(P, P^1, Y) = \frac{B_i e(P, P^1, Y)}{P_i} \tag{E.5b}$$

Application of Equation E.5b shows that its compensated demands and Marshallian demands are equal when prices are set at *new* levels.

Application to Observable Data

We now illustrate how estimates of ordinary demands (Equation E.2) give enough information to measure the theoretically correct benefits produced by a price change, as well as the consumer surplus approximation of those benefits. The demands can be estimated with standard econometric methods; usually the specified demands must be based on preferences that are limited by a budget constraint. For example, suppose income is \$50,000 per year, preferences for the first good, B_1 is 0.2, and the price of that good falls from an initial value of 100 to a new value of 50. The accompanying table shows the Marshallian and Hicksian demanded quantities of one good, the two welfare measures, and their consumer surplus approximation.

Conditions	P	Y	B_1	X_1	$X_1^*(P^0)$	$X_1^*(P^1)$	CV	CS	EV
Old	100	50,000	0.2	100	100	115	6,472	6,931	7,435
New	50	50,000	0.2	200	174	200			

The *CV* measures the maximum willingness to pay for a welfare-improving policy, while the *EV* measures the minimum willingness to accept payment that would take the place of a welfare-improving policy. The *CS* has no theoretical interpretation, but it is bound by the *CV* and *EV*.

Appendix F
Incentive Compatibility in Contingent Valuation Studies

Because incentive compatibility poses such a threat to the credibility of contingent valuation surveys, it is important to design incentive-compatible mechanisms to reveal respondents' true value of a resource. The challenge is to design the payment mechanism so that respondents face no incentives to lie. If a person receives a greater benefit by lying than by telling the truth, the person cannot be expected to routinely tell the truth. One way to encourage telling the truth is by structuring payment incentives carefully. For example, the following payment mechanism avoids incentives to lie by encouraging each individual respondent to compare the benefits he receives from a public good, *R,* to his costs[1]:

> *R* will be supplied only if the sum of stated WTPs over *all group members* exceeds the costs of supply. And, if *R* is supplied, each person in the group will be charged a price for their share of *R.* That price is equal to the difference between the cost supplying *R* and the sum of stated WTPs over all *other* group members (that is, the sum of willingness to pay (WTP) over all people *excluding* the respondent).

So think of the incentives facing any given respondent *j* in a group of people, all of whom are asked about their WTP for a public program *R* that will be supplied to all group members. Each person *j* knows that for *R* to be provided, the total WTP over everybody must exceed the cost.

$$WTP_j + \sum_{i \neq j} WTP_i > C(R) \tag{F.1}$$

Moreover, if this equation holds, *R* will be supplied and the price charged to *j* (tax) will be

$$T_j = C(R) - \sum_{i \neq j} WTP_i \tag{F.2}$$

What is attractive about this payment mechanism is that it puts each respondent in the driver's seat to influence the decision to supply *R.* Each *j*th person sees himself as making the decisive vote on the program. If this equation holds true,

$$\sum_{i \neq j} WTP_i < C(R) \tag{F.3}$$

which means if there is inadequate demand without person *j*, then a sufficiently high stated WTP_j will push the total bid over the top so the program is supplied. It will also create for *j* a tax bill equal to the difference between the sum of others' *WTP* and total cost. Look closer at the incentives this payment scheme provides: The benefit to *j* of producing a falsely high bid is that he can increase the probability of having the program built. But there is also a cost to *j* of overbidding that increases the probability of getting the project built. Remember, the project's successful completion brings

[1] Russell, C. S. 2001. *Applying Economics to the Environment.* New York: Oxford University Press.

Table F–1 Individual's payoff matrix for various contingent valuation bidding strategies[a]

Bidding strategy		Sum of others' bids							
	Bid	20	19	18	17	16	15	14	Worst outcome (lowest column) for row's bid
Bid very low	1	3	2	0	0	0	0	0	0
Bid low	2	3	2	1	0	0	0	0	0
Bid accurately	3	3	2	1	0	0	0	0	0
Bid high	4	3	2	1	0	−1	0	0	−1
Bid very high	5	3	2	1	0	−1	−2	0	−2

[a]Assumes individual's value of the public good is \$3 and that he finances others' shortfalls.

a large tax bill due to *j*, which will occur if the other bids are low compared to costs. The respondent sees both benefits and costs of overbidding, so there is no clear advantage to overbidding. In fact, overbidding is a poor strategy for maximizing net benefits. Consider the following data: $C(R)$ = cost = \$20 and WTP_j = *j*'s willingness to pay for R = \$3. The tax sharing rule (Equation F.2) indicates that any *j*th person has the payoff matrix shown in Table F–1, which contains numbers for five bids—one accurate (\$3) and two others on either side: very low (\$1), low (\$2), very high (\$5), and high (\$4). Each *j*th person is ignorant of the sum of others' bids. Person *j*'s net benefit for any decision consists of two parts, whether or not *R* is provided and the tax bill. If *R* is provided, *j*'s WTP is \$3, otherwise WTP is \$0. The tax bill depends on the sum of other bids and ranges from \$14 to \$20.

Assume that *j* uses a bidding strategy of minimizing the maximum loss in the far-right column. This strategy results in *j* being indifferent between underbidding and bidding accurately. But *j* finds it is a bad strategy to overbid, because overbidding leads to a higher probability of building *R* and being charged a high tax when the sum of other bids is low. Overbidding is a poor strategy because overbidding produces a higher net loss than bidding honestly, not having it built, and facing no tax charge.

Appendix G

CRITIQUES OF CONTINGENT VALUATION

The contingent valuation method (CVM) is used to measure monetary benefits of environmental improvements by asking people what they believe their benefits are. This certainly is a direct way to find the sought-after nugget, but is it really gold? The CVM has been criticized for many reasons, and many of these criticisms are compelling (Arrow et al., 1993). Before describing the details of these criticisms, it is important to remember that these debates take on considerable importance because of the difficulty of directly testing the results of CVM studies.

Rarely are there market-produced values to test against CVM-produced values. One way to deal with this difficulty of verifying CVM-produced values is to construct experiments in which an opportunity is created to pay for environmental improvements. Then, the results of estimated willingness to pay (WTP) can be compared with results from actual markets when the opportunity is made available to the same sample or an equivalent sample. Few such experiments have been tried. One, credited to Seip and Strand (2000), used CVM to estimate willingness to pay for membership in a Norwegian environmental organization, then compared that estimate with actual responses when some of the same respondents were presented with an opportunity to contribute real cash. The authors found that self-reported WTP was significantly greater than willingness to pay real cash. A study by Duffield and Patterson (1991) valued the environmental service of maintaining stream flow in two Montana fishing streams. The streams provided spawning grounds for two rare species of fish, which were illegal to catch, so there was no use value associated with the fish. So nonuse value was likely the primary motivation for survey respondents. One of two similar samples was asked about hypothetical willingness to contribute to the Montana Nature Conservancy, which would use the revenues to maintain sufficient stream flow to protect the fish. The other group was offered an opportunity to contribute real cash to the same organization for the same purpose. The authors found that the willingness to contribute was significantly higher when the contribution was hypothetical than when it required an immediate cash payment. Analysis by List (2001) examined methods to reduce inflated respondent values resulting from the hypothetical nature of the valuation exercise. Studies suggest that the CVM technique is likely to overstate real willingness to pay. For many economists, the most basic criticism of the CVM or any other stated preference method is that it is hypothetical. Cameron (1992) and Azevedo, Herriges, and Kling (2003) investigated inconsistencies between stated preferences and revealed preferences.

VIOLATIONS OF RATIONAL CHOICE

Some of the estimated values produced by CVM studies have been criticized because they are not consistent with the requirements of rational choice. This critique raises two questions: (1) What does rational choice require that values look like? (2) Why should outcomes be consistent with rational choice? Rationality requires consistency among choices an individual makes. For example, suppose an individual chooses a given basket of purchases at one set of prices and income.

Then, if one price falls and there are no other changes, the individual would buy more of that good and she would be better off. Likewise, an individual's preferences for public goods, like water quality, would be expected to show that same kind of consistency. Rationality also says that more of something regarded as desirable is better. This means that an individual's WTP for an environmental improvement increases with increased supplies of that good, but at a decreasing rate. Furthermore, if incremental WTP for greater amounts declines with the amount already available, it would not be expected to decline abruptly.

Well-known investigations have cast doubt on the consistency of CVM values with rationality, because willingness to pay did not increase with increased quantities of the good. For example, Nobel laureate Daniel Kahneman (1986) found that WTP for restoring all lakes in Ontario was only slightly higher than the WTP for cleaning up lakes in a single region. This kind of evidence was also found by Kahneman and Knetsch (1992), Desvousges et al. (1992), and Diamond and Hausman (1994). Desvousges et al. found that respondents' average WTP for a policy to protect 2,000 migratory birds from dying from oil spills was as large as that for preventing 20,000 or 200,000 birds from dying from the same cause. Willingness-to-pay values consistent with rational choice would show a diminishing marginal WTP for additional protection. That is, we would expect the added benefit from 2,000 birds to 3,000 birds protected to be larger than from 200,000 to 201,000 birds protected. But a drop in marginal willingness to pay to zero, particularly when the WTP for the first 2,000 birds was a large value, strains the bounds of credibility as expressions of consistent, rational choices.

You might ask this question: Why not just take the respondents' answers at face value, then hand the WTP numbers over to policymakers? There are two answers. One is that we do not know what a WTP response means without an underlying assumption of consumer rationality. Rationality requirements impose a constraint on the possible values in a benefit-cost analysis, without which the benefit numbers are meaningless and cannot be compared to costs on a level playing field. The second answer is that without consumers' choices being governed and limited by real prices and a finite budget constraint, there are no objective counterparts to verify the values.

Implausible Values

The CVM approach is typically used to measure the value of some program to avoid, restore, or reverse environmental damage, whether it be toxic wastes, a lost wilderness, or loss of visibility near an important natural area. In each case, a high percentage of respondents typically express zero willingness to pay but the *average* WTP over the whole sample is usually more than a dollar, often in the range of \$20 to \$50 and sometimes more. With 110 million households in the United States, alone, many survey responses are apt to produce very large dollar totals, often exceeding \$1 billion. Some CVM critics argue that the implausibility of such large sums cast doubt on the CVM method.

One can imagine many kinds of environmental damage: dead animals, aquifer contamination in different places, acid rain produced by several power plants, and the like. Would the typical person in a typical household really be willing to pay \$50, \$20, or even \$5 to prevent every one of those incidents? Not likely, many say, because the total WTP for all the environmental improvement programs could become a larger part of one's income or maybe even exceed it. Even if the CVM gives plausible estimates of any single program in isolation, its critics believe it produces inflated values when large numbers of environmental programs or program elements are added up. If individual CVM respondents are not reminded to decide what public or private goods they plan to substitute for the environmental improvement in question, their responses to CVM survey questions risk significant inflation.

Budget Constraint Violations

Even if CVM survey respondents believe the hypothetical referendum questions are plausible and relevant, they risk offering a WTP response for the resource with little consideration of how they plan to finance it. Respondents have been known to say they are willing to pay several thousand dollars for an environmental program, even though they are not wealthy and probably cannot afford the amount and would not actually pay the money even if they could find it. For these reasons, follow-up questions may be able to produce more plausible values: What other consumption would they plan to give up? How many extra hours would they plan to work to earn income to pay for that stated value? What contributions to other causes would they forego? For example, suppose a respondent reveals a WTP of $50 for a project to protect the critical habitat of an endangered species. However, if asked what current or planned expenditures she expects to give up to pay for the program, she might instead reevaluate her response and revise it downward to $10. A budget constraint places an upper limit on a realistic WTP for a single environmental improvement. Large numbers of such improvements competing for consumers' budgets mean that the WTP for any one in isolation is smaller as the respondent is asked to support larger numbers of them. For this reason, CVM critics believe that the value numbers are unrealistically high. Many of the newer CVM surveys remind respondents of the economic constraints within which WTP decisions are made to ensure more credible and plausible valuations. Loomis (1994 and 1996) and Brown, Ajzen, and Hrubes (2003) provide good examples.

Ignorance of the Program

For CVM surveys to elicit meaningful information about WTP for environmental services, survey respondents must understand precisely what service they are being asked to value or vote for, and their WTP response must value that service and not something else. All too often, CVM surveys are vague about the program, service, or project being valued. A related problem is that people are asked to value services they have never purchased and never thought of purchasing. Therefore, people may not know how to place a value on them. Lack of clarity or lack of experience casts doubt on WTP estimates produced by the survey.

Consider the following example: Suppose policymakers want to know the value of a program that would safeguard against toxic chemical spills into an aquifer. People's WTP responses would be expected to depend on how long it takes for the chemical to degrade naturally into the aquifer, how cheaply the spill could be cleaned up after it occurred, and the human health impacts of such a spill before it became benign. Without detailed information on these important facts, it is hard to expect that even well-informed, educated, and disciplined respondents could place accurate values on programs to protect aquifer contamination. Moreover, even if respondents understood all those facts, they might not believe them and even if they believed the data they might not incorporate all that into their answers. For example, respondents who believed the survey questions overstated the seriousness of effects would offer a lower WTP for the program based on their optimism that nature could solve the problem on its own. People who were pessimistic about nature's solutions to the problem would report a falsely high WTP for the program to prevent the contamination. In any case, respondents need an accurate and detailed description of the environmental service provided by the program, and they need to believe the survey's description and act on it, before WTP responses can provide information useful for policymakers to design economically efficient environmental programs.

Relevant Population Not Sampled

Major environmental policy actions produce benefits that potentially affect large populations, possibly in several countries. For actions that affect global climate change, the relevant population is everybody living in current generations as well as in many future generations. However, many environmental policies are carried out by people other than those who are influenced by the policy. For example, in the United States, environmental damage lawsuits are brought by trustees on behalf of a legally well-defined group. Those trustees define a limited population for which economic damages are measured, even though many outside that defined group may also suffer damages.

FEEL-GOOD EFFECTS

Many people like the idea of making commitments to worthy social causes such as environmental programs. This motivation is often uncovered in CVM surveys and has led critics to conclude that individuals' responses to CVM questions serve the same function as charitable contributions. Respondents wish to support an organization and they like to feel the "warm glow" inside that comes from donating to worthy social causes. When this occurs, CVM responses are not accurate estimates of WTP; rather, they indicate approval for the environmental program in question. A discrepancy may occur between attitude and behavior. People may be reflecting attitude toward supporting an environmental issue rather than behavior reflecting a WTP for a particular environmental service. If this warm-glow effect occurs on a survey, the stated value is not measuring preferences for the program but pleasure in supporting a good cause. List et al. (2004) examined the role of social isolation on stated values. Social isolation (i.e., a respondent knowing her stated values are not identifiable) may increase the private inner satisfaction of knowing she contributes to a worthy cause without fanfare.

OTHER BIASES

Bias occurs when the expected value of a parameter estimate is not equal to the parameter. For example, if the true mean willingness to pay for an environmental improvement is $7 per household per year, and the chosen method produces an average value of $10, there is an upward bias of $3. Bias can be a problem when estimating values of environmental programs for a benefit-cost analysis that informs policy decisions. People can make rational and informed choices about their valuations of goods in most market situations, but it may be harder to do so for public environmental services. So there is a risk that CVM-measured values for public programs may not be equal to what people are actually prepared to pay. Several problems tied to the CVM itself risk producing bias in results coming from CVM surveys: (1) commitment bias, (2) payment vehicle bias, (3) order effect bias, (4) embedding effect bias, (5) sample bias, and (6) starting point bias.

1. **Commitment bias.** Many marketing studies have found that survey respondents overstate their willingness to purchase a product described and/or shown to them, especially if the respondent makes no commitment. For example, the surveyor shows the respondent a cherry-flavored wine cooler packaged in an attractive bottle with gold foil, offers the respondent a taste, then asks if she would buy one four-pack per week at a price of $8 per pack. The respondent answers yes but never buys the product when it appears at the grocery store. The survey response may simply have been a strategic response, that is, the respondent's strategy of achieving a personal goal having little to do with the survey. It might also show a bias that comes from respondents who believe a product has value simply because they are being asked about it. It is easy for this bias to occur because the respondent commits no time or money. For example, one study found that respondents' WTP to prevent a particular future oil spill was several hundred times smaller after they were asked to value several other public goods, than when they valued the avoided spill as a stand-alone service item (Kemp and Maxwell, 1993).

2. **Payment vehicle bias.** CVM studies typically include a payment vehicle for the respondent, which states the method by which the hypothetical payment will be charged. Several payment vehicles have been used at one time or another: increased utility bills, taxes paid into a fund earmarked for the environmental service, increased income or sales taxes, and a higher price for the service itself. The choice of payment vehicle can make a difference in the estimated WTP, so there is a potential for a payment vehicle bias. Results of a CVM study may produce more credibility if the payment vehicle is specified to the respondent along with reminders that payments come at the expense of expenditures on other goods. For example, respondents who said they would vote for a public program

priced at $X could be asked where that expenditure of $X will be reduced in their existing budget (Berrerns et al., 2000). Jorgensen and Syme (2000) describe some approaches for dealing with protest responses, that is respondents who protest some of the valuation exercise. To increase the credibility of CVM surveys, researchers often select a payment vehicle that is close to the actual one that would be used to finance the proposed environmental program.

3. Order effect bias. Survey respondents' stated value of an environmental improvement can be affected significantly by the order in which questions are posed. Consider, for example, a study that asks respondents to value preserving a rare mammal *M,* then a rare bird *B;* another survey asks to protect *B* first, then *M.* Many researchers have found that *M*'s values are much lower when the *M* question is asked after the *B* question. The dependence of value on order of presentation might be explained by an income effect, a substitution effect, or a combination. If a respondent states a WTP of $20 to pay for the first environmental service in a sequence of several, then that person has less income to spend on the second and later elements in the sequence. However, income is not reduced much by paying for the first element, so the income effect in isolation should not produce a great reduction in WTP from one value element to the next. Substitution effects, for example between the rare mammal and the rare bird, could be large enough to matter, especially in measuring the total value of the package of elements. Respondents may have preferences for which one environmental improvement is a strong substitute for another.

When people's indifference curves show strong substitution, the added value of any element after the first is already provided could be as low as zero. A policy producing a package of environmental improvements could easily have zero marginal value after the first element is already provided. For example, if a resident of city *A* is willing to pay $100 per year for clean air in city *A* and then is offered a program that preserves clean air at a national park *B,* the value of *B*'s clean air, given that *A* already gets clean air, could be as low as zero. If the two are perfect substitutes, then either clean air in city *A* or clean air over park *B* is worth $100 *in isolation,* but the incremental benefit provided by supplying either is zero *after* the first is provided. Still hotly debated is the question of whether substitution effects could produce the high level of order inconsistency that has cropped up in many CVM studies.

4. Embedding effects bias. Microeconomic theory tells us that an individual's total benefits produced by greater quantities of an environmental service should increase at a decreasing rate. If CVM respondents' stated values are only slightly higher for large changes in the amount of the good offered than for small changes, then theory tells us we should question the validity of their responses. Despite this warning, much empirical research shows that respondents often fail to distinguish between small and large quantities in their valuations of an environmental service when the different quantities are embedded in one another. Embedding effects occur when environmental services are valued much lower when they occur as a part of a package of services than when valued on their own. As stated above, Desvousges et al. (1992) found that different samples of respondents value 2,000 migratory birds approximately the same as they value 200,000 birds. It is unlikely that declining marginal utility can explain all or even most of the absence of what is expected to be a much wider variability on total value for such different quantities of environmental services. Diamond and Hausman (1994) also raised questions about the credibility of CVM estimates in connection with embedding effects. The presence of these effects can produce widely different stated WTP numbers for a single environmental service, depending on which element was embedded inside another package of environmental improvements. As a result, there are few straightforward methods to select any one CVM as correct.

5. Sampling bias. Survey research elicits information about a known population using a small sample drawn from the population. Typically, the sample is selected by a probability mechanism that produces a random sample. In a random sample, each individual member of the population has a known and measured probability of being drawn from the population. For CVM studies, the relevant population targeted is all people with standing affected by the policy. For many programs, is who is affected is obvious, but often for environmental problems, that is a tough call to make. The greater the spread between people who bear the cost of a program and those who receive its benefits, the harder it is to choose the right population for sampling. Those who will directly consume the service in question are users, but so too is any individual who would suffer direct negative costs for which they would be willing to pay to avoid. For example, nearby residents of a hunting area who dislike gun noise are influenced by a program that improves wildlife habitat. Potential (not current) users should also be included, which could mean future generations for programs like long-term climate change. Even people who make no active use of a particular beach, river, bay, or other environmental resource might derive satisfaction from its mere existence, even if they have no intention of making direct use of it. This concept is sometimes known as nonuse values or passive-use value. In concept, nonuse values should be included as a part of total benefits. For environmental goods, CVM surveys that sample passive users often produce significantly higher WTP estimates than those that include only use benefits. However, much debate remains on appropriate sampling methods for CVM surveys to estimate passive-use values.

6. Starting point bias. An important problem arises in CVM when starting point values are presented to respondents. The dichotomous choice question format (Arrow et al., 1993) is intended to reduce starting point bias. Still, some have argued that responses to dichotomous choice questions are strongly influenced by starting point bias, because respondents may very well act on cues of resource value presented.

References

Arrow, K., R. Solow, P. Portney, E. Leamer, R. Radner, and H. Schuman. 1993. Report of the NOAA panel on contingent valuation. *Federal Register.* 58 (10): 4601–4614.

Azevedo, C. D., J. A. Herriges, and C. L. Kling. 2003. Combining revealed and stated preferences: Consistency tests and their interpretations. *American Journal of Agricultural Economics.* 85 (3): 525–537.

Berrens, R. P., A. K. Bohara, C. L. Silva, D. Brookshire, and M. McKee. 2000. Contingent values for New Mexico instream flows: With tests of scope, group-size reminder and temporal reliability. *Journal of Environmental Management.* 58 (1): 73–90.

Brown, T. C., I. Ajzen, and D. Hrubes. 2003. Further tests of entreaties to avoid hypothetical bias in referendum contingent valuation. *Journal of Environmental Economics and Management.* 46: 353–361.

Cameron, T. 1992. Combining contingent valuation and travel cost data for the valuation of nonmarket goods. *Land Economics.* 68: 302–317

Desvousges, W. H., F. R. Johnson, R. W. Dunford, K. J. Boyle, S. P. Hudson, and K. N. Wilson. 1992. Measuring Natural Resource Damages with Contingent Valuation: An Experimental Evaluation of Accuracy. Research Triangle Park, NC. Research Triangle Institute.

Diamond, P., and J. Hausman. 1994. Contingent valuation: Is some number better than no number. *Journal of Economic Perspectives.* 8: 45–64

Duffield, W. J., and D. A. Patterson. 1991. Field Testing Existence Values: An Instream Flow Trust Fund for Montana Rivers. Paper read at the American Economic Association, New Orleans, LA, January 3–6.

Jorgensen, B. S., and G. J. Syme. 2000. Protest responses and willingness to pay: Attitude toward paying for stormwater pollution abatement. *Ecological Economics.* 33 (2): 251–265.

Kahneman, D. 1986. Comments. In *Valuing Environmental Goods,* edited by R. G. Cummings, D. S. Brookshire, and W. D. Schulze. Totowa, NJ: Rowman and Allanheld.

Kahneman, D., and J. L. Knetsch. 1992. Valuing public goods: The purchase of moral satisfaction. *Journal of Environmental Economics and Management.* 22 (1): 57–70.

Kemp, M. A., and C. Maxwell. 1993. Exploring a budget context for contingent valuation estimates. In *Contingent Valuation: A Critical Assessment,* edited by J. A. Hausman. Amsterdam: North Holland.

List, J. A. 2001. Do explicit warnings eliminate the hypothetical bias in elicitation procedures? Evidence from field auctions for sportscards. *American Economic Review.* 91 (5): 1498–1507.

List, J. A., R. P. Berrens, A. K. Bohara, and J. Kerkvliet. 2004. Examining the role of social isolation on stated preferences. *American Economic Review.* 94 (3): 741–752.

Loomis, J., A. Gonzales-Caban, and R. Gregory. 1994. Do reminders of substitutes and budget constraints influence contingent valuation estimates. *Land Economics.* 70: 499–506.

Loomis, J., and D. White. 1996. Economic benefits of rare and endangered species: Summary and meta-analysis. *Ecological Economics,* 18 (3): 197–206.

Seip, K., and J. Strand. 2000. Willingness to Pay for Environmental Goods in Norway: A Contingent Valuation Study with Real Payment. SAF Center for Applied Research, Department of Economics, University of Oslo.

Appendix H

Infinite Time Horizons and Forest Rotation Analysis

Analyzing the net present value of a stream of revenues and costs for an infinite number of forest rotations is an old problem, first analyzed by the German forester Faustman, in the nineteenth century. The problem is defined by a forester who plants a stand of trees, waits t periods to harvest, then plants the stand again. The process is repeated an infinite number of times. The aim is to identify how long the period t should be (i.e., the optimal rotation), to maximize the net present value of income over the infinite number of rotations. To make this decision, the forester needs to measure the opportunity cost of delaying the sequence of harvests from period t to period $t + 1$, knowing that the delay from t to $t + 1$ delays all future rotations, and income produced by those rotations, by one year. The opportunity cost of future income delayed must be considered in the decision to delay a harvest on the current rotation by a year. Finding this cost requires computing the net present value of an infinite number of rotations.

Revenue

Revenue from selling timber, called stumpage value (SV) is defined by the stumpage price (P) times the volume of harvested timber sold (Q)

$$SV = PQ \tag{H.1}$$

Depending on the kind of timber sold and local market conditions, stumpage price may be a constant, independent of the stand's volume ($P = P_0$); or price may increase with volume because larger trees are often worth more per unit volume ($P = P(Q)$) as higher valued uses (e.g., sawtimber and veneer logs) now become possible. Whichever of these two patterns the stumpage price has, assume it has the same pattern whether applied to the first, second, or last rotation. That is, even if price increases with the stand's total volume, the impact of volume on price is identical in the current and in all future rotations.

The standard present value formula can be applied to find the present value of revenue (PVR) of an infinitely repeating number of rotations, in which each rotation has a nominal undiscounted stumpage value, from Equation H.1, identical in periods t, $2t$, $3t$, and so on:

$$PVR = \frac{SV}{(1 + r)^t} + \frac{SV}{(1 + r)^{2t}} + \frac{SV}{(1 + r)^{3t}} + \cdots + \frac{SV}{(1 + r)^{Nt}} \tag{H.2}$$

For example, if the forester harvests in year $t = 9$, then the rotation's length is 9 years. Assuming this cycle repeats itself, revenues accrue in year $t = 9$, year $2t = 18$, year $3t = 27$ and so forth. Equation H.2 shows that the present value of this future stream of revenues is found by discounting back to the present those future years' harvests. Equation H.2 is based on a cycle of revenues from a sequence of identical, repeated stand harvests. It allows for either constant prices or prices that increase with the stand's volume. It does not allow for prices that change over time due to factors like changes in timber scarcity, for this would require a different stumpage value in each rotation.

A standard mathematical trick is used to calculate the sum of an infinite series by multiplying both sides of Equation H.2 by the denominator of the first term on its right-hand side, which gives

$$PVR(1 + r)^t = SV + \frac{SV}{(1 + r)^t} + \frac{SV}{(1 + r)^{2t}} + \cdots + \frac{SV}{(1 + r)^{(N-1)t}} \quad \text{(H.3)}$$

The last term in Equation H.2 approaches zero as the number of rotations, *N*, approaches infinity. So the left- and right-hand sides of Equation H.3 can be subtracted from Equation H.2 to produce

$$PVR\,(1 + r)^t - PVR = SV \quad \text{(H.4)}$$

Equation H.4 can be used to solve for the present value of an infinite number of rotations in which each single rotation produces a nominal stumpage value in period *t* equal to *SV:*

$$PVR = \frac{SV}{(1 + r)^t - 1} \quad \text{(H.5)}$$

Cost

Two important costs are harvesting and planting costs. Harvesting costs, *HC*s, are incurred in the harvest year, defined above as period *t*. Because *HC* increases with the volume of timber harvested, it can be expressed as

$$HC = hQ \quad \text{(H.6)}$$

where *h* is the harvest cost per unit volume and *Q* remains the volume harvested in year *t*.

Each rotation's harvest cost (*HC*) is incurred in the same year (*t*) as its revenues (*SV*). So the same mathematical trick used earlier lets us substitute *HC* into Equation H.5 for *SV.* This substitution lets us find the present value of an infinitely repeating series of future harvest costs, defined as *PVHC:*

$$PVHC = \frac{HC}{(1 + r)^t - 1} \quad \text{(H.7)}$$

Planting costs (*PC*) have an important difference from harvesting costs. They are a fixed cost, do not vary with harvest volume, and are incurred in the first year of a rotation. So for each future rotation, *PC* can be expressed as

$$PC = PC_0 \quad \text{(H.8)}$$

which states that planting costs do not increase with timber volume harvested or with the rotation length. Because the forester incurs *PC* at the *beginning* of each rotation, not at the end, *PC* occurs *t* periods earlier than *HC*. The present value of these planting costs, *PVPC,* can be expressed as

$$PVPC = \frac{PC}{(1 + r)^0} + \frac{PC}{(1 + r)^t} + \frac{PC}{(1 + r)^{2t}} + \cdots + \frac{PC}{(1 + r)^{(N-1)t}} \quad \text{(H.9)}$$

where planting costs start in period *0* and end in period $N - 1$ after *N* rotations. A similar algebraic manipulation combining Equations H.2 and H.3 to produce H.5 shows that *PVPC* from Equation H.9 can be expressed as

$$PVPC = \frac{PC}{1 - (1 + r)^{-t}} \quad \text{(H.10)}$$

Equation H.10 looks slightly different from Equation H.5 or H.7 because planting costs take place *t* periods earlier than revenues or harvesting costs.

NET PRESENT VALUE

Net present value of the forester's income is the sum of all discounted revenues minus discounted planting and harvesting costs. It is Equation H.5 minus the sum of Equations H.7 and H.10 which can be expressed as

$$NPV = \frac{SV - HC}{(1 + r)^t - 1} - \frac{PC}{1 - (1 + r)^{-t}} \qquad \text{(H.11)}$$

Equation H.11 lets us compute the net present value of an infinitely repeating set of three terms: stumpage value; harvesting costs incurred in periods t, $2t$, $3t$, and so on; and planting costs incurred in period 0, t, $2t$, and so forth.

Appendix I

DYNAMIC EFFICIENCY FOR A DEPLETABLE RESOURCE

This appendix presents a calculus-based treatment of how a social-welfare-maximizing government or profit-maximizing producer forced to act as a perfect competitor allocates a fixed depletable energy resource, such as oil, over several time periods. The analysis involves the solution to a constrained optimization problem: Assume there are only 50 barrels of reserves and that the demand functions for each period ($t = 0, 1$) are given by the linear equation $P_t = 50 - 0.5\,Q_t$.

The goal is to maximize the present value of net benefits from two periods, the sum of consumer and producer surplus, which is the total area under each period's demand minus total production costs. The equation for that objective is $TB(Q_0) - C(Q_0) + [1/(1+r)][TB(Q_1) - C(Q_1)]$. (Notice that the second term is deflated by a one-period discount factor. It is maximized subject to the constraint that $Q_0 + Q_1 = 50$. This constrained optimization aim is accomplished by maximizing the Lagrangian expression:

$$L = TB(Q_0) - C(Q_0) + [1/(1+r)][TB(Q_1) - C(Q_1)] - \lambda(Q_0 + Q_1 - 50)$$

where $TB(\,)$ is the area beneath each period's demand and $C(\,)$ is each period's total cost. The first order conditions for an interior maximum requires differentiating this equation with respect to each of the three unknown variables: Q_0, Q_1, and λ. That differentiation produces the following three equations:

$$\partial L/\partial Q_0 = P_0 - MC_0 - \lambda = 0 \tag{I.1}$$

$$\partial L/\partial Q_1 = [1/(1+r)][P_1 - MC_1] - \lambda = 0 \tag{I.2}$$

$$\partial L/\partial \lambda = Q_0 + Q_1 - 50 = 0 \tag{I.3}$$

Remember that the derivative of total gross benefits, TB, in each period is marginal benefit, P. Solving Equations I.1 and I.2 for the two unknowns, output in each of the two periods, Q_0 and Q_1, produces

$$P_0 - MC_0 = (1/1.1)(P_1 - MC_1) \tag{I.3a}$$

If marginal cost is zero for both periods, $MC_0 = MC_1 = 0$. Substitute in the demand equation for the first two periods, for the terms P_0 and P_1:

$$50 - 0.5Q_0 = (1/1.1)(50 - 0.5Q_1) \tag{I.4a}$$

Combining Equation I.4a with Equation I.3, which describes the total supply available to allocate between the periods,

$$Q_0 + Q_1 = 50 \tag{I.4b}$$

produces two linear equations in the two unknowns, Q_0 and Q_1. Solving those two equations using your favorite method for solving a system of linear equations produces the optimal output levels of oil for the two periods:

$$Q_0^* = 28.6 \qquad (I.5)$$
$$Q_1^* = 21.4$$

Inserting those quantities of oil back into each period's demand equation lets you compute the prices associated with those optimal outputs:

$$P_0^* = 50 - 0.50(28.6) = 35.70 \qquad (I.6)$$
$$P_1^* = 50 - 0.50(21.4) = 39.30$$

Since the system of Equations I.4a and I.4b is linear, this simple two-period problem can be generalized to as many periods as desired using standard matrix inversion methods available in some spreadsheets for solving linear systems. All you need to know is each period's demand function, the number of periods, the discount rate, and total reserves. The system's linearity comes from the assumption of linear demands and constant marginal cost. The system can be solved even with nonlinear demands and rising marginal costs, but more advanced methods are required.

We can also solve for the unknown, λ, which is interpreted as the marginal economic value (shadow price) associated with discovering one extra barrel of reserves (increasing supply from 50 to 51). Inserting the now known values of P_0 (35.70) and P_1 (39.30) into either Equation I.1 or I.2 produces:

$$\lambda = P_0 - MC_0 = (1/1.1)(P_1 - MC_1) = 35.70 \qquad (I.7)$$

The shadow price, λ, equals 35.70 when reserves equal 50 barrels and the discount rate is 0.10 but it changes with changes in the supply of reserves and the discount rate. If you write out the size of reserves and the discount rate as separate parameters (numbers that can vary) in the two-equation system, you will find that Equation I.7 generalizes to a shadow price for oil under a wider range of conditions: $\lambda = \lambda$ (reserves, discount rate. Larger reserves reduce the shadow price, thus producing a downward-sloping marginal benefits schedule as a function of the size of reserves. A lower discount rate raises it.

Equation I.7 says that the first extra barrel of reserves discovered increases the present value of net income by \$35.70. Furthermore, if the cost of finding that extra barrel is less than \$35.70, there is a positive net income from investing in its discovery.

Glossary

Abatement cost. Cost of controlling the level of emissions released into the environment, or the cost of reducing ambient concentrations.
Accounting stance. Point of view from which benefits and costs are measured (e.g., national, regional, or local).
Acid rain. Mix of sulfuric and nitric acids with other airborne particles that falls as dry or wet deposits.
Age structure effect. Changes in the age distribution caused by population growth.
Aquaculture. Human-controlled fish rearing and harvesting.
Aquifer. An underground rock formation that stores and yields significant quantities of water.
Arbitrage. Process of buying a good or resource in one market and simultaneously selling it in another market at a better price. It is an economic force that eliminates price differences, which helps markets perform a valuable social service.
Asymmetric information. *See* information asymmetry.
Average cost pricing. Prices charged for resource use based on average costs.
Average consumer surplus. Total area under a demand curve divided by the quantity demanded at a specified price.
Average fixed cost. Total fixed cost divided by number of units of output.
Average revenue product. Total revenue divided by total resources used to produce a good or harvest a resource.
Average total cost. Total cost divided by number of units of output.
Average variable cost. Total variable cost divided by number of units of output.
Average willingness to pay. Total willingness to pay for a given number of units of a resource divided by that number of units.
Averting expenditure method. Estimation of benefits of environmental improvements as the change in spending on goods that substitute for a cleaner environment.
Banking. A program that harnesses the power of the market in which firms are allowed to store emission reduction credits or allowances for subsequent use or sale.
Benefit-cost analysis. Analysis of gains and losses of a proposed action or program that attempts to quantify benefits and costs in order to choose decisions for which benefits exceed costs by the largest amount.
Benefit-cost ratio. Ratio of the present value of benefits to the present value of costs associated with a particular proposed policy or program. If the ratio is less than one, it displaces more resources than it produces.
Benefits transfer. Transfer of a measured benefit of an environmental improvement from one context or geographic location to another one where a study has not yet occurred.
Bequest value. Willingness to pay for the opportunity to make gifts to future generations.
Biocentric. Term used for designing policies whose goals are centered on biology and ecology.
Biochemical oxygen demand. Amount of oxygen used per unit time in the biochemical oxidation of organic compounds.
Biodiversity. The variety of species, their genetic variability, and the variety of ecosystems they occupy.
BTU tax. A per unit charge based on the energy content of fuel.
Bubble policy. Transferable permit program that permits existing pollution sources to use emission reduction credits to help satisfy emission standards.
Burden of proof. In law based on evidence, it is the requirement of proving facts in dispute on an issue raised between the parties in a lawsuit. In criminal law, the burden of proof required by the prosecutor is to prove the guilt of the accused beyond a reasonable doubt. In a civil trial, a plaintiff must prove his or her case by a preponderance of evidence, which translates to a greater than 50% probability that all the presented facts necessary to win a judgment are true.
Bycatch. Fish unintentionally caught as a by-product harvesting a targeted species.
Cap-and-trade program. A system of regulations in which overall pollution emissions are limited and in which the regulatory authority assigns tradeable quota shares to polluters.
Capital. Any human-produced resource yielding satisfaction to its owner; includes durable goods, financial assets, land, and natural resources.
Capitalize. To find the present value of or to discount a stream of future benefits or costs.

Carbon sinks. Absorbers of carbon dioxide, such as forests and oceans.

Carbon tax. A policy proposal that would limit climate change by placing a per unit emissions tax on all carbon-emitting sources.

Carrying capacity. The population a given habitat can sustain indefinitely.

Cartel. A collusive agreement among producers to restrict production and raise prices. The group acts like a monopolist to share the gains from a monopolist's behavior.

Cash crop. Agricultural commodity that can be directly sold for money, as opposed, for example, to a crop raised for subsistence.

Catchment. Land area from which all rainfall flows into a given river.

Certainty equivalent. Monetary amount that, if received with certainty, would produce the same satisfaction as a risky benefit.

Certainty equivalent present value. A risky cash flow discounted with the risk-free discount rate.

Ceteris paribus. Latin expression that means other things that could change are held constant. For example, in a demand curve, factors other than price that affect demand, like incomes, are held constant.

Characteristic waste. Hazardous waste identified as exhibiting characteristics that suggest a significant risk.

Chlorofluorocarbons (CFCs). A family of chemicals for which the preponderance of evidence suggests that they contribute to ozone depletion.

Choke price. Maximum price anyone is willing to pay for any unit of a resource. At prices exceeding the choke price, resource demand is zero.

Coase theorem. A theory named after Nobel laureate Ronald Coase that suggests that in the absence of transaction costs, an efficient allocation of a scarce resource results through bargaining by affected individuals regardless of resource ownership property assigned to the various individuals.

Cobweb model. A model of time lags between supply and demand in which naive expectations on the part of suppliers leads to continual price oscillations. Often thought to apply to agricultural markets.

Command-and-control approach. A policy that directly regulates polluters through the use of rules or standards.

Common pool resource. Resources such as groundwater basins, rivers, marine fisheries, and community forests that have two characteristics: (1) it is hard to exclude several users from appropriating from the resource, and (2) the resources used by one user block use by others.

Common property regimes. A property rights system in which resources are managed by group decisions.

Common property resource. A resource for which property rights are shared among owners.

Comparative advantage. A principle that states people and nations specialize in those activities for which their productivity advantage is greatest or their productivity disadvantage is least compared with their trading partners.

Comparative risk analysis. Decision rule based on evaluating relative risk.

Competition. Rivalry among firms that compete for customers, a shorthand term for perfect competition. A market situation in which the number of sellers and buyers is large and the products offered by sellers are homogeneous. No firm can affect the price, so each firm faces a horizontal demand curve.

Competitive equilibrium. A condition in which marginal private benefit (MPB) equals marginal private cost (MPC) or where marginal profit (Mπ) equals zero for all goods and resources.

Composition of demand effect. Demand shifts brought about by changes in the relative input prices. For example, rising costs of virgin ores combined with stable prices for recycled inputs reduces the price of firms' products that use recycled inputs heavily, thereby making those products more desirable to buyers.

Compound interest. Interest earnings that occur on the original principle as well as on previous interest earnings.

Congestion pricing. Charging higher prices during peak hours to reduce vehicle traffic and related air pollution to promote use of public transport in those periods.

Conjunctive use. Joint use, development, and management of surface and groundwater.

Conservation easement. A purchased right to use land that permits the current owner to earn an income by protecting the land for purposes like outdoor recreation, open space, wildlife habitat, or ecosystem preservation.

Consumer price index (CPI). An index of the cost of living that measures inflation for a typical market basket of consumer goods, in which the weight assigned to each commodity is the share of spending on that commodity by consumers in a given time period. The index is set to 100 for a specified base year. The annual change in the CPI is the inflation for consumer goods, so a CPI of 200 says prices have doubled since the base year.

Consumer surplus. The amount by which total benefits produced by a good, service, or policy exceeds what consumers pay.

Consumption. Total spending by individuals or a nation on consumer goods during a given period.

Contingent choice. A method for measuring values of environmental improvements in which a respondent is asked to rank various combinations of environmental services and payments charged for those services.

Contingent ranking. A resource valuation technique that asks respondents to rank alternative situations or policies involving different proposed levels of an environmental amenity.

Contingent valuation method. Use of survey to value unpriced goods and services by asking users the maximum amount they would be willing to pay for use or minimum compensation they would require to willingly give up use.

Cost-effective abatement. Allocating abatement across polluting sources such that the marginal cost of further abatement for each source is equal.

Cost-effective policy. A policy that meets an objective of displacing the fewest amount of economic resources.

Cost-effectiveness analysis. Analysis of measures to meet program goals at minimum cost.

Criteria pollutants. In the United States, conventional air pollutant with ambient standards set by the federal Environmental Protection Agency (EPA). Examples are sulfur oxides, lead, particulate matter, carbon monoxide, nitrogen dioxide, and ozone.

Crop. An agricultural commodity sold directly for money. It is sometimes contrasted with a crop raised for a family's subsistence.

Current reserves. Geologically known stock of a mineral (e.g., oil) that can profitably be extracted at current prices.

Dam. An engineered barrier built to hold back water and raise its level, forming a reservoir to prevent flooding, reduce the costs of drought, supply water, generate power, or produce wildlife habitat.

Damage function method. A way to value environmental improvements by modeling the relationship between a contaminant and its observed effects, which permits analysts to estimate damage reductions arising from policy.

***De minimis* risk.** Latin expression meaning a negligible level of risk for which further reductions fail to justify the costs of doing so.

Deadweight loss. Loss produced by failure of economic efficiency to occur, typically caused by institutions or policies that block prices from being equal to marginal costs.

Debt-nature swap. Purchase and cancellation of a developing country's debt in exchange for environmentally related action on the part of that nation.

Decentralized markets. Decisions that result from profit-motivated transactions, which are typically contrasted with centralized allocation decisions made by government.

Declining block pricing structure. Pricing scheme for water or power in which the per unit price of different blocks decline as use increases.

Degradable pollutant. Pollutant that breaks into constituent harmless parts in water or air.

Demand. One or more individuals are willing to pay for a certain number of units of a good or service offered through a market process. Demand combines preference with ability to pay.

Demand curve. A graph showing the quantity of a commodity or service consumers buy as the price varies, holding other things equal. Typically, price is measured on the vertical axis and quantity on the horizontal.

Demand management. Policies that reduce shortages by reducing the use of a resource. Typically compared to supply management, in which policies reduce shortages by increasing supply.

Demographic transition. A theory that relates the stages of economic development to population growth rates through four stages.

Derby. A race for fish that occurs when a total allowable catch is set and fishers race with each other to harvest fish before the catch quota is met and the fishing season ends. Derbies are undesirable because they depress fish prices and are often dangerous to fishers.

Derived demand. The demand for a resource that is derived from consumer demand for final products produced by that resource. Thus, timber demand is derived from demand for housing, furniture, and other wood products.

Desalination. Process for removing salts from brackish or saline water or seawater. It tends to be expensive and economically infeasible except where substitute water is scarce or absent.

Deterrence. Process by which a risk-neutral firm is deterred from violating the law when the expected penalty exceeds the cost savings or revenue gains from violation.

Diminishing marginal returns. A law of production that states that the added output from successive increases in an input eventually diminishes.

Diminishing marginal utility. In consuming any good or service, the principle that, beyond some point, consuming more of any one commodity causes marginal utility to decline.

Direct compliance costs. The cost of environmental regulation includes direct and indirect costs. Direct costs include pollution abatement and costs of firms, consumers, and government, as well as opportunity costs attributed directly to regulation.

Discount factor. The term $1/(1 + r)^t$, where r is the discount rate and t is the number of periods.

Discount rate. Rate used to convert a stream of benefits or costs into its present value.

Discounting. Process of finding present value of a stated amount of income in which a future value is reduced to arrive at the present value. The value today of income to be received at a future date is less than its future value because the potential to earn interest on that income today is foregone when it is used today.

Dissipates resource rents. See *rent dissipation.*

Dose-response relationship. A relationship that describes the association between exposure to a contaminant and the observed response. Useful for formulating pollution control policies.

Dynamic economic efficiency. A program, policy, or proposal that generates the largest present discounted value of net benefits among all competitors.

Dynamically efficient sustained yield. Sustained yield that produces the highest present value of net benefits among all its competitors.

Ecolabel. Product label designed to inform consumers of the environmental cost of the goods.

Econometrics. A branch of economics that combines statistical methods with economic theory to test economic theories, estimate relationships, or conduct policy analysis.

Economic efficiency. Designed policies, programs, and actions for which total benefits exceed total costs by the greatest amount.

Economic theory of fertility. Theory stating that couples make rational choices about family size based on the perceived incremental benefit versus incremental cost of an additional child.

Economics. The study of how scarce resources are allocated among competing uses.

Economy of scale. Situation that occurs when the average cost of producing a unit of a good or a service declines as more and more are produced; occurs when a production function is such that when all inputs are doubled, output more than doubles.

Ecosystem service. Flow of materials, energy, and information from natural capital stocks that combine with manufactured and human capital to increase human welfare.

Efficient pricing. An organization of prices, typically when prices are set equal to marginal costs, that promotes an economically efficient allocation of resources.

Effluent allowance. Tradeable permit issued to a polluting source that assigns it the right to release effluents. By harnessing the power of the market, these allowances can reduce the total cost of complying with pollution standards.

Effluent charge. Fee or tax charged on the emission of a pollutant.

Elasticity. Responsiveness of one variable brought about by a percentage change in another. The elasticity of Y with respect to X is measured by the percentage change in Y divided by the percentage change in X.

Elasticity of substitution. A measure of the extent to which two inputs substitute for each another in production. A high elasticity of substitution between an environmentally benign input and a dangerous one means that a small price increase in the dangerous input inexpensively induces producers to substitute the benign input.

Embedding. Process that occurs when willingness-to-pay responses for a particular good (e.g., protecting a mountain stream) are approximately equal to the willingness-to-pay responses for a more inclusive good (e.g., protecting a wilderness area that includes the stream). The presence of embedding makes it difficult to isolate the economic value of any single environmental policy element, thereby making it difficult to know which elements by themselves have benefits exceeding costs.

Emission charge. A fee imposed directly on actual discharge of pollution into air or water.

Emission permit. Policy instrument for tackling environmental externalities in which the permit holder has the right to emit a set quantity of a pollutant. When the permits are tradeable, pollutants can be reduced at a lower cost than under command-and-control regulations.

Emission standard. Legal limit set on the quantity of a pollutant any single source may emit.

Emission trading. Use of market forces to minimize the cost of reducing emissions in various pollution reduction programs. An example is the U.S. Environmental Protection Agency's Clean Air Markets program (*http://www.epa.gov/airmarket/index.html*), in which a program of tradeable allowances for sulfur dioxide emissions allowed utilities to achieve their emission reduction targets more cheaply than without the program.

Emission trading program. Began in 1976 in the United States, state-controlled programs designed and implemented in cooperation with the U.S. Environmental Protection Agency.

Emissions banking. Program through which a source accumulates emission reduction credits if it reduces emissions more than required by law and deposits these credits.

Environmental economics. Application of economic principles to protect the environment.

Environmental equity. A standard based on the fairness of the environmental risk burden across segments of society, generations, or regions.

Environmental policy analysis. Evaluation of environmental policy using standards such as economic efficiency, equity, and sustainability.

Environmental risk. The probability that environmental damage will take place from exposure to some environmental hazard.

Environmental tax reform. The use of market taxing incentives to achieve environmental goals.

Epidemiology. Study of the distribution and determinant of diseases and injuries in human populations. It has five components: (1) a hunch that some factor may influence occurrence of disease; (2) formulate a hypothesis concerning influence of factor on disease occurrence; (3) test the hypothesis to see if there is a statistical association between factor and disease occurrence; (4) assess validity of association by seeing if the observed association really exists; (5) make a conclusion.

Equilibrium. Condition in which an economic variable like price or quantity is at rest with no tendency to change. For a business, it is the point at which it maximizes profit by setting marginal revenue equal to marginal cost, which blocks economic incentives to change. For a consumer, it is where the consumer's willingness to substitute one commodity for another equals the commodity price ratio.

Equimarginal principle. A principle that states you maximize the total benefit from using any scarce resource if you allocate it so that the last unit of the resource used in each activity brings the same added benefit (e.g., profit, utility, or other output) in each of its uses.

Equity. Fairness; also that part of a firm's assets on which no debt is owed to creditors.

Evaporation. Loss of water moisture as a vapor, excluding transpiration.

Evapotranspiration. Combined losses of water produced by evaporation and transpiration.

Even-aged forest. Forest for which trees were established at nearly the same time and are about the same age.

Ex ante. Latin expression meaning planned actions.

Ex post. Latin expression meaning realized outcomes.

Exclusion principle. Concept used to distinguish public and private goods. One person's use of a private good excludes another's, but use of a public good excludes nobody.

Existence value. Consumers' willingness to pay for the assurance that an environmental asset is protected, even though they have no plans to use or consume it.

Expectation. Belief people hold about uncertain future economic outcomes. Expectations are rational if they are, on average, correct in light of available information. They are adaptive if people assume that the future will be no different from the past but adapt their expectations based on past forecast errors.

Expected damages saved. The sum of all possible values for damages saved by a policy; each value is multiplied by its probability of occurence.

Expected value. Average value of a variable in some known population; mathematically measured as the algebraic sum of the potential values multiplied by each probability of occurrence. From this definition we can quantify an expected profit, present value of net benefit, or revenue. In cases where the value of some decision depends on which of several outcomes takes place, the expected value of the decision is the sum over all possible outcomes of each outcome occurring multiplied by the value that would result for each outcome.

Expenditure function. The minimum monetary expenditure needed to sustain a constant utility level in the face of any prices or environmental qualities the consumer confronts. If prices and qualities do not change from their baseline level, the expenditure function equals actual money income.

Exposure analysis. A method of analysis that involves estimating the quantity of the pollutant that people breathe, drink, absorb through skin, or are otherwise exposed to in a period of time.

External diseconomy. A firm's or consumer's actions impose uncompensated costs on third parties for whom no market transaction occurs. Pollution is an example. Same as a negative externality, spillover cost, or unpriced negative side effect.

External economy. A firm's or consumer's actions impose uncompensated benefits on third parties for whom no market transaction occurs. For example, your neighbor avoids turning a wetland into a parking lot; the wetland serves as a recharge area for your water supply. Same as a positive externality or spillover benefit.

Externality. Positive externalities are uncompensated benefits received by third parties as a consequence of a buyer-seller transaction. For example, when one homeowner buys a front yard floodlight, neighbors may benefit from increased safety at night. Negative externalities are uncompensated costs borne by third parties as a consequence of a buyer-seller transaction. For example, when firms can avoid expensive cleanup by continuing to pollute, they create an external cost shared by many others.

Extractive reserve. Forested area set aside for harvesting latex, nuts, fruits, oils, and other products as a way to create sustainable economic benefits from the rainforest without logging it.

Factor of production. Input used in production; includes equipment, machinery, labor, land, water, minerals, and the natural environment.

Female availability effect. Reduced labor force participation by females resulting from an increase in the population growth rate. An important concept to measure for nations that seek to alter their population growth rates.

Fertility rate. The average number of live births per lifetime produced by a woman in a nation. This rate has been falling worldwide over the past 100 years and is below the replacement level in many European nations.

Final good. A good or service produced for final human use (e.g., drinking water) and not used as an input for further production (e.g., irrigation water used to produce food).

Firm. Essential producing unit in a capitalistic economic system. The firm hires various factors of production to produce and sell outputs. Policymakers who wish to influence employment of inputs or production of outputs are typically more effective if they understand the firm's motivations.

Fishing effort. Use of fishing inputs (vessel, gear, labor) that is often measured as the dollar value of total inputs. Fishing effort and maximum sustained yield are often measured when setting up individual transferable quota programs in attempts to sustain commercial fisheries.

Fixed cost. Cost a firm incurs even if its output for the period is zero. These costs do not vary with output.

Flat rate pricing. Unit pricing scheme that charges the same price for each additional unit of waste. This kind of pricing scheme results in a zero marginal cost for additional use.

Flow. A variable whose value is defined between two points in time, such as income per month or cubic feet per second of water flowing in a stream. *See* stock.

Flow pollutant. Pollutant for which the natural environment has some capacity to absorb.

Free rider. A person or nation who enjoys the benefits of a public good without contributing to the costs of maintaining it. Free-rider incentives discourage participation in supply of costly public goods. For example, any country has an incentive to avoid enacting policies to contribute to controlling global warming, because it is cheaper to free ride on the steps taken by other nations.

Fully regulated forest. A forest that produces an even flow of timber forever.

Future value. The value of any income or wealth to a specified future date accumulated with the force of compound interest. The future value of a financial asset or cost is a current value multiplied by $(1 + r)^t$, where r = interest rate and t is the number of years in the future.

General equilibrium. State of balance for the economy as a whole in which prices of all goods and factors of production are at a level in which all markets are in equilibrium. At these prices, commodity producers and sellers of factors want to supply just what buyers want to buy. Partial equilibrium refers to equilibrium in a single market. Empirical general equilibrium models, called computable general equilibrium models, are expensive to build and hard to verify.

Gini coefficient. A measure of the degree of inequality in income or owned assets such as land. Values range from 0.0 (complete equality) to 1.0 (complete inequality).

Global warming. A condition caused by sunlight reaching the earth's surface and radiating back to the atmosphere where its absorption by greenhouse gasses warms the earth's surface.

Government failure. Government actions that block economic efficiency from occurring.

Green power. Power generated from renewable energy sources, such a wind and solar polar, geothermal power, hydropower, and various forms of biomass.

Green-adjusted NDP. National income measurement that accounts for renewable and nonrenewable resource depletion as well as for environmental damages from pollution.

Greenhouse abatement cost function. Costs incurred by policies that slow greenhouse warming.

Greenhouse damage function. Costs society bears from incurring damages produced by a changing climate.

Greenhouse gas. An atmospheric gas that limits heat from radiating back into space, which has a warming effect on the earth's atmosphere. Examples are carbon dioxide and methane.

Groundwater. Fresh water that lies beneath the earth's surface, typically in aquifers.

Hazard identification. Scientific analysis to determine the presence of a relationship between a pollutant (a hazard) and any adverse effects, such as those on human health.

Hedonic property study. Valuation method in which the economic value of an environmental amenity is measured by differences in the property values exposed to different levels of the amenity. For example, the value of policies that reduce airport noise from *X* to *Y* is measured by property value reductions for homes with airport noise *X* compared with similar property located farther from the airport, which currently receive noise level *Y*.

Hedonic wage valuation. Method of analysis that measures the value of an environmental amenity from differences in the values of wages paid to workers exposed to different amenity levels.

Hicks-Kaldor compensation principle. A principle that states that a policy improves society's overall welfare if the gainers could compensate the losers and still be better off. It does not require that actual compensation be paid.

High-grading. Discarding low-value fish or minerals in favor of high-valued ones in order to increase the income derived from a harvest quota or other capacity limit.

Hotelling's rule. Profit or net benefits maximization from a depletable resource requires that the discounted value is equal in all periods for price minus marginal cost.

Human capital. Stock of knowledge, skills, and talents of the workforce.

Hydrologic cycle. Natural movement of water from the atmosphere to the surface, beneath the ground, and back into the atmosphere.

Hypothetical bias. Biased answers from survey respondents that have the potential to arise if responses are based on hypothetical rather than actual choices.

Income. Flow of payments per unit time, including wages, dividends, interest, and rents that are made to a person, group, or to a nation.

Income elasticity. Quantification of the responsiveness of a good's sales to changes in consumer income. It is measured as the percentage change in quantity of a commodity purchased in response to a 1% change in income. Elasticities larger than one are said to be income elastic.

Increasing block pricing structure. Pricing scheme in which the per unit price of different blocks of water, power, or other commodities increases as water use increases.

Indifference curve. A graph showing combinations of consumption for two goods. All points on one indifference curve show combinations of two goods whose consumption provides equal levels of consumer satisfaction. The indifference curve is bowed in to the origin, showing that greater amounts of good *X* are required for a single unit of good *Y* sacrificed when there is less of *Y* and more of *X* in the mix.

Indifference map. A family of indifference curves. Higher levels of satisfaction are shown by indifference curves lying farther to the right. The indifference map combined with the consumer's budget constraint is used to derive the consumer's demand curve.

Individual transferable quotas. A method for protecting and sustaining a fishery and the income derived from it by limiting the number of fish caught. Shares of a total allowable catch are allocated to individual fishers. Initial quota allocations are typically based on historical catches or some other politically acceptable method. Quotas can be traded among fishers, and a competitive quota market is expected to allocate quota to the most economically efficient fish harvesters.

Inelastic. If the absolute value of the price elasticity on a demand or supply curve is less than one in some range, the curve is "inelastic" in that range. When demand is price inelastic, it means that a reduction in price produces such a small increase in quantity demanded that total revenues are reduced.

Information asymmetry. One party to a transaction has more or better information than the other party. Typically it is the seller that knows more about the product than the buyer; however; it is possible for the reverse to be true.

Information bias. A problem that may occur when respondents to contingent valuation surveys are asked to value environmental attributes for which they have little knowledge.

Input-output table. A table showing the monetary flows of sales and purchases among all sectors in an economy. The table can be expressed as absolute money flows or as coefficients that lie between zero and one. Tables have been built for local, regional, national, and global economies.

Intellectual capital. Stock of useful knowledge; also called the state of technology.

Interest rate. Price paid to lender for borrowing money or other resources for a period of time. If the interest rate is 5% per year, then $50 is paid for the use of $1,000 per year.

Intermediate good. Good or service that has gone through some manufacturing process but is not yet a final good available for sale to final buyers. Irrigation water is an intermediate good.

Internal rate of return. Discount rate at which the present value of a known stream of net benefits equals zero. A higher internal rate of return is typically associated with a more productive project or program. This method of project appraisal can be a useful way to rank programs when a nation's capital markets are too immature to produce a well-defined interest rate.

Invisible hand. A concept assigned to the famous eighteenth-century British economist Adam Smith that refers to the efficiency in which a well-functioning competitive market coordinates the complex and interdependent allocation of scarce resources in an economy without the hand of economic planners. This theory holds each person pursues his or her own self-interest and a market system composed of such people works to the benefit of everybody, as though an invisible hand orchestrated the process.

Involuntary risk. A risk beyond a person's control which is not the result of a willful act.

Joint and several liability. A legal doctrine used to assign the cost of cleaning up toxic sites. It states that any group of parties responsible for contaminating the site can be assessed the entire cost, regardless of the size of each individual contribution.

Joint cost. A cost ascribed to producing several outputs together—the cost cannot be attributed to any single output. A joint cost exists if deleting any one of several outputs has no effect on the cost.

Joint implementation. An institution in which nations work jointly to reduce the greenhouse gas emissions efficiently. It establishes the foundation for a system of tradeable carbon permits.

Law of demand. An economic law that says the quantity demanded of a good will be greater at a lower price than its quantity demanded at a higher price.

Law of diminishing marginal returns. Situation in which increasing output in the short run requires that the firm add increasingly more variable inputs such as labor or water.

Law of one price. Tendency for the price of any traded good to become equal in all areas, except for differences caused by shipping costs.

Law of supply. A direct relationship between price and quantity supplied of a good, that is, more output will be voluntarily supplied by profit-making producers only when the price increases.

Liability law. Rule for assigning liability to individuals or firms that could prevent environmental damage. Also, a rule that awards compensation by the courts from an injurer to an injured party after damage has occurred.

Long run. Period of time over which full adjustment can take place. For an individual firm, it is the period of time over which all inputs, including the size of the capital stock, are variable.

Malthusian Theory of Population. A theory expressed by the eighteenth-century classical economist Thomas Malthus that states that the natural tendency of population is to increase geometrically (1,2,4,8,16. . .), while the food supply tends to increase only arithmetically (1,2,3,4. . .). Food production per person therefore tends to decrease over time, and through starvation, reduced population growth. The theory has been found to be untrue in developed countries.

Marginal abatement cost. Added opportunity cost of reducing greenhouse gasses or other pollutants by one unit (typically 1%).

Marginal benefit. Added benefit of one more unit.

Marginal cost. Addition to total cost when output is increased by one more unit.

Marginal cost pricing. Setting a resource's price use based on the marginal opportunity cost associated with its use. This pricing scheme typically promotes economic efficiency. For example, charging each water user a price equal to the marginal cost of use saves the need for expensive new water purification facilities and environmental damages associated with that new water development.

Marginal external cost. Increase in total external costs paid by society as pollution damages that occur as a result of a single unit increase in a firm's output.

Marginal extraction cost. Cost of mining one more unit of resource.

Marginal net benefit. Marginal benefit minus marginal cost.

Marginal opportunity cost. Added opportunity cost of supplying the incremental unit of output measured by what it displaces.

Marginal private cost. Increase in total private cost resulting from a one-unit increase in a firm's output.

Marginal product. Added product resulting from one more unit of a given input, such as labor or water, when all other inputs are fixed.

Marginal revenue. Difference in total revenue derived from selling $n+1$ rather than n units. In a perfectly competitive market, marginal revenue is the same as price because the added output does not affect price. For a monopoly seller, marginal revenue is less than price because the price must be reduced on all previous units in order to sell one more unit of output.

Marginal revenue product. Marginal revenue times marginal product. It is the added total revenue from hiring one more unit of input and from selling the added units of output produced.

Marginal social benefit. The sum of marginal private benefit and marginal external benefit.

Marginal social cost. The increase in total cost borne by society that occurs as a consequence of a one-unit increase in output produced by a firm. It is the sum of marginal private cost and marginal external cost.

Marginal willingness to pay. Amount of money an individual is willing to pay for the last unit of a good or service produced by a proposed policy.

Market. A place, mechanism, or institution in which buyers and sellers communicate and exchange goods or services after agreeing on the price and terms and conditions of sale. Markets are an important institution for allocating scarce resources.

Market approach. An incentive-based policy that encourages conservation practices or pollution reduction strategies.

Market economy. An economic system in which resource allocation decisions are guided by prices, which come from the interaction of demand and supply that originates from voluntary purchasing and production decisions of private consumers and producers.

Market equilibrium. Balance achieved when a good's price is set so that the quantity demanded by buyers just equals the quantity supplied by sellers, which results in neither a surplus nor shortage.

Market failure. A failure of the price and market system in which an institution or policy blocks an economically efficient allocation of resources. Examples include externalities, common property resources, and public goods.

Maximum sustained yield. Maximum number or quantity of resource units that can be harvested without damaging the productive capacity of the resource stock.

Mean annual increment. Cumulative volume of a forest stand at any given period divided by the length of time the stand has existed. Measured in units of volume per year.

Microeconomic theory of fertility. A theory that attributes fertility differences among couples, cultures, and nations to the economic environment and associated incentives influencing childbearing decisions.

Minimum sustainable population. *See* minimum viable population.

Minimum viable population. Population level below which growth is negative, which leads to extinction.

Model. Simplified representation of a process, situation, or object; typically used in a formal framework to represent the essential features of a complex system through a small number of simplifying relations. Often represented by way of graphs or mathematical equations.

Monopoly. A market structure in which a commodity is sold by a single firm. When a policy increases the output of a product sold by a monopoly (e.g., the policy increases electric power output because of a dam built), its product price equals its marginal benefit to society but exceeds its marginal revenue.

Monopsony. A market structure in which a resource is bought by a single firm. When a policy increases the quantity of a resource bought by a monopsony (e.g., the policy increases sugar output but there is only one sugar buyer), the resource price equals the marginal resource cost but is less than the marginal benefit it produces for society.

Multiple use forestry. Management of a forest for several purposes. It can mean many uses on each acre, a single dominant use on different land parcels, different uses over time, or a combination.

National income account. Account that is used to measure total income produced by a nation's economy.

Natural capital. Stock of natural resources, together with the components and structural ecosystem relations that function jointly as the foundation for life.

Natural equilibrium. Stock levels of a renewable resource that persist if there are no outside influences.

Natural resource economics. Use of economic principles to improve the adequacy of natural resources.

Negative externality. An external effect that generates costs to a third party.

Negative feedback loop. Reaction that tends to return a process to equilibrium.

Negligence. A legal doctrine stating that the party responsible for contamination owes a duty to affected parties to exercise due caution. Failure to fulfill that due caution may lead a court to find that because of negligence by the responsible party, the injurer must pay compensation to the victim.

Net benefit. Excess of benefits over costs resulting from some action.

Net present value. Present value of future revenues (or benefits) minus present value of future costs associated with a project or proposal.

Netting. A marketable permit program to limit air pollution in which firms deciding to modify or expand existing operations may avoid new source review requirements if the resulting increase in emissions is less than a prespecified amount. Netting allows emission trading among points within a source for the same type of pollutant, such that any emission increase due to a modification is offset by a reduction from another point within that same source.

Nominal income. Money income expressed in the national currency (e.g., dollars) for the year in which it actually occurs (includes inflation).

Nonattainment region. A region in which pollution concentrations fail to meet existing ambient requirements.

Nonexcludability. Characteristic of a public good that makes it impossible to prevent others from sharing in the benefits of consumption even if they fail to pay their share.

Nonmarket. Not traded in the market for a price.

Nonpoint-source pollutant. A source of air or water pollution that cannot be identified accurately and degrades the environment in a way that is difficult to detect over a broad area.

Nonrenewable resource. A class of natural resources having the characteristic that the overall stock cannot replenish itself within the human time frame.

Nonrival. Characteristic use of public good such that one person's consumption does not preclude that of another.

Nonuse values. Resource value that arises from motivations other than personal use; sometimes called passive use values.

Occam's razor. Principle stated by English philosopher William of Occam which says the simplest theory that explains the observed facts should be selected when several theories are available. For any given phenomenon, Occam's razor eliminates concepts, variables, or constructs not required to explain it.

Offset plan. Institution developed for nonattainment areas that encourages emission trading such that releases from one source are more than offset by reduced emissions achieved by other sources.

Old growth. Standing timber that is suitably old and for which net annual volume growth is negligible.

Open access. Condition resulting from a lack of a property rights system that would otherwise limit access to a resource.

Opportunity cost. Value of the best alternative displaced by carrying out a given decision or policy. It can be expressed in nonmoney terms; for example, one opportunity cost of an intensively managed timber production area can be recreation visitor days displaced. The opportunity cost of establishing a wilderness area is the potential net income from commercial products displaced. Opportunity cost is an especially useful concept for estimating impacts of conserving or depleting unpriced goods such as environmental quality or workplace safety.

Optimal rotation. Tree harvest age at which net present value of a forest is maximized; may include timber and nontimber values.

Optimum. In a choice process, the point at which some output is maximized (e.g., satisfaction, income, present value of net benefits, profit, visitor days). The verb is *optimize;* the adjective is *optimal.* The word *optimum* can be used as a noun and as an adjective.

Option value. Consumers' willingness to pay for the opportunity to use some resource in the future, even if they do not use it now.

Ozone layer. Ozone existing in the stratosphere that protects Earth from harmful effects of ultraviolet radiation.

Ozone-depleting gas. Global pollutant that destroys the stratospheric ozone layer. Examples are chlorofluorocarbons and halons.

Pareto optimum. An economically efficient resource allocation, named after Italian economist Vilfredo Pareto. This means an allocation for which no resource reallocation could make anyone better off without a sacrifice, that is, without reducing at least one person's welfare.

Payback period. Number of years required to recover the initial capital invested in a project.

Peak-load pricing. Charging resource users during peak-use periods the higher cost of supplying resources. The higher charge during the peak period covers the cost of expanded capacity, because the need for expansion is caused by increased peak-load demands.

Peak period. Times of high resource demand (e.g., the demand for electricity during the hottest part of summer when air conditioning is in heavy use).

Photovoltaic technology. A family of technologies that converts light directly into electricity. It is attractive because it is renewable and environmentally benign.

Physical capital. Plants, machines, infrastructure, and buildings—all of which come from devoting part of each period's total output to investment.

Pigouvian tax. A per unit tax on a good or resource for which the use generates a negative externality. The intent is to reduce the output of the good or use of the resource.

Point-source pollutant. Any single identifiable source of pollution, such as a fixed location or facility, from which contaminants are discharged. They are easier to regulate and assign responsibility to than nonpoint sources. Most municipal and industrial sources are point sources.

Pollution. Human-produced emissions that exceed an ecosystem's capacity to assimilate harmlessly.

Pollution abatement. Reducing, eliminating, or disposing of unwanted emissions that are harmful to health or the natural environment.

Precautionary principle. A principle that states precautionary measures should be taken when evidence suggests an activity generates costly or dangerous harms, even if there is considerable uncertainty about the quantity or the mechanism of those harms.

Present value. A future value can be discounted back to the present value by dividing it by $(1 + r)^t$, where r = interest rate and t is years into the future.

Present value analysis. A method that discounts a future value into its corresponding present value that accounts for the discount rate.

Present value benefits. Time-adjusted magnitude of added benefits associated with an environmental policy proposal.

Present value cost. Time-adjusted magnitude of added costs associated with an environmental policy proposal.

Present value of net benefits. Difference between present value of benefits and present value of costs. It is used to determine the economic feasibility of a decision, based on its size and sign.

Price elasticity of demand. Responsiveness of buyers to a price change; measures the percentage change in quantity bought divided by the percentage change in price. Demand is elastic if the elasticity exceeds one, and inelastic if elasticity is less than one.

Price elasticity of supply. Like price elasticity of demand, but measures the responsiveness of suppliers to a price change; measures the percentage change in quantity supplied divided by the percentage change in price.

Prior appropriation doctrine. Principle that the right to use water is assigned to the person who first diverts water from the stream and puts it to beneficial use.

Private good. A commodity that has the characteristics of rivalry in consumption and excludability. *See* public good.

Private marginal cost. Cost that is borne by the producer of producing an additional unit of the resource or good.

Producer surplus. The extra income received by efficient firms above the minimum required to produce current output. For an industry of firms producing the same product, producer surplus is measured as the area above the industry supply curve and below the price line.

Production function. A mathematical or graphic function showing the amount of output that can be potentially achieved with given inputs. It can be applied to a single firm, a sector, a region, or a nation.

Profit maximization. A degree of profit achieved at the output level where marginal revenue equals marginal cost ($MR = MC$), or where $M\pi = 0$.

Property rights. A bundle of entitlements that define the owner's rights, privileges, and limitations for use of the resource.

Public choice. Analysis that predicts politicians' behavior in voting for government programs based on the theory that politicians are self-interested welfare-maximizers with an objective function that includes current and discounted future income, reelection, power, and control.

Public good. A good or service with these two characteristics: (1) it can be used by several people simultaneously and (2) use by one does not reduce available supply for others. Typically contrasted with a private good, which can only be used by one person, and its supply to one displaces its use by others.

Quantitative risk assessment. Quantitative measures for hazard identification, exposure assessment, dose response, and risk characterization.

Rational choice. A selection that is consistent with the objectives and preferences of those making the decision, given the available information.

Rational policy analysis. Analysis that considers all relevant alternatives, identifies and evaluates all consequences that would follow from the adoption of each alternative, and selects the alternative and its associated consequences that would be preferable in terms of society's most valued ends.

Reasonable use. Under a riparian law system, any water use that avoids imposing significant damage to other landowners that border a watercourse.

Regulated forest. A forest with equal quantities of even-aged timber in each age class, so that equal harvests could continue into the indefinite future.

Renewable resources. Resources that can be replenished over time, either naturally or through human action.

Rent dissipation. When current use rates of a renewable resource exceed the dynamically efficient use rate.

Rent seeking. Use of scarce resources for lobbying and other political activities motivated toward acquiring increased benefits for special interest groups through protective regulation or legislation.

Replacement. Level of total fertility that is compatible with a stationary population. The number is typically slightly larger than 2.0.

Riparian law. A system of water rights that allocates the right to use water to the landowner adjacent to the water.

Risk. Variability in an expected flow of cash or benefits. The greater the variation, the greater the risk. Generally risk refers to the potential for loss.

Risk assessment. Evaluation of the risk posed to health or environment by a hazard.

Risk aversion. A situation in which a decision maker dislikes greater variation or risk in an expected cash flow.

Risk-benefit analysis. Method of making decisions that evaluates the risks of an environmental hazard in conjunction with benefits of failing to regulate it.

Risk characterization. Analysis of risk to illustrate how human populations or communities are affected by hazardous human activity.

Risk management. Process by which a decision maker evaluates and selects from alternative actions for coping with environmental risk.

Risk neutral. A condition in which a person is indifferent between receiving an outcome of *X* for certain and taking a risky bet with an expected value equal to *X*.

Risk premium. Percentage points that a risk-averse investor adds to the risk-free discount rate to account for risk in future revenues. This premium represents compensation for a willingness to undertake some risk.

Rotation. Age at which timber is harvested; years between planting and harvesting.

Safe minimum standard. Achieved by avoiding the physical conditions brought about by human action that would make it uneconomical to halt and reverse depletion of a resource.

Scarcity. Situation in which demand exceeds supply at a zero price.

Secondary effect. Change in income resulting when a natural resource project buys from or sells to related businesses.

Sensitivity analysis. Numerical analysis that tests the sensitivity of a known outcome to changing values of an important variable. For example, you might test how the net present value of a project or program changes as you adjust the discount rate.

Severance tax. A tax imposed on a mineral's extraction.

Shifter. Anything other than price that shifts a demand or supply curve to the left or right (e.g., per capita income, which can shift the a commodity's demand curve to the right).

Short run. Period of time in which not all factors of production can fully adjust (e.g., the size of the capital stock, which cannot be adjusted in the short run).

Shortage. Demand for a commodity exceeding supply that occurs because the price is beneath its equilibrium.

Social discount rate. Discount rate used for public project evaluations based on the social opportunity cost of funds.

Social marginal cost. Cost of producing one more unit of the resource that is paid for by society at large. Includes private marginal costs added to external marginal costs.

Social opportunity cost. Society's highest-valued losses resulting from the use of resources for a given, specified period.

Social rate of time preference. A concept developed by economist Stephen Marglin that is defined as the rate of discount at which society is collectively willing to give gifts to future generations. It is an expression of the electorate concerning the rate of discount they wish to apply to future costs and benefits in evaluating government projects that provide services for future generations.

Social welfare. Total net benefits produced by a program or policy; often measured as the algebraic sum of consumer surplus and producer surplus.

Spillover. Another term for externality.

Stand. A group of standing trees.

Starting point bias. Type of bias that may arise if a respondent to a contingent valuation survey checks off an answer from a preset list of possibilities, and the respondent's answers vary according to the range specified by the instrument.

Static economic efficiency. An economic standard for selecting among competing project or policy decisions when time is not important. A decision satisfies this standard if its choice results in maximizing net benefits.

Stewardship. A sense of duty for preserving a natural resource or environment for future generations.

Stock. A variable whose value is defined at a single instant (e.g., water in a reservoir or personal wealth). *See* flow.

Stock pollutant. Pollutant that builds up in the natural environment because the environment has little capacity to absorb it.

Stocking. Establishing quantity of trees or livestock or other renewable resource on an area, especially in connection with achieving some management objective.

Strict liability. Legal standard that identifies individuals responsible for damages even without proving negligence.

Stumpage. Standing timber.

Stumpage value. Estimated amount that buyers would pay for standing timber for immediate harvest. When expressed as an amount per unit volume, it is described as stumpage price.

Sulfur allowance auction. In the United States, an annual auction that requires utilities to place a proportion of sulfur allowances up for sale each year; proceeds are returned to the utilities. Because the government receives no revenue, it is called a zero-revenue auction. It guarantees a continued supply of permits and provides price information.

Sulfur allowance program. A transferable sulfur permit program directed to electric utilities; the aim is to reduce sulfur emissions by 10 million tons from 1980 levels.

Sunk costs. Costs that have been incurred and cannot be reduced even if an activity ceases.

Supply. A concept that refers to the quantities of a commodity that producers will supply per unit of time at varying prices. A supply curve is a mathematical function and the graph is commonly drawn with price on the y-axis and quantity on the x-axis.

Sustainable development. Methods for satisfying the present generation's needs without compromising the ability of future generations to meet their own needs.

Sustainable yield. Resource harvest levels maintainable forever.

Threshold. Level of exposure to a hazard up to which there is no response. After the threshold, a response occurs.

Timber. Trees for purposes of producing commercial wood products.

Time path. Sequence of decisions made or outcomes resulting at several future time periods.

Total net benefit. Total benefit minus total cost.

Total profit. Total revenue minus total costs.

Tradeable carbon permits. A proposed system of transferable permits whose aim is to limit climate change by limiting carbon emissions and allocating transferable quotas to nations.

Trade-off analysis. Process in which changing one output alters other outputs. Typically, as one output is increased another output decreases.

Tragedy of the commons. A concept developed by Garrett Hardin that describes excessive use of a common-pool resource under open access property rights.

Transactions cost. Cost incurred in attempting to carry out a transaction. In buying a piece of real estate, costs could include payments to the broker for arranging the sale, to the bank for one-time fees, and to government for necessary forms.

Transferable emission permit system. A system in which permits defined by the right to emit known amounts of emissions can be transferred. Policymakers can use permits to control or limit pollutants while receiving industry support.

Transpiration. Use of water by plants and trees.

Travel cost model. A method of analysis in which differential travel costs from varying travel distances are used to estimate the recreational value of an unpriced environmental resource.

Use value. Satisfaction people receive from directly using some dimension of the environment.

User cost. Opportunity cost created by scarcity of an exhaustible resource. For example, when a unit of an exhaustible resource is used, it can never be used again. User cost is the highest net benefit that could have been earned had the unit been used in any other time period.

Utilitarianism. An ethical system requiring that public policy result in the greatest satisfaction and/or least displaced satisfaction for the people affected. For a utilitarian environmental policy to be ethically correct, the utility gained by that action should exceed the utility lost when summed over all affected members of society. Basing policy actions on this philosophy supports any program promoting economic benefits exceeding costs to the relevant affected population.

Utility. Principle that evaluates actions according to their ability to increase an individual's satisfaction. Measured by human satisfaction or well-being.

Virtual water. Water saved by reducing exports of products that consume water.

Voluntary risk. A risk deliberately assumed or taken on by an individual.

Water right. A property right to use, in accord with its priority, a specified part of the stream flow of a river or stream, by reason of its appropriation.

Weak complementarity. Two conditions that are required before the benefits of an environmental improvement can be measured as the changed area beneath a demand function: (1) there is a finite price at which the demand for the service is zero under its highest environmental quality being considered by the policy; (2) there are no benefits from a quality improvement at that price.

Weak global scarcity hypothesis. Hypothesis under which production (typically of food) keeps up with population growth, but the supply curve is sufficiently steep that food prices increase faster than other prices. This results in a problem of food affordability rather than physical availability.

Weak sustainability. A concept of sustainability based on economic growth models. A central requirement is that reductions in natural capital can be compensated for by increasing other forms of capital (physical, intellectual, human).

Willingness to pay. The conventional measure of the economic value of a policy or program that is used in benefit-cost analysis. It is measured as the maximum money amount an individual or group is willing to pay for a given good, service, or amenity.

With and without principle. A means of determining the benefit from any program by measuring the net present value (NPV) of the project *with* the program minus the NPV *without* the program.

Index

Acid rain program, 10
Africa. *See also specific country*
DDT in, 512
girls' education, 413
ivory trade, 53
rangelands, drought on, 230
sustaining development in poor cities, 556
Agriculture. *See also* Food
biotechnology in, 319–320
complexity of decisions, 301
computer literacy and, 320
environmental impact, 312–315
environmental protection, 317–318
fertilizer, 313–314
global warming, addressing, 319
irrigation, 313
multifunctionality, 315
pesticides, 314
property rights and, 323
small-farm friendly policies, 319
soil erosion, 312–313
water use efficiency, 318–319
wildlife habitats and, 314
Air pollution, 6. *See also* Pollution
Clean Air Act, economic efficiency of, 60
clean air *versus* affordable cars, 16
Alaska's Prudhoe Bay, 20
Analysis
benefits and costs. *See* Benefit-cost analysis (BCA)
climate change, 421–427
cost-effectiveness analysis, 115–118
environmental impact analysis, 119
government abuses of, 88
importance of, 9–10
Anasazi people, 2–3
Animal rights, 32
Animal *versus* human welfare, 542
Annual increment (AI), 192
Asia. *See also specific country*
Central Asian, pollution control incentives, 485
Southeast Asian water rights, 543
Avoided costs method, 175–176
Benefit-cost analysis (BCA), 93
advantages of, 119–120, 122–123
alternatives to, 115–119, 121–122
benefits by period, 103
benefits per unit, 101–102
climate change, 426–427
costs by period, 103–104
costs per unit, 102
decision rules, 108
design, implementation, and review programs, 94–95
economic principles of, 95–100
employment impacts, 114–115
environmental protection, 29
equimarginal principle, 97–99
goals of, 95–97
greenhouse warming, 426–427
incremental analysis, 97
inflation, 115
input quantity, 102
limits of, 120–121, 122
measuring benefits, 108–109
alternative cost saved, 110–111
marginal and nonmarginal changes, 109–110
mathematical methods, 110
user surveys, 111
measuring costs, 111–112
net benefits by period, 104
net present value, 104–105
calculation of, 105–106
postponing projects or regulations, 100
sensitivity analysis, 106–108
steps to, 101–108
output quantity, 102
postponing projects or regulations, 100
purpose of, 93–94
risk management, 516–518
risky programs, 115
scope of, 97
scoping, 101
secondary effects, 113–114
sequencing, 100
time estimates, 112–113
U.S. environmental programs, 100
what is, 93
why measure benefits/costs, 93
with and without principle, 99
Berlin Mandate, 435–437
Birth control. *See* Fertility regulation
Brazilian climate change negotiation, 440
British thermal units (BTUs), 358
Brundtland report, 9
Buenos Aires Plan of Action, 439
Bycatch, 350

California
power meltdown, 387
RECLAIM program, 6
water resources, 3
Cap-and-trade programs, 50
Carbon monoxide in Korea, 515
Carbon taxes, 428
Central Valley Project (CVP), 3
CERCLA, 5–6
Children
cost of, 408–412
demand for, 407
fertility regulation. *See* Fertility regulation
mortality, 412
women's roles in reducing, 414
Chilean environmental regulations, 467
China
energy programs, 563
food famine of 1958–1961, 324
one-child policy, 19, 409
population programs, 9
water pollution policy, 450
water programs, 292
Chloroflurocarbons (CFCs), 5
Clean Air Act, 16
economic efficiency of, 60
Climate change, 418
adjustment, 428
benefit-cost analysis, 426–427
carbon taxes, 428, 430–431
control of, 145
economic analysis, 421–427
economic impacts, 422

emission trading, 431–433
Kyoto Protocol, 438
engineering *versus* behavioral approaches, 429
gasoline taxes, 430–431
greenhouse gases. *See* Global warming
international agreements, 434–442
mitigation, 428
New Hampshire's experience, 429
outcomes, probabilities, and impacts, 419
policy choices, 427–431
prevention, 428, 430–437
public perception, 422
regional adaptive capacity and vulnerability, 420–421
scientific evidence, 419–421
technological change, role of, 421–422
United Nations Framework Convention on Climate Change, 434–435
Closed-ended choice, 178
Coase theorem, 73–75
Cobb-Douglas preference ordering, 577
Cod wars, 352
Common property, 79
conservation appeals, 83
private property rights, managing through, 84–85
public action, managing through, 83–84
Thailand' fisheries, 343
tragedy of the commons, 79–83
avoidance of, 87
Brundtland report, 9
when it works, 85–86
Compensated demands, 578
Compensating variation (CV), 155–158
Competitive markets, 36–37
Competitive supply, 55–58
Conference on the Human Environment, 566
Conservation easements, 214–215
Conservation of energy, 389–392
Consumer surplus, 573–574
approximation, 158–160
and producer surplus, 61–62
Contingent choice, 178–179
Contingent valuation method (CVM), 176–178
critiques of, 581–587
hypothetical *versus* actual cash payment, 183
Cost-effectiveness analysis, 115–118, 576

DDT in Africa, 512
Deep ecology, 31
Definitions, 593–603
Demand equation, 577
Demand for goods, 37
change in demand, 40–42
demand curves, 38
equilibrium supply. *See* Equilibrium demand and supply
law of demand, 37–38
market demand, 38–39
mathematics, 570–572
quality demanded, change in, 40
water demand management, 274–278, 287
Demand function for a private good, 160–161
Demand schedules, 157–158
Development, sustainable. *See* Sustainable development
Dichotomous choice, 179–180
Discount rate, 127
climate change, control of, 145
energy, 368–369
environmental policy, 135–136
environmental protection and, 140
government borrowing, monetary cost of, 137
net present value, calculation of, 128–130
environmental policy, 135–136
private investments, 134–135
tables of discounted values, 130–134
for saving lives, 132
search for, 136–146
social opportunity cost of capital, 137–139
social rate of time preference. *See* Social rate of time preference
tables, 130–134
Doc's theory of overfishing, 334
Drinking water, 4–5. *See also* Water
environmental justice in regulations, 535
poverty and, 290
Drought
Africa's rangelands, 230
Anasazi people, 2–3
water pricing, 284–285
Dynamic efficiency, for a depletable resource, 591–592

Ecofeminism, 32
Ecolabeling, 50
Ecolabel programs, 564
Economically optimum stocking rate, 230–231
optimum input use, 231–233
optimum supply response, 234–235
shifts in supply, 235–238
Economic efficiency, 25, 59
benefit-cost analysis, 95–96
Clean Air Act, 60
fisheries model, 337–338
of perfect competition, 58–63
pollution control, 448–459, 462
rangelands, 227–228
risk(s), 496–499
water, 265–269
Economic theory, 35
competitive markets, 36–37
demand. *See* Demand for goods
supply. *See* Supply of goods
Ecotourism, 217–218
Education, 412
African girls, 413
food education for women, 322
Egalitarianism, 30
Egyptian climate change, 441
Electricity, 370–371. *See also* Energy
Elitism, 30
Emissions
charges, 474–480
efficient levels, 457–459
permits, 6
Emission trading, 480–482
climate change, 431–433, 438
Kyoto Protocol, 438
Empire building, 88
Employees. *See* Labor force
Empowerment, 7
Endangered species
food *versus* endangered fish, 17–18
livestock *versus* endangered wolves, 16–17
timber *versus,* 15, 204–205
wildlife *versus* endangered wolves, 16–17
Endangered Species Act, 251–252
Energy, 356–357
affordable energy *versus* functioning ecosystems, 15
backstop fuel, presence of, 367–368
Brazilian challenges, 379
Chinese sustainable energy programs, 563
conservation, 389–392
discount rate, 368–369
electricity, 370–371
environmental adders, 383
future outlook, 377–380
green power, 388
Hotelling Rule, 363–366
Japan's photovoltaics, 379
marginal cost pricing, 381–383
market allocations, 360
economic concepts, 360–369
energy sources, 369–371
future outlook, 377–380
OPEC cartel, 372–377
nuclear energy, 370
OPEC cartel, 372–377
policy
need for, 358–359
objectives, 359–360
pollution taxes, 383–384
nd poverty, 358
power *versus* recreation, 18
profit maximization, 360–361
renewable energy
Brazilian challenges, 373
California's power meltdown, 387
conservation, 389–392
fossil fuels *versus,* 369–370
Germany, promoting in, 385
green power, 388
India, promoting in, 386
Japan's photovoltaics, 379
promotion of, 384–387
standards, 387–388
voluntary action, 388–389
resource base, size of, 366–367
sources of, 369–371
taxes, 383–384

use, measuring, 357–358
user costs, 361–363
versus wildlife, 20
Environmental impact analysis, 119
Environmental justice, 526–527
benefits and costs, 534–538
drinking water regulations, 535
evidence of, 540–543
hazardous wastes TSDF sites, 530–534
human *versus* animal welfare, 542
measuring, 538–540
in Nepal's community forestry, 537
principles of, 527
risk assessment, 529–530
South Asian water rights, 543
in the United States, 533
waste treatment facility sites, 536
what is, 527–529
Environmental policy. *See* Policy
Environmental protection
agriculture, 317–318
benefit-cost analysis, 29
benefits of, 27
cost-effectiveness analysis, 29
costs of, 27–28
discount rates and, 140
how much is necessary, 28–29
Environmental risk. *See* Risk(s)
Environmental values
compensating variation, 155–158
consumer surplus approximation, 158–160
demand function for a private good, 160–161
demand schedules, 157–158
economic valuation, 152
aggregation, and social welfare, 162–165
compensating variation, 155–158
consumer surplus approximation, 158–160
demand function for a private good, 160–161
demand schedules, 157–158
equivalent variation, 155–158
expenditure function, 153–154
price changes, 162
quality changes, 162
equivalent variation, 155–158
expenditure function, 153–154
information, uses of, 151
measurement methods, 166
avoided costs method, 175–176
benefit transfer method, 182–184
contingent valuation method, 176–182, 183
hedonic pricing, 169–171
market prices, 166–167
productivity, 168–169
travel cost method, 171–175
monetary measures of, 151
nature of, 153
policy, support of, 165
price changes, 162
quality changes, 162
relevance of, 151
United States *versus* Europe, 152
use *versus* nonuse values, 153
Equilibrium demand and supply, 46
changes in supply, demand and equilibrium, 49, 51–52
equilibrium, 48
inverse price/quantity relationship, 52–54
market price and quantity, 48–49
price as a rationing method, 49
shortages, 48
surpluses, 46–47
Equimarginal principle, 97–99
forests and forestry, 207–209
Equivalent variation (EV), 155–158
Erosion, 312–313
Ethiopian rangelands, 243
Europe's fisheries, 341
Existence value, 153
Expenditure function(s), 153–154, 577
Externalities, 71
Coase theorem, 73–75
ethics *versus*, 27
negative externalities, 71–72
positive externalities, 72–73
water charges, 282
Extractive reserves, 215–216
Exxon Valdez oil spill, 8, 15

Family planning. *See* Fertility regulation
Family size
China's one-child policy, 19, 409
potential, 410
unconstrained optimal, 409
Federal Lands Policy and Management Act, 252
Fertility regulation
China's one-child policy, 19, 409
motivation for, 410
Fertilizer, 313–314
Finland's forest policy, 202
Fish
endangered fish *versus* food, 17–18
fisheries. *See* Fisheries
Fisheries
biological model, 335–337
cod wars, 352
economically efficient model, 337–338
enforcement issues, 351–353
Europe's fisheries, 341
expanded territorial waters, 351
fish farming, 342–343
fish ranching, 342–343
individual transferable quotas, 347–351
joint owners, 340–342
limited entry regulations, 347–351
market allocations, 339
joint owners, 340–342
sole owner, 339–340
market solutions, 342–343
models
biological model, 335–337
economically efficient model, 337–338
open access regulations, 344–347
overfishing, 333–334
regulations
enforcement issues, 351–353
expanded territorial waters, 351
limited entry regulations, 347–351
open access regulations, 344–347
sole owner, 339–340
Southeast Asian fishers, 353
Thailand, 343
Fish farming, 342–343
Fish ranching, 342–343
Food, 296
agriculture
decisions, complexity of, 301
factor-factor relations, 304–306
factor-product relations, 306–308
goals of farmers, 301–302
output-price relations, 309–311
product-product relations, 302–304
technology and production, 308–309
Chinese Famine of 1958–1961, 324
distribution problems, 297–300
education of women, 322
emergency relief, 316
endangered fish *versus*, 17–18
forests, sustainable use of, 220–221
genetically modified, 509
government policies, 323–327
Green Revolution, 318
economics of, 300
sustainable food policy challenges, 299
health risks, 323
macroeconomic policies, 327
market outcomes, 301–315
outcomes
desirable, 300–301
institutional design for improving, 315–328
market, 301–315
population reduction and, 327–328
production measures
intermediate-term, 316–323
long-term, 323–328
short-term, 316
production mechanisms, 297
Russia's food security, 325
Food stamp programs, 321
Forest ecology, 190–192
Forest economics, 192–193
single rotation, 193–198
sustained management, 198–200
Forests and forestry, 189–190
benefits and costs
distribution of, 210–211
linking to decisions, 209–210
community management, 216–217
conservation easements, 214–215
economics. *See* Forest economics
ecotourism, 217–218
equimarginal principle, 207–209
extractive reserves, 215–216
Finland's policy, 202
food security and, 220–221

gender issues, 218–220
global warming, mitigation of, 218
green certification, 213–214
India, reducing poverty in, 216
infinite time horizons and rotation analysis, 588–590
legislation, 211–212
management
community-based, 216–217
multiple-use, 200–204
single-use, 190–200
sustained management, 198–200
multiple-use management, 200–204
Nepal forestry, 203, 537
Russian policy, 193
single-use management, 190–200
stewardship incentives, 220
sustained management, 198–200, 221
timber. *See* Timber
tradeoff analysis, 204–207
tropical forests
economic incentives to sustain, 219
underevaluation of, 201
Frontiers, 76

Gasoline taxes, 430–431
G-77/China, 439
Gender. *See* Women
Genetically modified food, 509
Geneva Declaration, 437
Germany's promotion of renewable energy, 385
Global warming. *See also* Climate change
agriculture and, 319
benefit cost analysis, 426–427
Berlin Mandate, 435–437
Brazilian negotiations, 440
Buenos Aires Plan of Action, 439
current costs *versus* future benefits, 19
deforestation and, 218
G-77/China, 439
Geneva Declaration, 437
greenhouse gases, controlling
benefits of, 425
costs of, 425–426
greenhouse gas treaty, 442
international agreements, 434–442
Kyoto Protocol, 437–439
Lebanon's cedar trees, 2
Marrakesh conference, 439–441
potential policy responses, 434
Umbrella Group, 439
United Nations Framework Convention on Climate Change, 434–435
Glossary, 593–603
Goods
demand for. *See* Demand for goods
public. *See* Public goods
supply of. *See* Supply of goods
Governments, 63–64
borrowing, monetary cost of, 137
failures, 86–88
social rate of time preference, 140–143

Grazing. *See* Rangelands
Green-adjusted NDP, 553–554
Green certification, 213–214
Greenhouse effect. *See* Climate change; Global warming
Greenhouse gas (GHG) treaty, 442
Green Revolution, 318
economics of, 300
sustainable food policy challenges, 299

Harvesting costs (HCs), 195, 584
Hazardous waste treatment, storage, and disposal facilities (TSDFs), 530–534
Hedonic pricing, 169–171
Hicksian demand function, 578
Himalayan rangelands, 229
Hotelling Rule, 363–366
Human-centered goals, 30
versus nature-centered goals, 32–33
Human rights
versus living space, 19
water, right to, 265

Incentive compatibility, in contingent valuation studies, 579–580
India
forestry, reducing poverty through, 217
pollution control, 460
renewable energy, promotion of, 386
sustainable environmental policies, 562
Individual transferable fish quotas, 347–351
Indonesian water pricing policy, 279
Inflation, 115
Informational contributions, 95
Institutions
efficient institutions, 70
food outcomes, 315–328
governments. *See* Governments
market. *See* Market system
rangeland development, 244–245
risk-management. *See* Risk-management institutions
social, 54–55. *See also* Governments; Market system
sustainable development, 561–566
Insurance. *See* Risk insurance
Interdependent markets, 62
Interstate water compacts, 288
Iranian ranchers, 234
Irrigation, 313
ISO 14000, 565–566
Ivory trade, in Africa, 53

Japan's photovoltaics, 379
Johannesburg Summit, 567
Justice. *See* Environmental justice

Klamath Basin conflict, 17–18
Korean carbon monoxide risks, 515
Kyoto Protocol, 437–439
sustainable development, 567

Labor force
age structure, 406
impacts, benefit-cost analysis, 114–115
output per worker, 405–406
Land ethic, 31
Law of demand, 37–38
Law of supply, 42–43
Lebanon's cedar trees, 2
Livestock
endangered wolves *versus,* 16–17
rangelands. *See also* Rangelands
economically optimum stocking rate, 230–238
grazing and forage dynamics, 240
overgrazing and profit motive, 240–241
U.S. grazing fees, 245–251
riparian ecosystems *versus,* 18–19
Living space, *versus* human rights, 19

Malawi water programs, 560
Marginal benefit, 573
cooperation with marginal cost, 60–61
Marginal cost pricing, 55–58, 63
energy, 381–383
water, 281–285
politically workable, 290–291
Market demand, 38–39
Market failures, 70
common property. *See* Common property
externalities. *See* Externalities
public goods. *See* Public goods
Market system, 22–26
competitive firm's supply curve, 58
competitive markets, 36–37
competitive supply, 55–58
demand for goods. *See* Demand for goods
failures. *See* Market failures
interdependent markets, 62
perfect competition, economic efficiency of, 58–63
supply of goods. *See* Supply of goods
Marrakesh conference, 439–441
Mean annual increment (MAI), 191
Middle East water policy, 260
Mining, *versus* public lands recreation, 18
Monteverde Cloud Forest Preserve, 217–218
Multiple Use Sustained Yield Act, 211

National Environmental Policy Act, 212
National Forest Management Act, 212
Natural resource economics, 1–2
Natural resource issues, 2–3
Nature-centered goals, 30–32
versus human-centered goals, 32–33
Negative externalities, 71–72
Nepal forestry, 203, 537
Net benefit, 28
Net present value (npv), 104–105, 590
calculation of, 105–106
discount rate, calculation from, 128–130
environmental policy, 135–136

private investments, 134–135
tables of discounted values, 130–134
postponing projects or regulations, 100
private investments, 134–135
sensitivity analysis, 106–108
steps to, 101–108
New Hampshire's climate change experience, 429
Nicaraguan natural disasters, sustaining development after, 558
Nuclear energy, 370

OPEC cartel, 372, 374
competitive supply response, 375–376
income elasticity, 377
shared interests, 374–375
Open-ended choice, 178
Option value, 153
Ozone layer, depletion of, 5. *See also* Global warming

Pacific yew tree, 20
Passive-use values, 153
Pesticides, 314
Pigouvian taxes, 84
Planting costs (PCs), 589
Policy
agenda for, 10–11
choices. *See* Policy choices
design of, 11
discount rate, use of, 135–136
energy policy
need for, 358–359
objectives, 359–360
environmental value, support of, 165
evaluation of, 12
food, 325–327
goals, 30–33
markets *versus* governments, 54–64
goods and means, 26–29
implementation of, 11–12
Policy choices
affordable energy *versus* functioning ecosystems, 15
clean air *versus* affordable cars, 16
climate change, 427–431
endangered fish *versus* food, 17–18
endangered species *versus* timber, 204–205
endangered wolves *versus* livestock and wildlife, 16–17
energy *versus* wildlife, 20
global warming control costs *versus* future benefits, 19
living space *versus* human rights, 19
market system and, 22–26
mining *versus* public lands recreation, 18
opportunities, displacement of, 20–21
population control, 411–412
power *versus* recreation, 18
riparian ecosystems *versus* livestock, 18–19
short-run human benefits *versus* long-run environmental damages, 20
timber *versus* endangered species, 15, 204–205
timber *versus* recreation, 205–207
Pollution, 445
abatement
costs, 452–456
efficient abatement, 456–459
absorptive capacity, 446
air. *See* Air pollution
assignment of property rights, 464–466
Chilean regulations, 467
China's water pollution policy, 450
command-and-control approach, 466–473
Common law, 462–464
continuous *versus* episodic, 447–448
controlling
Central Asian, economic incentives, 485
command-and-control approach, 466–473
decentralized policies, 461–466
economic efficiency, 448–459
equity in, 459–460
in India, 460
market incentives, use of, 473–482
sustainability, 460–461
damages, 448–452
decentralized policies, 461–466
desirable outcome criteria, 448–461
emission charges, 474–480
emission trading. *See* Emission trading
enforcement of laws, 482–486
environmental sustainability, 460–461
India, controlling in, 460
liability laws, 462
Common law, 462–464
efficiency of, 462
enforcement of, 482–486
statutory law, 464
local *versus* regional *versus* global, 447
market incentives, controlling with, 473–482
occurrence, 447–448
origin, 447
outcomes
desirable outcome criteria, 448–461
unregulated markets, 461
physical bases of, 446–448
point *versus* nonpoint sources, 447
property rights assignments, 464–466
rights to no pollution, 465–466
rights to pollute, 465
rules or standards, 466
Chilean regulations, 467
cost-effectiveness, 468–471
economics of, 466–471
incentives produced by, 471–473
uniformity, 468
spatial scale, 447
statutory law, 464
stock *versus* flow, 446
taxes, 383–384
water. *See* Water pollution
Poor persons
energy and poverty, 358
food stamp programs, 321
India forestry, reducing through, 217
water and, 289, 290
Population growth, 396–397
China's programs, 9
one-child policy, 19, 409
controlling
fertility regulation, *below*
policy choices, 411–412
whose welfare counts, 404
cost of children, 408–412
demand for children, 407
and demography, 399–400
economic growth
effect of, 401–403
effect on, 403–406
emerging questions, 413–415
environmental impact, 397–399
family size
China's one-child policy, 19, 409
potential, 410
unconstrained optimal, 409
fertility regulation
China's one-child policy, 19, 409
motivation for, 410
policy options, 411–412
price of, 410–411
food, population reduction and, 327–328
forecasting, 398
labor force
age structure, 406
output per worker, 405–406
macroeconomic forces, 400–406
microeconomic forces, 407–412
policy evaluation, 398–399
structural analysis, 398
Positive externalities, 72–73
Poverty. *See* Poor persons
Power. *See* Energy
Present value, 128. *See also* Net present value (npv)
Price(s)
changes in, valuing, 162
and competitive supply, 55–58
demand for goods. *See* Demand for goods
environmental value, measurement of
hedonic pricing, 169–171
market prices, 166–167
farming output-price relations, 309–311
marginal cost. *See* Marginal cost pricing
OPEC cartel's stability, 377
rangeland forage, 249–251
rangelands, 243–244
supply of goods. *See* Supply of goods
sustainable development, 555
underpriced resources, 25–26
water. *See* Water
PRI formula, 246–247
Productivity, 168–169
Programming methods, to value public actions, 575

Property rights, 24
and agriculture, 323
assignment of, 464–466
common property. *See* Common property
water, 289
Protection. *See* Environmental protection
Public action
common property, management of, 83–84
valuation using programming
methods, 575
Public goods, 75–76
excludability, 77
preferences for, 77–78
rivalry, 77
Public lands recreation. *See* Recreation

Random utility models (RUMs), 174
Rangelands, 225–226
Africa's rangelands, drought on, 230
economically optimum stocking rate,
230–238
economic efficiency, 227–228
economic value
of forage, 238–239
of range improvements, 239
emerging issues, 252–253
equity, 228
Ethiopian rangelands, 243
forage pricing schemes, 249–251
grazing
and forage dynamics, 240
overgrazing, profit motive and, 240–241
U.S. grazing fees, 245–251
Himalayan rangelands, 229
improvements, economic value of, 239
infrastructure development, 245
institutional development, 244–245
Iranian ranchers, 234
legislation, 251–252
livestock
economically optimum stocking rate,
230–238
grazing and forage dynamics, 240
overgrazing and profit motive, 240–241
U.S. grazing fees, 245–251
market incentive policies, 242–243
multiple uses for, 241
outcomes
desirable, 227–228
unregulated market, 228–241
pricing policies, 243–244
regulatory policies, 244
scarcity and competition, 226–227
sustainability, 228
Himalayan rangelands, 229
Rational ignorance, 87
RECLAIM program, 6
Recreation
mining *versus,* 18
timber *versus,* 205–207
Renewable energy
Brazilian challenges, 373
California's power meltdown, 387
conservation, 389–392
fossil fuels *versus,* 369–370
Germany, promoting in, 385
green power, 388
India, promoting in, 386
Japan's photovoltaics, 379
promotion of, 384–387
standards, 387–388
voluntary action, 388–389
Rent seeking, 87
Rio Summit, 566–567
Risk(s), 491–492
assessment. *See* Risk assessment
carbon monoxide in Korea, 515
causality, 493
choices under, 501
complexity, 492–493
DDT in Africa, 512
economic efficiency, 496–499
equity, 499–500
expected utility theory, 497–498
"human capital" valuation method, 496
insurability, 504
insurance, 501–504
legislation, 511
management, 507–518
risk-management institutions, 518–523
nature of, 492–495
outcomes
desirable, 495–500
unregulated market, 500–504
perceived *versus* objective, 494–495
risk-management institutions, 518–523
sustainability, 500
unacceptable risk, 494
valuation methods, 496
expected damages saved, 498–499
expected utility theory, 497–498
human capital, 496
using value information, 499
willingness-to-pay, 496–497
voluntary *versus* involuntary, 493–494
"willingness-to-pay" valuation method,
496–497
Risk assessment, 505
dose-response analysis, 505–506
environmental justice, 529–530
exposure analysis, 506–507
genetically modified food, 509
hazard identification, 505
list characterization, 507
synthesis, 507
uses and abuses, 508
Risk insurance, 501
adverse selection, 503–504
contract enforceability, 504
limited loss exposure, 503
moral hazard, 503
premium measurability, 502
risk pooling, 502
verifiable loss, 501
Risk management, 507–509
acceptable risk, 509–510
benefit-cost analysis, 516–518
economic analysis, 510, 512–518
Risk-management institutions, 518
consensus, mediation, and dialogue, 522
joint and several liability, 521
liability, 518–521
market-based incentives, 521–522
monitoring, research, and surveillance, 522
negligence *versus* strict liability, 519–521
regulations *versus* liability, 521
risk communication, 522–523
Risky programs, 115
River basin transboundary
management, 287
River basin withdrawals, 261–262
Russia
food security and garden plots, 325
forest policy, 193

Saving lives, discount rate for, 132
Shortages, 468
Social contract views of society, 528
Social institutions, 54–55. *See also*
Governments; Market system
Social opportunity cost of capital,
137–139
Social rate of time preference, 139–140
calculation and application, 145–146
collective burden, willingness to bear, 141
government's role, 142–143
intergenerational responsibility, 140–141
long-term discounting, 144–146
Soil erosion, 312–313
South Asia
Southeast Asian fishers, 353
water rights, 543
Spillover benefits, 26
Spillover costs, 26
Studying economics, importance of, 6–9
Superfund, 5–6
Supply of goods, 42
change in supply, 44–46
competitive supply, 55–58
equilibrium supply. *See* Equilibrium
demand and supply
law of supply, 42–43
mathematics, 570–572
quantity supplied, change in, 44
supply curves, 43–44
competitive firms, 58
water supply management, 271–274, 287
Surpluses
consumer surplus. *See* Consumer surplus
equilibrium demand and supply, 46–47
Sustainable development, 546
in Africa's poor cities, 556
Chinese energy programs, 563
coevolutionary development, 552
conditions for sustainability, 554–555
Conference on the Human
Environment, 566
consumption per capita,
nondeclining, 548
critical natural capital, 556–557
definitions of sustainability, 548–553
ecolabel programs, 564

ecosystem stability and resilience, 551
extended producer responsibility, 562–563
food policy challenges, 299
forestry management, 198–200, 221
global actions, 566–567
green-adjusted NDP, 553–554
Hartwick-Solow Rule, 555–556
in India, 562
institutional design, 561–566
instruments for implementing sustainability, 555–560
irreversibility, avoiding, 558–559
ISO 14000, 565–566
Johannesburg Summit, 567
Kyoto Protocol, 567
Malawi water programs, 560
measuring progress, 553–555
Nicaraguan natural disasters, 558
nondeclining consumption per capita, 548
nondeclining natural capital stock, 550–551
policy applications, 560–567
prices, 555
production opportunities, 548–549
rangelands, 228
Himalayan rangelands, 229
reasons for sustainability, 547–548
resource use, 550
Rio Summit, 566–567
risk avoidance, 551–552
safe minimum standard, 557–558
species existence, 552–553
steady-state principles, 559–560
Tanzanian sustainable exports, 565
tax reform, 563–564
water equity and sustainability, 269–270
what to sustain, 547–553
World Commission on Environment and Development, 566

Tanzanian sustainable exports, 565
Taxes
carbon taxes, 428, 430–431
as costs of production, 46
Pigouvian taxes, 84
pollution taxes, 383–384
as public gifts, 142
sustainable development, 563–564
Taxol, 20
Taylor Grazing Act, 251
Technology
agriculture
biotechnology in, 319–320
technology and production, 308–309
and climate change, 421–422
improvements in, 45
Thailand's fisheries, 343
Theory of economics. *See* Economic theory
Timber
versus endangered species, 15, 204–205
forest management, 190–200
versus recreation, 205–207
Total benefit, 574
Toxic wastes, 5–6
Tragedy of the commons, 79–83
avoidance of, 87
Brundtland report, 9
Travel cost models (TCMs), 171
individual models, 174
random utility models, 174
zonal models, 171–174
Tropical forests
economic incentives to sustain, 219
underevaluation of, 201

Umbrella Group, 439
Underpriced resources, 25–26
United Nations Framework Convention on Climate Change (UNFCCC), 434–435
U.S. Pacific Yew Act, 20
Use values, 153
Utilitarianism, 30, 527

Values
environmental. *See* Environmental values
individual values, and social welfare, 152–165
Virtual water, 288–289

Wastes
hazardous wastes TSDF sites, 530–534
pollution. *See* Pollution
toxic, 5–6
treatment facility sites, 536
Water, 258–263
agricultural use, 313, 318–319
California's ultimate resource, 3
China's water programs, 292
connection charges, 281
decentralized management, 289–290
demand management, 274–278
integrated supply/demand management, 287
distribution, 273
for drinking. *See* Drinking water
drought. *See* Drought
economic efficiency, 265–269
equity and sustainability, 269–270
externality charges, 282
flows, allocation of, 266–268
geographically differentiated rates, 282
groundwater law, 271
human rights, 265
interbasin transfer, 271–272
interstate compacts, 288
laws, 270–271
long-run pricing, 281
management, 264
decentralized management, 289–290
demand management, 274–278, 287
integrated supply/demand management, 287
river basin transboundary management, 287
supply management, 271–274, 287
marginal cost pricing, 281–285
politically workable, 290–291
markets, 282–283
Middle Eastern policy, 260
outcomes
actual, 270–280
desirable, 264–270
peak-load pricing, 283–284
price of, 277–278
Indonesia, 279
marginal cost pricing, 281–285, 290–291
peace and, 268
two-tiered pricing, 285–287
Zimbabwe, 282
property and, 289, 290
recycling, 273–274
regulations, 279–280
riparian ecosystems *versus* livestock, 18–19
river basin transboundary management, 287
river basin withdrawals, 261–262
shareholder participation, 291
South Asian water rights, 543
stocks, 269
storage, 272
supply management, 271–274
integrated supply/demand management, 287
surface water law, 270–271
transfers, 274–277
treatment, 272–273
two-tiered pricing, 285–287
use patterns, 263–264
virtual, 288–289
Water pollution. *See also* Pollution
acid rain program, 10
China's policy, 450
Wealth of Nations, The, 15, 24
Welfare measures, 578
Wilderness Act, 211–212
Wildlife
agricultural impact, 314
endangered. *See* Endangered species
energy *versus,* 20
Zambia wildlife management, 80
With and without principle, 99
Wolves, endangered *versus* livestock and wildlife, 16–17
Women
African girls' education, 413
child mortality, roles in reducing, 414
food education, 322
forestry issues, 218–220
status of, 412
World Commission on Environment and Development, 566

Yew tree, 20

Zambia wildlife management, 80
Zimbabwe water prices, 282
Zonal models, 171–174